Statistics and Data Analysis in Geology

Third Edition

John C. Davis

Kansas Geological Survey
The University of Kansas

John Wiley & Sons

New York · Chichester · Brisbane · Toronto · Singapore

ASSOCIATE EDITOR   Mark Gerber
MARKETING MANAGER   Kevin Molloy
PROGRAM COORDINATOR   Denise Powell
PRODUCTION EDITOR   Brienna Berger
DESIGNER   Madelyn Lesure
COVER PHOTO   Bill Bachmann/Photo Researchers

This book was printed and bound by Courier. The cover was printed by Phoenix Color.

ISBN 0-471-17275-8

Library of Congress Cataloging in Publication Data:
Davis, John C.
      Statistics and data analysis in geology—3rd ed.
      Includes bibliographies and index.
      1.   Geology—Data processing. 2. Geology—
      Statistical methods. I. Title

QE48.8.D38   2002   550'.72   85-12331

Printed in the United States of America

10 9 8 7 6 5 4 3 2 1

Preface

My original motivation for writing this book, back in 1973, was very simple. Teaching the techniques of data analysis to engineers and natural scientists, both university students and industry practitioners, would be easier, I reasoned, if I had a suitable textbook. It was. By 1986 when I revised *Statistics and Data Analysis in Geology* for its second edition, technology had progressed to the point that personal computers were almost commonplace and every young geologist was expected to have at least some familiarity with computing and analysis of data. This was a time of transition when personal computers offered the freedom of access and ease of use missing in the centralized mainframe environment, but these PC's lacked the power and speed necessary for many geological applications. In the intervening years since the appearance of the second edition, computing technology has evolved with almost unbelievable speed. I now have on my desktop a small crystalline cube, a "supercomputer" capable of outperforming devices that existed a decade ago at only a few sites in the world.

Although computing tools have advanced rapidly, our skills as educators have not kept pace. Almost all undergraduate students in the natural sciences and engineering, including the Earth sciences, are required to take classes in mathematics, statistics, data analysis, and computing. Graduate students, as a matter of course, are expected to have proficiency in these areas. Unfortunately, Earth science students voice an almost universal complaint: material taught in such courses is not relevant to their studies. In part this criticism reflects a certain mental rigidity present in some young minds that refuse to make an effort to stretch their imaginations. But it also reflects, in part, the absence of anything quantitative in many geology courses.

It is not surprising when students protest, "Why should I study this dull and boring topic when the material is never used in my field?" In an attempt to contribute to the solution of this educational impasse, I've made a major change in this edition of my book. The text now includes numerous geological data sets that illustrate how specific computational procedures can be applied to problems in the Earth sciences. In addition, each chapter ends with a set of exercises of greater or lesser complexity that the student can address using methods discussed in the text. It should be noted that there is no "teacher's manual" containing correct answers. Like most real-world situations, there may be more than one solution to a problem. An answer may depend upon how a question is framed. Acknowledging that no students, not even graduate assistants, like to do drudge work such as data entry, I've provided all of the data for examples and exercises as digital files on the World Wide Web. Thus, while there may be many excuses for failing to work an exercise, entering data incorrectly should not be one of them!

We have already noted that computing technology has changed enormously during the 28 years this book has been in print. Computers are no longer made that can read floppy disks and double-sided diskettes are being phased out by optical disks. We can be sure that computer technology will continue to evolve at a dizzying pace; to provide some degree of security from obsolescence, the data files are available on the World Wide Web at two sites, one maintained by John Wiley & Sons and the other by the Kansas Geological Survey. The WWW addresses are

http://www.wiley.com/college/davis

and

http://www.kgs.ku.edu/Mathgeo/Books/Stat/index.html

In addition to the downloadable files from the 3rd edition of *Statistics and Data Analysis in Geology*, you may also find additional data sets and exercises at this site as they are made available from time to time.

The basic arrangement of topics covered in the book is retained from earlier editions, progressing from background information to the analysis of geological sequences, then maps, and finally to multivariate observations. The discussion of elementary probability theory in Chapter 2 has been revised in recognition of the unfortunate fact that fundamentals of probability often are passed over in introductory courses in favor of a cookbook recitation of elementary statistical tests. These tests are also included here, but because probability forms the basis for almost all data analysis procedures and a thorough grounding in the concepts of probability is essential to understanding statistics, this introductory section has been expanded. The discussion of nonparametric methods introduced in the 2nd edition has been expanded because geologic data, particularly data collected in the field, seldom satisfy the distribution assumptions of classical parametric statistics. The effects of closure, which results in unwarranted relationships between variables when they are forced to sum to a constant value, are examined in detail. Geological measurements such as geochemical, petrographic, and petrophysical analyses, grain-size distributions—in fact, any set of values expressed as percentages—constitute compositional data and are subject to closure effects. The statistical transformations proposed by John Aitchison to overcome these problems are discussed at length.

In the 2nd edition, I revised the discussion of eigenvalues and eigenvectors because these topics had proved to be difficult for students. They are still difficult, so their treatment in the chapter on matrix algebra has been rewritten and a new section on singular value decomposition and the relationship between R- and Q-mode factor methods has been added to the final chapter on multivariate analysis.

The central role of geostatistics and regionalized variable theory in the study of the spatial behavior of geological and other properties is now firmly established. With the help of Ricardo Olea, I have completely revised the discussion of the many varieties of kriging and provide a series of simple demonstrations to illustrate how geostatistical methodologies work. I also have revised the section on contour mapping to reflect modern practices.

A discussion of fractals has been added, not because fractals have demonstrated any particular utility in geological investigations, but because they seem to hold a promise for the future. On a more prosaic topic, the section on regression has been expanded to include several variants that have special significance in the Earth sciences. To make room for these and other discussions, some subjects that proved to be of limited utility in geologic research have been deleted. Moving most tables to the WWW sites has made additional room in the text.

Because this is not a reference book, references are not emphasized. Citations are made to more specialized or advanced texts that I have found to contain especially lucid discussions of the points in question rather than to the most definitive or original sources. Those who wish to pursue a topic in depth will find ample references to the literature in the books I have included; those that simply want an elaboration on some point will probably find the books in Suggested Readings adequate for their needs.

I am fortunate to have enjoyed the help and encouragement of many people in the creation and evolution of this book throughout its several editions. The list of those who provided technical reviews and critical comments over the years reads like a "Who's Who" of mathematical geology and includes, in alphabetical order, Frits Agterberg, Dave Best, Paul Brockington, Jim Campbell, Ted Chang, Felix Chayes, Frank Ethridge, Je-an Fang, Colin Ferguson, John Griffiths, Jan Harff, Günther Hausberger, Ute Herzfeld, George Koch, Michael McCullagh, Gerry Middleton, Vera Pawlowsky, Floyd Preston, Nick Rock, Robert Sampson, Paul Switzer, Keith Turner, Leopold Weber, and Zhou Di. In addition, there have been dozens of others who have called or written to clarify a specific point or to bring an error to my attention, or to suggest ways in which the text could be improved. To all of these people, named and unnamed, I owe my deepest appreciation.

My esteem for my two mentors, Dan Merriam and John Harbaugh, was expressed in my dedication to the second edition of this book. My debt to these dear friends and colleagues remains as large as ever. However, those to whom I owe the greatest debt of gratitude for help with this 3rd edition are my associates and co-workers at the Kansas Geological Survey, particularly Ricardo Olea, John Doveton, and David Collins, who have provided examples, data, and exercises, and who have patiently reviewed specific topics with me in order to clarify my thoughts and to help me correct my misconceptions and errors. Ricardo has been my guide through the sometimes controversial field of geostatistics, and John has generously shared the store of instructional material and student exercises that he has patiently assembled through years of teaching petrophysics.

Most especially, I must acknowledge the assistance of Geoff Bohling, who volunteered to shoulder the burden of reading every word in the manuscript, working each example and exercise, and checking all of the computations and tables. Geoff created many of the statistical tables in the Appendix from the basic equations of distributions, and all of the calculations in the text have benefited from his careful checking and verification. Of course, any errors that remain are the responsibility of the author alone, but I would be remiss if I did not acknowledge that the number of such remaining errors would be far greater if it were not for Geoff's careful scrutiny.

I would also like to note that I have benefited from the nurturing environment of the Kansas Geological Survey (KGS) at The University of Kansas. KU has provided an intellectual greenhouse in which mathematical geology has flourished for over 30 years. I especially wish to acknowledge the support and encouragement of two previous directors of the Kansas Geological Survey, Bill Hambleton and Lee Gerhard, who recognized the importance of geology's quantitative aspects. Bill had the foresight to realize that the massive, expensive mainframe dinosaurs of computing in the 1960's would evolve into the compact, indispensable personal tools of every working geologist, and his vision kept the KGS at the forefront of computer applications. Mathematical geology advances, as does all of science, by the cumulative efforts of individuals throughout the world who share a common interest and who have learned that methodologies created in one part of the globe will find important applications elsewhere. Aware of this synergistic process, Lee encouraged visits and exchanges with the world's leaders in mathematical geology and its related disciplines, creating a heady ferment of intellectual activity that remains unique. It was with their support and encouragement that I have been able to write the three editions of this book.

My final expression of gratitude is the deepest and is owed to my editor, layout designer, proofreader, typesetter, reviewer, critic, companion, and source of inspiration—Jo Anne DeGraffenreid, without whose tireless efforts this edition would never have been completed. She carefully polished my words, refined my grammar, and detected obscure passages, insisting that I rewrite them until they were understandable. She checked the illustrations and equations for consistency in style and format, designed the layout, selected the book type, and in a Herculean effort, set the entire manuscript in camera-ready form using the TEX typesetting language. Most importantly, she encouraged me throughout the process of seemingly never-ending revision, and took me home and poured for me a generous libation when I despaired of ever laying this albatross to rest. To her I dedicate this book.

John C. Davis
Lawrence, KS

"...when you can measure what you are speaking about and express it in numbers, you know something about it; but when you cannot express it in numbers, your knowledge is of a meagre and unsatisfactory kind; it may be the beginning of knowledge, but you have scarcely in your thoughts advanced to the state of science, whatever the matter may be."

— *Lord Kelvin*

CONTENTS

Page

Preface v

1. Introduction 1
The Book and the Course it Follows 3
Statistics in Geology 6
Measurement Systems 7
A False Feeling of Security 9
Selected Readings 10

2. Elementary Statistics 11
Probability 11
Continuous Random Variables 25
Statistics 29
Summary Statistics 34
Joint Variation of Two Variables 40
Induced Correlations 46
Logratio Transformation 50
Comparing Normal Populations 55
Central Limits Theorem 58
Testing the Mean 60
P-Values 64
Significance 65
Confidence Limits 66
The t-Distribution 68
  Degrees of freedom 69
  Confidence intervals based on t 72
  A test of the equality of two sample means 72
  The t-test of correlation 74
The F-Distribution 75
  F-test of equality of variances 76
  Analysis of variance 78
  Fixed, random, and mixed effects 83
  Two-way analysis of variance 84

Contents

Nested design in analysis of variance 88
The χ^2 Distribution 92
Goodness-of-fit test 93
The Logarithmic and Other Transformations 97
Other transformations 102
Nonparametric Methods 102
Mann–Whitney test 103
Kruskal–Wallis test 105
Nonparametric correlation 105
Kolmogorov–Smirnov tests 107
Exercises 112
Selected Readings 119

3. Matrix Algebra 123
The Matrix 123
Elementary Matrix Operations 125
Matrix Multiplication 127
Inversion and Solution of Simultaneous Equations 132
Determinants 136
Eigenvalues and Eigenvectors 141
Eigenvalues 141
Eigenvectors 150
Exercises 153
Selected Readings 157

4. Analysis of Sequences of Data 159
Geologic Measurements in Sequences 159
Interpolation Procedures 163
Markov Chains 168
Embedded Markov chains 173
Series of Events 178
Runs Tests 185
Least-Squares Methods and Regression Analysis 191
Confidence belts around a regression 200
Calibration 204
Curvilinear regression 207

Reduced major axis and related regressions 214
Structural analysis and orthogonal regression 218
Regression through the origin 220
Logarithmic transformations in regression 221
Weighted regression 224
Looking at residuals 227
Splines 228
Segmenting Sequences 234
Zonation 234
Seriation 239
Autocorrelation 243
Cross-correlation 248
Cross-correlation and stratigraphic correlation 254
Semivariograms 254
Modeling the semivariogram 261
Alternatives to the semivariogram 264
Spectral Analysis 266
A quick review of trigonometry 266
Harmonic analysis 268
The continuous spectrum 275
Exercises 278
Selected Readings 288

5. Spatial Analysis 293

Geologic Maps, Conventional and Otherwise 293
Systematic Patterns of Search 295
Distribution of Points 299
Uniform density 300
Random patterns 302
Clustered patterns 307
Nearest-neighbor analysis 310
Distribution of Lines 313
Analysis of Directional Data 316
Testing hypotheses about circular directional data 322
Test for randomness 322
Test for a specified trend 325
Test of goodness of fit 326
Testing the equality of two sets of directional vectors 326
Spherical Distributions 330

Contents

Matrix representation of vectors 334

Displaying spherical data 338

Testing hypotheses about spherical directional data 341

    A test of randomness 341

Fractal Analysis 342

Ruler procedure 343

Grid-cell procedure 346

Spectral procedures 351

Higher dimensional fractals 353

Shape 355

Fourier measurements of shape 359

Spatial Analysis by ANOVA 366

Computer Contouring 370

Contouring by triangulation 374

Contouring by gridding 380

Problems in contour mapping 391

Extensions of contour mapping 394

Trend Surfaces 397

Statistical tests of trends 407

Two trend-surface models 412

Pitfalls 414

Kriging 416

Simple kriging 418

Ordinary kriging 420

Universal kriging 428

    Calculating the drift 433

    An example 435

Block kriging 437

Exercises 443

Selected Readings 452

6. Analysis of Multivariate Data 461

Multiple Regression 462

Discriminant Functions 471

Tests of significance 477

Multivariate Extensions of Elementary Statistics 479

Equality of two vector means 483

Equality of variance–covariance matrices 484

Cluster Analysis 487

Introduction to Eigenvector Methods, Including Factor Analysis 500
  Eckart–Young theorem 502
Principal Component Analysis 509
  Closure effects on principal components 523
R-Mode Factor Analysis 526
  Factor rotation 533
  Maximum likelihood factor analysis 538
Q-Mode Factor Analysis 540
  A word about closure 546
Principal Coordinates Analysis 548
Correspondence Analysis 552
Multidimensional Scaling 560
Simultaneous *R*- and *Q*-Mode Analysis 566
Multigroup Discriminant Functions 572
Canonical Correlation 577
Exercises 584
Selected Readings 594

Appendix 601

Table A.1. Cumulative probabilities for the standardized normal
    distribution 601

Table A.2. Critical values of t for ν degrees of freedom and
    selected levels of significance 602

Table A.3. Critical values of F for ν_1 and ν_2 degrees of freedom
    and selected levels of significance 603

Table A.4. Critical values of χ^2 for ν degrees of freedom and
    selected levels of significance 607

Table A.5. Probabilities of occurrence of specified values
    of the Mann–Whitney W_x test statistic 608

Table A.6. Critical values of Spearman's ρ for testing the significance
    of a rank correlation 613

Table A.7. Critical values of D in the Kolmogorov–Smirnov
    goodness-of-fit test 614

Table A.8. Critical values of the Lilliefors test statistic, T, for
    testing goodness-of-fit to a normal distribution 617

Table A.9. Maximum likelihood estimates of the concentration
    parameter κ for calculated values of $\overline{R}$ 618

Contents

Table A.10. Critical values of $\overline{R}$ for Rayleigh's test for the presence
of a preferred trend 619

Table A.11. Critical values of $\overline{R}$ for the test of uniformity of
a spherical distribution 620

Index 621

Chapter 1
Introduction

Mathematical methods have been employed by a few geologists since the earliest days of the profession. For example, mining geologists and engineers have used samples to calculate tonnages and estimate ore tenor for centuries. As Fisher pointed out (1953, p. 3), Lyell's subdivision of the Tertiary on the basis of the relative abundance of modern marine organisms is a statistical procedure. Sedimentary petrologists have regarded grain-size and shape measurements as important sources of sedimentological information since the beginning of the last century. The hybrid Earth sciences of geochemistry, geophysics, and geohydrology require a firm background in mathematics, although their procedures are primarily derived from the non-geological parent. Similarly, mineralogists and crystallographers utilize mathematical techniques derived from physical and analytical chemistry.

Although these topics are of undeniable importance to specialized disciplines, they are not the subject of this book. Since the spread of computers throughout universities and corporations in the late 1950's, geologists have been increasingly attracted to mathematical methods of data analysis. These methods have been borrowed from all scientific and engineering disciplines and applied to every facet of Earth science; it is these more general techniques that are our concern. Geology itself is responsible for some of the advances, most notably in the area of mapping and spatial analysis. However, our science has benefited more than it has contributed to the exchange of quantitative techniques.

The petroleum industry has been among the largest nongovernment users of computers in the United States, and is also the largest employer of geologists. It is not unexpected that a tremendous interest in geomathematical techniques has developed in petroleum companies, nor that this interest has spread back into the academic world, resulting in an increasing emphasis on computer languages and mathematical skills in the training of geologists. Unfortunately, there is no broad heritage of mathematical analysis in geology—adequate educational programs have been established only in scattered institutions, through the efforts of a handful of people.

Many older geologists have been caught short in the computer revolution. Educated in a tradition that emphasized the qualitative and descriptive at the expense of the quantitative and analytical, these Earth scientists are inadequately prepared in mathematics and distrustful of statistics. Even so, members of the profession quickly grasped the potential importance of procedures that computers now make so readily available. Many institutions, both commercial and public, provide extensive libraries of computer programs that will implement geomathematical applications. Software and data are widely distributed over the World Wide Web through organizations such as the International Association for Mathematical Geology (http://www.iamg.org/). The temptation is strong, perhaps irresistible, to utilize these computer programs, even though the user may not clearly understand the underlying principles on which the programs are based.

The development and explosive proliferation of personal computers has accelerated this trend. In the quarter-century since the first appearance of this book, computers have progressed from mainframes of ponderous dimensions (but minuscule capacity) to small cubes that perch on the corner of a desk and contain the power of a supercomputer. Any geologist can buy an inexpensive computer for personal use that will perform more computations faster than the largest mainframe computers that served entire corporations and universities only a few short years ago. For many geologists, a personal computer has replaced a small army of secretaries, draftsmen, and bookkeepers. However, these ubiquitous plastic boxes with their colorful screens seem to promise much more than just word-processing and spreadsheet calculations—if only geologists knew how to put them to use in their professional work.

This book is designed to help alleviate the difficulties of geologists who feel that they can gain from a quantitative approach to their research, but are inadequately prepared by training or experience. Ideally, of course, these people should receive formal instruction in probability, statistics, numerical analysis, and programming; then they should study under a qualified geomathematician. Such an ideal is unrealistic for all but a few fortunate individuals. Most must make their way as best they can, reading, questioning, and educating themselves by trial and error. The path followed by the unschooled is not an orderly progression through topics laid out in curriculum-wise fashion. The novice proceeds backwards, attracted first to those methods that seem to offer the greatest help in the research, exploration, or operational problems being addressed. Later the self-taught amateur fills in gaps in his or her background and attempts to master the precepts of the techniques that have been applied. This unsatisfactory and even dangerous method of education, comparable perhaps to a physician learning by on-the-job training, is one many people seem destined to follow. The aim of this book is to introduce organization into the self-educational process, and guide the impatient neophyte rapidly through the necessary initial steps to a glittering algorithmic Grail. Along the way, readers will be exposed to those less glamorous topics that constitute the foundations upon which geomathematical procedures are built.

This book is also designed to aid another type of geologist-in-training—the student who has taken or is taking courses in statistics and programming. Such curriculum requirements are now nearly ubiquitous in universities throughout the world. Unfortunately, these topics are frequently taught by persons who have little knowledge of geology or any appreciation for the types of problems faced by Earth scientists. The relevance of these courses to the geologist's primary field is often obscure. A feeling of skepticism may be compounded by the absence of mathematical applications in geology courses. Many faculty members in the Earth sciences received their formal education prior to the current emphasis on geomathematical methodology, and consequently are untrained in the quantitative subjects their students are required to master. These teachers may find it difficult to demonstrate the relevance of mathematical topics. In this book, the student will find not only generalized developments of computational techniques, but also numerous examples of their applications in geology and a library of problem sets for the exercises that are included. Of course, it is my hope that both the student and the instructor will find something of interest in this book that will help promote the widening common ground we refer to as geomathematics.

The Book and the Course it Follows

Readers are entitled to know at the onset where a book will lead and how the author has arranged the journey. Because the author has made certain assumptions about the background, training, interests, and abilities of the audience, it is also necessary that readers know what is expected of them. This book is about quantitative methods for the analysis of geologic data—the area of Earth science which some call **geomathematics** and others call **mathematical geology**. Also included is an introduction to geostatistics, a subspecialty that has grown into an entire branch of applied statistics.

The orientation of the book is methodological, or "how-to-do-it." Theory is not emphasized for several reasons. Most geologists tend to be pragmatists, and are far more interested in results than in theory. Many useful procedures are *ad hoc* and have no adequate theoretical background at present. Methods which are theoretically developed often are based on statistical assumptions so restrictive that the procedures are not strictly valid for geologic data. Although elementary probability is discussed and many statistical tests described, the detailed development of statistical and geostatistical theory has been left to others.

Because the most complex analytical procedure is built up of a series of relatively simple mathematical manipulations, our emphasis is on operations. These operations are most easily expressed in matrix algebra, so we will study this subject, illustrating the operations with geological examples.

The first edition of this text (published in 1973) devoted a chapter to the FORTRAN computer language and most procedures in that edition were accompanied by short program listings in FORTRAN. When the second edition appeared in 1986, FORTRAN no longer dominated scientific programming and computer centers maintained extensive libraries of statistical and mathematical routines written in many computer languages. Large statistical packages implemented almost every procedure described in the text, so program listings were no longer necessary. Now at the time of this third edition, there are many easy-to-use interactive programs to perform almost any desired statistical calculation; these programs have graphical interfaces and run on personal computers. In addition, there are inexpensive, specialized programs for geostatistics, for analysis of compositional data, and for other "nonstandard" procedures of interest to Earth scientists. Some of these are distributed free or at nominal cost as "shareware." Computation is no longer among the major problems facing researchers today; they must be concerned, rather, with interpretation and the appropriateness of their approach. As a consequence, this third edition contains many more worked examples and also includes an extensive library of problem sets accessible over the Internet.

The discussion in the following chapters begins with the basic topics of probability and elementary statistics, including the special steps necessary to analyze compositional data, or variables such as chemical analyses and grain-size categories that sum to a constant. The next topic is matrix algebra. Then we will consider the analysis of various types of geologic data that have been classified arbitrarily into three categories: (1) data in which the sequence of observations is important, (2) data in which the two-dimensional relationships between observations are important, and (3) multivariate data in which order and location of the observations are not considered.

The first category contains all classes of problems in which data have been collected along a continuum, either of time or distance. It includes time series, calculation of semivariograms, analysis of stratigraphic sections, and the interpretation of chart recordings such as well logs. The second category includes problems in which spatial coordinates or geographic locations of samples are important, *i.e.*, studies of shape and orientation, contour mapping, trend-surface analysis, geostatistics including kriging, and similar endeavors. The final category is concerned with clustering, classification, and the examination of interrelations among variables in which sample locations on a map or traverse are not considered. Paleontological, mineralogical, and geochemical data often are of this type.

The topics proceed from simple to complex. However, each successive topic is built upon its predecessors, so aspects of multiple regression, covered in Chapter 6, have been discussed in trend analysis (Chapter 5), which has in turn been preceded by curvilinear regression (Chapter 4). The basic mathematical procedure involved has been described under the solution of simultaneous equations (Chapter 3), and the statistical basis of regression has first been discussed in Chapter 2. Other techniques are similarly developed.

The first topic in the book is elementary statistics. The final topic is canonical correlation. These two subjects are separated by a wide gulf that would require several years to bridge following a typical course of study. Obviously, we cannot cover this span in a single book without omitting a tremendous amount of material. What has been sacrificed are all but the rudiments of statistical theory associated with each of the techniques, the details of all mathematical operations except those that are absolutely essential, and all the embellishments and refinements that typically are added to the basic procedures. What has been retained are the fundamental algorithms involved in each analysis, discussions of the relations between quantitative techniques and example applications to geologic problems, and references to sources for additional details.

My contention is that a quantitative approach to geology can yield a fruitful return to the investigator; not so much, perhaps, by "proving" a geological hypothesis or demonstrating its validity, but by gaining insights from the critical examination of phenomena that is prerequisite to any quantitative procedure. Numerical analysis requires that collection of data be carefully controlled, with consideration given to extraneous influences. As a consequence, the investigator may acquire a closer familiarity with the objects of study than could otherwise be attained. Certainly a paleontologist who has made careful measurements on a large collection of randomly selected fossil specimens has a far greater and more accurate understanding of the natural variation of these organisms than does the paleontologist who relies on informal examination. The rigor and objectivity required by quantitative methodologies can compensate in part for insight and experience which otherwise must be gained by many years of work. At the same time, the discipline necessary to perform quantitative research will hasten the growth and maturity of the scientist.

The measurement and analysis of data may lead to interpretations that are not obvious or apparent when other means of investigation are used. Multivariate methods, for example, may reveal clusterings of objects that are at variance with accepted classifications, or may show relationships between variables where none were expected. These findings require explanation. Sometimes a plausible explanation cannot be found; but in other instances, new theories may be suggested which would otherwise have been overlooked.

Perhaps the greatest worth of quantitative methodologies lies not in their capability to demonstrate what is true, but rather in their ability to expose what is false. Quantitative techniques can reveal the insufficiency of data, the tenuousness of assumptions, the paucity of information contained in most geologic studies. Unfortunately, upon careful and dispassionate analysis, many geological interpretations deteriorate into a collection of guesses and hunches based on very little data, of which most are of a contradictory or inconclusive nature.

If geology were an experimental science like chemistry or physics—in which observations can be verified by any competent worker—controversy and conflict might disappear. However, geologists are practitioners of an observational science, and the rigorous application of quantitative methods often reveals us for the imperfect observers that we are. Indeed, a decline into scientific skepticism is one of the dangers that often traps geomathematicians. These workers are often characterized by a suspicious and iconoclastic attitude toward geological platitudes. Sadly it must be confessed that such cynicism is often justified. Geologists are trained to see patterns and structure in nature. Geomathematical methods provide the objectivity necessary to avoid creating these patterns when they may exist only in the scientist's desire for order.

Statistics in Geology

All of the techniques of quantitative geology discussed in this book can be regarded as statistical procedures, or perhaps "quasi-statistical" or "proto-statistical" procedures. Some are sufficiently well developed to be used in rigorous tests of statistical hypotheses. Other procedures are *ad hoc*; results from their application must be judged on utilitarian rather than theoretical grounds. Unfortunately, there is no adequate general theory about the nature of geological populations, although geology can boast of some original contributions to the subject, such as the theory of regionalized variables. However, like statistical tests, geomathematical techniques are based on the premise that information about a phenomenon can be deduced from an examination of a small sample collected from a vastly larger set of potential observations on the phenomenon.

Consider subsurface structure mapping for petroleum exploration. Data are derived from scattered boreholes that pierce successive stratigraphic horizons. The elevation of the top of a horizon measured in one of these holes constitutes a single observation. Obviously, an infinite number of measurements of the top of this horizon could be made if we drilled unlimited numbers of holes. This cannot be done; we are restricted to those holes which have actually been drilled, and perhaps to a few additional test holes whose drilling we can authorize. From these data we must deduce as best we can the configuration of the top of the horizon between boreholes. The problem is analogous to statistical analysis; but unlike the classical statistician, we cannot design the pattern of holes or control the manner in which the data were obtained. However, we can use quantitative mapping techniques that are either closely related to statistical procedures or rely on novel statistical concepts. Even though traditional forms of statistical tests may be beyond our grasp, the basic underlying concepts are the same.

In contrast, we might consider mine development and production. For years mining geologists and engineers have carefully designed sampling schemes and drilling plans and subjected their observations to statistical analyses. A veritable blizzard of publications has been issued on mine sampling. Several elaborate statistical distributions have been proposed to account for the variation in mine values, providing a theoretical basis for formal statistical tests. When geologists can control the means of obtaining samples, they are quick to exploit the opportunity. The success of mining geologists and engineers in the assessment of mineral deposits testifies to the power of these methods.

Unfortunately, most geologists must collect their observations where they can. Logs of oil wells have been made at too great a cost to ignore merely because the well locations do not fit into a predesigned sampling plan. Paleontologists must be content with the fossils they can glean from the outcrop; those buried in the subsurface are forever beyond their reach. Rock specimens can be collected from the tops of batholiths in exposures along canyon walls, but examples from the roots of these same bodies are hopelessly deep in the Earth. The problem is seldom too much data in one place. Rather, it is too little data elsewhere. Our observations of the Earth are too precious to discard lightly. We must attempt to wring from them what knowledge we can, recognizing the bias and imperfections of that knowledge.

Many publications on the design of statistical experiments and sampling plans have appeared. Notable among these is the geological text by Griffiths (1967), which is in large part concerned with the effect sampling has on the outcome of statistical tests. Although Griffiths' examples are drawn from sedimentary petrology, the methods are equally applicable to other problems in the Earth sciences. The book represents a rigorous, formal approach to the interpretation of geologic phenomena using statistical methods. Griffiths' book, unfortunately now out of print, is especially commended to those who wish to perform experiments in geology and can exercise strict control over their sampling procedures. In this text we will concern ourselves with those less tractable situations where the sample design (either by chance or misfortune) is beyond our control. However, be warned that an uncontrolled experiment (*i.e.*, one in which the investigator has no influence over where or how observations are taken) usually takes us outside the realm of classical statistics. This is the area of "quasi-statistics" or "proto-statistics," where the assumptions of formal statistics cannot safely be made. Here, the well-developed formal tests of hypotheses do not exist, and the best we can hope from our procedures is guidance in what ultimately must be a human judgment.

Measurement Systems

A quantitative approach to geology requires something more profound than a headlong rush into the field armed with a personal computer. Because the conclusions reached in a quantitative study will be based at least in part on inferences drawn from measurements, the geologist must be aware of the nature of the number systems in which the measurements are made. Not only must the Earth scientist understand the geological significance of the recorded variables, the mathematical significance of the measurement scales used must also be understood. This topic is more complex than it might seem at first glance. Detailed discussions and references can be found in Stevens (1946), the book edited by Churchman and Ratoosh (1959) and, from a geologist's point of view, in Griffiths (1960).

A *measurement* is a numerical value assigned to an observation which reflects the magnitude or amount of some characteristic. The manner in which numerical values are assigned determines the *scale of measurement*, and this in turn determines the type of analyses that can be made of the data. There are four measurement scales, each more rigorously defined than its predecessor, and each containing greater information. The first two are the nominal scale and the ordinal scale, in which observations are simply classified into mutually exclusive categories. The final two scales, the interval and ratio, are those we ordinarily think of as "measurements" because they involve determination of the magnitudes of an attribute.

The *nominal scale* of measurement consists of a classification of observations into mutually exclusive categories of equal rank. These categories may be identified by names, such as "red," "green," and "blue," by labels such as "A," "B," and "C," by symbols such as ☀, ◇, and ●, or by numbers. However, numbers are used only as identifiers. There can be no connotation that 2 is "twice as much" as 1, or that 5 is "greater than" 4. Binary-state variables are a special type of nominal data in which symbolic tags such as 1 and 0, "yes" and "no," or "on" and "off" indicate the presence or absence of a condition, feature, or organism. The classification of fossils as to type is an example of nominal measurement. Identification of one fossil as a brachiopod and another as a crinoid implies nothing about the relative importance or magnitude of the two.

The number of observations occurring in each state of a nominal system can be counted, and certain nonparametric tests can be performed on nominal data. A classic example we will consider at length is the occurrence of heads or tails in a coin-flipping experiment. Heads and tails constitute two categories of a nominal scale, and our data will consist of the number of observations that fall into them. A geologic equivalent of this problem consists of the appearance of feldspar and quartz grains along a traverse across a thin section. Quartz and feldspar form mutually exclusive categories that cannot be meaningfully ranked in any way.

Sometimes observations can be ranked in a hierarchy of states. Mohs' hardness scale is a classic example of a ranked or ***ordinal scale***. Although the minerals on the scale, which extends from one to ten, increase in hardness with higher rank, the steps between successive states are not equal. The difference in absolute hardness between diamond (rank ten) and corundum (rank nine) is greater than the entire range of hardness from one to nine. Similarly, metamorphic rocks may be ranked along a scale of metamorphic grade, which reflects the intensity of alteration. However, the steps between grades do not represent a uniform progression of temperature and pressure.

As with the nominal scale, a quantitative analysis of ordinal measurements is restricted primarily to counting observations in the various states. However, we can also consider the manner in which different ordinal classes succeed one another. This is done, for example, by determining if states tend to be followed an unusual number of times by greater or lesser states on the ordinal scale.

The ***interval scale*** is so named because the length of successive intervals is a constant. The most commonly cited example of an interval scale is that of temperature. The increase in temperature between 10° and 20° C is exactly the same as the increase between 110° and 120° C. However, an interval scale has no natural zero, or point where the magnitude is nonexistent. Thus, we can have negative temperatures that are less than zero. The starting point for the Celsius (centigrade) scale was *arbitrarily set* at a point coinciding with the freezing point of water, whereas the starting point on the Fahrenheit scale was chosen as the lowest temperature reached by an equal mixture of snow and salt. To convert from one interval scale to another, we must perform two operations: a multiplication to change the scale, and an addition or subtraction to shift the arbitrary origin.

Ratio scales have not only equal increments between steps, but also a true zero point. Measurements of length are of this type. A 2-in. long shell is twice the length of a 1-in. shell. A shell with zero length does not exist, because it has no length at all. It is generally agreed that "negative lengths" are not possible. To convert from one ratio scale to another, such as from inches to centimeters, we must only perform the single operation of multiplication.

Ratio scales are the highest form of measurement. All types of mathematical and statistical operations may be performed with them. Although interval scales in theory convey less information than ratio scales, for most purposes the two can be used in the same manner. Almost all geological data consist of continuously distributed measurements made on ratio or interval scales, because these include the basic physical properties of length, volume, mass, and the like. In subsequent chapters, we will not distinguish between the two measurement scales, and they may occur intermixed in the same problem. An example occurs in trend-surface analysis where an independent variable may be measured on a ratio scale while the geographic coordinates are on an interval scale, because the coordinate grid has an arbitrary origin.

A False Feeling of Security

Perhaps this chapter should be concluded with a precautionary note. If you pursue the following topics, you will become involved with mathematical methods that have a certain aura of exactitude, that express relationships with apparent precision, and that are implemented on devices that have a popular reputation for infallibility. Computers can be used very effectively as devices of intimidation. The presentation of masses of numbers, all expressed to eight decimal places, overwhelms the minds of many people and numbs their natural skepticism. A geologic report couched in mathematical jargon and filled with computer output usually will bluff all but a few critics, and those who understand and comment often do so in equally obtuse terms. Hence, both the report and criticism pass over the heads of most of the intended audience. The greatest danger, however, is to researchers themselves. If they fall sway to their own computers, they may cease to critically examine their data and the interpretative methods. Hypnotized by numbers, he or she may be led to the most ludicrous conclusions, totally blind to any reality beyond the computer screen. Keep in mind the little phrase posted on the wall of every computation center: "GIGO—Garbage In, Garbage Out."

The first chapter in the first edition of this book began and ended with quotations; these were repeated in the second edition. I have no reason to remove them now, as they are as relevant today as they were then. An anonymous critic left the following rhyme on my desk almost 30 years ago. It remains posted on my wall to this day.

What could be cuter
Than to feed a computer
With wrong information
But naïve expectation
To obtain with precision
A Napoleonic decision?

— *Major Alexander P. de Seversky*

SELECTED READINGS

Churchman, C.W., and P. Ratoosh [Eds.], 1959, *Measurement: Definitions and Theories*: John Wiley & Sons, Inc., New York, 274 pp.

Fisher, R.A., 1953, The expansion of statistics: *Jour. Royal Statistical Soc.*, Series A, v. 116, p. 1-6.

Griffiths, J.C., 1960, Some aspects of measurement in the geosciences: *Mineral Industries,* v. 29, no. 4, Pennsylvania State Univ., p. 1, 4, 5, 8.

Griffiths, J.C., 1967, *Scientific Method in Analysis of Sediments*: McGraw-Hill, Inc., New York, 508 pp.

Stevens, S.S., 1946, On the theory of scales of measurement: *Science,* v. 103, p. 677-680.

Chapter 2
Elementary Statistics

Geologists' direct observations of our world are confined to the outer part of the Earth's crust, yet they must attempt to understand the nature of the Earth's core and mantle and the deeper parts of the crust. Furthermore, the processes that modify the Earth, such as mountain building and continental evolution, are generally beyond the geologists' capabilities for direct manipulation. No other scientists, with the exception of astronomers, are more removed from the bulk of their study material and less able to experiment on their subject.

Geology, to a major extent, remains a science that is principally concerned with observation. Because geologists depend heavily on observations, particularly observations in which there is a large portion of uncertainty, statistics should play an important role in their research. Although the term "statistics" once referred simply to the collection of numerical facts such as baseball scores, it has come to include the analysis of data, and especially the uncertainty associated with such data. Statistical problems, whether perceived or not, occur wherever there are elements of chance. Geologists need to be conscious of these problems, and of some of the statistical tools that are available to help solve the problems.

Probability

Although many descriptions and definitions of statistics have been written, it perhaps may be best considered as the determination of the probable from the possible. In any circumstance, there are a variety (sometimes an infinity) of possible outcomes. All these have an associated probability that describes their frequency of occurrence. From an analysis of probabilities associated with events, future behavior or past states of the object or event under study may be estimated.

All of us have an intuitive concept of probability. For example, if asked to guess whether it will rain tomorrow, most of us would reply with some confidence that rain is likely or unlikely, or perhaps in rare circumstances, that it is certain to rain, or certain not to rain. An alternative way of expressing our estimate would be to use a numerical scale, as for example a percentage scale. If we state that the chance of rain tomorrow is 30%, then we imply that the chance of it not raining is 70%.

Scientists usually express probability as an arbitrary number ranging from 0 to 1, or an equivalent percentage ranging from 0 to 100%. If we say that the probability of rain tomorrow is 0, we imply that we are absolutely certain that it will not rain. If, on the other hand, we state that the probability of rain is 1, we are absolutely certain that it will. Probability, expressed in this form, pertains to the likelihood of an event. Absolute certainty is expressed at the ends of this scale, 0 and 1, with different degrees of uncertainty in between. For example, if we rate the probability of rain tomorrow as 1/2 (and therefore of no rain as 1/2), we express our view with a maximum degree of uncertainty; the likelihood of rain is equal to that of no rain. If we rate the probability of rain as 3/4 (1/4 probability of no rain), we express a smaller degree of uncertainty, for we imply that it is three times as likely to rain as it is not to rain.

Our estimates of the likelihood of rain may be based on many different factors, including a subjective "feeling" about the matter. We may utilize the past behavior of a phenomenon such as the weather to provide insight into its probable future behavior. This "relative frequency" approach to probability is intuitively appealing to geologists, because the concept is closely akin to uniformitarianism. Other methods of defining and arriving at probabilities may be more appropriate in certain circumstances. In carefully prescribed games of chance, the probabilities attached to a specific outcome can be calculated exactly by combinatorial mathematics; we will use this concept of probability in our initial discussions because of its relative simplicity. An entire branch of statistics treats probabilities as subjective expressions of the "degree of belief" that a particular outcome will occur. We must rely on the subjective opinions of experts when considering such questions as the probability of failure of a new machine for which there is no past history of performance. The subjective approach is widely used (although seldom admitted to) in the assessment of the risks associated with petroleum and mineral exploration, where relative-frequency based estimates of geologic conditions and events are difficult to obtain (Harbaugh, Davis, and Wendebourg, 1995). The implications contained in various concepts of probability are discussed in books by von Mises (1981) and Fisher (1973). Fortunately, the mathematical manipulations of probabilities are identical regardless of the source of the probabilities.

The chance of rain is a discrete probability; it either will or will not rain. A classic example of discrete probability, used almost universally in statistics texts, pertains to the outcome of the toss of an unbiased coin. A single toss has two outcomes, heads or tails. Each is equally likely, so the probability of obtaining a head is 1/2. This does not imply that every other toss will be a head, but rather that, in the long run, heads will appear one-half of the time. Coin tossing is, then, a clear-cut example of discrete probability. The event has two states and must occupy one or the other; except for the vanishingly small possibility that the coin will land precisely on edge, it must come up either heads or tails.

An interesting series of probabilities can be formed based on coin tossing. If the probability of obtaining heads is 1/2, the probability of obtaining two heads in a row is $1/2 \cdot 1/2 = 1/4$. Perhaps we are interested in knowing the probabilities of obtaining three heads in a row; this will be $1/2 \cdot 1/2 \cdot 1/2 = 1/8$. The logic behind this progression is simple. On the first toss, our chances are 1/2 of obtaining a head. If we do, our chances of obtaining a second head are again 1/2, because the second toss is not dependent in any way on the first. Likewise, the third toss is independent of the two preceding tosses, and has an associated probability of 1/2 for heads. So, we have "one-half of one-half of one-half" of a chance of getting all three heads.

Suppose instead that we are interested in the probability of obtaining only one head in three tosses. All possible outcomes, denoting heads as H and tails as T, are:

$$
\begin{array}{lll}
\text{HHH} & \text{HTH} & \text{TTT} \\
\text{HHT} & \text{THH} & \text{[THT]} \\
\text{[HTT]} & \text{[TTH]} &
\end{array}
$$

Bracketed combinations are those that satisfy our requirements that they contain only one head. Because there are eight possible combinations, the probability of getting only one head in three tosses is 3/8.

What we have found is the number of possible **combinations** of three things (either heads or tails), taken one item at a time. This can be generalized to the number of possible combinations of n items taken r at a time. Symbolically, this is represented as $\binom{n}{r}$.

It can be demonstrated that the number of possible combinations of n items, taken r items at a time, is

$$
\binom{n}{r} = \frac{n!}{r!(n-r)!} \tag{2.1}
$$

The exclamation points stand for **factorial** and mean that the number preceding the exclamation point is multiplied by the number less one, then by the number less two, and so on:

$$
n! = n \cdot (n-1) \cdot (n-2) \cdot (n-3) \cdot \ldots \cdot \tag{2.2}
$$

The value of 3! is $3 \cdot 2 \cdot 1 = 6$. In our coin-flipping problem,

$$
\binom{3}{1} = \frac{3!}{1!(3-1)!} = \frac{3 \cdot 2 \cdot 1}{1 \cdot (2 \cdot 1)} = \frac{6}{2} = 3
$$

That is, there are three possible combinations that will contain one head. By this equation, how many possible combinations are there that contain exactly two heads?

$$
\binom{3}{2} = \frac{3!}{2!(3-2)!} = \frac{3 \cdot 2 \cdot 1}{2 \cdot 1(1)} = \frac{6}{2} = 3
$$

$$
\begin{array}{lll}
\text{HHH} & \text{[HTH]} & \text{TTT} \\
\text{[HHT]} & \text{[THH]} & \text{THT} \\
\text{HTT} & \text{TTH} &
\end{array}
$$

These combinations are bracketed above in our collection of possible outcomes.

Next, how many possible combinations of three tosses contain exactly three heads?

$$
\binom{3}{3} = \frac{3!}{3!(3-3)!} = \frac{3 \cdot 2 \cdot 1}{3 \cdot 2 \cdot 1(1)} = 1
$$

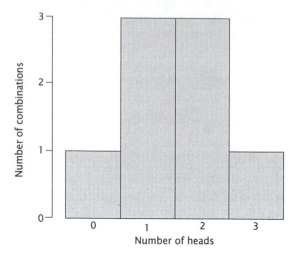

Figure 2–1. Bar graph showing the number of different ways to obtain a specified number of heads in three flips of a coin.

Note that 0! is defined as being one, not zero. Finally, the remaining possibility is the number of combinations that contain no heads:

$$\binom{3}{0} = \frac{3!}{0!(3-0)!} = \frac{3 \cdot 2 \cdot 1}{1(3 \cdot 2 \cdot 1)} = 1$$

Thus, with three flips of a coin, there is one way we can get no heads, three ways we can get one head, three ways we can get two heads, and one way we can get all heads. This can be shown in the form of a bar graph as in **Figure 2–1**.

We can count the number of total possible combinations, which is eight, and convert the frequencies of occurrence into probabilities. That is, the probability of getting no heads in three flips is one correct combination [TTT] out of eight possible, or 1/8. Our histogram now can be redrawn and expressed in probabilities, giving the discrete probability distribution shown in **Figure 2–2**. The total area under the distribution is 8/8, or 1. We are thus certain of getting some combination on the three tosses; the shape of the distribution function describes the likelihood of getting any specific combination. The coin-flipping experiment has four characteristics:

1. There are only two possible outcomes (call them "success" and "failure") for each trial or flip.
2. Each trial is independent of all others.
3. The probability of a success does not change from trial to trial.
4. The trials are performed a fixed number of times.

The probability distribution that governs experiments such as this is called the **binomial distribution**. Among its geological applications, it may be used to forecast the probability of success in a program of drilling for oil or gas. The four characteristics listed above must be assumed to be true; such assumptions seem most reasonable when applied to "wildcat" exploration in relatively virgin basins. Hence, the binomial distribution often is used to predict the outcomes of drilling programs in frontier areas and offshore concessions.

Under the assumptions of the binomial distribution, each wildcat must be classified as either a discovery ("success") or a dry hole ("failure"). Successive wildcats

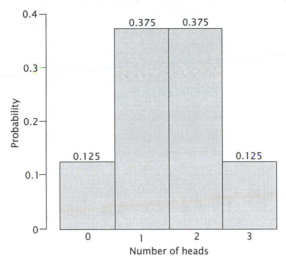

Figure 2–2. Discrete distribution giving the probability of obtaining specified numbers of heads in three flips of a coin.

are presumed to be independent; that is, success or failure of one hole will not influence the outcome of the next hole. (This assumption is difficult to justify in most circumstances, as a discovery usually will affect the selection of subsequent drilling sites. A protracted succession of dry holes will also cause a shift in an exploration program.) The probability of a discovery is assumed to remain unchanged. (This assumption is reasonable at the initiation of exploration, but becomes increasingly tenuous during later phases when a large proportion of the fields in a basin have been discovered.) Finally, the binomial is appropriate when a fixed number of holes will be drilled during an exploratory program, or during a single time period (perhaps a budget cycle) for which the forecast is being made.

The probability p that a wildcat hole will discover oil or gas can be estimated using industry-wide success ratios that have been observed during drilling in similar regions, using the success ratio of the particular company making the evaluation, or simply by making a subjective "guess." From p, the binomial model can be developed as it relates to exploratory drilling in the following steps:

1. The probability that a hole will result in a discovery is p.
2. Therefore, the probability that a hole will be dry is $1 - p$.
3. The probability that n successive wildcats will all be dry is

$$P = (1 - p)^n$$

4. The probability that the nth hole drilled will be a discovery but the preceding $(n - 1)$ holes will all be dry is

$$P = (1 - p)^{n-1} p$$

5. The probability of one discovery in a series of n wildcat holes is

$$P = n(1 - p)^{n-1} p$$

since the discovery can occur on *any* of the n wildcats.

6. The probability that $(n - r)$ dry holes will be drilled, followed by r discoveries, is

$$P = (1 - p)^{n-r} p^r$$

15

7. However, the $(n - r)$ dry holes and the r discoveries may be arranged in $\binom{n}{r}$ combinations or, equivalently, in $n!/(n-r)!r!$ different ways. So, the probability that r discoveries will be made in a drilling program of n wildcats is

$$P = \frac{n!}{(n-r)!\,r!}(1-p)^{n-r}p^r \tag{2.3}$$

This is an expression of the binomial distribution, and gives the probability that r successes will occur in n trials, when the probability of success in a single trial is p.

The binomial equation can be solved to determine the probability of occurrence of any particular combination of successes and failures, for any desired number of trials and any specified probability. These probabilities have already been computed and tabulated for many combinations of n, r, and p. Using either the equation or published tables such as those in Hald (1952), many interesting questions can be investigated. For example, suppose we wish to develop the probabilities associated with a five-hole exploration program in a virgin basin where the success ratio is anticipated to be about 10%. What is the probability that the entire exploration program will be a total failure, with no discoveries? Such an outcome is called "gambler's ruin" for obvious reasons, and the binomial expression has the terms

$$n = 5$$
$$r = 0$$
$$p = 0.10$$

$$P = \binom{5}{0} \cdot 0.10^0 \cdot (1 - 0.10)^5$$

$$= \frac{5!}{5!\,0!} \cdot 1 \cdot 0.90^5$$

$$= 1 \cdot 1 \cdot 0.59 = 0.59$$

The probability that no discoveries will result from the exploratory effort is almost 60%.

If only one hole is a discovery, it may pay off the costs of the entire exploration effort. What is the probability that one well will come in during the five-hole exploration campaign?

$$P = \binom{5}{1} \cdot 0.10^1 \cdot (1 - 0.10)^4$$

$$= \frac{5!}{4!\,1!} \cdot 0.10 \cdot 0.90^4$$

$$= 5 \cdot 0.10 \cdot 0.656 = 0.328$$

Using either the binomial equation or a table of the binomial distribution, the probabilities associated with all possible outcomes of the five-hole drilling program can be found. These are shown in **Figure 2–3**.

Other discrete probability distributions can be developed for those experimental situations where the basic assumptions are different. Suppose, for example, an

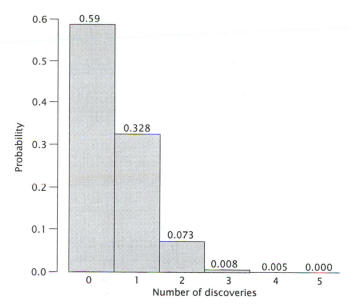

Figure 2–3. Discrete distribution giving the probability of making n discoveries in a five-hole drilling program when the success ratio (probability of a discovery) is 10%.

exploration company is determined to discover two new fields in a virgin basin it is prospecting, and will drill as many holes as required to achieve its goal. We can investigate the probability that it will require $2, 3, 4, \ldots$, up to n exploratory holes before two discoveries are made. The same conditions that govern the binomial distribution may be assumed, except that the number of "trials" is not fixed.

The probability distribution that governs such an experiment is called the **negative binomial**, and its development is very similar to that of the binomial distribution. As in that example, p is the probability of a discovery and r is the number of "successes" or discovery wells. However, n, the number of trials, is not specified. Instead, we wish to find the probability that x dry holes will be drilled before r discoveries are made. The negative binomial has the form

$$P = \binom{r + x - 1}{x}(1 - p)^x p^r \tag{2.4}$$

Note the similarity between this equation and Equation (2.3); the term $r + x - 1$ appears because the last hole drilled in a sequence must be the rth success. Expanding Equation (2.4) gives

$$P = \frac{(r + x - 1)!}{(r - 1)!\,x!}(1 - p)^x p^r \tag{2.5}$$

If the regional success ratio is assumed to be 10%, the probability that a two-hole exploration program will meet the company's goal of two discoveries can be calculated:

$$P = \frac{(2 + 0 - 1)!}{(2 - 1)!\,0!} \cdot (1 - 0.10)^0 \cdot 0.10^2$$

$$= \frac{1!}{1!\,0!} \cdot 0.90^0 \cdot 0.10^2$$

$$= 1 \cdot 1 \cdot 0.01 = 0.01$$

17

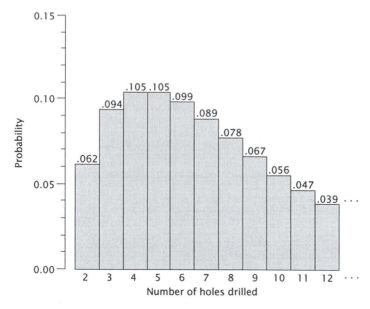

Figure 2–4. Discrete distribution for exactly two successes in a drilling program of n exploratory holes when the probability of a discovery is 25%.

The probabilities attached to other drilling programs having different numbers of holes or probabilities of success can be found in a similar way. The possibility that five holes will be required to achieve two successes when the regional success ratio is 25% is

$$P = \frac{(2 + 3 - 1)!}{(2 - 1)!3!} \cdot (1 - 0.25)^3 \cdot 0.25^2$$

$$= \frac{24}{1 \cdot 6} \cdot 0.422 \cdot 0.062 = 0.105$$

We can calculate the probabilities attached to a succession of possible outcomes and plot the results in the form of a distribution, just as we have done previously. **Figure 2–4** is a negative binomial probability distribution for a drilling program where the probability of a discovery on any hole is 25% and the drilling program will continue until exactly two discoveries have been made. Obviously, this distribution must start at two, since this is the minimum number of holes that might be required, and continues without limit (in the event of extremely bad luck!); we show the distribution only up to 12 holes.

The probabilities calculated are low because they relate to the likelihood of obtaining two successes and exactly x dry holes. It may be more useful to consider the distribution of the probability that more than x dry holes must be drilled before the goal of r discoveries is achieved. This is found by first calculating the negative binomial distribution in **cumulative form** in which each successive probability is added to the preceding probabilities; the cumulative distribution gives the probability that the goal of two successes will be achieved in $(x + r)$ or fewer holes as shown in **Figure 2–5**. If we subtract each of these probabilities from 1.0 we obtain the desired probability distribution (**Fig. 2–6**). The negative binomial will appear again in Chapter 5, as it constitutes an important model for the distribution of points in space.

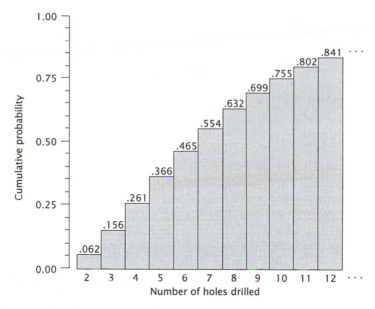

Figure 2–5. Discrete distribution giving the cumulative probability that two discoveries will be made by or before a specified hole when the probability of a discovery is 25%.

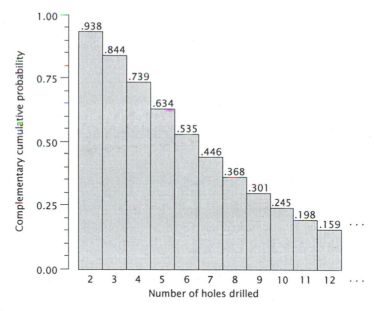

Figure 2–6. Discrete distribution giving the probability that more than a specified number of holes must be drilled to make two discoveries when the probability of a discovery is 25%.

There are other discrete probability distributions that apply to experimental situations similar to those appropriate for the binomial. These include the ***Poisson distribution***, which can be used instead of the binomial when p, the probability of success, is very small. The Poisson distribution will be discussed in Chapter 4, where it will be applied to the analysis of rare, random events in time (such as earthquakes or volcanic eruptions), and in Chapter 5, where it will serve as a model for objects located randomly in space. The **geometric distribution** is a special case of the negative binomial, appropriate when interest is focused on the number of trials prior to the initial success. The **multinomial distribution** is an extension of the binomial where more than two mutually exclusive outcomes are possible. These topics are extensively developed in most books on probability theory, such as those by Parzen (1960) or Ash (1970).

An important characteristic of all of the discrete probability distributions just discussed is that the probability of success remains constant from trial to trial. Statisticians discuss simple experiments called **sampling with replacement** in which this assumption holds strictly true. A typical experiment would involve an urn filled with red and white balls; if a ball is selected at random, the probability it will be red is equal to the proportion of red balls originally in the urn. If the ball is then returned to the urn, the proportions of the two colors remain unchanged, and the probability of drawing a red ball on a second trial remains unchanged as well. The probability also will remain approximately constant if there are a very large number of balls in the urn, even if those selected are not returned, because their removal causes an infinitesimal change in the proportions among those remaining. This latter condition usually is assumed to prevail in many geological situations where discrete probability distributions are applied. In our binomial probability example, the "urn" consists of the geologic basin where exploration is occurring, and the red and white balls correspond to undiscovered reservoirs and barren areas. As long as the number of undrilled locations is large, and the number of prospects that have been drilled (and hence "removed from the urn") is small, the assumption of constant probability of discovery seems reasonable. However, if a sampling experiment is performed with a small number of colored balls initially in the urn and those taken from the urn are not returned, the probabilities obviously change with each draw. Such an experiment is called **sampling without replacement**, and is governed by the discrete **hypergeometric distribution**. Geologic problems where its use is appropriate are not common, but McCray (1975) presents an example from geophysical exploration for petroleum.

In some circumstances it is possible to know the size of the population within which discoveries will be made. Suppose an offshore concession contains ten well-defined seismic features that seem to represent structures caused by movement of salt at depth. From experience in nearby offshore tracts, it is believed that about 40% of such seismic features will prove to be productive structures. Because of budgetary limitations, it is not possible to drill all of the features in the current exploration program. The hypergeometric distribution can be used to estimate the probabilities that specified numbers of discoveries will be made if only some of the identified prospects are drilled.

The binomial distribution is not appropriate for this problem because the probability of a discovery changes with each exploratory hole. If there are four reservoirs distributed among the ten seismic features, the discovery of one reservoir increases the odds against finding another because there are fewer remaining to be discovered. Conversely, drilling a dry hole on a seismic feature increases the probability that the remaining untested features will prove productive, because one nonproductive feature has been eliminated from the population.

Calculating the hypergeometric probability consists simply of finding all of the possible combinations of producing and dry features within the population, and then enumerating those combinations that yield the desired number of discoveries.

The probability of making x discoveries in a drilling program of n holes, when sampling from a population of N prospects of which S are believed to contain reservoirs, is

$$P = \frac{\binom{S}{x}\binom{N-S}{n-x}}{\binom{N}{n}} \tag{2.6}$$

This is the number of combinations of the reservoirs taken by the number of discoveries, times the number of combinations of barren anomalies taken by the number of dry holes, all divided by the number of combinations of all the prospects taken by the total number of holes in the drilling program.

The hypergeometric probability distribution can be applied to our offshore concession that contains ten seismic features, of which four are likely to be structures containing reservoirs. Unfortunately, we cannot know in advance of drilling which four of the ten features will prove productive. If the current season's exploration budget permits the drilling of only four of the prospects, we can determine the probabilities attached to the various possible outcomes.

What is the probability that the drilling program will be a total failure, with no discoveries among the four features tested?

$$P = \frac{\binom{4}{0}\binom{6}{4}}{\binom{10}{4}} = \frac{1 \cdot 15}{210} = 0.071$$

The probability of gambler's ruin is approximately 7%. What is the probability that one discovery will be made?

$$P = \frac{\binom{4}{1}\binom{6}{3}}{\binom{10}{4}} = \frac{4 \cdot 20}{210} = 0.381$$

The probability that one discovery will be made is 38%.

A histogram can be prepared which shows the probabilities attached to all possible outcomes in this exploration situation (**Fig. 2–7**). Note that the probability of some success is $(1.00 - 0.07)$, or 93%.

The preceding examples have addressed situations where there are only two possible outcomes: a hole is dry, or oil is discovered. If oil is found, the well cannot be dry, and *vice versa*. Events in which the occurrence of one outcome precludes the occurrence of the other outcome are said to be **mutually exclusive**. The probability that one event or the other happens is the sum of their separate probabilities; that is, *p* (*discovery* **or** *dry hole*) = *p* (*discovery*) + *p* (*dry hole*). This is called the **additive rule of probability**.

Events are not necessarily mutually exclusive. For example, we may be drilling an exploratory hole for oil or gas in anticipation of hitting a porous reservoir sandstone in what we have interpreted as an anticlinal structure from seismic data. The

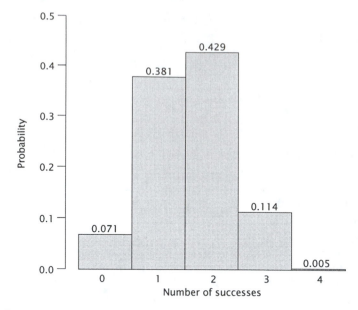

Figure 2–7. Discrete distribution for the probability of n discoveries in drilling four out of ten prospects when four prospects contain oil.

two outcomes, *hit porous sandstone* and *drill into an anticline*, are not mutually exclusive as we hope that both can occur simultaneously. Since the presence of a sandstone is governed by factors that operated at the time of deposition, and since the occurrence of an anticlinal fold is presumed to be related to tectonic conditions at a later time, the two outcomes are unrelated, or **independent**. If two events are not mutually exclusive but *are* independent, the **joint probability** that they will occur simultaneously is the product of their separate probabilities of occurrence. That is, p (*hit sandstone **and** drill anticline*) $=$ p (*hit sandstone*) $\times$ p (*drill anticline*). This is the **multiplicative rule of probability**.

Two events may be related in some way, so that the outcome of one is dependent in part on the outcome of the other. The joint probability of such events is said to be **conditional**. Such events are extremely important in geology, because we may be able to observe one event directly, but the other event is hidden. If the two are conditional, the occurrence of the observable event tells us something about the likely state of the hidden event. For example, the upward movement of magma in chambers beneath a volcano such as Mt. St. Helens in Washington is believed to cause a harmonic tremor, a particular type of earthquake. We cannot directly observe an active magma chamber, but we can observe and record the seismic activity associated with a volcano. If a conditional relationship exists between these two events, the occurrence of harmonic tremors may help predict eruptions. If p (*tremor*) is the probability that a harmonic tremor occurs and p (*eruption*) is the probability of a subsequent volcanic eruption, then p (*tremor **and** eruption*) $\neq$ p (*tremor*) $\times$ p (*eruption*) if the two events have a conditional relationship.

The conditional probability that an eruption will occur, given that harmonic tremors have been recorded, is denoted p (*eruption* | *tremor*). In this instance the conditional probability of an eruption is greater than the unconditional probability, or p (*eruption*), which is simply the probability that an eruption will occur without any knowledge of other events. Other conditional probabilities may be lower than the corresponding unconditional probabilities (the probability of finding a fossil, given that the terrain is igneous, is much lower than the unconditional probability of finding a fossil). Obviously, geologists exploit conditional probabilities in all phases of their work, whether this is done consciously or not.

The relationship between conditional and unconditional probabilities can be expressed by **Bayes' theorem**, named for Thomas Bayes, an eighteenth century English clergyman who investigated the manner in which probabilities change as more information becomes available. Bayes' basic equation is:

$$p(A, B) = p(B|A)p(A) \tag{2.7}$$

which states that $p(A, B)$, the **joint probability** that both events A and B occur, is equal to the probability that B will occur given that A has already occurred, times the probability that A will occur. $p(B|A)$ is a conditional probability because it expresses the probability that B will occur conditional upon the circumstance that A has already occurred. If events A and B are related (or dependent), the fact that A has already transpired tells us something about the likelihood that B will then occur. Conversely, it is also true that

$$p(A, B) = p(A|B)p(B)$$

Therefore, the two can be equated, giving

$$p(B|A)p(A) = p(A|B)p(B)$$

which may be rewritten as

$$p(B|A) = \frac{p(A|B)p(B)}{p(A)} \tag{2.8}$$

This is a most useful relationship, because sometimes we know one form of conditional probability but are interested in the other. For example, we may determine that mining districts often are characterized by the presence of abnormal geomagnetic fields. However, we are more interested in the converse, which is the probability that an area will prove to be mineralized, conditional upon the presence of a magnetic anomaly. We can gather estimates of the conditional probability $p(anomaly \mid mineralization)$ and the unconditional probability $p(mineralization)$ from studies of known mining districts, but it may be more difficult to directly estimate $p(mineralization \mid anomaly)$ because this would require the examination of geomagnetic anomalies that may not yet have been prospected:

If there is an all-inclusive number of events B_i that are conditionally related to event A, the probability that event A will occur is simply the sum of the conditional probabilities $p(A|B_i)$ times the probabilities that the events B_i occur. That is,

$$p(A) = \sum_{i=1}^{n} p(A|B_i)p(B_i) \tag{2.9}$$

If Equation (2.9) is substituted for $p(A)$ in Bayes' theorem, as given in Equation (2.8), we have the more general equation

$$p(B_i|A) = \frac{p(A|B_i)\, p(B_i)}{\sum_{i=1}^{n} p(A|B_i)\, p(B_i)} \tag{2.10}$$

23

A simple example involving two possible prior events, B_1 and B_2, will illustrate the use of Bayes' theorem. A fragment of a hitherto unknown species of mosasaur has been found in a stream bed in western Kansas, and a vertebrate paleontologist would like to send a student field party out to search for more complete remains. Unfortunately, the source of the fragment cannot be identified with certainty because the fossil was found below the junction of two dry stream tributaries. The drainage basin of the larger stream contains about 18 mi^2, while the basin drained by the smaller stream includes only about 10 mi^2. On the basis of just this information alone, we might postulate that the probability that the fragment came from one of the drainage basins is proportional to the area of the basin, or

$$p(B_1) = \frac{18}{28} = 0.64$$

$$p(B_2) = \frac{10}{28} = 0.36$$

However, an examination of a geologic report and map of the region discloses the additional information that about 35% of the outcropping Cretaceous rocks in the larger basin are marine, while almost 80% of the outcropping Cretaceous rocks in the smaller basin are marine. We may therefore postulate the conditional probability that, given a fossil is derived from basin B_i, it will be a marine fossil, as proportional to the percentage of the Cretaceous outcrop area in the basin that is marine, or for basin B_1

$$p(A|B_1) = 0.35$$

and for basin B_2

$$p(A|B_2) = 0.80$$

Using these probabilities and Bayes' theorem, we can assess the conditional probability that the fossil fragment came from basin B_1, given that the fossil is marine.

$$p(B_1|A) = \frac{p(A|B_1)\,p(B_1)}{p(A|B_1)\,p(B_1) + p(A|B_2)\,p(B_2)}$$

$$= \frac{(0.35)\,(0.64)}{(0.35)\,(0.64) + (0.80)\,(0.36)}$$

$$= 0.44$$

Similarly, the probability that the fossil came from the smaller basin is

$$p(B_2|A) = \frac{p(A|B_2)\,p(B_2)}{p(A|B_1)\,p(B_1) + p(A|B_2)\,p(B_2)}$$

$$= \frac{(0.80)\,(0.36)}{(0.35)\,(0.64) + (0.80)\,(0.36)}$$

$$= 0.56$$

Fortunately for the students who must search the area, it seems somewhat more likely that the fragment of marine fossil mosasaur came from the smaller basin than from the larger. However, the differences in probability are very small and, of course, depend upon the reasonableness of the assumptions used to estimate the probabilities.

Continuous Random Variables

To introduce the next topic we must return briefly to the binomial distribution. **Figure 2–2** shows the probability distribution for all possible numbers of heads in three flips of a coin. A similar experiment could be performed that would involve a much larger number of trials. **Figure 2–8**, for example, gives the probabilities associated with obtaining specified numbers of "successes" (or heads) in ten flips of a coin, and **Figure 2–9** shows the probability distribution that describes outcomes from an experiment involving 50 flips of a coin. All of the probabilities were obtained either from binomial tables or calculated using the binomial equation.

In each of these experiments, we have enumerated all possible numbers of heads that we could obtain, from zero up to three, to ten, or to 50. No other combinations of heads and tails can occur. Therefore, the sum of all the probabilities within each experiment must total 1.00, because we are absolutely certain to obtain a result from among those enumerated. We can conveniently represent this by setting the areas underneath histograms in **Figures 2–8** and **2–9** equal to 1.00, as was done in the histogram of **Figure 2–2**. The greater number of coin tosses can be accommodated only by making the histogram bars ever more narrow, and the histogram becomes increasingly like a smooth and continuous curve. We can imagine an ultimate experiment involving flips of an infinite number of coins, yielding a histogram having an infinite number of bars of infinitesimal width. Then, the histogram would be a continuous curve, and the horizontal axis would represent a continuous, rather than discrete, variable.

In the coin-tossing experiment, we are dealing with discrete outcomes—that is, specific combinations of heads and tails. In most experimental work, however, the possible outcomes are not discrete. Rather, there is an infinite continuum of possible results that might be obtained. The range of possible outcomes may be finite and in fact quite limited, but within the range the exact result that may appear cannot be predicted. Such events are called ***continuous random variables***. Suppose, for example, we measure the length of the hinge line on a brachiopod and find it to be 6 mm long. However, if we perform our measurement using a binocular microscope, we may obtain a length of 6.2 mm, by using an optical comparator we may measure 6.23 mm, and with a scanning electron microscope, 6.231 mm. A continuous variate can, in theory, be infinitely refined, which implies that we can always find a difference between two measurements, if we conduct the measurements at a fine enough scale. The corollary of this statement is that every outcome on a continuous scale of measurement is unique, and that the probability of obtaining a specific, exact result must be zero!

If this is true, it would seem impossible to define probability on the basis of relative frequencies of occurrence. However, even though it is impossible to observe a number of outcomes that are, for example, exactly 6.000...000 mm, it is entirely feasible to obtain a set of measurements that fall within an interval around this value. Even though the individual measurements are not precisely identical, they are sufficiently close that we can regard them as belonging to the same class. In effect, we divide the continuous scale into discrete segments, and can then count the number of events that occur within each interval. The narrower the class boundaries, the fewer the number of occurrences within the classes, and the lower the estimates of the probabilities of occurrence.

When dealing with discrete events, we are counting—a process that usually can be done with absolute precision. Continuous variables, however, must be measured

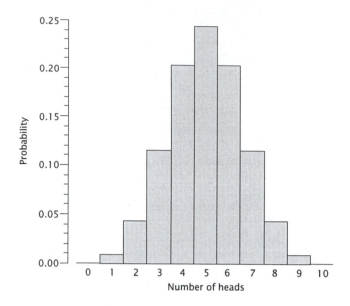

Figure 2–8. Discrete distribution giving the probability of obtaining specified numbers of heads in ten flips of a coin.

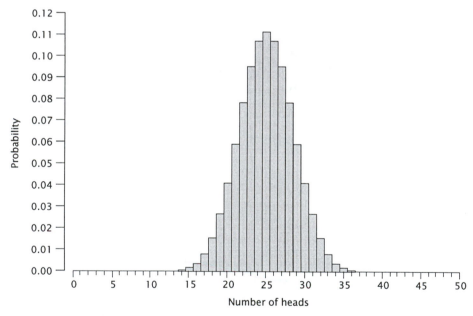

Figure 2–9. Discrete distribution giving the probability of obtaining specified numbers of heads in 50 flips of a coin.

by some physical procedure, and these inherently are limited in both their accuracy and precision. Repeated measurements made on the same object will display small differences whose magnitude may reflect both natural variation in the object, variation in the measurement process, and variation inadvertently caused by the person making the measurements. A single, exact, "true" value cannot be determined;

rather, we will observe a continuous distribution of possible values. This is a fundamental characteristic of a continuous random variable.

To further illustrate the nature of a continuous random variable, we can consider the problem of performing permeability tests on core samples. Permeabilities are determined by measuring the time required to force a certain amount of fluid, under standardized conditions, through a piece of rock. Suppose one test indicates a permeability of 108 md (millidarcies). Is this the "true" permeability of the sample? A second test run on the same specimen may yield a permeability of 93 md, and a third test may register 112 md. The permeability that is recorded on the instruments during any given run is affected by conditions which inevitably vary within the instrument from test to test, vagaries of flow and turbulence that occur within the sample, and inconsistencies in the performance of the test by the operator. No single test can be taken as an exactly correct measure of the true permeability. The various sources of fluctuation combine to yield a continuously random variable, which we are sampling by making repeated measurements.

Variation induced into measurements by inaccuracy of instrumentation is most apparent when repeated measurements are made on a single object or a test is repeated without change. This variation is called *experimental error*. In contrast, variation may occur between members of a set if measurements or experiments are performed on a series of test objects. This is usually the variation that is of scientific interest. Sometimes the two types of variations are hopelessly mixed together, or *confounded*, and the experimenter cannot determine what portion of the variability is due to variation between his test objects and what is due to error.

Rather than a single piece of rock, suppose we have a sizable length of core taken from a borehole through a sandstone body. We want to determine the permeability of the sandstone, but obviously cannot put 20 ft of core into our permeability apparatus. Instead, we cut small plugs from the larger core at intervals and determine the permeability of each. The variation we see is due in part to differences between the test plugs, but also results from differences in experimental conditions. Devising methods to estimate the magnitude of different sources of variation is one of the major tasks of statistics.

Repeated measurements on large samples drawn from natural populations may produce a characteristic frequency distribution. Most values are clustered around some central value, and the frequency of occurrence declines away from this central point. A graph of the distribution (**Fig. 2–10**) appears bell-shaped, and is called a *normal distribution*. It often is assumed that random variables are normally distributed, and many statistical tests are based on this supposition.

As with all frequency distributions, we may define the total area underneath the normal curve as being equal to 1.00 (or if we wish, as 100%), so we can calculate the probability directly from the curve. You should note the similarity of the bell-shaped continuous curve shown in **Figure 2–10** to the histogram of the binomial distribution in **Figure 2–9**. However, in **Figure 2–10** there is an infinite number of subdivisions along the horizontal axis so the probability of obtaining one exact, specific event is essentially zero. Instead, we consider the probability of obtaining a result within a specified range. This probability is proportional to the area of the frequency curve bounded by these limits. If our specified range is wide, we are more likely to observe an event within them; if the range is extremely narrow, observing an event is extremely unlikely.

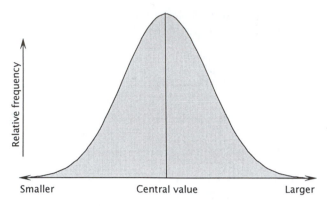

Figure 2–10. Plot of the normal frequency distribution.

Two terms have been introduced in preceding paragraphs without definition. These are "population" and "sample," two important concepts in statistics. A ***population*** consists of a well-defined set (either finite or infinite) of elements. Commonly, these elements are measurements of a specific nature made on items of a specified type. A ***sample*** is a subset of elements taken from a population. A finite population might consist of all oil wells drilled in Kansas in 1963. An example of an infinite geologic population might be all possible thin sections of the Tensleep Sandstone, or all possible shut-in tests on a well. Note in the latter example that the population includes not only the limited number of tests that have been run, but also all possible tests that could be run. Tests that actually were performed may be regarded as a sample of all potential tests.

Geologists typically attach a different meaning to the noun, "sample," than do statisticians. A geological sample, such as a "hand sample" of a rock, a "cuttings sample" from a well, or a "grab sample" or "channel sample" from a mine face, is a physical specimen and when represented by a quantitative or qualitative value would be called an observation or event by a statistician. What a statistician describes as a sample would likely be called a "collection" or "suite of samples" by a geologist. In this book, we will always use the noun "sample" in the statistical sense, meaning a set of observations taken from a population. The verb, "to sample," has essentially the same meaning for both geologists and statisticians and means the act of taking observations.

There are several practical reasons why we might wish to take samples. Many populations are infinite or so vast that it is only possible to examine a subset. Sometimes the measurements we make, such as chemical analyses, require the destruction of the material. By sampling, only a small part of the population is destroyed. Most geological populations extend deep into the Earth and are not accessible in their entirety. Finally, even if it were possible to observe an entire population, it might be more efficient to sample. There is always a point beyond which the increase in information gained from additional observations is not worth the increase in the cost of obtaining them.

Although all populations exhibit diversity, there is no real population whose elements vary without limit. Because any population has characteristic properties and the variation of its constituent members is limited, it is possible to select a relatively small, random sample that can adequately portray the traits of the population.

If observations with certain characteristics are systematically excluded from the sample, deliberately or inadvertently, the sample is said to be *biased*. Suppose, for example, we are interested in the porosity of a particular sandstone unit. If we exclude all loose and crumbly rocks from our sample because their porosity is difficult to measure, we will alter the results of the study. It is likely that the range of porosities will be truncated at the high end, biasing the sample toward low values and giving an erroneously low estimate of the variation in porosity within the unit.

Samples should be drawn from populations in a random manner. This means that each item in the population has an equal opportunity to be included in the sample. A random sample will be unbiased, and as the sample size is increased, will provide an increasingly refined picture of the nature of the population. Unfortunately, obtaining a truly random sample may be impractical, as in the situation of sampling a geologic unit that is partially buried. Samples within the unit at depth do not have the same opportunity of being chosen as samples at outcrops. The problems of sampling in such circumstances are complex; some of the references at the end of this chapter discuss the effects of various sampling schemes and the relative merits of different sampling designs. However, many geologic problems involve the analysis of data collected without prior design. The interpretation of subsurface structure from drill-hole data is a prominent example.

Statistics

Distributions have certain characteristics, such as their midpoint; measures indicating the amount of "spread"; and measures of symmetry of the distribution. These characteristics are known as *parameters* if they describe populations, and *statistics* if they refer to samples. Statistics may be used to estimate parameters of parent populations and to test hypotheses about populations.

Although summary statistics are important, sometimes we can learn more by examining the distribution of the observations as shown on different plots and graphs. A familiar form of display is the *histogram*, a bar chart in which a continuous variable is divided into discrete categories and the number or proportion of observations that fall into each category is represented by the areas of the corresponding bars. (As we have already seen, histograms are useful for showing discrete distributions but now we are interested in their application to continuous variables.) Usually the limits of categories are chosen so all of the histogram intervals will be the same width, so the heights of the bars also are proportional to the numbers of observations within the categories represented by the bars. If the vertical scale on the bar chart reads in number of observations, the graphic is called a *frequency histogram*. If the number of observations in each category are divided by the total number of observations, the scale reads in percent and the bar chart is a *relative frequency histogram*. Since a histogram covers the entire range of observations, the sum of the areas of all the bars will represent either the total number of observations or 100%. If the observations have been selected in an unbiased, representative manner, the sample histogram can be considered an approximation of the underlying probability distribution.

The appearance of a histogram is strongly affected by our choice of the number of categories and the starting value of the first category, especially if the sample contains only a few observations. Dividing the data into a small number of categories increases the average number in each and the histogram will be relatively reproducible with repeated sampling. Unfortunately, such a histogram will contain little detail and may not be particularly informative. Increasing the number of categories reveals more details of the distribution, but because each category will contain fewer observations, the histogram will be less stable. The choice of origin for histogram categories also may influence the shape of the histogram. Interactive software allows the user to dynamically vary the width of the histogram intervals and move the origin, so alternatives can be easily evaluated. **Figure 2–11** shows four different histograms representing 125 airborne measurements of total radiation, recorded on the Istrian peninsula of Croatia. The data are contained in file CROATRAD.TXT at the Web sites (see Preface). If you have access to an interactive statistics package, you can experiment with these data to see the effects of changing the size and origin of the histogram categories. Examples shown in **Figure 2–11** are only a few of the possible histograms that could be constructed from these data.

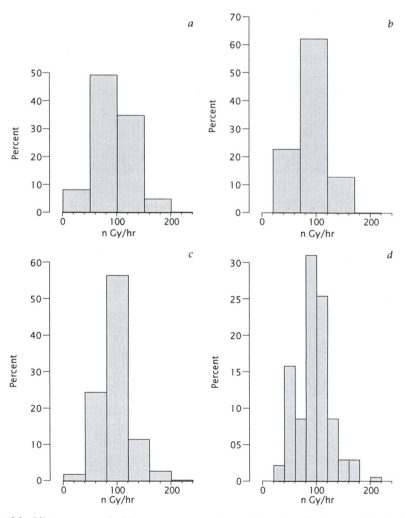

Figure 2–11. Histograms of airborne measurements of total radiation on the Istrian peninsula of Croatia, shown with different class intervals or histogram origins.

An alternative to a histogram is to show the data in the form of a **cumulative plot**. We will illustrate the relation of this graphic to a conventional histogram

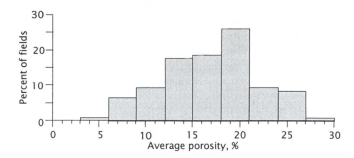

Figure 2–12. Histogram of field-wide average porosities of oil fields producing from the "D" and "J" sands in the Denver–Julesburg Basin of Colorado. Vertical axis is compressed for comparison with Figure 2–13.

using observations in file DJPOR.TXT, which gives the field-wide average porosities for 105 oil fields producing from the Cretaceous "D" and "J" sands in the Denver–Julesburg Basin of eastern Colorado. **Figure 2–12** is a histogram of these data in which the vertical axis is compressed for easier comparison with **Figure 2–13**, where each successive histogram bar begins at the top of the preceding bar. In effect, we have stacked the histogram bars so that the successive categories show the cumulative numbers or proportions of observations.

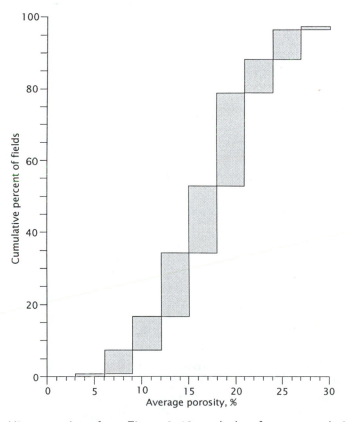

Figure 2–13. Histogram bars from Figure 2–12 stacked to form a cumulative distribution.

The great advantage of plotting data in cumulative form, however, comes about because we can show the individual observations directly, and avoid the loss of resolution that comes from grouping the observations into categories for a histogram. To do this, we must first rank the observations from smallest to largest, divide each observation's rank by the number of observations to convert it into a fraction, then multiply by 100 to express it as a percentile. That is,

$$\text{percentile of } x_i = 100 \left(\frac{\text{rank of } x_i}{n} \right) \tag{2.11}$$

where n is the number of observations. By graphing the percentile of each observation versus its value, we form a cumulative plot (**Fig. 2–14**). Note that both the cumulative histogram and the cumulative plot have a characteristic ogive form.

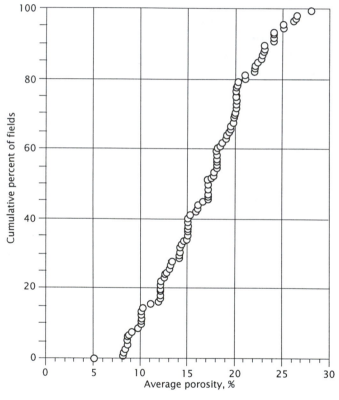

Figure 2–14. Cumulative plot of individual porosity measurements used to construct Figures 2–12 and 2–13.

Successive divisions of a distribution are called *quantiles*. If we rank all observations in a sample and then divide the ranks into 100 equal-sized categories, each category is a *percentile*. Suppose our sample contains 300 observations; the three smallest values constitute the first percentile. Each category is called a *decile* if the ranked sample is divided into ten equal categories, and a *quartile* if it is divided into four equal categories. Certain divisions of a distribution such as the 5th and 95th percentiles, the 25th and 75th percentiles (also called the 1st and 3rd quartiles), and the 50th percentile (also called the 5th decile, the 2nd quartile, or the median) are considered especially diagnostic and are indicated on the graphic plots we will consider next.

Box-and-whisker plots were devised by John Tukey (1977) to more effectively show the essential aspects of a sample distribution. There are many variants of the box-and-whisker plot, but all are graphs that show the spread of the central 50% of a distribution by a box whose lower limit is set at the first quartile and whose upper limit is set at the third quartile. The 50th percentile (second quartile or median) usually is indicated by a line across the box. The mean, or arithmetic average of the observations, may also be indicated by an asterisk or diamond. "Whiskers" are lines that extend from the ends of the box, usually to the 5th and 95th percentiles. Observations lying beyond these extremes may be shown as dots. **Figure 2–15** shows a histogram and several alternative box-and-whisker plots produced by several popular commercial programs. The data are 125 airborne measurements of radiation emitted by ^{137}Cs, recorded on the Istrian peninsula of Croatia. This component of total radiation (see **Fig. 2–11**) reflects fallout from the Chernobyl reactor accident in the Soviet Union during April of 1986. The data are given in file CROATRAD.TXT.

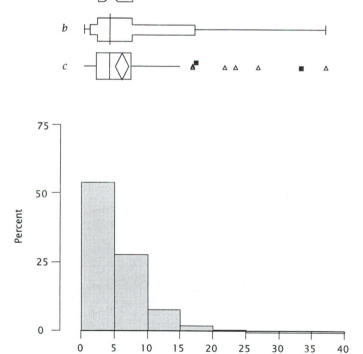

Figure 2–15. Histogram and alternative forms of box-and-whisker plots of airborne measurements of ^{137}Cs radiation recorded on the Istrian peninsula of Croatia.

Summary Statistics

The most obvious measure of a population or sample is some type of average value. Several measures exist, but only a few are used in practice. The *mode* is the value that occurs with the greatest frequency. In an asymmetric distribution such as that shown in **Figure 2–16**, the mode is the highest point on the frequency curve. The *median* is the value midway in the frequency distribution. In **Figure 2–16**, one-half of the area below the distribution curve is to the right of the median, one-half is to the left. The median is the 50th percentile, the 5th decile, or the 2nd quartile. The *mean* is another word for the arithmetic average, and is defined as the sum of all observations divided by the number of observations. The *geometric mean* is the *n*th root of the products of the *n* observations, or equivalently, the exponential of the arithmetic mean of the logarithms of the observations. In asymmetric frequency curves, the median lies between the mean and the mode. In symmetric curves such as the normal distribution, the mean, median, and mode coincide.

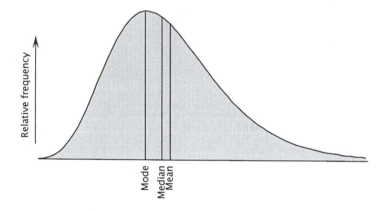

Figure 2–16. Asymmetric distribution showing relative positions of mean, median, and mode.

Certain symbols traditionally have been assigned to measures of distribution curves. Generally, the symbols for population distributions are Greek letters, and those for sample distributions are Roman. The sample mean, for example, is designated $\overline{X}$ and the population mean is μ (mu). A common objective in an investigation is to estimate some parameter of a population. A statistic we compute based on a sample taken from the population is used as an estimator of the desired parameter. The use of Greek and Roman symbols serves to emphasize the difference between parameters and the equivalent statistics.

The sample mean has two highly desirable properties that make it more useful as an estimator of the average or central value of a population than either the sample median or mode. First, the sample mean is an unbiased estimate of the population mean. (A (sample) statistic is an unbiased estimate of the equivalent (population) parameter if the average value of the statistic, from a large series of samples, is equal to the parameter. Second, it can be demonstrated that, for symmetrical distributions such as the normal, the sample mean tends to be closer to the population mean than any other unbiased estimate (such as the median) based on the same sample. This is equivalent to saying that sample means are less variable

Table 2–1. Chromium content of an Upper Pennsylvanian shale from Kansas.

Replicate	Cr (ppm)
1	205
2	255
3	195
4	220
5	235
TOTAL =	1110
MEAN =	1110/5 = 222

than sample medians, hence they are more efficient in estimating the population parameter.

In geochemical analyses, it is common practice to make multiple determinations, or **replicates**, of a single sample. The most nearly correct analytical value is taken to be the mean of the determinations. **Table 2–1** lists five values for chromium, in parts per million (ppm), obtained by spectrographic analysis of replicate splits of a Pennsylvanian shale specimen from southeastern Kansas. The table shows the steps in calculating the mean, whose equation is simply

$$\overline{X} = \frac{\sum_{i=1}^{n} x_i}{n} \tag{2.12}$$

Another characteristic of a distribution curve is the spread or dispersion about the mean. Various measures of this property have been suggested, but only two are used to any extent. One is the **variance**, and the other is the square root of the variance, called the **standard deviation**. Variance may be regarded as the average squared deviation of all possible observations from the population mean, and is defined by the equation

$$\sigma^2 = \frac{\sum_{i=1}^{n}(x_i - \mu)^2}{n} \tag{2.13}$$

The variance of a population, σ^2, is given by this equation. The variance of a sample is denoted by the symbol s^2. If the observations $x_1, x_2, \ldots, x_n$ are a random sample from a normal distribution, s^2 is an efficient estimate of σ^2.

The reason for using the average of squared deviations may not be obvious. It may seem, perhaps, more logical to define variability as simply the average of deviations from the mean, but a few simple trials will demonstrate that this value will always equal zero. That is,

$$\frac{\sum_{i=1}^{n} x_i - \overline{X}}{n} = 0 \tag{2.14}$$

Another choice might be the average absolute deviation from the mean, or **mean deviation**, MD:

$$MD = \frac{\sum_{i=1}^{n} |x_i - \overline{X}|}{n} \tag{2.15}$$

The vertical bars denote that the absolute value (*i.e.*, without sign) of the enclosed quantity is taken. However, the mean deviation is less efficient than the sample variance. If we take repeated samples, the mean deviations will be more variable than variances calculated from the same samples. Although not intuitively obvious, the variance has properties that make it far more useful than other measures of scatter.

Because variance is the average squared deviation from the mean, its units are the square of the units of the original measurements. A granite, for example, may have feldspar phenocrysts whose longest axes have an average length of 13.2 mm and a variance of 2.0 mm^2. Many people may find themselves reluctant to regard areas as an appropriate measurement unit for the dispersion of lengths! Fortunately, in most instances where we are concerned with variance, it is standardized or converted to a form independent of the measurement units. This is a topic discussed in greater detail elsewhere in this chapter.

To provide a statistic that describes dispersion or spread of data around the mean, and is in the units of measurement of the data, we can calculate the ***standard deviation***. This is defined simply as the square root of variance and is symbolically written as σ for the population parameter and s for the sample statistic. In equation form,

$$\sigma = \sqrt{\sigma^2} = \sqrt{\frac{\sum_{i=1}^{n}(x_i - \mu)^2}{n}} \qquad (2.16)$$

A small standard deviation indicates that observations are clustered tightly around a central value. Conversely, a large standard deviation indicates that values are scattered widely about the mean and the tendency for central clustering is weak. This is illustrated in **Figure 2–17**, which shows two symmetric frequency curves having different standard deviations. Curve *a* represents the percent oil saturation (s_o) measured in cores from the producing zone of a northeastern Oklahoma oil field. Curve *b* is the same type of data from a field in West Texas. The mean oil saturation differs in the two fields, but the major difference between the curves reflects the fact that the Texas field has a much greater variation in oil saturation.

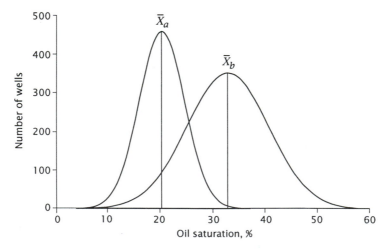

Figure 2–17. Distribution of percent oil saturation (s_o) measured on cores from a field (*a*) in northeastern Oklahoma and (*b*) in west Texas.

A most useful property of normal distributions is that areas under the curve, within any specified range, can be precisely calculated and expressed in terms of standard deviations from the mean. For example, slightly over two-thirds (68.27%) of observations will fall within one standard deviation on either side of the mean of a normal distribution. Approximately 95% of all observations are included within the interval from +2 to −2 standard deviations, and more than 99% are covered by the interval lying three standard deviations on both sides of the mean. This is illustrated in **Figure 2–18.**

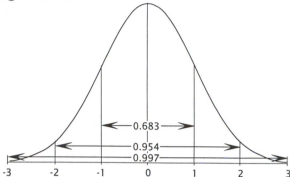

Figure 2–18. Areas enclosed by successive standard deviations of the standard normal distribution.

The distribution of measured oil saturations in cores from the northeastern Oklahoma field (**Fig. 2–17,** curve a) has a mean of 20.1% s_0 and a standard deviation of 4.3% s_0. If we assume that the distribution is normal, we would expect about two-thirds of the cores tested to have oil saturations between about 16% s_0 and 24% s_0. Examination of the original data shows that there are 1145 cores having saturations within this range, or about 68% of the data. Only 101 cores, or about 6% of the total number of observations, have saturations outside the 2σ range; that is, oil saturations less than 12% s_0 or more than 29% s_0.

Equation (2.13) is called the definitional equation of variance. This equation is not often used for hand calculation, involving as it does n subtractions, n multiplications, and n summations. Instead, a formula suitable for computation with a calculator is used which is algebraically equivalent but easier to perform. This equation is

$$s^2 = \frac{\sum_{i=1}^{n} x_i^2 - n\overline{X}^2}{n-1} \tag{2.17}$$

or alternatively,

$$s^2 = \frac{n\sum_{i=1}^{n} x_i^2 - \left(\sum_{i=1}^{n} x_i\right)^2}{n(n-1)} \tag{2.18}$$

On hand calculators, $\sum x_i$ and $\sum x_i^2$ can be found simultaneously, thus reducing the number of operations by n. However, this formula requires subtracting two quantities, $\sum x_i^2$ and $(\sum x_i)^2$, and both may be very large and very nearly the same. Problems can arise if significant digits are truncated during this operation, so it is better to use the definitional equation to calculate variance in a computer program.

To compute variances and standard deviations, we generate intermediate quantities which can be used directly in many techniques we will discuss in following chapters. The **uncorrected sum of squares** is simply $\sum x_i^2$; the **corrected sum of squares** (SS) is defined as

$$SS = \sum_{i=1}^{n} \left(x_i - \overline{X}\right)^2 \tag{2.19}$$

or, in the braically equivalent hand computational form,

$$SS = \sum_{i=1}^{n} x_i^2 - \frac{\left(\sum_{i=1}^{n} x_i\right)^2}{n} \tag{2.20}$$

Once this quantity is obtained, the variance can be found by division by $n - 1$.

$$s^2 = \frac{SS}{n-1} = \frac{\sum_{i=1}^{n} x_i^2 - \left(\sum_{i=1}^{n} x_i\right)^2 / n}{n-1} = \frac{n \sum_{i=1}^{n} x_i^2 - \left(\sum_{i=1}^{n} x_i\right)^2}{n(n-1)} \tag{2.21}$$

Introduction of the quantity $(n - 1)$, which also appears in Equations (2.17), (2.18), and (2.21), requires some explanation. Variance is defined as the average squared deviation from the mean. However, when we sample, we do not know the population mean, μ, rather we estimate it by the sample mean, $\overline{X}$. The sample mean is calculated in a manner that minimizes the squared deviations about it. In other words, the operation $\overline{X} = \sum x_i / n$ produces a value $\overline{X}$ for which $\sum (x_i - \overline{X})^2$ is the minimum of all possible values that could be selected. Because of this property of the sample mean, it tends to underestimate variance when used in Equation (2.13). That is, $s^2 = \sum (x_i - \overline{X})^2 / n$ is a biased estimator of $\sigma^2 = \sum (x_i - \mu)^2 / n$. In order to correct for bias, we reduce the denominator of the variance equation to $n - 1$, producing a larger but more realistic estimate of the variance.

The computation of these quantities can be illustrated with the geochemical data on chromium in shale replicates from **Table 2–1**. Rewriting the table to include a column of squares gives **Table 2–2**. Assuming that the analytical values are distributed approximately normally, we would expect about two-thirds of the readings to lie between 198 and 246 ppm. Examination of the table will show that three of the five values, or 60%, do indeed fall in this range.

Table 2–2. Sums of squares and variance computations for data in Table 2–1.

x	x^2
205	42,025
255	65,025
195	38,025
220	48,400
235	55,225
$\sum x_i = 1110$	$\sum x_i^2 = 248,700$

$$(\sum x_i)^2 = 1,232,100$$

$$SS = 248,700 - \frac{1,242,100}{5} = 2280$$

$$s^2 = \frac{2280}{4} = 570$$

$$s = \sqrt{570} = 23.88$$

In computing the sums of squares for these data, figures having seven decimal places are created. As noted earlier, this tendency to produce extremely large numbers during computation may lead to problems if a computer carries too few significant digits. It also may lead to problems on output if format fields are not large enough to contain the numbers to be printed.

Persons who have not performed statistical analyses usually find it difficult to obtain a "feel" for the numerical value of a variance or standard deviation. Is a variance of 10 large or small? What is the meaning of a standard deviation of 23? The way to interpret both variance and standard deviation is not to attach a significance to a specific numerical value, but to compare one variance to another. The sample having the largest variance or standard deviation has the greater spread among the values of the observations, provided all the measurements were made in the same units.

The dispersion in a variable is sometimes given by the *coefficient of variation*, C_v, which is a dimensionless measure of variability expressed as a fraction of the mean.

$$C_v = \frac{s}{\overline{X}} \tag{2.22}$$

It is not uncommon for variability in a sample to be related to the overall magnitude of the sample; in such instances the coefficient of variation will remain relatively constant. The coefficient of variation is not widely used by statisticians, although it is popular with reservoir geologists and engineers. Ball and others (1997) provide a number of examples of the use of C_v in the characterization of reservoir heterogeneity.

In many geological problems, more than one variable is measured on each observational unit. Examples include a series of measurements made on a collection of corals, results from a sequence of tests run on a group of wells, or sedimentary parameters of a collection of sandstones. The data from such problems can conveniently be arranged in an $n \times m$ array, where n is the number of observations and m is the number of variables measured. The complete analyses from which the data for **Table 2–1** were taken, for example, contain 17 variables. If we consider three of these, the trace amounts of nickel, vanadium, and chromium, we have the data array shown in **Table 2–3**.

Table 2–3. Chromium, nickel, and vanadium in an Upper Pennsylvanian shale from Kansas.

	Cr (ppm)	Ni (ppm)	V (ppm)
	205	130	180
	255	165	215
	195	100	135
	220	135	200
	235	145	205
TOTALS =	1110	675	935
MEANS =	222	135	187

Each column can be summed and a mean value and standard deviation computed. However, the different variables may not be independent, but rather there may be some form of conditional relationship between them. It is important that we be able to assess the nature and strengths of these conditional relationships, just as it was important to assess the conditional probabilities of occurrences of discrete events.

Joint Variation of Two Variables

Computational procedures used to calculate the variance of a single property can be extended to calculation of a measure of the mutual variability of a pair of properties. This measure, called the **covariance**, is the joint variation of two variables about their common mean. This relation is illustrated in **Figure 2–19**, which shows the form of the probability surface created from two normal curves. Both x_1 and x_2 have probability curves similar to that shown in **Figure 2–18**. Just as the variance measures the spread of values around the central point as illustrated in **Figure 2–18**, the covariance measures the distribution of values around a joint mean.

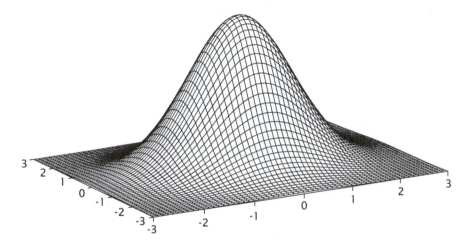

Figure 2–19. Joint probability distribution of two independent normal distributions. Both x_1 and x_2 are normally distributed.

To calculate covariance, we first must calculate a quantity analogous to the sum of squares. This is called the **corrected sum of products** (*SP*) and is defined by

$$SP_{jk} = \sum_{i=1}^{n} \left(x_{ij} - \overline{X}_j \right) \left(x_{ik} - \overline{X}_k \right) \tag{2.23}$$

In this notation, x_{ij} is the ith measurement of variable j, and x_{ik} is the ith measurement of variable k. The symbol SP_{jk} is the sum of products between variables j and k. In computational form, this becomes

$$SP_{jk} = \sum_{i=1}^{n} \left(x_{ij} x_{ik} \right) - \frac{\sum_{i=1}^{n} x_{ij} \sum_{i=1}^{n} x_{ik}}{n} \tag{2.24}$$

The quantity $\sum \left(x_{ij} x_{ik} \right)$ is called the **uncorrected sum of products**. The relationship of SP_{jk} to the sum of squares can easily be seen if we consider j and k to be the same. Then,

$$SP_{jj} = \sum_{i=1}^{n} \left(x_{ij} x_{ij} \right) - \frac{\sum_{i=1}^{n} x_{ij} \sum_{i=1}^{n} x_{ij}}{n} = \frac{n \sum_{i=1}^{n} x_{ij}^2 - \left(\sum_{i=1}^{n} x_{ij} \right)^2}{n} = SS_j \tag{2.25}$$

If we compute sums of products and sums of squares for all possible combinations of our three variables in **Table 2–3**, we can arrange the results in a 3 × 3 array of the form:

$$
\begin{array}{c c c c}
 & \text{Cr} & \text{Ni} & \text{V} \\
\text{Cr} & SS_{Cr} & SP_{Cr,Ni} & SP_{Cr,V} \\
\text{Ni} & SP_{Ni,Cr} & SS_{Ni} & SP_{Ni,V} \\
\text{V} & SP_{V,Cr} & SP_{V,Ni} & SS_{V}
\end{array}
$$

It should be apparent that some of the entries are duplicates; for example, the sum of products for vanadium and nickel is the same as the sum of products for nickel and vanadium. This can be generalized to $SP_{jk} = SP_{kj}$. This feature will be of help to us in subsequent chapters.

Just as variance was calculated by dividing SS by $(n-1)$, we calculate covariance by dividing SP by $(n-1)$.

$$
\text{cov}_{jk} = \frac{SP_{jk}}{n-1} = \frac{\sum_{i=1}^{n} x_{ij}x_{ik} - \left(\sum_{i=1}^{n} x_{ij}\sum_{i=1}^{n} x_{ik}/n\right)}{n-1}
$$

$$
= \frac{n\sum_{i=1}^{n} x_{ij}x_{ik} - \sum_{i=1}^{n} x_{ij}\sum_{i=1}^{n} x_{ik}}{n(n-1)} \tag{2.26}
$$

Returning to the replicate geochemical analyses in **Table 2–3**, we now can calculate the covariances between the three elements. Referring to chromium and nickel as x_1 and x_2, respectively, we can calculate the entries in **Table 2–4**. We now know the variance of x_1 (chromium) and the covariance of x_1 and x_2 (chromium and nickel). Calculate the variance of x_2 (nickel) following Equation (2.21) and complete the 2×2 array:

$$
\begin{array}{c c c}
 & x_1 & x_2 \\
\text{Chromium} & 570 & 537.5 \\
\text{Nickel} & 537.5 & s_2^2
\end{array}
$$

Table 2–4. Computation of covariance between chromium (x_1) and nickel (x_2).

x_1^2	x_1	$x_1 x_2$	x_2	x_2^2
42,025	205	26,650	130	16,900
65,025	255	42,075	165	27,225
38,025	195	19,500	100	10,000
48,400	220	29,700	135	18,225
55,225	235	34,075	145	21,025
$\sum x_1^2 = 248,700$	$\sum x_1 = 1110$	$\sum x_1 x_2 = 152,000$	$\sum x_2 = 675$	$\sum x_2^2 = 93,375$

$$
SP_{1,2} = 152,000 - \frac{(1,110)(675)}{5} = 2150
$$

$$
\text{cov}_{1,2} = \frac{2150}{4} = 537.5
$$

Three additional quantities remain to be calculated in order to complete our analysis of the geochemical data of **Table 2–3**. These are the covariances between chromium and vanadium (cov_{13}) and nickel and vanadium (cov_{23}) and the variance

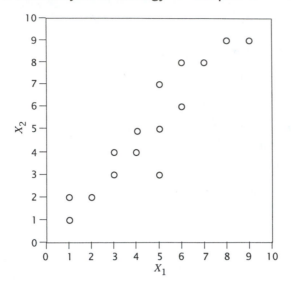

Figure 2–20. Scatter diagram of two variables with high covariance. Variance of $x_1 = 5.7$, variance of $x_2 = 7.1$, covariance = 5.9.

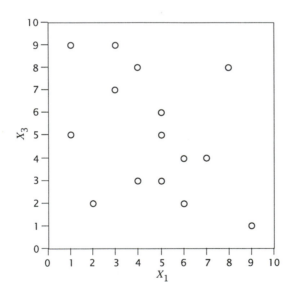

Figure 2–21. Scatter diagram of two variables with low covariance. Variance of $x_1 = 5.7$, variance of $x_3 = 7.1$, covariance = -2.3.

of vanadium (x_3). Compute the quantity (cov$_{13}$) following the procedure used in **Table 2–4**.

Figure **2–20** is a scatter diagram of two variables that are closely related and have a relatively high covariance. The two variables plotted in **Figure 2–21** have the same means and variances as those in **Figure 2–20**, but are independent of each other and have relatively low covariance. Interpretation of covariance values must proceed in the same manner as an interpretation of variances; individual values are not too meaningful because they are dependent upon the units of measurement.

In order to estimate the degree of interrelation between variables in a manner not influenced by measurement units, we must calculate a <u>dimensionless</u> measure of joint variation. The **correlation coefficient**, r_{jk}, is the ratio of the covariance of two variables to the product of their standard deviations. The units of covariance are the products of the units of the two variables, and these also are the units of the standard deviations in the denominator, so all the units cancel out and r_{jk} is a unitless ratio:

$$r_{jk} = \frac{\text{cov}_{jk}}{s_j s_k} \tag{2.27}$$

The covariance of two variables may equal but cannot exceed the product of the standard deviations of the variables, so correlation ranges from +1 to −1. A correlation of +1 indicates a perfect direct relationship between two variables; a correlation of −1 indicates that one variable changes inversely with relation to the other. Between the two extremes is a spectrum of less-than-perfect relationships, including zero, which indicates the lack of any sort of linear relationship at all.

Figure 2–22 shows correlations between 20 observations of the set of 11 artificial variables listed in file CORREL.TXT. In **Figure 2–22 a**, a strong correlation between variables x_1 and x_2 is evident; the correlation coefficient is +0.99. A less pronounced correlation between variables x_3 and x_4 is shown in **Figure 2–22 b**; the correlation coefficient is +0.83. Variables x_5 and x_6 in **Figure 2–22 c** were selected from a random number table, so they have no relation to one another and their correlation is only +0.07. Variables x_7 and x_3 have a negative correlation of −0.65 and are shown in **Figure 2–22 d**. This illustrates that values of one negatively correlated variable decrease as the other variable increases. An interesting extreme is shown in **Figure 2–22 e**: variable x_9 is invariant, that is, its values do not change. Attempts to calculate the correlation coefficient will encounter division by zero; the correlation is undefined in this situation. In the example shown in **Figure 2–22 f**, there is an obvious interdependency between variables x_{10} and x_{11}, as they define points that lie on a circle. The relationship between x_{10} and x_{11} may be expressed as

$$x_{11} = \sqrt{a^2 - x_{10}^2}$$

assuming the center of the circle lies at the origin. The radius of the circle is equal to the coefficient a. However, if the correlation between x_{10} and x_{11} is computed, it will be zero. This is because the correlation coefficient is an expression of the **linear** relationship between the two variables, and the circular relationship shown is not linear. There are many possible nonlinear relationships that may exist between two variables; the correlation coefficient is not a satisfactory measure of the strength of such relationships.

When using calculators, the <u>sample correlation coefficient</u>, r_{jk}, between variables x_j and x_k is commonly found using the equation

$$r_{jk} = \frac{SP_{jk}}{\sqrt{SS_j SS_k}}$$

$$= \frac{\sum_{i=1}^{n} x_{ij} x_{ik} - \left(\sum_{i=1}^{n} x_{ij} \sum_{i=1}^{n} x_{ik} \right) / n}{\sqrt{\left(\sum_{i=1}^{n} x_{ij}^2 - \left(\sum_{i=1}^{n} x_{ij} \right)^2 / n \right) \left(\sum_{i=1}^{n} x_{ik}^2 - \left(\sum_{i=1}^{n} x_{ik} \right)^2 / n \right)}} \tag{2.28}$$

but with computers, using the definitional equations for the sums of products and sums of squares reduces the risk that an erroneous value will result from numerical

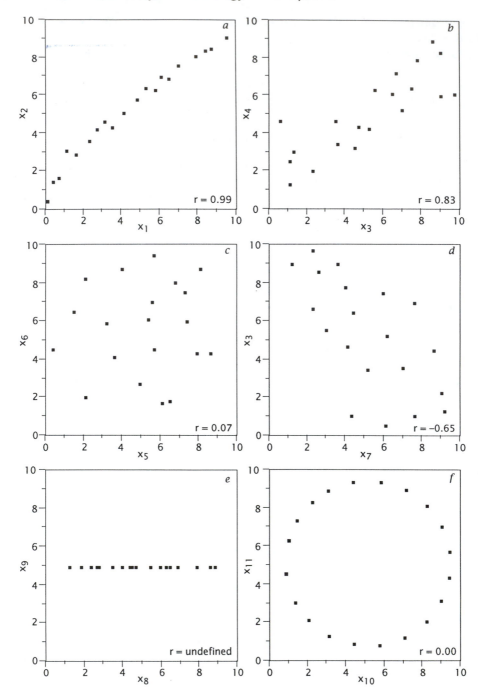

Figure 2–22. Scatter diagrams between pairs of variables showing different correlations.

truncation. Truncation is apt to be troublesome when the original variables contain many digits, especially if the variables are highly correlated.

If r_{jk} measures the linear relationship between two variables, it should be possible to compute the line of dependence between them. This leads into the branch of statistics called regression analysis, which is especially important to geologists because most techniques for surface fitting are special applications of regression. An extensive discussion of these topics will be deferred until later, and we will content ourselves at this point with the computation of r_{jk}.

Table 2–5. Lengths and widths of shells of the brachiopod, *Composita.*

Length (mm)	Width (mm)
18.4	15.4
16.9	15.1
13.6	10.9
11.4	9.7
7.8	7.4
6.3	5.3

Table 2–6. Sums of squares, cross products, and correlation between data in Table 2–5.

x_1^2	x_1	$x_1 x_2$	x_2	x_2^2
338.56	18.4	283.36	15.4	237.16
285.61	16.9	255.19	15.1	228.01
184.96	13.6	148.24	10.9	118.81
129.96	11.4	110.58	9.7	94.09
60.84	7.8	57.72	7.4	54.76
36.69	6.3	33.39	5.3	28.09
$\sum x_1^2 = \overline{1039.62}$	$\sum x_1 = \overline{74.4}$	$\sum x_1 x_2 = \overline{888.48}$	$\sum x_2 = \overline{63.8}$	$\sum x_2^2 = \overline{760.92}$

$$SP_{1,2} = (888.48) - \frac{(74.4)(63.8)}{6} = 97.37$$

$$\text{cov}_{1,2} = \frac{97.37}{5} = 19.47$$

$$SS_1 = (1039.62) - \frac{(74.4)^2}{6} = 117.06$$

$$SS_2 = (760.92) - \frac{(63.8)^2}{6} = 82.51$$

$$s_1^2 = \frac{117.06}{5} = 23.41 \quad s_1 = \sqrt{23.41} = 4.84$$

$$s_2^2 = \frac{82.51}{5} = 16.50 \quad s_2 = \sqrt{16.50} = 4.06$$

$$r_{1,2} = \frac{19.47}{(4.84)(4.06)} = 0.991$$

Biological characteristics may be highly correlated within a group of organisms, because measurements are related to the overall size of individuals. For example, consider the data in **Table 2–5**, giving lengths and widths of shells of the brachiopod, *Composita.* It should be apparent from the observations that there is a strong correspondence between the two measures. Just how strong this correspondence is may be evaluated by the correlation coefficient. To find the correlation between the two measurements, it is necessary to either create columns of squares and cross products, or columns of the differences between the variables and their means. The former is done in **Table 2–6**, where length is x_1 and width is x_2. The resulting correlation of 0.99 is extremely high, and confirms our suspicions that there is a direct relationship between shell length and width in the organisms. Such extremely high correlations are not always found; in fact, it may be a problem to determine if any real correlation exists at all. This is a subject we will discuss again.

Induced Correlations ← *may be hard to tell if induced or actual relationship*

Some correlations between variables do not reflect the relationships between them, but are induced by an operation or transformation that has been performed on the variables. Two unrelated random variables are expected to have a correlation of zero. However, certain operations on the variables may lead to a correlation other than zero even though there is no linear relationship between them. Correlations that do exist may be changed or even reversed by such operations.

Suppose pebbles are randomly selected from a shingle beach and three orthogonal axes are measured on each pebble. No attempt is made to measure the longest or shortest axes of the pebbles in any particular order. We might suppose that the measurements will be correlated, because a large pebble will most likely have large values for all three axes and, conversely, a small pebble will have small measurements for all three axes.

Table 2–7. Axial lengths, in cm, of pebbles collected on a shingle beach; axes are listed in order of measurement.

Pebble	Axis 1	Axis 2	Axis 3
a	3	7	8
b	16	5	8
c	10	12	9
d	13	5	12
e	14	16	5
f	9	8	14
g	16	13	13
h	6	3	11
i	9	15	9
j	13	10	9
TOTALS	109	94	98
MEANS	10.9	9.4	9.8
CORRELATIONS	$r_{1,2} = 0.279$	$r_{1,3} = -0.021$	$r_{2,3} = -0.349$

Table 2–7 lists the measurements made on a collection of pebbles and the correlations between the variables. The data are also shown in the form of scatter diagrams in the upper part of **Figure 2–23**. However, if the axes are now assigned in a conventional manner (that is, the longest axis of a pebble is, by definition, the *a*-axis; the shortest axis is the *c*-axis; and the intermediate axis is the *b*-axis), the ordering causes a change in the correlations (**Table 2–8**). This is especially apparent on the scatter diagrams shown in the lower part of **Figure 2–23**, as the definitions force all of the points to plot below the 45° diagonal dashed line. Because of this,

Table 2–8. Axial lengths, in cm, of pebbles collected on a shingle beach; measurements are sorted into longest (a), intermediate (b), and shortest (c) axes.

Pebble	a-Axis	b-Axis	c-Axis
a	8	7	3
b	16	8	5
c	12	10	9
d	13	12	5
e	16	14	5
f	14	9	8
g	16	13	13
h	11	6	3
i	15	9	9
j	13	10	9
TOTALS	134	98	69
MEANS	13.4	9.8	6.9
CORRELATIONS	$r_{a,b} = 0.597$	$r_{a,c} = 0.499$	$r_{b,c} = 0.467$

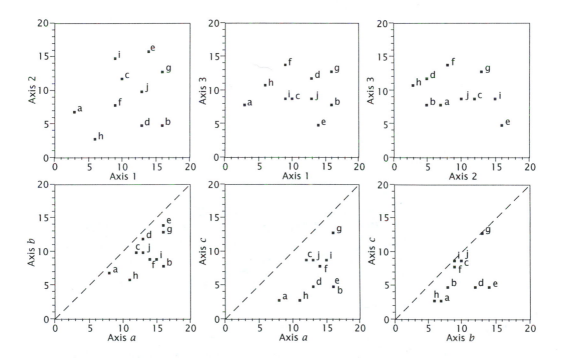

Figure 2–23. Scatter diagrams of axial length, in cm, of pebbles collected on a shingle beach. Top: Original measurements collected in random order. Bottom: Measurements sorted into a-, b-, and c-axes, which causes all points to plot below diagonals of diagrams.

there must always be a positive correlation between any pair of axes, or between the ratios of two axes and the third axis (for example, between b/a versus c).

If variables are expressed as ratios of other variables, spurious correlations may be created. This is especially troublesome in the geochemistry of igneous and metamorphic rocks, where element ratios are sometimes used for deducing the evolution of a fractionating melt. Rollinson (1993) provides a well-argued critique of this practice, and especially the use of so-called "Pearce element ratio diagrams," which are x-y plots of ratios of cation concentrations in which the ratios on both axes have a common denominator. File SCHELLER.TXT contains 54 analyses for major elements of granites from the Schellerhau pluton in the Erzgebirge Mountains of Germany. A plot of Na_2O versus K_2O shows a moderate negative correlation between the two oxides. When sodium and potassium are converted to cation percents, then used as the numerators in ratios formed with titanium, the two elements have a strong positive correlation (**Fig. 2–24**).

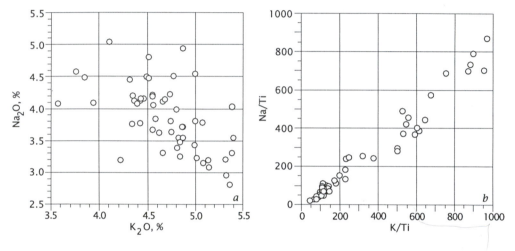

Figure 2–24. Scatter plots of (*a*) Na_2O and K_2O concentrations in granites from the Schellerhau pluton of Germany, and (*b*) after recasting concentrations of Na and K as ratios with Ti.

A ***closed data set*** is one in which all variables measured on an individual add to a fixed total such as 1.00 or 100%, which means the individual variables are proportions of a whole. These data also are referred to as ***compositional data***. Because the sum of the variables is a fixed number, an increase in the proportion of one variable can only occur at the expense of other variables. This creates the most troublesome form of induced correlation, the ***spurious negative correlations*** that appear as a result of closure.

In an ***open data set*** in which the measurements are not expressed as proportions, two linearly independent variables will have a correlation that is not significantly different from zero. If an open data set is closed by converting the measurements to proportions, apparently significant negative correlations will appear even though the original open data consisted entirely of independent variables. In the special case of a three-variable closed data set, the correlations between the closed variables are determined solely by the variances according to the following relationship:

$$r_{1,2} = \frac{s_3^2 - \left(s_1^2 + s_2^2\right)}{2 s_1 s_2} \tag{2.29}$$

48

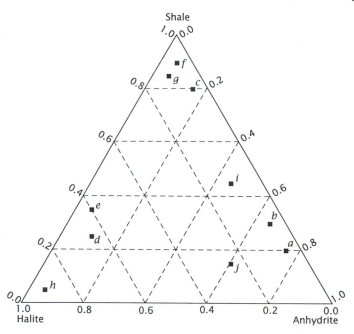

Figure 2–25. Triangular diagram of the halite–anhydrite–shale compositional system. Points represent average composition of 5-ft segments of the Wellington Formation (Permian) measured in a well in central Kansas.

Such intercorrelations are inherent in any geologic data that are plotted on triangular diagrams, such as sandstone-shale-limestone compositional triangles or ternary phase diagrams. These inverse relationships result solely from the fact that as the proportion of one constituent increases, the proportions of the other two constituents must decrease.

A triangular diagram of the halite-anhydrite-shale sedimentary rock compositional system is given in **Figure 2–25.** The points plotted represent the estimated lithologic proportions in 5-ft intervals of the Hutchinson Salt member of the Permian Wellington Formation in a borehole in central Kansas. The compositions were calculated from the responses of gamma-ray, neutron, density, and sonic logging tools used to measure the petrophysical properties of the interval. The test well was drilled to investigate a potential disposal site for radioactive wastes.

Table 2–9 gives the composition of the ten intervals plotted on **Figure 2–25,** estimated to 5%. The table also gives the variances of the three mineralogical components, and the correlations calculated from these variances. Notice that the covariances are not necessary to calculate the correlations, as these are predetermined by the variances and the effect of closure.

Because constant-sum data are so prevalent in geology, many attempts have been made to devise ways of assessing the statistical significance of their intercorrelations. Koch and Link (1980, volume II, chapter 11) discuss a number of special statistical techniques for analyzing closed data and Chayes (1971) has written a book on the closure problem; unfortunately, his proposed statistical procedures have proved ineffective (Kork, 1977, Aitchison, 1981). Three fundamental classes of problems arise in the analysis of compositional data. First, the observed correlations derive both from interrelations between the variables and from the mathematical constraint of closure. It is impossible to determine how much of the observed

Table 2–9. Lithologic composition, estimated to 5%, of 5-ft intervals in the Permian Wellington Formation of central Kansas; estimates based on petrophysical measurements made by well-logging tools.

Interval	Anhydrite	Shale	Halite
a	75	20	5
b	65	30	5
c	15	80	5
d	10	25	65
e	5	35	60
f	5	90	5
g	5	85	10
h	5	5	90
i	45	45	10
j	60	15	25
TOTALS	290	430	280
MEANS	29	43	28
VARIANCES	832.22	962.22	1001.11
STD. DEVIATIONS	28.85	31.02	31.64

$$r_{1,2} = \frac{1001.11 - (832.22 + 962.22)}{2 \cdot 28.85 \cdot 31.02} = \frac{-793.47}{1789.85} - 0.44$$

$$r_{1,3} = \frac{962.22 - (832.22 + 1001.11)}{2 \cdot 28.85 \cdot 31.64} = \frac{-871.11}{1787.66} - 0.48$$

$$r_{2,3} = \frac{832.22 - (962.22 + 1001.11)}{2 \cdot 31.02 \cdot 31.64} = \frac{-1131.11}{1962.95} - 0.58$$

correlation is "real," and how much is a mathematical artifice. Second, if we extract a subset of variables (called a **subcomposition**) and recalculate their proportions, the correlations between variables will change in an unpredictable manner. For this reason, relationships perceived on a ternary diagram of a subcomposition may be totally at odds with those we would infer from the correlations between variables in the complete data set. Third, once we know $m - 1$ of the variables, the mth variable must be the difference between their sum and 100%. That is, the true dimensionality of the data set is at most one less than the number of variables.

Logratio Transformation

Aitchison (1986) provides an extensive treatment of closure and offers a number of transformations that convert compositional variables into forms that can be analyzed by conventional statistical techniques. He justifies his procedure by noting that the absolute magnitudes of compositional variables are not directly meaningful, since they are really ratios whose denominators are the sum of all the constituents. Instead, we are interested in the relative magnitudes of some constituents as compared to others. This being the case, we should look at the ratios between constituents. The ratios are not constrained to the interval 0 to 1 (or 0 to 100%, or 0 to $1,000,000$ ppm), but can assume any positive value. Aitchison further notes that it is difficult to compute variances and covariances of ratios, but that the process can be simplified by the use of logarithms. After taking logarithms of the ratios, the

Table 2–10. Chemical compositions, in oxide percents, of 11 pyroxenes. R_2O_3 is the sum of Al_2O_3, TiO_2, Fe_2O_3, and Cr_2O_3.

SiO_2	R_2O_3	FeO	MgO	CaO
57.75	1.87	3.65	36.50	0.23
49.97	4.20	28.64	15.75	1.44
46.48	1.33	47.21	3.53	1.45
54.99	4.70	1.59	17.27	21.45
48.72	1.89	26.85	1.07	21.47
52.85	5.02	5.71	16.48	19.94
47.58	5.67	21.73	7.42	17.60
50.11	4.67	18.90	16.33	9.99
49.79	8.79	26.33	6.99	8.10
53.06	1.73	17.52	23.61	4.08
50.51	3.53	29.21	12.89	3.86

resulting transformed values are free to vary over the entire range of real numbers, from $-\infty$ to $+\infty$.

As Rollinson (1993) points out, Aitchison's logratio procedure and ratio correlation would seem to be similar, but this is not the case. Correlations between ratios are likely to be misleading, as seen in **Figure 2–24**, but the covariances of logratio variables can be interpreted in a meaningful manner. Depending upon the nature of the study, Aitchison provides three alternative forms of his transformation.

For simple descriptive purposes, Aitchison advocates the use of a ***compositional variation array*** in which every compositional variable forms a ratio with every other variable. Because of the large number of ratios that could be formed, this would seem to be a formidable task. Fortunately, algebraic identities greatly reduce the number of computations. First, we must find the ***logratio variances***, $s^2_{j/k}$

$$s^2_{j/k} = \text{var}\left(\ln\frac{x_j}{x_k}\right) \tag{2.30}$$

where the variances are calculated in the usual manner (Eq. 2.17), except that the variables x_i are now the logratio variables x_j/x_k. Because

$$\ln\frac{a}{b} = -\ln\frac{b}{a}$$

the variance of $\ln x_j/x_k$ will be equal to the variance of $\ln x_k/x_j$, so we only have to find variances for half the possible ratios. Furthermore, since the log of 1 is zero, the variances along the diagonal must also be zero. That is,

$$s^2_{k/k} = 0$$

Table 2–10 lists the analyses of 11 pyroxenes, each composed of five oxides (R_2O_3 is a composite variable consisting of $Al_2O_3 + TiO_2 + Fe_2O_3 + Cr_2O_3$). The ten unique logratio-transformed variables that can be created as ratios of the original variables are shown in **Table 2–11**. Means and variances of these logratio variables form the compositional variation array given in **Table 2–12**.

Table 2–11. Ten unique logratio transforms of pyroxene compositional variables listed in Table 2–10.

$\frac{SiO_2}{R_2O_3}$	$\frac{SiO_2}{FeO}$	$\frac{SiO_2}{MgO}$	$\frac{SiO_2}{CaO}$	$\frac{R_2O_3}{FeO}$	$\frac{R_2O_3}{MgO}$	$\frac{R_2O_3}{CaO}$	$\frac{FeO}{MgO}$	$\frac{FeO}{CaO}$	$\frac{MgO}{CaO}$
3.430	2.761	0.459	5.526	−0.669	−2.971	2.096	−2.306	2.764	5.067
2.476	0.557	1.155	3.547	−1.920	−1.322	1.070	0.598	2.990	2.392
3.554	−0.016	2.578	3.467	−3.569	−0.976	−0.086	2.593	3.483	0.890
2.460	3.543	1.158	0.941	1.084	−1.301	−1.518	−2.385	−2.602	−0.217
3.250	0.596	3.818	0.819	−2.654	0.569	−2.430	3.223	0.224	−2.999
2.354	2.225	1.165	0.975	−0.129	−1.189	−1.379	−1.060	−1.251	−0.191
2.127	0.784	1.858	0.995	−1.344	−0.269	−1.133	1.0745	0.211	−0.864
2.373	0.975	1.121	1.613	−1.398	−1.252	−0.760	0.146	0.636	0.491
1.734	0.637	1.963	1.816	−1.097	0.229	0.082	1.326	1.179	−0.147
3.423	1.108	0.810	2.565	−2.315	−2.614	−0.858	−0.298	1.457	1.755
2.661	0.548	1.366	2.572	−2.113	−1.295	−0.089	0.818	2.024	1.206

Table 2–12. Compositional variation array of logratio-transformed variables in Table 2–11. Upper diagonal contains variances, lower diagonal contains means.

	SiO_2	R_2O_3	FeO	MgO	CaO
SiO_2		0.3682	1.2187	0.8868	2.1653
R_2O_3	2.7129		1.6262	1.1202	1.5991
FeO	1.2471	−1.4658		3.2009	3.3923
MgO	1.5865	−1.1264	0.3393		4.1854
CaO	2.2578	−0.4552	1.0106	0.6713	

If $s^2_{j/k}$ is near zero, this indicates there is little relative variation between x_j and x_k. If $s^2_{j/k}$ is greater than 1, there is a large relative variation between x_j and x_k, as between MgO and CaO. If $\overline{X}_{j/k}$ is positive, x_j tends to be larger on average than x_k. If $\overline{X}_{j/k}$ is positive but smaller than $\sqrt{s^2_{j/k}}$, then x_j is larger than x_k for a substantial number of the observations. If $\overline{X}_{j/k}$ is positive and exceeds $\sqrt{s^2_{j/k}}$, then the observations of x_j are consistently and persistently larger than those of x_k. If $\overline{X}_{j/k}$ is negative, x_k tends to exceed x_j, and there is a set of relationships complementary to those just listed.

Although the computational variation array is a relatively simple way of exploring relationships among compositional variables, it does not contain true covariances. We can determine such an array by finding all possible ratios of the original compositional variables, then computing the covariances between the logarithms of these ratios. The logratio covariances are defined by Equation (2.26) for covariance, in which variables x_j and x_p have been replaced with logratio variables

$\ln(x_{j/k})$ and $\ln(x_{p/q})$.

$$\text{cov}_{j/k,p/q} = \text{cov}\left\{\ln(x_j/x_p),\ \ln(x_k/x_q)\right\}$$

$$= \frac{\sum_{i=1}^{n} \ln(x_{ij}/x_{ip}) \ln(x_{ik}/x_{iq}) - \left(\sum_{i=1}^{n} \ln(x_{ij}/x_{ip}) \sum_{i=1}^{n} \ln(x_{ik}/x_{iq})/n\right)}{n-1} \quad (2.31)$$

You can appreciate that this may result in a very large collection of covariances! For example, the five original variables in the pyroxene data given above will form a 25×25 array of possible logratio combinations. Fortunately, many of these are equivalent because of algebraic identities; for example, all covariances in which one of the ratios is 1 will vanish, and covariances involving reciprocal ratios will be identical. Aitchison demonstrates that we can choose any one of the original variables (call the variable selected x_d). We then form all possible logratios of the original variables in which we use x_d as the denominator; there will be $m-1$ of these transformed variables (for the pyroxene data, there will be four logratio variables in which the denominator is x_d). Finally, we calculate the covariances between all possible pairs of these logratios; there will be $1/2\,(m-1)\,(m-2)$ unique covariances and they can be arranged in a **logratio covariance array** whose elements are $\text{cov}_{j/d,\,p/d}$.

$$\text{cov}_{j/d,\,p/d} = \text{cov}\left\{\ln(x_j/x_d),\ \ln(x_k/x_d)\right\} \quad (2.32)$$

These covariances are appropriate for use in analysis of variance, detection of outliers, and linear discriminant analysis, among other applications. All possible logratio covariances can be created from this subset because of the relationship

$$\text{cov}_{j/k,\,p/q} = \text{cov}_{j/d,\,p/d} + \text{cov}_{k/d,\,q/d} - \text{cov}_{j/d,\,q/d} - \text{cov}_{k/d,\,p/d} \quad (2.33)$$

thus demonstrating that all the information on interrelations between the original compositional variables are contained in the covariances between any subset of logratio variables that share a common denominator. Note that it doesn't matter which original variable is selected to be the denominator of the logratios; the entire collection of logratio covariances can be recreated using any of them. The five-variable set of pyroxene analyses yields the 4×4 set of logratio covariances given in **Table 2–13**. All of the logratios were formed using SiO_2 as the denominator variable, x_d.

Table 2–13. Logratio covariances using $Si O_2$ as denominator of ratios.

	R_2O_3/SiO_2	FeO/SiO_2	MgO/SiO_2	CaO/SiO_2
R_2O_3/SiO_2	0.3682	−0.0196	0.0674	0.4672
FeO/SiO_2	−0.0196	1.2188	−0.5477	−0.0041
MgO/SiO_2	0.0674	−0.5477	0.8867	−0.5667
CaO/SiO_2	0.4672	−0.0041	−0.5667	2.1653

Aitchison shows that the logratio covariances also can be calculated from the logratio variances by

$$\text{cov}_{j/k,\,p/q} = \frac{1}{2}\left(s_{j/q}^2 + s_{k/p}^2 - s_{j/k}^2 - s_{p/q}^2\right) \quad (2.34)$$

Even though the set of logratio covariances contains the information necessary to determine all of the remaining possible logratio covariances, the fact that it does not directly express one of the variables (the one chosen to be the denominator) may prove troubling in some applications. Methods involving eigenvector extraction, such as principal component, factor, and canonical analyses, would seem to require a set of covariances in which all variables are directly expressed. For these applications, Aitchison proposes the use of the **centered logratio covariance** in which the divisors of the logratios are the geometric means of the original compositional variables. A geometric mean, g_n, is computed for each of the n rows of the data matrix. The geometric mean is discussed more thoroughly later in this chapter; here we merely note that it is given in the current context by

$$g_n = \left(\prod_{j=1}^{m} x_j \right)^{\frac{1}{m}} = \exp\left[\frac{1}{m} \sum_{j=1}^{m} \ln x_j \right] \tag{2.35}$$

It is the mth root of the product of all components, or more usually, the exponential of the arithmetic average of the logarithms of all the components. Once we have obtained the g_n's, we can use them as the denominators instead of x_d to form logratios of all the variables. The geometric means act just as the variable x_d

$$\text{cov}_{j/g, p/g} = \text{cov}\left\{ \ln\left(x_j / g_j \right), \ln\left(x_k / g_k \right) \right\} \tag{2.36}$$

However, it is not necessary to compute the g_n's, because a centered logratio data matrix can be created by first taking the logarithms of all the data values. Then, the arithmetic averages of the logarithmic values in each row are found and subtracted from every value in the row. All the rows in this transformed data matrix will now have a mean of zero, and the grand mean of the matrix will be zero as well. We now proceed to compute the covariances between the columns of the transformed data matrix in the usual manner. Note that we again have as many logratio variables as there are compositional variables, since none of the original variables are now relegated to the role of divisor. This reintroduces certain mathematical problems that had been relieved when the dimensionality of the data set was reduced by using one variable as a common denominator; we will consider these problems in Chapter 6. **Table 2–14** contains the 5×5 centered logratio covariance array calculated for the pyroxene analyses.

Table 2–14. Centered logratio covariances for pyroxene analyses.

	$\ln\left(\frac{SiO_2}{g}\right)$	$\ln\left(\frac{R_2O_3}{g}\right)$	$\ln\left(\frac{FeO}{g}\right)$	$\ln\left(\frac{MgO}{g}\right)$	$\ln\left(\frac{CaO}{g}\right)$
$\ln\left(SiO_2/g\right)$	0.00409	−0.00174	−0.05691	0.04619	−0.02665
$\ln\left(R_2O_3/g\right)$	−0.00174	0.36066	−0.08234	0.10777	0.43474
$\ln\left(FeO/g\right)$	−0.05691	−0.08234	1.10086	−0.56250	−0.09180
$\ln\left(MgO/g\right)$	0.04619	0.10777	−0.56250	0.97505	−0.55122
$\ln\left(CaO/g\right)$	−0.02665	0.43474	−0.09180	−0.55122	2.10791

Aitchison's transformation removes spurious negative correlations between compositional variables, but in some circumstances there are difficulties in its application. Because the transformation uses logarithms, it cannot be applied if some values are zero or missing. There are three possible solutions. You may argue that "zero amounts" are simply below the detection limit of the analytical device, and replace zeros with very small values that are smaller than the detection limit or possible rounding error. If this is not acceptable, variables having zero or missing values can be amalgamated with other variables. Possibly the least satisfying solution is to add a constant to all original values prior to logratio transformation.

Comparing Normal Populations   *Important stuff to know!*

Before proceeding, let us return for a moment to frequency distribution curves, specifically to the normal distribution. If we could measure the lengths of a very large collection of *Composita*, rather than only the sample of six presented in **Table 2–5**, we would find that a frequency diagram would look something like that shown in **Figure 2–26**. The mean, say 14.2 mm, would have the greatest frequency, and progressively larger and smaller specimens would occur with decreasing frequency. Approximately two-thirds of the shell lengths would occur within a standard deviation, perhaps 4.7 mm, of the mean. Now consider the width measurements we made while examining this very large collection of *Composita*. The width frequency distribution is similar in form to the length distribution, but the mean and standard deviation are different. It might look, for example, like **Figure 2–27**, with a mean of 10.3 mm and a standard deviation of 3.6 mm.

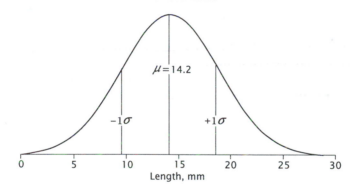

Figure 2–26. Hypothetical frequency distribution of the population of *Composita* lengths.

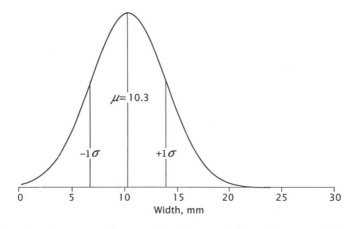

Figure 2–27. Hypothetical frequency distribution of the population of *Composita* widths.

Can we compare the two distributions with each other? They are measured in the same units, which makes the problem easier than if we wished to compare shell length to shell weight. We can draw the two distributions on the same millimeter scale, giving **Figure 2–28**.

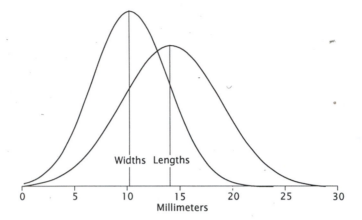

Figure 2–28. Frequency distributions of *Composita* lengths and widths redrawn on the same millimeter scale.

It might be simpler to compare the two if they were centered about the same mean. We could center them about a common mean by subtracting enough from all observations of one population (or adding to the values of the other population) to move the means until they coincide. Instead, let us subtract the mean from each observation in each of the two populations. That is, the new measurements are $x_i' = x_i - \overline{X}$. This will move each of the distributions along the millimeter scale until they are centered about zero, which is the mean of both transformed distributions. This is shown in **Figure 2–29**.

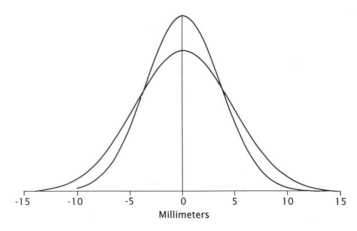

Figure 2–29. Frequency distributions of *Composita* lengths and widths after both have been adjusted to have zero mean.

Unfortunately, we are dependent upon the millimeter scale of the frequency distributions. This is no great problem with lengths and widths, but if we wish to compare the distributions to one describing weights of shells, we cannot do it. Is there an additional transformation that can be made which will make our normal distributions independent of measurement units? One extremely useful transformation is called **standardization**, and will result in new values for the individuals that not only have zero mean but are measured in units of standard deviations. This is done simply by subtracting the mean of the distribution from each observation and dividing by the standard deviation of the distribution. The new variable, z_i, has what is called the **standard normal form**.

$$z_i = \frac{x_i - \overline{X}}{s}$$

mean = 0 measured in units of standard deviation (2.37)

Now, our frequency curves of different *Composita* measurements are identical, and have the form shown in **Figure 2–30**. The characteristics of the standard normal distribution are extremely well known, and tables of the areas under specified segments of the curve are available in almost all statistics books. Remember that the areas are directly expressible as probabilities. With the use of a table such as Appendix Table **A.1**, we can find the probability of encountering a sample, by random selection from a normal population, whose measurement falls within a specified range. We must know, however, the variance or standard deviation of the population.

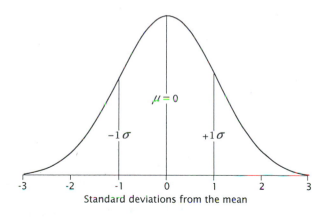

Figure 2–30. *Composita* frequency distributions after standardizing lengths and widths to have zero mean and unit standard deviation.

Let us make the unrealistic assumption that we have examined the entire population of *Composita*, so we know that the mean and standard deviation of their lengths is 14.2 mm and 4.7 mm, respectively. What is the probability of finding, by chance, a specimen shorter than 3 mm? To find the answer, we must convert 3 mm to units of standard deviation, and then examine Appendix Table **A.1**.

$$z = \frac{3.0 - 14.2}{4.7} = -2.38 \quad = Z \ value$$

The probability of finding a *Composita* smaller than -2.4 standard deviations is the cumulative probability to this point; from Appendix Table **A.1**, we can see that it is 0.0082, which is very small indeed. Now, what is the probability of finding a specimen of *Composita* longer than 20 mm? Again, converting to standard normal form:

$$z = \frac{20.0 - 14.2}{4.7} = 1.23$$

Because the total area under the normal distribution curve is 1.00, the probability of obtaining a measurement of 1.2 standard deviations or greater than the mean is the same as 1.0 minus the cumulative probability of obtaining anything smaller. Stated in another way,

$$p\,(1.2 \text{ or larger}) = 1.0 - p\,(1.2 \text{ or smaller})$$

Appendix Table **A.1** will give us the cumulative probability up to 1.2, which is 0.8849. Therefore, the probability of finding a *Composita* longer than 20 mm is

$$1.0000 - 0.8849 = 0.1151$$

or slightly greater than one chance out of ten. Next we can compute the probability of finding, at random, a specimen of *Composita* whose length falls in the size range from 15 to 20 mm:

$$\text{for 15 mm} \qquad z = \frac{15.0 - 14.2}{4.7} = 0.17$$

$$\text{for 20 mm} \qquad z = \frac{20.0 - 14.2}{4.7} = 1.23$$

$$p\,(1.23 \text{ or less}) = 0.8849$$
$$p\,(0.17 \text{ or less}) = \underline{0.5793}$$
$$p\,(\text{between } 0.17 \text{ and } 1.23) = 0.3056$$

Approximately one out of three specimens will fall in this size range.

Central Limits Theorem

In this example, we have assumed that we are drawing samples from a normally distributed population. Unfortunately, we ordinarily do not know what shape the population distribution may have, and may even suspect that it is distinctly non-normal. This does not mean that the normal distribution is of no use, however, because of a remarkable property called the **central limits theorem**. This states that if sets of random samples are taken from any population, and the means calculated for these samples, the sample means will tend to be normally distributed. The tendency toward normality becomes more pronounced for samples of larger size.

The central limits theorem may not seem reasonable at first glance; it is difficult to see why the means of samples should form a normal distribution if the samples were collected from a population quite different in form. However, a simulation experiment will show that the theorem does indeed hold true. Suppose we sample from a parent population having the distinctly nonnormal U-shape shown in **Figure 2–31**. Most of the individual observations in a sample will come from the two extremes of the distribution, which contain the bulk of the population. When these values are averaged together to form a mean, high values will tend to be counterbalanced by low values, resulting in a mean near the center of the distribution.

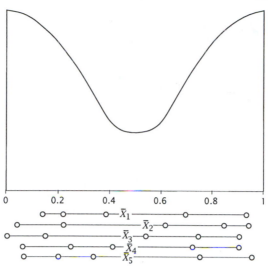

Figure 2–31. Five samples of five observations drawn randomly from a U-shaped distribution. Means of samples are shown by $\overline{X}$.

Only in the very rare circumstance when all of our randomly selected observations happen to come from either the high end or the low end will we calculate a mean that differs much from the central value.

Note that the sample means are clustered near the central value of the distribution in **Figure 2–31**. If this experiment were repeated a thousand or more times, we would find that the sample means would plot as the familiar, bell-shaped normal curve. Essentially the same results would be obtained if we began with almost any other form of original distribution, as shown in **Figure 2–32**, adapted from Lapin (1982). Since the distribution of sample means tends to be normal, it may be described by only two statistics—its mean and variance. Both theoretical and empirical studies show that the mean of the sample means is equal to the population mean. That is,

$$\overline{X}_{\overline{X}} = \mu$$

[handwritten: mean of sample = population mean]

The variance of the sample means is equal to the variance of the population divided by the size of the samples collected, or

$$s_{\overline{X}}^2 = \frac{\sigma^2}{n}$$

The standard deviation of the sample means is the square root of this quantity, and is called the **standard error of the estimate of the mean,** or simply the **standard error.** It describes the variability that can be expected in the means of samples taken by repeated random selection from the same population. The standard error is

$$s_e = \sqrt{\frac{\sigma^2}{n}} = \sigma\sqrt{\frac{1}{n}} \qquad (2.38)$$

59

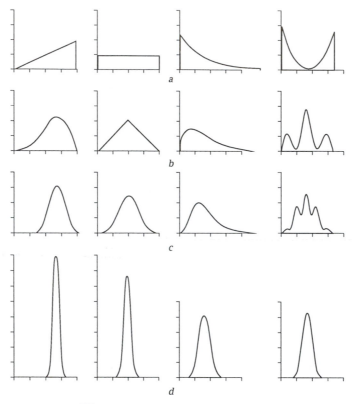

Figure 2–32. Distributions of $\overline{X}$ for large numbers of samples of size n, selected at random from nonnormal populations. Central limits theorem insures that distribution of $\overline{X}$ tends toward normal distribution as n increases. Parent populations from which samples are drawn are shown in (*a*). Distributions of $\overline{X}$ are shown for (*b*) samples of size $n = 2$, (*c*) samples of size $n = 4$, and (*d*) samples of size $n = 25$. After Lapin (1982).

Testing the Mean

The central limits theorem allows us to formulate statistical tests based on the characteristics of the normal curve, and to apply these tests even in circumstances in which the population sampled is not normally distributed. Suppose our paleontologist who has been examining *Composita* is presented with a large slab covered with brachiopods. These fossils look like *Composita*, but are extremely large, the average length for six specimens being 20.0 mm. Recall that we "know" that the mean and standard deviation of the population of *Composita* is 14.2 mm and 4.7 mm, respectively. Is it possible that the new sample of brachiopods is drawn from this population?

We can determine the difference between the mean of our new sample and the population mean. This difference can then be compared to the variation we expect to see in the means of samples selected at random from a specified population. This variation is given by the standard error, and is a function of both the variance of the population and the size of the sample.

The comparison between the difference in means and the standard error can be made in the following way:

$$z = \frac{\overline{X} - \mu}{s_e} = \frac{\overline{X} - \mu}{\sigma\sqrt{1/n}} \tag{2.39}$$

Note that the test statistic is calculated in a manner exactly equivalent to that used to convert a variable to standardized form (see Eq. 2.37). The test statistic, z, is normally distributed with a mean of zero and a standard deviation of one, if the sample mean was indeed drawn from the hypothesized population. If z is excessively large, we will tend to conclude that the sample was not taken from this population. A formal decision, however, requires that we establish a consistent procedure for evaluating the test statistic.

The initial step in statistical testing is the posing of an appropriate hypothesis or statement about the variable in question. Ordinarily, this is done in the form of a **null hypothesis**, symbolized H_0, which is a hypothesis of no difference. We may, as in the present example, speculate that a given sample of observations has been drawn from a parent population which has a certain specified mean. The null hypothesis is expressed in the form *there is no difference*

$$H_0 : \mu_1 = \mu_2 \tag{2.40}$$

which states that the mean, μ_1, of the parent population from which the sample was drawn is equal to (or is not different from) the mean of a population having a specified mean μ_2.

In our example, we are hypothesizing that the mean of the population from which the slab of brachiopods was taken is the same as the mean of the *Composita* population.

Having posed the null hypothesis, an **alternative hypothesis** must be given. An appropriate alternative in this situation might be

$$H_1 : \mu_1 \neq \mu_2 \tag{2.41}$$

stating that the mean of the population from which the sample was drawn does not equal the specified population mean. We now can devise procedures to test the hypothesis, with specified levels of probability of correctness. If the two parent populations are not the same, we must conclude that the slab of fossils was not drawn from the *Composita* population, but from the population of some other genus.

Once a null hypothesis and alternative are expressed, we can make a decision to either accept or reject the null hypothesis on the basis of our statistical test. There are also two possible states of the null hypothesis; it may be true or false. This combination produces four possible outcomes, of which two are correct and two incorrect. The possibilities can be graphically illustrated:

	Hypothesis is correct	Hypothesis is incorrect
Hypothesis accepted	Correct decision	Type II error, β
Hypothesis rejected	Type I error, α	Correct decision

Either acceptance of a true hypothesis or rejection of a false hypothesis will result in a correct decision. If a null hypothesis is rejected when it is in fact true, a type I error has been committed. Conversely, if an erroneous hypothesis is accepted, a type II error is committed. In terms of our example, the illustration above may be redrawn:

	Actuality	
Hypothesis	Slab is *Composita*	Slab is not *Composita*
μ of slab $=$ μ of *Composita*	Correct decision	Type II error, β
μ of slab $\neq$ μ of *Composita*	Type I error, α	Correct decision

Here, "μ of slab" refers, of course, to the mean of the population from which the slab came.

In standard statistical procedures, the probability of committing a type I error is called the ***level of significance*** and is denoted by α; this probability must be specified before running the test. In order to minimize the possibility of committing a type II error, we express the null hypothesis with the intention that it will be rejected. If we do reject the null hypothesis, there is no possibility of committing a type II error, and the probability of a type I error is known because it has been specified. If, however, the test fails to reject the null hypothesis, the probability of a mistake in the form of a type II error remains. This probability, called β, generally is not known. Thus, if we reject the hypothesis of equality, we can state that the two parent populations have different means. The probability that this statement is incorrect and that our decision is an error is equal to α. On the other hand, if H_0 is not rejected, the statement that the means of the two populations are equal is accompanied by an unknown probability, β, of a mistake.

The logic of statistical tests is based on the premise that the null hypothesis and its alternative are mutually exclusive and all inclusive. The null hypothesis is an explicit statement; therefore, the alternative must be general. If H_0 is rejected, we are stating that the specific relationship described by the null hypothesis does not exist. Rather, the true relationship is contained somewhere in the infinite realm of possibilities encompassed by the alternative. We cannot determine what the true relationship is; we can only state what it is not. Some statisticians express the possible outcomes of statistical tests as "rejection of the null hypothesis" versus "failure to reject." Failure to reject, containing as it does an unassessed probability of error, is not equivalent to acceptance. Statistical tests, in a sense, do not tell you what is, but only what is not.

Returning to the null hypothesis and alternative given in Equations (2.40) and (2.41), suppose that it is decided that a probability level of a type I error of 5% would be appropriate. In other words, we are willing to risk rejecting a correct hypothesis five times out of 100 trials.

We are assuming that we know the variance of the population against which we are checking. Our paleontologist has established that the variance of *Composita* lengths is 22.1 (recall that the standard deviation was 4.7). We now may set up a formal statistical test in the following manner.

1. The hypothesis and alternative:
$$H_0 : \mu_1 = \mu_0$$
$$H_1 : \mu_1 \neq \mu_0$$

2. The level of significance: $\alpha = 0.05$

3. The test statistic:

$$z = \frac{\overline{X} - \mu_0}{\sigma\sqrt{1/n}}$$

(2.42)

The test statistic, z, has a frequency distribution that is a standardized normal distribution, provided the observations in the sample were selected randomly from a normal population whose variance is known. We have specified that we are willing to reject the hypothesis of equality of means when they actually are equal one time out of 20; that is, we will accept a 5% risk of a type I error. On the standardized normal distribution curve, therefore, we wish to determine the extreme regions that contain 5% of the area of the curve. This part of the probability curve is called the **area of rejection** or the **critical region**. If the computed value of the test statistic falls into this area, we will reject the null hypothesis.

Because the alternative is simply one of nonequality, the hypothesis will be rejected if the test statistic is either too large or too small. That is, there are three possible true situations: $\mu_1 = \mu_0$, $\mu_1 > \mu_0$, or $\mu_1 < \mu_0$. We are not interested in distinguishing between the latter two possibilities. The critical region, therefore, occupies the extremities of the probability distribution and each subregion contains 2.5% of the total area of the curve.

The rationale of this particular test may be summarized as follows: We know the characteristics of the normal curve; these have been derived from theoretical considerations and their use has been empirically justified. The central limits theorem tells us that the means of samples will be normally distributed. Since we can find the standard error, we know what percentage of sample means will occupy various size ranges (for example, we know that two-thirds of the sample means will occur within one standard deviation of the population mean). If sample means are drawn from this population without bias (that is, by random selection of their constituent observations), the probabilities of obtaining a sample mean from within a specific range of the distribution curve is equal to the area within that portion of the curve. If a mean is drawn from a region of significantly low probability, we conclude that our sample was not obtained from the population specified by the null hypothesis, and we reject it. However, there is a finite possibility, equal to the area of rejection, that we did by chance obtain a sample that yielded a mean from this extreme region of the population.

Working through the *Composita* example, the outline takes the following form:

1.
$$H_0 : \mu_{\text{slab}} = 14.2 \text{ mm}$$
$$H_1 : \mu_{\text{slab}} \neq 14.2 \text{ mm}$$

2. $\alpha = 0.05$

3.
$$z = \frac{20.0 - 14.2}{4.7\sqrt{1/6}} = 3.023$$

We are prepared to reject the hypothesis of equality of means if the sample mean is either too large or too small. This leads to a two-tailed test, diagrammed in **Figure 2–33**.

The critical region, which we have decided should contain 5% of the area of the normal distribution, is split into two equal parts, each containing 2.5% of the total area. If the computed value of z falls into the left-hand region, the sample came from a population having a smaller mean than our known population. Conversely,

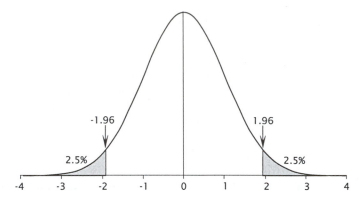

Figure 2–33. Normal distribution curve with two critical regions (shown by shading) which contain a total of 5% of the area under the curve.

if z falls into the right-hand region, the mean of the sample's parent population is larger than the mean of the known population. From Appendix Table **A.1**, we find that approximately 2.5% of the area of the curve is to the left of a ***critical value*** of $z = -1.9$, and 97.5% (100% − 2.5% = 97.5%) is to the left of a critical value of $z = +1.9$. The computed test value of $z = 3.0$ exceeds 1.9, so we conclude that the means of the two populations are not equal, and the collection of fossils must represent some genus other than *Composita*. It is important to note the assumptions that have been made in the application of this test. The normal test assumes:

1. The sample of brachiopods was selected randomly.

2. The population of lengths of *Composita* is known to be normally distributed.

3. The variance in lengths of *Composita* is known to be 22.1 mm.

If in a particular instance any of these test assumptions seem unwarranted, the results of the test are suspect. We must then look for an alternative procedure whose assumptions are more realistic.

P-Values

The development of computer programs for statistical analysis has led to alternatives to the classical procedure of specifying a fixed level of significance and then either accepting or rejecting the null hypothesis, depending upon the outcome of a statistical test. Although the classical testing approach imposes discipline on the investigator, who might otherwise be tempted to fudge the criterion for rejection, it also forces a specific significance level on others who might have quite different opinions about the acceptable risk. This rigidity can be avoided by the use of a ***p-value***, defined as the smallest level of significance at which the null hypothesis would be rejected for a specific test. It is the probability, assuming that H_0 is correct, that the observed (or a more extreme) test result could be obtained by random chance.

In the *Composita* example just discussed, we calculated a test value of $z = 3.02$ which was significant at the 5% level in a two-tailed test, leading to rejection of the null hypothesis. From the table of the standardized normal distribution in Appendix Table **A.1**, we see that the area to the right of $z = 3.023$ is $(1.000 - 0.9987) = 0.0013$. Since our test was two-tailed, the *p*-value associated with this test is twice

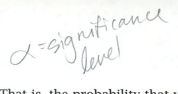
α = significance level

this amount, or $p = 0.0026$. That is, the probability that we would obtain a sample of six specimens of *Composita* whose mean length was 20.0 mm by drawing at random from a population whose mean and standard deviation are 14.2 and 4.7 is less than three in a thousand.

It is easy to determine if a test result falls into the critical region. Simply compare the p-value with the specified α-level; if the p-value is smaller, the null hypothesis should be rejected. It's not necessary to consult a table of critical values.

Significance

Before continuing with additional statistical tests, a few comments on the choice of *levels of significance* may be helpful. Most statistics texts, particularly those concerned with agricultural statistics or industrial quality control, repeatedly use significance levels of one in 20 ($\alpha = 0.05$) or one in 100 ($\alpha = 0.01$) in their examples and exercises. This practice may suggest that there is something particularly important about these specific levels, but this is not the case. Setting the level of significance is a responsibility of the researcher, who must decide what risk of rejecting a true hypothesis is appropriate. In geology, we often deal with circumstances of great uncertainty, and it may be unrealistic to demand that a statistical test produce a decision that may be in error only one time in 100, or even one time in 20. If extremely stringent levels of significance are set, we may find that our null hypothesis can never be rejected, and we always need greater and greater amounts of data, which we may not be able to produce. By setting more modest levels of significance, we may be able to come to conclusions more frequently, even though the possibility that these conclusions are erroneous may be high by comparison to standards in other areas.

Figure 2–34 illustrates the effect of setting different levels of significance for a hypothetical statistical test of petroleum prospects. We may imagine that a company has found some quantitative variable that can be measured prior to drilling to indicate whether or not a prospect may be productive. The company applies a statistical test to the variable to decide if a prospect should be drilled or abandoned. The null hypothesis states that a prospect comes from the population of barren localities; the alternative is that it comes from the population of nonbarren, or producing, localities.

If a conventional level of significance such as $\alpha = 0.05$ is set as in **Figure 2–34 a**, very few prospects will be found to differ from the barren null population. Those determined to be different will almost assuredly result in discoveries when drilled. The company will achieve a very high success ratio, but will drop many leases that may subsequently prove to be productive when drilled by more adventuresome explorationists. In summary, the company will seldom drill, will seldom fail, and will leave many reservoirs unfound.

In contrast, a level of significance such as $\alpha = 0.40$ might be set, as in **Figure 2–34 b**. Then, many prospects will be identified as drillable, but the failure rate will be much higher. With this decision criterion, the company will drill often, fail often, and will leave much less oil undiscovered.

The oil industry regards the consequences of failing to drill locations where oil exists (a type II error) as much more dire than drilling dry holes (a type I error). This is because the financial reward of a single large discovery may offset the cost of tens or even hundreds of dry holes. The industry's wildcat success ratio is about

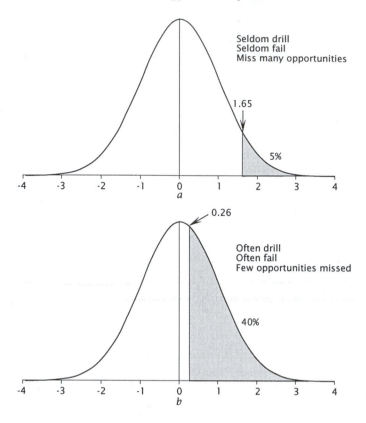

Figure 2–34. Distribution of test statistic with critical region specified for rejection of hypothesis that prospect is barren. (*a*) Critical region set at $\alpha = 0.05$. (*b*) Critical region set at $\alpha = 0.40$.

10% in the United States. If these wells were being drilled on the basis of statistical tests, this would represent a level of significance of almost $\alpha = 0.90$!

This is perhaps an extreme example, but it does illustrate the point that the level of significance must be set at a value appropriate for the particular circumstances of the test. The α-level should be based on an assessment of the consequences of making a type I error. These consequences may be tangible and involve loss of money, time, or even lives, or they may be intangible and involve damage to professional reputation or personal pride. To preserve intellectual honesty, the investigator must decide at the outset the amount of risk that he or she is willing to assume, and set the level of significance accordingly. Selecting a level of significance after a test has been run and the results are known is shameless gerrymandering, and the levels set may reflect the investigator's desire to accept or reject a hypothesis rather than a dispassionate evaluation of the risks involved.

Confidence Limits

In the z-test just described, we considered the probability that a sample with an observed mean $\overline{X}$ could have been drawn at random from the population of a continuous variable whose mean and variance were known. Sometimes we don't have a preconceived idea about the value of the population mean; instead, we would like

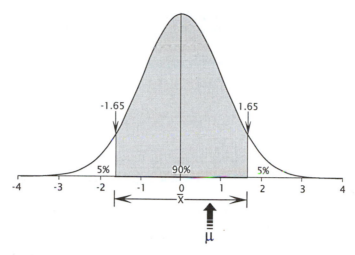

Figure 2–35. Distribution of standard error of the mean. Shaded interval contains 90% of area under curve and defines confidence interval around mean. Population mean μ is captured by this interval with 90% probability.

to estimate what the population mean might be, based on the characteristics of the sample. Because the variable is continuous, the probability that any specific point estimate we make will be correct must be zero. Therefore, we must express our estimate in the form of an interval, which we can do by considering the standard error, s_e, associated with $\overline{X}$.

A *confidence interval* is an interval of plausible values that a population parameter might assume, based on the value of a statistic which estimates that parameter. It is constructed so that the true value of the parameter will be contained within the interval with specified probability. The width of the interval conveys information about how precisely the value of the parameter can be estimated.

We can construct a confidence interval for the mean by using the standard error of the mean, which describes the variation in our estimates of the mean that we expect if we take samples repeatedly. Since this distribution is normal (because of the central limits theorem), we know the proportions of areas under the curve that correspond to different distances away from its center. We can turn these distances into limits that will enclose the true population mean, with an associated probability of successful capture. **Figure 2–35** shows a distribution of the standard error of the mean. The shaded interval extends from -1.65 standard deviations to $+1.65$ standard deviations from the sample mean, $\overline{X}$, and contains 90% of the area under the curve. Therefore, the limits

$$\overline{X} \pm 1.65 \cdot s_e \tag{2.43}$$

will contain the true mean of the population 90% of the time. That is, if we calculate a series of such intervals, each based on a random sample of the same size, we expect that over the long term only one out of ten intervals will fail to include the true (but unknown) population mean, μ.

The standard error is dependent on the size of the sample, so the confidence interval will decrease with increasing n. Conversely, if we specify a higher degree of confidence, the interval will increase. This conforms with our commonsense expectations; we can increase the probability that our interval captures the true mean by making the interval larger, or by increasing the size of our sample.

Equation (2.39) can be rearranged into an interval of the form

$$\left(\overline{X} - z \cdot s_e\right) < \mu < \left(\overline{X} + z \cdot s_e\right)$$

or

$$\left(\overline{X} - z \cdot \frac{\sigma}{\sqrt{n}}\right) < \mu < \left(\overline{X} + z \cdot \frac{\sigma}{\sqrt{n}}\right) \tag{2.44}$$

We can find a 90% confidence interval for the mean length of *Composita*, based on the sample of six observations given in **Table 2–5**. The critical value, z, corresponds to the standardized normal score for a two-tailed test at the desired level of significance, and can be obtained from Appendix Table **A.1**. For a 90% confidence interval, we want the z-scores corresponding to 5% and 95% cumulative probability (*i.e.*, the z-values that define 5% critical regions in the upper and lower tails).

$$\overline{X} = 12.4$$

$$s_e = \frac{4.7}{\sqrt{6}} = 1.92$$

$$\overline{X} - (1.65 \cdot 1.92) < \mu < \overline{X} + (1.65 \cdot 1.92)$$
$$12.4 - 3.17 < \mu < 12.4 + 3.17$$
$$9.2 < \mu < 15.6$$

The interval 9.2 to 15.6 does, indeed, capture the true population mean of 14.2 mm.

The t-Distribution

An assumption made in the test given in Equation (2.39) is seldom true in practical applications. We rarely know the parameters of a population. We have not examined the entire population of *Composita* and it is obviously impossible to do so. Because μ and σ are not known, the best that can be done is to estimate them from samples. An amount of uncertainty is associated with the estimates themselves, so decisions based upon estimates cannot be as precise as those based on known parameters.

The uncertainty introduced into estimates derived from samples can be accounted for by using a probability distribution which has a wider "spread" than the normal distribution. One distribution of this type is called the **t-distribution**, which is similar to the normal distribution, but dependent upon the size of sample taken. A typical t-distribution curve is shown in **Figure 2–36**. The exact shape of the curve changes according to the number of observations in the sample being used to estimate the population. When the number of observations in the sample is infinite, the t-distribution and the normal distribution are identical. In fact, when the number of observations exceeds about 30, the two distributions become so similar that we can use z rather than t, even though we do not know the population parameters. This threshold of sample size separates **small-sample statistics**, where we must compensate for sampling uncertainties, from **large-sample statistics**, where we can use the normal distribution as an approximation.

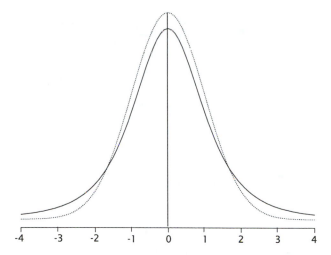

Figure 2–36. Comparison of t-distribution of four degrees of freedom (solid line) to normal distribution (dotted line).

Degrees of freedom

In tests based on samples, we must estimate a number of population parameters in order to calculate the test statistic. It seems intuitively unwise to both estimate the parameters and perform the test from the same set of data without somehow compensating for the double use of observations. This is done by considering a quantity called ***degrees of freedom***, which may be defined as the number of observations in a sample minus the number of parameters estimated from the sample. In other words, the degrees of freedom are the number of observations in excess of those necessary to estimate the parameters of the distribution. Degrees of freedom are symbolically indicated by the Greek letter v (nu) and are always positive integers.

As an example, consider **Table 2–15**, which represents the calculation of the mean and variance of the five replicate measurements of chromium in a shale sample. The mean is estimated from the five independent observations and so has five degrees of freedom. The variance is estimated as the average of the squared deviations from the mean, or the five squared differences, $(x_i - \overline{X})^2$. However, note that if we determine four of these differences, we automatically know the fifth, since

$$x_5 = 5\overline{X} - (x_1 + x_2 + x_3 + x_4)$$

$$x_5 = (5 \cdot 222) - (205 + 255 + 195 + 220) = 235$$

Therefore, there are only four independent items of information from which to estimate the variance.

Unfortunately, the concept of degrees of freedom is seldom explained in beginning statistical texts, but rather is presented as an apparently arbitrary quantity such as $n - 1$. An excellent general discussion of degrees of freedom, in both a physical as well as statistical context, is given by Walker (1940). We will briefly discuss the reasons for the various degrees of freedom that are associated with different statistical tests as these arise.

Table 2–15. Computation of variance of five chromium replicates.

Cr	$Cr - \overline{X}$	$(Cr - \overline{X})^2$
205	−17	289
255	33	1089
195	−27	729
220	−2	4
235	13	169
TOTALS = 1110	0	2280

MEAN = 1110/5 = 222

VARIANCE = 2280/4 = 570

Tables of the t-distribution (and other sample-based distributions) are used in exactly the same manner as tables of the cumulative standard normal distribution, except that two entries are necessary to find a probability in the table. The two entries are the desired level of significance (α, the probability of a type I error) and the degrees of freedom (ν). Appendix Table **A.2** is an abbreviated set of the t-statistic; more extensive tables are contained in most of the references listed at the end of this chapter.

Tests called **t-tests** and based on the t-probability distribution are useful for establishing the likelihood that a given sample could be a member of a population with specified characteristics, or for testing hypotheses about the equivalency of two samples. These are the types of problems discussed in introductory statistics courses and are fundamental in experimental sciences and quality-control fields.

Table 2–16. Porosity measurements of ten cores of Tensleep Sandstone, Pennsylvanian, from the Bighorn Basin of Wyoming.

Core Number	Porosity (%)	Core Number	Porosity (%)
01	13	06	29
02	17	07	18
03	15	08	27
04	23	09	20
05	27	10	24

TOTAL = 213

MEAN = 21.3

$s^2 = 30.46$ $s = 5.52$ $s_e = 0.57$

For example, we may wish to test the hypothesis that a suite of Tensleep Sandstone cores, listed in **Table 2–16**, came from a parent population having an average porosity of more than 18%. Assuming the cores were randomly collected from a normal population, the t-statistic may be computed by

$$t = \frac{\overline{X} - \mu_0}{s_e} = \frac{\overline{X} - \mu_0}{s\sqrt{1/n}} \tag{2.45}$$

where

$$\overline{X} = \text{mean of the sample}$$
$$\mu_0 = \text{hypothesized mean of population (18\%)}$$
$$n = \text{number of observations in the sample}$$
$$s = \text{standard deviation of observations}$$
$$s_e = \text{standard error of mean}$$

Note that the test is essentially identical to Equation (2.38) except that we must estimate the standard error by $s_e = s\sqrt{1/n}$ rather than by $\sigma\sqrt{1/n}$ since we do not know the population variance. In formal statistical terms, we are testing the hypothesis

$$H_0 : \mu_1 \leq \mu_0$$

against the alternative

$$H_1 : \mu_1 > \mu_0$$

The first hypothesis states that the mean of the population from which the sample was drawn is equal to or less than a mean of 18%. The alternative hypothesis states that the parent population of the sample has an average porosity greater than 18%.

Two items must be specified to obtain the critical value of t from Appendix Table **A.2**. These are the desired level of significance and the degrees of freedom. In this particular test, one parameter is assumed (μ_0) and another is estimated (σ is estimated by s, the sample standard deviation). Therefore, in the sample of ten porosity measurements, there are nine degrees of freedom. We are interested in rejecting the null hypothesis only if the mean porosity significantly exceeds 18%; therefore, the critical region occurs only at high values of the test statistic, as shown in **Figure 2–37**. Such a test is called **one-tailed**, because the critical region occupies only one extreme of the test distribution. If we wish to test this hypothesis with the probability of rejecting it when it is true only one time in 20 ($\alpha = 0.05$), the computed value of t must exceed 1.83 for a one-tailed test. The statistical test is given in the same manner as before:

1. $H_0 : \mu_1 \leq 18\%$
    $H_1 : \mu_1 > 18\%$

2. $\alpha = 0.05$

3. $t = \dfrac{21.3 - 18.0}{5.52\sqrt{1/10}} = 1.89$

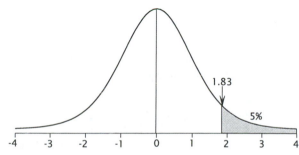

Figure 2–37. Student's t-distribution for nine degrees of freedom with one critical region (shown by shading) that contains 5% of the area under the curve. The critical value of $t = 1.83$.

The computed value of 1.89 exceeds the table value of t for nine degrees of freedom and the 5% ($\alpha = 0.05$) level of significance, and so lies in the critical region or region of rejection. On this basis we can reject the null hypothesis, leaving us with the alternative that the porosity of the population from which the Tensleep Sandstone sample was taken is indeed greater than 18%. If the computed value of t had been less than 1.83, we could only say that there is nothing in the sample to suggest that the population mean is greater than 18%. Note that we have not said that the mean is less than 18%, but only that there is no basis for suggesting that it is greater. As stated before, this indecisiveness is a consequence of the manner in which statistical tests are formulated. They can demonstrate, with specified probabilities, what things are not, but they cannot stipulate what they are.

Confidence intervals based on t

We can determine a confidence interval around a small-sample mean that will capture the true population mean a specified proportion of the time by modifying the large-sample confidence interval given in Equation (2.43). The interval is based on critical values of the t-distribution rather than the standard normal distribution. It is calculated as

$$\overline{X} \pm t_{\alpha/2,n-1} \cdot s_e \tag{2.46}$$

Here, n is the size of the sample and $t_{\alpha/2,n-1}$ is the critical value of t for a two-tailed test with critical regions containing $\alpha/2$ percent of the area under the curve in each tail. The standard error is estimated as $s/\sqrt{n}$, so the confidence interval can be written out as

$$\left(\overline{X} - t_{\alpha/2,n-1}\frac{s}{\sqrt{n}}\right) < \mu < \left(\overline{X} + t_{\alpha/2,n-1}\frac{s}{\sqrt{n}}\right) \tag{2.47}$$

We can find a 95% confidence interval for mean porosity of the Tensleep Sandstone, based on the sample of size 10 we examined above.

$$\left(21.3 - 2.262 \cdot \frac{5.52}{\sqrt{10}}\right) < \mu < \left(21.3 + 2.262 \cdot \frac{5.52}{\sqrt{10}}\right)$$

$$(21.3 - 3.948) < \mu < (21.3 + 3.948)$$

$$17.35 < \mu < 25.25$$

Ninety-five percent of the intervals calculated in this manner from samples of size 10 will capture the true mean porosity. On the basis of this one sample, we expect the mean porosity to be somewhere between 17.35% and 25.25%. The probability that the mean porosity actually falls somewhere outside this range is 5%. Note that since confidence intervals are based on two-sided critical regions, the critical value of t is different than the critical value of t in the one-tailed test of the preceding section, even though that test also had a 5% level of significance.

A test of the equality of two sample means

From another area in Wyoming, ten additional measurements of core porosity in the Tensleep Sandstone have been obtained. These are listed in **Table 2–17**. Are the means of the two sample collections the same? This is a somewhat different problem than the one previously considered. We now wish to compare statistics

Table 2–17. Porosity measurements on ten cores of Tensleep Sandstone, Pennsylvanian, from the Wind River Basin of Wyoming.

Core Number	Porosity (%)	Core Number	Porosity (%)
11	15	16	26
12	10	17	24
13	15	18	18
14	23	19	19
15	18	20	21
		TOTAL =	$\overline{189}$
		MEAN =	18.9
	$s^2 = 23.21$	$s = 04.82$	

of two samples against one another, rather than against a proposed population parameter. The appropriate test is again a t-test, but it is calculated in a more elaborate manner.

The hypothesis we are testing is

$$H_0 : \mu_1 = \mu_2$$

which states that the mean of the population from which the first sample was drawn is the same as the mean of the parent population of the second sample. This hypothesis is posed against the alternative

$$H_1 : \mu_1 \neq \mu_2$$

that the two population means are not equal. Again, we must specify a level of significance, say 10% ($\alpha = 0.10$). Our test statistic now has the form

$$t = \frac{\overline{X}_1 - \overline{X}_2}{s_e} \tag{2.48}$$

The standard error of the mean must be based on the characteristics of both samples, so we must generalize s_e to

$$s_e = s_p \sqrt{\frac{1}{n_1} + \frac{1}{n_2}} \tag{2.49}$$

Here, s_p is a **pooled estimate** of the standard deviation, found by combining the sample variances of the two data sets

$$s_p^2 = \frac{(n_1 - 1)\, s_1^2 + (n_2 - 1)\, s_2^2}{n_1 + n_2 - 2} \tag{2.50}$$

The subscripts refer, respectively, to sample 1 from the Bighorn Basin and sample 2 from the Wind River Basin. The process of pooling the two sample variances costs an additional degree of freedom, since two parameters $\left(\sigma_1^2 \text{ and } \sigma_2^2\right)$ must be estimated. The degrees of freedom for the t-test of equivalency given in Equation

(2.48) is therefore $v = n_1 + n_2 - 2$. Is the difference between the two sample means significant at the 10% level?

$$s_p^2 = \frac{9\,(30.46) + 9\,(23.21)}{10 + 10 - 2} = \frac{483.03}{18} = 26.84$$

$$s_p = 5.18$$

$$t = \frac{21.3 - 18.9}{5.18\sqrt{1/10 + 1/10}} = \frac{2.4}{2.32} = 1.03$$

Because the table values of t for a two-tailed test with 18 degrees of freedom and 10% level of significance (5% in each tail) are -1.73 and $+1.73$, the computed test value does not fall into either critical region, and the null hypothesis cannot be rejected. (Remember the critical region contains 10% of the area under the t-distribution curve.) We must conclude that there is no evidence to suggest that the two samples came from populations having different means.

Three assumptions are necessary to perform this test. One is that both samples were selected at random. The second is that the populations from which the samples were drawn are normally distributed. The third assumption is that the variances of the two populations are equal. The first assumption may be difficult to justify in many geologic problems, and may be a serious source of error if the samples are strongly and systematically biased (as they might be if the porosities were measured only on cores from producing zones of oil fields). A population may be tested for normality, but departures from normality seldom are a problem because of the central limits theorem, provided the sample is fairly large. The third assumption, equality of the variances of the two groups, is critical. Fortunately, this assumption can easily be checked and approximate t-tests are available if the variances of the two samples prove to be significantly different.

The t-test of correlation

Earlier, we introduced the correlation coefficient as a standardized measure of the linear relationship between two variables. However, we did not consider the question of the statistical significance of a given correlation coefficient. The sample correlation, r, is an estimate of the parameter ρ (rho), which expresses the relationship between two variables of a population. Provided both variables are normally distributed and the observations are chosen at random from the population, we can test the significance of r.

The most useful test is of the hypothesis and alternative

$$H_0 : \rho = 0$$
$$H_1 : \rho \neq 0$$

That is, we determine whether the observed sample correlation is significantly different from zero. The null hypothesis states that the two variables are independent, and that any nonzero value for r has arisen simply because of the vagaries of random sampling from a population whose variables are uncorrelated. A t-test for the significance of r is given by

$$t = \frac{r\sqrt{n-2}}{\sqrt{1-r^2}} \tag{2.51}$$

which has $n - 2$ degrees of freedom.

As an example, we may test the significance of the correlations we have measured between axial lengths of pebbles from a shingle beach, using the data in **Table 2–8**. The first correlation, between the a- and b-axes, is $r_{ab} = 0.597$, based upon ten pairs of measurements. The test statistic is therefore

$$t = \frac{0.597\sqrt{10 - 2}}{\sqrt{1 - 0.597^2}} = \frac{1.688}{0.802} = 2.10$$

The critical value for t with eight degrees of freedom and a 10% level of significance is $t = 1.860$. Remember that the test is two-sided, as r may be significantly greater or smaller (negative) than zero, so our area of rejection must be split into upper and lower regions. Because the test statistic falls into the upper critical region, we must conclude that there is a real correlation between the lengths of the longest and the intermediate axes of the beach pebbles.

For the other two correlations in the data set of **Table 2–8**, we have $r_{ac} = 0.499$ and $r_{bc} = 0.467$. The corresponding test values are

$$t = \frac{0.499\sqrt{10 - 2}}{\sqrt{1 - 0.499^2}} = \frac{1.411}{0.866} = 1.629$$

$$t = \frac{0.467\sqrt{10 - 2}}{\sqrt{1 - 0.467^2}} = \frac{1.321}{0.884} = 1.494$$

The critical value remains the same, so we see that neither of these two correlations is significantly different from zero. In other words, the observed correlations between the longest and shortest axes and between the intermediate and shortest axes could arise merely by chance in a random sample of ten pebbles, if the variables were completely independent of one another.

The F-Distribution

Tests to determine equality of variances are based on a probability distribution called the **F-distribution**, which is named in honor of the distinguished statistician, Sir Ronald Fisher. This is the theoretical distribution of values that would be expected by randomly sampling from a normal population and calculating, for all possible pairs of sample variances, the ratios

$$F = \frac{s_1^2}{s_2^2} \tag{2.52}$$

It seems reasonable that the sample variances will range more from trial to trial if the number of observations used in their calculation is small. Therefore, the shape of the F-distribution would be expected to change with changes in sample size. This returns us to the idea of degrees of freedom, except in this situation the F-distribution is dependent upon two values of v, one associated with each variance in the ratio. Also, it should be apparent that the distribution cannot be negative, because it is the ratio of two positive numbers. If the samples are large, the average of the ratios should be close to 1.0.

F-test of equality of variances

Because the F-distribution describes the probabilities of obtaining specified ratios of sample variances drawn from the same population, an **F-test** can be used to check the equality of the variances we obtain in statistical sampling. We may hypothesize that two samples are drawn from populations having equal variances. After computing the F-ratio, we then can ascertain the probability of obtaining, by chance, that specific value from two samples from one normal population. If it is unlikely that such a ratio could be obtained, we regard this as indicating that the samples came from different populations having different variances.

For any pair of variances, two ratios can be computed, s_1^2/s_2^2 and s_2^2/s_1^2; one ratio will be a fraction and the other will be greater than one. Logically, we should use a two-tailed test and reject the hypothesis of equality of the two variances if the ratio falls into either critical region. However, the F-distribution is constrained on the left by zero and has a long tail to the right, so the left-hand critical region is narrow while the right-hand critical region is very wide. If we always place the larger variance in the numerator, the ratio will always be greater than 1.0 and the calculated test statistic will always fall on the right. We then can test for significance using a one-tailed critical region on the right side of the distribution. This is shown in **Figure 2–38**, a typical F-distribution curve in which the critical region or area of rejection has been shaded.

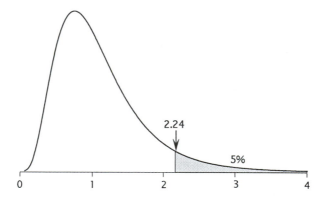

Figure 2–38. A typical F-distribution with $\nu_1 = 10$ and $\nu_2 = 25$ degrees of freedom; critical region (shown by shading) contains 5% of the area under the curve. Critical value of $F = 2.24$.

As an example of an elementary application of the F-test, consider a comparison between the two sample sets of porosity measurements on the Tensleep Sandstone. We are interested in determining if the variation in porosity is the same in the two areas. For our purposes, we will be content with a level of significance of 5%. That is, we are willing to run the risk of concluding that the porosities are different when actually they are the same one time out of every 20 trials.

The variances of the two samples may be computed by Equation (2.21). Then, the F-ratio between the two may be calculated by Equation (2.52), where s_1^2 is the larger variance and s_2^2 is the smaller. We now are testing the hypothesis

$$H_0 : \sigma_1^2 = \sigma_2^2$$

against

$$H_1 : \sigma_1^2 > \sigma_2^2$$

The null hypothesis states that the parent populations of the two samples have equal variances; the alternative hypothesis states that they do not. Degrees of freedom associated with this test are $(n_1 - 1)$ for v_1, and $(n_2 - 1)$ for v_2. The critical value of F with $v_1 = 9$ and $v_2 = 9$ degrees of freedom and a level of significance of 5% ($\alpha = 0.05$) can be found from Appendix Table A.3; that value is 3.18.

The value of F calculated from Equation (2.52) will fall into one of two areas. If the calculated value of F exceeds 3.18, the null hypothesis is rejected and we conclude that the variation in porosity is not the same in the two groups. If the calculated value is less than 3.18, we would have no evidence for concluding that the variances are different.

In most practical situations, we have no knowledge of the parameters of the population except for estimates made from samples. In comparing two samples, it is appropriate to first determine if their variances are statistically equivalent. If they appear to be equal, and the samples have been selected without bias from a naturally occurring population, it is probably safe to proceed to additional statistical tests.

Table 2-18. Concentration of microparticles in meltwater.

Concentration (ppb)			
Antarctica, $n = 16$		Greenland, $n = 18$	
3.7	0.6	3.7	1.6
2.0	1.4	7.8	2.4
1.3	4.4	1.9	1.3
3.9	3.2	2.0	2.6
0.2	1.7	1.1	3.7
1.4	2.1	1.3	2.2
4.2	4.2	1.9	1.8
4.9	3.5	3.7	1.2
		3.4	0.8

As an example, consider the following problem: Snow and ice collected from permanently frozen parts of the Earth contain small quantities of micron-sized bits of dust called microparticles. Individual grains range in size from about 0.5 to 3.0μ; these have been injected into the atmosphere by many agents, including volcanic eruptions, dust storms, and micrometeorite infall. The particles are so small they would remain suspended indefinitely, but they are scrubbed out by snow because they serve as nuclei for tiny ice crystals. The ice in turn is incorporated into the permanent snowfields of polar regions. It has been postulated that the concentration of microparticles in snow should be uniform over the Earth, because of mixing of the atmosphere and the manner by which microparticles are removed from the air. The theory, if true, has significance for the advisability of atmospheric testing of nuclear weapons, so two suites of snow samples have been collected carefully from the Greenland ice cap and from Antarctica. Under controlled conditions, the snow has been melted and the quantity of contained microparticles determined by an electron particle classifier. The volume concentration in parts per billion of microparticles in melted snow is given in **Table 2-18**. Do the two samples appear to be drawn from the same population, and do your conclusions tend to substantiate or refute the idea of atmospheric homogeneity?

Assuming the samples have been collected without bias and the distribution of microparticles is normal throughout the snowfields, the first step is to test the equality of variances in the two sample sets. This is done using Equation (2.52). The hypothesis and alternative are

$$H_0 : \sigma_1^2 = \sigma_2^2$$
$$H_1 : \sigma_1^2 > \sigma_2^2$$

If the variances are not significantly different, the next step in the procedure is to test equality of means. The appropriate test is Equation (2.48). For obvious reasons, the level of significance attached to this test cannot be higher than the significance attached to the test of equality of variances. The appropriate hypothesis and alternative are

$$H_0 : \mu_1 = \mu_2$$

$$H_1 : \mu_1 \neq \mu_2$$

because there is no reason to suppose that one region should have a larger mean than the other. If the variances and means cannot be distinguished (that is, the null hypotheses cannot be rejected), there is no statistical evidence to suggest that microparticle concentrations in the two areas are derived from different populations. On the other hand, if either test rejects its null hypothesis, the question of atmospheric homogeneity is in serious doubt. [If Equation (2.52) rejects the hypothesis of equality of variances, Equation (2.48) cannot be applied. Approximate tests, such as one described in Devore (1995), are available for testing equality of means from samples of unequal variances, but they would serve little purpose in this problem.]

Analysis of variance

Up to this point we have only considered techniques for comparing two samples, yet many problems involve several sets of observations. For example, suppose we have obtained five pieces of calcite-cemented sandstone. Each of the five rocks in our collection is lithologically somewhat different; one has conspicuously coarser sand grains, another appears to contain some clay, a third is slightly ferruginous, and so on. We wish to determine if the carbonate content is the same in each. We can consider this as a problem in the branch of statistics called ***analysis of variance.***

In general, techniques in this field involve separating the total variance in a collection of measurements into various components or sources. The tests of equality operate by simultaneously considering both differences in means and in variances.

A possible experimental approach to the problem would be to break each rock into a number of fragments and determine the carbonate content of each fragment by measuring its weight loss after acid treatment. Each fragment is called a ***replicate***. The purpose of breaking the original rock into replicates is to determine the variability within weight determinations on each specimen. Obviously, if the variation within replicates of a single specimen is large compared to the differences between specimens, the differences will be difficult to detect.

Suppose we break the original rocks into six fragments and analyze each. The variation we see arises from several causes; variation of composition within the original rock specimens, inadvertent differences in the manner the replicates were treated (perhaps the residue from one acid treatment was washed more vigorously than others), differences in weighing the residue (the replicates may retain different amounts of moisture, or the balance may vary in its response because of temperature changes during the day, *etc.*), and other subtle influences. These sources of variation all combine to produce what is known as *experimental error*, or the variation not accounted for by differences between the rocks.

To avoid the possibility of introducing a systematic error into the statistical analysis, the replicates must be treated in random order. This is known as *randomizing* the observations. The need for this step is apparent if there is some factor which consistently changes during the time of the experiment, perhaps continued drying of the acidified replicates as they await their turn to be weighed. If we weigh all six replicates of sample 1, then all replicates of sample 2, and so on, a greater weight loss may be recorded for the last weighed replicates simply because they dried over a longer period of time. One way to avoid this potential problem is to sequentially number each replicate and assign them to the analytical process according to a random number table. In fact, if the treatment proceeds in stages, it's best to assign each replicate randomly to each step of the treatment. Then the various sources of error are mixed up, or *confounded*, over all of the replicates, rather than being concentrated in a few.

Determining the equivalency of the five rock samples can be done by a technique called the *one-way analysis of variance*, because only one source of possible variation is analyzed. The hypothesis and alternative are

$$H_0 : \mu_1 = \mu_2 = \mu_3 = \mu_4 = \mu_5$$
$$H_1 : \text{at least one mean is different}$$

Certain assumptions are necessary to perform a test to choose between the two hypotheses. The assumptions are (*a*) each set of replicates represents a random sample from different populations, (*b*) each parent population is normally distributed, and (*c*) each parent population has the same variance.

The data for our problem are given in **Table 2–19** and at the Web sites in file ONEOVA.TXT. In the one-way analysis, total variance of the data set is broken into two parts—variance within each set of replicates and variance among the rocks. Statisticians have developed a formalized procedure for analysis of variance which is contained within an *ANOVA* (*AN*alysis *O*f *VA*riance) table. This lists the sources of variation, a column of corrected sums of squares resulting from the various

Table 2–19. Carbonate cement in five sandstone samples with replicates; numbers in brackets signify order of analysis.

| Replicate | Carbonate Cement (%) | | | | |
	Sample 1	Sample 2	Sample 3	Sample 4	Sample 5
1	19.2 [11]	18.7 [04]	12.5 [28]	20.3 [23]	19.9 [21]
2	18.7 [08]	14.3 [19]	14.3 [16]	22.5 [30]	24.3 [06]
3	21.3 [09]	20.2 [14]	8.7 [20]	17.6 [24]	17.6 [18]
4	16.5 [17]	17.6 [07]	11.4 [29]	18.4 [03]	20.2 [22]
5	17.3 [26]	19.3 [05]	9.5 [27]	15.9 [13]	18.4 [12]
6	22.4 [15]	16.1 [25]	16.5 [01]	19.0 [02]	19.1 [10]

sources, degrees of freedom associated with each, a column called *mean squares* which are nothing more than the sample-based estimates of the variances, and the *F*-test value. The design of an ANOVA table appropriate for our problem is outlined in general form in the box below:

Source of Variation	Sum of Squares	Degrees of Freedom	Mean Squares	*F*-Test
Among Rocks	SS_A	$m-1$	MS_A	MS_A/MS_E
Within Replications ("Error")	SS_E	$N-m$	MS_E	
Total Variation	SS_T	$N-1$		

The symbology of analysis of variance differs from author to author; we will use a notation that is a compromise. The total variation in the data is given by the *total sum of squares*, SS_T, or the sum of the squared differences between the observations and the overall or *grand mean*, $\overline{\overline{X}}$

$$SS_T = \sum_{j=1}^{m} \sum_{i=1}^{n} \left(x_{ij} - \overline{\overline{X}}\right)^2 = \sum_{j=1}^{m} \sum_{i=1}^{n} x_{ij}^2 - \frac{\left(\sum_{j=1}^{m} \sum_{i=1}^{n} x_{ij}\right)^2}{N} \qquad (2.53)$$

where m = number of samples and n = number of replicates per sample. In this symbology, x_{ij} is the ith replicate of the jth sample. The double summation indicates that we first sum down each of the columns containing the n replicates, then sum the m column totals. The total number of observations, N, is equal to the number of replicates per sample times the number of samples, or $N = n \times m$. The final term in the equation is referred to as the *correction term* and appears in other equations as well.

Similarly, the variation that is attributable to the differences between the samples can be measured by the sum of the squared differences between the sample means and the grand mean. This is the *sum of squares among* the samples, SS_A

$$SS_A = \sum_{j=1}^{m} \left(\overline{X}_j - \overline{\overline{X}}\right)^2 = \sum_{j=1}^{m} \frac{\left(\sum_{i=1}^{n} x_{ij}\right)^2}{n} - \frac{\left(\sum_{j=1}^{m} \sum_{i=1}^{n} x_{ij}\right)^2}{N} \qquad (2.54)$$

which means that we sum the replicates within each sample $\left(\sum_{i=1}^{n} x_{ij}\right)$, square each of the totals, divide by the number of replicates in each sample (n), sum the resulting figures for all samples, and finally, subtract the correction term.

The second source of variation is that within samples, which pools the sums of squared differences of the individual replicates from the means of their sample. This is referred to as the *sum of squares within*, or the *error sum of squares*, SS_E

$$SS_E = \sum_{j=1}^{m} \left(\sum_{i=1}^{n} x_{ij} - \overline{X}_j\right)^2$$

$$= \sum_{j=1}^{m} \sum_{i=1}^{n} x_{ij}^2 - \sum_{j=1}^{m} \frac{\left(\sum_{i=1}^{n} x_{ij}\right)^2}{n} \qquad (2.55)$$

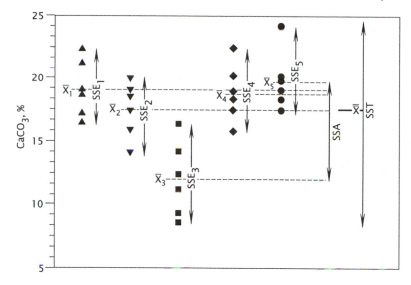

Figure 2–39. Sources of variation for calcium carbonate cement within five sandstone samples, each having six replicate analyses.

The variation that the different sums of squares express is graphically illustrated in **Figure 2–39**. Note that the first term of Equation (2.55) is the first term of the equation for SS_T (Eq. 2.53) and the last term is the first term in the equation of SS_A (Eq. 2.54). Therefore, we can find SS_E by the operation

$$SS_E = SS_T - SS_A \qquad (2.56)$$

If we divide the sums of squares by their degrees of freedom, we get estimates of the variance. In effect, the ANOVA procedure divides the total variance into its components and allows us to determine which of these are significant. The number of degrees of freedom for variance in the total data set is $N - 1$. Degrees of freedom for variance between the samples is $m - 1$, because we are estimating this variance from the means of the replicates of each sample. The difference between the two is the degrees of freedom attributable to the error variance, or variance within the samples.

Working through this problem will help clarify the analysis of variance procedure. First, the total variance of the data in **Table 2–19** must be found. Using Equation (2.53) for SS_T, we get

$$SS_T = 383.79$$

Next, we must determine the variance of the average values for the five samples. Following Equation (2.54), we first sum all five columns, square the totals, divide each by six, and subtract the correction term. This gives us the sum of squares among the samples, or

$$SS_A = 237.42$$

Finally, we subtract SS_A from SS_T to give the sum of squares within the samples, *i.e.*, the sum of squares attributable to error

$$SS_E = 146.37$$

Total degrees of freedom are $N - 1$, or 29. Because we are estimating the variance between the samples from five measurements (the five column means), degrees of freedom for SS_A are $m - 1$, or 4. The leftover degrees of freedom must be associated with the leftover sums of squares, or the error measure. The difference in degrees of freedom is $N - m$ or 25. Now the quantities SS_T, SS_A, and SS_E are corrected sums of squares and must be divided by the appropriate degrees of freedom to give variances (or **mean squares**, which is simply another name for the same thing).

$$\text{Total Variance} = \frac{SS_T}{N - 1} = \frac{383.79}{29} = 13.23$$

$$\text{Variance Between} = MS_A = \frac{SS_A}{m - 1} = \frac{237.42}{4} = 59.35$$

$$\text{Error Variance} = MS_E = \frac{SS_E}{N - m} = \frac{146.37}{25} = 5.85$$

The rationale in analysis of variance may be clearer if we consider the extreme situation where replicates are all identical. Then, the means of the columns will be the same as the entries within the columns and the variance calculated by considering all observations will be the same as that based only on the column means. In other words, the error measure will vanish, there being no variance unaccounted for by the differences within the replicates. Such an unlikely result would indicate that the original samples are indeed different, each set of replicates having been drawn from separate populations having zero variances.

Our example is less extreme. Calculating the F-test, we obtain the following critical value. Having chosen a critical region based on our selection of the desired level of significance and the degrees of freedom, we can now reject or accept the hypothesis:

$$F = \frac{MS_A}{MS_E} = \frac{59.35}{5.85} = 10.14$$

In the jargon of applied statistics, a source of variation to be investigated may be called a "treatment" or a "factor," and the different categories are called "levels" of the treatment or factor. The actual values measured are called "responses." These terms obviously are derived from agricultural applications, where many of the techniques of analysis of variance were originally developed. In our example, the rock specimens constitute the treatment (there is only one treatment in a one-way analysis), the five different rocks are the levels, and the carbonate measurements are the responses.

The one-way analysis of variance is appropriate if we wish to test the hypothesis that a number of populations, represented by samples or different levels of a treatment, are identical. However, we must be able to randomly select responses within the levels and make measurements in random order. This may be a severe restriction in certain situations and may lead to analyses in which too much information about the variance is lost. For example, it may be suspected that a certain step in the measurement technique is causing an increase in variance. Using a one-way model, we cannot assess the magnitude of the introduced variance because it is mixed with other sources of variance in the SS_E sum of squares. However, more elaborate statistical designs may allow us to isolate this cause of variation and estimate its magnitude.

Fixed, random, and mixed effects

Analyses of variance may be formulated according to different "models," depending upon the way observations have been collected and the intent of the analysis. In a *fixed-effects model* (or Model I), the specific levels of a treatment are the only ones of interest, and any statistical conclusions are valid only for these specific levels. In a sense, a fixed-effects model is an extension of an ordinary t-test of the difference in means of two samples, where the conclusion applies only to the two samples being compared. Any extrapolation of the results from a fixed-effects model to a broader spectrum of levels must be based on judgment, not on statistical inference. The one-way ANOVA we have just concluded follows a fixed-effects model, and the test result has no implications beyond the five rock samples.

In contrast, the levels of treatment in a *random-effects model* (or Model II) have been chosen in a manner so they represent a broader set of possible treatment levels. The specific treatment levels used are of interest only insofar as they are representative of all levels of treatment. The purpose of the analysis is to estimate the relative contributions of different sources of variation (the treatments) to the total variance.

A *mixed-effects model* incorporates some fixed-treatment levels and has other treatments whose levels are random. Krumbein and Graybill (1965) have a concise discussion of the different implications of these models in geological situations. Eisenhart (1947) offered the initial definitions of these models, which are extensively discussed in most books on design of experiments and linear models, such as Morrison (1983) and Neter, Wasserman, and Kutner (1985).

For our purposes, the distinctions between fixed-, mixed-, and random-effects ANOVA models are not computationally important, although they may have a critical bearing on the way we present our conclusions. The result of a fixed-effects analysis is relevant only for the specific set of observations being tested, whereas the result of a random-effects analysis has implications beyond the test material itself.

We can briefly examine a one-way ANOVA in which the random-effects model is appropriate. The instructor in a large introductory petroleum engineering laboratory course was concerned about the ability of students to perform standard laboratory procedures. To determine if students differ significantly in their abilities, he randomly selected five class members and had each student measure the porosities of four core plugs cut from a single core. The plugs were measured in random order using a single porosimeter. The measured porosities are listed in **Table 2–20** and in file RESENG.TXT. Are the differences between students significant?

This is a random-effects model because the experimental subjects, the students, were chosen at random from a larger population and the inference that students do, or do not, differ significantly in their ability to perform this particular measurement can be applied to the entire class. The experimental objective was to determine if differences between students account for a significant proportion of the variation in laboratory performance, and not to draw contrasts between the specific students participating in the exercise. Under the random-effects model, the results are applicable to the entire population of students in the class, but any

Table 2–20. Porosities, in percent, of four core plugs cut from a single sandstone core, as measured by five randomly chosen students in a petroleum engineering laboratory course.

Treatment	Porosities (%)				Totals $(\sum_{i=1}^{n} x_{ij})$	Means $(\sum_{i=1}^{n} x_{ij}/n)$
Student 1	11	18	24	15	68	17
Student 2	26	25	22	11	84	21
Student 3	13	19	22	10	64	16
Student 4	26	21	19	22	88	22
Student 5	19	27	28	22	96	24
			TOTALS		400	20

extrapolation beyond this particular class, either to past or future classes or to similar classes at other universities, must be based on judgment.

Two-way analysis of variance

Many tests, some of elaborate design, are described in texts on analysis of variance and design of experiments. Excellent descriptions of some designs most useful for geologists are contained in the books by Griffiths (1967) and Krumbein and Graybill (1965). Here we must limit our consideration to a few additional examples and their statistical designs.

The St. Peter Sandstone of Ordovician age is a remarkably pure orthoquartzite which occurs in the upper Mississippi River Valley. Because the sand grains of this unit are so well rounded and sorted, the St. Peter is unusually homogeneous in character. For this reason, petroleum reservoirs in the St. Peter Sandstone should react during pumping in a manner that could be closely predicted by theoretical models of reservoir behavior, even though these are based on idealized conditions. Deviations from reality in model behavior might indicate erroneous assumptions in the model's structure.

A small oil field in southern Illinois appeared ideally suited for examining the coincidence of model behavior to actual production performance. Carefully documented development and production data were available for the field. However, before performing an extensive analysis of the reservoir's behavior, it seemed prudent to examine the fundamental assumption that St. Peter Sandstone is essentially homogeneous in its properties. From the collection of drill cores taken during development of the field, ten were randomly selected for analysis. Each core was sliced at a random position and two 1-in. "plugs" were cut, one horizontally across the core and the other vertically, down the core. Using a liquid permeameter, the rates of fluid flow through the plugs were measured. Measurements from the vertical core plugs express flow perpendicular to bedding; measurements from the horizontal core plugs express flow in a plane parallel to bedding. Using the flow rates, permeabilities in millidarcies were calculated. The 20 permeability values are listed in **Table 2–21** and are contained in file STPETER.TXT. From these 20 responses, we will attempt to determine two things; (*a*) whether there are significant differences in permeability with location in the field (*i.e.*, between wells), and (*b*) if there are significant differences between horizontal and vertical permeabilities.

Table 2–21. Directional permeability of randomly selected cores of St. Peter Sandstone, Illinois.

DIrectional Permeability (md)			
Vertical	Horizontal	Vertical	Horizontal
1037	1124	928	943
936	960	1108	1165
842	921	821	803
1121	1202	797	792
1043	1028	949	1004

The problem may be considered as a *two-way analysis of variance*. We will be concerned with two main sources of variation; those arising because of differences between the cores, and those arising because of differences in the direction of flow in the permeability measurements. A third source of variation is the leftover, residual, or error variance, corresponding to variance within replicates in the one-way ANOVA. In this example, we will consider two hypotheses:

$$H_0 : \mu_{\text{well 1}} = \mu_{\text{well 2}} = \cdots = \mu_{\text{well 10}}$$
$$H_0 : \mu_{\text{vertical}} = \mu_{\text{horizontal}}$$

The corresponding alternative hypotheses are that at least one well is different, and that vertical and horizontal permeabilities are not the same. With one exception, this problem is much like the one-way ANOVA on carbonate determinations which we just worked. We have made two permeability measurements on each randomly selected piece of core. However, the two measurements are not replicates taken in some random fashion, but are of a distinctly different nature. Usually we arrange the data from such an experiment in an array in which the columns might be the permeabilities measured using horizontal or vertical core plugs; the rows would represent the ten wells. The manner in which the permeability measurements were taken constitutes a *treatment*, implying that the numbers generated using one level of treatment may be fundamentally different from those resulting from other levels of treatment even though the samples used are the same. Because the measurements are not completely randomized but are instead separated according to the treatment level, the data can be analyzed for differences between the columns (*i.e.*, between vertical and horizontal flow) as well as differences between rows (*i.e.*, between wells). Variance due to differences in treatment has been separated by the statistical design.

Four fundamental assumptions about the nature of the parent population are made in performing this test: (*a*) each combination of treatment and object is a random sample drawn from different populations, (*b*) each parent population is normal, (*c*) each parent population has the same variance, and (*d*) there is no interaction between different treatments and different samples. The last assumption is a statement that a particular combination of treatment and sample will not produce a greater variance than treatments and samples in other combinations. If we performed the ANOVA using replications (*i.e.*, more than one permeability measurement on each direction/core combination), we could assess interaction, but in this simple design we must assume that it does not occur. If interactions do exist, they cannot be detected by this experimental design, and their presence will invalidate the test results. Interaction between parameters erroneously assumed to be independent has led more than one researcher to grief. A good introduction to the effects of interactions is contained in Hicks and Turner (1999).

The appropriate ANOVA for two-way analysis without replications is given below. SS_T is calculated as in Equation (2.53); SS_A, the sum of squares attributable to differences between the levels of treatment or factor A, is calculated according to Equation (2.54). SS_B is the sum of squares within treatment B and is calculated by

$$SS_B = \sum_{i=1}^{n} \frac{\left(\sum_{j=1}^{m} x_{ij}\right)^2}{m} - \frac{\left(\sum_{j=1}^{m}\sum_{i=1}^{n} x_{ij}\right)^2}{N} \qquad (2.57)$$

where m = number of samples and n = number of levels in the second treatment or factor. The error sum of squares, SS_E, is found by

$$SS_E = SS_T - (SS_A + SS_B) \qquad (2.58)$$

From the new equations, you can see that SS_B is a measure of the variance of the treatments as determined from averages of the samples within each treatment. The error sum of squares is reduced by the amount assigned to this source of variation. Symbolic conventions are the same as in the one-way ANOVA, except that n is now the number of levels in the second treatment rather than replications.

Source of Variation	Sum of Squares	Degrees of Freedom	Mean Squares	F-Tests
Among Treatment A	SS_A	$m - 1$	MS_A	MS_A / MS_E [a]
Among Treatment B	SS_B	$n - 1$	MS_B	MS_B / MS_E [b]
Error	SS_E	$(m - 1)(n - 1)$	MS_E	
Total Variation	SS_T	$N - 1$		

[a] Test of significance of differences between levels of treatment A.
[b] Test of significance of differences between levels of treatment B.

After selecting the desired level of significance for the two hypotheses, we can use this statistical design to test the permeability data in **Table 2–21**. The questions to be answered are (*a*) is there a significant difference in permeabilities over the field, and (*b*) is there a significant difference in vertical and horizontal permeabilities? By extending the same calculations that we used in the one-way procedure, we can find the additional terms MS_B and the new error term MS_E. The results are contained in the ANOVA table that follows. There are highly significant differences in permeability between the wells, and a somewhat less significant difference between vertical and horizontal permeabilities. These results suggest that the St. Peter Sandstone cannot be regarded as geographically homogenous in permeability,

Source of Variation	Sum of Squares	Degrees of Freedom	Mean Squares	F-Tests
Among Wells	331632.45	9	36848.05	20.07[a]
Among Directions	14204.45	1	14204.45	7.74[b]
Error	16522.05	9	1835.8	
Total Variation	362358.95	19		

[a] $p < 0.0001$.
[b] $p = 0.0213$.

at least at the scale of the spacing between wells in a field. Neither is the reservoir rock homogenous with respect to the direction of flow.

It may be instructive to compare one- and two-way ANOVA's to see if there are advantages to using the more complicated statistical model. If we ignore the difference between vertical and horizontal measures of permeability, we can consider the St. Peter data to consist of two replicate measurements from each of ten wells. This is the resulting one-way ANOVA table:

Source of Variation	Sum of Squares	Degrees of Freedom	Mean Squares	F-Test
Among Wells	331632.45	9	36848.05	11.99[a]
Error	30726.50	10	3072.65	
Total Variation	362358.95	19		

[a] $p < 0.0003$.

Note that SS_E is larger for the one-way analysis than for the two-way design, because variation due to the difference between vertical and horizontal permeability measurements is confounded with random variation. Because SS_A is unchanged, the F-ratio for the simpler model must be lower and somewhat less significant. In this example, either ANOVA procedure detects highly significant differences between wells, but in other situations significant differences may not be found unless more than one treatment effect is considered. This does not mean that a more complicated statistical model is always better. If a treatment is included in a model when in fact it does not contribute significantly to total variation, the error sum of squares will not be reduced but the degrees of freedom for error will be smaller. This will inflate MS_E and reduce the value of the test statistic, F. The significance of additional treatments or factors can be tested, but this topic will be deferred until Chapter 5 where it will be discussed with regression.

Nested design in analysis of variance

A conventional two-way analysis of variance assumes that observations can be assigned randomly to combinations of the two factors or treatments. Sometimes this is not possible because levels of one factor occur within, or are subsets of, levels of the second factor. Such a design is called **nested** or **hierarchical**, and the analysis of variance must be modified to account for the lack of randomization. Nested designs often are required in geology, because measurements are constrained by the available sampling localities. We will illustrate a nested design from sedimentary petrology.

Five cores have been taken from different wells drilled in a limestone reservoir; the mean pore size of voids in the rock is to be determined by microscopic examination. A vertical thin section is cut from each core and examined using a petrographic microscope with a traversing stage. Four parallel traverses are made on each thin section and the chord distance across the first four pores encountered along each traverse is measured in millimeters. A "tight pore" whose apparent chord length is less than 1 mm is recorded as zero. The data consist of 80 measurements of pore chord distances, recorded on sets of four traverses, with each set of traverses nested inside one of the five thin sections. The data are contained in file NESTED.TXT, where the first variable is the measured chord length, the second variable is the number of the traverse (from 1 to 20), and the third variable is the number of the thin section (from 1 to 5). Traverses 1 through 4 are nested within slide 1, traverses 5 through 8 are nested within slide 2, and so on.

The ANOVA table for a nested two-way design has the following appearance:

Source of Variation	Sum of Squares	Degrees of Freedom	Mean Squares	F-Tests
Treatment A	SS_A	$a-1$	MS_A	MS_A/MS_E
Treatment B within A	$SS_{B(A)}$	$a(b-1)$	$MS_{B(A)}$	$MS_{B(A)}/MS_E$
Error	SS_E	$ab(n-1)$	MS_E	
Total Variation	SS_T	$abn-1$		

The total sum of squares, SS_T, is defined in the conventional manner; that is, as the sum of the squared differences between all measurements and the grand mean of all measurements. In our example, SS_T is the sum of the 80 squared differences between all measurements on all traverses on all thin sections, and the average value. Because the grand mean must be estimated, the total degrees of freedom are $80 - 1 = 79$.

$$SS_T = \sum_{i=1}^{a}\sum_{j=1}^{b}\sum_{k=1}^{n}\left(x_{ijk}-\overline{\overline{X}}\right)^2 = \sum_{i=1}^{a}\sum_{j=1}^{b}\sum_{k=1}^{n}x_{ijk}^2 - \frac{\left(\sum_{i=1}^{a}\sum_{j=1}^{b}\sum_{k=1}^{n}x_{ijk}\right)^2}{abn} \qquad (2.59)$$

SS_A is the sum of squares attributable to differences between levels of the main treatment, or in our example, between thin sections. Usually, this is the variation of interest. SS_A is defined as the sum of the squared differences between the means of each level of treatment A and the grand mean; in our example, the sum of squared differences between the grand mean and the averages of all measurements on each thin section. The degrees of freedom are $5 - 1 = 4$.

$$SS_A = bn \sum_{i=1}^{n} \left(\overline{X}_{i..} - \overline{\overline{X}}\right)^2 = \frac{\sum_{i=1}^{a} \left(\sum_{j=1}^{b}\sum_{k=1}^{n} x_{ijk}\right)^2}{bn} - \frac{\left(\sum_{i=1}^{a}\sum_{j=1}^{b}\sum_{k=1}^{n} x_{ijk}\right)^2}{abn} \qquad (2.60)$$

$SS_{B(A)}$ is the sum of squares of treatment B within levels of treatment A, and is a measure of the variation in traverses that are nested (or confined) within thin sections. The differences between the means of each of the four traverses on a slide and the overall mean for the slide are found, resulting in $4 \times 5 = 20$ differences which are squared and summed. Since means must be estimated for five slides, there are $20 - 5 = 15$ degrees of freedom for $SS_{B(A)}$.

$$SS_{B(A)} = n \sum_{i=1}^{a}\sum_{j=1}^{b} \left(\overline{X}_{ij.} - \overline{X}_{i..}\right)^2 = \frac{\sum_{i=1}^{a}\sum_{j=1}^{b}\left(\sum_{k=1}^{n} x_{ijk}\right)^2}{n} - \frac{\sum_{i=1}^{a}\left(\sum_{j=1}^{b}\sum_{k=1}^{n} x_{ijk}\right)^2}{bn} \qquad (2.61)$$

The error sum of squares, SS_E, can be found by difference, because $SS_T = SS_A + SS_{B(A)} + SS_E$, or directly as the sum of the squared differences between each observation and the mean of the nested level of B within A in which the observation occurs. Each of the four pore chord measurements along a traverse is compared to the mean of that traverse for all the traverses and all the slides, resulting in $4 \times 4 \times 5 = 80$ values that are squared and summed. Because the means of 20 traverses must be estimated, SS_E has $80 - 20 = 60$ degrees of freedom.

$$SS_E = \sum_{i=1}^{a}\sum_{j=1}^{b}\sum_{k=1}^{n}\left(x_{ijk} - \overline{X}_{ij.}\right)^2 = \sum_{i=1}^{a}\sum_{j=1}^{b}\sum_{k=1}^{n} x_{ijk}^2 - \frac{\sum_{i=1}^{a}\sum_{j=1}^{b}\left(\sum_{k=1}^{n} x_{ijk}\right)^2}{n} \qquad (2.62)$$

The completed ANOVA for testing the significance of differences in pore chord diameters is given below. The nested ANOVA indicates that the pore chord lengths are essentially the same for all thin sections, so the thin sections can be treated as coming from a single population. The weakly significant ($p = 0.0625$) difference between traverses comes almost entirely from the traverses across the second thin section, which is exceptionally variable.

Source of Variation	Sum of Squares	Degrees of Freedom	Mean Squares	F-Tests
Thin Sections	45.075	4	11.269	1.053^a
Traverses w/i Thin Sections	282.875	15	18.858	1.7625^b
Error	642.0	60	10.70	
Total Variation	969.95	79		

$^a p = 0.3876$ (not significant).
$^b p = 0.0625$.

If this analysis is erroneously done as a two-way ANOVA with replications, the differences between thin sections and between traverses do not even approach statistical significance. The ability of the ANOVA to detect differences is severely reduced because traverses are compared across all thin sections, rather than being compared only to other traverses on the same thin section. Variance attributable to differences between traverses within thin sections contributes instead to the error variance. The following two-way ANOVA table shows the consequences of incorrectly considering traverses to be an independent second treatment; significant differences between traverses are no longer detected.

Source of Variation	Sum of Squares	Degrees of Freedom	Mean Squares	F-Tests
Thin Sections (Factor A)	45.075	4	11.269	0.9247^a
Traverses (Factor B)	23.04	1	23.04	1.8905^b
Error	901.835	74	12.187	
Total Variation	969.95	79		

$^a p = 0.4544$ (not significant).
$^b p = 0.1733$ (not significant).

The ANOVA designs we have examined are said to be **balanced**; this means that there are identical numbers of observations in each level of each treatment. There are advantages to a balanced design; the standard errors at each level reflect only the differences between observations within the levels and not different numbers of observations. It is simpler to pool variances based on equal-sized samples, so computations are somewhat easier. However, when using statistical computer packages, the computational effort is not an important consideration and unbalanced designs can easily be analyzed. This may have a tremendous advantage for geologists, because often the nature of a geological problem precludes designing a balanced statistical experiment.

To illustrate the application of an unbalanced, multi-way analysis of variance, we will assess the importance of different influences on measurements of groundwater elevation in observation wells in western Kansas. Results of the study will be used to improve the accuracy and precision of the water-level monitoring effort. Field parties use a steel tape, lowered down a well bore, to determine the depth to the water's surface in selected wells. Measurements are made in midwinter, when most wells (particularly those used for irrigation) are not pumping. Many factors could influence the measurements, including the tapes themselves, the experience of the persons making the readings, the condition of the wells, and the aquifers being tapped by the wells. To assess these (and other) possible contributors, 48 wells were selected at random out of a population of 562 observation wells; each well in the sample was measured at least twice, by different field crews. The resulting set of 112 measurements are contained in file WATER.TXT. To eliminate the effects of geographic location and land surface elevation, the depth to water has been subtracted from the depth measured in the same well in the previous year. This annual difference in water-table elevation is the response variable in the file. Each measurement is accompanied by a well ID number, the initials of the person making the measurement, a binary (two-state) variable that indicates if the measuring tape had a weight attached to its lower end, a nominal code indicating the use of the well (irrigation, household, stock watering, or unused), and a nominal code identifying the aquifer or combination of aquifers from which the well draws. We can construct a statistical model of the annual change in depth to water as a function of these variables, and then estimate how much each factor or treatment contributes to the total variation. This results in an ANOVA of four treatments with unequal numbers of levels.

Source of Variation	Sum of Squares	Degrees of Freedom	Mean Squares	F-Tests
Between Measurers	87.317	5	17.463	5.246^a
Weight on Tape?	76.048	1	76.048	22.845^b
Use of Well	113.907	3	37.969	11.406^b
Aquifer Pumped	85.182	3	28.394	8.529^b
Error	329.557	99	3.329	
Total Variation	692.011	111		

[a] $p = 0.0003$; [b] $p < 0.0001$.

Because each well was measured twice, we also could check for lack of fit of the model, or for interactions between different factors. However, there are not enough replicates to check all possible interactions, because we would run out of degrees of freedom if we attempted to assess them.

The ANOVA table shows that all treatments are contributing significantly to the variance in annual change in water-table elevation. Operator variance, or that contributed by different persons measuring the wells, comes mostly from one individual, DRL, whose measurements are both lower on average and more variable than measurements made by others. Measurements made using tapes without weights were significantly lower and more variable than measurements made using weighted tapes. Wells used for household water supply have significantly greater annual change in the depth to water than do wells used for other purposes, probably reflecting the fact that these wells are pumped year-round. Finally, there are significant differences in the response variable for wells that tap different aquifers.

The analysis suggests that some operators may need more training or better supervision, that tapes should always be equipped with weights, that home water supply wells should not be used for monitoring purposes, and that wells which draw from different aquifers probably should not be mixed together in an analysis. In this example, which was taken from a much larger study, all of the treatments proved to be making significant contributions to the total variance. The complete analysis found that other treatments were not significant sources of variance and probably could be ignored in future studies.

Analyses of variance are among the most widely used statistical tests, especially in fields such as quality control, product development, and clinical experimentation. Consequently, computer programs are readily available which will perform ANOVA's of almost any degree of complexity.

The χ^2 Distribution

We must now introduce another probability distribution based on the properties of a normal population. If samples of size n are taken from a normal population having a mean, μ, and standard deviation, σ, each observation within a sample can be standardized by Equation (2.37) to the standard normal form (rewritten here to include population parameters rather than sample statistics):

$$z_i = \frac{x_i - \mu}{\sigma} \tag{2.63}$$

If the standardized values of z_i are squared and summed, they form a new statistic which we can denote as $\sum z^2$. That is,

$$\sum z^2 = \sum_{i=1}^{n} \left(\frac{x_i - \mu}{\sigma} \right)^2 \tag{2.64}$$

Because this is a sample-based statistic, it will vary from sample to sample. If we draw all possible samples of size n from a normal population and plot the values of $\sum z^2$ they will form a χ^2 (chi-squared) distribution. The distribution of χ^2 is dependent upon degrees of freedom that are related to the size of the samples involved in its creation. A typical χ^2 curve is shown in **Figure 2–40**, and tables of χ^2 for various degrees of freedom and levels of significance are given in Appendix Table **A.4**. Note that the curve, like that of the F-distribution, goes from zero to

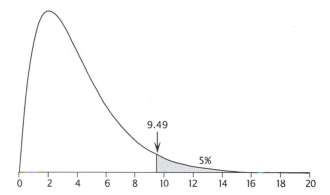

Figure 2–40. A typical χ^2 distribution for four degrees of freedom with critical region (shown by shading) containing 5% of area under the curve. Critical value of $\chi^2 = 9.49$.

positive values approaching infinity. This intuitive derivation of the χ^2 distribution is developed further in Li (1964, Chapter 7).

The great utility of the χ^2 distribution is that it can be used for tests of nominal and ordinal data. Up to this point, we have considered only tests of continuous (interval or ratio) variables. We now will examine some methods to treat count data such as the number of echinoids per unit area on the sea floor, the number of plagioclase crystals encountered on traverses across a thin section, and the number of grains within certain size classes of a disaggregated sandstone.

Goodness-of-fit test

A commonly encountered problem in elementary statistics is comparing a distribution of sample observations to some specified model distribution. As examples, a researcher may wish to apply statistical tests that assume the data are drawn from a population having certain characteristics, such as a normal or lognormal distribution. The frequency distribution of the sample may be compared to the hypothetical distribution to see if this assumption is warranted. In highway construction, allowable percentages of the various size classes of particles in aggregate may be contractually specified; a geologist or engineer may wish to know if the distribution of particle sizes in a natural gravel deposit or in the output from a rock crusher meets these specifications. In both of these problems, it is necessary to measure the correspondence between the form of two distributions, one estimated from a sample and the other assumed or specified. A probabilistic answer is needed to the question, "do the two distribution curves have the same shape?"

A similar problem arose in the course of a bottom-sampling project in Whitewater Bay, Florida, where 48 measurements of surface-water salinity were made. The measurements are listed in **Table 2–22** and in the file WHITE.TXT on the CD-ROM. Geographic considerations lead us to believe that the salinity variation in the sample area is random and should be normally distributed. If this hypothesis is true, it would imply a free mixing and interchange between open marine water and freshwater entering the bay. On the other hand, if some mechanism is operating which tends to keep fresh and saline waters separated in the bay, the distribution of salinity values would reflect this. It may in turn be possible to draw inferences about water circulation patterns and the expected distribution of bottom-sediment types.

Table 2–22. Measurements recorded at 48 stations in Whitewater Bay, Florida.

				Salinity (‰)					
46	53	58	60	60	49	59	48	46	78
37	58	46	46	47	48	42	50	63	48
62	49	47	36	40	39	61	43	53	42
59	60	52	34	40	36	67	44	40	
40	56	51	51	35	47	53	49	50	

We can test how well the distribution of sample values conforms to a normal distribution by a test procedure called the **goodness-of-fit** test. We hypothesize that the population of salinities from which our samples are drawn is normally distributed with an unknown mean, μ, and variance, σ^2. The alternative to this hypothesis is, of course, that the parent population is not normally distributed. A test statistic may be devised by dividing a standard normal distribution into a number of classes. The probability that a random observation from a standard normal distribution will fall into one of the classes is equal to the area under the curve within the class. From these probabilities, the number of observations that are expected in each class can be calculated. The expected frequency of occurrence within each class can be compared with the frequency of sample observations that actually fall within the classes. If the number in each class deviates significantly from that expected, it seems unlikely that the sample was drawn from a normal population. By use of the χ^2 distribution, we can attach probabilistic meaning to the words "significant" and "unlikely" in this statement.

The test statistic is calculated by the equation

$$\chi^2 = \sum_{j=1}^{k} \frac{\left(O_j - E_j\right)^2}{E_j} \tag{2.65}$$

where O_j is the number of observations within the jth class, and E_j is the number of observations expected in the jth class. There are k classes or intervals.

In this problem, the test statistic is computed by dividing a standard normal curve into a number of segments, such that each contains the same area and hence the same probability of occurrence. When our data are standardized, we expect that about the same number of standardized values will occur in each of the intervals. The number of observations actually falling into an interval is counted, the difference between this and the number we expect to find is determined, and the difference is squared. The squared difference is divided by the expected number and the values for all classes are summed. If this statistic falls into the critical region of the appropriate χ^2 distribution, the null hypothesis is rejected and we may conclude that the sample distribution does not follow a normal distribution.

The sampling distribution of χ^2 resembles **Figure 2–40**, although the exact shape depends upon the degrees of freedom. However, the degrees of freedom are not based on the number of observations in the same sense as in previous tests. In the χ^2 problem, our "observations" actually are the four categories of the sample distribution that we are comparing against the equivalent four categories of the standard normal curve. The number of degrees of freedom is (number of categories $= k$) $- 3$, or in our example, one. We lose one degree of freedom because the distributions are constrained to sum to a constant (1.0 or 100%), so when we determine the area within $k - 1$ of the categories, we automatically know the area of the final category. We lose an additional two degrees of freedom because we must estimate μ by $\overline{X}$ and σ^2 by s^2 in order to standardize the observations. The critical value of $\chi^2 = 2.71$ is that for a 10% level of significance ($\alpha = 0.10$) and one degree of freedom (taken from Appendix Table **A.4**).

Like the F-distribution, χ^2 is not centered around zero, but is entirely positive. Because the deviations of expected from observed frequencies in each class are squared, negative numbers do not appear. Consequently, χ^2 tests are always one-tailed, with their region of rejection on the right.

In the example, the normal distribution can be split into four classes of equal probability whose limits are $-\infty < z_i < -0.675$, $-0.675 < z_i < 0.000$, $0.000 < z_i < 0.675$, and $0.675 < z_i < \infty$. If the salinity measurements are distributed normally, approximately $48/4 = 12$ measurements should fall into each of the four intervals when the salinity values are standardized. The first step is to standardize the data by Equation (2.37). The Whitewater Bay samples have a mean of $\overline{X} = 49.54$ and a standard deviation of $s = 9.27$, so the individual observations are standardized by

$$z_i = \frac{x_i - \overline{X}}{s} = \frac{x_i - 49.54}{9.27}$$

The standardized scores are listed and sorted in **Table 2–23**, where they have been numbered according to the four classes into which they fall. Calculating the χ^2 test value, we obtain

$$
\begin{aligned}
\chi^2 &= \frac{(13 - 12)^2}{12} + \frac{(14 - 12)^2}{12} + \frac{(8 - 12)^2}{12} + \frac{(13 - 12)^2}{12} \\
&= \frac{1}{12} + \frac{4}{12} + \frac{16}{12} + \frac{1}{12} \\
&= \frac{22}{12} = 1.83
\end{aligned}
$$

The χ^2 statistic is less than the critical value of 2.71 for one degree of freedom and a 10% level of significance. Therefore, there is no evidence to suggest that the measurements of surface-water salinities are not normally distributed.

In effect, we have compared a four-category histogram of observations with a four-category histogram based on the normal distribution and found that they were indistinguishable. It may strike you that the comparison is rather crude, and perhaps more subtle distinctions could be detected if the comparison was based on histograms having greater numbers of categories. This is entirely correct, but we may encounter the problem of not enough observations to reliably estimate the proportions in a large number of categories. The rule of thumb is that all categories should have an expected frequency of five or greater; if the expected frequency is less than five, the category should be combined with another to achieve this minimum value. One degree of freedom is lost for every combination of categories. Applying the test of goodness of fit is a balance between increasing the number of categories (thus increasing the degrees of freedom and reducing the critical value), while simultaneously decreasing our ability to estimate the proportions in each category.

Table 2–23. Salinity (in ppt) at stations in Whitewater Bay, Florida, with standardized scores and membership in four categories.

No.	Salinity	Standard Scores	Four Classes	No.	Salinity	Standard Scores	Four Classes
46	78	3.070292	4	27	48	−0.16633	2
34	67	1.883532	4	36	48	−0.16633	2
42	63	1.451983	4	47	48	−0.16633	2
3	62	1.344096	4	13	47	−0.27421	2
33	61	1.236208	4	22	47	−0.27421	2
9	60	1.128321	4	30	47	−0.27421	2
16	60	1.128321	4	1	46	−0.3821	2
21	60	1.128321	4	12	46	−0.3821	2
4	59	1.020434	4	17	46	−0.3821	2
31	59	1.020434	4	41	46	−0.3821	2
7	58	0.912547	4	39	44	−0.59788	2
11	58	0.912547	4	38	43	−0.70576	1
10	56	0.696772	4	32	42	−0.81365	1
6	53	0.37311	3	48	42	−0.81365	1
35	53	0.37311	3	5	40	−1.02942	1
43	53	0.37311	3	23	40	−1.02942	1
14	52	0.265223	3	24	40	−1.02942	1
15	51	0.157336	3	44	40	−1.02942	1
20	51	0.157336	3	28	39	−1.13731	1
37	50	0.049448	3	2	37	−1.35309	1
45	50	0.049448	3	18	36	−1.46097	1
8	49	−0.05844	2	29	36	−1.46097	1
26	49	−0.05844	2	25	35	−1.56886	1
40	49	−0.05844	2	19	34	−1.67675	1

We can quickly assess the consequences of using an additional category in testing the distribution of Whitewater Bay salinities. A normal distribution can be divided into five equal categories whose limits are $-\infty < z_i < -0.840$, $-0.840 < z_i < -0.223$, $-0.223 < z_i < 0.223$, $0.223 < z_i < 0.840$, and $0.840 < z_i < \infty$. We expect 20% of the observations to fall into each of these categories, or $48 \times 0.2 = 9.6$ observations. Then,

$$\chi^2 = \frac{(10 - 9.6)^2}{9.6} + \frac{(11 - 9.6)^2}{9.6} + \frac{(10 - 9.6)^2}{9.6} + \frac{(5 - 9.6)^2}{9.6} + \frac{(12 - 9.6)^2}{9.6}$$

$$= \frac{0.16}{9.6} + \frac{1.96}{9.6} + \frac{0.16}{9.6} + \frac{21.16}{9.6} + \frac{5.76}{9.6} = \frac{29.2}{9.6} = 3.04$$

The critical value for χ^2 with two degrees of freedom and a 10% level of significance is $\chi^2 = 4.61$, so again the test statistic does not fall into the critical region. Even with a somewhat more detailed histogram, there is no evidence to indicate that the salinities are not normally distributed.

Of course, we are not restricted to considering only normal distributions with the χ^2 statistic. We may test the goodness of fit of a data set against *any* specified curve, such as a uniform, lognormal, exponential, or arbitrary distribution. The test procedure remains the same, although the degrees of freedom must be adjusted to account for the number of parameters estimated. Cochran (1952) provides an extensive discussion of these tests.

The Logarithmic and Other Transformations

Many geological variables very obviously do not follow a normal distribution. In geochemical surveys, for example, element concentrations typically form a highly skewed distribution, as shown in **Figure 2–41 a,** a histogram of lead measured in 274 soil samples collected on the Istrian peninsula of Croatia (the data are given in file CROPB.TXT). The concentration of selenium in plant material assayed during a geochemical reconnaissance or the concentrations of iodine detected in ground-water samples may form similar skewed distributions. The pattern, comprised of a minimum-value lower limit, a low-level "background" containing the bulk of the observations, a tail of decreasing numbers of observations having higher concentrations, and a few "anomalies" whose concentrations may exceed the background by orders of magnitude, is so ubiquitous that it has been called the "lognormal law" of geochemistry. Such distributions can be transformed to a more tractable shape simply by taking the logarithms of the concentration values, as shown in **Figure 2–41 b.**

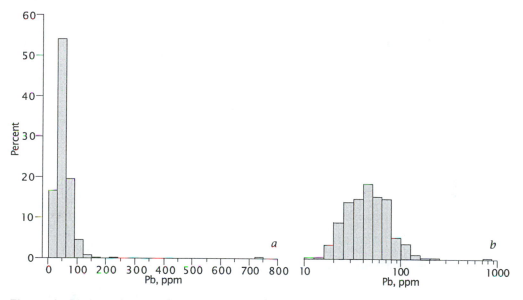

Figure 2–41. Histograms of Pb measured in soil samples collected on the Istrian peninsula of Croatia plotted on (*a*) an arithmetic scale and (*b*) a logarithmic scale.

The weight-percent, grain-size distribution of sediments is often pronouncedly skewed, and an entire system of size classification has been based on this fact (Krumbein and Pettijohn, 1938). The phi transformation consists of converting grain-size class limits in millimeters to base-2 logarithms, then changing their sign. The transformation was originally done to simplify calculations in pre-computer days, and its use persists because of tradition.

When the volumes of oil fields in a region are plotted, the result is a highly skewed distribution such as that shown as a cumulative probability graph in **Figure 2–42 a.** The data, for 1234 Cretaceous fields in the Denver–Julesburg Basin of Colorado, are contained in file DJBASIN.TXT. Field volume is expressed as the estimated ultimate production of oil from the fields, in thousands of barrels. Most

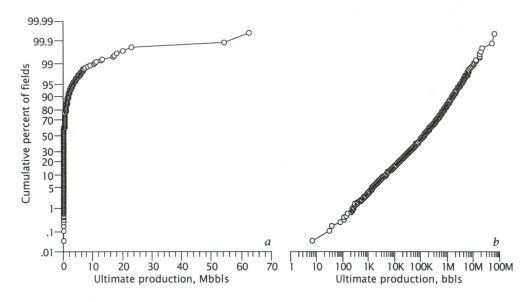

Figure 2–42. Cumulative distributions of ultimate production from Cretaceous oil fields in the Denver–Julesburg Basin, Colorado, shown on (*a*) an arithmetic scale and (*b*) a logarithmic scale.

fields are small, but there are decreasing numbers of much larger fields, and a few rare giants that greatly exceed all others in volume. Transforming the volumes by taking their logarithms results in a nearly straight-line plot on a cumulative probability graph (**Fig. 2–42 b**).

If the observations shown in **Figure 2–41a** and **Figure 2–42a** are converted to logarithmic form (that is, we use $y_i = \log x_i$ instead of x_i for each observation), we see that their distributions become nearly normal, or at least much more symmetrical (**Fig. 2–41 b** and **Fig. 2–42 b**). Variables whose logarithms follow normal distributions are said to be **lognormal**. Because of its common occurrence in geology, the lognormal distribution is extremely important. However, if we confine our attention to the transformed variable, y_i, rather than x_i itself, the properties of the lognormal distribution can be explained simply by reference to the normal distribution.

The mean and variance of a log transformed variable y_i are found in the usual way:

$$\overline{Y} = \frac{\Sigma y_i}{n}$$

and

$$s_y^2 = \frac{\Sigma \left(y_i - \overline{Y} \right)^2}{n - 1} \tag{2.66}$$

However, in terms of the original untransformed variable, x_i, the mean, $\overline{Y}$, corresponds to the nth root of the products of x_i, which is the **geometric mean**, g.

$$\exp \overline{Y} = g = \sqrt[n]{\prod x_i} \tag{2.67}$$

The symbol $\prod$ is analogous to Σ, except it means that all the elements in the indicated series are to be multiplied together, rather than added together. $\prod$ has limits exactly like those used with the summation symbol, $\sum$. Sometimes the limits are omitted if they are clearly implied in the expression. As an example,

$$\prod_{i=1}^{3} x_i, \quad \text{where } x_1 = 2, \ x_2 = 3, \ x_3 = 4$$

is

$$\prod x_i = 2 \times 3 \times 4 = 24$$

The variance of a logarithmically transformed variable is called the **geometric variance** and is equivalent to

$$s_g^2 = \exp s_y^2$$

$$= {}^{n-1}\sqrt{2 \prod_{i=1}^{n} \left(\frac{x_i}{\text{GM}} \right)} \tag{2.68}$$

In practice, of course, it is simplest to convert our observations into logarithms and then compute the mean and variance. If the geometric mean and variance are desired, they are found by taking the antilogs of $\overline{Y}$ and s_y^2. As long as we work with the data in its transformed state, all of the statistical procedures that are appropriate for ordinary variables are applicable to log-transformed variables.

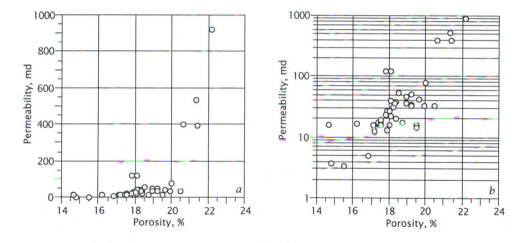

Figure 2–43. Porosity versus permeability measured on core plugs from wells in a field in eastern Oklahoma plotted on (*a*) arithmetic scales and (*b*) log-linear scales.

In addition to converting a skewed variable into a more symmetric form, log-arithmic transformation may also be useful in stabilizing the variance. **Figure 2–43a** is a plot of permeabilty versus porosity measured on 44 plugs cut from cores taken from wells in a gas field in eastern Oklahoma (the data are given in file PORPERM.TXT). As we would expect, cores that are more porous tend to be more permeable, but more-porous rocks also tend to be more variable in their permeability. By transforming permeabilty to log permeabilty and plotting the data on a log-linear graph (**Fig. 2–43b**), the variance in permeability becomes consistent and the relation between porosity and permeability is linearized.

If two variables having skewed distributions are cross plotted, the relationship between them can sometimes be made more apparent if both variables are logarithmically transformed. **Figure 2–44a** is a plot of the volumes of petroleum reservoirs

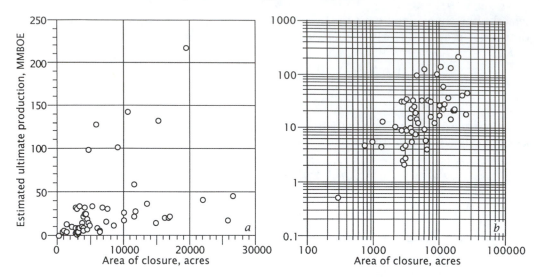

Figure 2–44. Field volumes versus areas of closure for petroleum reservoirs in salt structures in the Outer Continental Shelf of Texas and Louisiana, shown on (*a*) arithmetic scales and (*b*) log–log scales.

in an area of the Outer Continental Shelf of Texas and Louisiana plotted against the areas of closure of the salt structures in which the reservoirs are located. The data are taken from Table 4.1 of Harbaugh, Davis, and Wendebourg (1995) and are reproduced in file OCS.TXT. In general, there is a positive association between the two variables; that is, larger structures tend to contain larger reservoirs. However, the variation in reservoir volume also increases as the areas of structural closure increase, so that the variance tends to be proportional to the mean.

A logarithmic transformation may correct this condition, as can be seen in **Figure 2–44b.** Both reservoir volume and area of structural closure are transformed into their logarithms to produce a log–log plot. The variance of the logarithms of reservoir volumes remains almost constant for all values of the logarithms of the structural-closure areas.

The characteristics of the lognormal distribution are discussed at length in the monograph by Aitchison and Brown (1969) and in a geological context by Koch and Link (1980). The normal distribution is often explained as arising when repeated measurements of some fixed quantity are made. Each individual measurement is perturbed by many random errors that add (sometimes in one direction and sometimes in the opposite) to the measurement. Usually these random errors cancel one another, and the final measurement approaches the true value. But on rare occasions, most of the random errors will have the same sign and an extreme measurement occurs. The result is the familiar bell-shaped normal distribution of measurements.

A lognormal distribution may arise under the same circumstances if the random errors are multiplicative rather than additive. Most of the random perturbations, when multiplied together, will produce an intermediate product near the geometric mean. Rarely, by random chance, all of the perturbations will be very small and their product will be near zero. Equally rarely, all of the perturbations will be large and their product will be an extremely large value. The result of many random realizations will be a distribution that starts at zero, rises to a maximum, then slopes downward and extends to extremely large values.

Biological scientists often refer to the "law of proportionate effect," which states that the change in a variable at any step in a process is a random proportion of the previous value of the variable. Thus, for example, the probable change in size of colonies of microbes with time is proportional to the size of the colonies at the previous time. Large colonies will tend to expand (or shrink) to a much greater extent than will small colonies. Perhaps oil pools accumulated in the same manner, so that during initial hydrocarbon migration, larger accumulations tended to increase at a proportionately greater rate than did small accumulations. Such a process would result in a lognormal distribution.

Geologists may be more familiar with the "theory of breakage" that has been advanced to explain the lognormal distribution of particle sizes observed not only in natural sediments but also in the crushed material produced by mills and grinders. Suppose we begin with a collection of equal-sized particles, and then break each particle at random. In general, this will result in one smaller and one larger fragment for each original particle. If we then break each fragment again at random, the smaller fragment will produce still smaller pieces, while the larger fragment will tend once more to yield a larger and a smaller piece. If the process is repeated again and again, an extremely large number of very small particles will be produced, with a few "select" grains whose dimensions approach those of the original particles. In other words, the lognormal size distribution so commonly observed in sediments will result.

Although taking logarithms will make a positively skewed distribution more symmetical, the resulting transformed distribution is not necessarily lognormal. If we log transform the values for lead in file CROPB.TXT, or the oil-field volumes in file DJBASIN.TXT, and test them for normality, we will see that they do not follow a lognormal model perfectly (**Fig. 2–45**). This is neither surprising nor distressing, because most of the data assumed to follow a normal distribution actually deviate from normality in some manner. Provided the distributions (or their transforms) are reasonably symmetical and do not have excessive tails, statistical tests based on the assumption of normality will perform in an acceptable manner.

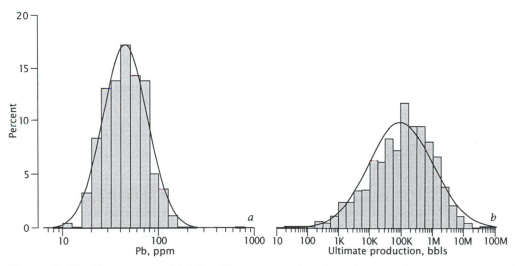

Figure 2–45. Histograms of (*a*) log Pb concentrations in soil on the Istrian peninsula of Croatia and (*b*) log oil-field volumes in the Denver–Julesburg Basin of Colorado with fitted normal distributions having the same means and variances.

Other transformations

Certain other transformations of x_i may make its distribution more nearly normal, or make its variance more consistent, or yield some other statistically beneficial results. Ordinarily, there is nothing at all sacred about our original scale of measurement, and we may feel free to change scales if it seems a useful thing to do. However, we must keep in mind that our statistical analysis is now testing a different hypothesis which pertains to the characteristics of the transformed variable, and not necessarily to the original variable. Also, we should be careful not to befuddle ourselves with transformations so exotic that we lose sight of the original nature of the geologic properties we are attempting to understand.

If our data consist of integer counts such as the number of discovery wells in a township or number of zircon grains in a set of thin sections, the numbers may tend to follow a Poisson distribution (discussed in Chapter 5). Rather than treating the data as discrete values, we can convert them to approximately normal form by the **square-root transformation.** That is, we replace every value x_i by $y_i = \sqrt{x_i}$. This transformation will make the variances more uniform and will tend to shorten the long tail of the Poisson distribution. If the observed values, x_i, are less than about ten, the transformation $y_i = \sqrt{x_i - 1/2}$ is preferable, especially if some of the counts are zero.

Robinson (1982) suggests the use of **power transforms** such as $y_i = x_i^2$, $y_i = x_i^3$, and so on, for the enhancement of petrophysical logs measured in boreholes. Powering a variable causes a much greater increase in larger values than in smaller values. If power transformation is followed by rescaling, so that the range $(y_{max} - y_{min})$ is the same as the original range of the data $(x_{max} - x_{min})$, the effect on a borehole log will be to accentuate variations in parts of the log where readings are high and subdue and flatten out parts of the log where readings are low. A power transformation has the same effect on a data distribution and may be used to correct a negative skew. However, power transforms may also tend to inflate the variance and make it nonuniform.

Negatively skewed distributions can sometimes be made approximately normal by an **arcsine transformation**, $y_i = \arcsin x_i$. The original variables should be scaled within the range of zero to 1.00. A different arcsine transformation, described in Chapter 4, can be used to convert binomial distributions to approximately normal form.

Nonparametric Methods

All of the preceding statistical techniques are **parametric**; that is, they assume that the calculated test values have distributions whose shapes are of known form. These test distributions (t, F, and χ^2) all describe the results of random sampling from normal populations, and are defined by equations that have only a few simple parameters. If we estimate these parameters, we can easily determine specified areas under the test distributions and, in turn, estimate probabilities of occurrence of the test results. The ease by which this can be done is one of the things that makes the parametric approach to statistical testing so appealing. Parametric tests have an underlying assumption of normality. However, we can invoke the central limits theorem to justify use of these tests even when the sampled population is not normal—provided our sample size is large and the population does not differ too much from normality. Sometimes, however, we must work with a small sample whose size cannot be increased and whose underlying population may be distinctly nonnormal. In such circumstances we cannot rely on the central limits theorem to justify use of parametric tests; we must turn to a category of alternative procedures called *nonparametric statistical tests*. Nonparametric tests use information of a lower rank, such as nominal or ordinal observations, rather than metric data required by conventional tests. No assumptions about the form of the parent population are required, hence the name "nonparametric." In general, a nonparametric test is less powerful than its parametric equivalent if the sampled population follows the assumed distribution. However, if it does not, the nonparametric test is usually more powerful.

Nonparametric tests are seldom covered in introductory statistics classes, and their use in geology has not been widespread. Rock (1988) reviews most of the uses of nonparametric statistics and provides extensive citations to applications in the Earth sciences and other fields. Cheeney (1983) emphasizes the use of nonparametric techniques in geology. There are many excellent specialized texts on nonparametric statistics, including Siegel and Castellan (1988), Noether (1991), Gibbons and Chakraborti (1992), and Conover (1999), among others. A concise introduction is given in a paperback monograph by Gibbons (1993a).

Mann–Whitney test

We have discussed how the t-distribution can be used to test for equality of the means of populations from which two samples have been drawn. A nonparametric equivalent, called the Mann–Whitney test, checks for equivalence of the medians of the two samples. Suppose we collect two samples of size n and m, where $m \leq n$, and wish to test the hypothesis that both came from the same population. We combine the two samples and sort them so the observations are arranged in order from smallest to largest. Each observation is assigned a rank such that the smallest is ranked 1, the next larger is ranked 2, and so on up to the largest observation, which is ranked $(n + m)$. If the two samples have been drawn randomly from the same population, we would expect that observations from one of the samples would be scattered more or less uniformly through the ranked sequence.

We will call x_i the ith observation in the smaller sample, and y_j the jth observation in the larger sample. The rank of an observation is denoted $R(x_i)$ or $R(y_j)$. If there are ties (two or more observations having the same values), we assign each of them a rank equal to their average rank. Next, we calculate the sums of the ranks of the two groups:

$$W_x = \sum_{i=1}^{m} R(x_i)$$
$$W_y = \sum_{j=1}^{n} R(y_j)$$

(2.69)

The test statistic is simply the sum of the ranks of the smaller group, or W_x. Exact probabilities of occurrence for specified values of W_x for small sample sizes (eight or fewer observations in each group) are given in Appendix Table **A.5**. The table, abridged from Siegel and Castellan (1988), lists exact probabilities that W_x will be equal to or less than a lower critical value, C_L, or equal to or greater than an upper critical value, C_U. For larger sample sizes, W_x is approximately normally distributed and can be tested using normal tables such as **A.1**. The expected mean and standard deviation of the statistic are

$$\overline{X} = nm/2$$
$$s = \sqrt{nm(n + m + 1)/12} \qquad (2.70)$$

which can be used to convert the test value W_x to standard normal form, or z-score, using Equation (2.37).

$$z = \frac{W_x - \overline{X}}{s}$$
$$= \frac{W_x - nm/2}{\sqrt{nm(n + m + 1)/12}} \qquad (2.71)$$

As an example of the use of the Mann–Whitney test, we may examine the data given in **Table 2–24**. In the U.S. Gulf Coast region, oil and gas are produced from structural traps associated with salt domes. Prospects can be identified by mapping subsurface horizons which are detected by their seismic reflections. Potential hydrocarbon traps include tilted fault blocks on the flanks of salt piercements and closed anticlines over the crests of salt domes. **Table 2–24** lists the areas of closure of prospects selected from marine seismic maps of two concessions off the Louisiana coast. We wish to know if the areal extents of prospects differ between the two areas.

Table 2–24. Areas of oil and gas prospects, in acres, on salt dome structures in two regions off the Louisiana coast; areas measured on maps based on marine seismic surveys.

Eastern Region			Western Region		
Prospect	Size	Rank	Prospect	Size	Rank
x_1	802	16	y_1	312	12
x_2	174	8	y_2	55	2
x_3	158	6	y_3	220	9
x_4	140	4	y_4	276	11
x_5	166	7	y_5	154	5
x_6	328	13	y_6	37	1
x_7	239	10	y_7	478	14
x_8	99	3	y_8	666	15

Table 2–24 also gives the ranks for pooled observations of prospect areas in the two concessions. The sum of the ranks of the prospects in the first area is $\sum_{i=1}^{m} R(x_i) = 67$ and the sum of the ranks of the prospects in the second area is $\sum_{j=1}^{n} R(y_j) = 69$. We will choose the smaller value, 67, as the test statistic whose probability of occurrence is read from Appendix Table **A.5**. Since both m and n are equal to 8, we determine from Table **A.5** that the probability of obtaining a test value of W_x that is equal to or less than the critical value of $C_L = 67$ by random chance is $p = 0.48$. This strongly suggests that the distributions of field areas are the same in the two regions, and that the differences in their rankings are not significant.

The Mann–Whitney test appears in slightly different forms under several names in the statistical literature. These include the Wilcoxon test, the Siegel–Tukey test, and Festinger's test. The Siegel–Tukey variant is especially interesting because it can be used to test the equivalency of variances in two samples, and thus is a nonparametric substitute for the simple F-test.

Kruskal–Wallis test

A nonparametric test of the equivalency of several samples has been devised by Kruskal and Wallis. In effect, this test is a nonparametric alternative to the one-way analysis of variance. The procedure is very similar to that used in the Mann–Whitney test; the observations from the k samples are combined or pooled and then ranked from smallest to largest. For each sample, the sum of the ranks is found:

$$R_k = \sum_{i=1}^{n_k} R\,(x_{ik}) \tag{2.72}$$

where $R\,(x_{ik})$ represents the rank of the ith observation in the kth sample. The total number of observations is $N = \sum_{j=1}^{k} n_k$, where n_k is the number of observations in the kth sample.

The null hypothesis that we wish to test is that all of the k populations from which the samples are taken have identical distributions. The alternative is that at least one of the populations has a different central value. We must assume that all observations have been collected randomly, and that the samples are independent of one another.

From the sum of the ranks, we can compute the Kruskal–Wallis H-statistic:

$$H = \frac{12}{N\,(N-1)} \sum_{j=1}^{k} \frac{[R_k - n_k\,(N+1)\,/2]^2}{n_k} \tag{2.73}$$

An algebraically equivalent form that is somewhat easier to use is

$$H = \frac{12}{N\,(N-1)} \sum_{j=1}^{k} \frac{R_k^2}{n_k} - 3\,(N+1) \tag{2.74}$$

Critical values of H have been tabulated only for sets of three samples, each consisting of up to five observations [see, for example, Hollander and Wolfe's (1999) Table A.7]. Fortunately, H is approximately distributed as χ^2 with $k-1$ degrees of freedom, so the test can readily be applied to larger problems. Rock (1988) provides a correction for H when there are tied ranks; the correction factor is negligible unless the ties are numerous.

Nonparametric correlation

In earlier sections, the calculation of the correlation coefficient has been described at length, and a parametric t-test for the significance of a sample correlation coefficient was given in Equation (2.51). However, there are many circumstances in which the conventional correlation coefficient (sometimes called the **Pearsonian product-moment correlation coefficient**) is not appropriate, yet some measure of relationship between variables is desirable.

105

The textural properties of sandstones usually are considered to reflect environmental conditions at the time of deposition. The high energy of a beach environment, for example, will cause winnowing and abrasion, resulting in well-sorted deposits of coarse, well-rounded grains. In low-energy environments, deposits are likely to consist of more poorly sorted material, with a finer grain size and more angular particles. Folk (1951) has defined **textural maturity** as the degree to which a sand is both well sorted and well rounded. These two properties are presumed to go hand-in-hand; sandstones that exhibit anomalous relationships (for example, a well-sorted angular sandstone) are said to show "textural inversion."

It would seem a simple matter to check the concept of textural maturity by collecting a suite of sandstone samples, measuring their textural properties, and then calculating a measure of the relationship between the properties. Unfortunately, roundness and degree of sorting usually are not measured on an interval or ratio scale; rather, ordinal scales are used. Sorting usually is expressed in ordinal classes as poor, moderate, or well sorted, while roundness is classed as angular, subangular, subrounded, or rounded. The ordinary correlation coefficient cannot be used to measure the strength of the relationship between roundness and sorting when expressed in these terms.

An alternative measure is **Spearman's rank correlation coefficient**, which, as the name suggests, expresses the similarity between two sets of rankings. If we make two sets of ordinal measurements on a number of objects, we can designate one of the sets as x and the other as y. We then rank each measurement and call the two sets of ranks $R(x_i)$ and $R(y_i)$. Spearman's coefficient measures the similarity between these two sets of ranks.

$$r' = 1 - \frac{6 \sum_{i=1}^{n} [R(x_i) - R(y_i)]^2}{n(n^2 - 1)} \tag{2.75}$$

The term inside the brackets of the numerator is simply the difference between the rank of property x and the rank of property y as measured on the ith object.

The textural properties of 12 reservoir sandstones are given in **Table 2–25**. Each of the sandstones was compared to every other sandstone in order to rank them from most poorly sorted to best sorted, and from most angular to most rounded. The rankings are given in the table. As in the Mann–Whitney test, tied ranks are given the average of the ranks that would have been assigned if there were no ties.

Spearman's rank correlation between the two variables is

$$r' = 1 - \frac{6 [162]}{12 (12^2 - 1)}$$

$$= 1 - \frac{972}{1716}$$

$$= 0.43$$

The rank correlation, r', is analogous to r in that it varies from $+1.0$ (perfect correspondence between the ranks) to -1.0 (perfect inverse relationship between the ranks). A rank correlation of $r' = 0.0$ indicates that the two sets of ranks are independent. A rank correlation of $r' = 0.43$ suggests there is a weak positive relationship between amount of grain roundness and degree of sorting.

The conventional t-test of the significance of a correlation coefficient cannot be applied to r', because the t-test assumes the sample is drawn from a bivariate

Table 2–25. Roundness and sorting of reservoir sandstones.

Formation	Age	Sorting[a]	Rank	Round-ness[b]	Rank	$[R(x_i) - R(y_i)]$	$[R(x_i) - R(y_i)]^2$
Lakota Ss.	Cretaceous	P	4	SR	11	−7	49
Berea Ss.	Miss.	W	10	SA	9	1	1
Boise Ss.	Pliocene	P	2	A	1	1	1
Big Clifty Ss.	Miss.	M	8	SA	4	4	16
Clear Creek Ss.	Penn.	M	6	SA	6	0	0
Bromide Ss.	Ordovician	W	9	SR	12	−3	9
Noxie Ss.	Penn.	P	3	SA	8	−5	25
Green River Fm.	Eocene	M	7	SA	3	4	16
Reagan Ss.	Cambrian	W	11	SA	7	4	16
Peru Ss.	Devonian	W	12	SR	10	2	4
Bartlesville Ss.	Penn.	M	5	A	2	3	9
Mt. Simon Ss.	Cambrian	P	1	SA	5	−4	16
						TOTAL	162

[a] Categories of sorting are P = poor, M = moderate, W = well sorted.
[b] Categories of roundness are A = angular, SA = subangular, SR = subrounded.

normal population. Fortunately, a table of critical values is available that allows Spearman's rank correlation coefficients to be tested directly; part of this table is reproduced as Appendix Table **A.6**. As in the t-test, the null hypothesis is that the two variables are independent, or that $\rho' = 0$. The most common alternative is $\rho' \neq 0$, so the test is two-tailed, with either very large positive or very large negative correlations leading to rejection. Suppose we decide that a significance level of $\alpha = 0.05$ would be appropriate to test the correlation between roundness and sorting. Then, the upper critical value would correspond to $1 - \alpha/2 = 0.975$, or 0.5804 for $n = 12$. The lower critical value, corresponding to $\alpha/2 = 0.025$, is −0.5804. Our computed correlation, $r' = 0.43$, does not fall beyond either of these limits, so we cannot reject the hypothesis that roundness and sorting are independent of one another. If there is a relationship between these two properties, our sample of 12 observations is not adequate to detect it at the 5% level of significance.

Kolmogorov–Smirnov tests

One extremely useful group of nonparametric procedures includes Kolmogorov–Smirnov tests, extensively discussed in a geological context by Cheeney (1983). Among their other applications, they can be used to test for goodness of fit, and thus are an alternative to the χ^2 methods discussed earlier. Although χ^2 goodness-of-fit tests also are nonparametric in the sense that they can be applied to observations following any kind of distribution, the Kolmogorov–Smirnov tests are superior in certain circumstances. Their most obvious advantage is that they compare cumulative distribution functions directly, so it is not necessary to group observations into arbitrary categories. For this reason, they are more sensitive than is the χ^2 test to deviations in the tails of distributions where frequencies are low and categories must be combined.

Cheeney provides an excellent introduction to the Kolmogorov–Smirnov procedure using measurements on basalts in Scotland. The columnar form that basalts often exhibit is believed to be due to shrinkage of the cooling lava toward uniformly

Table 2–26. Number of sides on basalt columns measured
at Yellowcraigs, Scotland (from Cheney, 1983).

Number of Sides	Frequency	Cumulative Frequency	Empirical Cumulative Distribution
3	1	1	0.03
4	3	4	0.12
5	8	12	0.36
6	15	27	0.82
7	4	31	0.94
8	1	32	0.97
9	0	32	0.97
10	1	33	1.00

spaced centers. If this is correct, the resulting fractures should form six-sided columns, since a network of face-centered hexagons is the most uniform possible pattern. **Table 2–26** gives the number of sides observed on 33 basalt columns measured at Yellowcraigs, east of Edinburgh. Certainly most of the columns have six sides, but they range in number from three sides to ten. We can test the hypothesis that basalt columns are six-sided by comparing their empirical cumulative distribution of number of sides against a theoretical cumulative distribution which states that all basalts have six sides. Both of these distributions are shown in **Figure 2–46**, which also illustrates graphically how the Kolmogorov–Smirnov procedure works. In general, we select a sample from some unknown population and wish to test its goodness of fit to a hypothetical model of a specific population. Both the sample and the hypothetical model are plotted together in cumulative form, each scaled so their cumulative sums are 1.0; that is, we express both as cumulative probability distributions. We then look for the greatest difference between the two. This maximum difference is the Kolmogorov–Smirnov statistic, D, or

$$D = \max |CDF - EDF| \tag{2.76}$$

Table 2–26 gives the frequencies of occurrence of basalt columns observed to have different numbers of sides, the cumulative frequencies, and the empirical cumulative distribution function (EDF) found as the successive cumulative frequencies divided by the total number of observations. The theoretical cumulative distribution (CDF) has only two steps; 0 for all numbers of sides less than six, and 1 for all numbers of sides equal to or greater than six. It is obvious from inspection that the greatest difference occurs for the class of five-sided basalt columns, where $D = |0.00 - 0.36| = 0.36$. Critical values of D are tabulated in Appendix Table **A.7**. For $n = 33$ degrees of freedom and a significance level of $\alpha = 0.95$, we find that the critical value is 0.208; our test statistic exceeds this value, so we must conclude that our sample of 33 Scottish basalt columns was *not* drawn from a theoretical population of six-sided columns. Apparently, the centers of contraction in the cooling basalt are not perfectly uniformly distributed.

Appendix Table **A.7** gives critical values for the Kolmogorov–Smirnov statistic, D, and can be used for either one-tailed or two-tailed hypotheses. The two-tailed null hypothesis, used above, states that classes of the distribution from which the

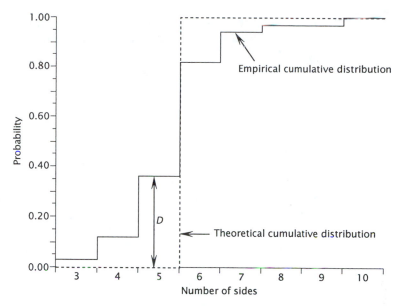

Figure 2–46. Kolmogorov–Smirnov procedure for testing goodness of fit of an empirical distribution of the number of sides on basalt columns at Yellowcraigs, Scotland (solid line), against a hypothetical model (dashed line) specifying that all columns have six sides.

sample is derived are *equal* to those of the hypothetical model for all values of x. The one-tailed null hypothesis states that all classes of the sample distribution are *equal to or less than* those of the hypothetical model (we use the maximum positive difference as the test statistic), or that all classes of the sample distribution are *equal to or greater than* those of the hypothetical model (we use the maximum negative difference as the test statistic). In most instances, we will use a two-tailed hypothesis and alternative.

Ordinarily, the Kolmogorov–Smirnov test is used when the hypothetical model can be completely specified. That is, the parameters of the distribution are known (or assumed) from information other than that contained within the sample itself. A variation devised by Lilliefors (1967) allows us to use the Kolmogorov–Smirnov procedure for testing the fit of a sample to a normal distribution with an unspecified mean and variance. This variation permits the Kolmogorov–Smirnov procedure to be used in a manner exactly analogous to the χ^2 procedure.

The mean, $\overline{X}$, and variance, s^2, of the sample are found in the usual manner, and the empirical cumulative distribution is determined from the ranked observations as in the ordinary Kolmogorov–Smirnov procedure. A normal cumulative probability distribution is then calculated that has the same mean and variance as the sample. The largest absolute difference between the normal cumulative distribution function and the empirical distribution function is found; this is the test statistic, T. The test statistic is compared to critical values of T using special tables such as Appendix Table **A.8**. Since the cumulative normal distribution is calculated using a mean and variance estimated from the sample, degrees of freedom have been lost. The Lilliefors table of critical values compensates for this loss of degrees of freedom.

In some discussions of the Lilliefors procedure (for example, Conover, 1999), it is emphasized that the observational data first must be converted to standardized form using the z-transformation:

$$z_i = \frac{x_i - \overline{X}}{s}$$

The standardized data are ranked and converted to an empirical cumulative distribution as before, and compared to a standard normal distribution in cumulative form. The standardizing step is a holdover from pre-computer days, when it was much easier to obtain appropriate values of the standardized normal distribution from tables such as Appendix Table **A.1**. The test value of T is identical, whether the two distributions are plotted along a z-scale or on the original scale of measurement, because T is the difference in the associated probabilities. This is shown in **Figure 2–47**, which uses both a standardized scale and a scale of original units.

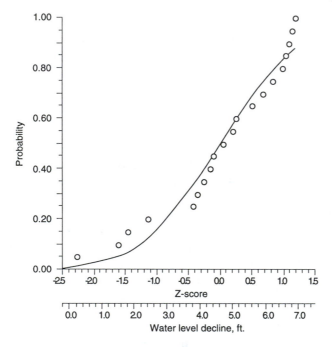

Figure 2–47. Lilliefors variation of Kolmogorov–Smirnov procedure testing goodness of fit of cumulative distribution of water-level declines in western Kansas observation wells (open dots) to a cumulative normal distribution (line).

In western Kansas, the depth to the water table is measured each year in observation wells in order to monitor any changes in groundwater levels due to irrigation pumping. **Table 2–27** contains measurements of the drawdown in depth to water recorded in 20 observation wells within a small area between 1997 and 1998. Although in general the water level is lower in 1998, we may hypothesize that the differences in water level are random and follow a normal distribution. The sample size is too small to test this hypothesis using a χ^2 test, but we may use the Lilliefors modification of the Kolmogorov–Smirnov procedure.

Table 2–27. Drawdown of water table (in feet) from 1997 to 1998 in 20 observation wells in an area of western Kansas

Drawdown	z-Score	Rank	Empirical CDF	Normal PDF	Difference
0.2	− 2.2869	1	0.05	0.0111	0.0389
1.5	− 1.6143	2	0.1	0.0532	0.0468
1.8	− 1.4591	3	0.15	0.0723	0.0777
2.4	− 1.1487	4	0.2	0.1254	0.0746
3.8	− 0.4243	5	0.25	0.3357	− 0.0857
3.9	− 0.3725	6	0.3	0.3547	− 0.0548
4.1	− 0.2691	7	0.35	0.3939	− 0.0439
4.3	− 0.1656	8	0.4	0.4342	− 0.0343
4.4	− 0.1138	9	0.45	0.4547	− 0.0047
4.7	0.0414	10	0.5	0.5165	− 0.0165
5	0.1966	11	0.55	0.5779	−0.0279
5.1	0.2484	12	0.6	0.5981	0.0019
5.6	0.5071	13	0.65	0.6939	− 0.0439
5.9	0.6623	14	0.7	0.7461	− 0.0461
6.2	0.8175	15	0.75	0.7932	− 0.0432
6.5	0.9727	16	0.8	0.8347	− 0.0347
6.6	1.0245	17	0.85	0.8472	0.0028
6.7	1.0762	18	0.9	0.8591	0.0409
6.8	1.1280	19	0.95	0.8703	0.0797
6.9	1.1797	20	1	0.8809	0.1191

Columns in **Table 2–27** show the 1998–1997 differences in water level, in both feet of drawdown and as standardized scores; the associated empirical probabilities; and the probabilities associated with the equivalent normal scores. The cumulative distributions are shown in **Figure 2–47**. The maximum difference between the empirical distribution and the normal distribution is $T = 0.119$, which can be compared to the critical values in Appendix Table **A.8** for $n = 20$; the critical value for $p = 0.90$ is 0.174. Since the test statistic is less than this amount, it cannot be demonstrated that the changes in water level do not follow a normal distribution.

We can also test the normality of salinity measurements from Whitewater Bay (**Table 2–22**) that were used to illustrate the χ^2 goodness-of-fit test. **Figure 2–48** shows the empirical cumulative distribution of salinity values compared to a cumulative probability distribution with the same mean and variance. The maximum absolute difference between the two curves occurs at $z_i = 0.37$, corresponding to sample no. 35, which has a salinity of 53 ppt. The critical values of the Lilliefors test statistic given in Appendix Table **A.8** extend only to $n = 30$; however, approximate values for larger n may be found by the formulas indicated on the table. For $n = 48$ and a level of significance of $\alpha = 0.10$, the critical value is $T = 0.116$. The computed test statistic is

$$T = |0.70 - 0.64| = 0.06$$

which does not fall into the critical region. Therefore, we cannot reject the null hypothesis that the samples were collected from a normally distributed population.

There are dozens, perhaps hundreds, of alternative nonparametric tests and their variations described in the statistical literature and in specialized texts such as Conover (1999) and Hollander and Wolfe (1999). Although not widely covered in

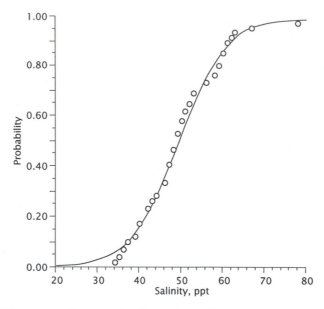

Figure 2–48. Kolmogorov–Smirnov fit of cum .iative distribution of Whitewater Bay salinity measurements (open dots) to a cumuiative normal distribution (line).

conventional introductory statistics texts, nonparametric techniques are beginning to find their way into mainstream books as the importance of statistical testing under less than ideal circumstances becomes increasingly recognized. Unfortunately, "less than ideal circumstances" are the norm rather than the exception in the Earth sciences, and the use of nonparametric techniques will (or should) become more widespread in the future. Rock's (1988) monograph contains a treasure trove of applications of nonparametric methods in the Earth sciences and should be consulted by any geologist seeking information on the available techniques and how they have been applied.

EXERCISES

At this point we must end our discussion of elementary statistical procedures—not because we have considered them in all of their aspects, but simply to allow space for other topics. The material presented is an extreme condensation of what typically is a one-semester course in introductory statistics, plus selected other topics from more advanced courses. We cannot hope to have done justice to the subject in the short space allotted to it here. However, you should now have some feel for the basic concepts of statistical testing and a passing familiarity with the jargon of the field.

This brief discussion of the testing of statistical hypotheses will serve as an introduction to more advanced tests and procedures presented in later chapters. This material by no means exhausts the potential of the discipline of statistics. As Wallis and Roberts (1960) state in their preface, "statistics is a lively and fascinating subject; but studying it is too often excruciatingly dull." By casting this branch of mathematics in the context of geological investigation, this book will hopefully provide the motivation and special interest required to enliven the subject for Earth scientists, and readers will further investigate the practical and rewarding field of statistics. The following exercises should help you hone your new (or resharpen your old) statistical skills.

Exercise 2.1

An offshore concession in the China Sea was evaluated by an oil company, whose geologists postulated three possible analogs for the habitat of petroleum in the concession: The geologic setting of the concession might be similar to (1) an oil province in Xian province where the exploration success ratio is 20%, or (2) the Gulf of Tonkin, which has a success ratio of only 10%, or, (3) it might be most similar to the Java Sea, which has a highly favorable success ratio of 35%. The geologists subjectively assigned a likelihood of 0.70 that the concession would prove to be like Xian, 0.10 that it would be like the Gulf of Tonkin, and 0.20 that it would be like the Java Sea.

The first four wildcats drilled in the concession resulted in three discoveries. How should the likelihoods assigned to the three possible analog areas be revised in light of this outcome?

Exercise 2.2

A university mineralogy laboratory contains two beam balances used to determine specific gravity by the immersion method. The students suspect that the two balances are not equal in quality. To test this, they have repeatedly weighed a 150-gm weight on each balance, as recorded in the following table. Which is the better instrument?

Balance A	Balance B
150.2	150.1
150.3	149.5
149.5	150.2
150.0	150.1
150.4	149.7
149.6	150.4

Exercise 2.3

The West Lyons oil field was discovered in 1963 in Rice County, Kansas, and was originally estimated to contain 22 million barrels of oil. The reservoir is a sandstone of Pennsylvanian (Upper Carboniferous) age and has been cored in 94 wells in the field. File WLYONS.TXT contains core measurements of porosity and water saturation and the thickness of the reservoir for each of these wells. Assume that porosity is normally distributed and the parameters of the population can be estimated from the sample statistics. Then, answer the following questions.

What is the probability of measuring
1. A core porosity that is exactly 15%?
2. A core porosity less than 6%?

    3. A core porosity greater than 15%?

    4. A core porosity between 15% and 16%?

Production from the West Lyons oil field is now declining and operators are considering the application of a proprietary enhanced recovery procedure (called PERP) to stimulate recovery. The PERP method works best on sandstone reservoirs whose average porosity is 15% or greater. Assume the population of suitable reservoirs is normally distributed with a mean porosity of 15% and a standard deviation of 5%. Could the random taking of 94 cores from such a population result in the distribution of porosity measurements we observe for the West Lyons field?

The production geologist in charge of the West Lyons field has speculated that the characteristics of the field vary with reservoir thickness, perhaps reflecting differences in depositional environment and hence grain sorting and packing. The distribution of reservoir thickness does appear bimodal, with a distinction between wells containing more than 30 ft of reservoir sandstone and those that contain less. Are core porosities measured where the reservoir sand is over 30 ft thick significantly different from those measured where the sandstone is thinner?

Exercise 2.4

File CELTIC.TXT contains measurements of the mean size of grains taken from 24 bottom samples collected from the La Chapelle bank in the Celtic Sea off the northeastern coast of Great Britain. The first column (Quartz) in the file gives the mean grain size (in microns) for quartz particles, and the second column (Carbonate) gives the mean grain size for carbonate particles. The carbonate particles are mostly shell fragments. Since the bank is a hydrodynamic feature created by shelf-edge current flow, quartz and carbonate grains must be in hydraulic equilibrium. The variable *Quartz* can be regarded as having been collected at random from off the La Chapelle bank and is a sample from the population of all possible quartz mean grain sizes within the bank. The variable *Carbonate* also is a random sample of mean grain sizes, but drawn from a population of carbonate mean grain sizes. Are the two populations the same?

Exercise 2.5

The chemical composition of a mineral often depends upon thermodynamic conditions in the environment in which it formed. Zeolites are complex silicates that form in sediments at relatively low temperatures, either during diagenesis or low-grade metamorphism; they also may form as hydrothermal vein deposits. File HEULAND.TXT gives the percent potassium analyzed in 92 heulandites, a common zeolite. Each specimen has been classified as either sedimentary (S) or hydrothermal (H) in origin. Is the difference in potassium content significant?

Exercise 2.6

In the classic study (Winsauer and others, 1952) that led to the "Humble Equation" relating porosity and readings on resistivity logs of wells, many characteristics of the pore geometry and rock properties of sandstones were examined. Among these were the permeabilty and the cementation factor, the exponent m in Archie's equation. In spite of the importance of Archie's equation in petrophysics, it is based on a surprisingly small number of sandstones that were analyzed in this original study. The data are given in file HUMBLE.TXT.

Calculate the mean and median of the cementation factor, m. Are the two measures of centrality similar? Now calculate the mean and median of permeability and compare these two measures. What can you deduce about the relative shapes of the two distributions?

Exercise 2.7

Rangely oil field in Colorado produces from the Pennsylvanian (Upper Carboniferous) Weber Sandstone. Because of the long history of production and water injection, and the discontinuous nature of permeable zones in the reservoir, pressures vary erratically both vertically and horizontally within the field. Maintaining efficient production under secondary recovery requires that these variations be mapped to indicate areas where increased flooding is required.

In order to collect *in situ* pressure data, a wireline formation tester is used to take fluid samples and pressure measurements at different depths within the borehole. The formation tester collects two fluid samples of each tested zone. However, the results of the two tests may not be identical, in part because the device may not seat at exactly the same spot in the borehole, and because the second test may require excessive time to equilibrate. Formation tests are run in conjunction with porosity logs that are used to select the high porosity zones to be examined. Data from the Rangely field are given in file RANGELY.TXT.

The file contains the well I.D. number, the test number (1 or 2), depth in feet, porosity in percent, pressure in psi, and the pressure build-up time in minutes. Test the hypothesis that the pressures recorded during the collection of the two sets of measurements are statistically identical.

Exercise 2.8

A petroleum research laboratory has evaluated two alternative procedures for measuring core porosity in sandstone core plugs. The conventional approach requires extracting any hydrocarbons with a solvent, drying the core plugs, weighing them, saturating the plugs with water, then weighing them again. A quicker procedure involves determining the bulk volume of the core plug by mercury displacement, disaggregating the sandstone, and then determining the grain volume by displacement of a light organic liquid. Because porosity determinations must be run on cores with widely differing characteristics, a selection of five diverse cores was used for a comparative study of the two procedures. The data are listed in file TWOWAY.TXT, taken from a study originally reported by Pyle and Sherborne (1939) and used as a statistical example by Griffiths (1967). The data file contains three columns and ten rows. The first column, labeled Porosity, contains the porosity measurements, the second column, labeled Procedure, contains an index equal to 1 for the traditional approach or 2 for the new technique; the third column, labeled Sandstone, contains numbers 1 through 5, indicating the five different sandstone cores.

Calculate a two-way ANOVA for these data, using *Procedure* as levels of one treatment and *Sandstone* as levels of the other. Is the difference in porosity measured by the two techniques signficantly different? Do the cores from different sandstones have significantly different porosities? Note that by blocking the differences between the sandstones, the *F*-test of the difference between procedures is more sensitive than if a fully randomized analysis (a *t*-test of equality of means) had been performed. Confirm this by comparing the probability associated with the *F*-test to the probability for the results of a *t*-test.

Exercise 2.9

In the Precambrian Agua Caliente Formation of northern Mexico, the relationship between porosity and percent quartz grains can be estimated from the nine measurements listed in file AGUACAL.TXT. Calculate the correlation between the two variables and examine a cross plot of the data. Outliers can have an extreme effect on bivariate relationships. A simple and effective check of correlation is to calculate both Pearsonian and Spearman's correlation coefficients. If the two are radically different, you should suspect either the presence of significant outliers, or a significant nonlinearity in the relationship between the two variables. Compute the Pearsonian correlation coefficient, then rank both sets of measurements and examine the Spearman's correlation coefficient to see if such problems are evident in these data.

Exercise 2.10

The original lunar crust has been highly fragmented and remelted because of meteorite impacts over billions of years. However, specimens of anorthositic rocks brought back from several landing sites in the lunar highlands are thought to be relics of an initially uniform primal lunar crust. In igneous melts, P_2O_5 is restricted to late residual magmatic liquids; if these rocks were derived from an initially homogeneous primal melt, the concentration of P_2O_5 should be essentially the same even though the specimens were collected at widely different localities. File MOON-CRST.TXT lists analyses for P_2O_5 measured on 45 specimens of lunar anorthosites collected at four landing sites in the lunar highlands. The analyses were made by two different laboratories. Are there significant differences in P_2O_5 concentrations between the different sites? Are your findings affected by differences in analyses made by the two laboratories?

Exercise 2.11

From the Mediterranean Sea floor, the Italian Navy collected a deep-sea core measuring 478 cm in length. The core was split and grain-size analyses were made of 51 intervals using three different measurement techniques. The percentage of sand was determined for all the splits by sieving, but the percentages of silt-sized and clay-sized material were measured in different manners, using a commercial particle-size classifier widely employed by the paint and cosmetics industries, a laboratory-built photoextinction sedimentometer, and the classical Andreasen's pipette method. File ITALNAVY.TXT contains the resulting data. The columns are

(1) Depth of sediment, in centimeters
(2) Clay percent determined by particle-size classifier
(3) Clay percent determined by photoextinction sedimentometer
(4) Clay percent determined by Andreasen's pipette method
(5) Silt percent determined by particle-size classifier
(6) Silt percent determined by photoextinction sedimentometer
(7) Silt percent determined by Andreasen's pipette method
(8) Sand percent determined by seiving.

We would like to determine if the three methods give equivalent results. A quick answer can be obtained by looking at the distributions obtained by the three measurement techniques for clay and for silt. But visual inspection will not provide a statistical test of the significance of the differences we may see. A one-way ANOVA will compare the means and variances of the three sets of measurements for either the silt or clay fractions. A two-way ANOVA can be used if we also wish to consider the silt and clay fractions simultaneously.

Some statistical programs insist that all of the response variables for different treatments must be contained in a single column, with the different treatments identified by labels in another column. A relatively easy way to do this is to re-arrange the data table in ITALNAVY.TXT so that the three columns for clay (or silt) are stacked on top of one another to form a single column, accompanied by a new column that contains the appropriate label for each entry. A two-way analysis will require stacking the silt and clay measurements together.

Are the three methods of sediment analysis equivalent? Do the procedures differ in the same sense for silt and clay? Is there a relationship between sediment composition and depth in the core?

Exercise 2.12

The landscape of northwestern England is marked by numerous drumlins, which form distinctive patterns of contours on topographic maps ("basket-of-eggs topography"). Students in a geomorphology class have collected information on drumlin fields in five locations: near Appleby, in Eden Valley, around Carlisle, at Tyne Gap, and in the Solway Lowlands. The five locations can be regarded as having been randomly selected from the drumlin region of northwestern England. For each site, 20 drumlins were selected at random on topographic maps and their long and short axes measured in the field. The measurements were used to calculate "elongation values" given in file DRUMLIN.TXT. Are there significant differences in elongation of drumlins at the various locations? What does this imply about the shapes of drumlins throughout the region?

Exercise 2.13

Major oxides were analyzed on 54 Schellerhau Granite samples collected in the Erzgebirge Mountains of eastern Germany. File SCHELLER.TXT contains values for these oxides plus loss on ignition (LOI); each analysis sums to 100%, within analytical and rounding error. Compute correlations between all pairs of the oxide constituents of the granites, but note that because of closure effects, the validity of the correlations are suspect. Use Aitchison's procedures to transform the data to logratios having SiO_2 as the common denominator and calculate correlations between pairs of the logratio variables. How has the pattern of correlations been changed?

Exercise 2.14

A scientific commission in Europe has attempted to standardize chemical analyses made for environmental purposes. As part of this effort, four different procedures for analysis of cobalt were evaluated at five different laboratories. Each laboratory was randomly assigned a ground sample which was split eight ways. The splits were randomly assigned to undergo the four analytical procedures so that each procedure was used to analyze two splits. Note that the experimental design is nested, because the splits were made at each laboratory. The data are given in file COBALT.TXT. Determine if there are significant differences between the results obtained by the laboratories and between the experimental procedures.

Suppose the ground sample had been split originally into 40 parts, and a random selection of eight splits sent to each laboratory for analysis. How would this alter the experimental design? Assume the data in file COBALT.TXT result from this scenario. Does this change your findings?

Exercise 2.15

An environmental study team is investigating possible contamination from a phosphate plant in an intermontane basin of a western U.S. state. The prevailing winds in the basin blow consistently in the same direction and there is concern that airborne particulate fluorides have been carried over pastures where it may have contaminated grass that is being consumed by grazing cattle. File COWURINE.TXT contains analyses of urinary fluoride concentrations, in ppm, collected from ten cows in a downwind pasture and six cows in a similar pasture that is several miles distant from the plant in an upwind direction (the samples were collected by the team's biologists, not by geologists). Because of the small sample size and lack of information on the form of the distribution of the variable, a nonparametric test seems appropriate to determine if the cattle in the downwind pasture show evidence of greater fluoride excretion than cattle in the upwind pasture. Do the data suggest significant differences?

Exercise 2.16

There is a relationship between topographic elevation and intensity of Bouger gravity measurements. Higher topographic elevations are indicative of a thicker crust and hence lower gravity because the crust tends to be in isostatic equilibrium and to displace mantle material of higher density. However, the relationship between elevation and gravity is imperfect, since intensity is also influenced by compositional variations in the crust and mantle, both below and adjacent to the location where the gravity is being measured. File BOUGER.TXT contains 166 Bouger gravity measurements in milligals along a traverse in southwestern Kansas. The file also contains the longitude and latitude in decimal degrees and the ground-surface elevation in meters at each of the measurement stations. Determine the correlation between Bouger gravity and topographic elevation for these data. Is the relationship statistically significant?

Exercise 2.17

A shallow core drilled through Paleocene strata on Spitzbergen Island in the Arctic Ocean north of Norway has been analyzed for the presence of light hydrocarbons (file SPTZBRGN.TXT). Because the interval is in the permafrost zone, researchers speculate that light hydrocarbons generated in the lower siltstone were diffusing upward through the pore water to the overlying sandstone when they were frozen in place. Elsewhere, such light hydrocarbons would have continued their movement, and this phase of primary migration would not have been recognized. Among the variables measured are total organic carbon (TOC) in percent and Rock-Eval pyrolysis hydrogen index in mg/g organic carbon (the Rock-Eval instrument, widely used in the petroleum industry, is a pyrolitic chromatograph specifically designed for source-rock evaluation). Is there a relationship between TOC and the hydrogen index? Do the values for these two variables differ significantly in the siltstone and the overlying sandstone?

SELECTED READINGS

Aitchison, J., 1981, A new approach to null correlations of proportions: *Mathematical Geology,* v. 13, no. 2, p. 175–189.

Aitchison, J., 1986, *The Statistical Analysis of Compositional Data:* John Wiley & Sons, Inc., New York, 416 pp. *The basic reference on analysis of constant-sum data. Most examples are geologic.*

Aitchison, J., and J.A.C. Brown, 1969, *The Lognormal Distribution: With Special Reference to Its Uses in Economics:* Cambridge Univ. Press, Cambridge, U.K., 176 pp.

Ash, R.B., 1970, *Basic Probability Theory:* John Wiley & Sons, Inc., New York, 337 pp. *Contains an extensive discussion of the various philosophical bases of statistical analysis and probability.*

Ball, L.D., P.W.M. Corbett, J.L. Jensen, and J.J.M. Lewis, 1997, The role of geology in the behavior and choice of permeability predictors: *SPE Formation Evaluation,* v. 12, no. 1, p. 32–39.

Chayes, F., 1971, *Ratio Correlation: A Manual for Students of Petrology and Geochemistry:* Univ. Chicago Press, Chicago, Ill., 99 pp. *An extensive discussion of the statistical behavior of ratio and other closed forms of data. The examples are from geochemistry.*

Cheeney, R.F., 1983, *Statistical Methods in Geology for Field and Lab Decisions:* Allen & Unwin Ltd., London, 169 pp. *Four chapters of this book are devoted to nonparametric statistical tests, which the author advocates because they are easy to compute in the field.*

Cochran, W.G., 1952, The χ^2 test of goodness-of-fit: *Annals of Mathematical Statistics,* v. 3, p. 315–345.

Conover, W.J., 1999, *Practical Nonparametric Statistics,* 3rd ed.: John Wiley & Sons, Inc., New York, 584 pp. *Most of the nonparametric tests mentioned in this chapter are discussed at length in this book. It contains a particularly thorough treatment of runs tests.*

Devore, J.L., 1995, *Probability and Statistics for Engineering and the Sciences,* 4th ed.: Wadsworth Publ. Co., Inc., Belmont, Calif., 743 pp., diskette.

Duckworth, W.E., 1968, *Statistical Techniques in Technological Research*: Methuen & Co., London, 303 pp. *One of the most readable statistics books designed for the scientist who wishes to apply statistical techniques. Emphasis is on experimental designs, some of which are rather sophisticated. However, the excellent presentation manages to smooth out the going as much as possible.*

Eisenhart, C., 1947, The assumptions underlying the analysis of variance: *Biometrics,* v. 3, no. 1, p. 1–21.

Fisher, R.A., 1973, *Statistical Methods and Scientific Inferences,* Revised ed.: Hafner Press, New York, 175 pp. *A very readable discussion of the evolution of statistical thought. Chapters 2 and 5 illustrate the basis, and some of the consequences, of different concepts of probability.*

Folk, R.L., 1951, Stages of textural maturity in sedimentary rocks: *Jour. Sedimentary Petrology,* v. 21, p. 127–130.

Freund, J.E., and F.J. Williams, 1991, *Dictionary/Outline of Basic Statistics*: Dover Publications, Inc., New York, 195 pp. *An essential requirement for every student or applier of statistics. Available as an inexpensive paperback, this is a well-written dictionary of most statistical terms, compilation of common (and some uncommon) statistical formulas and tests, and set of statistical tables.*

Gibbons, J.D., 1993a, *Nonparametric Statistics, An Introduction*: Quantitative Applications in the Social Sciences, Series 07-90, Sage Publications, Inc., Newbury Park, Calif., 87 pp. *A compact summation of the major elements of nonparametric statistical tests.*

Gibbons, J.D., 1993b, *Nonparametric Measures of Association*: Quantitative Applications in the Social Sciences, Series 07-91, Sage Publications, Inc., Newbury Park, Calif., 97 pp.

Gibbons, J.D., and S. Chakraborti, 1992, *Nonparametric Statistical Inference,* 3rd ed.: Marcel Dekker, New York, 544 pp.

Griffiths, J.C., 1967, *Scientific Method in Analysis of Sediments*: McGraw–Hill, Inc., New York, 508 pp. *Chapters 13 through 22 (approximately the second half of the book) actually are an introductory text on applied statistics. The book, unfortunately now out of print, is especially valuable for its consideration of sampling problems and emphasis on correct statistical design.*

Guenther, W.C, 1973, *Concepts of Statistical Inference,* 2nd ed.: McGraw–Hill, Inc., New York, 512 pp. *A concise introductory statistics text for those who want the maximum amount of information from a minimum of reading. Covers a wide spectrum of topics not usually found in an introductory text.*

Hahn, G.J., and W.Q. Meeker, 1991, *Statistical Intervals: A Guide for Practitioners*: John Wiley & Sons, Inc., New York, 392 pp. *An encyclopedia of information on all sorts of statistical intervals, including confidence intervals.*

Hald, A., 1952, *Statistical Tables and Formulas*: John Wiley & Sons, Inc., New York, 97 pp.

Harbaugh, J.W., J.C. Davis, and J. Wendebourg, 1995, *Computing Risk for Oil Prospects: Principles and Programs*: Pergamon Press, Oxford, 452 pp., 2 diskettes.

Hicks, C.R., and K.V. Turner, Jr., 1999, *Fundamental Concepts in the Design of Experiments*, 5th ed.: Oxford Univ. Press, New York, 565 pp. *An intermediate text on experimental design and analysis of variance with a graphic explanation of interaction in two-way designs, including examples.*

Hollander, M., and D.A. Wolfe, 1999, *Nonparametric Statistical Methods*, 2nd ed.: John Wiley & Sons, Inc., New York, 787 pp.

Koch, G.S., Jr., and R.F. Link, 1980, *Statistical Analysis of Geological Data*: Dover Publications, Inc., New York, 850 pp. *A one-volume paperback reprint of the original two-volume edition, this contains a detailed discussion of statistical analysis of geologic data, especially mine-assay values. Includes an extensive treatment of sample designs, the use of analysis of variance methods, and multiple regression. A number of specialized topics and examples are considered in detail.*

Kork, J.O., 1977, Examination of the Chayes-Kruskal procedure for testing correlations between proportions: *Mathematical Geology*, v. 9, no. 6, p. 543–562.

Krumbein, W.C., and F.A. Graybill, 1965, *An Introduction to Statistical Models in Geology*: McGraw–Hill, Inc., New York, 475 pp. *It is essential that any serious practitioner of geostatistics master this classic reference, now out of print. Concentrate first on Chapters 6 through 10.*

Krumbein, W.C., and F.J. Pettijohn, 1938, *Manual of Sedimentary Petrography*: Appleton–Century–Crofts, Inc., New York, 549 pp.

Lapin, L.L., 1982, *Statistics for Modern Business Decisions*, 3rd ed.: Harcourt Brace Jovanovich, Inc., New York, 887 pp. *Chapter 7 contains results of a simulation experiment to demonstrate the central limits theorem.*

Li, J.C.R., 1964, *Statistical Inference*, Vol. 1: Edwards Bros., Inc., Ann Arbor, Mich., 658 pp. *The first volume of an encyclopedic coverage of elementary statistics. The approach is largely intuitive. The discussion of the χ^2 distribution in Chapter 7 is especially helpful.*

Lilliefors, H.W., 1967, On the Kolmogorov–Smirnov test for normality with mean and variance unknown: *Jour. American Statistical Assoc.*, v. 62, p. 399–402.

McCray, A.W., 1975, *Petroleum Evaluations and Economic Decisions*: Prentice Hall, Inc., Englewood Cliffs, N.J., 448 pp. *Chapter 3 is an extensive discussion of probability distributions that arise in petroleum exploration.*

Morrison, D.F., 1983, *Applied Linear Statistical Methods*: Prentice Hall, Inc., Englewood Cliffs, N.J., 562 pp.

Neter, J., W. Wasserman, and M.H. Kutner, 1985, *Applied Linear Statistical Models*, 2nd ed.: Richard D. Irwin, Inc., Homewood, Ill., 1127 pp.

Noether, G.E., 1991, *Introduction to Statistics: The Nonparametric Way*: Springer-Verlag, New York, 414 pp.

Parzen, E., 1960, *Modern Probability Theory and Its Applications*: John Wiley & Sons, Inc., New York, 464 pp.

Pyle, H.C., and J.E. Sherborne, 1939, Core analysis: *Trans. AIME Petrol. Tech.*, v. 132, p. 33–61.

Reyment, R.A., 1971, *Introduction to Quantitative Paleoecology*: Elsevier Publ. Co., Amsterdam, 227 pp. *An extremely readable book on the application of elementary statistics to problems in ecology and paleoecology. Chapter 5 covers life tables, a topic we have not considered in this text. Many nonparametric methods are also described.*

Robinson, J.E., 1982, *Computer Applications in Petroleum Geology*: Hutchinson Ross Publ. Co., Stroudsburg, Pa., 164 pp. *An inexpensive paperback, with emphasis on mapping.*

Rock, N.M.S., 1988, *Numerical Geology*: Springer-Verlag, Berlin, 427 pp. *Combines concise discussions of methods with extensive citations to the geological literature through the 1980's. Includes a 49-p. glossary of terms and a 51-p. bibliography of mathematical geology.*

Rollinson, H.R., 1993, *Using Geochemical Data: Evaluation, Presentation, Interpretation*: Longman Scientific & Technical, Harlow, Essex, U.K., 352 pp. *Contains a very good discussion of compositional data and the dangers of attempting to interpret patterns in plots of ratios of constant-sum variables.*

Saager, R., and A.J. Sinclair, 1974, Factor analysis of stream sediment geochemical data from the Mount Nansen area, Yukon Territory, Canada: *Mineralogica Deposita*, v. 9, p. 243–252.

Siegel, S., and N.J. Castellan, Jr., 1988, *Nonparametric Statistics for the Behavioral Sciences, 2nd ed.*: McGraw-Hill, Inc., New York, 399 pp. *A revised edition of the first applied book on nonparametric statistics.*

Tukey, J.W., 1977, *Exploratory Data Analysis*: Addison-Wesley Publ. Co., Inc., Reading, Mass., 688 pp.

von Mises, R., 1981, *Probability, Statistics, and Truth*: Dover Publications, Inc., New York, 244 pp. *Although dated, this translation of a 1939 German classic is a basic introduction to the philosophy of statistics. The first three chapters are especially pertinent to the definition of probability.*

Walker, H.M., 1940, Degrees of freedom: *Jour. Educational Psychology*, v. 31, p. 253–269. *An excellent tutorial on the concept of degrees of freedom.*

Wallis, W.A., and H.V. Roberts, 1960, *Statistics, A New Approach*: Free Press, Glencoe, Ill., 646 pp. *An entertaining introduction to statistics, with many examples, stories, and anecdotes about the origin of statistical procedures and their applications. For those who prefer narratives to mathematics.*

Winsauer, W.O., H.M. Shearin, Jr., P.H. Masson, and M. Williams, 1952, Resistivity of brine-saturated sands in relation to pore geometry: *Bull. Am. Assoc. Petroleum Geologists*, v. 36, p. 253–277. *Source of the "Humble Equation" in petrophysics.*

Chapter 3
Matrix Algebra

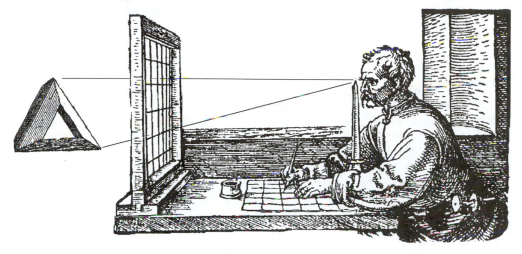

This chapter is devoted to matrix algebra. Most of the methods we will discuss in subsequent chapters are based on matrix manipulations, especially as performed by computers. In this chapter, we will examine the mathematical operations that underlie such techniques as trend-surface analysis, principal components, and discriminant functions. These techniques are almost impossible to apply without the help of computers, because the calculations are complicated and must be performed repetitively. However, with matrix algebra we can express the basic principles involved in a manner that is succinct and easily understood. Once you master the rudiments of matrix algebra, you will be able to see the fundamental structure within the complex procedures we will examine later.

Most geologists probably have not taken a course in matrix algebra. This is unfortunate; the subject is not difficult and is probably one of the most useful tools in mathematics. College courses in matrix algebra usually are sprinkled liberally with theorems and their proofs. Such an approach is certainly beyond the scope of this short chapter, so we will confine ourselves to those topics pertinent to techniques that we will utilize later. Rather than giving derivations and proofs, the material will be presented by examples.

The Matrix

A *matrix* is a rectangular array of numbers, exactly the same as a table of data. In matrix algebra, the array is considered to be a single entity rather than a collection of individual values and is operated upon as a unit. This results in a great simplification of the statement of complicated procedures and relationships. Individual numbers within a matrix are called the *elements* of the matrix and are identified by subscripts. The first subscript specifies the row in which the element occurs and the second specifies the column. The individual elements of a matrix may be measurements of variables, variances or covariances, sums of observations, terms in a series of simultaneous equations or, in fact, any set of numbers.

As an example, in Chapter 2 you were asked to compute the variances and covariances of trace-element data given in **Table 2–3**. Your answers can be arranged in the form of the matrix below.

$$\begin{bmatrix} \text{var}_{Cr} & \text{cov}_{Cr\,Ni} & \text{cov}_{Cr\,V} \\ \text{cov}_{Ni\,Cr} & \text{var}_{Ni} & \text{cov}_{Ni\,V} \\ \text{cov}_{V\,Cr} & \text{cov}_{V\,Ni} & \text{var}_{Ni} \end{bmatrix} = \begin{bmatrix} 570 & 537.5 & 663.75 \\ 537.5 & 562.5 & 718.75 \\ 663.75 & 718.75 & 1007.5 \end{bmatrix}$$

We can designate a matrix (perhaps containing values of several variables) symbolically by capital letters such as [X], **X**, (X), or ‖X‖. In a change from earlier editions of this book, we will adopt the commonly used boldface notation for matrices. Individual entries in a matrix, or its elements, are indicated by subscripted italic lowercase letters such as x_{ij}. Particularly in older books, you may encounter different conventions for denoting individual elements of a matrix. The symbol x_{ij} is the element in the ith row and the jth column of matrix **X**. For example, if **X** is the 3×3 matrix

$$\mathbf{X} = \begin{bmatrix} 1 & 4 & 7 \\ 2 & 5 & 8 \\ 3 & 6 & 9 \end{bmatrix}$$

x_{33} is 9, x_{13} is 7, x_{21} is 2, and so on. The **order** of a matrix is an expression of its size, in the sense of the number of rows and/or the number of columns it contains. So, the order of **X**, above, is 3. If the number of rows equals the number of columns, the matrix is **square**. Entries in a square matrix whose subscripts are equal (*i.e.*, $i = j$) are called the **diagonal elements**, and they lie on the **principal diagonal** or **major diagonal** of the matrix. In the matrix of trace-element variances and covariances, the variances lie on the diagonal and the off-diagonal elements are the covariances. The diagonal elements in the matrix above are 1, 5, and 9. Although data arrays usually are in the form of rectangular matrices, often we will create square matrices from them by calculating their variances and covariances or other summary statistics. Many useful operations that can be performed on square matrices are not possible with nonsquare matrices. However, two forms of nonsquare matrices are especially important; these are the **vectors**, $1 \times m$ (row vector) and $m \times 1$ (column vector).

Certain square matrices have special importance and are designated by name. A **symmetric matrix** is a square matrix in which all observations $x_{ij} = x_{ji}$, as for example

$$\begin{bmatrix} 1 & 2 & 3 \\ 2 & 4 & 5 \\ 3 & 5 & 6 \end{bmatrix}$$

The variance–covariance matrix of trace elements given above is another example of a square matrix that is symmetrical about the diagonal.

A **diagonal matrix** is a square, symmetric matrix in which all the off-diagonal elements are 0. If all of the diagonal elements of a diagonal matrix are equal, the matrix is a **scalar matrix**. Finally, a scalar matrix whose diagonal elements are equal to 1 is called an **identity matrix** or **unit matrix**. An identity matrix is almost always indicated by **I**:

$$I = \begin{bmatrix} 1 & 0 & 0 \\ 0 & 1 & 0 \\ 0 & 0 & 1 \end{bmatrix}$$

Elementary Matrix Operations

Addition and subtraction of matrices obey the rules of algebra of ordinary numbers, with one important additional characteristic. The two matrices being added or subtracted must be of the same order; that is, they must have the same number of rows and columns.

To perform the operation $C = A + B$, every element of **A** is added to its corresponding element in **B**. If the matrices are not of the same order, there will be leftover elements, and the operation cannot be completed. Subtraction, such as $C = A - B$, proceeds in exactly the same manner, with every element of **B** subtracted from its corresponding element in **A**.

Table 3–1. Bentonite production in Wyoming, 1964.

District	Clay (100,000 tons)		
	Drilling Mud	Foundry Clay	Miscellaneous
Eastern	105	63	5
Montana Border	218	80	2
Central	220	76	1

As an illustration, **Table 3–1** contains 1964 production figures for bentonite from three mining districts in Wyoming. Three major grades of clay were produced: clay for drilling mud; foundry clay; and a miscellaneous category that includes cattle feed binder, drug and cosmetic uses, and pottery clay. These data can be expressed in a 3×3 matrix, **A**:

$$A = \begin{bmatrix} 105 & 63 & 5 \\ 218 & 80 & 2 \\ 220 & 76 & 1 \end{bmatrix}$$

Production figures for the following year may be expressed in the same manner, giving the matrix **B**:

$$B = \begin{bmatrix} 84 & 102 & 4 \\ 240 & 121 & 1 \\ 302 & 28 & 0 \end{bmatrix}$$

Total production for the 2 years in the three districts is the sum, **C**, of the the matrices **A** and **B**:

$$\begin{array}{ccccc} A & + & B & = & C \end{array}$$

$$\begin{bmatrix} 105 & 63 & 5 \\ 218 & 80 & 2 \\ 220 & 76 & 1 \end{bmatrix} + \begin{bmatrix} 84 & 102 & 4 \\ 240 & 121 & 1 \\ 302 & 28 & 0 \end{bmatrix} = \begin{bmatrix} 189 & 165 & 9 \\ 458 & 201 & 3 \\ 522 & 104 & 1 \end{bmatrix}$$

Similarly, the change in production can be found by subtracting:

$$
\begin{array}{ccccc}
\mathbf{B} & - & \mathbf{A} & = & \mathbf{D}
\end{array}
$$

$$
\begin{bmatrix}
84 & 102 & 4 \\
240 & 121 & 1 \\
302 & 28 & 0
\end{bmatrix}
-
\begin{bmatrix}
105 & 63 & 5 \\
218 & 80 & 2 \\
220 & 76 & 1
\end{bmatrix}
=
\begin{bmatrix}
-21 & 39 & -1 \\
22 & 41 & -1 \\
82 & -48 & -1
\end{bmatrix}
$$

Note that **A** was subtracted from **B** simply to show increases in production as positive values.

As in ordinary algebra, $\mathbf{A} + \mathbf{B} = \mathbf{B} + \mathbf{A}$, and $(\mathbf{A} + \mathbf{B}) + \mathbf{C} = \mathbf{A} + (\mathbf{B} + \mathbf{C})$, provided all are $n \times m$ matrices. The order of subtraction is, of course, mandatory.

Transposition is a matrix operation in which rows become columns and columns become rows. Each element x_{ij} becomes the element x_{ji} in the transpose. The operation is indicated symbolically by $\mathbf{X}^T$ or by $\mathbf{X}'$. So,

$$
\mathbf{X} =
\begin{bmatrix}
1 & 4 \\
2 & 5 \\
3 & 6
\end{bmatrix}
\qquad
\mathbf{X}^T =
\begin{bmatrix}
1 & 2 & 3 \\
4 & 5 & 6
\end{bmatrix}
$$

Note that the first row has become the first column of the transpose, and the second row has become the second column. In some of the calculations we will consider later, a row vector, **A**, becomes a column vector, $\mathbf{A}^T$, when transposed, and *vice versa*. The row and column vectors

$$
\mathbf{A} =
\begin{bmatrix}
1 & 2 & 3 & 4
\end{bmatrix}
\qquad
\mathbf{A}^T =
\begin{bmatrix}
1 \\
2 \\
3 \\
4
\end{bmatrix}
$$

are the transpose of each other.

A matrix may be *multiplied by a constant* by multiplying each element in the matrix by the constant. For example

$$
3 \times
\begin{bmatrix}
1 & 4 \\
2 & 5 \\
3 & 6
\end{bmatrix}
=
\begin{bmatrix}
3 & 12 \\
6 & 15 \\
9 & 18
\end{bmatrix}
$$

Strictly speaking, a matrix cannot be divided by a constant, but we can perform an equivalent operation. If we multiply a matrix by a value equal to the inverse of a constant, we obtain the same numerical result as if we divided each element of the matrix by the constant. The inverse of the constant, c, is indicated by c^{-1}, which represents $1/c$.

Table 3–2. Measurements of axes of pebbles (in inches) collected from glacial till.

		Axis	
Sample	a	b	c
1	3.4	2.2	1.8
2	4.6	4.3	4.2
3	5.4	4.7	4.7
4	3.9	2.8	2.3
5	5.1	4.9	3.8

As a simple example, consider **Table 3–2,** which contains measurements of the a-, b-, and c-axes of chert pebbles collected in a glacial till. The measurements were recorded in inches and we wish to convert them to millimeters. If the data are expressed in the form of the matrix **E**, we may multiply **E** by the constant 25.4 to obtain a matrix containing the measurements in millimeters:

$$25.4 \quad \times \quad \mathbf{E} \quad = \quad \mathbf{M}$$

$$25.4 \times \begin{bmatrix} 3.4 & 2.2 & 1.8 \\ 4.6 & 4.3 & 4.2 \\ 5.4 & 4.7 & 4.7 \\ 3.9 & 2.8 & 2.3 \\ 5.1 & 4.9 & 3.8 \end{bmatrix} = \begin{bmatrix} 86.36 & 55.88 & 45.72 \\ 116.84 & 109.22 & 106.68 \\ 137.16 & 119.38 & 119.38 \\ 99.06 & 71.12 & 58.42 \\ 129.54 & 124.46 & 96.52 \end{bmatrix}$$

Matrix Multiplication

Recall the coin-flipping problem from Chapter 2, where we considered the probability of obtaining a succession of heads if the probability of heads on one flip was $1/2$. The probability that we would get three heads in a row was $1/2 \times 1/2 \times 1/2$, or $1/2^3$. We can develop an equivalent set of probabilities for lithologies encountered in a stratigraphic section. Suppose we have measured an outcrop and identified the units as sandstone, shale, or limestone. At every foot, the rock type can be categorized and the type immediately above noted. We would eventually build a matrix of frequencies similar to that below. This is called a *transition frequency matrix* and tells us, for example, that sandstone is followed by shale 18 times, but followed by limestone only 2 times. Similarly, limestone follows shale 41 times, succeeds itself 51 times, but follows sandstone only 2 times:

		To		
		Sandstone	Shale	Limestone
	Sandstone	59	18	2
From	Shale	14	86	41
	Limestone	4	34	51

We can convert these frequencies to probabilities by dividing each element in a row by the total of the row. This will give the *transition probability matrix* shown below, from which the probability of proceeding from one state to another can be assessed. This subject will be considered in detail in a later chapter, where its use in time-series analysis will be examined. Now, however, we are interested in the matrix of probabilities, which is analogous to the single probability associated with the flip of a coin:

		To		
		Sandstone	Shale	Limestone
	Sandstone	0.74	0.23	0.03
From	Shale	0.10	0.61	0.29
	Limestone	0.05	0.38	0.57

Just as we can find the probability of producing a string of heads in a coin-flipping experiment by powering the probability associated with a single flip, we can determine the probability of attaining specified states at successive intervals by powering the transition probability matrix. That is, the probability matrix, **P**, after n steps through the succession is equal to $\mathbf{P}^n$. The nth power of a matrix is simply the matrix times itself n times. To perform this operation, however, we must know the special procedures of matrix multiplication.

The simplest form of multiplication involves two square matrices, **A** and **B**, of equal size, producing the product matrix, **C**. An easy method of performing this operation is to arrange the matrices in the following manner:

$$\begin{bmatrix} b_{11} & b_{12} & b_{13} \\ b_{21} & b_{22} & b_{23} \\ b_{31} & b_{32} & b_{33} \end{bmatrix}$$

$$\begin{bmatrix} a_{11} & a_{12} & a_{13} \\ a_{21} & a_{22} & a_{23} \\ a_{31} & a_{32} & a_{33} \end{bmatrix} \begin{bmatrix} c_{11} & c_{12} & c_{13} \\ c_{21} & c_{22} & c_{23} \\ c_{31} & c_{32} & c_{33} \end{bmatrix}$$

To obtain the value of an element c_{ij}, multiply each element of row i of **A**, starting at the left, by each element of column j of **B**, starting at the top. All the products are summed to obtain the c_{ij} element of the answer. The steps in multiplication are demonstrated below on the two matrices,

$$\begin{bmatrix} 1 & 4 & 7 \\ 2 & 5 & 8 \\ 3 & 6 & 9 \end{bmatrix} \times \begin{bmatrix} 1 & 2 & 3 \\ 3 & 4 & 5 \\ 5 & 6 & 7 \end{bmatrix}$$

First, multiply a_{11} by $b_{11} = 1$,

$$\begin{bmatrix} ① & 4 & 7 \\ 2 & 5 & 8 \\ 3 & 6 & 9 \end{bmatrix} \times \begin{bmatrix} ① & 2 & 3 \\ 3 & 4 & 5 \\ 5 & 6 & 7 \end{bmatrix}$$

Then, a_{12} by $b_{21} = 12$,

$$\begin{bmatrix} 1 & ④ & 7 \\ 2 & 5 & 8 \\ 3 & 6 & 9 \end{bmatrix} \times \begin{bmatrix} 1 & 2 & 3 \\ ③ & 4 & 5 \\ 5 & 6 & 7 \end{bmatrix}$$

Finally, a_{13} by $b_{31} = 35$,

$$\begin{bmatrix} 1 & 4 & ⑦ \\ 2 & 5 & 8 \\ 3 & 6 & 9 \end{bmatrix} \times \begin{bmatrix} 1 & 2 & 3 \\ 3 & 4 & 5 \\ ⑤ & 6 & 7 \end{bmatrix}$$

The entry c_{11} is the sum of these three values, $1 + 12 + 35 = 48$. These steps can be summarized in the diagram below. Note that each entry c_{ij} in the product matrix results from multiplying and summing the products of elements in the ith row of matrix **A** by elements in the jth column of matrix **B**.

To find element c_{11} 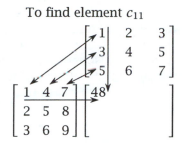   To find element c_{32}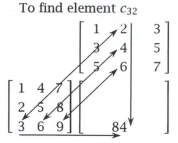

The completed matrix multiplication has the appearance

$$
\begin{array}{c}
\begin{bmatrix} 1 & 2 & 3 \\ 3 & 4 & 5 \\ 5 & 6 & 7 \end{bmatrix} \\
\begin{bmatrix} 1 & 4 & 7 \\ 2 & 5 & 8 \\ 3 & 6 & 9 \end{bmatrix} \begin{bmatrix} 48 & 60 & 72 \\ 57 & 72 & 87 \\ 66 & 84 & 102 \end{bmatrix}
\end{array}
$$

In general, if the order of multiplication is reversed to $\mathbf{B} \times \mathbf{A} = \mathbf{C}$, a different answer will be obtained:

$$
\begin{array}{c}
\begin{bmatrix} 1 & 4 & 7 \\ 2 & 5 & 8 \\ 3 & 6 & 9 \end{bmatrix} \\
\begin{bmatrix} 1 & 2 & 3 \\ 3 & 4 & 5 \\ 5 & 6 & 7 \end{bmatrix} \begin{bmatrix} 14 & 32 & 50 \\ 26 & 62 & 98 \\ 38 & 92 & 146 \end{bmatrix}
\end{array}
$$

In the operation $\mathbf{A} \times \mathbf{B} = \mathbf{C}$, the matrix $\mathbf{B}$ is said to be **_premultiplied_** by $\mathbf{A}$. Similarly, the matrix $\mathbf{A}$ can be said to be **_postmultiplied_** by $\mathbf{B}$. This is simply a verbal way of specifying the order of multiplication.

If two square matrices are multiplied, the product is a square matrix of the same size. However, if an $m \times n$ matrix is multiplied by an $n \times r$ matrix, the result is an $m \times r$ matrix. That is, the product matrix has the same number of rows as the premultiplier matrix on the left and the same number of columns as the postmultiplier matrix on the right. For example, premultiplying a 3×2 matrix by a 5×3 matrix results in a 5×2 matrix:

$$
\begin{bmatrix} 1 & 2 & 4 \\ 2 & 1 & 2 \\ 3 & 1 & 1 \\ 2 & 3 & 1 \\ 1 & 2 & 0 \end{bmatrix} \times \begin{bmatrix} 3 & 4 \\ 2 & 1 \\ 0 & 1 \end{bmatrix} = \begin{bmatrix} 7 & 10 \\ 8 & 11 \\ 11 & 14 \\ 12 & 12 \\ 7 & 6 \end{bmatrix}
$$

However, the 3×2 matrix cannot be postmultiplied by the 5×3 matrix because the number of columns (two) in the left matrix would not equal the number of rows (five) in the right matrix.

Multiplying a matrix by its transpose results in a square, symmetric matrix product whose size is determined by the order of multiplication. Typically, a data array consists of n rows and m columns, where n is much larger than m. If such an array is premultiplied by its transpose, the **minor product matrix** will be $m \times m$:

$$\begin{bmatrix} 1 & 2 & 3 \\ 4 & 5 & 6 \end{bmatrix} \times \begin{bmatrix} 1 & 4 \\ 2 & 5 \\ 3 & 6 \end{bmatrix} = \begin{bmatrix} 14 & 32 \\ 32 & 77 \end{bmatrix}$$

But reversing the order of multiplication yields the $n \times n$ **major product matrix**:

$$\begin{bmatrix} 1 & 4 \\ 2 & 5 \\ 3 & 6 \end{bmatrix} \times \begin{bmatrix} 1 & 2 & 3 \\ 4 & 5 & 6 \end{bmatrix} = \begin{bmatrix} 17 & 22 & 27 \\ 22 & 29 & 36 \\ 27 & 36 & 45 \end{bmatrix}$$

The equation for the general case of matrix multiplication is

$$c_{ij} = \sum_{k=1}^{n} a_{ik} b_{kj} \tag{3.1}$$

In a series of multiplications, the sequence in which the multiplications are accomplished is not mandatory if the arrangement is not changed. That is,

$$\mathbf{A} \times \mathbf{B} \times \mathbf{C} = (\mathbf{A} \times \mathbf{B}) \times \mathbf{C} = \mathbf{A} \times (\mathbf{B} \times \mathbf{C})$$

Because powering is simply a series of multiplications, a square matrix can be raised to a power. So,

$$\mathbf{A}^2 = \mathbf{A} \times \mathbf{A}$$

and

$$\mathbf{A}^3 = \mathbf{A}^2 \times \mathbf{A} = \mathbf{A} \times \mathbf{A} \times \mathbf{A}$$

Note that nonsquare matrices cannot be powered, because the number of rows and columns of a rectangular matrix would not accord if the matrix were multiplied by itself.

As an example, we can power the array of transition probabilities discussed at the first of this section. In matrix form,

$$\mathbf{T} = \begin{bmatrix} 0.74 & 0.23 & 0.03 \\ 0.10 & 0.61 & 0.29 \\ 0.05 & 0.38 & 0.57 \end{bmatrix}$$

So,

$$\mathbf{T}^2 = \begin{bmatrix} 0.572 & 0.322 & 0.106 \\ 0.150 & 0.505 & 0.345 \\ 0.104 & 0.460 & 0.437 \end{bmatrix}$$

and

$$\mathbf{T}^3 = \begin{bmatrix} 0.461 & 0.368 & 0.171 \\ 0.178 & 0.474 & 0.348 \\ 0.144 & 0.470 & 0.385 \end{bmatrix}$$

If we continue to power the transition probability matrix, it converges to a stable configuration (called the *stationary probability matrix*) in which each column of the matrix is a constant. These are the proportions of the specific lithologies represented by the columns. In this example, the proportions are 23% sandstone, 45% shale, and 32% limestone. We can see that the columns are converging on these values at the 10th power of **T**:

$$\mathbf{T}^{10} = \begin{bmatrix} 0.248 & 0.443 & 0.309 \\ 0.230 & 0.449 & 0.321 \\ 0.228 & 0.450 & 0.322 \end{bmatrix}$$

Square matrices also can be raised to a fractional power, most commonly to the one-half power. This is equivalent to finding the square root of the matrix. That is, $\mathbf{A}^{1/2}$ is a matrix, **X**, whose square is **A**:

$$\mathbf{A}^{1/2} = \mathbf{X}$$
$$\mathbf{X}^2 = \mathbf{X} \times \mathbf{X} = \mathbf{A}$$

Finding fractional powers of matrices can be computationally troublesome. Fortunately, in the applications we will consider, we will only need to find the fractional powers of diagonal matrices, which have special properties that make it easy to raise them to a fractional power. If we raise the diagonal matrix **A** to the one-half power, the result is a diagonal matrix whose nonzero elements are equal to the square roots of the equivalent elements in **A**. For example, if **A** is 3×3,

$$\begin{bmatrix} a_{11} & 0 & 0 \\ 0 & a_{22} & 0 \\ 0 & 0 & a_{33} \end{bmatrix}^{1/2} = \begin{bmatrix} \sqrt{a_{11}} & 0 & 0 \\ 0 & \sqrt{a_{22}} & 0 \\ 0 & 0 & \sqrt{a_{33}} \end{bmatrix}$$

As we defined it earlier, the identity matrix is a special diagonal matrix in which the diagonal terms are all equal to 1. The identity matrix has an extremely useful property; if a matrix is multiplied by an identity matrix, the resulting product is exactly the same as the initial matrix:

$$\mathbf{A} \quad \times \quad \mathbf{I} \quad = \quad \mathbf{A}$$

$$\begin{bmatrix} 1 & 4 & 7 \\ 2 & 5 & 8 \\ 3 & 6 & 9 \end{bmatrix} \times \begin{bmatrix} 1 & 0 & 0 \\ 0 & 1 & 0 \\ 0 & 0 & 1 \end{bmatrix} = \begin{bmatrix} 1 & 4 & 7 \\ 2 & 5 & 8 \\ 3 & 6 & 9 \end{bmatrix}$$

Thus, the identity matrix corresponds to the 1 of ordinary multiplication. This property is especially important in operations in the following sections.

Inversion and Solution of Simultaneous Equations

Division of one matrix by another, in the sense of ordinary algebraic division, cannot be performed. However, by utilizing the rules of matrix multiplication, an operation can be performed that is equivalent to solving the equation

$$A \times X = B$$

for the unknown matrix, X, when the elements of A and B are known. This is one of the most important techniques in matrix algebra, and it is essential for the solution of simultaneous equations such as those of trend-surface analysis and discriminant functions. The techniques of matrix inversion will be encountered again and again in the next chapters of this book.

The equation given above is solved by finding the inverse of matrix A. The *inverse matrix* (or *reciprocal* matrix) A^{-1} is one that satisfies the relationship $A \times A^{-1} = I$. If both sides of a matrix equation are multiplied by A^{-1}, the matrix A is effectively removed from the left side. At the same time, B is converted into a quantity that is the value of the unknown matrix X. The matrix A must be a square matrix. Beginning with

$$A \times X = B$$

premultiply both sides by the inverse of A, or A^{-1}:

$$A^{-1} \times A \times X = A^{-1} \times B$$

Since $A^{-1} \times A = I$ and $I \times X = X$, the equation reduces to

$$X = A^{-1} \times B \tag{3.2}$$

Thus, the problem of division by a matrix reduces to one of finding a matrix that satisfies the reciprocal relationship. In some situations, an inverse cannot be found because division by zero is encountered during the inversion process. A matrix with no inverse is called a *singular matrix*, and presents problems beyond the scope of this chapter.

The inversion procedure may be illustrated by solving the following pair of simultaneous equations in matrix form. The unknown coefficients are $x_1 = 2$ and $x_2 = 3$. We will attempt to recover them by a process of matrix inversion and multiplication:

$$4x_1 + 10x_2 = 38$$
$$10x_1 + 30x_2 = 110$$

This is a set of equations of the general type

$$A X = B$$

where A is a matrix of coefficients, X is a column vector of unknowns, and B is a column vector of right-hand sides of the equations. In the specific set of equations given above, we have

$$\begin{bmatrix} 4 & 10 \\ 10 & 30 \end{bmatrix} \times \begin{bmatrix} x_1 \\ x_2 \end{bmatrix} = \begin{bmatrix} 38 \\ 110 \end{bmatrix}$$

To solve the equation, the matrix A will be inverted and B will be multiplied by A^{-1} to give the solution for X.

It may not be apparent why the set of simultaneous equations can be set into the matrix form shown. You can satisfy yourself on this point, however, by multiplying the two terms, $\mathbf{A}\mathbf{X}$, to obtain the left-hand side of the simultaneous equation set:

$$\begin{bmatrix} 4 & 10 \\ 10 & 30 \end{bmatrix} \times \begin{bmatrix} x_1 \\ x_2 \end{bmatrix} = \begin{bmatrix} 4x_1 + 10x_2 \\ 10x_1 + 30x_2 \end{bmatrix}$$

Working through this multiplication, you will see that all of the terms are associated with the proper coefficients. By the rules of matrix multiplication,

$$\begin{bmatrix} 4 & 10 \\ 10 & 30 \end{bmatrix} \begin{bmatrix} x_1 \\ x_2 \end{bmatrix} = \begin{bmatrix} 4x_1 + 10x_2 \\ \end{bmatrix}$$

Then, multiplying the bottom row,

$$\begin{bmatrix} 4 & 10 \\ 10 & 30 \end{bmatrix} \begin{bmatrix} x_1 \\ x_2 \end{bmatrix} = \begin{bmatrix} \\ 10x_1 + 30x_2 \end{bmatrix}$$

We will solve the simultaneous equation set by first inverting the term $\mathbf{A}$. Place the $\mathbf{A}$ matrix beside an identity matrix, $\mathbf{I}$, and perform all operations simultaneously on both matrices. The purpose of each operation is to convert the diagonal elements of $\mathbf{A}$ to ones and the off-diagonal elements to zeros. This is done by dividing rows of the matrix by constants and subtracting (or adding) rows of the matrix from other rows:

1. $\begin{bmatrix} 4 & 10 \\ 10 & 30 \end{bmatrix} \begin{bmatrix} 1 & 0 \\ 0 & 1 \end{bmatrix}$ The matrix $\mathbf{A}$ is placed beside an identity matrix, $\mathbf{I}$;

2. $\begin{bmatrix} 1 & 2.5 \\ 10 & 30 \end{bmatrix} \begin{bmatrix} 0.25 & 0 \\ 0 & 1 \end{bmatrix}$ row one is divided by 4, the first element in the row, to produce 1 at a_{11};

3. $\begin{bmatrix} 1 & 2.5 \\ 0 & 5 \end{bmatrix} \begin{bmatrix} 0.25 & 0 \\ -2.5 & 1 \end{bmatrix}$ 10 times row one is subtracted from row two to reduce a_{21} to 0;

4. $\begin{bmatrix} 1 & 2.5 \\ 0 & 1 \end{bmatrix} \begin{bmatrix} 0.25 & 0 \\ -0.5 & 0.2 \end{bmatrix}$ row two is divided by 5 to give 1 at a_{22}, and

5. $\begin{bmatrix} 1 & 0 \\ 0 & 1 \end{bmatrix} \begin{bmatrix} 1.5 & -0.5 \\ -0.5 & 0.2 \end{bmatrix}$ 2.5 times row two is subtracted from row one to reduce the final off-diagonal element to 0.

The matrix is now inverted. Work may be checked by multiplying the original matrix $\mathbf{A}$ by the inverted matrix, $\mathbf{A}^{-1}$, which should yield the identity matrix

$$\begin{bmatrix} 1.5 & -0.5 \\ -0.5 & 0.2 \end{bmatrix} \times \begin{bmatrix} 4 & 10 \\ 10 & 30 \end{bmatrix} = \begin{bmatrix} 1 & 0 \\ 0 & 1 \end{bmatrix}$$

Because

$$\mathbf{A}^{-1}\mathbf{A} = \mathbf{I}$$

the following identities hold:

$$\mathbf{A}^{-1}\mathbf{A}\mathbf{X} = \mathbf{A}^{-1}\mathbf{B}$$
$$\mathbf{I}\mathbf{X} = \mathbf{A}^{-1}\mathbf{B}$$
$$\mathbf{X} = \mathbf{A}^{-1}\mathbf{B}$$

By postmultiplying the inverted matrix $\mathbf{A}^{-1}$ by the matrix $\mathbf{B}$, the unknown matrix, $\mathbf{X}$, is solved,

$$\begin{array}{cccc} \mathbf{A}^{-1} & \times & \mathbf{B} & = & \mathbf{X} \end{array}$$

$$\begin{bmatrix} 1.5 & -0.5 \\ -0.5 & 0.2 \end{bmatrix} \times \begin{bmatrix} 38 \\ 110 \end{bmatrix} = \begin{bmatrix} 2 \\ 3 \end{bmatrix}$$

The column vector contains the unknown coefficients which we find to be equal to $x_1 = 2$ and $x_2 = 3$. You will recall that it was stated that these were the coefficients originally in the equation set, so we have recovered the proper values.

As an additional example of the solution of simultaneous equations by matrix inversion, we can set the equations below into matrix form and solve for x_1 and x_2 by inversion,

$$2x_1 + x_2 = 4$$
$$3x_1 + 4x_2 = 1$$

The steps in the inversion process can be written out briefly:

1. $\begin{bmatrix} 2 & 1 \\ 3 & 4 \end{bmatrix} \times \begin{bmatrix} x_1 \\ x_2 \end{bmatrix} = \begin{bmatrix} 4 \\ 1 \end{bmatrix}$

2. $\begin{bmatrix} 2 & 1 \\ 3 & 4 \end{bmatrix}^{-1} = \begin{bmatrix} 4/5 & -1/5 \\ -3/5 & 2/5 \end{bmatrix}$

3. $\begin{bmatrix} 4/5 & -1/5 \\ -3/5 & 2/5 \end{bmatrix} \times \begin{bmatrix} 4 \\ 1 \end{bmatrix} = \begin{bmatrix} 3 \\ -2 \end{bmatrix}$

Therefore, the unknown coefficients are $x_1 = 3$ and $x_2 = -2$.

It may be noted that the procedure just described is almost exactly the same as the classical algebraic method of solving two simultaneous equations. In fact, the solution of simultaneous equations is probably the most important application of matrix inversion. The advantage of matrix manipulation over the "try it and see" approach of ordinary algebra is that it is more amenable to computer programming. Almost all of the techniques described in subsequent chapters of this book involve the solution of sets of simultaneous equations. These can be expressed conveniently in the form of matrix equations and solved in the manner just described.

Matrix inversion can, of course, be applied to square matrices of any size, and not just the 2×2 examples we have investigated so far. Demonstrate this to yourself by inverting the 3×3 matrix below:

$$\begin{bmatrix} 1 & 2 & 3 \\ 2 & 6 & 5 \\ 3 & 5 & 6 \end{bmatrix}$$

If we need the inverse of a diagonal matrix, the problem is much simpler. The inverse of a diagonal matrix is simply another diagonal matrix whose nonzero elements are the reciprocals of the corresponding elements of the original matrix. Considering the 3×3 matrix, **A**,

$$\begin{bmatrix} a_{11} & 0 & 0 \\ 0 & a_{22} & 0 \\ 0 & 0 & a_{33} \end{bmatrix}^{-1} = \begin{bmatrix} 1/a_{11} & 0 & 0 \\ 0 & 1/a_{22} & 0 \\ 0 & 0 & 1/a_{33} \end{bmatrix}$$

Certain combinations of otherwise complicated operations become very simple when the matrices involved are diagonal matrices. For example, consider the multiplication

$$\mathbf{A}^{-1}\mathbf{A}^{1/2} = \mathbf{A}^{-1/2}$$

If **A** is 3×3, the product is

$$\begin{bmatrix} a_{11} & 0 & 0 \\ 0 & a_{22} & 0 \\ 0 & 0 & a_{33} \end{bmatrix}^{-1/2} = \begin{bmatrix} 1/\sqrt{a_{11}} & 0 & 0 \\ 0 & 1/\sqrt{a_{22}} & 0 \\ 0 & 0 & 1/\sqrt{a_{33}} \end{bmatrix}$$

In some applications, the inverse may not be required, but only the solutions to a set of simultaneous equations. In the handworked example, we wanted the values of the matrix **X** in the equation

$$\begin{bmatrix} 4 & 10 \\ 10 & 30 \end{bmatrix} \times \begin{bmatrix} x_1 \\ x_2 \end{bmatrix} = \begin{bmatrix} 38 \\ 110 \end{bmatrix}$$

To find this, we inverted **A** and then postmultiplied $\mathbf{A}^{-1}$ by **B** to give **X**. We could have instead found **X** directly by operating on **B** as **A** was transformed into an identity matrix. To do this, we would utilize what is called an **augmented matrix** that has one more column than it has rows. The column vector, **B**, then occupies the $(n + 1)$ column of the matrix, and the remaining $(n \times n)$ part is inverted. Repeating the same problem:

1. $\begin{bmatrix} 4 & 10 & | & 38 \\ 10 & 30 & | & 110 \end{bmatrix}$ Matrices **A** and **B** are combined in an $n \times (n + 1)$ matrix.

2. $\begin{bmatrix} 1.0 & 2.5 & | & 9.5 \\ 1.0 & 3.0 & | & 11.0 \end{bmatrix}$ Row one is divided by 4 and row two is divided by 10.

3. $\begin{bmatrix} 1.0 & 2.5 & | & 9.5 \\ 0.0 & 0.5 & | & 1.5 \end{bmatrix}$ Row one is subtracted from row two.

4. $\begin{bmatrix} 1.0 & 0 & | & 2.0 \\ 0.0 & 0.5 & | & 1.5 \end{bmatrix}$ Row two is multiplied by 5 and the product is subtracted from row one.

5. $\begin{bmatrix} 1.0 & 0.0 & | & 2.0 \\ 0.0 & 1.0 & | & 3.0 \end{bmatrix}$ Row two is divided by 0.5.

So, the $(n + 1)$ column of the augmented matrix contains the solution to the simultaneous equation set, and our original matrix has been replaced by an identity matrix.

Few mathematical procedures have received the attention given to matrix inversion. Dozens of methods have been devised to solve sets of simultaneous equations, and hundreds of programmed versions exist. Some are especially tailored to deal with special types of matrices, such as those containing many zero elements (such matrices are called *sparse*) or possessing certain types of symmetry. Numerical computation packages for personal computers, such as MATHEMATICA® and MATLAB®, contain alternative algorithms that can be used to calculate the inverse of matrices. Some of these procedures, such as singular value decomposition (SVD), will find approximate inverses even when exact solutions do not exist.

Determinants

Before discussing our final topic, which is eigenvalues and eigenvectors and how they are obtained, we must examine an additional property of a square matrix called the ***determinant***. A determinant is a single number extracted from a square matrix by a series of operations, and is symbolically represented by det $\mathbf{A}$, $|\mathbf{A}|$, or by

$$\begin{vmatrix} a_{11} & a_{12} \\ a_{21} & a_{22} \end{vmatrix}$$

It is defined as the sum of $n!$ terms of the form

$$(-1)^k a_{1i_1} a_{2i_2} \dots a_{ni_n} \tag{3.3}$$

where n is the number of rows (or columns) in the matrix, the subscripts $i_1, i_2, \dots, i_n$ are equal to $1, 2, \dots, n$, taken in any order, and k is the number of exchanges of two elements necessary to place the i subscripts in the order $1, 2, \dots, n$. Each term contains one element from each row and each column. The process of obtaining a determinant from a square matrix is called ***evaluating the determinant***.

We begin the process of evaluating the determinant by selecting one element from each row of the matrix to form a term or combination of elements. The elements in a term are selected in order from row $1, 2, \dots, n$, but each combination can contain only one element from each column. For example, we might select the combination $a_{12} a_{21} a_{33}$ from a 3×3 matrix. Note that the method of selection places the elements in proper order according to their first, or row, subscript. The term contains one and only one element from each row and each column. We must find all possible combinations of elements that can be formed in this way. If a matrix is $n \times n$, there will be $n!$ combinations which contain one element from each row and column, and whose first subscripts are in the order $1, 2, \dots, n$.

Since the order of multiplication of a series of numbers makes no difference in the product, that is, $a_{11} a_{22} a_{33} = a_{22} a_{11} a_{33} = a_{33} a_{22} a_{11}$ and so on, we can rearrange our combinations without changing the result. We wish to rearrange each combination until the second, or column, subscript of each element is in proper numerical order. The rearranging may be performed by swapping any two adjacent elements. As the operation is performed, we must keep track of the number of exchanges or transpositions necessary to get the second subscript in the correct order. If an even number of transpositions is required (*i.e.*, 0, 2, 4, 6, *etc.*), the product is given a positive sign. If an odd number of transpositions is necessary (1, 3, 5, 7, *etc.*), the product is negative.

In a 2×2 matrix

$$\begin{bmatrix} a_{11} & a_{12} \\ a_{21} & a_{22} \end{bmatrix}$$

we can find two combinations of elements that contain one and only one element from each row and each column. These are $a_{11}a_{22}$ and $a_{12}a_{21}$.

The second subscripts in $a_{11}a_{22}$ are in correct numerical order and no rearranging is necessary. The number of transpositions is zero, so the sign of the product is positive. However, $a_{12}a_{21}$ must be rearranged to $a_{21}a_{12}$ before the second subscripts are in numerical order. This requires one transposition, so the product is negative. The determinant of a 2×2 matrix is therefore

$$\begin{vmatrix} a_{11} & a_{12} \\ a_{21} & a_{22} \end{vmatrix} = +a_{11}a_{22} - a_{12}a_{21}$$

For a numerical example, we will consider the matrix

$$\begin{bmatrix} 2 & 1 \\ 4 & 3 \end{bmatrix}$$

The determinant is

$$\begin{vmatrix} 2 & 1 \\ 4 & 3 \end{vmatrix} = +(2 \times 3) - (1 \times 4) = 2$$

Next, let us consider a more complex example, a 3×3 determinant:

$$\begin{vmatrix} a_{11} & a_{12} & a_{13} \\ a_{21} & a_{22} & a_{23} \\ a_{31} & a_{32} & a_{33} \end{vmatrix}$$

There are $3!$, or $3 \times 2 \times 1 = 6$, combinations of elements in a 3×3 matrix that contain one element from each row and column and whose first subscripts are in the order $1, 2, 3$. Start with the top row and pick an entry from each row. Be sure to choose in order from the first row, second row, third row, ... nth row, with no more than one entry from each column. All possible combinations that satisfy these conditions in a 3×3 matrix are

$$\begin{array}{cc} a_{11}a_{22}a_{33} & a_{11}a_{23}a_{32} \\ a_{12}a_{23}a_{31} & a_{12}a_{21}a_{33} \\ a_{13}a_{21}a_{32} & a_{13}a_{22}a_{31} \end{array}$$

To determine the signs of each of these terms, we must see how many transpositions are necessary to get the second subscripts in the order $1, 2, 3$. For $a_{11}a_{22}a_{33}$, no transpositions are necessary, so $k = 0$ and the term is positive. Transpositions for the others and the resulting signs are given below:

$$a_{11}\,a_{23}a_{32} = a_{11}\,a_{32}\,a_{23} \qquad\qquad k = 1 \quad \text{sign} = -$$

$$a_{12}\,a_{23}a_{31} = a_{12}a_{31}\,a_{23} = a_{31}\,a_{12}\,a_{23} \qquad k = 2 \quad \text{sign} = +$$

$$a_{12}a_{21}\,a_{33} = a_{21}\,a_{12}\,a_{33} \qquad\qquad k = 1 \quad \text{sign} = -$$

$$a_{13}a_{21}\,a_{32} = a_{21}\,a_{13}a_{32} = a_{21}\,a_{32}\,a_{13} \qquad k = 2 \quad \text{sign} = +$$

$$a_{13}\,a_{22}a_{31} = a_{13}a_{31}\,a_{22} = a_{31}\,a_{13}a_{22} = a_{31}\,a_{22}\,a_{13} \qquad k = 3 \quad \text{sign} = -$$

Thus, there are three negative and three positive terms in the determinant. Summing according to the signs just found yields a single number, which is

$$+ a_{11}a_{22}a_{33} - a_{11}a_{23}a_{32} + a_{12}a_{23}a_{31} - a_{12}a_{21}a_{33} + a_{13}a_{21}a_{32} - a_{13}a_{22}a_{31}$$

We can now try a matrix of real values:

$$\begin{vmatrix} 4 & 3 & 2 \\ 2 & 4 & 1 \\ 1 & 0 & 3 \end{vmatrix}$$

The six terms possible are

$$(4 \times 4 \times 3) = 48$$
$$(4 \times 1 \times 0) = 0$$
$$(3 \times 1 \times 1) = 3$$
$$(3 \times 2 \times 3) = 18$$
$$(2 \times 2 \times 0) = 0$$
$$(2 \times 4 \times 1) = 8$$

The first, third, and fifth of these require an even number of transpositions for proper arrangement of the second subscript and so are positive. The others require an odd number of transpositions and are therefore negative. Summing, we have

$$\det \mathbf{A} = +48 - 0 + 3 - 18 + 0 - 8 = 25$$

This method of evaluating a determinant is described by Pettofrezzo (1978). A more conventional approach (see, for example, Anton and Rorres, 1994) uses what is called the "method of cofactors," but the two can be shown to be equivalent.

We now have at our command a system for reducing a square matrix into its determinant, but no clear grasp of what a determinant "really is." Determinants arise in many ways, but they appear most conspicuously during the solution of sets of simultaneous equations. You may not have noticed them, however, because they have been hidden in the inversion process we have been using.

Consider the set of equations:

$$a_{11}x_1 + a_{12}x_2 = b_1$$
$$a_{21}x_1 + a_{22}x_2 = b_2$$

Expressed in matrix form, this becomes

$$\begin{bmatrix} a_{11} & a_{12} \\ a_{21} & a_{22} \end{bmatrix} \begin{bmatrix} x_1 \\ x_2 \end{bmatrix} = \begin{bmatrix} b_1 \\ b_2 \end{bmatrix}$$

and we have discussed how the vector of unknown x's can be solved by matrix inversion. However, with algebraic rearrangement, the unknowns also can be found by the equations

$$x_1 = \frac{b_1 a_{22} - a_{12} b_2}{a_{11} a_{22} - a_{12} a_{21}}$$

and

$$x_2 = \frac{a_{11} b_2 - b_1 a_{21}}{a_{11} a_{22} - a_{12} a_{21}}$$

You will note that the denominators are the same for both unknowns. They also are the determinants of the matrix **A**. That is,

$$|\mathbf{A}| = \begin{vmatrix} a_{11} & a_{12} \\ a_{21} & a_{22} \end{vmatrix} = a_{11}a_{22} - a_{12}a_{21}$$

Furthermore, the numerators can be expressed as determinants. For the equation of x_1, the numerator is the determinant of the matrix

$$|\mathbf{B}\,\mathbf{A}_{i2}| = \begin{vmatrix} b_1 & a_{12} \\ b_2 & a_{22} \end{vmatrix} = b_1 a_{22} - b_2 a_{21}$$

and for x_2, it is the determinant of

$$|\mathbf{A}_{1i}\,\mathbf{B}| = \begin{vmatrix} a_{11} & b_1 \\ a_{21} & b_2 \end{vmatrix} = a_{11}b_2 - a_{21}b_1$$

This procedure can be generalized to any set of simultaneous equations and provides one common method for their solution. This procedure for solving equations is called **Cramer's rule**. The rule states that the solution for any unknown x_i in a set of simultaneous equations is equal to the ratio of the two determinants. The denominator is the determinant of the coefficients (in our example, the a's). The numerator is the same determinant except that the ith column is replaced by the vector of right-hand terms (the vector of b's). Let us check the rule with an example used before:

$$\begin{bmatrix} 4 & 10 \\ 10 & 30 \end{bmatrix} \times \begin{bmatrix} x_1 \\ x_2 \end{bmatrix} = \begin{bmatrix} 38 \\ 110 \end{bmatrix}$$

The denominators of the ratios for both unknown coefficients are the same:

$$\begin{vmatrix} 4 & 10 \\ 10 & 30 \end{vmatrix} = (4 \times 30) - (10 \times 10) = 20$$

The numerator of x_1 is the determinant

$$\begin{vmatrix} 38 & 10 \\ 110 & 30 \end{vmatrix} = (38 \times 30) - (110 \times 10) = 40$$

so $x_1 = 40/20 = 2$. For x_2, the numerator is the determinant

$$\begin{vmatrix} 4 & 38 \\ 10 & 110 \end{vmatrix} = (4 \times 110) - (10 \times 38) = 60$$

so $x_2 = 60/20 = 3$. These are the same unknowns we recovered by matrix inversion.

The determinant of an arbitrary square matrix such as the 3×3 example above may be a positive value, a negative value, or zero. If the matrix is symmetric (the variety of matrix we will encounter most often), its determinant cannot be negative. However, the distinction between a positive determinant and a zero determinant is very important because a matrix whose determinant is zero cannot be inverted by ordinary methods. That is, the matrix will be singular.

What circumstances will lead to singularity? The condition indicates that two or more rows (or columns) of the matrix are linear combinations or linear transformations of other rows; that is, the values in some rows (or columns) are dependent on values in other rows. For example, the determinant

$$\begin{vmatrix} 1 & 2 & 3 \\ 4 & 5 & 6 \\ 2 & 4 & 6 \end{vmatrix} = 0$$

is zero because the third row of the matrix is simply twice the first row. Similarly, the determinant

$$\begin{vmatrix} 1 & 2 & 3 \\ 4 & 5 & 6 \\ 5 & 7 & 9 \end{vmatrix} = 0$$

is zero because the third row is the sum of rows one and two. Of course, in real problems the source of singularity usually is not so obvious. Consider the data in file BALTIC.TXT, which gives the weight-percent of sand in five successive size fractions, measured on bottom samples collected in an area of the Baltic Sea. We can calculate correlations between the five sand size categories and place the results in a square, symmetric correlation matrix:

$$\begin{bmatrix} 1 & 0.243 & -0.301 & 0.096 & -0.261 \\ 0.243 & 1 & -0.969 & -0.562 & -0.422 \\ -0.301 & -0.969 & 1 & 0.340 & 0.253 \\ 0.096 & -0.562 & 0.340 & 1 & 0.691 \\ -0.261 & -0.422 & 0.253 & 0.691 & 1 \end{bmatrix}$$

It is not obvious that this matrix should be singular with a zero determinant, yet it is. The linear dependence comes about because the weight-percentages in the five size categories sum to 100 for each observation, so there are induced negative correlations between the size categories. (Actually, because of rounding during computations, you may compute a correlation matrix that is not exactly singular. Depending upon the numerical precision of the computer program, rather than exactly 0, you may observe a very small determinant such as −0.0002. A matrix with a determinant near zero is said to be ***ill-conditioned***.)

Finally, there is another special case of interest. An identity matrix has a determinant equal to 1.0. If several variables are completely independent of each other, their correlations will be near zero and they will form a correlation matrix that approximates an identity matrix. The determinant of such a matrix will be close to one, and its logarithm will be close to zero; this is the basis for one test of independence between variables.

Eigenvalues and Eigenvectors

The topic we will consider next usually is regarded as one of the most difficult topics in matrix algebra, the determination of eigenvalues and eigenvectors (also called "latent" and "proper" values and vectors). The difficulty is not in their calculation, which is cumbersome but no more so than many other mathematical procedures. Rather, difficulties arise in developing a "feel" for the meaning of these quantities, especially in an intuitive sense. Unfortunately, many textbooks provide no help in this regard, placing their discussions in strictly mathematical terms that may be difficult for nonmathematicians to interpret.

A lucid discussion and geometric interpretation of eigenvectors and eigenvalues was prepared by Peter Gould for the benefit of geography students at Pennsylvania State University. The following discussion leans heavily on his prepared notes and a subsequent article (Gould, 1967). We will consider a real matrix of coordinates of points in space and interpret the eigenvalues and associated functions as geometric properties of the arrangement of these points. This approach limits us, of course, to small matrices, but the insights gained can be extrapolated to larger systems even though hand computation becomes impractical. In this regard, it may be noted that we are entering a realm where the computational powers of even the largest computers may be inadequate to solve real problems.

Eigenvalues

Having worked through determinants, we can use them to develop eigenvalues. Consider a hypothetical set of simultaneous equations expressed in the following matrix form:

$$\mathbf{A}\mathbf{X} = \lambda \mathbf{X} \tag{3.4}$$

This equation states that the matrix of coefficients (the a_{ij}'s) times the vector of unknowns (the x_i's) is equal to some constant (λ) times the unknown vector itself. The problem is the same as in the solution of the simultaneous equation set

$$\mathbf{A}\mathbf{X} = \mathbf{B}$$

except now

$$\mathbf{B} = \lambda \mathbf{X}$$

Our concern is to find values of λ that satisfy this relationship. Equation (3.4) can be rewritten in the form

$$(\mathbf{A} - \lambda \mathbf{I})\mathbf{X} = 0 \tag{3.5}$$

where $\lambda \mathbf{I}$ is nothing more than an identity matrix (of the same size as $\mathbf{A}$) times the quantity λ. That is,

$$\lambda \mathbf{I} = \begin{bmatrix} \lambda & 0 & 0 \\ 0 & \lambda & 0 \\ 0 & 0 & \lambda \end{bmatrix}$$

for a 3×3 matrix. Written in conventional form, the equivalent of the three simultaneous equations is

$$\begin{aligned} (a_{11} - \lambda)\, x_1 + a_{12}x_2 + a_{13}x_3 &= 0 \\ a_{21}x_1 + (a_{22} - \lambda)\, x_2 + a_{23}x_3 &= 0 \\ a_{31}x_1 + a_{32}x_2 + (a_{33} - \lambda)\, x_3 &= 0 \end{aligned} \tag{3.6}$$

Let us assume that there are solutions to these equations other than the trivial case where all the unknown x's $= 0$. Look back at Cramer's rule for the solution of simultaneous equations, in which the unknowns are expressed as the ratio of two determinants. Because the numerator in our present example would contain a column of zeros, the determinant of the numerator also will be zero. That is, the solution for the **X** vector is

$$\mathbf{X} = \frac{0}{|\mathbf{A}|}$$

Rewriting, this becomes

$$|\mathbf{A}|\,\mathbf{X} = 0 \qquad\qquad (3.7)$$

If the vector **X** is not zero, it follows that the determinant of the matrix **A** must be zero, or

$$|\mathbf{A} - \lambda\,\mathbf{I}| = \begin{vmatrix} a_{11} - \lambda & a_{12} & a_{13} \\ a_{21} & a_{22} - \lambda & a_{23} \\ a_{31} & a_{32} & a_{33} - \lambda \end{vmatrix} = 0 \qquad\qquad (3.8)$$

We usually know the coefficients, a_{ij}, of the matrix, so this equation can be used to determine the values of $\lambda\,\mathbf{I}$ that satisfy all of these various conditions. This is done by expanding the determinant to yield a polynomial equation. Looking first at a 2×2 determinant,

$$\begin{vmatrix} a_{11} - \lambda & a_{12} \\ a_{21} & a_{22} - \lambda \end{vmatrix} = 0$$

Expanding gives

$$(a_{11} - \lambda)\,(a_{22} - \lambda) - a_{21}a_{12} = 0$$

Multiplying out the first term,

$$(a_{11} - \lambda)\,(a_{22} - \lambda) = (a_{11}a_{22}) - (a_{11}\lambda) - (a_{22}\lambda) + \lambda^2$$

Thus we have

$$(a_{11}a_{22}) - (a_{21}a_{12}) - (a_{11}\lambda) - (a_{22}\lambda) + \lambda^2 = 0$$

Because we know the various values of the elements a_{ij}, we can collect all of these terms together in the form of an equation such as

$$\lambda^2 + \alpha_1\lambda + \alpha_2 = 0 \qquad\qquad (3.9)$$

where the α's represent the sum of the numerical values of the appropriate a_{ij}'s. You should recognize that this is a quadratic equation of the general form

$$ax^2 + bx + c = 0$$

which can be solved for the unknown terms by factoring. The general solution to a quadratic equation is

$$x = \frac{-b \pm \sqrt{b^2 - 4ac}}{2a} \qquad\qquad (3.10)$$

If this seems unfamiliar, review the sections in an elementary algebra book that deal with factoring and quadratic equations. Now, we can try the procedures just outlined to find the eigenvalues of the 2×2 matrix:

$$\mathbf{A} = \begin{bmatrix} 17 & -6 \\ 45 & -16 \end{bmatrix}$$

First, we must set the matrix in the form

$$\mathbf{A} - \lambda \mathbf{I} = \begin{bmatrix} 17 - \lambda & -6 \\ 45 & -16 - \lambda \end{bmatrix}$$

Equating the determinant to zero,

$$\begin{vmatrix} 17 - \lambda & -6 \\ 45 & -16 - \lambda \end{vmatrix} = 0$$

we can expand the determinant

$$\begin{vmatrix} 17 - \lambda & -6 \\ 45 & -16 - \lambda \end{vmatrix} = (17 - \lambda)(-16 - \lambda) - (-6)(45) = 0$$

Multiplying out gives

$$-272 - 17\lambda + 16\lambda + \lambda^2 + 270 = 0$$

which can be collected to give

$$\lambda^2 - \lambda - 2 = 0$$

This can be factored into

$$(\lambda - 2)(\lambda + 1) = 0$$

So, the two eigenvalues associated with the matrix $\mathbf{A}$ are

$$\lambda_1 = +2 \qquad \lambda_2 = -1$$

This example was deliberately chosen for ease in factoring. We can try a somewhat more difficult example by using the set of simultaneous equations we solved earlier. This is the 2×2 matrix:

$$\mathbf{A} = \begin{bmatrix} 4 & 10 \\ 10 & 30 \end{bmatrix}$$

Repeating the sequence of steps yields the determinant

$$\begin{vmatrix} 4 - \lambda & 10 \\ 10 & 30 - \lambda \end{vmatrix} = 0$$

which is then expanded into

$$\begin{vmatrix} 4 - \lambda & 10 \\ 10 & 30 - \lambda \end{vmatrix} = (4 - \lambda)(30 - \lambda) - 100 = 0$$

or

$$\lambda^2 - 34\lambda + 20 = 0$$

There are no obvious factors in the quadratic equation, so we must apply the rule for a general solution:

$$x = \frac{-b \pm \sqrt{b^2 - 4ac}}{2a} = \lambda = \frac{-(-34) \pm \sqrt{-34^2 - 4 \times 1 \times 20}}{2 \times 1} = \frac{34 \pm \sqrt{1076}}{2}$$

$$\lambda_1 = 33.4 \qquad \lambda_2 = 0.6$$

We can check our work by substituting the eigenvalues back into the determinant to see if it is equal to zero, within the error introduced by round-off:

$$\begin{vmatrix} 4 - 33.4 & 10 \\ 10 & 30 - 33.4 \end{vmatrix} = (-29.4)(-3.4) - (10)(10) = -0.04$$

and

$$\begin{vmatrix} 4 - 0.6 & 10 \\ 10 & 30 - 0.6 \end{vmatrix} = (3.4)(29.4) - (10)(10) = -0.04$$

So, the eigenvalues we have found are correct within two decimal places.

Before we leave the computation of eigenvalues of 2×2 matrices, we should consider one additional complication that may arise. Suppose we want the eigenvalues of the matrix

$$\mathbf{A} = \begin{bmatrix} 2 & 4 \\ -6 & 3 \end{bmatrix}$$

Expressed as a determinant equal to zero, we have

$$\begin{vmatrix} 2 - \lambda & 4 \\ -6 & 3 - \lambda \end{vmatrix} = 0$$

which expands to

$$\begin{vmatrix} 2 - \lambda & 4 \\ -6 & 3 - \lambda \end{vmatrix} = (2 - \lambda)(3 - \lambda) + 24 = 0$$

or

$$\lambda^2 - 5\lambda + 30 = 0$$

The roots of this equation are

$$\lambda_1, \lambda_2 = \frac{5 \pm \sqrt{25 - 120}}{2}$$

But this leads to equations involving the square roots of negative numbers:

$$\lambda_1 = \frac{5 + \sqrt{-95}}{2} = 2.5 + 4.9i$$

$$\lambda_2 = \frac{5 - \sqrt{-95}}{2} = 2.5 - 4.9i$$

These are complex numbers, containing both real parts and imaginary parts which include the imaginary number, $i = \sqrt{-1}$. Fortunately, a symmetric matrix always yields real eigenvalues, and most of our computations involving eigenvalues and eigenvectors will utilize covariance, correlation, or similarity matrices which are always symmetrical.

Next, we will consider the eigenvalues of the third-order matrix:

$$\begin{bmatrix} 20 & -4 & 8 \\ -40 & 8 & -20 \\ -60 & 12 & -26 \end{bmatrix}$$

The determinant of the matrix is set to zero, giving

$$\begin{vmatrix} 20 - \lambda & -4 & 8 \\ -40 & 8 - \lambda & -20 \\ -60 & 12 & -26 - \lambda \end{vmatrix} = 0$$

Expanding out the determinant and combining terms yields

$$-\lambda^3 + 2\lambda^2 + 8\lambda = 0$$

This is a cubic equation having three roots that must be found. In this instance, the polynomial can be factored into

$$(\lambda - 4)\,(\lambda - 0)\,(\lambda + 2) = 0$$

and the roots are directly obtainable:

$$\lambda_1 = +4 \qquad \lambda_2 = 0 \qquad \lambda_3 = -2$$

Although the techniques we have been using are extendible to any size matrix, finding the roots of large polynomial equations can be an arduous task. Usually, eigenvalues are not found by solution of a polynomial equation, but rather by matrix manipulation methods that involve refinement of a successive series of approximations to the eigenvalues. These methods are practical only because of the great computational speed of digital computers. Utilizing this speed, a researcher can compress literally a lifetime of trial solutions and refinements into a few minutes.

We can now define another measure of the "size" of a square matrix. The *rank* of a square matrix is the number of independent rows (or columns) in the matrix and is equal to the number of nonzero eigenvalues that can be extracted from the matrix. A nonsingular matrix has as many nonzero eigenvalues as there are rows or columns in the matrix, so its rank is equal to its order. A singular matrix has one or more rows or columns that are dependent on other rows or columns, and consequently will have one or more zero eigenvalues; its rank will be less than its order.

Now that we have an idea of the manipulations that produce eigenvalues, we may try to get some insight into their nature. The rows of a matrix can be regarded as the coordinates of points in m-dimensional space. If we restrict our consideration to 2×2 matrices, we can represent this space as an illustration on a page and can view matrix operations geometrically.

Table 3–3. Concentrations of selected elements (in ppm) measured in soil samples collected in vineyards and associated terraces on the Istrian peninsula of Croatia.

Cr	Cu	Mg	V	Zn
125	25	6936	114	194
205	33	5368	143	212
171	25	5006	90	272
62	157	3600	59	129
137	88	3220	130	123
234	185	7450	162	264
270	52	4400	205	155
179	322	5000	150	135
113	29	8600	98	114
65	400	4000	60	40
80	225	2000	90	130
35	230	1000	100	50
176	30	3100	160	100
90	164	5000	105	105
52	200	9000	60	170
98	29	3100	89	87
130	59	7100	112	147
158	28	6400	143	133
69	30	7900	109	103
108	30	2300	136	84

We will use a series of 2×2 matrices calculated from data that might arise in an environmental study. **Table 3–3** lists trace-element concentrations for five elements measured on 20 soil samples collected in vineyards and adjacent terraces on the Istrian peninsula of Croatia (the data are contained in the file ISTRIA.TXT). For centuries, the growers have treated their grapes with "blue galicia," or copper sulfate, to prevent fungus. As a consequence, the soil is enriched in copper and other metals that are present as impurities in the crude sulfate compound.

Using the matrix operations we have already discussed, we will construct a matrix containing correlations between the concentrations of the different metals. The data in **Table 3–3** can be regarded as a 20×5 matrix, **M**. Define a row vector **V** having 20 elements, each equal to 1.0. The matrix multiplication, **VM**, will yield a five-element row vector containing the column totals of **M**. If we premultiply this row vector by 1/20, it will contain the means of each of the five columns.

We can now subtract the means from each observation to convert the data into deviations. By premultiplying the vector of means by the transpose of **V**, we create a 20×5 matrix in which every row is the same as the vector of means. Subtracting this matrix from **M** yields **D**, the data in the form of deviations from their means:

$$\mathbf{D} = \mathbf{M} - \mathbf{V}^{\mathrm{T}} n^{-1} \mathbf{V} \mathbf{M}$$

Here, n is the number of rows in **M** (*i.e.*, the number of observations) and n^{-1} is the inverse of n, or 1/20.

Premultiplying **D** by its transpose will yield a square 5×5 matrix whose individual entries are the sums of squares (along the diagonal) and cross products of the five elements, corrected for their means. If we divide a corrected sum of squares by $n - 1$ we obtain the variance, and if we divide a corrected sum of products by $n - 1$ we obtain the covariance. These are the elements of the covariance matrix, S, which we can compute by

$$S = (n - 1)^{-1}D^TD$$

A subset of S could serve our purposes (and the covariance matrix often is used in multivariate statistics), but the relationships will be clearer if we use the correlation matrix, R. Correlations are simply covariances of standardized variables; that is, observations from which the means have been removed and then divided by the standard deviation. In matrix D, the means have already been removed. We can, in effect, divide by the appropriate standard deviations if we create a 5×5 matrix, C, whose diagonal elements are the square roots of the variances found on the diagonal of S, and whose off-diagonal elements are all 0.0. If we invert C and premultiply by D, each element of D will be divided by the standard deviation of its column. Call the result U, a 20×5 matrix of standardized values;

$$U = DC^{-1}$$

We can calculate the correlation matrix by repeating the procedure we used to find S, substituting U for D:

$$R = (n - 1)^{-1}U^TU$$

$$R = \begin{bmatrix} 1 & -0.312 & 0.141 & 0.85 & 0.595 \\ -0.312 & 1 & -0.201 & -0.33 & -0.28 \\ 0.141 & -0.201 & 1 & -0.029 & 0.456 \\ 0.85 & -0.33 & -0.029 & 1 & 0.242 \\ 0.595 & -0.28 & 0.456 & 0.242 & 1 \end{bmatrix}$$

To graphically illustrate matrix relationships, we must confine ourselves to 2×2 matrices, which we can extract from R. Copper and zinc are recorded in the second and fifth columns of M, and so their correlations are the elements $r_{i,j}$ whose subscripts are 2 and 5:

$$R_{cu,zn} = \begin{bmatrix} r_{2,2} & r_{2,5} \\ r_{5,2} & r_{5,5} \end{bmatrix} = \begin{bmatrix} 1 & -0.28 \\ -0.28 & 1 \end{bmatrix}$$

If we regard the rows as vectors in X and Y, we can plot each row as the tip of a vector that extends from the origin. In **Figure 3–1**, the tip of each vector is indicated by an open circle, labeled with its coordinates. The ends of the two vectors lie on an ellipse whose center is at the origin of the coordinate system and which just encloses the tips of the vectors. The eigenvalues of the 2×2 matrix $R_{cu,zn}$ represent the magnitudes, or lengths, of the major and minor semiaxes of the ellipse. In this example, the eigenvalues are

$$\lambda_1 = 1.28 \qquad \lambda_2 = 0.72$$

Gould refers to the relative lengths of the semiaxes as a measure of the "stretchability" of the enclosing ellipse. The semiaxes are shown by arrows on **Figure 3–1**. The first eigenvalue represents the major semiaxis whose length from center to

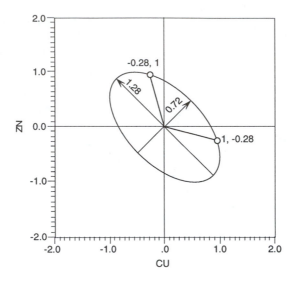

Figure 3–1. Ellipse defined by rows in matrix of correlations between copper and zinc. Eigenvectors of matrix correspond to principal semiaxes (arrows) of ellipse.

edge of the ellipse is 1.28 units. The second eigenvalue represents the length of the minor semiaxis, which is 0.72 units.

If the two vectors are closer together, the ratio between the semiaxes of the enclosing ellipse will change. For example, chromium and vanadium have very similar behavior in the vineyard soil samples, leading to a high correlation between the two. Their correlations are given by elements in the first and fourth rows and columns of **R**:

$$\mathbf{R}_{cr,v} = \begin{bmatrix} r_{1,1} & r_{1,4} \\ r_{4,1} & r_{4,4} \end{bmatrix} = \begin{bmatrix} 1 & 0.85 \\ 0.85 & 1 \end{bmatrix}$$

The rows of $\mathbf{R}_{cr,v}$ are plotted as vectors in **Figure 3–2**. The eigenvalues of this 2×2 matrix are

$$\lambda_1 = 1.85 \qquad \lambda_2 = 0.15$$

which define one very long major semiaxis and a short minor semiaxis. At the limit, we can imagine that two variables might behave in an identical fashion. Then, their rows in **R** would be so similar that they would be identical and the plotted vectors would coincide. That is,

$$\mathbf{R}_{x,y} = \begin{bmatrix} 1 & 1 \\ 1 & 1 \end{bmatrix}$$

The enclosing ellipse would collapse to a straight line of semiaxis length $\lambda_1 = 2$ and a minor semiaxis of $\lambda_2 = 0$.

At the opposite extreme, two variables which are completely unrelated will have a correlation of near zero. Magnesium and vanadium show such behavior in the vineyard samples. They are represented by elements in the third and fourth rows and columns of **R**, and are shown plotted as vectors in **Figure 3–3**.

$$\mathbf{R}_{mg,v} = \begin{bmatrix} r_{3,3} & r_{3,4} \\ r_{4,3} & r_{4,4} \end{bmatrix} = \begin{bmatrix} 1 & -0.029 \\ -0.029 & 1 \end{bmatrix}$$

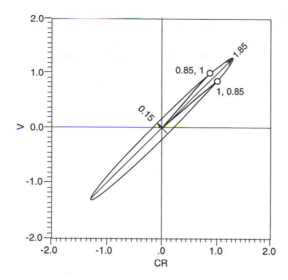

Figure 3–2. Elongated ellipse defined by rows in matrix of correlations between chromium and vanadium, which are highly correlated.

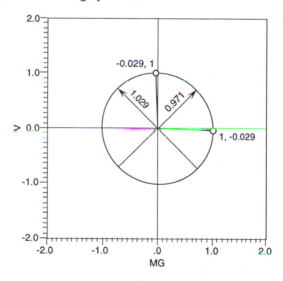

Figure 3–3. Nearly circular ellipse defined by rows in matrix of correlations between magnesium and vanadium, which have a correlation approaching zero.

The two eigenvalues of this matrix are

$$\lambda_1 = 1.029 \qquad \lambda_2 = 0.971$$

which are almost identical in size. As we can see, they define the major and minor semiaxes of an ellipse that is almost a circle, and both the semiaxes and the vectors are essentially radii. By definition, the axes of the ellipse are at right angles to each other, and the two plotted vectors also are almost orthogonal.

Some final notes on eigenvalues: You'll notice that the correlation matrices we've graphed are square, symmetrical about their diagonals, composed of real elements (that is, no imaginary numbers), and that the largest numbers in every row are on the diagonal. As a consequence of these special conditions, the eigenvalues will always be real numbers that are equal to or greater than zero. As you can verify by checking these examples, the sum of the eigenvalues of a matrix is always equal to the sum of the diagonal elements, or the *trace*, of the original matrix. In a correlation matrix, the diagonal elements are all equal to one, so the trace is simply the number of variables. The product of the eigenvalues will be equal to the determinant of the original matrix. Most (but not all) of the eigenvalue operations we will consider later will be applied to correlation or covariance matrices, so these special results will hold true in most instances. The methods just developed can be extended directly to $n \times n$ matrices, although the procedure becomes increasingly cumbersome with larger matrices.

Eigenvectors

We can examine the correlation matrices we calculated for the Istrian vineyard data to gain some insight into the geometrical nature of eigenvectors. First, consider the 2×2 matrix

$$\mathbf{R}_{cu,zn} = \begin{bmatrix} 1 & -0.28 \\ -0.28 & 1 \end{bmatrix}$$

with eigenvalues

$$\lambda_1 = 1.28 \qquad \lambda_2 = 0.72$$

Substituting the first eigenvalue into the original matrix gives

$$\begin{bmatrix} 1 - 1.28 & -0.28 \\ -0.28 & 1 - 1.28 \end{bmatrix} = \begin{bmatrix} -0.28 & -0.28 \\ -0.28 & -0.28 \end{bmatrix}$$

whose solution is the eigenvector

$$\begin{bmatrix} x_1 \\ x_2 \end{bmatrix} = \begin{bmatrix} -1 \\ 1 \end{bmatrix}$$

In **Figure 3–1**, we can interpret this eigenvector as the slope of the major semi-axis of the enclosing ellipse. If we regard the elements of the eigenvector as coordinates, the first eigenvector defines an axis which extends from the center of the ellipse into the second quadrant at an angle of 135°. The length is equal to the first eigenvalue, or 1.28.

Turning to the second eigenvalue, $\lambda_2 = 0.72$, the equation set is

$$\begin{bmatrix} 1 - 0.72 & -0.28 \\ -0.28 & 1 - 0.72 \end{bmatrix} = \begin{bmatrix} 0.28 & -0.28 \\ -0.28 & 0.28 \end{bmatrix}$$

whose solution gives the second eigenvector:

$$\begin{bmatrix} x_1 \\ x_2 \end{bmatrix} = \begin{bmatrix} 1 \\ 1 \end{bmatrix}$$

In **Figure 3–1**, this will plot as the vector direction $1/1 = 45°$, perpendicular to the major semiaxis of the ellipse. Its magnitude or length is 0.72.

We can determine the eigenvalues for the matrix of correlations between chromium and vanadium in a similar fashion. The matrix is

$$\mathbf{R}_{cr,v} = \begin{bmatrix} 1 & 0.85 \\ 0.85 & 1 \end{bmatrix}$$

with eigenvalues

$$\lambda_1 = 1.85 \qquad \lambda_2 = 0.15$$

The first eigenvector is

$$\begin{bmatrix} 1 - 1.85 & 0.85 \\ 0.85 & 1 - 1.85 \end{bmatrix} = \begin{bmatrix} -0.85 & 0.85 \\ 0.85 & -0.85 \end{bmatrix}$$

$$\begin{bmatrix} x_1 \\ x_2 \end{bmatrix} = \begin{bmatrix} 1 \\ 1 \end{bmatrix}$$

which defines a line having a slope of 45°. This axis bisects the angle between the two points and the center of the ellipse in **Figure 3–2**. The magnitude of the major semiaxis is equal to 1.85, the first eigenvalue of $\mathbf{R}_{cr,v}$. Similarly, we can show that the eigenvector associated with the second eigenvalue is

$$\begin{bmatrix} 1 - 0.15 & 0.85 \\ 0.85 & 1 - 0.15 \end{bmatrix} = \begin{bmatrix} 0.85 & 0.85 \\ 0.85 & 0.85 \end{bmatrix}$$

$$\begin{bmatrix} x_1 \\ x_2 \end{bmatrix} = \begin{bmatrix} 1 \\ -1 \end{bmatrix}$$

This procedure can be applied to the matrix $\mathbf{R}_{mg,v}$ and the eigenvectors found will again define directions of 135° and 45°, as shown in **Figure 3–3**. By now you no doubt suspect that the eigenvectors of 2×2 symmetric matrices will always lie at these specific angles, and this is indeed the case. The eigenvectors of real, symmetric matrices are always orthogonal, or at right angles to each other. This is not true of eigenvectors of matrices in general, but only of symmetric matrices. In addition, the eigenvectors of two-dimensional symmetric matrices are additionally constrained to orientations that are multiples of 45°. Incidentally, if two vectors, **A** and **B**, are orthogonal, then $\mathbf{A}^T\mathbf{B} = 0$.

Eigenvalue and eigenvector techniques are directly extendible to larger matrices, even though the operations become tedious. As an example, we will consider the full 5×5 correlation matrix **R** for trace metals from Istrian vineyard soils. The five eigenvalues of this matrix are

$$\Lambda = \begin{bmatrix} 2.453 & 1.233 & 0.789 & 0.465 & 0.061 \end{bmatrix}$$

and their associated eigenvectors are

$$\mathbf{V}_1 = \begin{bmatrix} 0.585 \\ -0.363 \\ 0.244 \\ 0.498 \\ 0.469 \end{bmatrix} \mathbf{V}_2 = \begin{bmatrix} -0.248 \\ -0.075 \\ 0.736 \\ -0.490 \\ 0.389 \end{bmatrix} \mathbf{V}_3 = \begin{bmatrix} 0.259 \\ 0.951 \\ 0.056 \\ 0.052 \\ 0.300 \end{bmatrix} \mathbf{V}_4 = \begin{bmatrix} -0.014 \\ -0.149 \\ -0.628 \\ -0.398 \\ 0.652 \end{bmatrix} \mathbf{V}_5 = \begin{bmatrix} -0.727 \\ 0.062 \\ -0.023 \\ 0.593 \\ 0.339 \end{bmatrix}$$

Each eigenvector can be regarded as a set of coordinates in five-dimensional space that defines the "direction" of a semiaxis of a hyperellipsoid. The length of each semiaxis is given by the corresponding eigenvalue. The first semiaxis is twice as long as the second, which is almost twice the length of the third. The fourth axis is very short, and the fifth axis is almost nonexistent; the hyperellipse defined by the correlation matrix, **R**, is really only a three-dimensional disk embedded in a space of five dimensions.

The slope of a line drawn from the origin of a graph through a point is defined by the ratio between the two coordinates of the point, and not by the actual magnitudes of the coordinates. Similarly, the absolute magnitudes of the elements in eigenvectors are not significant, only the ratios between the elements. An eigenvector can be scaled by multiplying by any arbitrary constant, and it will still define the same direction in multidimensional space. Different computer programs may return different eigenvectors for the same matrix; the eigenvectors simply have been scaled in different ways. Most programs ***normalize***, or scale each eigenvector so the sum of the squares of each element in a vector will be equal to 1.0. Others scale each eigenvector so the sum of its elements will be equal to its eigenvalue. Although such results appear to be different, the ratios between pairs of elements in the eigenvectors remain the same, and the vectors they define point in the same "direction." Also, you may note that the pattern of signs on the elements of the eigenvectors seems to be different for two otherwise identical sets of eigenvectors. This merely means that one set of vectors has been multiplied by (-1), reversing its "direction" but not changing its orientation in multivariate space.

Increasingly, computer programs for multivariate analysis employ alternative techniques for obtaining eigenvalues and eigenvectors. Rather than reducing a rectangular data matrix to a symmetrical, square correlation or covariance matrix and then extracting the desired eigenvalues and eigenvectors as we have done, these programs obtain results directly from the data matrix by ***singular value decomposition*** (SVD). An excellent description of SVD is given by Jackson (1991); Press and others (1992) provide a more compact presentation, as well as computer program listings. We will delay a discussion of this procedure until Chapter 6, where we can provide a motivation for our interest. Now, we merely note that an $n \times m$ rectangular matrix, **X**, can be decomposed into three other matrices:

$$\mathbf{X} = \mathbf{W}\mathbf{\Lambda}^{1/2}\mathbf{V}^\mathbf{T}$$

where **W** contains the eigenvectors of the major product matrix, $\mathbf{X}\mathbf{X}^\mathbf{T}$. **V** contains the eigenvectors of the minor product matrix, $\mathbf{X}^\mathbf{T}\mathbf{X}$, and $\mathbf{\Lambda}$ is an $m \times m$ diagonal matrix whose diagonal elements are the eigenvalues of either $\mathbf{X}\mathbf{X}^\mathbf{T}$ or $\mathbf{X}^\mathbf{T}\mathbf{X}$ (they will be identical except that $\mathbf{X}^\mathbf{T}\mathbf{X}$ will have $n - m$ extra eigenvalues, all equal to zero).

If you have worked through the small examples in this chapter, you can readily appreciate that the computational labor involved in dealing with large matrices can be formidable, even though the underlying, individual mathematical steps are simple. A modest data set such as ISTRIA.TXT will present a challenge to those who attempt to analyze the data by hand. Fortunately, there are many powerful computational tools available at modest cost (at least for student versions), and they run on almost any type of personal computer. A numerical computation package such as MATLAB®, Mathcad®, or MATHEMATICA®, and even some statistical packages, such as S-PLUS© , will provide all of the mathematical computation power you are likely to need for applications in the Earth sciences. We have attempted to present, in as painless a manner as possible, the rudiments of beginning matrix algebra. As stated at the conclusion of Chapter 2, statistics is too large a subject to be covered in one chapter, or even one book. Matrix algebra also is an impossibly large subject to encompass in these few pages. However, you should now have some insight into matrix methods that will enable you to understand the computational basis of techniques we will cover in the remainder of this book.

EXERCISES

Exercise 3.1

File BHTEMP.TXT contains 15 bottomhole temperatures (BHT's) measured in the Mississippian interval in wells in eastern Kansas. The measurements are in degrees Fahrenheit. Convert the vector of temperatures to degrees Celsius using matrix algebra.

Exercise 3.2

The following two matrices are defined:

$$\mathbf{A} = \begin{bmatrix} 1 & 2 \\ -2 & 0 \end{bmatrix} \qquad \mathbf{B} = \begin{bmatrix} -3 & 2 \\ -2 & -4 \end{bmatrix}$$

Compute the matrix products, $\mathbf{A}\mathbf{B}$ and $\mathbf{B}\mathbf{A}$. Two matrices which exhibit the property that will be apparent are said to be *commutative*. Demonstrate that for commutative matrices, $\mathbf{A}^{-1}\mathbf{B}^{-1} = (\mathbf{A}\mathbf{B})^{-1}$.

Consider the following two matrices,

$$\mathbf{C} = \begin{bmatrix} 2 & 1 & 0 \\ 3 & 4 & 0 \\ 0 & 0 & 2 \end{bmatrix} \qquad \mathbf{D} = \begin{bmatrix} 1 & -1 & 3 \\ 7 & 1 & 2 \\ 5 & 0 & 1 \end{bmatrix}$$

Compare the determinant, $|\mathbf{C}\mathbf{D}|$, of the matrix product to the product, $|\mathbf{C}| \cdot |\mathbf{D}|$, of the determinants of the two matrices. The result you obtain is general. Determine if $|\mathbf{C}| + |\mathbf{D}| = |\mathbf{C} + \mathbf{D}|$. This result also is general. For the matrices $\mathbf{C}$ and $\mathbf{D}$, demonstrate that $(\mathbf{C}\mathbf{D})^{\mathbf{T}} = \mathbf{D}^{\mathbf{T}}\mathbf{C}^{\mathbf{T}}$. Using matrix $\mathbf{C}$, show that $(\mathbf{C}^{-1})^{\mathbf{T}} = (\mathbf{C}^{\mathbf{T}})^{-1}$.

Exercise 3.3

File MAGNETIT.TXT contains the proportions of olivine, magnetite, and anorthite estimated by point-counting thin sections from 15 hand specimens collected at a magnetite deposit in the Laramie Range of Wyoming. The specific gravity is 3.34 for olivine, 2.76 for anorthite, and 5.20 for magnetite. Using matrix algebra, estimate the specific gravity of the 15 samples.

Exercise 3.4

Coordinates can be rotated by a matrix multiplication in which the premultiplier is a 2×2 matrix of sines and cosines of the angle of rotation,

$$\begin{bmatrix} \cos\theta & \sin\theta \\ -\sin\theta & \cos\theta \end{bmatrix}$$

where θ is the desired angle of rotation. Data in file PROSPECT.TXT were taken from a surveyor's notebook describing the outline of a gold prospect in central Idaho. Coordinates are given in meters from an arbitrary origin at the southwest corner of the property and were measured relative to magnetic north. The magnetic declination in this area is $18°30'$ east of true north. Convert the surveyor's measurements to coordinates relative to true north.

Exercise 3.5

Petrophysical well logs are strip charts made after the drilling of a well by lowering a sonde down the hole and recording physical properties versus depth in the well. Measurements include various electrical and sonic characteristics of the rocks, and both natural and induced radioactivity. The measured values reflect the composition of the rocks and the fluids in the pore space.

File KANSALT.TXT contains data for depths between 980 and 1030 ft below the surface in A.E.C. Test Hole No. 2, drilled in 1970 in Rice County, Kansas. At this depth, the well penetrated the Hutchinson Salt member of the Permian Wellington Formation, which was under investigation as a possible nuclear waste disposal site. The Wellington Formation is composed entirely of varying proportions of halite, anhydrite, and shale. Pure samples of these end members have distinct physical properties, so appropriate log responses can be used to estimate the relative amounts of halite, anhydrite, or shale at every foot within the Wellington Formation. A more detailed discussion of these data is given in Doveton (1986).

Table 3–4. Physical properties measured on pure samples of halite, anhydrite, and "shale" (clay minerals). From Gearhart-Owen (1975).

	Halite	Anhydrite	Shale
Apparent grain density (ρ_b), g/cc	2.03	2.98	2.43
Sonic transit time (Δ_t), μsec/ft	67	50	113

Two useful petrophysical properties are the apparent density (in grams per cubic centimeter) as measured by gamma-ray absorption and sonic transit time (in microseconds per foot). Laboratory-determined values for pure halite, anhydrite, and shale are given in **Table 3–4**. The apparent density and the sonic transmission time of a mixture of these three constituents can be calculated as the sum of the products of the densities and transit times for pure constituents times the proportions of the constituents. That is,

$$\rho_b = 2.03V_h + 2.98V_a + 2.43V_{sh}$$

$$\Delta_t = 67V_h + 50V_a + 113V_{sh}$$

where V_h, V_a, and V_{sh} are the proportions of halite, anhydrite, and shale. However, we want to reverse these equations, and for given values of ρ_b and Δ_t that we read from the well logs, estimate the proportions of the three constituents of the rock. Since three unknowns must be estimated, it seems we will require three equations and, hence, measurements of three log properties. However, because the proportions of halite, anhydrite, and shale must sum to one, we can use this constraint to provide the necessary third equation.

$$1 = V_h + V_a + V_{sh}$$

The three equations can be set into matrix form as

$$\mathbf{L} = \mathbf{CV}$$

$$\begin{bmatrix} \rho_b \\ \Delta_t \\ 1 \end{bmatrix} = \begin{bmatrix} 2.03 & 2.98 & 2.43 \\ 67 & 50 & 113 \\ 1 & 1 & 1 \end{bmatrix} \begin{bmatrix} V_h \\ V_a \\ V_{sh} \end{bmatrix}$$

However, what we really want to do is solve for $\mathbf{V}$, given values of $\mathbf{L}$ taken from the well logs. This means that $\mathbf{C}$ must be moved to the other side of the equal sign, which we can do by multiplying both sides of the equation by its inverse, $\mathbf{C}^{-1}$. Then,

$$\begin{bmatrix} 2.03 & 2.98 & 2.43 \\ 67 & 50 & 113 \\ 1 & 1 & 1 \end{bmatrix}^{-1} \begin{bmatrix} \rho_b \\ \Delta_t \\ 1 \end{bmatrix} = \begin{bmatrix} V_h \\ V_a \\ V_{sh} \end{bmatrix}$$

Perform the necessary matrix inversion and multiplications to determine the proportions of halite, anhydrite, and shale in the 50-ft interval of the Hutchinson Salt. Plot the record of lithologic compositions in the form of a lithologic strip log. Ten of these estimates have been used in Chapter 2 (**Table 2.9**) to demonstrate the effects of closure on the calculation of correlations among closed variables.

[Hint: $\mathbf{L}$, as given in file KANSALT.TXT, is a 2×50 matrix of ρ_b and Δ_t log responses. It must be converted to a 3×50 matrix by adding a column of 1's in order for the dimensions of the matrix multiplication to be correct. What does this column of 1's represent?]

Exercise 3.6

The state of stress in the subsurface can be represented in a 3×3 matrix, Σ, whose diagonal elements represent normal stresses and whose off-diagonal elements represent shear stresses. The meanings of the nine elements of the stress matrix can be seen by imagining a cube in a Cartesian coordinate system in which the X-axis points to the east, the Y-axis points to the north, and the Z-axis points up. The symbol σ_{xx} represents the normal stress directed onto the east or west face of the cube; it will be a positive value if the stress is compressional and a negative value if the stress is tensional. There is a similar meaning for σ_{yy} and σ_{zz}. The symbol σ_{xy} represents the shear stress on the east or west face of the cube, acting parallel to the Y-axis. A shear stress is positive if the compressional or tensional component agrees in sign with the direction of force. That is, both components of shear point in a positive coordinate direction, or both components point in a negative coordinate direction. Otherwise, the shear stress is negative. In order for the cube to be in rotational equilibrium, shear stresses on adjacent faces must balance; so, for example, $\sigma_{xy} = \sigma_{yx}$. This means that the stress matrix is symmetric about the diagonal:

$$\begin{bmatrix} \sigma_{xx} & \sigma_{xy} & \sigma_{xz} \\ \sigma_{xy} & \sigma_{yy} & \sigma_{yz} \\ \sigma_{xz} & \sigma_{yz} & \sigma_{zz} \end{bmatrix}$$

Turcotte and Schubert (1982) provide a more detailed discussion of stress in the subsurface and the measurement of stress components.

By finding the eigenvalues and eigenvectors of the 3×3 stress matrix, we can rotate the imaginary cube into a coordinate system in which all the shear stresses will be zero. The eigenvalues represent the magnitudes of the three orthogonal stresses. Their associated eigenvectors point in the directions of the stresses. The largest eigenvalue, λ_1, represents the maximum normal stress and the smallest, λ_3, represents the minimum normal stress. The maximum shear stress is given by $(\lambda_1 - \lambda_3)/2$ and occurs along a plane oriented perpendicular to a line that bisects the angle between the directions of maximum and minimum normal stress (that is, between the first and third eigenvectors). In a homogenous, isotropic material, failure (*i.e.*, faulting) will tend to occur along this plane. The orientation of this plane can be determined from the elements of the first eigenvector. In the conventional notation used by geologists, the strike of the first eigenvector is $\tan^{-1}(v_{12}/v_{11})$ and its dip is

$$\tan^{-1}\left(v_{13}\Big/\sqrt{v_{11}^2 + v_{12}^2}\right)$$

(Here, v_{ij} refers to the jth element of the ith eigenvector.) The strike and dip of the second and third eigenvectors can be found in the same manner.

Three-dimensional stress measurements have been made in a pillar in a deep mine, yielding the following stress matrix:

$$\begin{bmatrix} 61.2 & 4.1 & -8.2 \\ 4.1 & 51.5 & -3.0 \\ -8.2 & -3.0 & 32.3 \end{bmatrix}$$

The data are given in megapascals (MPa) and were recorded by strain gauges placed so the measurements have the same orientation as our imaginary cube (X increasing to the east, Y to the north, and Z increasing upward). Find the principal stresses and their associated directions. What is the maximum shear stress and what is the strike and dip of the plane on which this stress occurs?

SELECTED READINGS

Anton, H., and C. Rorres, 1994, *Elementary Linear Algebra,* 7th ed., Applications Version: John Wiley & Sons, Inc., New York, 800 pp. *A computationally oriented text on matrix algebra. Diskettes contain examples and exercises.*

Buchanan, J.L., and P.R. Turner, 1992, *Numerical Methods and Analysis*: McGraw-Hill, Inc., New York, 751 pp.

Davis, P.J., 1984, *The Mathematics of Matrices*: R.E. Krieger Publ. Co., Malabar, Fla., 368 pp. *Reprint of a classic. A highly readable text on matrix algebra with a minimum of mathematical jargon and a maximum of examples and applications.*

Doveton, J.H., 1986, *Log Analysis of Subsurface Geology: Concepts and Computer Methods*: John Wiley & Sons, Inc., New York, 273 pp. *Chapter 6 discusses matrix algebra techniques for resolving rock composition from well log responses, including the Hutchinson Salt (file KANSALT.TXT) exercise.*

Ferguson, J., 1988, *Mathematics in Geology*: Allen & Unwin Ltd., London, 299 pp. *Chapters 6 and 7 treat matrix algebra and its application to geological problems.*

Gearhart-Owen, 1975, *Formation Evaluation Data Handbook*: Gerhard-Owen Industries, Inc., Fort Worth, Texas, 240 pp.

Golub, G.H., and C.F. Van Loan, 1996, *Matrix Computations,* 3rd ed.: Johns Hopkins Univ. Press, Baltimore, Md., 694 pp.

Gould, P., 1967, On the geographic interpretation of eigenvalues: An initial exploration: *Trans. Inst. British Geographers,* No. 42, p. 53–86. *An intuitive look at eigenvalues and vectors by geometric analogy. Part of this chapter is derived from this excellent exposition, written originally for students.*

Jackson, J.E., 1991, *A User's Guide to Principal Components*: John Wiley & Sons, Inc., New York, 569 pp. *Appendices A and B are a concise summary of matrix algebra. Chapter 10 discusses singular value decomposition.*

Jensen, J.A., and J.H. Rowland, 1975, *Methods of Computation: The Linear Approach to Numerical Analysis*: Scott, Foresman and Co., Glenview, Ill., 303 pp.

Maron, M.J., and R.J. Lopez, 1991, *Numerical Analysis—A Practical Approach,* 3rd ed.: PWS-Kent Publ. Co., Boston, Mass., 743 pp. *Gives procedures and algorithms for matrix operations, especially different methods for inversion, solution of simultaneous equations, and extraction of eigenvalues.*

Ortega, J.M., 1990, *Numerical Analysis, a Second Course*: Society for Industrial and Applied Mathematics, Philadelphia, Pa., 201 pp. *A concise but complete text, issued as a paperback reprint by SIAM to "foster better understanding of applied mathematics."*

Pettofrezzo, A.J., 1978, *Matrices and Transformations*: Dover Publications, Inc., New York, 133 pp. *This paperback reprint of a classic text covers the traditional material for a one-semester matrix algebra course. It is liberally sprinkled with worked examples and problems.*

Press, W.H., S.A. Teukolsky, W.T. Vetterling, and B.P. Flannery, 1992, *Numerical Recipes: The Art of Scientific Computing,* 2nd ed.: Cambridge Univ. Press, Cambridge, U.K., 963 pp. *The "how-to" book of computer algorithms for numerical computation; contains succinct descriptions of eigenvalue techniques, including SVD. Available in several versions for different computer languages.*

Searle, S.R., 1982, *Matrix Algebra Useful for Statistics*: John Wiley & Sons, Inc., New York, 438 pp. *Examples and exercises are drawn from the biological sciences.*

Turcotte, D.L., and G. Schubert, 1982, *Geodynamics Applications of Continuum Physics to Geological Problems*: John Wiley & Sons, Inc., New York, 450 pp.

Wolfram, S., 1996, *The MATHEMATICA® Book*: Wolfram Media, Inc., Champaign, Ill., 1395 pp.

Chapter 4
Analysis of Sequences of Data

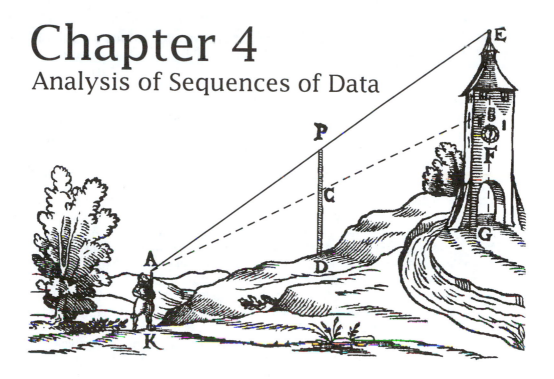

In this chapter we will consider ways of examining data that are characterized by their position along a single line. That is, they form a sequence, and the position at which a data point occurs within the sequence is important. Data sets of this type are common in geology, and include measured successions of lithologies, geochemical or mineralogical assays along traverses or drill holes, electric logs of oil wells, and chart recordings from instruments. Also in this general category are measurements separated by the flow of time, such as a sequence of water quality determinations at a river station, or the production history of a flowing gas well. Techniques for examining data having a single positional characteristic traditionally are considered part of the field of time-series analysis, although we will take the broader view that time and space relationships can be considered interchangeably.

Geologic Measurements in Sequences

Before proceeding to some geological examples and appropriate methods of examination, we must consider the nature of different types of sequences apt to be encountered by geologists. At one extreme, we may have a record which is quite precise, both in the variable which is measured and in the scale along which successive observations are located. Examples might include an electrical resistivity log from a borehole, or the production history of a commercial well. In the former, the variable is a measured attribute expressed in ohms (Ω) and the scale is measured in feet. In the latter example, the variable again is a measured attribute, barrels (bbl) of oil, and the scale is measured in days, months, or years. There are two important characteristics in either record. First, the variable being measured is expressed in units of an interval or ratio scale; 1000 bbl of oil is twice as large a quantity as 500 bbl, and a measurement of 10 Ω is ten times the resistance of 1 Ω. Second, the scales along which the data points are located also are expressed in units having magnitude. A depth of 3000 ft in a well is ten times a depth of 300 ft, and the decade between the years 1940 and 1950 has the same duration as the interval between 1950 and 1960. These may seem obvious or even trivial points to emphasize, but as we shall see, not all geologic sequences have such well-behaved characteristics.

At the opposite extreme, we can consider a stratigraphic sequence consisting of the lithologic states encountered in a sedimentary succession. Such a sequence might be a cyclothem of limestone–shale–limestone–shale–sandstone–coal–shale–limestone, from bottom to top. We are interested in the significance of the succession, but we cannot put a meaningful scale on the sequence itself. Obviously, the succession of lithologies represents changes that occurred through time, but we have no way of estimating the time scale involved. We could use thickness, but this may change dramatically from location to location even though the sequence is not altered. If thickness is considered, it may obscure our examination of the succession, which is the subject of our interest. Thus, the fact that limestone is the third state in the section and coal is the sixth has no significance that can be expressed numerically (that is, position 6 is not "twice" position 3). Likewise, the lithologic states of the units cannot be expressed on a numerical scale. We might code the sequences just given as $1 - 2 - 1 - 2 - 3 - 4 - 2 - 1$, where limestone is equated to 1, shale is 2, sandstone is 3, and coal is 4, but such a convention is purely arbitrary and expresses no meaningful relations between the states. It is obvious that this sequence poses different problems to the analyst than do the first examples.

There also are intermediate possibilities. For example, we may be interested in some measurable attribute contained in successive stages of a sequence. Perhaps we have measured the boron content of each lithologic unit in the cyclothem just discussed. We can utilize a distance scale of feet between samples and consider this a problem related to depth or distance. Alternatively, we can consider the relationship between the boron measurements and the sequence of states.

A closely related problem is the analysis of a sequence characterized by the presence or absence of some variable or variables at points along a line. We might be interested, for example, in the repeated recurrence of certain environment-dependent microfossils in the chips recovered during the drilling of a well. Another class of problems may be typified by the succession of mineral grains encountered on traverses across a thin section. In this case, we can use millimeters as a convenient spatial scale, but we have no way of evaluating whether olivine rates a higher number than plagioclase.

Data having the characteristic of being arranged along a continuum, either of time or space, often are referred to as forming a series, sequence, string, or chain. The nature of the data and the chain determine the questions that we can consider. Obviously, we cannot extract information about time intervals from stratigraphic succession data, because the time scale accompanying the succession is not known. We often substitute spatial scales for a time scale in stratigraphic problems, but our conclusions are no better than our fundamental assumptions about the length of time required to deposit the interval we have measured.

Table 4–1 is a classification of the various data-analysis techniques discussed in this chapter. We can consider two types of sequences. In the first, the distance between observations varies and must be specified for every point. In the second, the points are assumed to be equally and regularly spaced; the numerical value of the spacing does not enter into the analyses except as a constant. A subset of

Table 4–1. Techniques discussed in this chapter classified by the nature of the variable and its spacing along a line. Locations are explicit if X is specified for every Y; locations are implicit if X is implied by the order of observations.

Nature of Variables	Explicit Location in Time or Space	Implicit Location in Time or Space
Interval or Ratio Data	Interpolation Regression Splines	Zonation Seriation Autocorrelation Cross-correlation Semivariograms Periodograms Spectral Density
Nominal or Ordinal Data	Series of Events	Markov Chains Runs Tests

this category does not consider the spacing at all, and only the sequence of the observations is important.

The techniques also may be classified on the type of observations they require. Some necessitate interval or ratio observations; the variate must be measured on a scale and expressed in real numbers. Other methods accept nominal or ordinal data, and observations need only to be categorized in some fashion. In the methods discussed in this chapter, the classes are not ranked; that is, state A is not "greater" or "larger" in some sense than states B or C. Nominal data may be represented by integers, alphabetic characters, or symbols.

In the remainder of this chapter, we are going to examine the mathematical techniques required to analyze data in sequences. The methods described here do not exhaust the possibilities by any means. Rather, these are a collection of operations that have proved valuable in quantitative problem-solving in the Earth sciences, or that seem especially promising. Other methods may be more appropriate or powerful in specific situations or for certain data sets. However, a familiarity with the techniques discussed here will provide an introduction to a diverse field of analytical tools. Unfortunately, many of these methods were developed in scientific specialties alien to most geologists, and the description of an application in radar engineering, stock market analysis, speech therapy, or cell biology may be difficult to relate to a geologic problem. Some of the methods involve nonparametric statistics, and these are not widely considered in introductory statistics courses. Because of the general unfamiliarity of most Earth scientists with developments in the numerical analysis of data sequences, we have thought it best to present a potpourri of techniques and approaches. As you can see from **Table 4.1**, these cover a variety of sequences of different types, and are designed to answer different kinds of questions. None of the techniques can be considered exhaustively in this short space, but from the examples and applications presented, one or another may suggest themselves to the geologist with a problem to solve. The list of Selected Readings can then provide a discussion of a specific subject in more detail.

These methods provide answers to the following broad categories of questions: Are the observations random, or do they contain evidence of a trend or pattern? If a trend exists, what is its form? Can cycles or repetitions be detected and measured?

Can predictions or estimations be made from the data? Can variables be related or their effectiveness measured? Although such questions may not be explicitly posed in each of the following discussions, you should examine the nature of the methods and think about their applicability and the type of problems they may help solve. The sample problems are only suggestions from the many that could be used.

Geologists are concerned not only with the analysis of data in sequences, but also with the comparison of two or more sequences. An obvious example is stratigraphic correlation, either of measured sections or petrophysical well logs. A geologist's motive for numerical correlation may be a simple desire for speed, as in the production of geologic cross-sections from digitized logs stored in data banks. Alternatively, he may be faced with a correlation problem where the recognition of equivalency is beyond his ability. Subtle degrees of similarity, too slight for unaided detection, may provide the clues that will allow him to make a decision where none is otherwise possible. Numerical methods allow the geologist to consider many variables simultaneously, a powerful extension of his pattern-recognition facilities. Finally, because of the absolute invariance in operation of a computer program, mathematical correlation provides a challenge to the human interpreter. If a geologist's correlation disagrees with that established by computer, it is the geologist's responsibility to determine the reason for the discrepancy. The forced scrutiny may reveal complexities or biases not apparent during the initial examination. This is not to say that the geologist should unthinkingly bend his interpretation to conform with that of the computer. However, because modern programs for automatic correlation are increasingly able to mimic (and extend) the mental processes of a human interpreter, their output must be considered seriously.

Most techniques for comparing two or more sequences can be grouped into two broad categories. In the first of these, the data sequences are assumed to match at one position only, and we wish to determine the degree of similarity between the two sequences. An example is the comparison of an X-ray diffraction chart with a set of standards in an attempt to identify an unknown mineral. The chart and standards can be compared only in one position, where intensities at certain angles are compared to intensities of the standards at the same angles. Nothing is gained, for example, by comparing X-ray intensity at $20°2\theta$ with the intensity at $30°2\theta$ on another chart. Although the correspondence may be high, it is meaningless.

The fact that data such as these are in the form of sequences is irrelevant, because each data point is considered to be a separate and distinct variable. The intensity of diffracted radiation at $20°2\theta$ is one variable, and the intensity at $30°2\theta$ is another. We will consider methods for the comparisons of such sequences in greater detail in Chapter 6, when we discuss multivariate measures of similarity and problems of classification and discrimination. In this class of problems, an observation's location in a sequence merely serves to identify it as a specific variable, and its location has no other significance.

In contrast, some of the techniques we will discuss in this chapter regard data sequences as samples from a continuous string of possible observations. There is no *a priori* reason why one position of comparison should be better than any other. These methods of cross comparison superficially resemble the mental process of geologic correlation, but have the limitation that they assume the distance or time scales of the two sequences being compared are the same. In historic time series and sequences such as Holocene ice cores, this assumption is valid. In other circumstances such as stratigraphic correlation, equivalent thicknesses may not represent equivalent temporal intervals and the problem of cross comparison is much more complex.

As we emphasized in Chapter 1, the computer is a powerful tool for the analysis of complex problems. However, it is mindless and will accept unreasonable data and return nonsense answers without a qualm. A bundle of programs for analyzing sequences of data can readily be obtained from many sources. If you utilize these as a "black box" without understanding their operation and limitations, you may be led badly astray. It is our hope in this chapter that the discussions and examples will indicate the areas of appropriate application for each method, and that the programs you use are sufficiently straightforward so that their operation is clear. However, in the final analysis, the researcher must be his own guide. When confronted with a problem involving data along a sequence, you may ask yourself the following questions to aid in planning your research:

(a) What question(s) do I want to answer?
(b) What is the nature of my observations?
(c) What is the nature of the sequence in which the observations occur?

You may quickly discover that the answer to the first question requires that the second and third be answered in specific ways. Therefore, you avoid unnecessary work if these points are carefully thought out before your investigation begins. Otherwise, the manner in which you gather your data may predetermine the techniques that can be used for interpretation, and may seriously limit the scope of your investigation.

Interpolation Procedures

Many of the following techniques require data that are equally spaced; the observations must be taken at regular intervals on a traverse or line, or equally spaced through time. Of course, this often is not possible when dealing with natural phenomena over which you have little control. Many stratigraphic measurements, for example, are recorded bed-by-bed rather than foot-by-foot. This also may be true of analytical data from drill holes, or from samples collected on traverses across regions which are incompletely exposed. We must, therefore, estimate the variable under consideration at regularly spaced points from its values at irregular intervals. Estimation of regularly spaced points will also be considered in Chapter 5, when we discuss contouring of map data. Most contouring programs operate by creating a regular grid of control points estimated from irregularly spaced observations. The appearance and fidelity of the finished map is governed to a large extent by the fineness of the grid system and the algorithm used to estimate values at the grid intersections. We are now considering a one-dimensional analogy of this same problem.

The data in **Table 4–2** consist of analyses of the magnesium concentration in stream samples collected along a river. Because of the problems of accessibility, the samples were collected at irregular intervals up the winding stream channel. Sample localities were carefully noted on aerial photographs, and later the distances between samples were measured.

Although there are many methods whereby regularly spaced data might be estimated from these data, we will consider only two in detail. The first and most obvious technique consists of simple *linear interpolation* between data points to

Table 4–2. Measurements of magnesium concentration in stream water at 20 locations; distances are from stream mouth to sample locations.

Distance (m)	Magnesium (ppm)	Distance (m)	Magnesium (ppm)
0.0	6.44	11,098	2.86
1820	8.61	11,922	1.22
2542	5.24	12,530	1.09
2889	5.73	14,065	2.36
3460	3.81	14,937	2.24
4586	4.05	16,244	2.05
6020	2.95	17,632	2.23
6841	2.57	19,002	0.42
7232	3.37	20,860	0.87
10,903	3.84	22,471	1.26

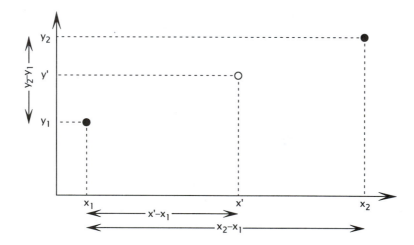

Figure 4–1. Linear interpolation between two data points along a sequence.

estimate intermediate points. This approach is illustrated in **Figure 4–1.** Assume y_1 and y_2 are observed values at points x_1 and x_2; we wish to estimate the value of y' at point x'. If we assume that a straight linear relation exists between sample points, intermediate values can be calculated from the geometric relationship

$$y' = \frac{(y_2 - y_1)(x' - x_1)}{x_2 - x_1} + y_1 \tag{4.1}$$

Expressed in other words, the difference between values of two adjacent points is assumed to be a function of the distance separating them. The value of a point halfway between two observations is exactly intermediate between the values of the two enclosing points. The nearer a point is to an observation, the closer its value is to that of the observation. The manganese values from stream samples listed in **Table 4–2** are shown in graphical form in **Figure 4–2 a**, and interpolated to regular 1000-m intervals in **Figure 4–2 b**.

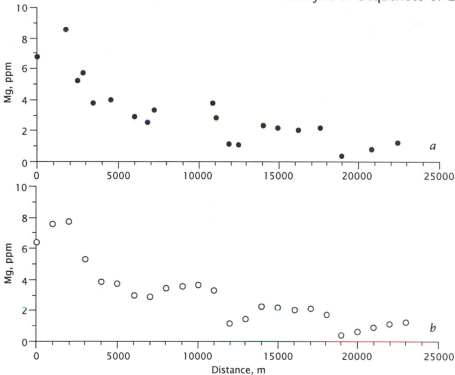

Figure 4–2. Magnesium concentration (parts per million) in water at 20 stream locations, measured in meters from stream mouth. (*a*) Original field measurements. (*b*) Values interpolated at 1000-m intervals.

Although linear interpolation is simple, it possesses certain drawbacks in many applications. If the number of equally spaced points is approximately the same as the number of original points, and the original points are somewhat uniformly spaced, the technique will give satisfactory results. However, if there are many more original points than interpolated points, most of the original data will be ignored because only two surrounding points determine an interpolated value. If the original data possess a large random component which causes values to fluctuate widely, interpolated points may also fluctuate unacceptably. Both of these objections may be met by techniques that consider more than two of the original values, perhaps by fitting a linear function that extends over several adjacent values. Wilkes (1966) devotes an entire chapter to various interpolation procedures.

If the original data are sparse and several values must be estimated between each pair of observations, linear interpolation will perform adequately, provided the idea of uniformity of slope between points is reasonable. In any problem where points are interpolated between observations, however, you must always remember that you cannot create data by estimation using any method. The validity of your result is controlled by the density of the original values and no amount of interpolation will allow refinement of the analysis beyond the limitations of the data. For example, we could estimate the magnesium content of the river at 500-m intervals, or even at every 5 m, but it is obvious that these new values would provide no additional information on the distribution of the metal in the stream.

We will next consider a method that produces equally spaced estimates of a variable and considers all observations between successive points of estimation.

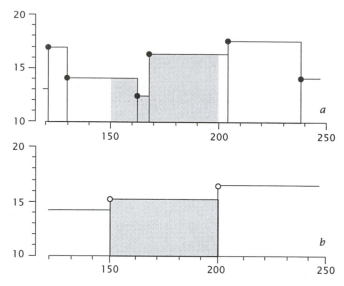

Figure 4–3. (*a*) Data sequence considered as a step function or "rectangular curve." (*b*) Equally spaced sequence created by rectangular integration. Shaded intervals in (*a*) and (*b*) have the same areas.

The technique is called ***rectangular integration***. If we regard the original data as a rectangular curve or step function in which the interval from one observation to the succeeding observation has a constant value, a data set might have the form shown in **Figure 4–3 a**. If we wish to create an equally spaced approximation to this distribution, we can generate another step function of rectangles of equal length whose areas equal the total areas of the original rectangles. This is shown graphically in **Figure 4–3 b**, with the resulting sequence of equally spaced values derived from the data in part **a**. The shaded area under the curve is the same in both illustrations. This procedure has the advantage of considering all data within an interval in estimating a point. Also, because the area under the estimated curve is equal to the area under the original curve, observations used in the estimation of a point are weighted proportionally to the length of interval they represent.

Calculation of an estimate by rectangular integration is easy in theory but presents a somewhat difficult programming challenge. Starting at one estimated point, the distance to the next observation must be calculated, multiplied by the magnitude of the observation to give the rectangular area, and the process repeated through all successive observations up to the next estimated point. That point is determined by summing the areas just found and dividing by the equally spaced interval to give the estimated value. The initial estimated point in a sequence is taken as the same as the first preceding data point.

An obvious difference in the two interpolation procedures is apparent when original data are sparse and more than one point must be estimated between two observations. Using linear interpolation, values will be created which lie on a straight line between two surrounding data points. In contrast, rectangular integration will create estimates that are equal to the first observation.

In the study of a metamorphic halo around an intrusive, a diamond-drill core was taken perpendicular to the intrusive wall. The entire core was split and all garnet crystals exposed on the split surface were removed, individually crushed, and

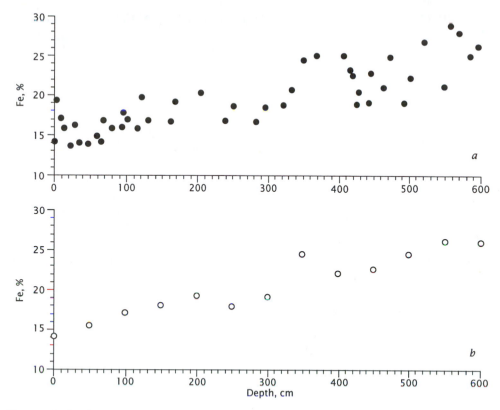

Figure 4–4. Iron content, in percent Fe_2O_3, of garnets taken from diamond-drill core through metamorphic halo. (*a*) Original measurements. (*b*) Values integrated to 50-cm intervals.

analyzed for iron content by a rapid spectrochemical method. Both the spacing between successive crystals and their iron content fluctuate through a wide range. Data from this core are shown in **Figure 4–4 a** and are given in file GARNETS.TXT. A generalized picture of compositional changes is desired, but the data seem too erratic for direct interpretation. As a preparatory step to further analysis, the data may be approximated by equally spaced estimates. The desired interval is 50 cm. Here we are presented with a situation that is different from the river data; observations are more abundant than estimates and we wish to preserve as much of the original information as possible. Rectangular integration seems more appropriate in this instance than linear interpolation. **Figure 4–4 b** shows the result of inter-polating iron concentration to 50-cm intervals by rectangular integration. It may be instructive to compare these results with those from linear interpolation and to compare both with the overlying original data to see how much detail is lost by the two approximation processes.

In geology, equal spacing procedures have been most widely used to pretreat stratigraphic data (measured sections, drilling-time logs, and similar records) prior to filtering or time-trend analysis. Time-series methods, such as autocorrelation and spectral analysis, require equally spaced data. Time-series techniques are inherently more powerful than other analytical methods for examining sequential data, and their use has become widespread. However, they require long strings of data, which has restricted their application to geophysics, well-log analysis, and the study of stratigraphic sequences and diamond-drill cores through ore deposits. Some work also has been done on mineral successions along traverses across thin sections. These applications will be considered in greater detail later in this chapter.

Markov Chains

In many geologic investigations, data sequences may be created that consist of ordered successions of mutually exclusive states. An example is a point-count traverse across a thin section, where the states are the minerals noted at succeeding points. Measured stratigraphic sections also have the form of series of lithologies, as may drill holes through zoned ore bodies where the rocks encountered are classified into different types of ore and gangue. Observations along a traverse may be taken at equally spaced intervals, as in point counting, or they may be taken wherever a change in state occurs, as is commonly done in the measurement of stratigraphic sections. In the first instance, we would expect runs of the same state; that is, several successive observations could conceivably fall in the same category. This obviously cannot happen if observations are taken only where states change.

Table 4–3. Stratigraphic succession shown in Figure 4–4 coded into four mutually exclusive states of sandstone (A), limestone (B), shale (C), and coal (D); observations taken at 1-ft intervals.

Top					
C	C	B	C	A	A
C	C	B	C	A	A
C	C	B	C	A	A
A	A	B	C	C	A
A	A	B	A	C	A
A	C	C	A	D	A
A	C	C	A	C	C
A	D	C	A	C	Bottom
A	D	B	A	D	
C	C	B	C	D	
C	C	C	C	C	

Sometimes we are interested in the nature of transitions from one state to another, rather than in the relative positions of states in the sequence. We can employ techniques that sacrifice all information about the position of observations within the succession, but that provide in return information on the tendency of one state to follow another. The data in **Table 4–3** represent the stratigraphic section shown in **Figure 4–5**, in which the sedimentary rock has been classified at successive points spaced 1 ft apart. The lithologies include four mutually exclusive states—sandstone, limestone, shale, and coal, arbitrarily designated A, B, C, and D, respectively. A 4×4 matrix can be constructed, showing the number of times a given rock type is succeeded, or overlain, by another. A matrix of this type is called a ***transition frequency matrix*** and is shown below. The measured stratigraphic section contains 63 observations, so there are $(n - 1) = 62$ transitions. The matrix is read "from rows to columns," meaning, for example, that a transition from state

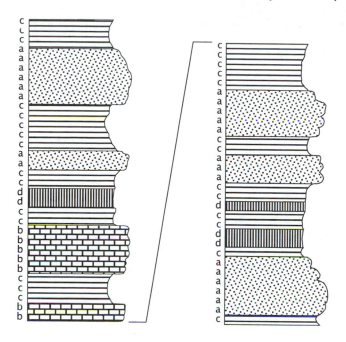

Figure 4–5. Measured stratigraphic column in which lithologies have been classified into four mutually exclusive states of sandstone (*a*), limestone (*b*), shale (*c*), and coal (*d*).

A to state C is counted as an entry in element $a_{1,3}$ of the matrix. That is, if we read from the row labeled A to the column labeled C, we see that we move from state A into state C five times in the sequence. Similarly, there are five transitions from state C to state A in the sequence; this number appears as the matrix element defined by row C and column A. The transition frequency matrix is a concise way of expressing the incidence of one state following another:

$$
\begin{array}{cc}
 & \begin{array}{cccc} & to & & \\ A & B & C & D \end{array} \\
from \begin{array}{c} A \\ B \\ C \\ D \end{array} & \left[\begin{array}{cccc} 18 & 0 & 5 & 0 \\ 0 & 5 & 2 & 0 \\ 5 & 2 & 18 & 3 \\ 0 & 0 & 3 & 2 \end{array}\right] \begin{array}{c} \text{Row} \\ \text{Totals} \\ 23 \\ 7 \\ 28 \\ 5 \end{array}
\end{array}
$$

Column Totals 23 7 28 5   63 Grand Total

Note that the row totals and the column totals will be the same, provided the section begins and ends with the same state; otherwise two rows and columns will differ by one. Also note that, unlike most matrices we have calculated before, the transition frequency matrix is asymmetric and in general $a_{i,j} \neq a_{j,i}$.

The tendency for one state to succeed another can be emphasized in the matrix by converting the frequencies to decimal fractions or percentages. If each element in the ith row is divided by the total of the ith row, the resulting fractions express the relative number of times state i is succeeded by the other states. In a probabilistic sense, these are estimates of the conditional probability, $p(j|i)$, the probability that state j will be the next state to occur, *given* that the present state is i. [We here introduce the unconventional but equivalent notation, $p\,(i \rightarrow j)$, which can be read as the probability that state i will be followed by state j. This alternative notation will be useful later.]

		to			Row Totals
	A	B	C	D	
A	0.78	0	0.22	0	1.00
B	0	0.71	0.29	0	1.00
C	0.18	0.07	0.64	0.11	1.00
D	0	0	0.60	0.40	1.00

(*from* labels the rows A, B, C, D)

Here, for example, we see that if we are in state C at one point, the probability is 64% that the lithology 1 ft up will also be state C. The probability is 18% that the lithology will be state A, 7% that it will be state B, and 11% that it will be state D. Since the four states are mutually exclusive and exhaustive, the lithology must be one of the four and so their sum, given as the row total, is 100%.

If we divide the row totals of the transition frequency matrix by the total number of transitions, we obtain the relative proportions of the four lithologies that are present in the section. This is called the **marginal** (or **fixed**) **probability vector**:

$$\begin{array}{c} A \\ B \\ C \\ D \end{array} \begin{bmatrix} 0.37 \\ 0.11 \\ 0.44 \\ 0.08 \end{bmatrix}$$

You will recall from Chapter 2 (Eq. 2.7) that the joint probability of two events, A and B, is

$$p(A, B) \;=\; p(B|A)p(A)$$

rearranging,

$$p(B|A) \;=\; \frac{p(B, A)}{p(A)}$$

So, the probability that state B will follow, or overlie, state A is the probability that both state A and B occur, divided by the probability that state A occurs. If the occurrence of states A and B are independent, or unconditional,

$$p(A, B) = p(A)\,p(B)$$

and

$$p(B|A) = \frac{p(A)\,p(B)}{p(A)} = p(B)$$

That is, the probability that state B will follow state A is simply the probability that state B occurs in the section, which is given by the appropriate element in the fixed probability vector. If the occurrences of all the states in the section are independent, the same relationship holds for all possible transitions; so, for example,

$$p(B|A) \;=\; p(B|B) \;=\; p(B|C) \;=\; p(B|D) \;=\; p(B)$$

This allows us to predict what the transition probability matrix should look like if the occurrence of a lithologic state at one point in the stratigraphic interval were completely independent of the lithology at the immediately underlying point. The expected transition probability matrix would consist of rows that were all identical to the fixed probability vector. For our stratigraphic example, this would appear as

$$
\begin{array}{c}
 & \text{to} & & \text{Row} \\
 & A \quad\; B \quad\;\; C \quad\;\; D & & \text{Totals}
\end{array}
$$

$$
from \quad
\begin{array}{c}
A \\
B \\
C \\
D
\end{array}
\left[
\begin{array}{cccc}
0.37 & 0.11 & 0.44 & 0.08 \\
0.37 & 0.11 & 0.44 & 0.08 \\
0.37 & 0.11 & 0.44 & 0.08 \\
0.37 & 0.11 & 0.44 & 0.08
\end{array}
\right]
\begin{array}{c}
1.00 \\
1.00 \\
1.00 \\
1.00
\end{array}
$$

We can compare this expected transition probability matrix to the transition probability matrix we actually observe to test the hypothesis that all lithologic states are independent of the immediately preceding states. This is done using a χ^2 test, first converting the probabilities to expected numbers of occurrences by multiplying each row by the corresponding total number of occurrences:

Expected Transition Probabilities					Totals	Expected Frequencies			
0.37	0.11	0.44	0.08		× 23 =	8.5	2.5	10.1	1.8
0.37	0.11	0.44	0.08		× 7 =	2.6	0.8	3.1	0.6
0.37	0.11	0.44	0.08		× 28 =	10.4	3.1	12.3	2.2
0.37	0.11	0.44	0.08		× 5 =	1.9	0.6	2.2	0.4

The χ^2 test is similar in form to the test equation (Eq. 2.65) described in Chapter 2. Each element in the transition frequency matrix constitutes a category, with both an observed and an expected number of transitions. These are compared by

$$
\chi^2 = \sum \frac{(O - E)^2}{E} \tag{4.2}
$$

where O is the observed number of transitions from one state to another, and E is the number of transitions expected if the successive states are independent. The test has $(m - 1)^2$ degrees of freedom, where m is the number of states (a degree of freedom is lost from each row because the probabilities in the rows sum to 1.00). As with other types of χ^2 tests, each category must have an expected frequency of at least five transitions. This is not the case in this example, but we can still make a conservative test of independence by calculating the test statistic using the four categories whose expected frequency is greater than five. The remaining categories can be combined until their expected frequencies exceed five.

The categories include the transitions $A \rightarrow A$, $A \rightarrow C$, $C \rightarrow A$, and $C \rightarrow C$. Combined categories can be formed of all elements in the B row, all elements in the D row, and the combination of transitions $A \rightarrow B$, $A \rightarrow D$, $C \rightarrow B$ and $C \rightarrow D$. The resulting χ^2 statistic is

$$
\chi^2 = \frac{(18 - 8.5)^2}{8.5} + \frac{(5 - 10.4)^2}{10.4} + \frac{(5 - 10.1)^2}{10.1} + \frac{(18 - 12.3)^2}{12.3}
$$

$$
+ \frac{(7 - 7.0)^2}{7.0} + \frac{(5 - 5.0)^2}{5.0} + \frac{(5 - 9.8)^2}{9.8}
$$

$$
= 20.99
$$

The critical value of χ^2 for nine degrees of freedom and a 5% level of significance is 16.92; the test value comfortably exceeds this, so we may conclude that the hypothesis of independence of successive states is not correct. There is a statistically significant tendency for certain states to be preferentially followed by certain other states.

A sequence in which the state at one point is partially dependent, in a probabilistic sense, on the preceding state is called a **Markov chain** (named after the Russian statistician, A.A. Markov). A sequence having the Markov property is intermediate between deterministic sequences and completely random sequences. Our stratigraphic section exhibits *first-order* Markov properties; that is, the statistical dependency exists between points and their immediate predecessors. Higher order Markov properties can exist as well. For example, a second-order Markov sequence exhibits a significant conditional relationship between points that are two steps apart.

From the transition probability matrix we can estimate what the lithology will be 2 ft (that is, two observations) above a given point. Suppose we start in limestone (state *B*). The following probabilities estimate the lithology to be encountered at the next point upward:

State *A (sandstone)*	0%
State *B (limestone)*	71%
State *C (shale)*	29%
State *D (coal)*	0%

Suppose the next point actually falls in a shale; we can then determine the probable lithology of the following point:

State *A (sandstone)*	18%
State *B (limestone)*	7%
State *C (shale)*	64%
State *D (coal)*	11%

So, the probability that the lithologic sequence will be *limestone* → *shale* → *limestone* is

$$p(B \rightarrow C) \times p(C \rightarrow B) = 29\% \times 7\% = 2\%$$

However, there is another way to reach the limestone state in two steps. The sequence *limestone* → *limestone* → *limestone* is also possible. The probability attached to this sequence is

$$p(B \rightarrow B) \times p(B \rightarrow B) = 71\% \times 71\% = 50\%$$

Since the other transitions *limestone* → *sandstone* and *limestone* → *coal* have zero probability, these two sequences are the only possible ones which lead from limestone and back again in two steps. The probability that the lithology two steps above a limestone will also be a limestone, regardless of the intervening lithology, is the sum of all possibilities. That is,

$$
\begin{aligned}
p(B \rightarrow A \rightarrow B) &= 0\% \\
p(B \rightarrow B \rightarrow B) &= 50\% \\
p(B \rightarrow C \rightarrow B) &= 2\% \\
p(B \rightarrow D \rightarrow B) &= 0\% \\
\text{Total} &= 52\%
\end{aligned}
$$

The same reasoning can be applied to determine the probability of any lithology two steps hence, from any starting lithology. However, all of the various sequences do not have to be worked out individually, because the process of multiplying and summing is exactly that used for matrix multiplication. If the transition probability matrix is multiplied by itself (that is, the matrix is squared), the result is the second-order transition probability matrix describing the second-order Markov properties of the succession:

$$\begin{bmatrix} 0.78 & 0 & 0.22 & 0 \\ 0 & 0.71 & 0.29 & 0 \\ 0.18 & 0.07 & 0.64 & 0.11 \\ 0 & 0 & 0.60 & 0.40 \end{bmatrix}^2 = \begin{bmatrix} 0.64 & 0.02 & 0.31 & 0.02 \\ 0.05 & 0.52 & 0.39 & 0.03 \\ 0.26 & 0.09 & 0.54 & 0.11 \\ 0.11 & 0.04 & 0.62 & 0.23 \end{bmatrix}$$

Note that the rows of the squared matrix also sum to 100%.

The existence of a significant second-order property can be checked in exactly the same manner as we checked for independence between successive states, by using a χ^2 test. If you repeat the test performed earlier, but using the second-order transition probability matrix, you should find that the sequence has no significant second-order properties.

We can estimate the probable state to be encountered at any step in the future simply by powering the transition probability matrix the appropriate number of times. If the matrix is raised to a sufficiently high power, it reaches a stable state in which the rows all become equal to the fixed probability vector, or in other words, becomes an independent transition probability matrix and will not change with additional powering.

You will note in the example that the highest transition probabilities are from one state to itself, particularly from sandstone to sandstone, from limestone to limestone, and from shale to shale. It is obvious that these transition probabilities are related to the thicknesses of the stratigraphic units being sampled and the distance between the sample points. For example, the frequencies along the main diagonal of the transition frequency matrix would be doubled while off-diagonal frequencies remained unchanged if observations were made every half-foot. This would greatly enhance the Markovian property, but in a specious manner. Selecting the appropriate distance between sampling points can be a vexing problem; if observations are too closely spaced, the transition matrix reflects mainly the thickness of the more massive stratigraphic units. If the spacing is too great, thin units may be entirely missed.

Embedded Markov chains

The difficulty of selecting an appropriate sampling interval can be avoided if observations are taken only when there is a change in state. A stratigraphic section, for example, would be recorded as a succession of beds, each one of a different lithology than the immediately preceding bed. **Table 4–4** contains the record of successive rock types penetrated by a well drilled in the Midland Valley of Scotland (these data are contained in file MIDLAND.TXT). The well was drilled through 1600 ft of Coal Measures of Carboniferous age, consisting of interbedded shales, siltstones, sandstones, and coal beds or root zones. These sediments are interpreted as having been deposited in a delta plain environment subject to repeated flooding, so we would expect that certain lithologies would occur in preferred relations to

Table 4–4. Successive lithologic states encountered in a drill hole through the Coal Measures in the Midland Valley of Scotland (after Doveton, 1971); mutually exclusive states are barren shale (*A*), shale with fossils of nonmarine bivalves (*B*), siltstone (*C*), sandstone (*D*), and coal or root zone (*E*); read across rows. Data are in file MIDLAND.TXT.

Top →

```
B E A E A D A C D C D C A B E A D C D C D C A E
D C A D C A E C D C B E A D C D C D C A B A E D
C A E C A D E A D A C A B E A D C A E C D C A B
A E A D E A D C E A C D C D C D C A B E A B A B
A B E A B A C A C A B A B E A C D C D C D C A C
B E A C A C B E C A D C A C D C E A C D A C D C
B A B E A C D C A B A B E A D A C E A D A D C A
E A C D A E A E A C D C E C A B C E C A D B E A
D C D E A D A C A B E A B A B E A B A B E C A C
D A E A C D C D C A C A C E A C D C D C A B E A
D E A C D C D E C D C E A C A E A C A E A C A B
C D A E A C D C E A C B E A C A E A D A B E A C
D E A D C A B E A D C D E A D C D A E A C D C A
D A E A D A D C A C E D A B D B A E A C A E C D
C D C D A E A E C D A B E A B E A E A C D E A D
A D E C D C A E A E A C D A E C D B E A D C D C
A D A B A B E A D B A E A  →  Bottom
```

others. The data are taken from one of a large number of wells studied by Doveton (1971).

The four-state transition frequency matrix for the section in the Scottish well is given below. One obvious difference between this matrix and the one we have considered previously is that all the diagonal terms must be zero, since a state cannot succeed itself. The transition probability matrix, computed by dividing each element of the transition frequency matrix by the appropriate row total, shares this same characteristic. Sequences in which transitions from a state to itself are not permitted are called **embedded Markov chains**, and their analysis presents special problems that have not always been appreciated by geologists studying stratigraphic records.

		A	*B*	*C*	*D*	*E*	Row Totals
	A	0	13	36	19	52	120
	B	29	0	5	4	0	38
from	*C*	35	2	0	45	12	94
	D	29	1	44	0	3	77
	E	26	23	9	9	0	67
Column Totals		119	39	94	77	67	397 Grand Total

to

The lithologic states have been coded as (*A*) unfossiliferous shale and mudstone, (*B*) shales containing nonmarine bivalves, (*C*) siltstone, (*D*) sandstone, and (*E*) coals and root zones. The corresponding transition probability matrix is

		to				Row Totals
	A	*B*	*C*	*D*	*E*	
A	0	0.11	0.30	0.16	0.43	1.00
B	0.76	0	0.13	0.11	0	1.00
from *C*	0.37	0.02	0	0.48	0.13	1.00
D	0.38	0.01	0.57	0	0.04	1.00
E	0.40	0.34	0.13	0.13	0	1.00

The marginal probability vector is

$$
\begin{matrix} A \\ B \\ C \\ D \\ E \end{matrix}
\begin{bmatrix} 0.30 \\ 0.10 \\ 0.24 \\ 0.19 \\ 0.17 \end{bmatrix}
$$

A χ^2 test, identical to Equation (4.2), can be used to check for the Markov property in an embedded sequence. This is done by comparing the observed transition frequency matrix to the matrix expected if successive states are independent. However, the fixed probability vector cannot be used to estimate the columns of the expected transition probability matrix. This would result in the expectation of transitions from a state to itself, which are forbidden. Rather, we must use a somewhat roundabout procedure to estimate the frequencies of transitions between independent states, subject to the constraint that states cannot succeed themselves. We begin by imagining that our sequence is actually a censored sample taken from an ordinary succession in which transitions from a state to itself can occur. The transition frequency matrix of this succession would look like the one we observe except that the diagonal elements would contain values other than zero. If we were to compute a transition probability matrix from this frequency matrix and then raise it to an appropriately high power, it would estimate the transition probability matrix of a sequence in which successive states were independent. If the diagonal elements were then discarded and the off-diagonal probabilities recalculated, the result would be the expected transition probability matrix for an embedded sequence whose states are independent.

How do we estimate the frequencies of transitions from each state to itself, when this information is not available? We do this by trial-and-error, searching for those values that, when inserted on the diagonal of the transition frequency matrix, do not change when the matrix is powered. The off-diagonal elements, however, will change until a stable configuration is reached, corresponding to the independent events model.

In practice it is not necessary to calculate the off-diagonal probabilities at all. We begin by assigning some arbitrarily large number, say 1000, to the diagonal positions of the observed transition frequency matrix. The fixed probability vector is found, by summing each row and dividing by the grand total, and then is used as an estimate of the transition probabilities along the diagonal. These probabilities are powered by squaring and multiplied by the grand total to obtain new estimates of the diagonal frequencies. These new estimates are inserted into the original transition frequency matrix and the process repeated. We can work through the first cycle of the procedure.

Step 1. Initial estimate of transition frequency matrix, with 1000 inserted in each diagonal position.

			to			
	A	*B*	*C*	*D*	*E*	Row Totals
A	1000	13	36	19	52	1120
B	29	1000	5	4	0	1038
from *C*	35	2	1000	45	12	1094
D	29	1	44	1000	3	1077
E	26	23	9	9	1000	1067
						5397 Grand Total

Step 2. Estimate of transition probabilities of diagonal elements, found by dividing row totals by grand total.

			to			
	A	*B*	*C*	*D*	*E*	Row Totals
A	0.208					0.208
B		0.192				0.192
from *C*			0.203			0.203
D				0.200		0.200
E					0.198	0.198

Step 3. Square the probabilities along the diagonal.

Step 4. Second estimate of transition frequency matrix using new diagonal elements calculated by multiplying probabilities on the diagonal by the grand total of 5397. Off-diagonal terms are the original observed frequencies. New row totals and grand total are then found

			to			
	A	*B*	*C*	*D*	*E*	Row Totals
A	232	13	36	19	52	352
B	29	199	5	4	0	237
from *C*	35	2	222	45	12	316
D	29	1	44	215	3	292
E	26	23	9	9	211	278
						1475 Grand Total

The process is repeated again and again, until the estimated transition frequencies along the diagonal do not change from time to time. This generally requires about 10 to 20 iterations, depending upon how closely the initial guesses were to the final, stable estimates. In this example, the estimates do not change after 10 iterations.

The final form of the transition frequency matrix with estimated diagonal frequencies is given below.

		to					
	A	*B*	*C*	*D*	*E*	Row Totals
A	66	13	36	19	52	186
B	29	3	5	4	0	41
from *C*	35	2	29	45	12	123
D	29	1	44	17	3	94
E	26	23	9	9	12	79
Column Totals	185	42	123	94	79	523 Grand Total

This matrix could be converted into an expected transition probability matrix of the hypothetical Markov sequence by dividing each element by the corresponding row total. However, such a matrix is of little interest because it pertains to the hypothetical sequence rather than the observed embedded sequence. The marginal totals are another matter, because they are required to compute the marginal probability vector:

$$
\begin{array}{cc}
A & \left[\,0.355\,\right] \\
B & 0.074 \\
C & 0.235 \\
D & 0.181 \\
E & \left[\,0.155\,\right]
\end{array}
$$

We may now calculate the expected probabilities and expected frequencies of a hypothetical sequence of independent states from the marginal probability vector. We are testing the hypothesis of independence between successive states by noting that, for example, if state A is independent of state B, then $p(A|B) = p(A)p(B)$. As $P(A)$ and $P(B)$ are given by the appropriate elements of the marginal probability vector, the estimated conditional probability that state A will follow state B is $p(A|B) = (0.355)(0.074) = 0.026$. The expected probabilities for all transitions are given below.

		to			
	A	*B*	*C*	*D*	*E*
A	0.125	0.026	0.083	0.064	0.055
B	0.026	0.006	0.017	0.013	0.012
from *C*	0.083	0.017	0.055	0.043	0.036
D	0.064	0.013	0.043	0.033	0.028
E	0.055	0.012	0.036	0.028	0.024

The expected frequencies are found by multiplying this matrix by the grand total, 523.

		to			
	A	*B*	*C*	*D*	*E*
A	65.5	13.6	43.5	33.5	28.8
B	13.6	3.1	8.9	6.8	6.3
from *C*	43.5	8.9	28.8	22.5	18.9
D	33.5	6.8	22.5	17.3	14.7
E	28.8	6.3	18.9	14.7	12.6

Note that the matrix is symmetrical and the diagonal elements remain unchanged, within the limits of rounding error. The off-diagonal elements are the expected frequencies of transitions within the embedded sequence, assuming independence between successive states. If the diagonal elements are stripped from the matrix, it may be compared directly to the observed transition frequency matrix because the row and column totals of the two are the same, again within rounding limits.

The comparison by χ^2 methods yields a test statistic of $\chi^2 = 172$. The test has $v = (m - 1)^2 - m$ degrees of freedom, where m is the number of states, or in this example, $v = 11$. The critical value of χ^2 for 11 degrees of freedom and an $\alpha = 0.05$ level of significance is 19.68, which is far exceeded by the test statistic. Therefore, we must conclude that successive lithologies encountered in the Scottish well are not independent, but rather exhibit a strong first-order Markovian property.

If tests determine that a sequence exhibits partial dependence between successive states, the structure of this dependence may be investigated further. Simple graphs of the most significant transitions may reveal repetitive patterns in the succession. Modified χ^2 procedures are available to test the significance of individual transition pairs. Some authors have found that the eigenvalues extracted from the transition probability matrix are useful indicators of cyclicity. (It should be noted, however, that extracting the eigenvectors from an asymmetric matrix such as the transition probability matrix may not be an easy task!) These topics will not be pursued further in this book; the interested reader should refer to the texts by Kemeny (1983) and Norris (1997), as well as the book on quantitative sedimentology by Schwarzacher (1975). Chi-square tests appropriate for embedded sequences are discussed by Goodman (1968). In a geological context, the articles by Doveton (1971) and Doveton and Skipper (1974), plus the comment by Türk (1979), are recommended.

Series of Events

An interesting type of time series we will now consider is called a *series of events*. Geological examples of this type of data sequence include the historical record of earthquake occurrences in California, the record of volcanic eruptions in the Mediterranean area, and the incidence of landslides in the Tetons. The characteristics of these series are (a) the events are distinguishable by when they occur in time; (b) the events are essentially instantaneous; and (c) the events are so infrequent that no two occur in the same time interval. A series of events is therefore nothing more than a sequence of the intervals between occurrences. Our data may consist of the duration between successive events, or the cumulative length of time over which the events occur. One form may be directly transformed into the other.

Series-of-events models may be appropriate for certain types of spatially distributed data. We might, for example, be interested in the occurrence of a rare mineral encountered sporadically on a traverse across a thin section or in the appearance of bentonite beds in a vertical succession of sedimentary rocks. Justification for applying series-of-events models to spatial data may be tenuous, however, and depends on the assumption that the spatial sequence has been created at a constant rate. This assumption probably is reasonable in the first example, but the second requires that we assume that the sedimentation rate remained constant through the series.

The historic record of eruptions of the volcano Aso in Kyushu, Japan, has been kept since 1229 (Kuno, 1962), and is given in **Table 4–5** and file ASO.TXT. Aso is

Table 4–5. Years of eruptions of the volcano Aso for the period 1229-1962.

1229	1376	1583	1780	1927
1239	1377	1584	1804	1928
1240	1387	1587	1806	1929
1265	1388	1598	1814	1931
1269	1434	1611	1815	1932
1270	1438	1612	1826	1933
1272	1473	1613	1827	1934
1273	1485	1620	1828	1935
1274	1505	1631	1829	1938
1281	1506	1637	1830	1949
1286	1522	1649	1854	1950
1305	1533	1668	1872	1951
1324	1542	1675	1874	1953
1331	1558	1683	1884	1954
1335	1562	1691	1894	1955
1340	1563	1708	1897	1956
1346	1564	1709	1906	1957
1369	1576	1765	1916	1958
1375	1582	1772	1920	1962

a complex stratovolcano, but all historic eruptions have been explosive, ejecting ash of andesitic composition. Although the ancient monastic records contain an indication of the relative violence and duration of some eruptions, for all practical purposes we must regard the record as one of indistinguishable instantaneous explosive events. Analysis of volcanic histories may shed some light on the nature of eruptive mechanisms and can even lead to physical models of the structure of volcanoes (Wickman, 1966). Of course, we would also hope that such studies might lead to predictive tools to forecast future eruptions.

Studies of series of events may have several objectives. Usually, an investigator is interested in the **mean rate of occurrence**, or number of events per interval of time. In addition, it may be necessary to examine the series in more detail, in order to estimate any pattern that may exist in the events. This additional information can be used to determine the precision of the estimate of the rate of occurrence, to assess the appropriateness of the sampling scheme, to detect a trend, and to detect other systematic features of the series.

Because series of events are very simple, in the sense that they consist of nominal occurrences (presence–absence), simple analytical techniques may prove to be the most effective. Cox and Lewis (1966) described a variety of graphical tools that are useful in examining series of events. These are illustrated using the data on the eruptions of Aso from **Table 4–5**.

A cumulative plot of the total number of events (n_t) to have occurred at or before time t, against time t, is given in **Figure 4–6**. This plot is especially good for showing changes in the average rate of occurrence. The slope of a straight line connecting any two points on the cumulative plot is the average number of events per unit of time for the interval between the two points.

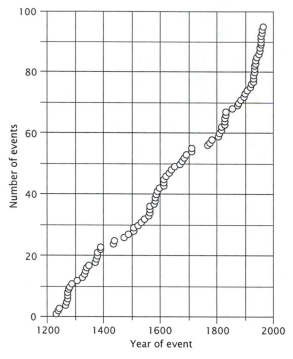

Figure 4–6. Cumulative number of eruptions of the Japanese volcano Aso plotted against years of eruptions.

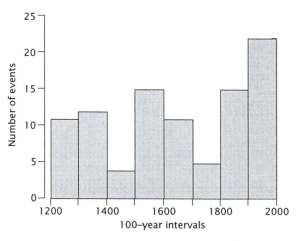

Figure 4–7. Histogram of number of eruptions of the Japanese volcano Aso occurring in successive 100-yr intervals.

A histogram of the number of events occurring in successive equal intervals of time is given in **Figure 4–7.** This histogram directly indicates local periods of fluctuation from the average rate of occurrence. The pattern shown by the histogram is sensitive to the length of the chosen intervals, so more than one histogram may be useful in examining a series.

The ***empirical survivor function*** is obtained by plotting the percent "survivors," or Y = proportion of time intervals longer than X, against X = length of time interval. The function estimates the probability that an event has not occurred before time X. In **Figure 4–8**, the points represent the percentage of intervals between eruptions which are longer than the specified number of years. If events occur randomly in time, the survivor function will be exponential in form.

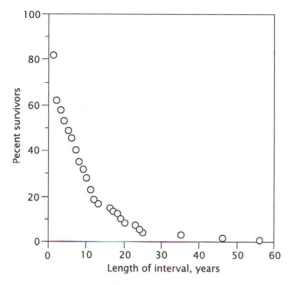

Figure 4–8. Empirical survivor function for the Japanese volcano Aso. The vertical axis gives the percent of intervals between eruptions that are longer than a specified duration, versus the duration in years along the horizontal axis.

This same function can be plotted in logarithmic form, as log Y against X. The **log empirical survivor function** is especially good for showing departures from randomness, which appear as deviations from the straight-line form of the plot (**Fig. 4–9**).

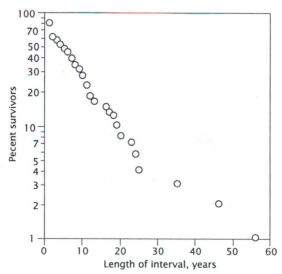

Figure 4–9. Log empirical survivor function of the Japanese volcano Aso. The vertical axis of Figure 4–8 is expressed in logarithmic form.

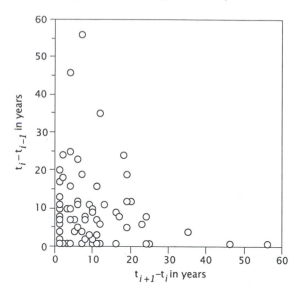

Figure 4–10. Serial correlation of durations between successive eruptions of the Japanese volcano Aso. Vertical axis is duration of quiet before the ith eruption, and horizontal axis is duration after the ith eruption.

A scatter diagram of the **serial correlation**, or first-order autocorrelation, of successive intervals between events is shown in **Figure 4–10.** The degree of correspondence between the length of an interval and the length of the immediately preceding interval is shown by plotting $x_i = t_{i+1} - t_i$ against $y_i = t_i - t_{i-1}$ where t_i is the time of occurrence of the ith event. This plot reveals any tendency for intervals to be followed by intervals of similar length. A scatter diagram with large dispersion and relatively high concentrations of points near the axes is typical of random series of events.

In most series-of-events studies, we hope that we can describe the basic features of the series in a way that will suggest a physical mechanism for the lengths of the intervals between occurrences. First we must consider the possibility of a trend in the data. We may check for a trend in two ways. A series may be subdivided into segments of equal length, provided each segment contains several observations. The numbers of events within each segment are taken to be observations located at the midpoints of the segments. A regression can then be run with these numbers as the dependent variable, y_i, and the locations of the midpoints of the segments as values of x_i. The slope coefficient of the regression can be tested by the ANOVA given later in **Table 4–9** (p. 197) to determine if it is significantly different from zero. The process is illustrated in **Figure 4–11.** Unfortunately, this test is not particularly efficient because degrees of freedom are lost when the series is divided into segments.

There are tests specifically designed to detect a trend in the rate of occurrence of events by comparing the midpoint of the sequence to its centroid. If the sequence is relatively uniform, the two will be very similar, but if there is a trend the centroid will be displaced in the direction of increasing rate of occurrence. If t_i is the time or distance from the start of the series to the ith event and N is the total number of events, we can calculate the centroid, S, by

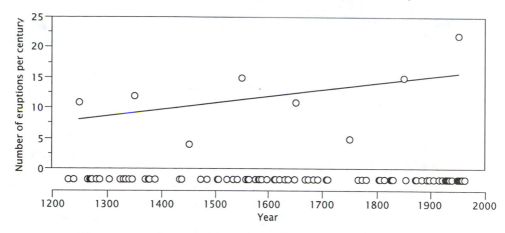

Figure 4–11. Eruptions of the Japanese volcano Aso considered as "instantaneous" happenings along a continuum of time. Cross plot shows number of eruptions per century versus midpoints of successive centuries. Fitted regression line estimates change in rate of occurrence.

$$S = \frac{\sum_{i=1}^{N} t_i}{N} \tag{4.3}$$

This statistic can in turn be used in Equation (4.4),

$$z = \frac{S - 1/2T}{T \big/ \sqrt{12N}} \tag{4.4}$$

where T is the total length of the series, z is the standardized normal variate, and the significance of the test result can be determined by normal tables such as Appendix Table **A.1**.

The test is very sensitive to changes in the rate of occurrence of events. Specifically, if the events are considered to be the result of a process

$$Y_t = e^{\alpha + \beta t} \tag{4.5}$$

the null hypothesis states that $\beta = 0$. You will recognize that the model is exponential; if β has any value other than zero, the rate of occurrence of Y_t will change with t. It is this possibility that we are testing.

If no trends are detected in the rate of occurrence, we may conclude that the series of events is stationary. We can next check to see if successive occurrences are independent. This can be done by computing the autocorrelation of the lengths between events. That is, we regard the intervals between events as a variable, X, located at equally spaced points. If the intervals are not independent, this will be expressed as a positive autocorrelation with a tendency for large values of x_i (long intervals between events) to be succeeded by large values; similarly, there will be a tendency for small values of x_i (short intervals) to be followed by other small values. We can compute autocorrelation coefficients for successive lags and test these for significance. Usually only the first few lags will be of interest. If the autocorrelation coefficients are not significantly different from zero, as tested by methods that will be developed later in this chapter, we can conclude that the events are occurring independently in time or space.

If we have established that the series is neither autocorrelated nor contains a trend, we may wish to test the possibility that the events are distributed according to a *Poisson distribution*. You will recall from Chapter 2 that the Poisson is a discrete probability distribution that can be regarded as the limiting case of the binomial when n, the number of trials, becomes very large, and p, the probability of success on any one trial, becomes very small. We can imagine that our time series is subdivided into n intervals of equal duration. If events occur randomly, the number of intervals that contain exactly $0, 1, 2, \ldots, x$ events will follow the binomial distribution. As we make the lengths of the intervals progressively shorter, n becomes progressively larger and the probabilities of occurrence decline. The binomial distribution becomes difficult to compute, but the Poisson can be readily used because it does not require either n or p directly. Instead, the product $np = \lambda$ is all that is needed, which is given by the *rate of occurrence* of events.

The Poisson probability model assumes that (a) the events occur independently, (b) the probability that an event occurs does not change with time, (c) the probability that an event will occur in an interval is proportional to the length of the interval, and (d) the probability of more than one event occurring at the same time is vanishingly small.

The equation for the Poisson distribution in this instance is

$$p(X) = e^{-\lambda}\lambda^X \big/ X! \tag{4.6}$$

Note that the rate of occurrence, λ, is the only parameter of the distribution. Typical Poisson frequency distributions are shown in **Figure 4–12**. The distribution is applicable to such problems as the rate that telephone calls come to a switchboard or the length of time between failures in a computer system. It seems reasonable that it also may apply to the series of geological events described at the beginning of this section. If we can determine that our series follows a Poisson distribution, we can use the characteristics of the distribution to make probabilistic forecasts of the series.

The Kolmogorov–Smirnov test provides a simple way to test the goodness of fit of a series of events to that expected from a Poisson distribution. First, the series must be converted to a cumulative form

$$y_i = \frac{t_i}{T}$$

where t_i is the time from the start of the series to the ith event, and T is the total length of the series. Three estimates can then be calculated

$$
\begin{aligned}
D^+ &= \sqrt{n}\max\left\{\frac{i}{n} - y_i\right\} \\
D^- &= \sqrt{n}\max\left\{y_i - \frac{i-1}{n}\right\} \\
D &= \max\left|D^+, D^-\right|
\end{aligned}
\tag{4.7}
$$

The first test is simply the maximum positive difference between the observed series and that expected from a Poisson, the second is the maximum negative difference, and the third is the larger of the absolute values of the two. The test statistic,

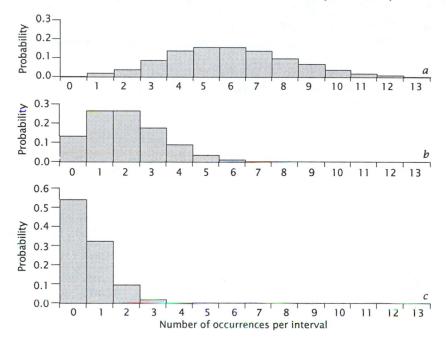

Figure 4–12. Poisson probability distributions with different rates of occurrence, λ, expressed as numbers of occurrences per interval. (*a*) $\lambda = 6.0$. (*b*) $\lambda = 2.0$. (*c*) $\lambda = 0.6$.

D, can be compared to two-tailed critical values given in Appendix Table **A.7.** If the statistic exceeds the critical value, the maximum deviation is larger than that expected in a sample collected at random from a Poisson distribution.

Runs Tests

The simplest type of sequence is a succession of observations arranged in order of occurrence, where the observations are two mutually exclusive categories or states. Consider a rock collector cracking open concretions in a search for fossils. The breaking of a concretion constitutes a trial, and each trial has two mutually exclusive outcomes: The concretion either contains a fossil or it does not. The sequence of successes and failures by the collector during the course of a day forms a special type of time series. We can experimentally create a similar succession by flipping pennies and noting the occurrence of heads or tails. The sequence generated might resemble this set of twenty trials:

H T H H T H T T T H T H T H T H H T T H H H

We intuitively expect, of course, that about ten heads will appear, and we can determine the probability of obtaining this (or any other) number of heads. Here we obtained 11 heads; assuming the coin is unbiased, the probability of obtaining this number in 20 trials is 0.16 or about one in six. We would expect similar trials to contain 9, 10, or 11 heads slightly more than one-third of the time. Results of this experiment follow the binomial distribution, discussed in Chapter 2.

One aspect that we have not considered, however, is the order in which the heads appear. We probably would regard a sequence such as

$$H\,H\,H\,H\,H\,H\,H\,H\,H\,H\,H\,T\,T\,T\,T\,T\,T\,T\,T\,T$$

as being very strange, although the probability of obtaining this many heads in 20 trials is the same as in the preceding example. At the other extreme, the regular alternation of heads and tails

$$H\,T\,H\,T\,H\,T\,H\,T\,H\,T\,H\,T\,H\,T\,H\,T\,H\,T\,H\,H$$

would also appear very unusual to us, although the probability of the number of heads is unchanged. What arouses our suspicions is not the proportion of heads but the order in which they appear. We assume that heads and tails will occur at random; in the two preceding examples, it seems very unlikely that they have.

We can test these sequences for randomness of occurrence by examining the number of runs. **Runs** are defined as uninterrupted sequences of the same state. The first set of trials contains 13 runs, the second only 2, and the third contains 19. Runs in the first sequence shown are underlined:

(Start)

H	T	HH	T	H	TTT	H	T	H	T	HH	TT	HHH	
1	1	3	4	5	6	7	8	9	10	11	12	13	(End)

We can calculate the probability that a given sequence of runs was created by the random occurrence of two states (heads and tails, in this example). This is done by enumerating all possible ways of arranging n_1 items of state 1 and n_2 items of state 2. The total number of runs in a sequence is denoted U; tables are available which give critical values of U for specified n_1, n_2, and level of significance, α. However, if n_1 and n_2 each exceed ten, the distribution of U can be closely approximated by a normal distribution, and we can use tables of the standard normal variate z for our statistical tests. The expected mean number of runs in a randomly generated sequence of n_1 items of state 1 and n_2 items of state 2 is

$$\overline{U} = \frac{2n_1 n_2}{n_1 + n_2} + 1 \tag{4.8}$$

The expected variance in the mean number of runs is

$$\sigma_{\overline{U}}^2 = \frac{2n_1 n_2 (2n_1 n_2 - n_1 - n_2)}{(n_1 + n_2)^2 (n_1 + n_2 - 1)} \tag{4.9}$$

By these equations, we can determine the mean number of runs and the standard error of the mean number of runs in all possible arrangements of n_1 and n_2 items. Having calculated these, we can create a z-test by Equation (4.10), where U is the observed number of runs:

$$z = \frac{U - \overline{U}}{\sigma_{\overline{U}}} \tag{4.10}$$

You will recognize that this is simply Equation (2.37) rewritten to include the runs statistics. We can formulate a variety of statistical hypotheses which can be tested with this statistic. For example, we may wish to see if a sequence contains more than the expected number of runs from a random arrangement; the null hypothesis and alternative are

$$H_0 : U \leq \overline{U}$$
$$H_1 : U > \overline{U}$$

and too many runs leads to rejection. The test is one-tailed. Conversely, we may wish to determine if the sequence contains an improbably low number of runs. The appropriate alternatives are

$$H_0 : U \geq \overline{U}$$
$$H_1 : U < \overline{U}$$

and too few runs will cause rejection of the null hypothesis. Again, the test is one-tailed. We may wish to reject either form of nonrandomness. A two-tailed test is appropriate, with hypotheses

$$H_0 : U = \overline{U}$$
$$H_1 : U \neq \overline{U}$$

We can work through the test procedure for the first series of coin flips and determine the likelihood of achieving this sequence by a random process. The null hypothesis states that there is no difference between the observed number of runs and the mean number of runs from random sequences of the same size. We will use a two-tailed test, and reject if there are too many or too few runs in the sequence. Therefore, the proper alternative is

$$H_1 : U \neq \overline{U}$$

Using a 5% ($\alpha = 0.05$) level of significance, our critical regions are bounded by -1.96 and $+1.96$. We first calculate the expected mean and standard deviation of runs for random sequences having n_1 heads ($n_1 = 11$) and n_2 tails ($n_2 = 9$):

$$\overline{U} = \frac{2 \cdot 11 \cdot 9}{11 + 9} + 1 = 10.9$$

$$\sigma_{\overline{U}}^2 = \frac{(2 \cdot 11 \cdot 9)(2 \cdot 11 \cdot 9 - 11 - 9)}{(9 + 11)^2 (9 + 11 - 1)} = 4.6$$

The test statistic is

$$z = \frac{U - \overline{U}}{\sigma_U} \approx \frac{13 - 10.9}{2.1} = 1.0$$

The number of runs in the sequence is one standard deviation from the mean of all runs possible in such a sequence, and does not fall within the critical region. Therefore, the number of runs does not suggest that the sequence is nonrandom. The other sequences, in contrast, yield very different test results. Because n_1 and n_2 are the same for all three sequences, $\overline{U}$ and σ_U also are the same. For the second sequence, the test statistic is

$$z = \frac{2 - 10.9}{2.1} = -4.2$$

and for the third,

$$z = \frac{19 - 10.9}{2.1} = 3.9$$

Both of these values lie within the critical region, and we would reject the hypothesis that they contain the number of runs expected in random sequences.

Geologic applications of this test may not be obvious, because we ordinarily must consider more than two states in a succession. Stratigraphic sections or traverses across thin sections, for example, usually include at least three states and these cannot be ranked in a meaningful way. We will consider ways that certain sequences can be reduced to dichotomous states, but first we will examine a geologic application of the runs test to a traverse through a two-state system.

Simple pegmatites originate by crystallization of the last, volatile-laden substances squeezed off from solidifying granitic magma. Their textures result from simultaneous crystallization of quartz and feldspar at the eutectic point. If the solidifying pegmatite is undisturbed, we might suppose that quartz and feldspar begin to appear at random locations within the cooling body. This situation may persist, with grains crystallizing at random, until the entire mass is solid. However, the presence of one crystal, perhaps feldspar, might stimulate the local crystallization of additional crystals of feldspar, eventually producing a patchwork texture. Alternatively, growth of a crystal of one state might locally deplete the magma of that constituent, retarding crystallization and resulting in a highly alternating mosaic of quartz and feldspar. A large slab of polished pegmatite used as a window ledge in the washroom of a geology building provides a way for students to investigate these alternative possibilities. The polished surface allows easy discrimination of adjacent grains, so a line drawn on the ledge produces a sequence through the quartz and feldspar grains in the pegmatite. The line on the polished slab may be regarded as a random sample of possible successions through the pegmatite body from which the slab was quarried. The quartz–feldspar sequence along the line is listed in **Table 4–6**. Our problem is to determine if the alternations between quartz and feldspar form a random pattern; if there is a systematic tendency for one state to succeed itself; or whether there is a tendency for one state to immediately succeed the other. Perform a runs test on this data and evaluate the three possibilities.

Table 4–6. Sequence of 100 feldspar (F) and quartz (Q) grains encountered along traverse through pegmatite.

(Start)	F Q Q F Q Q F F Q F Q F F F F F F F Q Q F Q F F F
	Q F F F F Q F F F Q Q F Q F Q Q Q F F F F F F Q F F
	F F F Q Q Q Q F F Q Q Q F F F F F F Q F Q F F F F
	F Q F Q F Q F F Q F F F F F Q F F F Q Q F Q F F Q (End)

We will now consider a related statistical procedure for examining what are called **runs up** and **runs down**. We are concerned, not with two distinct states, but whether an observation exceeds or is smaller than the preceding observation. **Figure 4–13** shows a typical sequence that can be analyzed by means of a runs test.

The segment *abc* is a run up, because each observation is larger than the preceding one; similarly, the segment *ghi* is a run down. Segment *cdef* is a run down even though the difference between *d* and *e* is zero. This is because the interval *de* lies between segments *cd* and *ef*, both of which run downward; therefore, the

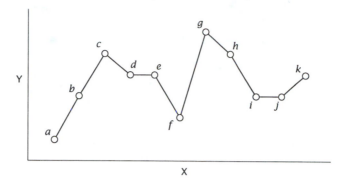

Figure 4–13. Sequence of data points to be analyzed by the method of runs up and down.

entire segment $cdef$ can be considered as a single downward run. The interval ij can be considered either as part of the run down ghi or the run up ijk, as the total number of runs remains the same in either case. In this example, we are assuming that the successive points have integer values. If the observations are expressions of magnitude, they ordinarily will contain fractional parts, and ties (two successive points with identical values) are unlikely.

By considering only differences in magnitude between successive points, we have reduced the data sequence to a string having only two states (or three, if ties occur). We can rewrite the sequence in **Figure 4–13** in the following form:

$$+ + + - 0 - + - - 0 +$$

Regarding the first zero as "−" gives a total of five runs, three of "+" and two of "−" (it makes no difference in the number of runs if we call the second zero "+" or "−"). We can now apply test procedures outlined for the case of sequences of two dissimilar items (Eqs. 4.8–4.10). We must have a large sample to utilize the normal approximation method presented here, but in most geologic problems, adequate numbers of samples will be available.

Table 4–7. Numbers of radiolarian tests per square centimeter in thin sections of siliceous shale.

(Bottom of section)																
1	2	3	2	3	5	7	9	9	11	10	12	7	4	3	2	3
	2	2	1	0	2	3	2	0	3	3	4	9	10	10	8	9	12
	10	12	14	22	17	19	14	4	2	1	0	0	8	14	16	27	(Top of section)

In the study of a silicified shale unit in the Rocky Mountains, it was noted that the rock contained unusual numbers of well-preserved radiolarian tests. Their presence in the silicified shale suggested a causal relationship, so a sequence of samples was collected at approximately equal intervals in an exposure through the unit. Thin sections were made of the samples and the number of radiolarian tests in a 10×10-mm area of the slides was counted. Data for 50 samples are given in **Table 4–7** and shown graphically in **Figure 4–14**. Does the abundance of

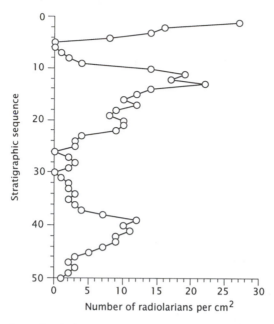

Figure 4–14. Number of radiolarian tests per square centimeter in thin sections of siliceous Mowry Shale.

radiolarians vary at random through the section? A computer program could be written that will perform the necessary calculations, but the programming effort probably exceeds the difficulty of computing the test statistic by hand.

In this procedure, observations are dichotomized by comparing their magnitudes to the preceding observations. Actually, runs tests may be applied to data dichotomized by any arbitrary scheme, provided the hypothesis being tested reflects the dichotomizing method. For example, a common test procedure is to dichotomize a series by subtracting each observation from the median of all observations, and testing the signs for randomness of runs about the median. We also can test the randomness of runs about the mean, and we will use this as a test of residuals from trends later in this chapter. Runs tests are another example of the nonparametric procedures introduced in Chapter 2.

There are a number of variants on the runs tests described here. Information about these tests may be found in texts on nonparametric statistics, such as Conover (1999, p. 122–142) and Siegel and Castellan (1988, section 4.5). Examples of the geologic application of runs tests are included in Miller and Kahn (1962, chapter 14) and Rock (1988, topic 16). Some investigators consider the length of the longest run as an indicator of nonrandomness, and others use the number of turning points, which are points in the sequence where the signs of successive observations change. In certain instances these tests may be more appropriate than the procedures described here. The runs-up-and-down test generally is regarded as the most powerful of the runs tests because it utilizes changes in magnitude of every point with respect to adjacent points. Other dichotomizing schemes reflect only changes with respect to a single value such as the median or mean.

Runs tests are appropriate when the cause of nonrandomness is the object of investigation. They test for a form of nonrandomness expressed by the presence of too few or too many runs, and do not identify overall trends. It should be emphasized that randomness itself cannot be proven, as the condition of random occurrence is implied in the null hypothesis. Rather, at specified levels of significance, we can demonstrate that the null hypothesis is incorrect and the sequence is therefore not random. Or we can fail to reject the null hypothesis, implying that we have failed to find any indication of nonrandomness. We will next consider procedures for detecting trends, or systematic changes in average value, and will find that runs tests may be used to good advantage in conjunction with these procedures.

Least-Squares Methods and Regression Analysis

In many types of problems, we are concerned not only with changes along a sequence, but are also interested in where these changes occur. To examine these problems, we must have a collection of measurements of a variable and also must know the locations of the measurement points. Both the variable and the scale along the sequence must be expressed in units having magnitude; it is not sufficient simply to know the order of succession of points. We are interested in the general tendency of the data in most of the examples we will now consider. This tendency will be used to interpolate between data points, extrapolate beyond the data sequence, infer the presence of trends, or estimate characteristics that may be of interest to the geologist. If certain assumptions can justifiably be made about the distribution of the populations from which the samples are collected, statistical tests called *regression analyses* can be performed.

It must be emphasized that we are now using the expression "sequence" in the broadest possible sense. Regression methods are useful for much more than the analysis of observations arranged in order in time or space; they can be used to analyze *any* bivariate data set when it is useful to consider one of the variables as a function of the other. It is as though one variable forms a scale along which observations of the other variable are located, and we want to examine the nature of changes in this variable as we move up or down the scale.

Table 4–8. Moisture content of core samples of Recent mud in Louisiana estuary.

Depth, ft	Moisture (g water/100 g dried solids)
0.0	124.0
5.0	78.0
10.0	54.0
15.0	35.0
20.0	30.0
25.0	21.0
30.0	22.0
35.0	18.0

The data in **Table 4–8** are the moisture contents of samples from a core through Recent marine muds accumulating in a small inlet on the U.S. Gulf Coast in eastern Louisiana. These data are also in file LOUISMUD.TXT. The measurements were made

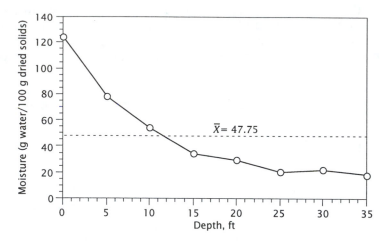

Figure 4–15. Plot of moisture content in grams water/100 grams dry weight versus depth below the sediment–water interface. Data collected from a core through Recent mud in Louisiana estuary.

by comparing the weight of a sample immediately after it was removed from the core barrel with its weight after forced drying. Moisture content is expressed as grams of water per 100 g of dried sediment. If we plot the measurements against depth, as in **Figure 4–15**, we can see that moisture content drops rapidly in the upper layers of sediment, but decreases slowly, if at all, in sediment near the bottom of the core. We will now consider various ways of representing the relationship implicit in these observations.

The value 47.75 indicated on **Figure 4–15** represents the mean moisture content of the samples. This value is the single number about which there is the smallest variance. That is, the sum of the squared deviations of moisture content is the smallest about this point. You will recall from Chapter 2 that if certain sampling precautions are observed, this value also is an unbiased and efficient estimator of the population mean, and is the "best guess," or predictor, of additional samples drawn from the same population. However, the mean is clearly not adequate to represent data in **Figure 4–15**. This is because the samples were taken sequentially, and hence are not independent. Rather than a point estimate, we need a line that will express the relation between moisture content and depth throughout the range of both variables. It seems intuitively reasonable to construct the line so that deviations from it are minimized in some way. One choice might be to minimize the sum of squared deviations from the line, reasoning by analogy to the mean. (The mean is the value about which the variance—and hence the sum of squared deviations—is smallest.) We could then construct a unique line about which the variance is a minimum. If the value of such a line were subtracted at the appropriate points from the corresponding observations, the collection of resulting numbers would have a mean of zero and smaller variance than the set of deviations that could be calculated from any other straight line through the data.

There are, however, several alternative ways that deviations from the fitted line can be defined and measured. For example, we might consider deviations in moisture content, deviations in depth, or some combination of the two. In **Figure 4–16**, the vertical lines represent the deviation of moisture content from a fitted line, or the differences between the observed amount of moisture at each depth and

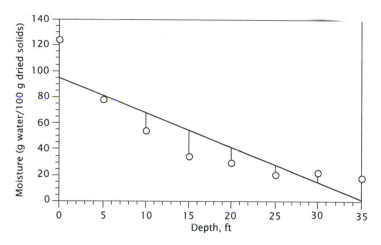

Figure 4–16. A line relating depth with moisture content, fitted by minimization of deviations in moisture content.

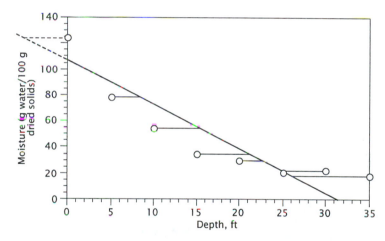

Figure 4–17. A line relating moisture content with depth, fitted by minimization of deviations in depth.

the moisture content predicted by a straight-line function of depth. In contrast, **Figure 4–17** shows horizontal lines representing the deviations in depth of each observation from a fitted line.

We also could consider deviations that are measured in a direction perpendicular to a fitted line, a topic we will examine later in this chapter and again in considerable detail in Chapter 6. It is possible to construct fitted lines in which any of these criteria of deviation are used; however, we should examine the implications of each in the light of our specific problem. If we minimize the deviations of moisture content from a fitted line, as in **Figure 4–16**, we are stating that the resulting line is the best estimator of moisture at specified depths. In contrast, if we minimize deviations in depth, as in **Figure 4–17**, we are estimating depth from moisture content. The use of a perpendicular criterion is an expression of the joint relationship of the two variables. In the particular class of problems we are examining in this chapter, time or distance is known and a second variable is distributed along this continuum. Therefore, the first alternative is appropriate. In other words, moisture content, Y, is a random variable, and depth, X, is fixed. Therefore, the problem is to predict Y from X. Other approaches will be considered in later sections of this book. Having agreed on the characteristics of the desired trend line, we now must define some terms. The variable being examined is the **dependent** or **regressed** variable, designated Y; individual observations of the dependent variable are indicated as y_i. Deviations of y_i from the fitted line will be minimized. The other variable is the **independent** or **regressor** variable and is denoted X, with individual observations, x_i. The fitted line will cross the Y-axis at a point b_0 (the intercept), and will have a slope, b_1. The equation of the line is

$$\hat{y}_i = b_0 + b_1 x_i \tag{4.11}$$

$\hat{y}_i$ is the estimated value of y_i at specified values of x_i. The deviations we are considering, therefore, are $\hat{y}_i - y_i$ and our problem becomes one of finding a method such that

$$\sum_{i=1}^{n} (\hat{y}_i - y_i)^2 = \text{minimum} \tag{4.12}$$

Derivation of the necessary technique requires differential calculus, so we will not consider a proof. Rather, we will present what are called the **normal equations** that give values of b_0 and b_1 defining a line having the desired characteristics:

$$\sum_{i=1}^{n} y_i = b_0 + b_1 \sum_{i=1}^{n} x_i \tag{4.13}$$

$$\sum_{i=1}^{n} x_i y_i = b_0 \sum_{i=1}^{n} x_i + b_1 \sum_{i=1}^{n} x_i^2 \tag{4.14}$$

Rewriting, we obtain

$$b_1 = \frac{\sum_{i=1}^{n} x_i y_i - \left(\sum_{i=1}^{n} x_i \sum_{i=1}^{n} y_i\right)/n}{\sum_{i=1}^{n} x_i^2 - \left(\sum_{i=1}^{n} x_i\right)^2 / n} = \frac{SP_{xy}}{SS_x} \tag{4.15}$$

and

$$b_0 = \frac{\sum_{i=1}^{n} y_i}{n} - b_1 \frac{\sum_{i=1}^{n} x_i}{n} = \overline{Y} - b_1 \overline{X} \tag{4.16}$$

We can use these equations to obtain the coefficients of the line, but you should recognize that Equation (4.13) and Equation (4.14) are a pair of simultaneous equations that can be solved by methods developed in Chapter 3. The two equations can be rewritten in matrix form:

$$\begin{bmatrix} n & \sum X \\ \sum X & \sum X^2 \end{bmatrix} \cdot \begin{bmatrix} b_0 \\ b_1 \end{bmatrix} = \begin{bmatrix} \sum Y \\ \sum XY \end{bmatrix} \tag{4.17}$$

Although there is hardly any advantage to the matrix expression in this simple case, matrix methods will be necessary to fit more complex lines. Therefore, we will solve the problem of moisture content versus depth by matrix algebra, and we will use this method throughout this chapter. The necessary quantities are $n = 8$, $\Sigma X = 140$, $\Sigma Y = 382$, $\Sigma XY = 3870$, and $\Sigma X^2 = 3500$. In matrix form,

$$\begin{bmatrix} 8 & 140 \\ 140 & 3500 \end{bmatrix} \cdot \begin{bmatrix} b_0 \\ b_1 \end{bmatrix} = \begin{bmatrix} 382 \\ 3870 \end{bmatrix}$$

Solving gives $b_0 = 94.67$ and $b_1 = -2.68$. We can use this line to estimate values of moisture content at various depths. The estimated values $\hat{y}_i$ at the sample points provide a measure of how well the least-squares line conforms to the raw data. If the line passed exactly through each sample point, $\hat{y}_i$ and y_i would correspond, and the sum of the squared deviations from the line would be zero. Of course, in this example they do not.

We can define three terms that express variation of the dependent variable. The first of these is measured by the *total sum of squares* (SS_T) of Y:

$$SS_T = \sum_{i=1}^{n} y_i^2 - \frac{\left(\sum_{i=1}^{n} y_i\right)^2}{n} = \sum_{i=1}^{n} \left(y_i - \overline{Y}\right)^2 \qquad (4.18)$$

This quantity, divided by $(n - 1)$, gives the variance of Y:

$$s_y^2 = \frac{SS_T}{n - 1} = \frac{n \sum_{i=1}^{n} y_i^2 - \left(\sum_{i=1}^{n} y_i\right)^2}{n \, (n - 1)} \qquad (4.19)$$

The second measure of variation is the *sum of squares due to regression* (SS_R):

$$SS_R = \sum_{i=1}^{n} \hat{y}_i^2 - \frac{\left(\sum_{i=1}^{n} \hat{y}_i\right)^2}{n} = \sum_{i=1}^{n} \left(\hat{y}_i - \overline{Y}\right)^2 \qquad (4.20)$$

As the expression on the right implies, the estimated values of $\hat{y}_i$ have the same mean as the original values. The sum of squares of these estimates, $\hat{y}_i$, provides a measure of the variation of the regression line around the mean. If $\hat{y}_i$ and y_i correspond for all observations, the sums of squares calculated by Equations (4.18) and (4.20) will be the same. Otherwise, the sum of squares due to regression will be smaller, and there will be "leftover" variation we can call the *sum of squares due to deviations* or the *error sum of squares* (SS_E):

$$SS_E = SS_T - SS_R \qquad (4.21)$$

This is a measure of the failure of the least-squares line to fit the data points. It can also be calculated by

$$SS_E = \sum_{i=1}^{n} \left(\hat{y}_i - y_i\right)^2 \qquad (4.22)$$

which is algebraically equivalent to Equation (4.21).

The *goodness of fit* of the line to the points can be defined by

$$R^2 = \frac{SS_R}{SS_T} \qquad (4.23)$$

If the line is a good estimator of the data, this ratio will be near unity (we will later discuss tests of "how good" this value is). Often R^2 is expressed as a percentage. The same terminology is commonly used in trend-surface analysis, which we will see is a direct extension of this method. A most useful relation is that the square root of goodness of fit is the *multiple correlation coefficient*, R:

$$R = \sqrt{R^2} = \sqrt{SS_R / SS_T} \qquad (4.24)$$

195

This definition is algebraically equivalent to the definition of the correlation coefficient given in Chapter 2:

$$r = \frac{SS_{xy}}{\sqrt{SS_x \cdot SS_y}} \tag{4.25}$$

In calculating the equation of the least-squares line between moisture content and depth, we have found all of the entries necessary to determine the various sums of squares, goodness of fit, and correlation. Compute the quantities SS_T, SS_R, SS_E, R^2, and R for the data in **Table 4–8** and in file LOUISMUD.TXT. (Note that the computer file also contains measurements from a second core hole. The two holes are identified by a 1 or 2 in the third column of the data matrix. At this time, work only with data from the first core hole.)

It is obvious from the appearance of the fitted line that a straight line does not closely approximate the data, even though the correlation is high. Poor fit may arise from several causes, including high variance in the dependent variable (excessive scatter in the data), or from the fitting of an inappropriate model. In this example, we would tend to suspect the latter, because the data points seem to approximate a curve rather than a straight line. We will investigate fitting curves after considering statistical tests that can be performed, provided the data meet certain requirements.

If y_i is a random variable observed at specified intervals of x_i, we can assume that our data follow the theoretical population model

$$y_i = \beta_0 + \beta_1 x_i + \varepsilon_i \tag{4.26}$$

where i represents successive observations. The quantity ε_i is a normally distributed random variable with mean zero and unknown variance, σ^2, independent of the value of y_i. That is, an observed value of y_i is assumed equal to the sum of a constant related to the mean (if both the y_i and x_i are converted to deviations around their means, β_0 vanishes), plus a linear function of x_i plus a random error or deviation, ε_i. This relationship is shown in **Figure 4–18**. At every point, a normal frequency distribution of possible values of y_i is assumed to exist around the line of regression. By the method of least squares, we can estimate the population regression parameters [the β's in the model given as Equation (4.26)] by sample regression coefficients [the b's in the computational equation (Eq. 4.11)]. If the assumptions we have just made are true, the least-squares method will give b's that are the most likely estimate of the regression parameters, and the line we compute will be closer to the true regression line than any other. If the linear equation we fit to the data is a true model of the regression, the variance of ε is equal to the variance around the regression line. On the other hand, if the model is not correct, the variance around the regression will be greater than the variance of ε.

We can use the sums of squares to compute variances that in turn can be used to test the two alternatives. In particular, SS_E can be used to estimate variance about the regression. We can estimate σ^2 adequately only if we have replicate measurements of y_i at each point x_i, because this is the only way we can obtain a measure of the variance in Y independently of the variance in X. However, SS_R will give an estimate of σ^2 when our model is correct and will estimate σ^2 plus the amount of bias if our model is incorrect. Using SS_R, an analysis of variance can be constructed that will lead to rejection if either the observations are too variable for reliable judgment or the postulated model is incorrect. **Table 4–9** shows the form of the analysis of variance.

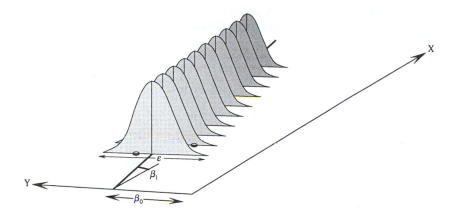

Figure 4–18. Components of the regression model $y_i = \beta_0 + \beta_1 x_i + \varepsilon_i$. Error is assumed to be normally distributed about the regression line.

Table 4–9. ANOVA for simple linear regression.

Source of Variation	Sum of Squares	Degrees of Freedom	Mean Squares	F-Test
Linear Regression	SS_R	1	MS_R	MS_R / MS_E
Deviation	SS_E	$n - 2$	MS_E	
Total Variation	SS_T	$n - 1$		

As in Chapter 2, mean squares are variance estimates made by dividing the appropriate sums of squares by their degrees of freedom. The MS_R has one degree of freedom because it is based on two "observations," the coefficients b_0 and b_1. The total variance has $(n - 1)$ degrees of freedom. Therefore, MS_E must have degrees of freedom equal to the difference between the two, or $(n - 1) - 1 = n - 2$. We can complete the ANOVA for the problem as we have done in **Table 4–10**. We are testing the hypothesis and alternative

$$H_0 : \beta_1 = 0$$
$$H_1 : \beta_1 \neq 0$$

The regression line is constrained to pass through the means, $\overline{X}$ and $\overline{Y}$. If the slope coefficient, β_1, is not significantly different from zero, this is equivalent to saying that the scatter in values of y_i about the regression line is no less than their scatter about $\overline{Y}$. We will use a 5% level of significance ($\alpha = 0.05$). The test statistic follows an F-distribution with degrees of freedom $\nu_1 = 1$ and $\nu_2 = 6$, so the critical region will consist of values exceeding $F = 5.99$. The computed test value falls well within the critical region, so we must reject the hypothesis that the variance about the regression line is no different than the variance in the observations. However, even

197

Table 4–10. Completed ANOVA for significance of regression of water content on depth.

Source of Variation	Sum of Squares	Degrees of Freedom	Mean Squares	F-Test
Linear Regression	7546.88	1	7546.88	23.07[a]
Deviation	1962.62	6	327.10	
Total Variation	9509.50	7		

[a]Significant at the $\alpha = 5\%$ level of significance.

though a significant linear trend exists in the data, a plot of the data suggests that we should be able to do better.

Fifty feet from the first core, a second core was taken through the mud sequence in the estuary. The water content of samples from the core can provide us with a set of replicates of y_i, allowing direct estimation of σ^2. We then can determine if the poor correlation between water content and depth is due to high variance or lack of fit to our model equation. Data from the second core are given in **Table 4–11** and also are included in file LOUISMUD.TXT, where they are identified by the number 2 in the column headed "Core Number." Plot these observations on a graph and compare their distribution to the observations from the first core hole.

Table 4–11. Moisture content of second core.

Depth, ft	Moisture (g water/100 g dried solids)
0.0	137.0
5.0	84.0
10.0	50.0
15.0	32.0
20.0	28.0
25.0	24.0
30.0	23.0
35.0	20.0

A regression can be computed using all of the observations from the two drill holes. We will calculate SS_T, SS_R, and SS_E in exactly the same manner as before except, of course, we now have twice as many observations. Because we have replicate measurements, we also can calculate a **sum of squares due to lack of fit** (SS_{LF}) and a **sum of squares due to pure error** (SS_{PE}). These divide the sum of squares of deviation into two parts. In the case of pairs of replicates we can find SS_{PE} by

$$SS_{PE} = 1/2 \sum_{i=1}^{n} (y_{i,1} - y_{i,2})^2 \qquad (4.27)$$

which has one degree of freedom for every replicated point. The remaining sum of squares, SS_{LF}, is found by subtraction, as are its degrees of freedom:

$$SS_{LF} = SS_E - SS_{PE} \qquad (4.28)$$

It is not necessary that we measure replicates of every point, but the analysis is more powerful if all are replicated. It is possible to use more than two replicates of y_i for each value of x_i, but the calculation of SS_{PE} becomes somewhat more complex. These and other modifications are described in books on regression such as that of Draper and Smith (1998) and will not be considered further here.

Table 4–12. ANOVA for simple linear regression with replicates. Number of observations of Y is n; number of points replicated is k.

Source of Variation	Sum of Squares	Degrees of Freedom	Mean Squares	F-Tests
Linear Regression	SS_R	1	MS_R	MS_R / MS_E [a]
Deviation	SS_E	$n - 2$	MS_E	
Lack of Fit	SS_{LF}	$(n - 2) - k$	MS_{LF}	
Pure Error	SS_{PE}	k	MS_{PE}	MS_{LF} / MS_{PE} [b]
Total Variation	SS_T	$n - 1$		

[a] Tests for goodness of fit.
[b] Tests for appropriateness of model.

Our modified analysis of variance table has the form shown in **Table 4–12**. Using the combined raw data from the two cores, complete the ANOVA and calculate SS_{PE} and SS_{LF}. The mean square of SS_{PE} is an estimate of $\sigma^2_{y|x}$, the variance around the regression line. It is found by

$$MS_{PE} = \frac{SS_{PE}}{k} \qquad (4.29)$$

where k is the number of data points that are replicated. In our case, we have taken replicates of all points, so k is equal to $n/2$, as half of the observations of y_i are replicates. We stated that SS_E was a measure of the variance around the regression plus any bias that might result from an inappropriate model, so the mean square of SS_{LF} is an estimate of the bias alone. We can test the appropriateness of the model equation by

$$F = \frac{MS_{LF}}{MS_{PE}} \qquad (4.30)$$

If the test value falls into the critical region, we may conclude that the model appears to be inadequate. If the test does not lead to rejection of the model, we can

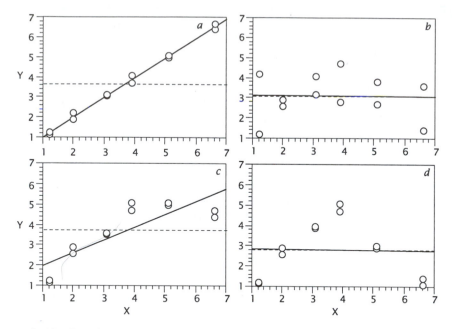

Figure 4–19. Possible straight-line regression situations. (*a*) Significant linear regression, no lack of fit. (*b*) Linear regression not significant, no lack of fit. (*c*) Significant linear regression, significant lack of fit. (*d*) Linear regression not significant, significant lack of fit.

combine both variance estimates, $MS_{LF} + MS_{PE} = MS_E$, and test for goodness of fit as we have done previously. Calculate this F-ratio by completing ANOVA **Table 4–12** and determine if the simple linear model is appropriate. The four possible outcomes for the two tests, one of the appropriateness of the model, and the other of goodness of fit, are graphically shown in **Figure 4–19.**

Confidence belts around a regression

Regression coefficients calculated from a sample are statistics that estimate population parameters, just as sample means and variances estimate population means and variances. The regression coefficients b_0 and b_1 are estimates of the underlying population regression parameters, β_0 and β_1. Until now, we have considered tests of significance of the regression coefficients, which are analogous to tests of sample means and variances which we examined in Chapter 2. We can also calculate a confidence interval around a regression coefficient, and can state that the true population coefficient lies within the confidence interval with specified probability. The confidence interval gives a range of plausible values for the regression parameter and has essentially the same form as confidence intervals about means or variances.

Like other confidence intervals, those around regression statistics are found by multiplying the standard error of the statistic by a multiple of a value from an appropriate standardized probability distribution. The value is a two-sided z- or t-score that encloses the proportion of the distribution that represents the desired level of probability.

First, let's calculate the confidence interval around the slope, or b_1, regression coefficient. The estimation variance of the slope coefficient is

$$\sigma_{b_1}^2 = \frac{\sigma^2}{\Sigma\left(x_i - \overline{X}\right)^2} \tag{4.31}$$

We do not know the population variance, σ^2, but it is estimated by MSE, so the standard error of the slope coefficient b_1 is the square root of

$$s_{b_1}^2 = \frac{MSE}{\Sigma\left(x_i - \overline{X}\right)^2} = \frac{MSE}{\Sigma x_i - \frac{(\Sigma x_i)^2}{n}} \tag{4.32}$$

Therefore, the confidence interval about b_1 is

$$b_1 \pm (t \text{ critical value}) \sqrt{\frac{MSE}{\Sigma\left(x_i - \overline{X}\right)^2}} \tag{4.33}$$

Since a confidence interval is two-sided, the critical value of t is one in which the critical region is split between the two tails of the distribution, or $t_{(1-\alpha/2, n-2)}$. Two degrees of freedom are lost since both the mean and variance must be estimated from the sample.

We can calculate confidence bands for the regression intercept coefficient, b_0, in much the same way. The estimation variance of the intercept is

$$\sigma_{b_0}^2 = \sigma^2 \frac{\Sigma x_i^2}{n \Sigma\left(x_i - \overline{X}\right)^2} = \sigma^2 \left[\frac{1}{n} + \frac{\overline{X}^2}{\Sigma\left(x_i - \overline{X}\right)^2}\right] \tag{4.34}$$

Again, we must estimate the population variance by MSE, so the standard error of the intercept coefficient b_0 is the square root of

$$s_{b_0}^2 = MSE \frac{\Sigma x_i^2}{n \Sigma\left(x_i - \overline{X}\right)^2} = MSE \left[\frac{1}{n} + \frac{\overline{X}^2}{\Sigma\left(x_i - \overline{X}\right)^2}\right] \tag{4.35}$$

The confidence interval about b_0 is

$$b_0 \pm (t \text{ critical value}) \sqrt{MSE \frac{\Sigma x_i^2}{n \Sigma\left(x_i - \overline{X}\right)^2}} \tag{4.36}$$

The critical value of t is again $t_{(1-\alpha/2, n-2)}$.

If the dependent variable, y_i, is normally distributed, then the sampling distributions of b_0 and b_1 will be exactly normal. If y_i is not normal, b_0 and b_1 will still be asymptotically normal and the confidence intervals will be approximately correct. For large samples, the t critical values in Equations (4.33) and (4.36) should be replaced with appropriate z-values from the standard normal distribution.

Confidence intervals can be fitted around both b_0 and b_1 simultaneously. The combination defines a **confidence band** or **belt**. This has the form of a quadratic space that is narrowest at the mean, $\overline{X}$, and flares outward toward both high and low values of x_i. The true population regression, defined by parameters β_0 and β_1, lies within the band with a probability of $p = 1 - \alpha$.

The regression estimates, $\hat{y}_i$, or points on the regression line corresponding to values x_i, are the local expected values of the population. In general, if other random samples were collected and used to estimate the regression coefficients, slightly different lines would be fitted and the values of $\hat{y}_i$ would be different. We can calculate a confidence band for the estimates $\hat{y}_i$, which will define a region in which we expect regression lines generated from different random samples to fall. This is equivalent to estimating confidence intervals simultaneously for b_0 and b_1.

The sampling variance for $\hat{y}_i$ is

$$\sigma_{\hat{y}_i}^2 = \sigma^2 \left[\frac{1}{n} + \frac{\left(x_i - \overline{X}\right)^2}{\sum \left(x_i - \overline{X}\right)^2} \right] \tag{4.37}$$

Since the population variance is unknown, we again estimate it by MS_E. This yields

$$s_{\hat{y}_i}^2 = MS_E \left[\frac{1}{n} + \frac{\left(x_i - \overline{X}\right)^2}{\sum \left(x_i - \overline{X}\right)^2} \right] \tag{4.38}$$

whose square root is the standard error of the estimate $\hat{y}_i$. The confidence interval around any estimated value is $\hat{y}_i \pm (t \text{ critical value}) \, s_{\hat{y}_i}$. A value of t is chosen for a two-tailed critical region and $n-2$ degrees of freedom. So, for every value x_i, we can plot values of $\hat{y}_i$ and its upper and lower confidence limits, $\hat{y}_i + (t \text{ critical value}) \, s_{\hat{y}_i}$ and $\hat{y}_i - (t \text{ critical value}) \, s_{\hat{y}_i}$. Connecting these limits produces a graphic display of the confidence belt.

Data in file BENIOFF.TXT are estimates of the depth to the Benioff zone at localities in island arcs around the world and analyses of the K_2O content of andesites collected from volcanoes that have erupted at the same localities (adapted from Hatherton and Dickinson, 1969). It is believed that the andesitic magmas originated by partial melting above the oceanic plate being subducted along the Benioff zone. There seems to be a relationship between composition and depth, as indicated by a cross plot of the two variables (**Fig. 4–20**). The correlation between variables is $r = 0.82$, which is highly significant. We can fit a regression using depth as the independent variable and percent K_2O as the dependent variable (*i.e.*, composition varies as a function of its depth of origin). The mean squared error, MS_E, of K_2O content is 0.1813, and the sum of the squared deviations of depth from mean depth is 1066.4. Using these variance estimates and the deviations of each individual x_i from $\overline{X}$, we can calculate the shaded 95% confidence band shown on **Figure 4–20**. If we were to repeatedly take 35 observations at random from this same population, we would expect the fitted regressions to fall outside the shaded band only 1 time in 20. The horizontal line on **Figure 4–20** represents the mean K_2O content, $\overline{Y} = 1.69$. Whether or not the line representing the mean falls within the confidence belt is graphically equivalent to a significance test of an hypothesis that the slope of the regression is zero; if the mean line falls outside, the hypothesis is rejected at the 95% significance level. If the mean line is enclosed by the confidence belt, the regression cannot be statistically distinguished from a constant equal to the mean, and changes in X have no apparent influence on values of Y.

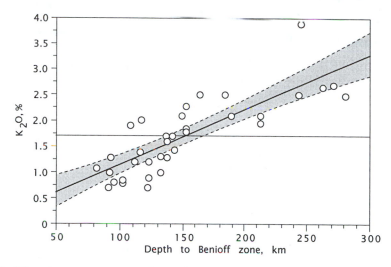

Figure 4–20. Linear regression of $K_2O\%$ in andesitic volcanics against depth to Benioff zone near source volcano. Shaded area within dashed lines represents 95% confidence belt around regression. Horizontal line is mean K_2O content of 1.69%.

Suppose we were to take repeated observations at some fixed value, x_i; their average would be an estimate of $\hat{y}_i$, and we would expect most such averages to fall within the shaded confidence band. But where would the individual observations fall? That is, can we predict the range of possible values that a new, randomly chosen observation of y_i might assume? Such a prediction must not only consider the scatter of observations about the sample-based regression line but also the uncertainty in the estimated regression itself (**Fig. 4–21**).

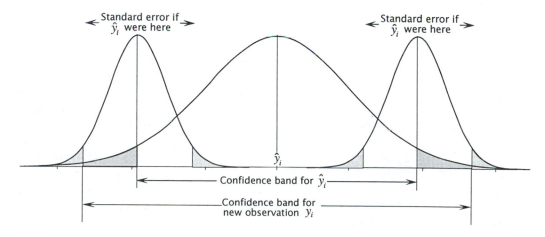

Figure 4–21. Probability distribution representing a confidence band around a regression estimate, $\hat{y}_i$, plus the standard errors of estimate surrounding $\hat{y}_i$, assuming it fell at an extreme limit of the confidence band. Their combination gives a confidence band for individual observations.

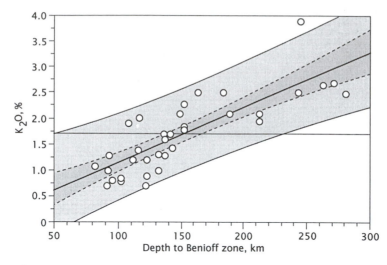

Figure 4–22. Linear regression of Figure 4–20 with added 95% confidence belt for individual observations shown by light gray shade within curved solid lines.

The sampling variance for a new observation can be estimated as

$$s_{y_{new}}^2 = MS_E \left[1 + \frac{1}{n} + \frac{\left(x_i - \overline{X}\right)^2}{\sum \left(x_i - \overline{X}\right)^2} \right] \tag{4.39}$$

which combines the sampling variance for $\hat{y}_i$ and the variance of Y. The confidence belt around a new observation is therefore $\hat{y}_i \pm$ (t critical value) $s_{y_{new}}$ where the critical value of t is $t_{1-\alpha/2,n-2}$.

We can calculate a 95% confidence belt for individual observations using data in the file BENIOFF.TXT. Such a belt is shown in **Figure 4–22**. Note that one out of the 35 observations lies outside the shaded belt, which is approximately what we expect at the specified level of significance.

Calibration

Sonic transit time, or time required for a compressional wave to move between two points, is measured in drill holes by sonic logging tools. Sound waves move faster in solid rock than through fluid-filled pores, so transit time increases with increasing porosity. File PRUDHOE.TXT contains measurements of sonic transit times recorded for a specific stratigraphic interval in 44 wells drilled in the Prudhoe Bay oil field on the Alaskan North Slope. Cores were taken of these corresponding intervals, and their porosities measured in the laboratory. We can determine the strength of the relationship between core porosity and sonic transit time by computing the regression between the two. This is shown as a fitted line on **Figure 4–23**. The correlation is $r = 0.88$. The ANOVA in **Table 4–13** confirms that the relationship is highly significant.

The regression $\Delta_t = 58.056 + 1.216\phi$ will allow us to predict sonic transit times from known core porosities with high precision. Unfortunately, such a relationship is not very useful, because the sonic log typically is run in wells for the purpose of estimating formation porosities; if core measurements are available, transit times

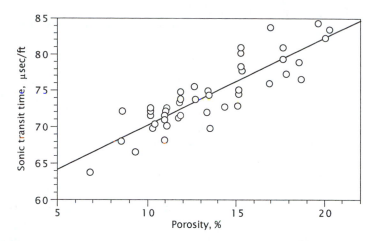

Figure 4–23. Sonic transit time measured by a sonic logging tool in wells drilled in the Prudhoe Bay field, Alaska, versus porosities measured on cores recovered from the same wells. Line is regression of sonic transit time on core porosity.

Table 4–13. ANOVA for the significance of linear regession of sonic transit time on core porosity in 44 wells in part of Prudhoe Bay field, Alaska.

Source of Variation	Sum of Squares	Degrees of Freedom	Mean Squares	F-Test
Linear Regression	731.549	1	731.549	
Deviation	223.172	42	5.314	137.67[a]
Total Variation	954.721	43		

[a]Highly significant ($F_{1,42} = 4.07$); $p < 0.0001$.

are not needed. What really is of interest is the reverse relationship, in which the porosities of uncored intervals are predicted from transit times recorded on well logs.

There are two possible approaches to this problem; which alternative is the most appropriate was at one time a subject of controversy among statisticians. The most obvious procedure is called **inverse regression** and simply consists of exchanging the roles of X and Y. Casting porosity as the dependent variable yields a regression equation, $\phi = -33.435 + 0.6302\Delta_t$. Unfortunately, the statistical uncertainty is now expressed in the core porosity for a specified sonic transit time, even though porosities measured on core samples represent the "truth." In effect, we are professing that the porosity in rocks results from the speed of transmission of sound waves, when we know that the converse is true.

The alternative approach is indirect, but both statistically and physically correct. The regression in which transit time is dependent upon core porosity is rewritten so that ϕ appears on the left side of the equation. The result is called a

calibration equation and has the general form

$$\hat{x}_i = (y_i - b_0)/b_1 \qquad\qquad (4.40)$$

For this example, the equation is $\hat{\phi} = (\Delta_t - 58.056)/1.216$. Using the calibration approach has several statistical advantages. The uncertainty in the calibration estimate is related to the sampling variance of the dependent variable, as it should be. The independent variable is treated as a known quantity, or fixed standard. It is used to establish the calibration line and regarded as having been set or measured without error.

Calibration is widely used in geochemistry and related areas where standards are used to create calibration curves for the analysis of specimens of unknown composition. Carbon-14 age dates have been calibrated against absolute ages obtained by counting tree rings. Analytical procedures such as X-ray fluorescence, chromatography, spectroscopy, and others depend upon calibrations achieved by recording the instrumental response of standard specimens of known composition; the proper application of statistical calibration has been a lively topic in the chemometrics literature.

If we calculate and plot confidence belts around the regression, we can use graphical methods to estimate the uncertainty when we use the regression as a calibration line to predict a value. **Figure 4–24** shows the regression of transit time on core porosity and the 95% confidence belt around the regression. Suppose we have a log response of $\Delta_t = 80\,\mu$ sec/ft and we want to predict the corresponding porosity: Draw a horizontal line at the level of the log response, intersecting the two confidence limits and the regression line. Drop perpendiculars from the points of intersection to the X-axis. The "best estimate" of the porosity is 18%, which is exactly the same value we would obtain by solving the calibration equation. The lower confidence limit about this estimate is 17.2% and the upper confidence limit is 18.9%. [Strictly speaking, these are not true confidence limits. Draper and Smith (1998) call them "fiducial limits" and provide equations for their estimation.]

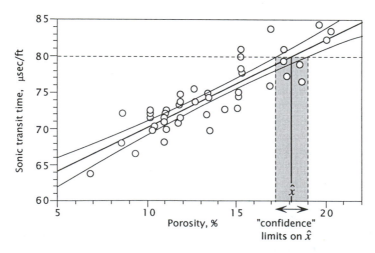

Figure 4–24. Regression used as calibration line to estimate porosity from sonic transit time read from well log. Curved lines are 95% confidence belt around regression; shaded interval is approximate "confidence" interval around back-calculated estimate of porosity.

This graphical technique will not work unless the relationship between X and Y is pronounced. The horizontal line may not intercept the upper and lower confidence limits if the slope of the regression line is low or the confidence band is very wide; or the line may intercept the lower band twice, producing an interval on X that does not include the back-calculated estimate itself. Either nonsensical result merely indicates that the relationship between X and Y is too poor to be used for calibration purposes.

Curvilinear regression

After calculating the F-test for the appropriateness of a regression, you may conclude that a straight line is inadequate. What you do next depends upon the objectives of the study and your knowledge (or beliefs) about the relationship between X and Y. On one hand, you may be able to make definite statements about the relation between the two variables. For example, we know from elementary physics that an object dropped from a height will accelerate at a constant rate of $9.8 \, \text{m/sec}^2$ due to the force of gravity. The distance the object will fall in a given length of time can be calculated and plotted (**Fig. 4–25**), and a parabola can justifiably be fitted to the data. On the other hand, we may know nothing about the physical relationship between the two variables X and Y (indeed, none may exist), but simply want a concise expression of one in terms of the other. Usually our problems lie between these two extremes; we suspect a causal relationship, but do not know its form. In the latter circumstance, we can fit an **approximating equation** to the data in the hope that it will shed some light on the underlying relationship, or at least describe the form of the X-Y relationship in a concise way. These equations are chosen because they are capable of approximating many classes of functions and commonly are used when the theoretical form of a function is not known.

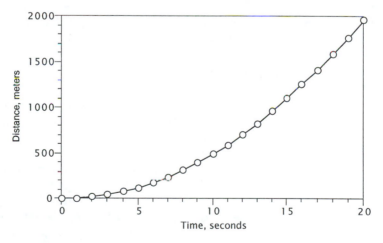

Figure 4–25. Theoretical plot of distance an object drops in free fall, versus time.

There are several approximating equations available, but the one most commonly used is a **polynomial expansion.** This is a summation of integer powers of the independent variable

$$y_i = b_0 + b_1 x_i + b_2 x_i^2 + b_3 x_i^3 + \ldots + b_m x_i^m \tag{4.41}$$

207

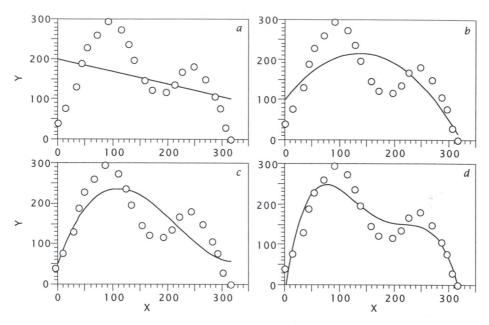

Figure 4–26. Polynomial regressions of increasing powers of X. (*a*) Linear or first-degree polynomial, $y_i = 200.3 - 0.3\,x_i$. (*b*) Quadratic or second-degree polynomial, $y_i = 102.1 + 1.7\,x_i - 0.006\,x_i^2$. (*c*) Cubic or third-degree polynomial, $y_i = 52.8 + 3.8\,x_i - 0.02\,x_i^2 + 0.00004\,x_i^3$. (*d*) Quartic or fourth-degree polynomial, $y_i = -5.9 + 8.5\,x_i - 0.09\,x_i^2 + 0.0004\,x_i^3 - 0.0000006\,x_i^4$.

An equation in which all terms are added together is called a ***linear function***, because the relationships between the dependent variable and any one of the independent variables graph as straight lines if the other variables are held constant. Expansion of our original equation by adding successive powers allows our original straight line to bend. One additional term (X^2) allows the line to reverse its slope; a second additional term (X^3) allows two changes in the direction of the slope, and so on. Increased flexibility means that the line can conform more closely to the data. In fact, when the number of additional powers reaches ($n - 1$), the line will go exactly through every data point. There is little purpose in computing such a line, because it is no more efficient than the original data. Hopefully, the essential aspects of the data array can be preserved with only a few terms in the polynomial equation. **Figure 4–26** shows typical polynomial curves incorporating various powers of X. The maximum power used in the polynomial defines the ***degree*** of the equation. That is, $y_i = b_0 + b_1 x_i + b_2 x_i^2 + b_3 x_i^3$ is a third-degree polynomial. So is $y_i = b_1 x_i^3$, because this is the same as stating that b_0, b_1, and b_2 are equal to zero. The polynomial equations are fitted to the observations by least-squares methods, and the process is called ***curve fitting***. If certain statistical assumptions are valid, the goodness of fit and appropriateness of the curve can be tested by extensions of the regression techniques we have just considered. The statistical procedures are grouped under ***curvilinear regression analysis***.

To fit a second-degree (or quadratic) curve to data, we must expand the normal equations to include additional terms. Equations (4.13) and (4.14), the two normal equations, become a set of three simultaneous equations:

$$\sum Y = b_0 n + b_1 \sum X + b_2 \sum X^2$$

$$\sum XY = b_0 \sum X + b_1 \sum X^2 + b_2 \sum X^3 \qquad (4.42)$$

$$\sum X^2 Y = b_0 \sum X^2 + b_1 \sum X^3 + b_2 \sum X^4$$

All summations are understood to extend over the observations from 1 to n. Rewriting these normal equations into matrix form gives

$$\begin{bmatrix} n & \sum X & \sum X^2 \\ \sum X & \sum X^2 & \sum X^3 \\ \sum X^2 & \sum X^3 & \sum X^4 \end{bmatrix} \cdot \begin{bmatrix} b_0 \\ b_1 \\ b_2 \end{bmatrix} = \begin{bmatrix} \sum Y \\ \sum XY \\ \sum X^2 Y \end{bmatrix} \qquad (4.43)$$

which can be solved readily by the matrix algebra procedure given in Chapter 3. Note that high powers of the independent variable are required in the equations.

The largest power in the matrix is twice the degree of the equation being fitted. This can be a major source of errors in computer programs that fit polynomials, because elements in the lower right part of the coefficient matrix may be many orders of magnitude greater than those in the upper left corner of the matrix. This may lead to round-off errors and loss of significance in critical digits, resulting in unstable or unreliable solutions to the simultaneous equations. Discussions of these problems are contained in books by Buchanan and Turner (1992) and Press and others (1992).

The structure of the coefficient matrix can be seen if we utilize a dummy variable, X^0, which we will define as being equal to 1 for every observation x_i. We can label the rows and columns in the matrix equation in the following manner:

$$\begin{array}{c} \\ \begin{array}{ccccccc} X^0 & X^1 & X^2 & X^3 & \cdots & X^m & \qquad b \qquad Y \end{array} \\ \begin{array}{c} X^0 \\ X^1 \\ X^2 \\ X^3 \\ \cdots \\ X^m \end{array} \begin{bmatrix} & & & & & \\ & & & & & \\ & & & & & \\ & & & & & \\ & & & & & \\ & & & & & \end{bmatrix} \cdot \begin{bmatrix} \\ \\ \\ \\ \\ \end{bmatrix} = \begin{bmatrix} \\ \\ \\ \\ \\ \end{bmatrix} \end{array} \qquad (4.44)$$

Entries within the body of the coefficient matrix and within the matrices of b coefficients and right-hand parts are sums of the cross products of the row and column labels. By our definition of X^0, the x_{11} entry becomes $\sum_{i=1}^{n} 1 \cdot 1 = n$, and the other entries in the top row are equal to 1 times the column label. Element x_{43} in the matrix is, for example, $\sum X^3 \cdot X^2 = \sum X^5$. Remember that multiplication of exponents consists of adding the powers. That is, $X^a \cdot X^b = X^{a+b}$.

To demonstrate the computations used in curvilinear regression, we can analyze the combined data from **Tables 4–8** and **4–11**, given in file LOUISMUD.TXT. We will work through the problem using a quadratic fit to demonstrate an additional procedure that tests to see if the increase in the degree of the polynomial has significantly improved the fit of the regression. The second-degree polynomial curve fitted to the data is shown in **Figure 4–27**. The regression equation is

$$y_i = \beta_0 + \beta_1 x_i + \beta_2 x_i^2$$
$$= 122.9 - 7.9 x_i + 0.1 x_i^2$$

209

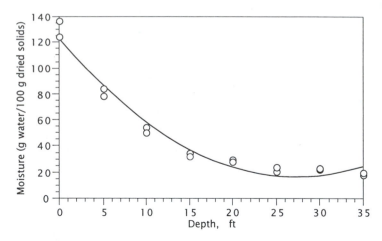

Figure 4–27. Second-degree polynomial regression fitted to moisture data from Tables 4–11 and 4–14 and file LOUISMUD.TXT.

Statistics necessary to perform an analysis of variance include:

$$
\begin{aligned}
SS_T &= 21{,}363.0 & SS_{LF} &= 563.8 \\
SS_R &= 20{,}673.2 & R^2 &= 0.97 \\
SS_E &= 689.8 & R &= 0.98 \\
SS_{PE} &= 126.0
\end{aligned}
$$

You will note that SS_T and SS_{PE} are the same as in the linear fit to this data, because they do not contain the estimated values, $\hat{Y}$. As we would expect, the more flexible quadratic equation fits the moisture-content data much more closely than does a straight line. The sum of squares due to deviations from the regression has been reduced from 5177.8 to 689.8. It seems obvious that this is a significant reduction, but this distinction is not always so apparent. The analysis of variance table can be expanded again to test this possibility. This new ANOVA is given in **Table 4–14**.

Table 4–14. ANOVA for significance of added terms in curvilinear regression.

Source of Variation	Sum of Squares	Degrees of Freedom	Mean Squares	F-Tests
Linear Regression	SS_{R1}	1	MS_{R1}	
Quadratic Regression	SS_{R2}	2	MS_{R2}	MS_{R2}/MS_E[a]
Added by Quadratic	SS_{R2-1}	1	MS_{R2-1}	MS_{R2-1}/MS_E[b]
Quadratic Deviation	SS_E	$n-3$	MS_E	
Total Variation	SS_T	$n-1$		

[a] Tests for significance of quadratic fit.
[b] Tests for significance of increase of quadratic over linear fit.

As you can see from the table, a sum of squares is created by subtracting the regression sum of squares for a linear fit (SS_{R1}) from the equivalent sum of squares for a quadratic fit (SS_{R2}). This new sum of squares is a measure of the increase in fit resulting from the additional regression term. In test b of the ANOVA table, this quantity, which we have designated SS_{R2-1}, is used to estimate the additional regression variance. Its significance is tested in exactly the same manner as the significance of the regression itself. If the resulting F-value falls in the critical region, the added term is making a significant contribution to the regression and should be retained. If the test is not significant, the additional power is not contributing to the regression. Note that test a may be significant when test b is not. This is because test a is actually examining the significance of the linear and quadratic terms combined; the linear fit may be highly significant but the quadratic contribution very low. Test a will then be significant because of the strength of the linear term alone. In different situations, we may find that either, neither, or both terms are significant.

You should also note that the correlation will always increase with the addition of terms. When the number of terms becomes equal to $(n-1)$, the correlation will equal 1.00, regardless of how wildly the data points are scattered. However, the tests outlined above may show that increases in correlation are not statistically meaningful. The F-ratio for significance of fit may decrease because the mean squares due to deviation (MS_E) may increase. This variance estimator is dependent in part on the number of observations used in its creation, or its degrees of freedom. These are being reduced constantly as we add coefficients to the regression equation. Remember from Chapter 2 that we lose one degree of freedom for each parameter we estimate, and the b's of the polynomial equation are estimates of β's, the population regression parameters.

This procedure for testing the significance of added terms can be extended to successively higher powers in the polynomial regression, provided the statistical assumptions are fulfilled. These tests can also be combined with the tests for lack of fit and pure error if replicates are available. **Table 4–15** is the completed ANOVA for a quadratic regression on the combined moisture-content data.

In some problems we are more concerned with the value of the intercept or the slope of the regression line than we are with estimates of points or deviations from these. As an example of a problem of this type, we will consider a stratigraphic succession from a long core through Lower Paleozoic rocks in eastern Oklahoma. In an attempt to interpret environmental conditions prevailing at the time of deposition, a geologist has studied a thick clastic unit cut by drilling. The unit consists of interbedded siltstones and sandstones believed to be offshore marine deposits. The geologist has postulated that the basin filled gradually, and as the shoreline advanced toward the location of the well, successively deposited sand layers should be increasingly thicker. The unit contains thousands of beds, and it would be extremely time-consuming to measure each of them. Instead, the thickness of a single sandstone bed was measured at successive 10-ft intervals. These measurements are listed in file OKLA.TXT and shown graphically in **Figure 4–28**. The geologist wishes to determine if the slope of individual bed thickness versus thickness of accumulated sediment is a real phenomenon.

Note that cumulative thickness, X, is measured at fixed locations, while individual bed thickness is a random variable; therefore, a regression model is appropriate.

Table 4–15. Completed ANOVA for significance of the quadratic regression of moisture content of sediment on depth of burial.

Source of Variation	Sum of Squares	Degrees of Freedom	Mean Squares	F-Tests
Linear Regression	16185.19	1	16185.19	
Quadratic Regression	20673.24	2	10336.62	$MS_{R2}/MS_E =$ 194.81 [a]
Added by Quadratic	4488.05	1	4488.05	$MS_{R2-1}/MS_E =$ 84.58 [b]
Quadratic Deviation	689.76	13	53.06	
Total Variation	21363.00	15		

[a] Quadratic regression is highly significant.
[b] Quadratic regression is a highly significant improvement over linear regression alone.

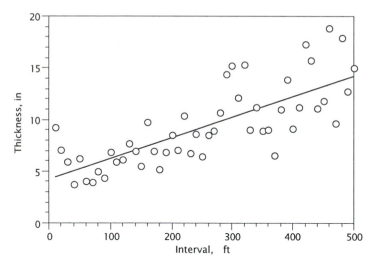

Figure 4–28. Thickness of sandstone beds at 10-ft intervals measured on a core through Lower Paleozoic sandstones in eastern Oklahoma. Line is fitted linear regression. Data are listed in file OKLA.TXT.

In other words, the geologist must test the hypothesis that the slope coefficient, b_1, of a linear regression is significantly greater than zero.

After we have calculated the regression equation $y_i = b_0 + b_1 x_i$ shown by the line in **Figure 4–28**, we can estimate the variance about the regression by MS_E. This can in turn be used to calculate the t-test statistic:

$$t = \frac{b_1}{\sqrt{MS_E/SS_x}} \tag{4.45}$$

The mean square due to deviation or error (MS_E) is equal to SS_E divided by $(n-2)$ degrees of freedom, as shown in **Table 4–12**. The corrected sums of squares of X, SS_x, is found by

$$SS_x = \sum_{i=1}^{n} x_i^2 - \frac{\left(\sum_{i=1}^{n} x_i\right)^2}{n} \tag{4.46}$$

This is a test of one of the hypotheses

1. $H_0 : \beta_1 = 0$
2. $H_0 : \beta_1 \leq 0$
3. $H_0 : \beta_1 \geq 0$

against their respective alternatives

1. $H_0 : \beta_1 \neq 0$
2. $H_0 : \beta_1 > 0$
3. $H_0 : \beta_1 < 0$

The first null hypothesis requires a two-tailed test, because either significantly positive or negative slopes lead to rejection. The other tests are one-tailed. Our geologist is only interested in establishing if bed thickness is increasing up the section; that is, if the slope of the regression between bed thickness and cumulative thickness is positive. Therefore, the second hypothesis and its alternative are the appropriate choices. The test will be one-tailed, with the critical region on the right.

We can quickly calculate the necessary entries for the test. Some of the numerical values required for this test are:

Regression equation: $y_i = 4.25 + 0.020x_i$

$SS_T = 730.94 \qquad SS_R = 425.18$
$SS_E = 305.76 \qquad SS_X = 1041250.00$
$R^2 = 0.58 \qquad R = 0.76$

There are 50 observations in the data set, so there are 48 degrees of freedom associated with SS_E. At a significance level of 5% ($\alpha = 0.05$), the critical value of t with $\nu = 46$ degrees of freedom is 1.68. The test statistic is

$$t = \frac{0.02}{\sqrt{6.37/1041250.00}} = 8.17$$

This value lies within the critical region, so we must reject the hypothesis that the slope of the regression is zero or negative. A small but definite increase in bed thickness occurs through the sequence.

The test just presented is a special case of

$$t = \frac{b_1 - \beta_1}{\sqrt{MS_E/SS_X}} \tag{4.47}$$

which tests the hypothesis that the regression slope is equal to some predefined value β_1. In the test in Equation (4.45), β_1 is zero. This test, or a version of it, is very important in time-series analysis. Time-series procedures are based on the assumption that there is no trend in the data, or that the slope of the regression on time (or distance) is zero. If a trend is present, it must be removed, or the analysis of the time series is invalid. A series having no significant linear trend is said to

213

be **stationary**. If a persistent trend or "drift" is present in the data, it is called **evolutionary** or **nonstationary**.

One of the assumptions of linear regression is that the variance is constant about the regression line. This assumption can be tested by examining the residuals from the fitted line. If the variance is constant, the residuals will form a more-or-less uniform band around the regression line. If there is a progressive change in the width of the band of deviations, the variance may not be constant. These two conditions are known by the somewhat terrifying names of **homoscedasticity** for constant variance, and **heteroscedasticity** for changing variance. A "quick and dirty" way to check the consistency of variance about a regression is to perform a linear regression on the absolute values of the deviations. A change in variance along the sequence will appear as a significant slope.

Another assumption made in regression analysis is that deviations or residuals from the regression are free from autocorrelation. **Autocorrelation** in this context means that residuals tend to occur as "clumps" of adjacent deviations on the same side of the regression line. The presence of large sequences of autocorrelated residuals may indicate that the regression model is inappropriate. It also happens that autocorrelated deviations may suggest phenomena of geologic interest. This subject will be pursued more thoroughly when we discuss trend-surface analysis, where autocorrelated positive residuals may be an indicator of economic potential in the form of oil or other mineral accumulations. Testing for autocorrelation is easy; we need only apply the runs test to the signs of the deviations from the regression, or we can use one of the procedures discussed in the section on autocorrelation.

Reduced major axis and related regressions

The regression procedures we have been discussing involve fitting a line to a collection of bivariate observations so that the squared deviations of one of the variables from the line is a minimum. If the deviations in the Y-direction are minimized, one set of linear regression coefficients is obtained, and if the deviations are minimized in the X-direction, a different set of coefficients for the inverse regression will result. When the two lines are plotted, as in **Figure 4–29**, they will cross at $\overline{X}, \overline{Y}$. The cosine of the angle between the two lines is numerically equal to the correlation between X and Y.

Sometimes physical circumstances leading to the observations dictate which variable should be considered dependent or else the purpose of the analysis clearly indicates which variable should be regressed onto the other. However, it may not be possible to rationally decide which variable should be X and which should be Y. This occurs, for example, in biometry where it may be useful to know the relationship between two sets of measurements such as the lengths and widths of shells, but it is not obvious which set of measurements should be expressed as a function of the other. Similar circumstances arise in petrophysics, where a common problem is relating measurements, such as sonic transit time and neutron density, made by different logging tools. Both measurements are subject to errors, and neither can be regarded as a function of the other, yet it is extremely useful to be able to cross plot the two variables and express their mutual relationship in some manner.

An appealing solution would be to fit a line that minimizes the deviations of the observations from the line in both the X- and Y-directions simultaneously. Such a line would split the difference between the regression lines of X on Y and Y on X.

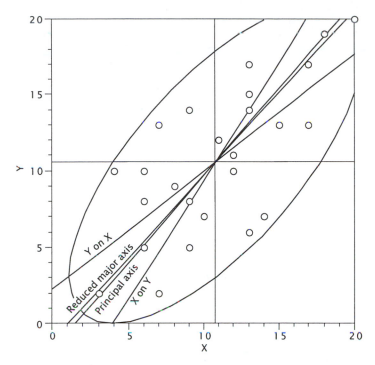

Figure 4–29. Scatter diagram of bivariate data from Table 6–19 (p. 532). Also shown are regression lines of Y on X and X on Y, reduced major axis line, and principal or major axis. Horizontal and vertical lines intersect at the bivariate mean.

It would conform more closely to the visual impression of the trend in the observations, and it would attribute the scatter of the data points to both variables, rather than assigning all of the deviations from the fitted line to a single variable.

There are alternative ways in which such a line could be defined. One method involves minimizing the squared deviations from the line in both the X- and Y-directions simultaneously. By the Pythagorean theorem, this is equivalent to minimizing the squared perpendicular deviations from the fitted line (**Fig. 4–30**). Such a line is called the **principal** or **major axis**, and can be found as the first eigenvector of a matrix containing the variances and covariance of X and Y. We will discuss computation of the principal axis at length in Chapter 6, under the heading Principal Component Analysis.

The second procedure minimizes the product of the deviations in both the X- and Y-directions. This in effect minimizes the sum of the areas of the triangles formed by the observations and the fitted line (**Fig. 4–30**), resulting in what is called the **reduced major axis**, often referred to simply as the "RMA line." Most articles on the reduced major axis have been published in the two journals *Biometrics* and *Biometrika*, reflecting the popularity of the technique among scientists concerned with growth in organisms. Although the properties of the reduced major axis have received scant attention from statisticians, they have been investigated by Kermack and Haldane (1950), Kruskal (1953), and Fuller (1987). A summary for geologists is given by Till (1974) and a more extensive treatment is contained in Miller and Kahn (1962). Doveton (1994) includes a detailed discussion for petrophysicists.

The reduced major axis is defined by an ordinary linear equation, $y_i = b_0 + b_1 x_i$, in which the b_0 coefficient represents the intercept and the b_1 coefficient represents

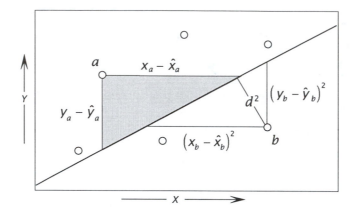

Figure 4–30. Criteria for fitting RMA and major axis lines. (*a*) Reduced major axis minimizes the product of deviations $(x_a - \hat{x}_a)(y_a - \hat{y}_a)$ from the fitted line, equivalent to minimizing the area of a triangle (shaded). (*b*) Principal axis minimizes the sum of the squared deviations $(x_b - \hat{x}_b)^2 + (y_b - \hat{y}_b)^2$, in effect minimizing the squared perpendicular deviations d^2.

Table 4–16. Sums of squares and other quantities for data listed in Table 6–19 and file BIVARIAT.TXT.

$$n = 25$$

$\sum X$	=	272	$\sum Y$	=	267
$\overline{X}$	=	10.88	$\overline{Y}$	=	10.68
s_X^2	=	20.277	s_Y^2	=	24.060
s_X	=	4.503	s_Y	=	4.905
SS_X	=	486.65	SS_Y	=	577.44

$$\text{cov}_{XY} = 15.585$$
$$SP_{XY} = 374.04$$
$$r = 0.706$$

the slope. The slope is defined as the ratio of the standard deviations of the two variables, X and Y, or

$$b_1 = s_y / s_x \qquad (4.48)$$

Since n is the same for both standard deviations, b_1 may be found by the equivalent equation

$$b_1 = \sqrt{\frac{SS_y}{SS_x}} \qquad (4.49)$$

The intercept of the reduced major axis is found in the conventional way, using Equation (4.16).

The computation of the reduced major axis can be demonstrated using the observations shown on **Figure 4–29** and listed in **Table 6–19** on p. 532, and contained in file BIVARIAT.TXT, which will also be used later to illustrate the calculation of principal components. Sums, sums of squares and cross products, means, variances, and covariances are given in **Table 4–16**. From these, we may first calculate the ordinary regression of Y on X and the inverse regression of X on Y. For the regression of Y on X:

$$b_1 = \frac{SP_{xy}}{SS_x} = \frac{374.04}{486.64} = 0.769$$

$$b_0 = \overline{Y} - b_1\overline{X} = 10.68 - 0.769\,(10.88) = 2.317$$

So, the regression equation is

$$\hat{y}_i = 2.32 + 0.77\,x_i$$

For the inverse regression of X on Y:

$$b_1 = \frac{SP_{xy}}{SS_y} = \frac{374.04}{577.44} = 0.648$$

$$b_0 = \overline{X} - b_1\overline{Y} = 10.88 - 0.648\,(10.68) = 3.962$$

yielding the inverse regression equation

$$\hat{x}_i = 3.96 + 0.65\,y_i$$

For the reduced major axis,

$$b_1 = \sqrt{\frac{SS_y}{SS_x}} = \sqrt{\frac{577.44}{486.64}} = 1.089$$

$$b_0 = \overline{Y} - b_1\overline{X} = 10.68 - 1.089\,(10.88) = -1.168$$

The equation of the RMA line is therefore

$$\hat{y}_i = -1.17 + 1.09\,x_i$$

For comparative purposes, the first eigenvector of the variance–covariance matrix of X and Y is

$$\begin{bmatrix} 0.663 \\ 0.749 \end{bmatrix}$$

which means that the eigenvector has a slope of 75 units in Y for 66 units in X, equivalent to a b_1 coefficient of 1.1297. The intercept is

$$b_0 = \overline{Y} - b_1\overline{X} = 10.68 - 1.130\,(10.88) = -1.611$$

The equation of the major axis can be written as

$$\hat{y}_i = -1.61 + 1.13\,x_i$$

In addition to the two regression lines, the major axis and reduced major axis are also shown on **Figure 4–29**. Note that the reduced major axis and the major axis are very similar to each other. The reduced major axis bisects the angle between the line of regression of Y on X and the line of regression of X on Y; the principal axis responds to the somewhat greater variance of Y by swinging to a slightly steeper angle.

The standard errors of both reduced major axis coefficients can be computed easily, and from them approximate tests of significance can be formulated. However, there are no equivalents to the more soundly based analyses of variance that can be performed in conventional regression. The standard error of the RMA slope is

$$se_{b_1} = b_1 \sqrt{\left(\frac{1 - r^2}{n} \right)} \tag{4.50}$$

The equivalency of the slopes b_1 and b_2 of two reduced major axis lines may be tested by

$$z = \frac{b_1 - b_2}{\sqrt{se_{b_1}^2 - se_{b_2}^2}} \tag{4.51}$$

which will be recognized as a variant of one of the elementary tests discussed in Chapter 2. The test statistic z is approximately normally distributed and its significance can be determined from a table of the standardized normal distribution.

The standard error of the intercept is

$$se_{b_0} = s_y \sqrt{\frac{1 - r^2}{n} \left(1 + \frac{\overline{X}^2}{s_x^2} \right)} \tag{4.52}$$

Equation (4.52) may be used to construct an approximate confidence interval around the computed value of b_0. Similarly, the standard error of the slope coefficient (Eq. 4.50) may be used to determine an approximate confidence interval around b_1. Other, essentially *ad hoc*, tests of the coefficients of the reduced major axis have been used. Because of the lack of theoretical underpinnings, however, the reduced major axis should be used primarily for descriptive purposes and not for tests of statistical significance.

Structural analysis and orthogonal regression

The relationship between ordinary regression, inverse regression, principal axes, and reduced major axes can be clarified by introducing another term, λ, which is equal to the ratio between the error variances of X and Y:

$$\lambda = \frac{\sigma_{e_y}^2}{\sigma_{e_x}^2} \tag{4.53}$$

Using this ratio, we can formulate a progression of line-fitting techniques which depend upon the relative error variances of the two variables being fitted. If $\lambda = \infty$, this means that X is measured without error and ordinary regression is appropriate. If $\lambda = 0$, then Y is measured without error and inverse regression is the proper choice. Between these two extremes, we may have $\lambda = 1$, which indicates that the two variables have the same error variances and the principal axis solution is appropriate. If the error variances are considered proportional to the variances of the variables themselves, their ratio will be equal to the ratio between the variances of X and Y, and the slope will be equal to the square root of this ratio, or the RMA solution. In addition, there is a complete range of possible intermediate ratios of the error variances. This methodological spectrum is referred to as **structural analysis** (Mark and Church, 1977; Kendall and Stuart, 1979; Fuller, 1987) or **orthogonal** *regression* (Carroll, Ruppert, and Stefanski, 1995). The approach requires that we know (or can reliably estimate) the error variances of the two variables and want to explicitly consider these variances in our analysis. The objective of a structural analysis is not to predict either expected values or individual observations (ordinary regression is appropriate for these purposes), but to estimate the numerical value of the β_1 coefficient. That is, to estimate the relationship between X and Y for comparison with some theoretical model.

Once we have established the ratio λ, we can use it to estimate β_1 from the corrected sums of squares and products of X and Y. The estimate is

$$b_1 = \frac{SS_y - \lambda SS_x + \sqrt{\left(SS_y - \lambda SS_x\right)^2 + 4\lambda SP_{xy}^2}}{2 SP_{xy}} \tag{4.54}$$

If we first fit an ordinary regression of X on Y and estimate the slope coefficient (which we will denote $b_{1(x|y)}$ to avoid confusion) and the correlation, r, we can calculate b_1 from these values:

$$b_1 = \frac{\left(\frac{b_{1(x|y)}}{r^2} - \lambda\right) + \sqrt{\left(\frac{b_{1(x|y)}}{r^2} - \lambda\right)^2 + 4\lambda b_{1(x|y)}^2}}{2 b_{1(x|y)}} \tag{4.55}$$

So, the problem is to estimate the error variance of X and of Y. If we have multiple measurements of x_i for given values of y_i, we can estimate the error variance of X as

$$s_{e_x}^2 = \frac{\sum\limits_{j=1}^{m} \left(x_{ij} - \overline{X}_i\right)^2}{m} \tag{4.56}$$

where the x_{ij} are m repeated measurements of X for a fixed value of y_i, and $\overline{X}_i$ is the average of the repetitions. The error variance of Y is estimated in a similar way, from the deviations of repeated measurements of y_i from their local mean.

Unfortunately, we seldom have the necessary repeated measurements to estimate the error variances, but sometimes we can draw on external information for their estimation. Other studies may provide estimates, or estimates may be based on accepted tolerance limits by assuming that the tolerance interval is equivalent to a 95% confidence band. Then, the error variance can be estimated as

$$s_{e_x}^2 \approx \left[(UTL_x - LTL_x)/1.96\right]^2 \tag{4.57}$$

Another possibility is to estimate λ as the ratio between the mean squared errors from ordinary regression and inverse regression; that is, by

$$\lambda = \frac{MS_{EY}}{MS_{EX}} \tag{4.58}$$

Since both these fitted regressions are certainly wrong (otherwise, why attempt to estimate λ?), the ratio is also incorrect, but may serve as a tentative estimate in the absence of better information.

Structural analysis should be used with caution, and only in those circumstances where the testing of a relational coefficient for conformity with a theoretical model is the objective, and the error properties of the variables are well understood. In most circumstances, when the purpose is to predict one variable from another, ordinary regression is the appropriate tool.

Regression through the origin

Sometimes we know that in a cross plot of one variable against another, when one variable goes to zero, the other must also be zero. For example, in a study of the distribution of pore sizes in a sedimentary rock, the pressure applied in a mercury porosimeter is plotted against the proportion of mercury injected into the pore space. If no pressure is applied, no mercury is injected, and it seems logical that any fitted line describing the relationship between pressure and injected mercury must pass through the origin of the graph. The fitted line would be defined by an equation with only one coefficient to be estimated:

$$y_i = \beta_1 x_i + \varepsilon_i \qquad (4.59)$$

The slope coefficient of a line constrained to go through the origin can be estimated by the normal equation

$$b_1 = \frac{\sum x_i y_i}{\sum x_i^2} \qquad (4.60)$$

Note that the terms in the normal equation are the raw sum of products of X and Y and the raw sum of squares of X, not the mean-corrected sums of squares and products used to estimate the slope in ordinary linear regression. This means that, in general, a regression through the origin will not pass through the bivariate mean. In turn, this implies that constraining the regression to pass through the origin introduces a bias, and may severely and adversely affect the goodness of fit (*i.e.*, the deviations from a constrained-fit line will be greater). We can see this in **Figure 4–31** (based on data in file HGCURVE.TXT), which shows injection pressures versus proportion of mercury injected into the pore space of a core sample of Jurassic limestone from an oil field in Florida. Both an ordinary regression of proportion of mercury injected on pressure and a fitted line constrained to pass through the origin are shown. Note that if an ordinary fitted regression does seem to pass through the origin, a confidence interval (Eq. 4.36) around b_0 should include zero.

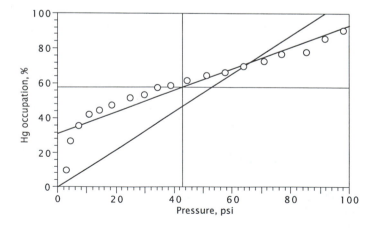

Figure 4–31. Injection pressure versus percentage of pore space occupied by mercury, from a mercury porosimeter study of a Jurassic limestone from Florida.

Fitting a regression through the origin often seems the logical thing to do because of the physical nature of the variables, and assuming linearity conveniently makes the equations more tractable. However, many relationships—especially those in engineering—only appear to be linear, while the actual phenomena behave in a far more complex manner. Fitting a straight line through the observations may give a completely misleading impression of the relationship between the variables being compared, and constraining the line to the origin may make the situation worse.

Logarithmic transformations in regression

In Chapter 2 we examined the lognormal and other distributions that are obtained when a variable is transformed, and noted that, after transformation, the data may resemble a normal distribution more closely than they did originally. Similarly, one or both variables in a regression may be transformed so that their joint distribution lies more closely along a straight line. This may provide an alternative to curvilinear regression or, as is often the case in engineering applications, transformation may be used to estimate a bivariate relationship when the distribution of one (or both) of the variables is highly skewed.

If only the independent variable X is transformed, the fundamental nature of ordinary regression is unchanged, because all of the variances are defined in terms of Y. However, if we transform Y, or both X and Y, the least-squares property of regression may be invalidated and a fitted regression may not exhibit the expected behavior.

Figure 4–32 shows a series of cross plots of hypothetical data. In a, the slope of the cloud of data points is lower at high values of X; by transforming X to its logarithm, the horizontal scale is compressed at the upper end and expanded at the lower, as in b. Since Y is unchanged, the properties of the regression are unaffected by the transformation. In c, the distribution of X is fairly uniform, but Y is skewed toward high values. A logarithmic transform of Y will reduce the extreme values, as in d. However, this changes the form of the regression model, from

$$y_i = b_0 + b_1 x_i$$

to

$$\log y_i = b_0 + b_1 x_i \tag{4.61}$$

or equivalently, to an exponential model

$$y_i = 10^{(b_0 + b_1 x_i)} \tag{4.62}$$

More seriously, although the deviations of log Y will be symmetrical about the fitted line and will sum to zero, and the sum of the squared deviations of log Y will be the smallest possible for any fitted line, these conditions do not hold for the deviations of Y itself. Positive deviations of Y (those above the fitted line) will be much greater than the negative deviations because of the logarithmic scaling. The fitted line (which is a curved line in linear space) will be biased toward low values and will tend to underestimate Y, and the underestimation will grow worse as the values of $\hat{Y}$ increase. We will address this problem later.

In e on **Figure 4–32**, both X and Y are skewed and plot along a sharply curved line. If logarithms of both variables are taken, it may succeed in straightening their

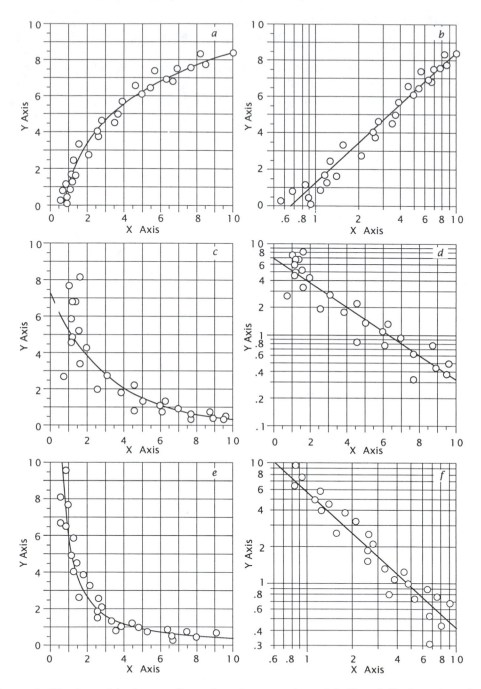

Figure 4–32. Logarithmic transformations in regression. (*a*) X and Y plotted on linear scales. (*b*) X in (*a*) transformed to log X. Regression is $y_i = 1.27 + 7.17 \log x_i$. (*c*) X and Y plotted on linear scales. (*d*) Y in (*c*) transformed to log Y. Regression is $\log y_i = 0.85 - 0.13 x_i$. (*e*) X and Y plotted on linear scales. (*f*) Both X and Y in (*e*) transformed to logarithms. Regression is $\log y_i = 0.74 - 0.12 \log x_i$.

relationship, as in the log–log plot shown in f. Again, because we have transformed the dependent variable Y, the regression model has been changed and the usual least-squares properties have been altered. The new model is

$$\log y_i = b_0 + b_1 \log x_i \qquad (4.63)$$

which is algebraically the same as the multiplicative model,

$$y_i = b_0 x_i^{b_1} \qquad (4.64)$$

The deviations of log Y from the fitted line are symmetric, although the deviations of Y are not. (This discussion is given in terms of common logarithms, however the same circumstances hold if natural logarithms are used, or even logarithms to base 2 or another base.)

Logarithms are used in many empirical relationships in engineering. This does not necessarily reflect a belief that the relationships are truly multiplicative or exponential, but is simply a holdover from pre-computer days when graphical methods were the only practical way to solve many problems. By plotting data on log–linear or log–log graph paper, complex patterns could be resolved into simple straight-line relationships and extrapolation became much easier. With the advent of personal computers, these graphical methods have been implemented as programs that fit lines by least squares, with little recognition of the underlying statistical assumptions.

Hydrologists have installed an automated data logger on an experimental well in an alluvial aquifer in the Kansas River valley to determine drawdown as the well is pumped. The data, collected at 5-sec intervals, are shown in **Figure 4–33**, and listed in file DRAWDOWN.TXT. This record exhibits a typical pattern of rapid initial drop in water level, with decline slowing to an approximately constant rate dependent upon the conductivity of the aquifer. If the data are transformed by taking the logarithms of time, an approximately linear pattern is revealed (**Fig. 4–34**). A regression can be fitted to these data to provide an estimate of the slope of the drawdown and, in turn, used to estimate the conductivity of the aquifer. **Table 4–17** is a combined ANOVA for the regression of drawdown on time, and on the logarithm of time. Although both regressions are significant, the regression on the logarithm of time is significantly better.

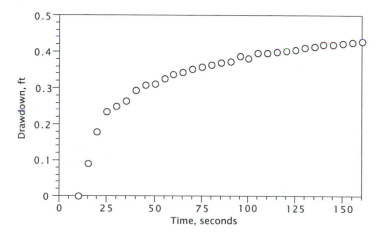

Figure 4–33. Drawdown curve for an observation well pumping from an alluvial aquifer in the Kansas River valley. Drop in water level recorded at 5-sec intervals by an automatic data logger after onset of pumping.

223

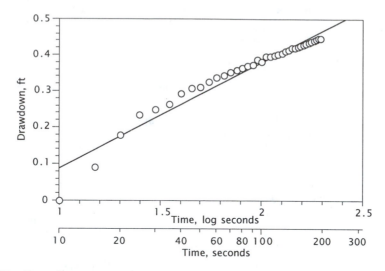

Figure 4–34. Drawdown curve plotted against log time, with fitted regression line. (In practice, early drawdown observations may be discarded.)

Table 4–17. ANOVA for logarithmic regression and linear regression of water-level decline on time.

Source of Variation	Sum of Squares	Degrees of Freedom	Mean Squares	F-Tests
Logarithmic Regression	0.345958	1	0.345958	
Logarithmic Deviation	0.018310	36	0.000509	680.19 [a]
Linear Regression	0.263491	1	0.263491	
Linear Deviation	0.100777	36	0.002799	94.13 [a] 5.50 [b]
Total Variation	0.364268	37		

[a] $p < 0.0001$.
[b] MS_{E2}/MS_{E1} tests equality of two error variances, $p < 0.001$.

Weighted regression

In ordinary regression, each observation has an equal influence on the position and orientation of the fitted line. However, some observations may be more important or more reliable than others, and should be given greater influence in the regression. This can be done by **weighting**, which simply involves multiplying each y_i by a quantity that reflects its relative importance. The normal equations for ordinary regression become

$$b_1 = \frac{\sum w_i \sum w_i x_i y_i - (\sum w_i x)(\sum w_i y_i)}{\sum w_i \sum w_i x_i^2 - (\sum w_i x_i)^2}$$

(4.65)

and

$$b_0 = \frac{\sum w_i y_i - b_1 \sum w_i x_i}{\sum w_i} \tag{4.66}$$

Note that if we set each weight equal to 1, the w_i vanish and Equations (4.65) and (4.66) become identical to the normal equations (Eqs. 4.15 and 4.16).

We often wish to compare one regression with another. This may be difficult if each regression has different weights assigned to variables. To avoid this problem we can scale the weights so their sum is equal to the number of observations. Each individual weight w_i is scaled by

$$\hat{w}_i = \frac{n w_i}{\sum w_i} \tag{4.67}$$

where n is the number of observations. Scaling the weights has no effect on statistics such as the correlation coefficient, r, or on the results of tests of significance, confidence levels, or probabilities, although the various sums of squares may be quite different.

An important application of weighted regression is to correct for the bias that occurs when the dependent variable, y_i, is transformed to $\log y_i$. Without correction, the fitted regression will underestimate y_i; the underestimation will be more severe as y_i becomes larger. In ordinary regression, the sum of the squared deviations from the fitted line (the residuals or "errors") is minimized.

$$\sum e_i^2 = \sum (y_i - \hat{y}_i)^2$$
$$= \sum [y_i - (b_0 + b_1 x_i)]^2 = \text{minimum}$$

If the dependent variable is transformed to $\log y_i$, the minimization is changed to

$$\sum \delta_i^2 = \sum (\log y_i - \log \hat{y}_i)^2$$
$$= \sum [\log y_i - \log(b_0 + b_1 x_i)]^2$$
$$= \sum (y_i - 10^{(b_0 + b_1 x_i)})^2 = \text{minimum}$$

The logarithmic transformation yields deviations (in linear space) approximately equal to $(\delta_i / y_i)^2$. Taking the logarithms of y_i has introduced a weighting equal to the inverse of the square of y_i; that is, we are minimizing the error relative to the magnitude of y_i^2, rather than the absolute error.

To adjust the regression so the observations have equal weights, we must apply a weighting equal to y_i^2. In scaled form, the weights are

$$\hat{w}_i = \frac{n y_i^2}{\sum y_i^2} \tag{4.68}$$

Calculating the regression with either raw weights (i.e., $w_i = y_i^2$) or standardized weights, $\hat{w}_i$, will not alter the fitted line or the goodness of fit.

File PORPERM.TXT contains 44 porosity and permeability measurements from a well drilled in eastern Oklahoma. **Figure 4–35 a** shows these data plotted as the logarithm of permeability against porosity. A fitted regression of log permeability is also shown and, as expected, the fitted line goes through the approximate center of the cloud of points and follows the trend of increasingly greater values of log

225

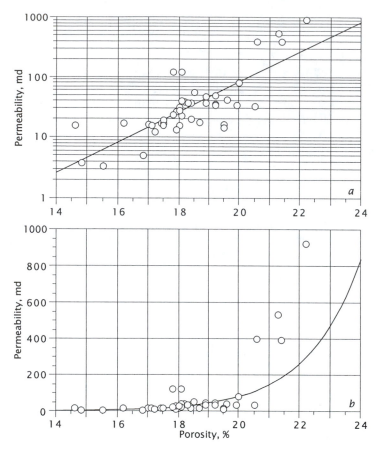

Figure 4–35. Permeability versus porosity measured on cores from a well in eastern Oklahoma. (*a*) Ordinary regression of log permeability on porosity. (*b*) Observations and
fitted regression from (*a*) transformed to linear scale of permeability.

permeability with higher porosities. However, when the Y-axis is transformed back
to a linear scale, as in **Figure 4–35 b**, we can see that the fitted regression (now in
the form of a curved line) underestimates all of the higher values of permeability.
If we determine the residuals ($\log \hat{y}_i - \log y_i$) in log space and take the antilogs of
the differences, we will see that these residuals are strongly skewed in a positive
direction and have a high variance ($s_e^2 = 14741.5$).

The fitted regression can be adjusted for the bias introduced by logarithmic
transformation by weighting each observation by $w_i = y_i^2$. A permeability/porosity
cross plot remains unchanged, but the fitted regression is altered by the weighting
(**Fig. 4–36 a**). The regression line no longer goes through the center of the cloud of
points, but instead is near the upper range of values. The slope of the fitted line
also is somewhat steeper. However, when the Y-axis is changed from a logarithmic
to a linear scale (**Fig. 4–36 b**), we see that the weighted regression closely fits the
higher values of permeability and porosity. Although the line overestimates low
values of porosity and permeability, these deviations are small in magnitude. If
we find the residuals ($\log \hat{y}_i - \log y_i$) and take their antilogs, we now find that the
residuals are nearly symmetrical and their variance is ($s_e^2 = 4394.6$), which is only
a third of the error variance of the unweighted regression.

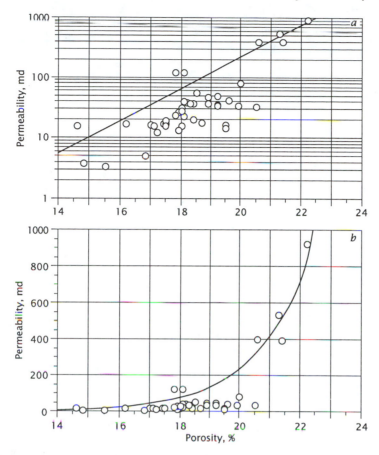

Figure 4–36. Permeability versus porosity measured on cores from a well in eastern Oklahoma. (*a*) Weighted regression of log permeability on porosity. (*b*) Observations and fitted weighted regression from (*a*) transformed to linear scale of permeability.

Draper and Smith (1998) provide a succinct discussion of weighted regression and least squares in matrix algebra formulation. Etnyre (1984a, b) gives an excellent two-part discussion of weighted regression with applications in well-log analysis.

Looking at residuals

The basic scatter diagram in regression is a cross plot of Y versus X. If we plot the regression line itself, it forms a cross plot of $\hat{Y}$ versus X. There are other cross plots that can be made that are useful for checking that the assumptions of regression are met. Among these are plots of the residuals $(\hat{Y} - Y)$ against X or the residuals $(\hat{Y} - Y)$ against the estimate $\hat{Y}$. Either of these will disclose systematic departures of the data from the fitted regression model (lack of fit) or systematic increases or decreases in variance (**Fig. 4–37**).

We also can plot the residuals against cumulative probability in order to determine if the residuals are normally distributed (**Fig. 4–38**). Although the least-squares procedure assures us that a fitted regression will retain its optimal properties even if the residuals are not normally distributed (provided they are symmetric), the calculation of confidence intervals and use of tests of significance assumes

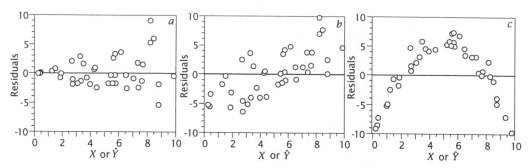

Figure 4–37. Plots of residuals from regression against $\hat{Y}$ or X. (*a*) Indicates variance of Y is not constant. Can possibly be corrected by $\log Y$ transformation or use of weighted regression. (*b*) Indicates systematic lack of fit, possibly caused by forcing a linear regression through the origin (no constant term). (*c*) Indicates systematic lack of fit caused by an inadequate regression model. Can possibly be corrected by use of polynomial regression or logarithmic transformation of X.

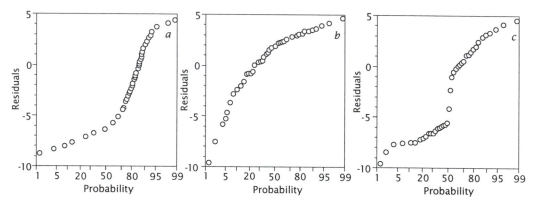

Figure 4–38. Cumulative probability plots of residuals from regression. (*a*) Indicates positively skewed residuals. Can possibly be corrected by logarithmic transformation of Y or by trimming extreme values. (*b*) Indicates negatively skewed residuals. Can possibly be corrected by use of an inverse log or square root transformation of Y. (*c*) Indicates bimodal distribution of residuals. Suggests that Y consists of a mixed sample from two or more populations.

normality of the residuals. The pattern of the cumulative probability distribution may suggest corrections or transformations that will improve the regression.

Splines

Some data may be conveniently thought of as strings of coordinate pairs. That is, the observations consist of measurements on two properties, and can be envisioned as defining a sequence of points in a two-dimensional space. For purposes of presentation or analysis, it may be desirable to connect these points with a smooth, continuous line. We can do this by the use of ***spline functions***.

Splines are one of a large class of piecewise functions that can be used to represent curves in two or three dimensions. The mathematical spline gains its name from a physical counterpart, the flexible drafting spline made from a narrow strip of wood or plastic that can be bent to conform to an irregular shape. A drafting spline is held by lead weights called "ducks," which fix the position of the spline at their points of attachment. Between the ducks, the spline flexes into a smooth, continuous form. A mathematical spline is similarly constrained at defined points, but between the points it flexes in a manner that results in a smoothly varying line.

Splines are not analytical functions, nor are they statistical models such as the polynomial regressions described earlier. Rather, they are purely arbitrary and devoid of any theoretical basis except that which defines the characteristics of the lines themselves. They are, however, extremely useful for interpolation and are important in software for generating computer displays. Interactive computer graphics systems are widely used for geological and geophysical modeling, and spline fitting plays an important role in this software.

Splines are piecewise polynomials that are constrained to have continuous derivatives at the joints between the pieces or segments. The most common spline consists of cubic polynomials, which are functions of the form

$$Y = \beta_1 + \beta_2 X + \beta_3 X^2 + \beta_4 X^3$$

The curve defined by a cubic polynomial can pass exactly through four points, but in order to fit a longer sequence it is necessary to use a succession of polynomial segments. To insure that there are no abrupt changes in slope or curvature between successive segments, the polynomial function is not fitted to four points, but only to two. This allows us to use additional constraints which will insure that the resulting spline has continuous first derivatives between segments (the slope of the line will be the same on either side of a joint) and continuous second derivatives (the rate of change in the slope of the line will not change across a joint). In general, a spline of degree m will have continuous derivatives across the points up to order $m - 1$.

Developing the spline equations requires knowledge of differential calculus, a skill not presumed for this book. Therefore, we will simply present the necessary equations in computational form and work through their application; those interested in their derivation are referred to the excellent introductory texts on computer graphics by Mortenson (1989) and Rogers and Adams (1990), and to the monograph on geologic applications of surface fitting techniques by Tipper (1979).

The mathematical notation used with spline functions is somewhat confusing, but can be clarified with the help of **Figure 4–39**, which shows a set of four observations connected by a piecewise spline function. The observations are plotted as points, and are symbolically indicated as P_i, with the understanding that P is actually a vector of Cartesian coordinates. That is, $P_i = x_i, y_i$. The intervals between successive points are referred to as **spans**; the chord or straight-line distance between two points is indicated as t_i with i assuming the value of the second point. A cubic spline function covers a single span or two points; the illustration, therefore, shows three successive splines, one from point P_1 to P_2, from P_2 to P_3, and from point P_3 to P_4.

In general form, the spline equation may be written as

$$\hat{P}_t = \beta_1 + \beta_2 t + \beta_3 t^2 + \beta_4 t^3 \tag{4.69}$$

which states that the coordinates of the spline at some distance t along a span are equal to a cubic polynomial function of t. To determine the coefficients, we must know the coordinates of the points, which define the ends of the splines, and the slopes of tangent lines at the points. In addition, we must specify end conditions that determine the behavior of the line in the first and last spans. The

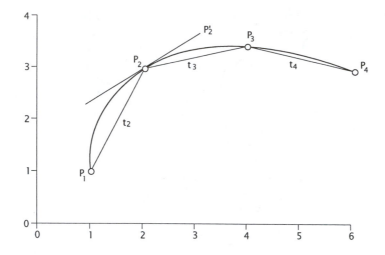

Figure 4–39. Four points connected by cubic spline function. Original observations are labeled P_i. Chord distances between points are t_i. Tangent to spline at interior point P_2 is indicated as P_2'.

point coordinates, of course, are given. The slopes, expressed as tangent vectors, must be determined. Various end conditions can be selected, depending upon the desired shape of the line at its terminus. We will consider only what are called *relaxed* or *natural* end conditions that do not require specifying tangent vectors at the end points.

To find the tangent vectors at the interior points (P_2 and P_3 of **Fig. 4–39**), we must solve a set of simultaneous equations of the form

$$\mathbf{M}\,\mathbf{P}' = \mathbf{B} \qquad (4.70)$$

where the unknown vector of $\mathbf{P}'$ coefficients are the desired tangents. The left-hand matrix that must be inverted is *tridiagonal*; that is, all elements are zero except for the diagonal elements and those immediately adjacent to the diagonal. For relaxed end conditions, $\mathbf{M}$ has the form of an n by n matrix:

$$\mathbf{M} = \begin{bmatrix} 1.0 & 0.5 & 0 & 0 & 0 & \cdots & 0 \\ t_3 & 2(t_2 + t_3) & t_2 & 0 & 0 & \cdots & 0 \\ 0 & t_4 & 2(t_3 + t_4) & t_3 & 0 & \cdots & 0 \\ 0 & 0 & t_5 & 2(t_4 + t_5) & t_4 & \cdots & 0 \\ 0 & 0 & 0 & t_6 & 2(t_5 + t_6) & \cdots & 0 \\ \vdots & \vdots & \vdots & \vdots & \vdots & \vdots & \vdots \\ 0 & 0 & 0 & 0 & 0 & 2 & 4 \end{bmatrix} \qquad (4.71)$$

The right-hand vector, B, has the form

$$B = \begin{bmatrix} \frac{3}{2t_2}(P_2 - P_1) \\ \frac{3}{t_2 t_3}[t_2^2(P_3 - P_2) + t_3^2(P_2 - P_1)] \\ \frac{3}{t_3 t_4}[t_3^2(P_4 - P_3) + t_4^2(P_3 - P_2)] \\ \frac{3}{t_4 t_5}[t_4^2(P_5 - P_4) + t_5^2(P_4 - P_3)] \\ \frac{3}{t_5 t_6}[t_5^2(P_6 - P_5) + t_6^2(P_5 - P_4)] \\ \vdots \\ \frac{6}{t_n}(P_n - P_{n-1}) \end{bmatrix} \quad (4.72)$$

The matrix equation is solved by inverting M and postmultiplying the inverse by B. Note that since the point coordinates P_i are double-valued (that is, each consists of a value for X and a value for Y), the right-hand matrix B is $n \times 2$, where n is the number of points to be included in the set of splines. Equation (4.72) shows the form of the terms in B. The first column of B is found by inserting the appropriate chord distances, t_k, and the X-coordinates of the observations. The second column is formed in the same manner, but using the Y-coordinates. Similarly, the solution matrix P' is also $n \times 2$. Each row of P' represents the slope of a tangent to the spline at the observation point, given in terms of X and Y.

To find the four β coefficients that define the kth spline (that is, the span between point P_k and point P_{k+1}), we set

$$\beta_1 = P_k$$
$$\beta_2 = P'_k$$
$$\beta_3 = \frac{3(P_{k+1} - P_k)}{t_{k+1}^2} - \frac{2P'_k}{t_{k+1}} - \frac{P'_{k+1}}{t_{k+1}} \quad (4.73)$$
$$\beta_4 = \frac{2(P_k - P_{k+1})}{t_{k+1}^3} + \frac{P'_k}{t_{k+1}^2} + \frac{P'_{k+1}}{t_{k+1}^2}$$

Finally, when the four coefficients are found for the kth span, points along the curve within this interval can be determined. The length of the chord between points k and $k + 1$ can be divided into a convenient number of parts and these successive distances inserted as t in Equation (4.69). This will provide a set of equally spaced coordinates that can be connected to form the spline curve. The process is repeated for each segment of the piecewise spline, using the appropriate slopes at the interior points, chord lengths, and point coordinates to find a new set of coefficients for each span.

To demonstrate the fitting of cubic splines, we will use four points shown in **Figure 4–40** whose coordinates are

$$P = \begin{bmatrix} 1 & 1 \\ 1 & 3 \\ 4 & 3 \\ 3 & 1 \end{bmatrix}$$

The chord lengths are $t_2 = 2.0$, $t_3 = 3.0$, and $t_4 = 2.236$. These distances are all that are required to form the matrix M, defined in Equation (4.71):

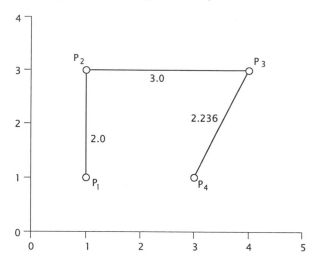

Figure 4–40. Four points to be fitted by cubic spline function. Chord lengths of spans between points are indicated.

$$\mathbf{M} = \begin{bmatrix} 1.0 & 0.5 & 0 & 0 \\ 3.0 & 10.0 & 2.0 & 0 \\ 0 & 2.236 & 10.472 & 3.0 \\ 0 & 0 & 2.0 & 4.0 \end{bmatrix}$$

The inverse of **M** is

$$\mathbf{M}^{-1} = \begin{bmatrix} 1.1875 & -0.0625 & 0.0139 & -0.0104 \\ -0.3749 & 0.1250 & -0.0279 & 0.0209 \\ 0.0934 & -0.0311 & 0.1184 & -0.0888 \\ -0.0467 & 0.0156 & -0.0592 & 0.2944 \end{bmatrix}$$

We must also determine the right-hand vector, **B**. The necessary information to find the elements of **B** consists of the chord lengths and the coordinates of the points. Since each point consists of two coordinates, the vector **B** has two columns, the first for X and the second for Y:

$$\mathbf{B} = \begin{bmatrix} \frac{3}{2\cdot2}(1-1) & \frac{3}{2\cdot2}(3-1) \\ \frac{3}{2\cdot3}[2^2(4-1)+3^2(1-1)] & \frac{3}{2\cdot3}[2^2(3-3)+3^2(3-1)] \\ \frac{3}{3\cdot2.236}[3^2(3-4)+2.236^2(4-1)] & \frac{3}{3\cdot2.236}[3^2(1-3)+2.236^2(3-3)] \\ \frac{6}{2.236}(3-4) & \frac{6}{2.236}(1-3) \end{bmatrix}$$

$$= \begin{bmatrix} 0 & 1.5 \\ 6 & 9 \\ 2.683 & -8.050 \\ -2.683 & -5.367 \end{bmatrix}$$

Multiplying **B** by $\mathbf{M}^{-1}$ yields **P'**:

$$\mathbf{P'} = \begin{bmatrix} -0.3097 & 1.2026 \\ 0.6187 & 0.6750 \\ 0.3723 & -0.6160 \\ -0.8552 & -1.0328 \end{bmatrix}$$

We now have all of the terms necessary to calculate the spline coefficients for each of the spans in our example. For the first span, we may substitute the appropriate values of t, $\mathbf{P}$, and $\mathbf{P}'$ into the set of equations in (4.73) to obtain:
For the X-coordinate,

$$\beta_1 = 1$$
$$\beta_2 = -0.3097$$
$$\beta_3 = \frac{3(1-1)}{2^2} - \frac{2(-0.3097)}{2} - \frac{0.6187}{2} = 0.0004$$
$$\beta_4 = \frac{2(1-1)}{2^3} + \frac{(-0.3097)}{2^2} + \frac{0.6187}{2^2} = 0.0773$$

For the Y-coordinate,

$$\beta_1 = 1$$
$$\beta_2 = 1.2026$$
$$\beta_3 = \frac{3(1-1)}{2^2} - \frac{2(1.2026)}{2} - \frac{0.6750}{2} = -0.0401$$
$$\beta_4 = \frac{2(1-3)}{2^3} + \frac{(1.2026)}{2^2} + \frac{0.6750}{2^2} = -0.0306$$

or

$$\mathbf{B} = \begin{bmatrix} 1 & 1 \\ -0.3097 & 1.2026 \\ 0.0004 & -0.0401 \\ 0.0773 & -0.0306 \end{bmatrix}$$

In a similar manner, we can determine the spline coefficients for spans 2 and 3. These are

$$\begin{bmatrix} 1 & 3 \\ 0.6187 & 0.6750 \\ 0.4634 & -0.2447 \\ -0.1121 & 0.0066 \end{bmatrix} \quad \begin{bmatrix} 4 & 3 \\ 0.3723 & -0.6160 \\ -0.5506 & -0.1872 \\ 0.0823 & 0.0286 \end{bmatrix}$$

Finally, we can use the spline coefficients to determine the coordinates of intermediate points on the splines between each observation. If we calculate a large number of such points and connect them with straight lines, the visual result will be an apparently smooth, continuous curve. This is the way in which a computer graphics system calculates and draws smoothly curving lines. For purposes of demonstration, we will content ourselves with only three intermediate points on each spline.

To find these intermediate points, we first divide each chord into four parts. The distances at $t_k/4$, $2t_k/4$, and $3t_k/4$ will define the values of t to be inserted into the spline equation. For the first spline, these distances are 0.5, 1.0, and 1.5. Inserting into Equation (4.69), first for X and then for Y,

$$\hat{P}_{0.5X} = 1 - 0.3097(0.5) + 0.0004(0.5^2) + 0.0773(0.5^2) = 0.8549$$
$$\hat{P}_{0.5Y} = 1 + 1.2020(0.5) - 0.0401(0.5^2) - 0.0306(0.5^2) = 1.5874$$

We may now compute the coordinates of the first spline at distances $t = 1.0$ and $t = 1.5$. These are

$$\text{for } t_{1.0}, \quad \begin{bmatrix} 0.7679 & 2.1319 \end{bmatrix}$$

$$\text{for } t_{1.5}, \quad \begin{bmatrix} 0.7969 & 2.6104 \end{bmatrix}$$

The process is now repeated for the second and third splines, yielding the sets of coordinates

$$\text{for spline 2,} \quad \begin{bmatrix} 1.6774 & 3.3714 \\ 2.5924 & 3.4841 \\ 3.4612 & 3.3548 \end{bmatrix}$$

$$\text{for spline 3,} \quad \begin{bmatrix} 4.0500 & 2.6020 \\ 3.8431 & 2.1163 \\ 3.4642 & 1.5722 \end{bmatrix}$$

These are shown plotted in **Figure 4–41**. Also shown is the smooth spline generated by calculating 30 intermediate points between each joint. Although the process of determining the spline coefficients is involved, once they have been found it is a relatively trivial matter to calculate as many points as desired along the curve.

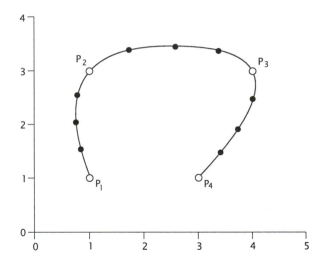

Figure 4–41. Smooth spline function consisting of 30 segments between each point of Figure 4–40. The three intermediate points on each spline calculated in text are shown as solid dots.

Segmenting Sequences

Zonation

The dividing of a sequence into relatively uniform segments, each of which is distinctive from adjacent segments, is referred to as **zonation**. Paleontologists, for example, may want to zone a stratigraphic sequence on the basis of consistent abundances of microfossils. Well logs may be subdivided into relatively uniform intervals that represent zones of constant lithology, corresponding to stratigraphic units. Airborne radiometric traverses may be subdivided into zones that can be interpreted as belts of uniform rock composition or mineralization.

There are basically two contrasting approaches to zonation. The simplest are local, interactive procedures that begin at one end of a sequence and progressively move to the other end, identifying abrupt changes in average value or steep gradients. The number of intervals that may be found is not predetermined. Such a "local boundary hunting" process is akin to filtering. It involves examining the elements of the sequence within a short interval, computing some value which characterizes that interval, then moving the interval of examination down the sequence by one position and repeating the process. Webster (1973) called the moving interval a "window" consisting of two parts, a segment from point $(i - h)$ on the sequence to point i, and another segment from point i to point $(i + h)$. The difference in average value within the two segments is calculated, perhaps scaled to account for the variance in the interval, and plotted against the position of the midpoint, i. When the difference exceeds some threshold value, a boundary is decreed.

Alternative difference measures have been suggested, depending upon the number of variables in the sequence and the assumptions made about the statistical nature of the sequence. If the log or traverse consists of only a single variable, the *generalized distance* may be calculated for the difference between the segments within the two halves of the window. This is the ratio of the squared difference between the means of the two segments to the pooled variance within the segments. We can denote the mean of the segment from x_{i-h} to x_i as $\overline{X}_1$ and its variance as s_1^2; the mean of the segment from x_i to x_{i+h} is $\overline{X}_2$ and the variance is s_2^2. Then, the generalized difference is

$$D^2 = \frac{\left(\overline{X}_1 - \overline{X}_2\right)^2}{s_1^2 + s_2^2} \tag{4.74}$$

Note that the pooled variance is simply the sum of the variances of the two segments, because both segments contain the same number of observations. Also note that the first and last h of the points in the sequence cannot be separated into different zones.

A plot of h versus D^2 will result in transformation of the original traverse into a new sequence in which zone boundaries appear as sharp spikes. **Figure 4-42a** shows values along a 6-km traverse in the upper Thames valley of England. The original multivariate measurements on soils have been condensed into a single variable by principal component analysis, which will be discussed extensively in Chapter 6. **Figure 4-42b** is a plot of D^2 along the same traverse. Abrupt changes in the first principal component result in conspicuous spikes in D^2, which correspond with boundaries between soil types. The illustration is adapted from Webster (1973).

Webster noted that the performance of the procedure depends upon the variability of the original sequence and the length of the moving window. A long window will average across small zones and may miss short intervals. However, it will subdue the erratic variation of a noisy original record. A short window is more sensitive and will identify small zones, but may create an irregular, uninterpretable plot of D^2. Generalizing D^2 to analyze a record consisting of multiple variables is straightforward and will be considered in more detail in Chapter 6. Here we merely note that there is more than one mean for each side of the split window, so the difference between the two segments is represented by a vector. Similarly, the pooled variance becomes a matrix of pooled variances and covariances, **S**. The operation equivalent to Equation (4.74) is given by the matrix equation

$$D^2 = (\overline{X}_1 - \overline{X}_2)' S^{-1} (\overline{X}_1 - \overline{X}_2) \tag{4.75}$$

Webster (1980) reports that in some instances D^2 is more indicative of boundaries if it is not corrected for the variance within the window. It then becomes simply

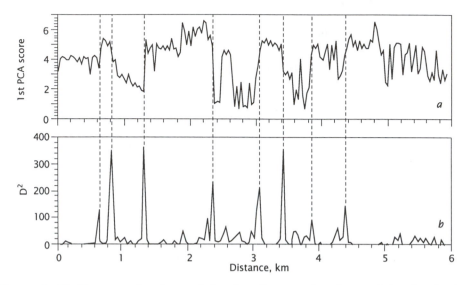

Figure 4–42. Transect showing variation in soil properties along a 6-km line in the upper Thames valley, England. (*a*) Variation in first principal component of 27 soil properties. (*b*) Values of D^2 along traverse. Maxima define boundaries shown on (*a*). After Webster (1973).

the squared Euclidean distance, E^2, equal to

$$E^2 = (\overline{X}_1 - \overline{X}_2)'(\overline{X}_1 - \overline{X}_2) \tag{4.76}$$

Webster also discusses the relative merits of standardizing the variables throughout the extent of the sequence, or converting the variables to principal components or canonical variates, both topics to be discussed in Chapter 6. It should be emphasized that local boundary hunting and all of its variations are *ad hoc* procedures whose merits, if any, lie in their success in producing a usable zonation.

One objection to local boundary hunting procedures is that they may find an inordinate number of boundaries, particularly within a highly variable part of the sequence. Global zonation is a different approach, using procedures that break the sequence into a specified number of segments which are as internally homogeneous as possible and as distinct as possible from adjacent segments. These procedures also are iterative, but consider the entire sequence at one time, not just the portion within a window.

One of the first, and still most practical, of these procedures was devised by Gill (1970), who used an iterative analysis of variance approach. First, the sequence is divided into two segments, a very short initial part, and the remainder of the sequence. The sum of squares within the segments, SS_W, is computed as

$$SS_W = \sum_{j=1}^{m} \sum_{i=1}^{n_j} \left(x_{ij} - \overline{X}_{\cdot j}\right)^2 \Big/ \sum_{j=1}^{m} n_j - m \tag{4.77}$$

where x_{ij} is the ith point within segment j, $\overline{X}_{\cdot j}$ is the mean of the jth segment, n_j is the number of points in the jth segment, and m is the number of segments. The sum of squares between segments, SS_B, is a measure of the variance of the segment means about $\overline{\overline{X}}_{..}$, the grand mean of the total sequence, or

$$SS_B = \sum_{j=1}^{m} \left(\overline{X}_{.j} - \overline{\overline{X}}_{..} \right) \Big/ m - 1 \qquad (4.78)$$

The partition between the two segments is moved along the sequence to successive positions and the two quantities SS_W and SS_B recomputed at each position. For every possible position of the boundary, the ratio

$$R = \frac{SS_B - SS_W}{SS_B} \qquad (4.79)$$

is calculated. The position corresponding to the maximum value of R is chosen as the location of the first zonal boundary.

Next, the two zones are themselves partitioned by repeating the process to insert an additional boundary which again maximizes the quantity R. The zonation is repeatedly run until the entire sequence is divided into the specified number of zones, or until the quantity R no longer increases with the addition of new boundaries.

Gill's procedure has been used as a way of automatically zoning digitized well logs. A similar procedure has been published by Hawkins and Merriam (1973, 1974) which uses global optimization, based on methods of dynamic programming. Their algorithm is iterative but also is recursive and takes advantage of Bellman's principle of optimality to insure that the final set of zone boundaries is the best possible of all sets of partitions that might have been chosen. With a nonrecursive procedure, it is always possible that the position selected as the best boundary between two zones is no longer the best when another boundary is inserted into one of the zones.

Hawkins and Merriam calculate a quantity that is the sum of the within-zone variances, equivalent to Gill's SS_W. If this quantity is computed for all possible partitions of the sequence into two segments, the result is a table of SS_W for $(n-1)$ possible locations for the first boundary. For each possible first partition, a new value of SS_W is then computed for all possible positions of a second boundary which would divide the sequence into three zones. By selecting the smallest value of SS_W for the second partition, the associated location for the first boundary is optimal, and will remain the best location no matter how many additional boundaries are inserted.

The process now iterates for the third cycle, and for every combination of the optimum first boundary with all possible second boundaries, all possible third boundary positions are found and their values of SS_W calculated. Selection of the smallest SS_W value then determines the optimum position for the second boundary. The process repeats again and again, until the specified number of boundaries has been found.

Because of the recursive nature of the algorithm, the final set of zones is guaranteed to have the smallest internal variance of any possible set of the same number of zones covering the total interval. This optimality is achieved at a high computational cost which formerly limited use of the process to relatively small sequences, a restriction now removed by faster computers.

Note that neither local boundary hunting procedures nor global zonation methods depend in any way on the distribution of the data, nor on sampling theory or other underlying assumptions. This is because zonation methods are not statistical; the objective of zonation is not to draw inferences about some population that is represented by a sample in the form of the observed sequence. The sequence itself is the object of interest, and questions of sampling and probability are not relevant. This is simultaneously a disadvantage and an advantage; there are no theoretical guides to a "best" zonation, but there are no constraints on what approaches may be tried (although some authors, such as Mehta, Radhakrishnan, and Srikanth, 1990, have attempted to cast zonation into a Bayesian framework). In this regard, zonation is very similar to cluster analysis, which will be examined in detail in Chapter 6. In fact, hierarchical cluster analysis can be modified to perform zonation of sequences. This simply involves introducing the constraint that only adjacent observations in a sequence can be clustered together.

In a well-log analysis program by Bohling and others (1998), a zonation procedure based on Ward's method of hierarchical cluster analysis has been implemented. The data consist of an $n \times m$ array of n observations of m variables along a traverse or well log; typically, n includes hundreds or even thousands of observations and m includes only a few different variables. The data are standardized so each variable has a mean of zero and a standard deviation of one. This is done so all variables have equal influence on the zonation. Prior to the initial iteration, every observation is considered to be an individual zone, so there are as many zones as there are points in the sequence. As a first step, we calculate a measure of total variance in the sequence,

$$SS_T = \sum_i^n x_{ij}'x_{ij} \tag{4.80}$$

(Because the observations have been standardized, they are already deviations from the grand mean.)

During the first iteration, the Euclidean distance between every observation at position i and its neighbor, $i + 1$, is determined by

$$E_i^2 = (x_{ij} - x_{i+1,j})'(x_{ij} - x_{i+1,j}) \tag{4.81}$$

This produces an array of $n - 1$ sequential differences. The sequence is examined to find the pair of points with the smallest difference; these are combined to form the start of the first composite zone. This zone will henceforth be treated as a single object defined by its vector mean

$$\overline{X}_k = \sum_i^{n_k} x_{ij} \tag{4.82}$$

where n_k is the number of adjacent observations in the kth zone; initially, this is simply two, observation i and observation $i + 1$. We also must calculate a measure of the variation within the zone

$$W_k = \sum_i^{n_k} (x_{ij} - \overline{X}_k)'(x_{ij} - \overline{X}_k) \tag{4.83}$$

If we sum this measure for all zones, we will obtain the sum of squares within zones, SS_W

$$SS_W = \sum_k^g W_k = \sum_k^g \sum_i^{n_k} (x_{ij} - \overline{X}_k)'(x_{ij} - \overline{X}_k) \tag{4.84}$$

In the second iteration, the vector of distances is examined again and the closest pair selected. This may be a new pair of observations which will form the start of a second composite zone, or it may consist of the initial composite zone plus a neighbor, in which case the initial zone simply expands. In either case, the measures of Equations (4.81) through (4.83) are recalculated and the process iterates for a third and subsequent times. At each iteration, the number of zones is decreased by one and the within-zones sum of squares increases.

Iteration continues until every point is included within a composite zone and all zones have been merged with their neighbors, resulting in a single zone that encompasses the entire sequence. SS_W will become equal to SS_T. At any step in the process, the relative amount of variance "explained" by the zonation can be expressed by the ratio

$$R^2 = \frac{SS_W}{SS_T} \tag{4.85}$$

given as a percentage.

At each iteration, a record is made of the zone membership of every observation in the sequence. The initial record consists simply of the integers 1 through n assigned in order from the first of the sequence to the last, representing the n initial "zones." The final record consists of the number 1 assigned to every observation, indicating that all observations are assigned to the same zone. Intermediate records show the assignment of observations to zones at each stage of zonation. It is up to the user to determine the appropriate number of zones and to select the corresponding stage of iteration for display.

Figure 4–43 shows part of the resistivity and density logs recorded through the Mississippian Lodgepole Formation in a well drilled in the Williston Basin of North Dakota. This well resulted in an unusually prolific discovery, and was interpreted as having penetrated a carbonate algal or "Waulsortian" mound. The data for the interval shown is contained in file LODGEPOL.TXT. The results of zonation by Bohling and other's (1998) clustering procedure are shown for 5, 10, 15, and 20 zones. These zones divide the reservoir into relatively homogeneous intervals based on petrophysical properties, and may serve as initial estimates of "flow units" for reservoir characterization. A plot of R^2 versus number of zones is shown in **Figure 4–44**; such plots may exhibit "natural breaks" which can be useful in deciding the appropriate number of zones for the interval. Breaks are not obvious in this example, and the appropriate number of zones must be based on external criteria.

Seriation

The terms *ordination* and *seriation* mean the putting of observations in some logical order, on the basis of their relative similarities. If the observations are characterized by multiple variables, this essentially means projecting them by some means onto a single line where their position is a logical expression of their place in the data set. This can be done by any of several techniques, such as principal component analysis and factor analysis, discussed in Chapter 6. Seriation has the additional connotation of chronological order; the term is widely used by archaeologists. Unfortunately, there is no guarantee that a sequence of observations arranged in order of similarity will also be arranged in a chronologically meaningful way.

The concepts of ordination and seriation have not been widely used in geology, except in applications of numerical taxonomy to paleontology (Sneath and Sokal, 1973). However, there is one area where the seriation concept seems useful, and that is in the geologic correlation of two stratigraphic sequences.

Two petrophysical well logs can be matched together on the basis of similarity in log response by "slotting," a dynamic programming procedure that shuffles the

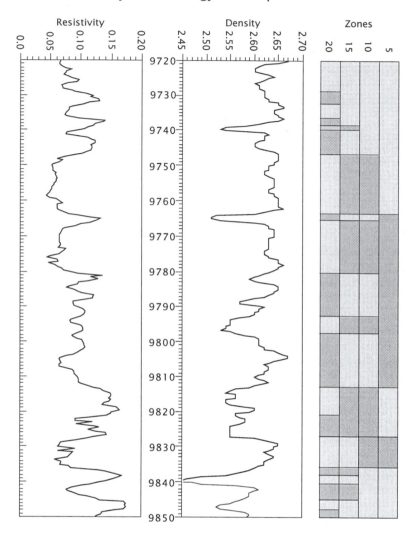

Figure 4–43. Well-log traces for the Lodgepole Formation (Mississippian) in the Dickinson State #74 well in Billings County, North Dakota. Resistivity, in ohm-meters, is plotted as $\sqrt{1/RT}$; bulk density is given in g/cc. Depths are in feet below surface elevation, recorded at 1/2-ft intervals. Shaded columns represent zonations into 20, 15, 10, and 5 zones.

two sections together like a deck of cards (Gordon and Reyment, 1979). Every point in one sequence is paired with the most similar point in the other sequence, subject to the constraint that stratigraphic order must be preserved in the two sections. In this way, true seriation is achieved, because the final arrangement is meaningful both lithologically and chronologically. Additional constraints can be introduced that will force specified points on the two sequences to match, or that will force a specified segment of one sequence to match with a point on the other. Thus, if marker beds are identified within two well logs being compared, these beds can be forced to correspond. The remainders of the logs are correlated on the basis of greatest similarity, subject to the constraints that correlation lines cannot cross, and that the marker beds *must* correlate.

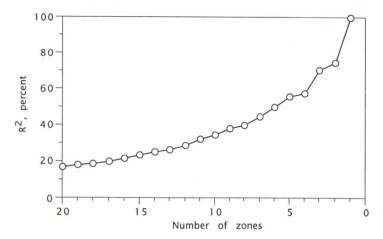

Figure 4–44. R^2 in percent of total sum of squares versus number of zones determined for the Lodgepole Formation in the Dickinson State #74 well.

The algorithm was published by Gordon and Reyment (1979). First, every point in the first well sequence is compared to every point in the second well sequence. If there are n observations on the first log and m on the second, this results in an $n \times m$ array of comparisons. A number of comparative measures could be used, but Gordon and Reyment (1979) use the simple dissimilarity measure

$$D_{j,k} = \sum_{l=1}^{p} \omega_l |u_{l,j} - v_{l,k}| \qquad (4.86)$$

where $u_{l,j}$ is the response for log variable l at depth j in the first well, and $v_{l,k}$ is the response for the same log variable at depth k in the second hole. Weights ω_l can be assigned to the different log variables if desired. The variables can be of any rank except nominal, although using integer values may result in ambiguous slotting because many dissimilarities may be indentical.

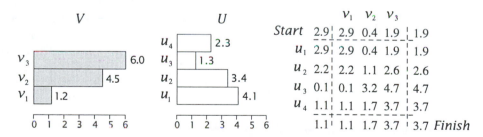

Figure 4–45. Artificial stratigraphic sections to be slotted together. Section V contains three intervals and section U contains four intervals. Characteristic measured on each interval ranges from 1.2 to 6.0. Matrix contains simple dissimilarity measures between all possible pairs of intervals in sections U and V.

The dynamic programming algorithm now seeks to trace a single path through this array, from the upper left corner to the lower right, such that the sum of the dissimilarities is a minimum. Stratigraphic order is preserved by requiring the search to move only down, or to the right. A small example is shown in **Figure 4–45**, slotting three intervals from one log with four intervals from another log. Note

241

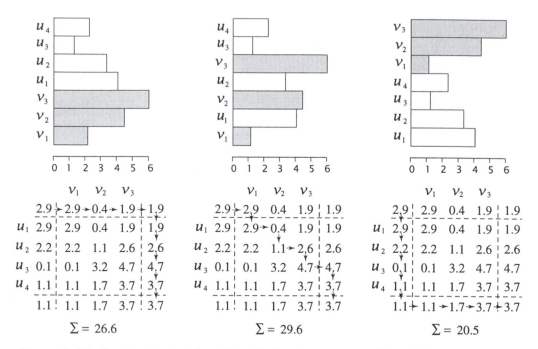

Figure 4–46. Results obtained by arbitrarily slotting sequences U and V together. Arrows in matrix show path of succession from bottom to top. Total dissimilarity is indicated by Σ.

that the outermost rows and columns of the matrix are repeated; this allows the algorithm to trace down one log or the other at the beginning and ending of the process. Some example paths through the matrix are shown in **Figure 4–46**, with the corresponding stratigraphic intervals slotted together in a single column.

To find the optimum path, a recursive procedure is used. Beginning at the upper left, the first interval can be either u_1 or v_1; the dissimilarity is the same, $2.9 + 2.9 = 5.8$, along either path. If u_1 is chosen, the next interval may be either u_2 with a total dissimilarity of $2.9 + 2.9 + 2.2 = 8.0$, or v_1 with a total dissimilarity of $2.9 + 2.9 + 2.9 = 8.7$. Alternatively, if v_1 is chosen as the first step, the second step could be either to v_2 or to u_1. The path to v_2 results in a total dissimilarity of $2.9 + 2.9 + 0.4 = 6.2$; the path to u_1 gives a dissimilarity of $2.9 + 2.9 + 2.9 = 8.7$. Of these alternatives, the path (Start) $\rightarrow v_1 \rightarrow v_2$ has the minimum dissimilarity, so the proper first step is to v_1.

Having set the initial point as v_1, with two possible second steps to either v_2 or u_1, the possible third steps must be examined. Again there are four possibilities: from v_2 to v_3, with a total dissimilarity of $2.9 + 2.9 + 0.4 + 1.9 = 8.1$; from v_2 to u_1 ($2.9 + 2.9 + 0.4 + 0.4 = 6.6$); from u_1 to u_2 ($2.9 + 2.9 + 2.9 + 2.2 = 10.9$); and from u_1 to v_2 ($2.9 + 2.9 + 2.9 + 0.4 = 9.1$). The smallest score occurs along the path (Start) $\rightarrow v_1 \rightarrow v_2 \rightarrow u_1$, so the optimal second step is to the point v_2. The procedure then iterates again, examining the outcomes of the four possible paths from v_2. The minimum-value path defines the optimum step after v_2. The process repeats until the lower left (Finish) point is reached. In the example, the optimum path is

$$(\text{Start}) \rightarrow v_1 \rightarrow v_2 \rightarrow u_1 \rightarrow u_2 \rightarrow u_3 \rightarrow u_4 \rightarrow v_3 \rightarrow (\text{Finish})$$

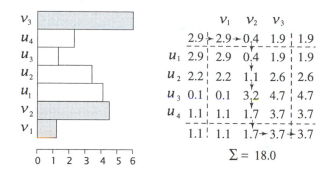

Figure 4–47. Optimal slotting of sequences U and V. Arrows in matrix indicate path of succession in slotted sequence. Total dissimilarity is only 18.0, the lowest possible for any succession.

with a total dissimilarity of 18.0. The matrix and the resulting slotted stratigraphic section are shown in **Figure 4–47.**

Because of its flexibility, slotting seems potentially to be a very powerful tool for correlation and the comparison of other types of sequences. Unfortunately, it is demanding of computer resources, even with the efficiency of dynamic programming, and the cost of slotting together long sequences may be prohibitive except for research purposes.

Autocorrelation

Figure 4–48 shows a gamma-ray log of part of the Pennsylvanian section measured in an oil well in northeastern Kansas; the log has been laid on its side to conserve space; the data are contained in file PAGELER.TXT. The interval penetrated consists of alternating limestones and shales. Because of the radiation emitted by potassium-40 in clay minerals, the shales are marked by relatively high log responses while the limestones are characterized by low radioactivity. This particular section has been noted for the existence of cyclothems, which are more or less regular repetitions of lithologies. A brief inspection of the log will show that the limestones do appear to be separated by shales of about the same thickness.

Repetitions, as well as other properties of a sequence, can be found by computing a measure of the self similarity of the sequence. That is, the sequence can be compared to itself at successive positions and the degree of similarity between the corresponding intervals computed. If every point in the sequence is compared successively to every other point, all positions of good correspondence will be detected, and also the degree of dissimilarity at other positions will be determined.

To perform this operation, the time series must have certain characteristics. It must consist of a sequence of observations of a variable, Y, measured at successive instants in time or at points in space. Each observation must be separated from the preceding observation by an interval of time or distance that is constant for the series. We indicate the position of an observation within the series by a subscript, such as y_t. It is therefore not necessary to explicitly consider the time or distance variable X, because it is implied by the subscript and can be found if needed by $x_t = \Delta t$, where Δ is the spacing between points. The entire time series contains n points and has a total length of $T = \Delta(n - 1)$.

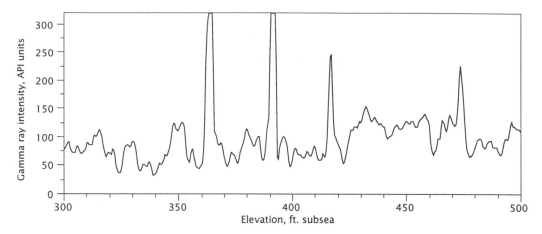

Figure 4–48. Gamma-ray log trace from Pageler No. 1 well, Wabaunsee County, Kansas. Logged interval includes upper Pennsylvanian rocks from 299.5 to 499.5 ft below sea level. Gamma-ray readings in API units. Responses greater than 320 API units are truncated. Recording interval is 0.5 ft.

The separation between any two points y_t and $y_{t+\tau}$ is referred to as a *lag* of length τ, where τ is the number of intervals between the points. It is the displacement between the time series and itself at a previous time or location. An analogy may be drawn between a time series and a chain. Each link in the chain corresponds to an observation in the series. If we lay two identical chains side by side and compare each link, we are making a cross comparison at lag 0. If we move one of the chains so the first link in one is matched to the second link in the other, then all of the other links are also offset by one. This position of comparison is lag 1. The chains can be offset by one more link, and the cross comparison will then be for lag 2, and so on.

The *autocovariance* for lag τ is the covariance between all observations y_t and lagged observations $y_{t+\tau}$. That is, the covariance is calculated between a series and itself displaced by a lag of length τ. The definitional equation of the autocovariance is

$$\text{cov}_\tau = \frac{1}{n-\tau} \sum_{t=1+\tau}^{n} y_t y_{t-\tau} - \overline{Y}_t \overline{Y}_{t-\tau} \qquad (4.87)$$

The autocovariance at lag 0 is simply the variance of the time series. If the series is very long and the lag τ is short, the mean of the series and that of the lagged series are essentially identical and Equation (4.87) can be simplified. However, if τ is an appreciable fraction of the length of the time series, the differences between the means become important. A computational equivalent of Equation (4.87) which also accounts for the loss of a degree of freedom is

$$\text{cov}_\tau = \frac{(n-\tau)\left(\sum_{t=1+\tau}^{n} y_t y_{t-\tau}\right) - \sum_{t=1+\tau}^{n} y_t \sum_{t=1+\tau}^{n} y_{t-\tau}}{(n-\tau)(n-\tau-1)} \qquad (4.88)$$

Conventionally, the autocovariance is calculated for lags from 0 to about $n/4$. The resulting values can be displayed as an *autocovariogram* or *autocovariance function*, which is a plot of autocovariance against lag. The units of autocovariance are the squares of the measurements of the time series; in our example, API units^2. This means that the autocovariance is sensitive to changes in the scale of the time

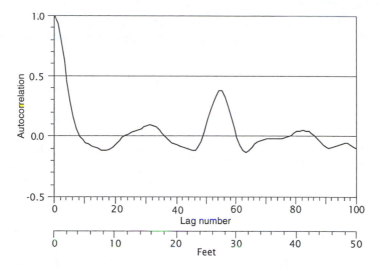

Figure 4–49. Autocorrelation function of gamma-ray log through the upper Pennsylvanian section in Pageler No. 1 well, Wabaunsee County, Kansas. Lag interval is 0.5 ft.

series, which makes it difficult to compare two autocovariograms. However, if the time series is standardized by subtracting the mean from each observation and dividing by the standard deviation, the series will be in units of standard deviation and the autocovariance will be in standardized form. As noted in Chapter 2, the covariance of a standardized variable is the correlation, and the same relationship is true for autocovariances.

Rather than first standardizing our time series, we may compute the autocorrelation directly by dividing the autocovariance by the variance of the time series. That is,

$$r_\tau = \frac{\text{cov}_\tau}{\text{var } Y} = \frac{[\sum_{t=1+\tau}^{n} y_t y_{t-\tau} - (n-\tau)\overline{Y}_t \overline{Y}_{t-\tau}]/(n-\tau-1)}{[\sum_{t=1}^{n}(y_t - \overline{Y})^2]/(n-1)} \tag{4.89}$$

The autocorrelation of a time series of finite length n and lag τ, which is some appreciable fraction of n, may be found by using Equation (4.88) as the numerator and the estimate $\sqrt{\text{var } y_t \text{ var } y_{t+\tau}}$ for the denominator. As in Equation (4.88), the limits of the summations extend from $t = 1 + \tau$ to n.

$$r_\tau = \frac{(n-\tau)\sum y_t y_{t-\tau} - (\sum y_t \sum y_{t-\tau})}{\sqrt{\left((n-\tau)\sum y_t^2 - (\sum y_t)^2\right)\left((n-\tau)\sum y_{t-\tau}^2 - (\sum y_{t-\tau})^2\right)}} \tag{4.90}$$

Figure 4–49 shows the autocorrelation function for the well log of **Figure 4–48**. The plot begins at a maximum value of 1.0 for lag $\tau = 0$, then drops and fluctuates around zero until it rises again to a peak at about lag 54, which corresponds to a spacing of about 27 ft, since the gamma-ray log was digitized at 0.5-ft intervals. This is the approximate vertical distance between successive limestone beds in the sequence shown in **Figure 4–48**.

Unfortunately, conflicts exist in time-series terminology. Some authors refer to the autocorrelation as a parameter of a population and use the term *serial correlation* as the equivalent statistic calculated from a sample. Others use serial correlation as meaning the correlation between two time series. Still others call this cross-correlation. We will use the terms "autocorrelation" and "cross-correlation" and will make no distinction in terminology between statistics and parameters.

It should be noted that the correlogram is actually two-sided, with a negative part that is the mirror image of the positive part. We can see how this arises if we return to our analogy of two chains. If one of the chains is designated A and the other B, and we successively move chain A ahead of chain B, we can regard the resulting autocorrelations as forming the positive part of the correlogram. But if we move chain B ahead of A and compute the autocorrelations, the relative movement of the two sequences is reversed, and the lags will be negative. As chains A and B are identical, it makes no difference which direction the two are shifted, because in practice we do not consider the negative part of the autocorrelogram; but, it does play a role in the development of the Fourier transform.

Correlograms help to reveal the characteristics of times series. This is done by comparing the correlogram of the time series to the correlograms of idealized models of the series, and determining how well they match. The simplest model that can be proposed for a time series is that successive observations are independent and are normally distributed. This means that there is no relationship between observations at one time, t, and observations at any other time, $t + \tau$. This is the behavior that we would expect if the time series had been generated by a random process. The expected autocorrelation for this model is $\rho_\tau = 0$ for all lags τ greater than zero. The theoretical correlogram will plot as a flat line through $r = 0$.

Other models assume some dependence between successive observations. For example, suppose a process generates random, normally distributed observations which are averaged in sets of w adjacent observations as they are created:

$$y_t = \sum_{t-w}^{t} z_t / w$$

This is a *moving average* model and will have a correlogram of the form

$$\rho_\tau = 1 - \tau / w$$

There are a wide variety of models of increasing complexity that can be proposed. A good introduction to these is given in Yule and Kendall (1969), and in a hydrological context by Yevjevich (1972). **Figure 4–50** illustrates how a complex time series can be built up by the combination of simpler elements. In **Figure 4–50a**, a regular sine wave and its correlogram are shown. The correlogram drops from +1.0 to 0 and then to −1.0 as the series moves out of phase with itself, or when peaks are matched with troughs. The correlation increases again until it reaches +1.0 when the signal is shifted exactly one wavelength. **Figure 4–50b** shows a signal created by the first model discussed, a sequence of random numbers. The correlogram drops immediately from +1.0 at 0 lag, then fluctuates slightly about 0. Both of these series are stationary; that is, no significant trend exists in the observations.

A nonstationary signal is shown in **Figure 4–50c**, where the observations increase steadily in value along the sequence. The correlogram shows steadily decreasing correlation. **Figure 4–50d** shows a combination of **Figure 4–50a** and **4–50b**, a sine wave with superimposed noise. Perfect autocorrelation does not occur except at 0 lag, but the periodic component of the time series is revealed by the peak in the correlogram following the second zero crossing. **Figure 4–50e** is a combination of **Figure 4–50a** through **4–50c**, a sine wave combined with a linear

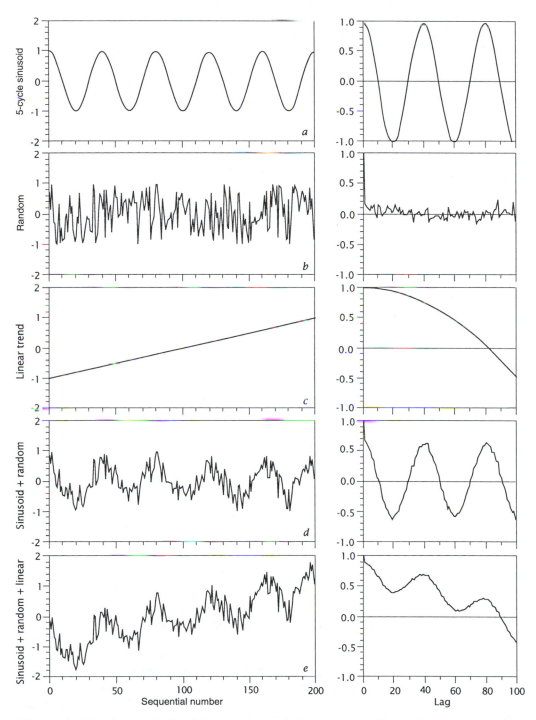

Figure 4–50. Some idealized time series and their autocorrelation functions. (*a*) Sine wave with wavelength of 20 units. (*b*) Sequence of random numbers, or "noise." (*c*) Sequence of linearly increasing numbers, or "trend." (*d*) Sine wave plus random noise [sequence (*a*) plus sequence (*b*)]. (*e*) Sine wave plus random noise plus linear trend [sequence (*a*) plus sequence (*b*) plus sequence (*c*)].

trend with superimposed noise. Note that the trend further reduces our ability to discern the periodic component in the signal.

As noted earlier, the expected or mean autocorrelation of a sequence of random numbers is zero. The expected variance in the autocorrelation of a random sequence at any lag τ is

$$\sigma_\tau^2 = 1/(n - \tau + 3) \tag{4.91}$$

These two parameters define the population of a random time series of a specified length n (Yule and Kendall, 1969, p. 639–641). You will recall from Chapter 2 that we can determine the probability of drawing a specific observation from a normal population having a known mean and variance by $z_i = (x_i - \mu)/\sigma$. Taking the square root of Equation (4.91) gives the expected standard deviation of autocorrelations, which can be used along with the expected mean of zero to give

$$z_\tau = \frac{r_\tau - 0}{\sqrt{1/(n - \tau + 3)}} = r_\tau \sqrt{n - \tau + 3} \tag{4.92}$$

This can be used as a conservative test of the hypothesis that an autocorrelation ρ_τ is zero, provided the length of the sequence, n, is large and the lag, τ, is small. "Large" and "small" are relative terms and difficult to define exactly. As a rule of thumb, n should exceed 50 and τ should not exceed $n/4$. (Some authors advocate more conservative limits, down to $n/10$ or less.) These restrictions are based on the fact that as the lag increases, r_τ is based on fewer and fewer observations. This not only causes an increase in the variance of r_τ but also results in an increasing violation of the assumption that the autocorrelation is a sample from an infinitely long time series. For these reasons, little importance should be attached to high autocorrelations at long lag intervals, unless the time series itself is many times larger.

Cross-correlation

If it is possible to compare a time series with itself at successive lags in order to detect dependencies through time, it would seem possible to compare two time series with each other in order to determine positions of pronounced correspondence. Two items of information may emerge from such a comparison: the strength of the relationship between the two series, and the lag or offset in time or distance between them at their position of maximum equivalence. The process of comparing two time series at successive lags is called ***cross-correlation***. The "zero lag" often is set where the origins of the two series are aligned; negative lags represent an arbitrary choice of the sense of movement of one sequence past the other. Since the two series are not identical, the cross-correlogram is not symmetric about its middle; lags in which series A leads series B differ from lags in which B leads A (**Fig. 4–51**). A complicating factor arises because series A may not necessarily be the same length as series B. In fact, one approach to "automated correlation" (in the geological sense of equating two stratigraphic sections) consists of moving a short, distinctive part of one stratigraphic interval past another entire stratigraphic section to determine the position of best match.

The equation for cross-correlation is the same as the ordinary linear correlation coefficient, and differs somewhat from the autocorrelation coefficient. If we designate the two series being compared as Y_1 and Y_2 and define n^* as the number

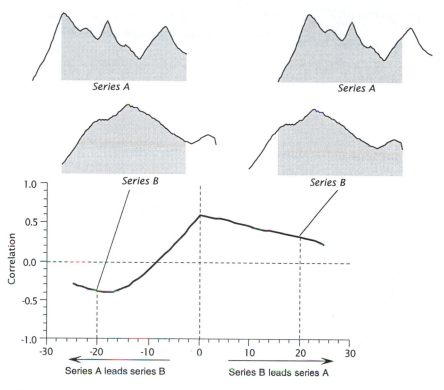

Figure 4–51. Cross-correlogram of two dissimilar series, A and B. Correlogram is asymmetric unless two series are identical. Shaded intervals indicate portions of A and B that are compared at two lag positions. Zero lag position indicates both series are aligned at their origins.

of overlapped positions between the two, the cross-correlation for match position m is

$$r_m = \frac{n^* \sum Y_1 Y_2 - \sum Y_1 \sum Y_2}{\sqrt{\left[n^* \sum Y_1^2 - (\sum Y_1)^2\right]\left[n^* \sum Y_2^2 - (\sum Y_2)^2\right]}} \tag{4.93}$$

or, equivalently,

$$r_m = \frac{\text{cov}_{1,2}}{s_1 s_2} \tag{4.94}$$

In this equation, $\text{cov}_{1,2}$ is the covariance between the overlapped portions of sequences Y_1 and Y_2, and s_1 and s_2 are the corresponding standard deviations. Note that the summations are understood to extend only over the segments of the two sequences which are overlapped at the match position. The match positions are numbered sequentially as shown in **Figure 4–51**, and the cross-correlogram is a plot of match position versus cross-correlation.

Because the summations in Equation (4.93) extend only over the overlapped segment, the denominator of the cross-correlation changes with m. In contrast, the denominator of the autocorrelation coefficient is based on the variance of the entire chain, and is considered to be constant for all lags. A variance derived from an entire sequence is more stable than an estimate derived from a shorter segment, and so is preferable. However, in cross-correlation we cannot expect the variances to be constant through the lengths of both chains, especially when one of the chains is short with respect to the other.

The significance of the cross-correlation coefficient can be assessed by the approximate test

$$t \approx r_m \sqrt{\frac{n^* - 2}{1 - r_m^2}} \tag{4.95}$$

which has $(n^* - 2)$ degrees of freedom. This test is derived from a test for the significance of the correlation between two samples drawn from normal populations. The null hypothesis states that the correlation between the two sequences at the specified match position is zero, or that which is expected if the two sequences are independent, random series.

Cross-correlation is used most appropriately to compare two series that may have a temporal dependency. As an example, we may analyze the data given in **Table 4–18** and in file ARSENAL.TXT. The Rocky Mountain Arsenal was a manufacturing plant that produced various noxious military compounds; it is located at Denver, Colorado, near the Front Range of the Rocky Mountains. Tremendous quantities of contaminated water were produced as a by-product of the weapons industry. In an attempt to dispose of this waste water, a deep injection well was drilled into basement rocks in 1961. Unfortunately, the well penetrated the shear zone of a major fault along the Rocky Mountain front, and there is evidence that the high-pressure injection of waste fluids served to lubricate and mobilize the fault. One of the variables given in **Table 4–18** is the month-by-month record of volume of water injected into the disposal well at the Rocky Mountain Arsenal over the 4-yr period the well was in use. The other variable is the number of earthquakes detected in Denver each month. In a study of the statistical relationship between these two time series, Bardwell (1970) plotted each in cumulative form and concluded there was a pronounced 3-mo lag between injection and earthquake incidence. Unfortunately, this method of visual comparison between two curves may be misleading because it requires arbitrary scaling to bring the two records into coincidence.

The two records are shown in graphic form in **Figures 4–52** and **4–53**. The cross-correlogram of these two time series is shown in **Figure 4–54**. Since in this example both series have a common origin and time scale, it is possible to express match position in terms of "positive" and "negative" lags from the position of initial coincidence. The position of greatest correspondence between the two series occurs at lag +1, when the number of earthquakes in a given month is compared to the volume of waste water injected one month previously. The correlation at this match position is $r = 0.62$. The second highest cross-correlation, $r = 0.60$, occurs at lag 2. The cross-correlation at zero lag is very slightly lower. This suggests that the tectonic response to operation of the injection well began very quickly and extended over an interval of one to two months.

Sometimes we may want to compare two time series that are periodic over the same interval. We can take advantage of the periodicity and compute a cross-correlogram in "circular" form. In effect, each time series is wrapped in a circle and the start of a series is connected to its end. Rather than imagining the two time series as chains that are moved by each other, we can picture them in the form of wheels. Each wheel has the same number of divisions around its rim, corresponding

Table 4–18. Waste injected by Rocky Mountain Arsenal and frequency of recorded earthquakes in the Denver, Colorado, area from March 1962 to October 1965 (after Bardwell, 1970).

Mo/Yr	Waste (MM gal)	Earth-quakes	Seq. Mo	Mo/Yr	Waste (MM gal)	Earth-quakes	Seq. Mo
Mar 1962	4.2	0	3	Jan 1964	0	5	25
Apr 1962	7.2	2	4	Feb 1964	0	2	26
May 1962	8.4	12	5	Mar1964	0	9	27
Jun 1962	8.0	35	6	Apr1964	0	9	28
Jul 1962	5.2	23	7	May 1964	0	2	29
Aug 1962	6.0	29	8	Jun 1964	0	4	30
Sep 1962	5.0	24	9	Jul 1964	0	4	31
Oct 1962	5.6	8	10	Aug 1964	0	5	32
Nov 1962	4.0	6	11	Sep 1964	0.6	2	33
Dec 1962	3.6	20	12	Oct 1964	1.8	14	34
Jan 1963	6.0	25	13	Nov 1964	2.4	2	35
Feb 1963	7.6	22	14	Dec 1964	2.0	7	36
Mar 1963	7.8	21	15	Jan 1965	2.0	1	37
Apr 1963	6.4	42	16	Feb 1965	1.7	30	38
May 1963	3.6	21	17	Mar 1965	1.6	9	39
Jun 1963	4.0	8	18	Apr 1965	3.6	19	40
Jul 1963	3.4	6	19	May 1965	4.0	11	41
Aug 1963	2.4	10	20	Jun 1965	6.4	38	42
Sep 1963	3.9	11	21	Jul 1965	8.9	62	43
Oct 1963	0	12	22	Aug 1965	5.4	48	44
Nov 1963	0	4	23	Sep 1965	6.4	87	45
Dec 1963	0	2	24	Oct 1965	3.8	5	46

to the n successive observations. If one wheel is rotated with respect to the other, cross comparisons can be made between the two at n different positions before repeating.

In the Chesapeake Bay of Maryland, as in many estuaries, there are complex changes in salinity caused by the mingling of freshwater with seawater during the diurnal tidal cycle. Fresh Chesapeake River water floats across the denser brine in the Bay; during periods of low tide, this water moves farther down the estuary. However, there is a counter-flow along the bottom that carries dense marine water up the Bay during the waning tide. **Table 4–19** gives salinity measurements (in parts per thousand) made at 1.5-hr intervals over a 24-hr period, for both surface water and bottom water (11-m depth average) at a collecting station offshore from Annapolis, Maryland. The records are shown in graphical form in **Figure 4–55**.

The cross-correlation function is shown in **Figure 4–56** and clearly indicates a lag of 2, representing a difference of 3 hr between the crest of saltwater invasion along the bottom and the maximum salinity in the surface waters. Because the records represent a 24-hr tidal cycle that repeats, the cross-correlations can be calculated in circular form. Therefore, cross-correlations can be found for up to 24 hr, or 16 1.5-hr lags.

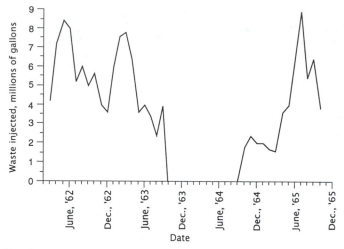

Figure 4–52. Quantity of liquid waste injected through Rocky Mountain Arsenal disposal well, in millions of gallons per month.

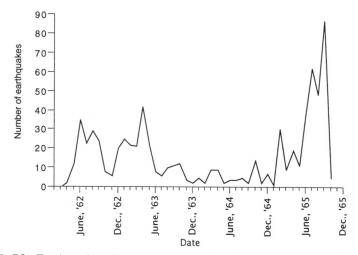

Figure 4–53. Earthquakes detected per month with epicenters near Denver, Colorado.

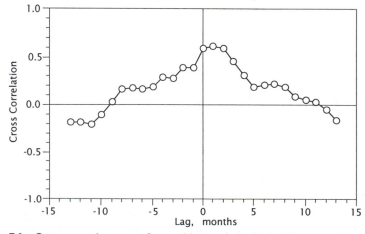

Figure 4–54. Cross-correlogram of monthly earthquake incidence and monthly volume of injected waste. Maximum correlation occurs at lag +1.

Table 4–19. Salinity of water in Chesapeake Bay at Station 11, offshore from Annapolis, Maryland, on July 3–4, 1927 (from Wells, Bailey, and Henderson, 1928).

	Time	State of Tide	Surface Salinity (ppt)	Bottom Salinity (ppt)
July 3	2:30 pm	1/4 ebb	6.97	11.10
	4:00 pm	1/2 ebb	6.20	11.54
	5:30 pm	Ebb	5.93	13.12
	7:00 pm	Ebb	6.32	13.52
	8:30 pm	1/4 flood	6.36	13.35
	10:00 pm	1/2 flood	6.72	12.83
	11:30 pm	3/4 flood	6.80	13.31
July 4	1:00 am	Flood	6.90	13.02
	2:30 pm	1/4 ebb	7.14	12.14
	4:00 pm	1/2 ebb	6.91	12.44
	5:30 pm	Ebb	6.76	12.60
	7:00 pm	Begin flood	6.74	12.79
	8:30 pm	3/8 flood	7.20	13.46
	10:00 pm	1/2 flood	7.45	12.33
	11:30 pm	3/4 flood	7.47	12.40
	1:00 pm	Flood	7.47	12.14

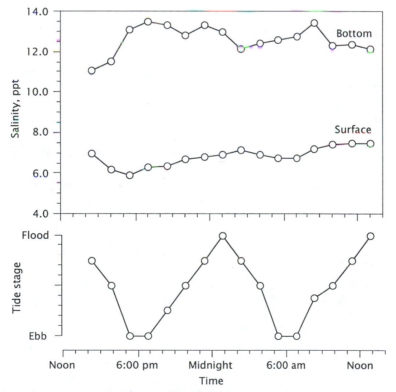

Figure 4–55. Salinities of bottom water and surface water in the Chesapeake Bay near Annapolis, Maryland, over one day. Lower curve is state of tide.

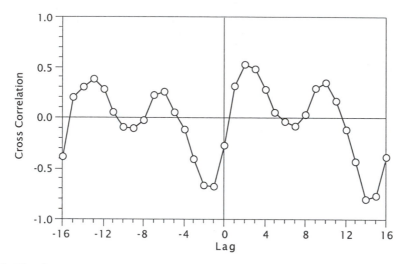

Figure 4–56. Cross-correlogram between salinities of bottom water and surface water in Chesapeake Bay. Maximum correspondence occurs at lag 2, representing a 3-hr time difference.

Cross-correlation and stratigraphic correlation

Geologists have repeatedly succumbed to the temptation to use cross-correlation procedures to match stratigraphic sections, thereby performing automated geological correlation. Although the literature on this topic is extensive, these efforts have met with a notable lack of success, except in special circumstances. The reasons for failure are not difficult to deduce. Cross-correlation presumes that the two sequences being compared are sampled at discrete, uniformly spaced points, and that the sampling interval in one succession is the same as in the other. Unfortunately, it is difficult to collect stratigraphic measurements that meet these requirements. In general, it is not possible to place sample points so they are equally spaced in geologic time, and if points are equally spaced in distance we are in effect assuming that the rates of sediment accumulation were constant throughout the time of deposition of the two sequences. If we consider stratigraphic sections to be analogous to the records produced by chart recorders, it is as though the recorders ran at different (and unknown) speeds at different times. To make matters worse, it is probable that the recorders were turned off entirely much of the time!

Semivariograms

The term *geostatistics* is widely applied to a special branch of applied statistics originally developed by Georges Matheron of the Centre de Morphologie Mathématique in Fontainebleau, France. Geostatistics was devised to treat problems that arise when conventional statistical theory is used in estimating changes in ore grade within a mine. However, because geostatistics is an abstract theory of statistical behavior, it is applicable to many circumstances in different areas of geology and other natural sciences.

A key concept of geostatistics is that of the *regionalized variable*, which has properties intermediate between a truly random variable and one that is completely deterministic. Typical regionalized variables are functions that describe natural phenomena that have geographic distributions, such as the elevation of the ground surface, changes in grade within an ore body, or the spontaneous electrical potential measured in a well by a logging tool. Unlike random variables, regionalized variables have continuity from point to point, but the changes in the variable are so complex that they cannot be described by any tractable deterministic function.

Even though a regionalized variable is spatially continuous, it is not usually possible to know its value everywhere. Instead, its values are known only through a sample of observations that are taken at specific locations. The size, shape, orientation, and spatial arrangement of these observations constitute the *support* of the regionalized variable, and the regionalized variable will have different characteristics if any of these are changed. For example, suppose we wish to determine the variation in ore grade in a disseminated molybdenum deposit. It seems likely that the answer we might obtain from analysis of 2-in. diamond-drill cores could be considerably different from what we might determine if we used the mill runs from muck car-sized volumes of ore. In both instances we might take exactly the same number of observations, and they might be centered on identical locations within the mine. However, the fact that one set of observations is based on volumes of a few cubic inches and the volumes in the second set are measured in cubic yards must inevitably affect the pattern of variation in ore grade we map through the mine. A principal concern of geostatistics is to relate the results obtained from one support (such as drill cores) to that obtained from another support (such as stope blocks).

Geostatistics involves estimating the form of a regionalized variable in one, two, or three dimensions. In the next chapter we will consider the estimation procedure, called *kriging*, more extensively. Now we will be concerned only with one of the basic statistical measures of geostatistics, the *semivariance*, which is used to express the rate of change of a regionalized variable along a specific orientation. Estimating the semivariance involves procedures similar to those of time-series analysis, hence the introduction of geostatistics at this point.

The semivariance is a measure of the degree of spatial dependence between observations along a specific support. For the sake of simplicity, we will assume the observations are point measurements of a property such as depth to a subsurface horizon. For computational tractability, we will further assume that the sample spacing is regular; that is, the observations are uniformly spaced along straight lines. If the spacing between observations along a line is some distance, Δ, the semivariance can be estimated for distances that are multiples of Δ:

$$\gamma_h = \sum_i^{n-h} (x_i - x_{i+h})^2 \Big/ 2n \qquad (4.96)$$

In this notation, x_i is a measurement of a regionalized variable, X, taken at location i, and x_{i+h} is another measurement taken h intervals away. We are therefore finding the sum of the squared differences between pairs of points separated by the distance, Δh. The number of points is n, so the number of comparisons between pairs of points is $n - h$. **Figure 4–57** illustrates the procedure; for $h = 1$, every point along a traverse is compared to its neighbor. For $h = 2$, every point is compared to a point two spaces away, and so forth.

If we calculate the semivariances for different values of h, we can plot the results in the form of a *semivariogram*, which is analogous to a correlogram. This data-based graphic is sometimes called an *experimental* semivariogram, to distinguish it from the *theoretical* semivariogram that characterizes the underlying

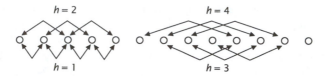

Figure 4–57. Traverse of equally spaced observations. For $h = 1$, semivariance is calculated from values at every location compared to values at adjacent locations. Semivariance for $h = 2$ is determined by comparison of values at every other location, for $h = 3$, locations compared are separated by two intervening locations, and for $h = 4$, locations compared are separated by three intervening locations. Semivariances for higher lags are determined in an analogous manner.

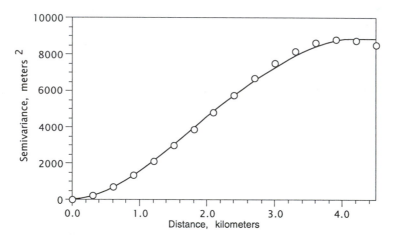

Figure 4–58. Semivariogram of depth in meters to a seismic reflecting horizon measured along a marine survey traverse in the Straits of Magellan, South America. Circles are values of experimental semivariance; line is fitted Gaussian model.

population. The latter usually is thought of as a smooth function represented by a model equation; the experimental semivariogram estimates its form. **Figure 4–58** shows a semivariogram for depth to a seismic reflecting horizon, as measured along the seismic profile shown in **Figure 4–59**. Note that when the distance between sample points is zero, the value at each point is being compared with itself. Hence, all the differences are zero, and the semivariance for y_0 is zero. If Δh is a small distance, the points being compared tend to be very similar, and the semivariance will be a small value. As the distance Δh is increased, the points being compared are less and less closely related to each other and their differences become larger, resulting in larger values of y_h. At some distance the points being compared are so far apart that they are not related to each other, and their squared differences become equal in magnitude to the variance around the average value. The semivariance no longer increases and the semivariogram develops a flat region called a *sill*. The distance at which the semivariance approaches the variance is referred to as the *range* of the regionalized variable, and defines a neighborhood within which all locations are related to one another.

For some arbitrary point in space, we can imagine the neighborhood as a symmetrical interval (or area or volume, depending on the number of dimensions) about the point. If the regionalized variable is stationary, or has the same average value

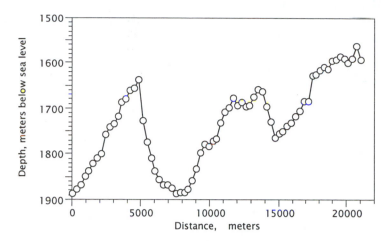

Figure 4–59. Subsea depths to reflecting horizon along an interpreted marine seismic profile in the Straits of Magellan, South America. Data are given in file MAGELLAN.TXT.

everywhere, any locations outside the interval are completely independent of the central point and cannot provide information about the value of the regionalized variable at that location. Within the neighborhood, however, the regionalized variable at all observation points is related to the regionalized variable at the central location and hence can be used to estimate its value. If we use a number of measurements made at locations within the neighborhood to estimate the value of the regionalized variable at the central location, the semivariogram provides the proper weightings to be assigned to each of these measurements.

We here digress to demonstrate a point that will be useful later. The semivariance is not only equal to the average of the squared differences between pairs of points spaced a distance Δh apart, it is also equal to the variance of these differences. That is, the semivariance can also be defined as

$$\gamma_h = \frac{\sum \left\{ (x_i - x_{i+h}) - \frac{\sum (x_i - x_{i+h})}{n} \right\}^2}{2n} \tag{4.97}$$

Note that the mean of the regionalized variable, x_i, is also the mean of the regionalized variable x_{i+h}, because these are the same observations, merely taken in a different order. That is,

$$\frac{\sum x_i}{n} = \frac{\sum x_{i+h}}{n}$$

Therefore, their difference must be zero:

$$\frac{\sum x_i}{n} - \frac{\sum x_{i+h}}{n} = 0$$

We can combine the summations

$$\frac{\sum x_i - \sum x_{i+h}}{n} = \frac{\sum (x_i - x_{i+h})}{n} = 0$$

Substituting into Equation (4.97), we see that the second term in the numerator is zero, so the equation is equal to Equation (4.96). Note that this relationship

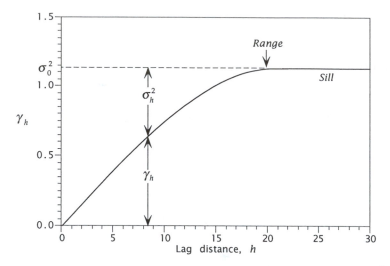

Figure 4–60. Relationship between semivariance, γ_h, and autocovariance, σ_h^2, for a stationary regionalized variable. σ_0^2 is the variance of the observations, or the autocovariance at lag 0. For values of h beyond the range, $\gamma_h = \sigma_0^2$.

is strictly true only if the regionalized variable is stationary. If the data are not stationary, the mean of the sequence changes with h and Equation (4.97) must be modified.

As you might expect, there are mathematical relationships between the semivariance and other statistics such as the autocovariance and the autocorrelation. If the regionalized variable is stationary, the semivariance for a distance is equal to the difference between the variance and the spatial autocovariance for the same distance (**Fig. 4–60**).

Unfortunately, it often happens that regionalized variables are not stationary, but rather exhibit changes in their average value from place to place. If we attempt to compute a semivariogram for such a variable, we will discover that it may not have the properties we have described. However, if we reexamine the definition of semivariance given in Equation (4.97), we note that it contains two parts, the first being the difference between pairs of points and the second being the average of these differences. If the regionalized variable is stationary, we have shown that the second part vanishes, but if it is not stationary, this average will have some value. In effect, the regionalized variable can be regarded as being composed of two parts, called the residual and the drift. The **drift** is the expected value of the regionalized variable at a point i, or computationally, a weighted average of all the points within the neighborhood around point i. The drift will have the form of a generalized approximation of the original regionalized variable. If the drift is subtracted from the regionalized variable, the **residuals**

$$R_i = x_i - \overline{X}_i$$

will themselves be a regionalized variable and will have local mean values equal to zero. In other words, the residuals will be stationary and it will be possible to compute their semivariogram.

Here we come to an awkwardly circular problem. The drift could be estimated if we knew the size of the neighborhood and the weights to be assigned points within the neighborhood. However, the weights can be calculated only if we know the semivariances corresponding to the distances between point i, the center of the neighborhood, and the various other points. Having once calculated the drift, it could be subtracted from the observed values to yield stationary residuals, which in turn could be used to estimate the neighborhood size and form of the semivari-ogram.

At this stage we must relax our definitional rigor and resort to trial and error. We first concede that we cannot determine the neighborhood in the sense that we have been using the term. Instead, the neighborhood is defined as a convenient but arbitrary interval within which we are reasonably confident that all locations are related to one another. Within this arbitrary neighborhood, we assume that the drift can be approximated by a simple expression such as

$$\overline{X}_0 = \sum b_1 x_i$$

or

$$\overline{X}_0 = \sum (b_1 x_i + b_2 x_i^2)$$

where the first represents a linear drift and the second a quadratic drift. The calculations involve all of the points within the arbitrary neighborhood, so there is an interrelation between neighborhood size, drift, and semivariogram for the residuals. If the neighborhood is large, the drift calculations will involve many points and the drift itself will be very smooth and gentle. Consequently, the residuals will tend to be more variable and their semivariogram will be complicated in form. Conversely, specification of a small neighborhood size will result in a more variable drift estimate, smaller residuals, and a simpler semivariogram.

Determining the b coefficients for the drift requires solving a set of simultaneous equations of somewhat foreboding complexity; they will be left to the section on kriging. The only variables in the equations are the semivariances corresponding to the different distances between point i and the other points within the neighborhood. However, we do not yet have a semivariogram from which to obtain the necessary semivariances. We must assume a reasonable form for the semivariogram and use it as a first approximation. It will be much easier to guess the form of a simple semivariogram, so this is an argument for using as small a neighborhood size as possible.

Next, the experimental estimates of the drift are subtracted from their corresponding observations to yield a set of experimental residuals. A semivariogram can be calculated for these residuals and its form compared to that of the semivariogram that was first assumed. If the assumptions that have been made are appropriate, the two will coincide, and we have successfully deduced the form of the drift and the semivariogram. Most likely they will differ, and we must try again.

The process of attempting to simultaneously find satisfactory representations of the semivariogram and drift expression is a major part of "structural analysis." It is to a certain extent an art, requiring experience, patience, and sometimes luck. The process is not altogether satisfying, because the conclusions are not unique; many combinations of drift, neighborhood, and semivariogram model may yield approximately equivalent results. This is especially apt to be true if the regionalized variable is erratic or we have only a short sequence. In such circumstances it may be difficult to tell when (or if) we have arrived at the proper combination of estimates.

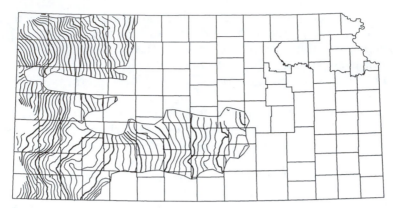

Figure 4–61. Contour map of elevation of water table in the High Plains aquifer in western Kansas, measured in winter, 1987–88. Semivariograms in Figure 4–62 are based on 327 observation wells in northwest corner of state. Water-table elevation ranges from approximately 1300 ft in the east to 3800 ft in the west; contour interval is 30 ft. After Olea (1999).

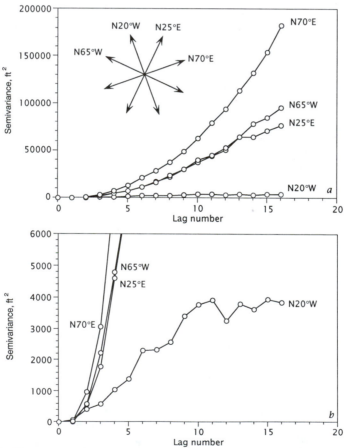

Figure 4–62. Experimental semivariograms of water-table elevations calculated in specific directions across the northwestern part of the High Plains aquifer of Kansas, shown in Figure 4–61. Orientations of traverses are indicated. Regional dip is approximately N70°E and regional strike is approximately N20°W. (*a*) Compressed y-axis scale. (*b*) Expanded y-axis scale showing details of semivariogram oriented N20°W.

In certain circumstances we can estimate the semivariogram in the presence of a drift without such an involved trial-and-error procedure. If the drift is reasonably consistent, an experimental semivariogram can be calculated in a direction that is drift-free, such as parallel to the regional strike. If we assume the regionalized variable would be isotropic if the drift were not present, this directional semivariogram can be used as an estimate of the drift-free semivariogram. **Figure 4–61** is a contour map of the elevation of the water table in the High Plains aquifer of Kansas, adapted from Olea (1999). The surface has a pronounced drift in the form of a linear slope to the east, roughly parallel to the topography. **Figure 4–62** shows several semivariograms calculated along traverses with different orientations across the northwestern part the aquifer; all show the effects of drift except for the semivariogram calculated along traverses oriented N20°W, parallel to the regional strike.

Modeling the semivariogram

In principle, the experimental semivariogram could be used directly to provide values for the estimation procedures we will discuss in the next chapter. However, the semivariogram is known only at discrete points representing distances Δh; in practice, semivariances may be required for any distance, whether it is a multiple of Δ or not. For this reason, the discrete experimental semivariogram must be modeled by a continuous function that can be evaluated for any desired distance. The model is an estimate of the form of the underlying population semivariogram of the regionalized variable. (There is a further mathematical constraint. The semivariogram model must be such that the associated spatial covariance model is *positive definite*—or more specifically, *nonnegative definite*—meaning that it cannot yield negative kriging estimation variances regardless of the spatial configuration of the data.) Fitting a model equation to an experimental semivariogram is a trial-and-error process, usually done by eye. Clark (1979) describes and gives examples of the manual process, while Jian, Olea, and Yu (1996) provide a program that fits a variety of models to the experimental semivariogram by weighted least squares.

Ideally, the model chosen to represent the semivariogram should begin at the origin, rise smoothly to some upper limit, then continue at a constant level. The *spherical model*, shown in **Figure 4–63 a**, has these properties. It is defined as

$$\gamma_h = \sigma_0^2 \left(\frac{3h}{2a} - \frac{h^3}{2a^3} \right) \tag{4.98}$$

for all distances up to the range of the semivariogram, a. Beyond the range, $\gamma_h = \sigma_0^2$. The spherical model usually is described as the ideal form of the semivariogram. Another model that is sometimes used is the *exponential* model:

$$\gamma_h = \sigma_0^2 \left(1 - e^{-3h/a} \right) \tag{4.99}$$

Figure 4–63 b shows an exponential model whose form should be compared to the spherical model fitted to the same data. The exponential never quite reaches the limiting value of the sill, but approaches it asymptotically. Also, the semivariance of the exponential model is lower than the spherical for all values of h less than the range.

The *linear* model is simpler than either the spherical or exponential, as it has only one parameter, the slope. The model has the form

$$\gamma_h = \alpha h \tag{4.100}$$

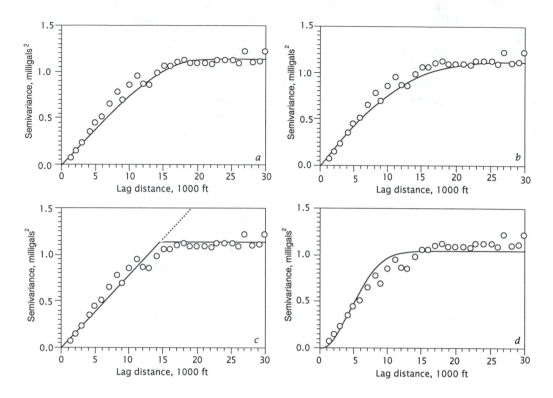

Figure 4–63. Models fitted to experimental semivariogram of raw gravity measurements (in milligals) in Elk County, Kansas. Direction of measurement is N20°E. (*a*) Spherical model. (*b*) Exponential model. (*c*) Linear model. (*d*) Gaussian model.

and plots as a straight line through the origin. Obviously, this model cannot have a sill, as it rises without limit. Sometimes the linear model is arbitrarily modified by inserting a sharp break at the sill value, as in **Figure 4–63 c**, so that

$$y_h = \alpha h \quad \text{for } h < a$$
$$y_h = \sigma_0^2 \quad \text{for } h \geq a$$

(4.101)

Armstrong and Jabin (1981), among others, have criticized the use of such a model, because the requirement for positive definiteness in the kriging equations means the semivariogram must be a continuous, smoothly varying function. However, provided the semivariogram is evaluated only for distances much less than the range, the linear model is a perfectly good approximation. This is apparent in **Figure 4–63**, where both the spherical and exponential models are almost coincident with a straight line near the origin. If the regionalized variable has been sampled at a sufficient density relative to the range, there will be no significant differences between estimates made assuming a linear model and those obtained using a spherical or other model.

A semivariogram tangent to the X-axis at the origin has a parabolic initial form, as in **Figure 4–63 d**, and is called a *Gaussian* model. This model indicates that the regionalized variable is exceptionally continuous and smooth over short distances. A truly random variable will have no continuity and its semivariogram will be a

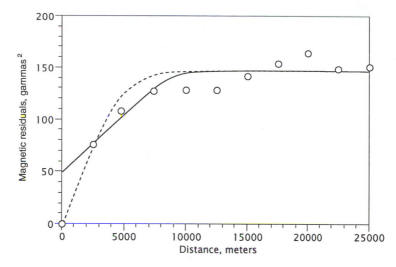

Figure 4–64. Model semivariograms fitted to experimental semivariogram of magnetic residuals in gammas, measured along a marine traverse in Straits of Magellan, South America. Dashed line is spherical model with no nugget effect. Solid line is spherical model with an assumed nugget effect of $y_0 = 50$. Original data after Olea (1977).

horizontal line equal to the variance. In some circumstances the semivariogram will appear not to go through the origin but rather will assume some nonzero value. This is referred to as a "nugget effect" and is shown in **Figure 4–64** where the nugget effect is combined with a spherical model. In theory, y_0 must equal zero; the nugget effect arises because the regionalized variable is either subject to sampling errors (*i.e.*, repeated measurements at the same location would yield different values) or is so erratic over a very short distance that the semivariogram goes from zero to the level of the nugget effect in a distance less than the sampling interval. A nugget effect can occur in combination with any semivariogram model.

Although in theory the calculation of semivariograms requires that observations be arranged at equally spaced points along parallel, straight traverses, this severely limits the circumstances in which they can be calculated. If we relax these requirements, we can determine semivariograms automatically, making the analysis easier and the resulting semivariograms more representative. First, we must use a search algorithm, such as those described in the section on contour mapping in Chapter 5, to locate "nearby" points around each observation. Then, we find the direction and distance from each observation to its neighboring points using simple geometry:

$$\phi = \arctan\left(\frac{y_2 - y_1}{x_2 - x_1}\right)$$

$$d = \sqrt{(y_2 - y_1)^2 + (x_2 - x_1)^2}$$

Here, x_1, y_1 are the coordinates of a point and x_2, y_2 are the coordinates of a neighboring point; ϕ is the orientation of a line drawn between the two points, and d is the length of the line. We also find the squared difference in value of the regionalized variable at the two points. We will have a collection of these squared differences, orientations, and distances for each observation. The distances and orientations are then sorted into regular classes, and the semivariance calculated

263

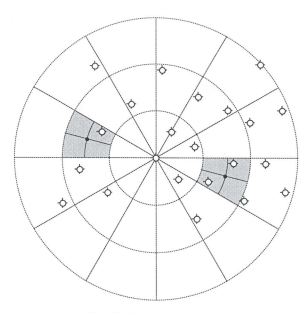

Figure 4–65. Pattern of radial "bins" placed over successive observations to identify neighbors to be used in calculating directional experimental semivariograms.

as though all of the points were located at the class centroids. **Figure 4–65** shows how this procedure places the neighboring points around each observation into "bins" for the purpose of calculating the semivariance for each combination of direction and distance. As an example, the shaded bins represent the lag $h = 2$ for a semivariogram oriented 15° south of east. The semivariance is calculated using all points that fall within these bins when the pattern is placed over each of the observations in turn. When performed for all bins in the pattern, the result is a series of directional semivariograms oriented 30° apart. Each semivariogram will have as many lags, h, as there are rings in the pattern. The directional semivariograms in **Figure 4–62** were computed using this algorithm.

Of course, assuming points lie along a straight line when in fact they deviate within angular limits, and assuming points are equally spaced when they actually are not introduces errors into the semivariogram. Hopefully, the increased error will be outweighed by the decrease in uncertainty that results from using a much larger number of observations. In addition, we gain the ability to assess anisotropy and to look for a drift-free direction if the regionalized variable is nonstationary.

Alternatives to the semivariogram

Although the preceding discussion has been cast entirely in terms of the semivariogram, other measures of spatial continuity are available. These include such variants as the **semimadogram**, a measure based on the absolute (as opposed to squared) differences between observations separated by lags h. The most important of these, however, is the **covariogram**, a plot of the covariances between all pairs of points which are a distance Δh apart. The equation for the spatial covariance (or autocovariance) for lag h is

$$C_h = \sum_i^{n-h} \left(x_i x_{i+h} - \overline{X}_i \overline{X}_{i+h} \right) \Big/ n - h \qquad (4.102)$$

where $\overline{X}_i$ is the mean of observations x_i and $\overline{X}_{i+h}$ is the mean of observations x_{i+h}. You will recognize that this equation is identical to Equation (4.87), but rewritten in the symbology commonly used in geostatistics. In principle, if the sequence is stationary and infinite (or at least very long compared to the lag), these two means will be the same and the correction term will be $\overline{X}^2$.

Most geostatistical discussions in the current literature are expressed in terms of spatial covariances, in part because this makes it easier to tie the topic into classical statistics. However, the semivariogram is the traditional measure of spatial variation originally developed within geostatistics, and for this reason it has been used here. All of the applications of semivariograms, including the modeling of spatial dependence and solution of kriging equations, can be performed using covariograms. In fact, the kriging equations may be somewhat easier to solve using covariances because zero elements (corresponding to γ_0) do not arise.

As **Figure 4–60** implies, it is easy to convert a semivariogram into a covariogram; all that is necessary is to subtract γ_h from a constant. In fact, Deutsch and Journel (1998), in their library of geostatistical routines, specify spatial continuity using semivariograms but then convert the semivariogram models to covariance models for solution of the kriging equations. **Figure 4–66** shows both a semivariogram and covariogram for a time series from the Eisenerz iron mine in Austria. The data from which these functions were calculated will be described more completely in the chapter exercises. They are automated measurements of the percent iron in samples from successive batches of ore that are moved at 1-hr intervals into a blending pile. The semivariogram and covariogram calculated from these stationary data are mirror images.

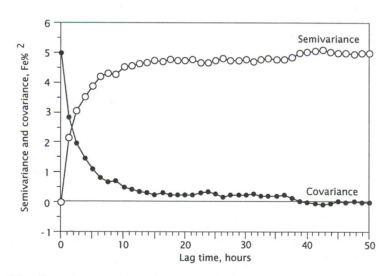

Figure 4–66. Mirror-image relationship between semivariogram and covariogram of stationary data. Functions calculated for time series of percent Fe in ore batches entering a blending pile at Eisenerz, Austria. Intervals between measurements are 1 hr in duration.

Spectral Analysis

Spectral analysis is known, in its various guises, as harmonic analysis, Fourier analysis, and frequency analysis. The roots of the method are interwoven with those of musical theory, reflecting a common concern with vibratory motion. Kepler, in the seventeenth century, applied the harmonic relationships he found in arithmetic, geometry, and music to the positions of the planets, and discovered the laws of planetary motion. However, it was Jean Baptiste Fourier (1768–1830) who provided proof that any continuous, single-valued function could be represented by a series of sinusoids, the relationship that now bears Fourier's name. Some of the terminology of spectral analysis was coined by electrical engineers for use in the analysis of electrical signals. Although an electrical signal is an energy wave that changes with time, engineers are accustomed to seeing the signal "frozen" on an oscilloscope. Engineering discussions of Fourier analysis always assume that the signal is time-variant; however, the fact that engineers examine the signal as a spatial phenomenon on an oscilloscope indicates that time and space can be considered equivalent. This is, of course, mathematically correct, and the unconscious ease with which electrical engineers make the transition should assure us that the two are conceptually interchangeable as well.

A *time series* is a sequence of observations taken at equal intervals of time, and a *spatial series* is a sequence of observations taken at equal intervals of distance; both can be treated, with some reservations, equivalently. (In the following discussion, you can read "spatial" for "time" where appropriate.) These sequences often are cyclic in their nature. Spectral analysis identifies the frequencies of cycles in a time series and estimates the amount of variation being contributed by each cycle. Spectral analysis can be approached in many ways. We will first consider it as a special type of regression that uses sine and cosine transformations. Later, we will examine other ways of developing spectral analysis.

A quick review of trigonometry

The definitions of the basic angular relationships of trigonometry:

$$\cos\theta = \frac{X}{V} \qquad \sin\theta = \frac{Y}{V} \qquad \tan\theta = \frac{Y}{X} = \frac{\sin\theta}{\cos\theta}$$

are illustrated by **Figure 4–67**. Note that the reference circle is labeled in degrees from 0° to 360°, and also in radians. A *radian* is defined as the angle at the center of a circle subtended by an arc whose length is equal to the radius of the circle, or V. The number of radians that can be fitted around the circumference of a circle is 2π, so one radian is equal to 57.296°. A rotating vector will pass through 360° or 2π radians when making one complete revolution; it will then pass through a second cycle that is indistinguishable from the first. That is, the trigonometric functions for a revolution and a half (3π radians or 540°) are identical to those for a half-revolution (1π radians or 180°). The values of the functions repeat again and again with successive revolutions.

If we graph values of either the sine or cosine of a rotating vector against the angle of the vector, the graph will be a sinusoidal wave form similar to that shown in **Figure 4–68**. The wave form repeats each time the vector makes a complete revolution. If V is taken to be a unit vector (*i.e.*, $V = 1$), the equation of the curve shown is

$$Y = \cos\theta$$

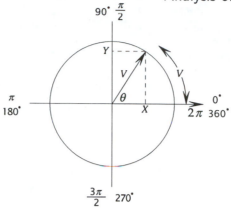

Figure 4–67. Basic trigonometric relationships expressed as functions of a vector V of unit length.

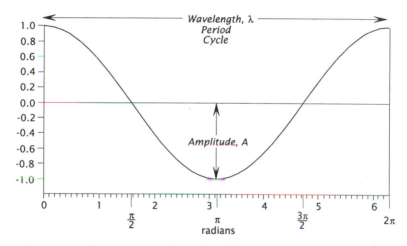

Figure 4–68. Terms applied to a regularly repeating cosine wave expressed in radians. Total length of curve is defined as equal to 2π radians.

where θ is given in radians. As shown, the wave form oscillates between $+1$ and -1, but this **amplitude** can be changed by multiplying $\cos \theta$ by any constant A.

$$Y = A \cos \theta$$

The distance between two similar points on successive wave forms is called the **wavelength, period,** or **cycle.** In the illustration, one wavelength is indicated as the distance from crest to crest. **Frequency** is the reciprocal of wavelength, being the number of wave forms, periods, or cycles that occur in some interval of time or distance.

The frequency, or number of waves in the basic interval, may be changed by multiplying θ by an integer, k:

$$Y = \cos(k\theta)$$

The cosine wave shown in **Figure 4–68** starts with the crest at zero radians, but the crest may be shifted either forward or backward by subtracting a **phase angle,** ϕ. The phase angle may assume any value between 0 and 2π.

$$Y = \cos(\theta - \phi)$$

267

Both the phase angle, ϕ, and the amplitude, A, should be subscripted because they are associated with a specific frequency, k, which is referred to as the **harmonic number**. Combining everything yields a general cosine equation for describing the form of any simple sinusoidal wave form:

$$Y_k = A_k \cos(k\theta - \phi_k)$$

This equation can be rewritten utilizing a trigonometric identity for the difference between two angles,

$$\cos(R - S) = \cos S \cos R + \sin S \sin R$$

to yield

$$Y_k = A_k \cos \phi_k \cos(k\theta) + A_k \sin \phi_k \sin(k\theta) \tag{4.103}$$

We can simplify this equation by defining two coefficients, $\alpha_k = A_k \cos \phi_k$ and $\beta_k = A_k \sin \phi_k$, giving

$$Y_k = \alpha_k \cos(k\theta) + \beta_k \sin(k\theta) \tag{4.104}$$

Harmonic analysis

Any time series, regardless of how complex in form (except that it must be continuous or without breaks, and there can be only one value of Y for each value of X), can be represented as the sum of a series of cosine wave forms. This is an expression of the **Fourier relationship**,

$$Y = \sum_k \left[\alpha_k \cos(k\theta) + \beta_k \sin(k\theta) \right] \tag{4.105}$$

Equation (4.105) is a linear equation (that is, all of the terms are added together) and resembles in form the polynomial regressions we considered earlier in this chapter. In the Fourier equation, the trigonometric terms $\cos(k\theta)$ and $\sin(k\theta)$ are equivalent to the power terms of a polynomial, such as X^2 and X^3. The α_k and β_k coefficients of these terms can be found by regression. However, if we attempt to estimate the coefficients by least-squares matrix methods, we quickly face insurmountably large computational problems if many harmonics are to be determined.

Note that a time series has the general form $Y = f(x)$ where x is measured in units of time or distance, but Equation (4.105) is expressed in terms of θ in radians, a measure of angle. The scale of x can be converted to radians by

$$\theta_j = \frac{2\pi x_j}{X} \tag{4.106}$$

where X is the total length of the time series. In effect, we have wrapped the time series in a circle, beginning at zero radians and ending at 2π radians. If the observations are equally spaced (which is almost always the case in time series), we can ignore the units of measurement of the x-scale entirely, and simply consider the number of observations, n, in the series:

$$\theta_j = \frac{2\pi j}{n} \tag{4.107}$$

where j is an integer ranging from 0 to $n-1$ (we use j as the index subscript rather than i to avoid later confusion with the imaginary number $i = \sqrt{-1}$). That is, the "location" of the first observation in the time series is θ_0 and the last is θ_{n-1}. The next observation in the series would be $\theta_n = 2\pi$, the start of the next cycle.

We can now express the Fourier relationship of Equation (4.105) as a regression model for $j = 0, \ldots, n$

$$y_j = \alpha_0 + \sum_{k=1}^{\frac{n-1}{2}} \left[\alpha_k \cos\left(2\pi j \frac{k}{n}\right) + \beta_k \sin\left(2\pi j \frac{k}{n}\right) \right] \tag{4.108}$$

The α's and the β's are regression coefficients to be estimated. The independent variables are the sine and cosine functions evaluated for the terms inside the parentheses. Note that k/n represents the frequency, the number of cycles per one unit of time or distance. The equation contains two unknowns for each k, of which there are $(n-1)/2$ values, so with the addition of the constant term, α_0, there is one coefficient for each observation in the series. The regression model will fit the data exactly.

The Fourier coefficients can be estimated by

$$\beta_k = \frac{2}{n} \sum_{j=0}^{n-1} y_j \sin\left(2\pi j \frac{k}{n}\right)$$

$$\tag{4.109}$$

$$\alpha_k = \frac{2}{n} \sum_{j=0}^{n-1} y_j \cos\left(2\pi j \frac{k}{n}\right)$$

Because of trigonometric relationships, the coefficient β_0 is always zero and α_0 simplifies to the mean of the time series, or

$$\alpha_0 = \frac{1}{n} \sum_{j=0}^{n-1} y_j \tag{4.110}$$

Once we have found the α_k and β_k coefficients, we can estimate the amplitude of the kth harmonic by

$$A_k = \left(\alpha_k^2 + \beta_k^2\right)^{1/2} \tag{4.111}$$

and its phase angle by

$$\phi_k = \arctan\left(\frac{\beta_k}{\alpha_k}\right) \tag{4.112}$$

Following the development in Christensen (1991), we can write the regression model in matrix form. First, we must define several vectors and matrices. The observations, y_j, in the time series form the $n \times 1$ column vector

$$\mathbf{Y} = [y_1, \ldots, y_n]'$$

The cosines of the successive points in the time series also form an $n \times 1$ column vector

$$\mathbf{C}_k = \left[\cos\left(2\pi 1 \frac{k}{n}\right), \cos\left(2\pi 2 \frac{k}{n}\right), \ldots, \cos\left(2\pi n \frac{k}{n}\right) \right]'$$

as do the sines,

$$\mathbf{S}_k = \left[\sin\left(2\pi 1\frac{k}{n}\right), \sin\left(2\pi 2\frac{k}{n}\right), \ldots, \sin\left(2\pi n\frac{k}{n}\right) \right]'$$

These can be combined into the $n \times 2$ matrix

$$\mathbf{Z}_k = [\mathbf{C}_k, \mathbf{S}_k]$$

There will be k such matrices, one for each harmonic number. We also can define a 2×1 vector of unknown coefficients,

$$\boldsymbol{Y}_k = [\alpha_k, \beta_k]'$$

Again, there will be k of these vectors, one for each harmonic. The regression model for an odd number of observations in the time series is

$$\mathbf{Y} = \alpha_0 + \sum_{k=1}^{\frac{n-1}{2}} \mathbf{Z}_k \boldsymbol{Y}_k \tag{4.113}$$

If n is even, the equation is slightly more complicated, as described in Christensen (1991). However, a time series containing an even number of points can always be made to contain an odd number simply by adding or deleting one observation from either end.

The vectors $\mathbf{C}_k$ and $\mathbf{S}_k$ are orthogonal, so the order in which the coefficients are estimated has no influence on the estimates. Also, the sum of squares associated with each coefficient is not influenced by other coefficients that may or may not be estimated. The sum of squares associated with the cosine coefficient $\mathbf{C}_k$ can be designated SSC_k and the sum of squares associated with the sine coefficient $\mathbf{S}_k$ can be designated SSS_k. The total sum of squares associated with frequency k/n is the sum of the two sums of squares for the two coefficients, $SSC_k + SSS_k$. Therefore, the mean square or variance associated with each frequency is simply

$$s_k^2 = \frac{SSC_k + SSS_k}{2} \tag{4.114}$$

In terms of the individual coefficients, this is equivalent to

$$s_k^2 = \frac{\alpha_k^2 + \beta_k^2}{2}$$

$$= \frac{A_k^2}{2} \tag{4.115}$$

or half the square of the amplitude.

Since the Fourier theorem holds that a time series can be regarded as the sum of many orthogonal sinusoidal functions or harmonics, the total variance of the time series must be the sum of the variances of these same harmonics. We can therefore express the variance of an individual harmonic k as a proportion of the total variance of the time series from which it has been derived. If we compute a large series of harmonics, the variances of the successive harmonics can be plotted in a **periodogram**, or as it is sometimes called, a **discrete power spectrum** or **line power spectrum**. The periodogram is defined only for integer values of k from 1

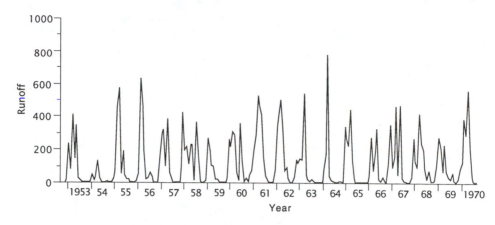

Figure 4–69. Eighteen-year record of monthly runoff of Cave Creek, in Kentucky, in hundredths of inches.

Table 4–20. Monthly runoff of Cave Creek, Kentucky, given in hundredths of inches. Data are collected during a "water year" beginning in October of the preceding calendar year and ending in September, in order to avoid breaking the sequence in the middle of the winter (after Haan, 1977).

Water year	Oct.	Nov.	Dec.	Jan.	Feb.	Mar.	Apr.	May	Jun.	Jul.	Aug.	Sep.
1953	2	5	19	240	86	416	147	354	31	18	7	1
1954	0	2	4	54	22	40	139	35	8	7	6	14
1955	2	4	30	73	463	579	59	197	55	24	28	3
1956	4	6	13	59	637	469	192	28	32	64	38	8
1957	7	10	172	308	325	103	392	68	24	6	5	2
1958	3	106	432	200	221	117	235	236	19	369	170	12
1959	6	9	17	270	195	112	102	24	24	5	4	2
1960	3	36	269	219	313	291	68	19	364	138	14	30
1961	12	52	79	204	295	532	476	414	159	48	18	4
1962	2	6	76	346	401	508	330	79	96	30	8	7
1963	39	141	124	150	146	548	52	25	14	29	11	3
1964	1	4	3	87	173	788	45	21	11	8	2	16
1965	15	7	347	276	230	449	146	31	8	5	1	2
1966	4	2	2	48	281	79	202	332	25	14	41	11
1967	7	119	357	97	161	466	50	476	33	14	15	7
1968	9	38	271	135	98	425	238	199	91	29	75	16
1969	14	22	112	278	216	73	237	74	40	27	66	17
1970	7	25	91	130	389	291	568	206	38	14	6	27

to $(n-1)/2$, and is simply a plot of variance or power versus harmonic number, k. The terms **variance** and **power** are synonymous; the latter arose in electrical engineering and is now widely used in all fields of engineering, signal analysis, and pattern recognition.

The time series shown in **Figure 4–69** represents the monthly flow of water in Cave Creek in Kentucky; data (from Haan, 1977) are listed in **Table 4–20** and are listed in sequential form in file CAVECREK.TXT. The periodogram shown in **Figure 4–70** has only a single pronounced spike at $k = 18$. Since the record covers the 18-yr

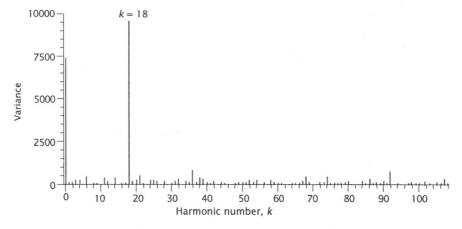

Figure 4–70. Periodogram of runoff data for Cave Creek, in Kentucky. Spike at harmonic number $k = 18$ represents an annual cycle, since the record in file CAVECRK.TXT contains monthly values for 18 yr.

interval from 1953 to 1970, the time series can be interpreted as being composed of an annual or seasonal cycle in water flow, plus random variation.

Note there is also a spike at $k = 0$. This variance is simply half the square of the α_0 coefficient, as the β_0 coefficient has vanished. As we have seen, α_0 is equivalent to the mean of the time series. The k_0 term corresponds to the variance of a cycle with an infinite wavelength; such a cycle would be a flat line throughout the extent of the time series. The variance is the square of this constant value, or the mean. If the mean outflow of Cave Creek had been subtracted from each value of the time series in **Table 4–20** before calculating the periodogram, the spike at $k = 0$ would not be present.

The variance estimates in the periodogram are subject to error from several sources. First, only the specific k/n harmonics can be calculated from a time series of length n, and different harmonics would result if n were changed. A true frequency present in the time series is unlikely to correspond exactly to one of the frequencies in the selection we calculate, especially if the observed time series is considered to be a discrete sample from a continuous process. If there is a periodicity in the time series, it may have a frequency that lies somewhere between $(k - 1)/n$ and k/n. We would expect the variance of this true frequency to be smeared out over the nearby frequencies that can be estimated from this particular time series.

If we want a measure of the importance of all of the frequencies in a neighborhood around k/n, it would seem reasonable to calculate an average of the variances of nearby frequencies. In doing so, we will create a smoothed approximation of the periodogram called the ***estimate of the spectral density***, or simply the ***spectral estimator***. Conventionally, the spectral estimator is shown in the form of a continuous plot of variance versus harmonic number rather than in the discrete form of a periodogram. Because our time series is regarded as a discrete sample of a continuous process, the underlying spectrum must also be continuous.

File ICECORE.TXT contains a time series that we do not anticipate to be strictly periodic. The series is a record of the ratios between isotopes ^{18}O and ^{16}O, as measured in specimens of ice from the GISP-2 core taken in the middle of the

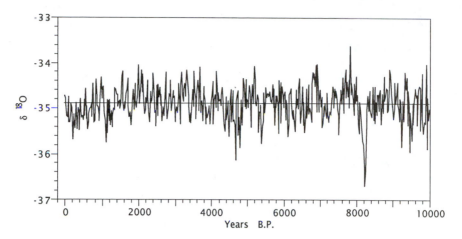

Figure 4–71. Variation in 20-yr average values of $\delta^{18}O$ measured on the GISP–2 ice core from Greenland. Observations are recorded in years before present (B.P.), where the "present" is defined as 1950. Record extends through the Holocene, from -30 B.P. to 10,000 B.P.

Greenland ice sheet. The ratio between the two oxygen isotopes in the atmosphere (called the $\delta^{18}O$ record) varies because the lighter ^{16}O evaporates more rapidly from seawater as the water temperature rises. The atmospheric $\delta^{18}O$ ratio is preserved in snow, which accumulates in annual layers like varves. The year in which any particular layer was deposited can be determined simply by counting the layers down from the surface (Lowe and Walker, 1997). Data in the file consist of $\delta^{18}O$ measurements averaged over 20-yr intervals, and covers the last 10,000 yr. This segment of the GISP-2 core is shown in **Figure 4–71**.

Note that from the past to the present, the values of $\delta^{18}O$ steadily decrease; this trend can be removed by fitting a linear regression of $\delta^{18}O$ against time, shown as a straight line in the figure. Values of Y are then replaced by their residuals, $(Y - \hat{Y})$. This process, called **detrending**, not only removes the trend, it also removes the mean.

The periodogram of the ice core data, computed on the residuals from the regression, is shown in **Figure 4–72**. It is much more erratic than the periodogram of the Cave Creek time series, suggesting that the ice core data are not dominated by regular cycles. There is an especially pronounced spike at $k = 11$, representing a cycle having a wavelength of about 909 yr. Because the data have been detrended, the variance at $k = 0$ (representing the mean of the series) is equal to zero and there is no spike as there is in **Figure 4–70**.

The smoothed estimate of the spectral density can be made by combining adjacent harmonics in the periodogram. Conventionally, this is done by **filtering**, which simply is a weighted moving average that extends over a small span of adjacent harmonics. The set of weights is called a **spectral window**, or filter. One commonly used window is the Tukey–Hanning filter,

$$\hat{s}_j^2 = \tfrac{1}{4}s_{j-1}^2 + \tfrac{1}{2}s_j^2 + \tfrac{1}{4}s_{j+1}^2 \tag{4.116}$$

Christensen (1991) suggests the use of a simple three-point moving average, which has been used to create the spectral density plot in **Figure 4–73**. Note that the spectrum still displays a single prominent peak, which now has been shifted to

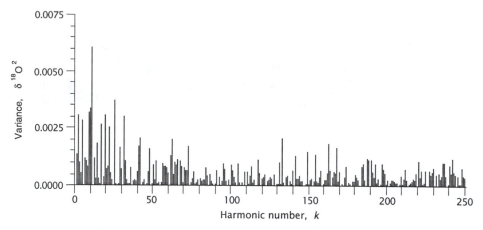

Figure 4–72. Periodogram of $\delta^{18}O$ values for the last 10,000 yr in the GISP–2 ice core from Greenland. Because data have been leveled, there is no spike at $k = 0$. Absence of pronounced spikes indicates record is stochastic, not periodic.

slightly lower frequencies. This peak represents a cyclic component having a wavelength of approximately 1000 yr. Possible causes of this component are conjectural; it has a frequency which is much higher than any of the Milankovitch orbital phenomena. Perhaps it reflects the approximate cycle time of feedback mechanisms that operate between the ocean and the atmosphere; these are believed to be the source of more pronounced oscillations of similar duration (the so-called "Dansgaard-Oeschger events") during the Weichselian (Wisconsinan) glacial period (Lowe and Walker, 1997).

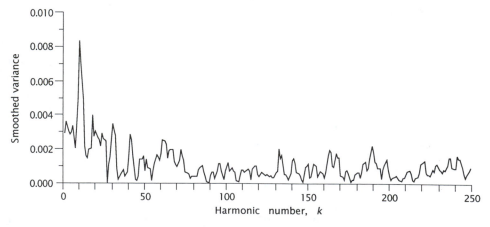

Figure 4–73. Spectral density of $\delta^{18}O$ values for the last 10,000 yr in the GISP–2 ice core from Greenland, estimated by smoothing periodogram in Figure 4–72. Largest peak represents a spectral component having a wavelength of about 1000 yr.

Error in the periodogram may also result from ***aliasing***, as illustrated in **Figure 4–74**. The highest frequency that can be estimated in the periodogram has a harmonic number of $(n-1)/2$. This is called the ***Nyquist frequency***, in which each wave form is defined by three successive observations and whose wavelength is equal to 2Δ, twice the sample spacing. The true sinusoidal wave form shown by a

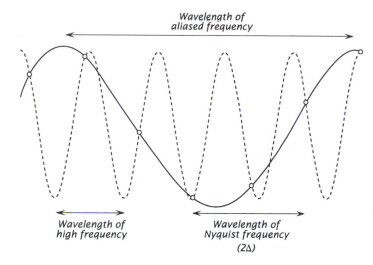

Wavelength of aliased frequency

Wavelength of high frequency

Wavelength of Nyquist frequency (2Δ)

Figure 4–74. High-frequency sinusoidal wave (dashed line) sampled at discrete points yields apparent lower frequency wave (solid line).

dashed line has a frequency that is higher than the Nyquist frequency as fixed by the sample spacing. Successive observations yield an apparent wave form shown by the solid line, or alias, whose frequency is only about one-fourth that of the actual signal. The variance of all wave forms whose frequencies are higher than the Nyquist frequency will be added to the variances of lower frequencies in the periodogram.

The continuous spectrum

Harmonic analysis and the construction of the periodogram or line spectrum is appropriate if the time series being investigated is truly periodic. There are few naturally occurring phenomena that possess true periodicity, except for those related to astronomical cycles such as monthly tides or seasonal changes. Most geologic time series are not periodic, but are stochastic. A sequence is *stochastic* if it can be characterized only by its statistical properties; this is in contrast to a deterministic sequence whose state can be predicted exactly from its coefficients.

Even though a time series may not contain truly periodic components, the methods of spectral analysis can provide valuable insights into the behavior of the process that has generated the sequence. Most time series can be regarded as continuous sequences, even though they ordinarily are sampled at discrete points. It is possible to calculate a *continuous spectrum* or *spectral density function* for such sequences, in which the variance of the time series is apportioned among a set of *frequency bands*. The continuous spectrum has the form of a continuous plot of variance versus frequency, and is analogous to a continuous probability distribution, with variances proportional to the areas under the spectral curve between limiting frequencies. The total area under the spectrum is equal to the total variance of the time series from which it is calculated. In contrast, the line spectrum of a periodic time series shows the variances attributable to defined individual frequencies.

The specific sequence of observations we wish to analyze can be considered a random sample from a large, perhaps infinite, set of such time series that could be produced by the process under study. The complete set of time series is called an *ensemble*, and is the population from which our particular sample has been taken.

A time series is said to be stationary (or time-invariant or nonevolutionary) if its properties do not change with time. A spatial series with the same characteristics is said to be *homogeneous*. If a time series is divided into small segments and the means of all these segments tend to be the same (and the same as the mean of the entire time series), it is referred to as being *first-order stationary* or stationary in the mean. If, in addition, the autocovariance changes only with lag and not with position along the time series, the series is *second-order stationary*. This is also called *weak stationarity*, or stationarity in the wider sense. If all higher moments are dependent only on lag and not on position, the series is *strongly stationary*, or possesses stationarity in the strict sense.

If a time series is not only strongly stationary, but all statistics are invariant from time series to time series within the ensemble, the ensemble is *ergodic*. Many statistical tests of time series assume ergodicity, just as univariate statistical tests may assume uniformity of variance. We may check for ergodicity if we have several time series that are all realizations of the same stochastic process. Commonly, however, we have only one time series available, and then no check is possible. We can test for stationarity in the mean and variance of a single time series by regression, or by dividing the series into segments and testing to see if the statistics for the segments are the same. If they are, the series is *self-stationary*, and ergodicity may be assumed. If the series is not self-stationary, the ensemble from which it is derived cannot be ergodic.

Sometimes a time series can be made stationary by *leveling* or *detrending*, which consists of subtracting a linear trend from the observations as was done to the $\delta^{18}O$ data. If the original series was characterized by a slow change in average value, the detrended series will have a stationary mean of zero.

For a stationary, stochastic time series that is continuous and sampled at discrete, equally spaced points, the continuous variance (or power) spectrum may be calculated by several methods. The most widely used approach involves calculating many values of the line spectrum by the Fast Fourier Transform (FFT) computer algorithm. These spectral values are then averaged across frequencies to produce a smoothed estimate of the continuous spectrum. The Fast Fourier Transform, as its name implies, is extremely rapid and requires only $n \log_2 n$ arithmetic operations rather than n^2 operations required by alternative procedures. In the FFT approach, the Fourier relationship must be expressed in complex form,

$$s_k^2 = \frac{1}{n} \sum_{j=-n/2}^{n/2} y_j e^{(-i2\pi jk/n)} \tag{4.117}$$

where i is the imaginary number $\sqrt{-1}$. Developing an FFT algorithm (there are several alternatives) using complex arithmetic requires mathematical and computational skills beyond those assumed in this book, so we will not pursue it further, except to note that most spectral analysis packages are based on an FFT algorithm. Typically, the time series must contain exactly N points where N is a power of 2, although there are procedures in which the length of the time series is based on the prime factors of the number of points. The largest prime factor determines how fast the algorithm runs—the smaller, the faster. An excellent introduction to the Fast Fourier Transform, including a discussion of complex arithmetic, is given in Rayner (1971), which also describes applications of the FFT in geography. Press and others (1992) provide a succinct discussion of the original FFT algorithm, as well as short descriptions of others. The first landmark article describing the Fast Fourier Transform algorithm as we now know it was published by Gentleman and Sande (1966), based upon the mathematics previously introduced by Cooley and Tukey (1965). Specialized books such as Nussbaumer (1982) and modern texts on time-series analysis (*e.g.*, Nerlove, Grether, and Carvalho, 1995; Bloomfield, 2000) treat the subject extensively.

In performing Fourier analysis, we have transformed our data from one domain to another. We began with observations in the form of values, y_j, at points in space or time, x_j. The succession of points forms a signal or wave form, defined by X- and Y-coordinates. The data, defined in this manner, are said to be in the time or spatial domain, depending upon whether X denotes points in time or distance. By determining the component frequencies in the signal, we have transformed the data to the frequency domain. A physical analogy can be drawn with the effect of a glass prism on sunlight, as shown in **Figure 4–75**. A beam of white light can be regarded as a complex wave form changing with time and composed of many colors (or wavelengths) of light. A prism acts as a frequency analyzer and separates the beam into its components, which appear as a rainbow display. Each colored band is separated from its neighbor by an amount proportional to the difference in their wavelengths or frequencies, and the intensity of each band is proportional to the contribution of that particular wavelength to the total intensity of the original beam. Examining the spectrum of a light source may tell us a great many things: the composition of the source, its temperature, the nature of the material through which the light passed, and so forth. In a similar fashion, examining the power spectrum of a data sequence may tell us a great deal about its nature and origin— information that may not be apparent in any other way.

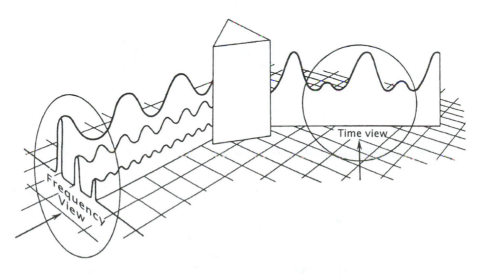

Figure 4–75. A prism acts as a frequency analyzer, transforming white light (time or spatial domain) into its constituent spectrum of colors (frequency domain). Courtesy of Hewlett–Packard Corp.

EXERCISES

Exercise 4.1

Both consistency of variance and autocorrelation of residuals are of interest in an analysis of a mining prospect in northern Quebec. A long trench has been cut by bulldozer across a gold prospect. Samples have been taken at intervals along the trench and the gold value of each sample has been determined by assay. A trend in the values seems apparent, and deviations from the regression seem more extreme near one end of the trench. This is suggestive of the "bonanza" characteristic of many gold deposits. Such deposits often have extremely low values through most of the mineralized zone, but contain occasional rich veins. In addition, large positive deviations seem to occur in clusters or runs, which also suggests that the trench transects mineralized veins. From the data given in file QUEBECAU.TXT and shown in **Figure 4–76**, determine the trend of gold values and examine the consistency of variance along the traverse. Use a runs test of the signs of the deviations to determine if they are distributed randomly about the regression. From the results of your analysis, does it seem possible that mineralized veins have been cut by the trench? Does extrapolation of gold values beyond the limits of the trench seem wise in view of the behavior of deviations from the regression?

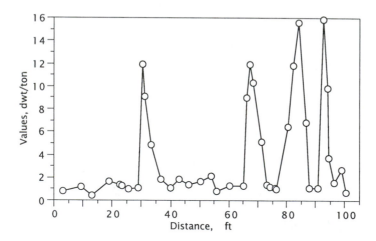

Figure 4–76. Gold assay values (in pennyweights/ton) of material collected in a prospect trench in northern Quebec. Data are listed in file QUEBECAU.TXT.

Exercise 4.2

The data shown in **Figure 4–77** (given in file GREENRIV.TXT) represent an unusual phenomenon in our science; the values are observations of a true time series from the geologic past. Only in very special circumstances are datable occurrences recorded in the rocks that allow us to establish a definitive time scale for a geologic sequence. The Eocene lake deposits of the Rocky Mountains consist of thinly laminated dolomitic oil shales hundreds of feet thick. It has been well established that the laminations are varves, or layered deposits caused by seasonal climatic changes in the lake basins. By measuring the thickness of these laminations, we can record annual changes in the rate of deposition through the lake's history. File GREENRIV.TXT contains the thickness in millimeters of a varved section deposited near the western shore of one of these major lakes in western Wyoming. We may attempt to answer several questions with these data. For example, was there a trend in the rate of deposition of dolomite through time, perhaps caused by a gradual climatic change? Is there evidence of cyclicity in the thickness of the laminae, possibly related to astronomical phenomena? Because the cyclicity we are seeking may have periods of many years (sunspot cycles, for example, last 11 yr), 101 observations have been made and are given in the file. Process the varve data and determine if significant trends or periodicities in varve thickness exist. Remember, the data must be stationary, so any significant linear trend must be removed prior to autocorrelation analysis.

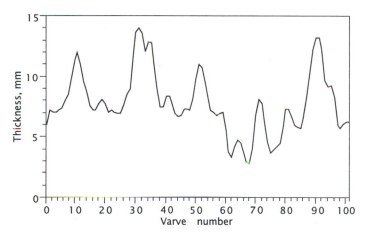

Figure 4–77. Thickness of varve laminae (in millimeters) plotted against cumulative number of varves from an arbitrary origin in a stratigraphic section of the Green River Shale in western Wyoming. Data are given in file GREENRIV.TXT.

Exercise 4.3

A large area of Precambrian anorthosite in the Laramie Range of Wyoming contains several bodies of magnetite. One of these is mined and the rock is crushed for use as high-density drilling mud. An igneous petrology class has visited the quarry and collected a suite of equally spaced samples along a traverse across the magnetite body. The samples vary, some are predominantly plagioclase, others mostly olivine, some almost entirely magnetite, and others are mixtures of the three minerals. Class members disagree as to whether a systematic variation in mineralogy occurs across the magnetite body. To test the possibility, the specific gravities of the samples were determined and are given in **Table 4–21** and shown in **Figure 4–78**.

Table 4–21. Specific gravities of specimens collected along a traverse across a magnetite body in the Laramie Range of Wyoming.

Specific Gravity	Sample Number	Specific Gravity	Sample Number
3.57	1.0	4.55	21.0
3.63	2.0	4.61	22.0
2.86	3.0	4.93	23.0
2.94	4.0	4.60	24.0
3.42	5.0	4.51	25.0
2.85	6.0	3.98	26.0
3.67	7.0	4.22	27.0
3.78	8.0	3.52	28.0
3.86	9.0	2.91	29.0
4.02	10.0	3.87	30.0
4.56	11.0	3.52	31.0
4.62	12.0	3.77	32.0
4.31	13.0	3.84	33.0
4.58	14.0	3.92	34.0
5.02	15.0	4.09	35.0
4.68	16.0	3.86	36.0
4.37	17.0	4.13	37.0
4.88	18.0	3.92	38.0
4.52	19.0	3.54	39.0
4.80	20.0		

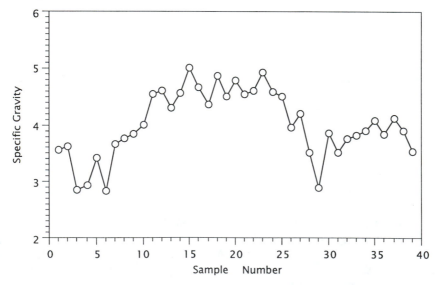

Figure 4–78. Variation in specific gravity of specimens of ultramafic rock collected along a traverse across an open-pit magnetite mine in the Laramie Range of Wyoming.

Is the variation in specific gravity along the traverse that which would be expected if the composition varied randomly about a central value?

Exercise 4.4

The Mowry Shale is a black, siliceous shale of Early Cretaceous age occurring in Colorado, Wyoming, and Montana. The interval is characterized by numerous bentonite beds, which were mined at several locations in Wyoming and Montana for drilling mud and foundry clay. Bentonite is composed almost entirely of montmorillonite, developed as an alteration product of rhyolitic or andesitic volcanic ash. **Table 4-22** and file MOWRY.TXT give the thickness between successive bentonite beds measured in an outcrop of Mowry Shale in Fremont County, Wyoming.

Table 4-22. Thickness, in feet, of intervals between successive bentonite beds in Cretaceous Mowry Shale in Fremont County, Wyoming.

Thickness of Intervals Between Bentonite Beds	Cumulative Height Above Base of Mowry Shale	Thickness of Intervals Between Bentonite Beds	CumulativeHeight Above Base of Mowry Shale
(Base of Mowry Shale)			
0	0	10	135
4	4	23	158
26	30	8	166
4	34	7	173
5	39	47	220
4	43	14	234
17	60	17	251
3	63	5	256
6	69	10	266
4	73	5	271
35	108	6	277
2	110	11	288
15	125	29	317
		(Top of Mowry Shale)	

A plot of thickness of interval versus distance above the base of the Mowry Shale is given in **Figure 4-79**. The bentonite beds represent ash falls from explosive eruptions of volcanoes in western Idaho. If it is assumed that the enclosing black shale was deposited at an approximately uniform rate, it may be possible to analyze this sequence of thicknesses as a series of events analogous to the historical series formed by the eruptions of Aso.

Test these data for a trend in the rate of occurrence. If none is observed, test for autocorrelation of successive intervals between events. Comment on (a) the possible effects of unequal rates of sedimentation of the black shale, and (b) the possibility that more than one volcano was active.

Exercise 4.5

The geothermal gradient can be determined by measuring the temperature in an observation well that has been left undisturbed sufficiently long so that the hole has come to thermal equilibrium with the enclosing rocks. Temperature measurements

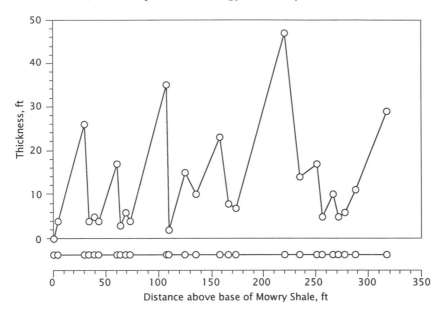

Figure 4–79. Thickness of bentonite beds in a section of Cretaceous Mowry Shale measured in Fremont County, Wyoming. Data are given in file MOWRY.TXT.

typically are made using a string of thermistors lowered into the hole. However, repeated measurements made over a period of time may exhibit temperature fluctuations that reflect the movement of fluids through convection cells that develop inside the bore hole. The durations of these fluctuations are determined by the viscosity of the fluid in the hole, the geothermal gradient, and the diameter of the hole. File TEMPER.TXT contains measurements taken at 5-min intervals at a depth of 180 m in the Foster No. 6 well in San Jacinto County, Texas (Gretener, 1981). The borehole fluid is freshwater, the interior diameter of the casing is 22 cm, and the local geothermal gradient is over $20°C/km$. The time series covers slightly less than 29 hr. Determine the spectrum to see if there is a significant periodicity to the temperature variations in the series. Remember to check for possible nonstationarity and to correct for a trend, if present. How would you explain a trend? If there are multiple peaks in the spectrum, can you offer an explanation for them?

Exercise 4.6

The presence of large amounts of sulfur in kerogen is believed to cause the generation of heavy oil at anomalously low temperatures because sulfur–carbon bonds are weaker than the hydrocarbon bonds of normal kerogen. This has led to the generation of unusual amounts of oil at very low thermal maturities in shales that were deposited in highly reducing environments, such as the Phosphoria, Woodford, and Monterey shales of California. File SULFUR.TXT contains API gravities and weight percent sulfur content of 48 oils from the Miocene Monterey Formation in Santa Barbara Channel and the offshore Santa Maria basin (Baskin and Peters, 1992). Increasing API gravity is considered to be an indication of increasing thermal maturity. What is the relationship between oil gravity and sulfur content in these oils, and does the relationship support the idea that sulfur content influences maturity?

Exercise 4.7

Many of the major rivers of the world have extensive deep-sea fans. The magnitudes of these fans presumably reflect the sediment load that has been transported by the river and the duration of time over which the river has been active. **Table 4–23** and file FANS.TXT list characteristics of 18 major river systems and their associated

Table 4–23. Modern river-fed, deep-sea fans.

Fan	Length, km	Volume, 1000 km³	Age, Ma	Deposition Rate, 10^6 cu m/yr	River Discharge, 10^6 MT/yr
Amazon	700	700	16.5	42	900
Columbia	310	27	1.0	27	13
Ganges	2800	4000	21	190	2180
Crati	16	0.001	0.006	0.15	2
Indus	1500	1000	16	66	480
St. Lawrence	1200	100	2	50	15
Magdalena	270	180	21	11.1	220
Mississippi	570	290	2.4	121	400
Nile	280	140	5.2	27	142
Fraser	260	9	0.6	15	20
Rhone	440	40	5.2	7.7	10
Valencia	160	6	5.2	1.3	7
Zaire	270	– – –	0.01	3	55
Cap Ferret*	75	13	2.0	6.5	
Delgada*	350	40	11.2	3.6	
La Jolla*	40	12	2.0	0.6	
Monterey*	300	64	24	2.7	
Navy*	40	0.08	0.4	0.2	

*Deep-sea fans in submarine canyons with no major inflowing river.

deep-sea fans (Wetzel, 1993). What is the relationship between deposition rate and size of submarine fan as expressed by its length? What is the relationship between deposition rate and sediment discharge? Are these relationships clearer if the data are transformed in some manner? The table contains data on five deep-sea fans which have no associated river source. What estimated sediment-discharge rate would be needed to produce these fans, if they were associated with rivers?

Exercise 4.8

Geomorphologists have held the concept that landscapes evolve to an equilibrium state in which erosion and transport of material are in balance. In maturity, the slope profiles of valley walls (where erosion occurs) should be related to the gradient of the stream in the valley (which controls removal of eroded material). The relationship is expressed as the slope ratio. **Table 4–24** contains means of valley wall slopes and stream channel gradients of second-order stream basins, measured on selected areas of mature topography (Strahler, 1950). Determine by linear regression the relationship (which may be more apparent if the slopes are logarithmically transformed) between the two sets of gradients. Which slope is logically the dependent variable?

Table 4–24. Mean channel and valley wall slopes, in degrees, for nine maturely dissected regions in the U.S.

Location	Channel Slope	Valley Wall Slope
Grant, LA	0.40065	1.9023
Grant, LA	0.90746	3.4267
Rappahannock, VA	1.8125	7.9292
Belmont, VA	2.3309	9.9339
Allen's Creek, IN	5.5518	17.6716
Hunter-Shandaken, NY	10.6780	31.4365
Mt. Gleason, CA	16.5837	32.2337
Petrified Forest, AZ	25.1157	54.5399
Perth Amboy, NJ	33.5416	48.1208

Exercise 4.9

On October 21, 1966, a colliery spoil heap in Aberfan, Wales, collapsed, burying 144 people. As a consequence of this disaster, the physical and chemical conditions of other abandoned colliery heaps in Great Britain, some over 100 years old, have been carefully examined. The data in file COLLIERY.TXT are shear strength measurements made on samples from the Brancepeth colliery tip in County Durham (Taylor, 1973). The triaxial principal stresses have been measured and resolved into the largest possible shear stress, $(\sigma_1 - \sigma_3)/2$, and the lowest possible shear stress, $(\sigma_1 + \sigma_3)/2$ (labeled Y-Axis and X-Axis in file COLLIERY.TXT). The sine of the angle of principal stress, ϕ, can be estimated as the slope of the best-fit line between these two variables. What are the coefficients of the line of best fit? Is there any significant difference between the slope as estimated by ordinary regression and by reduced major axis regression?

Exercise 4.10

File OREODONT.TXT contains measurements made on the fossil skulls of Oligocene oreodonts (Miller and Kahn, 1962). These pig-like mammals are very similar in appearance; paleontologists differentiate species primarily by the length of the face, the width of the skull, and the length and depth of the bulla, a depression below the opening of the ear. We will examine the relationship between two of the variables. Make a plot of the depth of the bulla versus length of the cheek-tooth row (variables d and b in file OREODONT.TXT). Mark each observation according to species, and construct bivariate ellipses around the resulting groups. Note the pattern formed by the groups. Can you fit a regression to the ungrouped observations so that the groups are arranged in a systematic manner? (This may require a polynomial regression or transformation of one or both variables.)

File OREODONT.TXT contains five columns of data. The first column, headed "Species," is a nominal identifier of species: DH = *Desmatochoerus hatcheri*, MC = *Merychoidodon culbertsoni*, MGL = *Megoreodon gigas loomisi*, OO = *Oreodontus osborni*, PM = *Prodesmatochoerus meeki*, P = *Psuedodesmatocherus* sp., S = *Subdesmatochoerus* sp. Column a is width of brain case. Column b is maximum length of cheek-tooth row. Column c is maximum length of bulla. Column d is maximum depth of bulla. All measurements are in millimeters.

Exercise 4.11

The Smackover Formation consists of dolomitized carbonates of Jurassic age. File SMACKOVR.TXT contains 46 measurements of porosity made by mercury injection on cores taken from wells in the Jay Field on the border between Alabama and Florida (Melas and Friedman, 1992). The file also contains porosities estimated from sonic logs from the same intervals as the cores. The calculations of sonic porosities are based on generalized mineralogical properties. It may be possible to improve log-based estimates by calibrating the log responses to core measurements. Determine a calibration curve that can be used to adjust well-log porosities, and estimate the "true" porosity (with an approximate 90% confidence interval) for (1) an interval whose log porosity is 10%, and (2) an interval with 15% log porosity.

Exercise 4.12

Figure 4–80 shows the percent clay in topsoil along a traverse taken at Sandford St. Martin, England (Webster and Cuanalo, 1975). Soil samples have been taken at 10-m intervals. The data are contained in file TOPSOIL.TXT. Calculate the experimental semivariogram and fit an appropriate model. What is the span, or range, of clay content, and what does this mean in terms of the relationship between observations? Does the semivariogram show a nugget effect, and if so, what is its magnitude?

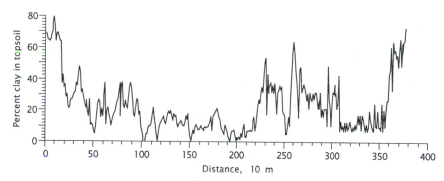

Figure 4–80. Clay, in percent, in topsoil along a traverse at Sandford St. Martin, England. Data are listed in TOPSOIL.TXT.

Exercise 4.13

Although we made discouraging remarks about attempts to perform lithostratigraphic correlation by computer, the special circumstances in this problem of geologic correlation not only make it appropriate to use cross-correlation, but it is about our only hope for success. **Table 4–25** contains measurements of varve thicknesses from a short sequence in the Green River Formation, an Eocene lake deposit in the Rocky Mountains. A similar but longer section has already been presented in **Figure 4–77** and given in file GREENRIV.TXT. These thinly laminated dolomitic oil shales are hundreds of feet thick. The two sections are only 10 mi apart, and presumably the shorter sequence is equivalent to some part of the larger. **Figure 4–77** shows the longer sequence and **Figure 4–81** shows the shorter sequence drawn to

Table 4–25. Thickness of successive varves in a stratigraphic section through the Green River oil shale; this section is located 10 mi north of section given in file GREENRIV.TXT.

(Top of section)						
10.8	11.7	11.0	9.9	9.8	9.9	10.0
10.0	10.2	10.8	11.3	12.0	13.5	15.6
15.0	13.4	14.6	13.0	10.3	9.4	9.0
10.1	10.3	9.2	9.1	9.0	9.2	10.7
10.8	8.9	9.4	10.6	12.6	14.2	12.3
11.1	11.0	(Bottom of section)				

the same scale. There are no marker beds or distinct features in this part of the monotonous varves, so we must base a correlation on the best match in thickness of individual laminations in the two sections. Compute the cross-correlogram and determine the position of best correlation between the two sections. How do you explain the other positions of lesser, but significant, match?

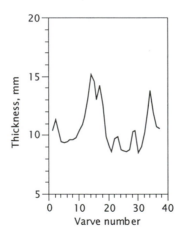

Figure 4–81. Thickness (mm) of varve laminae plotted against cumulative number of varves from arbitrary origin in short stratigraphic interval of Green River shale, western Wyoming; measured in outcrop about 10 mi from section shown in Figure 4–77.

Exercise 4.14

Some lunar basalts have thermal histories that are quite different than terrestrial basalts; they appear to have undergone impact melting and rapid recrystallization. Xenocrysts in lunar basalts have reaction rims whose dimensions reflect the difference in composition between the xenocryst and the enclosing melt and the length of time the rock was in a molten state. To understand details of the origin of basaltic breccias collected in the lunar highlands, experimental melts have been produced and cooled and the composition of xenocryst reaction rims measured by electron microprobe. A plot of composition versus distance from the edge of a xenocryst forms a "diffusion profile" whose shape approximates a differential equation for diffusion known as Fick's second law. File LUNARBAS.TXT contains electron microprobe measurements of mole percent fayalite (Fe_2SiO_4) determined at successive locations given as distances in microns from the edge of the xenocryst. Plot the data and fit an appropriate equation to the data by least squares. It may be necessary to use a quadratic model or a logarithmic transformation of a variable. The data are adapted from Sanford and Huebner (1980).

Exercise 4.15

The Eisenerz iron mine in Austria has been worked since the Middle Ages. The metamorphic deposit is composed of massive iron carbonate, mostly siderite, with admixed silicic materials. The mine runs continually, supplying ore to a large steel mill. To insure that the composition of the mill feed is appropriate, the output stream from the beneficiation plant at the Eisenerz mine is automatically sampled and analyzed. Depending upon its composition, the beneficiated ore is directed to one of three large storage piles. On the other side of the storage piles, material is removed and mixed in different proportions to obtain feed stock for the mill that has a nearly constant composition.

File EISENERZ.TXT contains a multivariate time series of the composition and tonnage of the output from the beneficiation plant for a 21-week period in 1996. The data represent 3524 average hourly measurements of Fe%, CaO%, SiO_2%, MgO%, MnO%, Al_2O_3%, and metric tonnes of beneficiated ore. The time series of Fe% was used to calculate the semivariogram shown in **Figure 4–66**, after leveling. Plots of the compositional variables versus time seem to contain periodicities, although the data are very noisy. Compute the spectra of the variables and examine them for common components. Do you anticipate any relationship between the spectra of Fe% and CaO%? Remember that the data may be nonstationary and some variables may require pretreatment before spectral analysis. These same data will be used again in Chapter 6 to examine their multivariate characteristics. The data are kindly provided by Voest-Alpine Erzberg Ges.m.b.H.

Exercise 4.16

The Cretaceous–Tertiary boundary seems marked by abrupt differences in faunal assemblages. Although the extinction of the dinosaurs has attracted the most popular and scientific attention, other organisms also experienced sudden changes in dominance. McKinney and his co-workers (1998) examined the relative importance of cyclostome and cheilostome bryozoans over time. They note that the number of species in a clade may reflect the group's diversity but does not necessarily indicate its dominance in the environment. As a novel measure of dominance, the authors weighed bryozoan fragments recovered from kilogram-sized samples of 70 marine continental shelf deposits that ranged in age from 140 million yr to the present. The data are given in file BRYOZOAN.TXT. The variables are age (millions of years B.P.) and proportion of bryozoan material that is of cyclostome origin (%). In their report, the authors show a cross plot of these two variables combined with a moving-average curve that suggests there is an abrupt jump in the proportion of cyclostome bryozoans after the Cretaceous–Tertiary boundary event. Analyzing data that are suspected to contain discontinuities is difficult; Yevjevich (1972) has an especially lucid discussion of such transients. A simple approach, appropriate for the data in BRYOZOAN.TXT, is to fit two successive linear models. The simplest is a straight-line regression of cyclostome proportion on age, which is equivalent to postulating that there is no discontinuity in the proportion of cyclostome bryozoans at the Cretaceous–Tertiary boundary. A more complicated model incorporates a jump at the boundary by fitting a two-part linear regression in which the pre-Tertiary data are fitted separately from the post-Cretaceous data. An ANOVA comparison of the residual mean squares will indicate if the two-part model is significantly better than a simpler linear model (remember that the two-part model requires four degrees of freedom).

Exercise 4.17

The velocity at which seismic waves travel is dependent on the density of the rocks traversed and their elastic properties. In turn, rock density depends upon composition or lithology. If sonic velocities can be measured and rock compositions are assumed, the geometry of rock bodies along a seismic profile can be modeled. To calibrate the modeling process, experimental measurements of sonic velocity and density have been made on very large collections of Earth materials; file SONIC.TXT contains one such set of measurements. These data include p-wave (compressional) seismic velocities in km/sec and rock densities in g/cm^3 for 357 igneous and metamorphic rocks (Schön, 1983). The rocks have been classified into eight lithologies (1—granite; 2—biotite and biotite–amphibole gneiss; 3—kyanite–garnet–biotite gneiss; 4—amphibolite and amphibole gneiss; 5—granulite; 6—diorite and gabbrodiorite; 7—gabbronorite; 8—ultrabasite) of increasing density. Plot these data and determine the relationship between density and p-wave velocity by regression. It may be necessary to use a polynomial regression or a transformation to obtain an adequate model. Use analysis of variance to test the statistical significance of the regressions. Do the individual lithologic categories exhibit the same relationship between velocity and density as seen in the entire suite of rocks?

SELECTED READINGS

Armstrong, M., and R. Jabin, 1981, Variogram models must be positive-definite: *Mathematical Geology,* v. 13, no. 5, p. 455–459.

Bardwell, G.E., 1970, Some statistical features of the relationship between Rocky Mountain Arsenal waste disposal and frequency of earthquakes: *Engineering Geology Case Histories,* No. 8, Geol. Soc. America, p. 33–37.

Baskin, D.K., and K.E. Peters, 1992, Early generation characteristics of a sulfur-rich Monterey kerogen: *Bull. Am. Assoc. Petroleum Geologists,* v. 76, p. 1–13.

Bloomfield, P., 2000, *Fourier Analysis of Time Series: An Introduction,* 2nd ed.: Wiley-Interscience Inc., New York, 261 pp.

Bohling, G., J.H. Doveton, W. Guy, W.L. Watney, and S. Bhattacharya, 1998, *Pfeffer 2.0 Manual*: Kansas Geol. Survey, Lawrence, Kansas, 164 pp.

Buchanan, J.L., and P.R. Turner, 1992, *Numerical Methods and Analysis*: McGraw-Hill, Inc., New York, 751 pp.

Carroll, R.J., D. Ruppert, and L.A. Stefanski, 1995, *Measurement Error in Nonlinear Models*: Chapman & Hall, New York, 305 pp. *Contains discussions of orthogonal regression, as well as more advanced topics in regression.*

Christensen, R., 1991, *Linear Models for Multivariate, Time Series, and Spatial Data*: Springer-Verlag, New York, 317 pp.

Clark, I., 1979, *Practical Geostatistics*: Applied Science Publishers, London, 129 pp. *This slim volume, although dated, remains one of the most readable books on geostatistics.*

Conover, W.J., 1999, *Practical Nonparametric Statistics, 3rd* ed.: John Wiley & Sons, Inc., New York, 584 pp.

Cooley, J.W., and J.W. Tukey, 1965, An algorithm for the machine computation of complex Fourier series: *Mathematical Computing*, v. 19, p. 297–301.

Cox, D.R., and P.A.W. Lewis, 1966, *The Statistical Analysis of Series of Events*: Methuen & Co., London, 285 pp.

Cressie, N.A.C., 1993, *Statistics for Spatial Data*, Rev. ed.: John Wiley & Sons, Inc., New York, 900 pp. *A formal treatment of all aspects of geostatistics, including semivariogram analysis.*

Deutsch, C.V., and A.G. Journel, 1998, *GSLIB, Geostatistical Software Library and User's Guide, 2nd* ed.: Oxford Univ. Press, New York, 369 pp., CD-ROM. *The accompanying CD-ROM contains FORTRAN routines for semivariogram analysis and modeling.*

Doveton, J.H., 1971, An application of Markov chain analysis to the Ayrshire Coal Measures succession: *Scottish Jour. Geology*, v. 7, p. 11–27.

Doveton, J.H., 1994, Geologic log analysis using computer methods: *Computer Applications in Geology*, No. 2: Am. Assoc. Petroleum Geologists, Tulsa, Okla., 169 pp. *Contains an accessible discussion of RMA and related models used in petrophysics.*

Doveton, J.H., and K. Skipper, 1974, Markov chain and substitutability analysis of turbidite succession, Cloridorme Formation (Middle Ordovician), Gaspé, Quebec: *Canadian Jour. Earth Sciences*, v. 11, p. 472–488.

Draper, N.R., and H. Smith, 1998, *Applied Regression Analysis, 3rd* ed.: John Wiley & Sons, Inc., New York, 706 pp. *An extremely complete reference on all aspects of regression analysis.*

Etnyre, L.M., 1984a, Practical application of weighted least squares methods to formation evaluation, Pt. I: The logarithmic transformation of non-linear data and selection of dependent variable: *The Log Analyst*, v. 25, no. 1, p. 11–21.

Etnyre, L.M., 1984b, Practical application of weighted least squares methods to formation evaluation, Pt. II: Evaluating the uncertainty in least squares: *The Log Analyst*, v. 25, no. 3, p. 11–20.

Fuller, W.A., 1987, *Measurement Error Models*: John Wiley & Sons, Inc., New York, 440 pp. *The first chapter contains a thorough although advanced discussion of regression when the error structure of the data is known.*

Gentleman, W.M., and G. Sande, 1966, *Fast Fourier Transforms—For Fun and Profit*: Bell Telephone Laboratories, Murray Hill, N.J., 65 pp.

Gill, D., 1970, Application of a statistical zonation method to reservoir evaluation and digitized-log analysis: *Bull. Am. Assoc. Petroleum Geologists,* v. 54, no. 5, p. 719–729.

Goodman, L.A., 1968, The analysis of cross-classified data: Independence, quasi-independence, and interactions in contingency tables with and without missing entries: *Jour. American Statistical Assoc.,* v. 63, p. 1091–1131.

Gordon, A.D., and R.A. Reyment, 1979, Slotting of borehole sequences: *Mathematical Geology,* v. 11, no. 3, p. 309–327.

Gretener, P.E., 1981, Geothermics: Using temperature in hydrocarbon exploration: *Education Course Note Series,* No. 17: Am. Assoc. Petroleum Geologists, Tulsa, Okla., 156 pp.

Haan, C.T., 1977, *Statistical Methods in Hydrology*: Iowa State Univ. Press, Ames, Iowa, 378 pp. *An excellent introduction to time series analysis of surface-water hydrologic data.*

Hatherton, T., and W.R. Dickinson, 1969, The relationship between andesitic volcanism and seismicity in Indonesia, the Lesser Antilles, and other island arcs: *Jour. Geophysical Research,* v. 74, no. 22, p. 5301–5310.

Hawkins, D.M., and D.F. Merriam, 1973, Optimal zonation of digitized sequential data: *Mathematical Geology,* v. 5, no. 4, p. 389–395.

Hawkins, D.M., and D.F. Merriam, 1974, Zonation of multivariate sequences of digitized geologic data: *Mathematical Geology,* v. 6, no. 3, p. 263–269.

Isaaks, E.H., and R.M. Srivastava, 1989, *Applied Geostatistics*: Oxford Univ. Press, New York, 561 pp. *Geostatistics taught through the medium of experiments with an exhaustive (partially artificial) data set. Chapter 7 is devoted to the semivariogram.*

Jian, X., R.A. Olea, and Y.-S. Yu, 1996, Semivariogram modeling by weighted least squares: *Computers & Geosciences,* v. 22, no. 4, p. 387–397.

Kemeny, J.G., 1983, *Finite Markov Chains*: Springer-Verlag, New York, 224 pp.

Kendall, M.G., and A. Stuart, 1979, *The Advanced Theory of Statistics,* Vol. 2. *Inference and Relationship,* 4th ed.: Charles Griffin & Co. Ltd., High Wycombe, U.K., 748 pp. *Includes a classic treatment of structural analysis.*

Kermack, K.A., and J.B.S. Haldane, 1950, Organic correlation and allometry: *Biometrika,* v. 37, p. 30–41.

Kruskal, W., 1953, On the uniqueness of the line of organic correlation: *Biometrics,* v. 9, p. 47–58.

Kuno, H., 1962, *Catalogue of the Active Volcanoes of the World Including Solfatara Fields,* Part XI: *Japan, Taiwan and Marianas*: International Volcanological Assoc., Naples, 332 pp.

Lowe, J.J., and M.J.C. Walker, 1997, *Reconstructing Quarternary Environments,* 2nd ed.: Addison Wesley Longman Ltd., Harlow, England, 446 pp.

Mark, D.M., and M. Church, 1977, On the misuse of regression in earth science: *Mathematical Geology,* v. 9, p. 63–75. *Contrasts ordinary regression and structural analysis in geological problems.*

McKinney, F.K., S. Lidgard, J.J. Sepkoski, Jr., and P.D. Taylor, 1998, Decoupled temporal patterns of evolution and ecology in two post-Paleozoic clades: *Science,* v. 281, p. 807–809.

Melas, F.F., and G.M. Friedman, 1992, Petrophysical characteristics of the Jurassic Smackover Formation, Jay Field, Conecuh Embayment, Alabama and Florida: *Bull. Am. Assoc. Petroleum Geologists,* v. 76, p. 81–100.

Mehta, C.H., S. Radhakrishnan, and G. Srikanth, 1990, Segmentation of well logs by maximum-likelihood estimation: *Mathematical Geology,* v. 22, no. 7, p. 853–869.

Miller, R.L., and J.S. Kahn, 1962, *Statistical Analysis in the Geological Sciences*: John Wiley & Sons, Inc., New York, 483 pp.

Mortenson, M.E., 1989, *Computer Graphics: An Introduction to the Mathematics and Geometry*: Industrial Press, New York, 381 pp.

Nerlove, M., D.M. Grether, and J.L. Carvalho, 1995, *Analysis of Economic Time Series, A Synthesis (Economic Theory, Econometrics, and Mathematical Economics),* 2nd ed.: Academic Press, New York, 468 pp.

Norris, J.R., 1997, *Markov Chains*: Cambridge Univ. Press, New York, 253 pp.

Nussbaumer, H.J., 1982, *Fast Fourier Transform and Convolution Algorithms,* 2nd ed.: Springer-Verlag, New York, 276 pp.

Olea, R.A., 1977, Measuring spatial dependence with semivariograms: *Series on Spatial Analysis,* No. 3, Kansas Geol. Survey, Lawrence, Kansas, 29 pp. *The data set, MAGELLAN.TXT, is taken from this publication.*

Olea, R.A., 1999, *Geostatistics for Engineers and Earth Scientists*: Kluwer Academic Publishers, Boston, Mass., 303 pp. *Chapter 5 covers semivariogram analysis.*

Press, W.H., S.A. Teukolsky, W.T. Vetterling, and B.P. Flannery, 1992, *Numerical Recipes: The Art of Scientific Computing,* 2nd ed.: Cambridge Univ. Press, Cambridge, U.K., 963 pp.

Rayner, J.N., 1971, *An Introduction to Spectral Analysis*: Pion Ltd., London, 174 pp.

Rock, N.M.S., 1988, *Numerical Geology*: Springer-Verlag, Berlin, 427 pp.

Rogers, D.F., and J.A. Adams, 1990, *Mathematical Elements for Computer Graphics,* 2nd ed.: McGraw–Hill Book Co., Boston, Mass., 611 pp.

Sanford, R.F., and J.S. Huebner, 1980, Model thermal history of 77115 and its implications for the origin of fragment-laden basalts, *in* Papike, J.J., and R.B. Merrill [Eds.], *Proceedings of a Conference on Lunar Highlands Crust*: Pergamon Press, New York, p. 253–269.

Schön, J., 1983, *Petrophysik-physikalische Eigenschaften von Gesteinen und Mineralien*: Akademie-Verlag, Berlin, 264 pp.

Schwarzacher, W., 1975, *Sedimentation Models and Quantitative Stratigraphy*: Elsevier Scientific Publ. Co., Amsterdam, 382 pp.

Siegel, S., and N.J. Castellan, Jr., 1988, *Nonparametric Statistics for the Behavioral Sciences,* 2nd ed.: McGraw–Hill, Inc., Boston, Mass., 399 pp.

Sneath, P.H.A., and R.R. Sokal, 1973, *Numerical Taxonomy: The Principles and Practice of Numerical Classification*: W.H. Freeman & Co., San Francisco, 573 pp.

Strahler, A.N., 1950, Equilibrium theory of erosional slopes approached by frequency distribution analysis, Pt. 1: *American Jour. Science,* v. 248, p. 673–696.

Taylor, R.K., 1973, Compositional and geotechnical characteristics of a 100-year-old colliery spoil heap: *Trans. Inst. Mining and Metallurgy,* Section A, p. 1–14.

Till, R., 1974, *Statistical Methods for the Earth Sciences: An Introduction*: John Wiley & Sons, Inc., New York, 154 pp.

Tipper, J., 1979, Surface modeling techniques: Series on Spatial Analysis, No. 4, Kansas Geol. Survey, Lawrence, Kansas, 108 pp. *A survey of graphics techniques from the field of computer-aided design (CAD) that may be useful in the Earth sciences.*

Türk, G., 1979, Transition analysis of structural sequences: Pt. I, Discussion: *Bull. Geol. Soc. America,* v. 90, p. 989–992.

Webster, R., 1973, Automatic soil-boundary location from transect data: *Mathematical Geology,* v. 5, no. 1, p. 27–37.

Webster, R., 1980, DIVIDE, A FORTRAN IV program for segmenting multivariate one-dimensional spatial series: *Computers & Geosciences,* v. 6, no. 1, p. 61–68.

Webster, R., and H.E. Cuanalo, 1975, Soil transect correlograms of North Oxfordshire and their interpretation: *Jour. Soil Science,* v. 26, p. 176–194.

Webster, R., and M.A. Oliver, 1990, *Statistical Methods in Soil and Land Resource Survey*: Oxford Univ. Press, Oxford, U.K., 316 pp. *Chapter 12, Spatial Dependence, discusses calculation and modeling of semivariograms as applied to soil transects.*

Wells, R.C., R.K. Bailey, and E.P. Henderson, 1928, Salinity of the water of Chesapeake Bay: U.S. Geol. Survey, Prof. Paper 154, p. 105–152.

Wetzel, A., 1993, The transfer of river load to deep-sea fans: A quantitative approach: *Bull. Am. Assoc. Petroleum Geologists,* v. 77, p. 1679–1692.

Wickman, F.E., 1966, Repose-period patterns of volcanoes: *Arkiv för Mineralogi och Geologi,* Bd. 4, p. 291–366.

Wilkes, M.V., 1966, *A Short Introduction to Numerical Analysis*: Cambridge Univ. Press, Cambridge, U.K., 76 pp.

Yevjevich, V., 1972, *Stochastic Processes in Hydrology*: Water Resources Publications, Fort Collins, Colo., 276 pp.

Yule, G.U., and M.G. Kendall, 1969, *An Introduction to the Theory of Statistics,* 14th ed.: Hafner Publ. Co., New York, 701 pp.

Chapter 5
Spatial Analysis

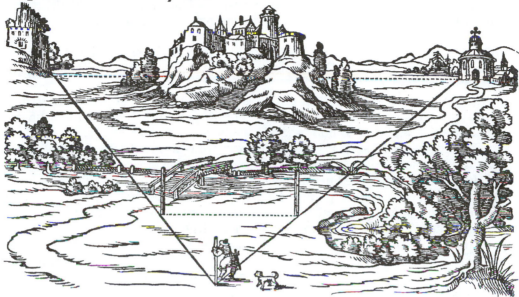

Although geologists study a three-dimensional world, their view of it is strongly two dimensional. This reflects in part the fact that the third dimension, depth, often is accessible to only a fraction of the extent of the other two spatial dimensions. Also, our thoughts are conditioned by the media in which we express them, and maps, photographs, and cross-sections are printed or drawn on flat sheets of paper. We may be interested in the geologic features exposed in a deep mine with successive levels, adits, and raises creating a complex three-dimensional net, yet we must reduce this network to flat projections in order to express our ideas concerning the relationships we see.

Geologic Maps, Conventional and Otherwise

Geologists are carefully trained to read, utilize, and create maps; probably no other group of scientists is as adept at expressing and envisioning dimensional relationships. Maps are compact and efficient means of expressing spatial relationships and details—they are as important to Earth scientists as the conventions for scales and notes are to the musician.

In this chapter, we will examine methods for analyzing features on what we loosely define as "maps": two-dimensional representations of areas. Usually the area is geographical (a quadrangle, mining district, country, *etc.*) and the map is a method for reducing very large-scale spatial relationships so they can be easily perceived. However, the representation may equally well be a "map" of a thin section or electron photomicrograph, where the relationships between features have been enlarged so they become visible. Maps, in this general definition, include traditional geologic and topographic maps and also aerial photographs, mine plans, peel prints, photomicrographs, and electron micrographs. In fact, any sort of two-dimensional spatial representation is included.

Among the topics we will consider that have obvious applications to fields as diverse as geophysics and microscopy is the probability of encountering an object with a systematic search across an area. We will look at the statistics of directional data in both two and three dimensions. Many natural phenomena are expressed as complicated patterns of lines and areas that can best be described as fractals, which we will touch upon. We will also look at ways of describing and comparing more conventional shapes of individual objects, ranging in size from islands to oil fields to microfossils.

Map relationships are almost always expressed in terms of points located on the map. We are concerned with distances between points, the density of points, and the values assigned to points. Most maps are estimates of continuous functions based on observations made at discrete points. An obvious example is the topographic map; although the contour lines are an expression of a continuous and unbroken surface, the lines are calculated from measurements taken at triangulation and survey control points. An even more obvious example is a structural contour map. We do not know that the structural surface is continuous, because we can observe it only at the locations where drill holes penetrate the surface. Nevertheless, we believe that it is continuous and we estimate its form from the measurements made at the wells, recognizing that our reconstruction is inaccurate and lacking in detail because we have no data between wells.

When mapping the surface geology of a desert region, we can stand at one locality where strike and dip have been measured and extend formation boundaries on our map with great assurance because we can see the contacts across the countryside. In regions of heavy vegetation or deep weathering, however, we must make do with scattered outcrops and poor exposures; the quality of the finished map reflects to a great extent the density of control points. Geologists should be intensely interested in the effects which control-point distributions have on maps, but few studies of this influence have been published. In fact, almost all studies of point distributions have been made by geographers. In this chapter, we will examine some of these procedures and consider their application to maps and also to such problems as the distribution of mineral grains in thin sections.

Geologists exercise their artistic talents as well as their geologic skills when they create contour maps. In some instances, the addition of geologic interpretation to the raw data contained in the observation points is a valuable enhancement of the map. Sometimes, however, geologic judgment becomes biased, and the subtle effects of personal opinion detract rather than add to the utility of a map. Computer contouring is totally consistent and provides a counterbalance to overly interpretative traditional mapping. Of course, subjective judgment is necessary in choosing an algorithm to perform mapping, but methods are available that allow a choice to be made between competing algorithms, based upon specified criteria. The principal motive behind the development of automatic contouring is economic, an attempt to utilize the petroleum industry's vast investment in stratigraphic data banks. Aside from this, one of the prime benefits of computerized mapping techniques may come from the attention they focus on the contouring process and the problems they reveal about map reliability. Contour mapping is the subject of one section in this chapter.

Trend-surface analysis is a popular numerical technique in geology. However, although it is widely applied, it is frequently misused. Therefore, we will discuss the problems of data-point distribution, lack of fit, computational "blowup," and inappropriate applications. Statistical tests are available for trend surfaces if they are to be used as multiple regressions; we will consider these tests and the assumptions prerequisite to their application.

The exchange between Earth scientists and statisticians has been mostly one way, with the notable exception of the expansion of the theory of regionalized variables. This theory, developed originally by Georges Matheron, a French mining engineer, describes the statistical behavior of spatial properties that are intermediate between purely random and completely deterministic phenomena. The most familiar application of the theory is in kriging, an estimation procedure important in mine evaluation, mapping, and other applications where values of a property must be estimated at specific geographic locations.

Two-dimensional methods are, for the most part, direct extensions of techniques discussed in Chapter 4. Trend-surface analysis is an offshoot of statistical regression; kriging is related to time-series analysis; contouring is an extension of interpolation procedures. We have simply enlarged the dimensionality of the subjects of our inquiries by considering a second (and in some cases a third) spatial variable. Of course, there are some applications and some analytical methods that are unique to map analysis. Other methods are a subset of more general multidimensional procedures. It is an indication of the importance of one- and two-dimensional problems in the Earth sciences that they have been included in individual chapters.

Systematic Patterns of Search

Most geologists devote their professional careers to the process of searching for something hidden. Usually the object of the search is an undiscovered oil field or an ore body, but for some it may be a flaw in a casting, a primate fossil in an excavation, or a thermal spring on the ocean's floor. Too often the search has been conducted haphazardly—the geologist wanders at random across the area of investigation like an old-time prospector following his burro. Increasingly, however, geologists and other Earth scientists are using systematic procedures to search, particularly when they must rely on instruments to detect their targets.

Most systematic searches are conducted along one or more sets of parallel lines. Ore bodies that are distinctively radioactive or magnetic are sought using airborne instruments carried along equally spaced parallel flight lines. Seismic surveys are laid out in regular sets of traverses. Satellite reconnaissance, by its very nature, consists of parallel orbital tracks.

The probabilities that targets will be detected by a search along a set of lines can be determined by geometrical considerations. Basically, the probability of discovery is related to the relative size of the target as compared to the spacing of the search pattern. The shape of the target and the arrangement of the lines of search also influence the probability. If the target is assumed to be elliptical and the search consists of parallel lines, the probability that a line will intersect a hidden target of specified size, regardless of where it occurs within the search area, can be calculated. These assumptions do not seem unreasonable for many exploratory surveys. Note that the probabilities relate only to intersecting a target with a line, and do not consider the problem of recognizing a target when it is hit.

McCammon (1977) gives the derivation of the geometric probabilities for circular and linear targets and parallel-line searches. His work is based mostly on the mathematical development of Kendall and Moran (1963). An older text by Uspensky (1937) derives the more general elliptical case used here.

Assume the target being sought is an ellipse whose dimensions are given by the major semiaxis a and minor semiaxis b. (If the target is circular, then $a = b = r$, the radius of the circle.) The search pattern consists of a series of parallel traverses spaced a distance D apart (**Fig. 5–1 a**). The probability that a target (smaller than the spacing between lines) will be intersected by a line is

$$p = \frac{P}{\pi D} \tag{5.1}$$

where P is the perimeter of the elliptical target. The equation for the perimeter of an ellipse is $P = 2\pi\sqrt{(a^2 + b^2)/2}$, where a and b are the major and minor semiaxes. Substituting,

$$p = \frac{2\pi\sqrt{\frac{a^2+b^2}{2}}}{\pi D} = \frac{2\sqrt{\frac{a^2+b^2}{2}}}{D} \tag{5.2}$$

We can define a quantity Q as the numerator of Equation (5.2); that is, $Q = 2\sqrt{(a^2 + b^2)/2}$. With this simplification, the probability of intersecting an elliptical target with one line in a set of parallel search lines can be written as

$$p = \frac{Q}{D} \tag{5.3}$$

In the specific case of a circular target, a and b are both equal to the radius, so Q can be replaced by twice the radius:

$$p = \frac{2r}{D} \tag{5.4}$$

At the other extreme, one axis of the ellipse may be so short that the target becomes a randomly oriented line. This geometric relationship is known as **Buffon's problem**, which specifies the probability that a needle of length ℓ, when dropped at random on a set of ruled lines having a spacing D, will fall across one of the lines. The probability is

$$p = \frac{2\ell}{\pi D} \tag{5.5}$$

where ℓ is the length of the target.

A similar geometric relationship, known as **Laplace's problem**, also pertains to the probabilities in systematic searches. Laplace's problem specifies the probability that a needle of length ℓ, when dropped on a board covered with a set of rectangles, will lie entirely within a single rectangle. A variant gives the probability that a coin tossed onto a chessboard will fall entirely within one square. In exploration, the complementary probabilities are of interest, *i.e.*, that a randomly located target will be intersected one or more times by a set of lines, such as seismic traverses, arranged in a rectangular grid (**Fig. 5–1 b**).

The general equation is

$$p = \frac{Q(D_1 + D_2 - Q)}{D_1 D_2} \tag{5.6}$$

where D_1 is the spacing between one set of parallel seismic traverses and D_2 is the spacing between the perpendicular set of traverses. In the specific instance of a

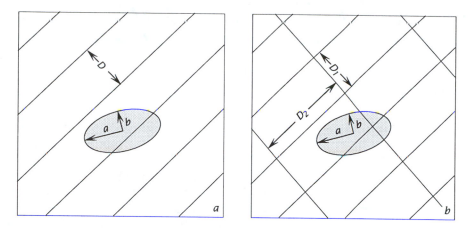

Figure 5–1. Search for an elliptical target with major semiaxis a and minor semiaxis b. (a) Using a parallel-line search of spacing D. (b) Using a grid search with spacing D_1 in one direction and D_2 in the perpendicular direction.

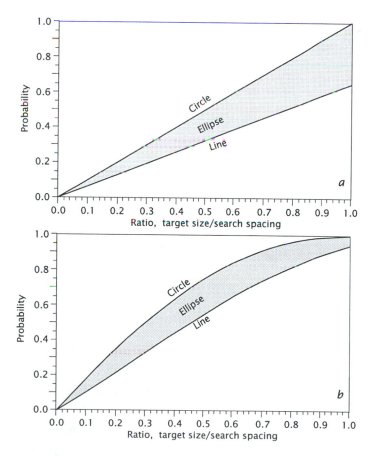

Figure 5–2. Probability of intersecting a target with a systematic pattern of search. Shape of target may range from a circle to a line; elliptical targets of various axial ratios fall in the shaded region. Horizontal axis is ratio (major dimension of target)/(spacing between search lines). (a) Parallel-line search pattern. (b) Square-grid search pattern. After McCammon (1977).

search in the pattern of a square grid, the equation simplifies to

$$p = \frac{Q}{D}\left(2 - \frac{Q}{D}\right)$$ (5.7)

Lambie (unpublished report, 1981) has pointed out that these equations for geometric probability are approximations of integral equations. Comparing exact probabilities found by numerical integration with those predicted by the approximation equations, he found that significant differences occur only for very elongate targets that are large with respect to spacing between search lines. Then, equations such as (5.3) and (5.6) may seriously overestimate the probabilities of detection.

The probabilities of intersecting a target, as calculated by the approximating equations, can be shown conveniently as graphs. McCammon (1977) presented such graphs in a particularly useful dimensionless form for various combinations of target shape and size relative to the spacing between the search lines. **Figure 5–2 a** gives the probability of detecting an elliptical target whose shape ranges from a circle to a line, using a search pattern of parallel lines. The relative size of the target is found by dividing the target's maximum dimension by the search line

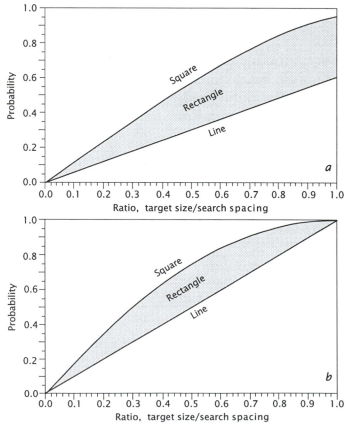

Figure 5–3. Probability of intersecting targets with regular search patterns ranging from squares to parallel lines. Rectangular search patterns with different ratios of D_1/D_2 fall in the shaded region. Horizontal axis is ratio (major dimension of target)/(minimum spacing between search lines). (*a*) Target is circular. (*b*) Target is a line. After McCammon (1977).

spacing. **Figure 5–2 b** is an equivalent graph for a search pattern consisting of a square grid of lines.

If the shape of the target is specified, the probabilities of intersection can be graphed for different patterns of search. **Figure 5–3 a**, for example, shows the probability of intersecting a circular target with search patterns ranging from a square grid, through rectangular grid patterns, to a parallel-line search. **Figure 5–3 b** is the equivalent graph for a line-shaped target. Between the two graphs, all possible shapes of elliptical targets and all possible patterns of search along two perpendicular sets of parallel lines are encompassed.

Distribution of Points

Geologists often are interested in the manner in which points are distributed on a two-dimensional surface or a map. The points may represent sample localities, oil wells, control points, or poles and projections on a stereonet. We may be concerned about the uniformity of control-point coverage, the distribution of point density, or the relation of one point to another. These are questions of intense interest to geographers as well as geologists, and the burgeoning field of locational analysis is devoted to these and similar problems. Although much of the attention of the geographer is focused on the distribution of shopping malls or public facilities, the methodologies are directly applicable to the study of natural phenomena as well.

The patterns of points on maps may be conveniently classified into three categories: regular, random, and aggregated or clustered. Examples of point distributions are shown in **Figure 5–4** and range from the most uniform possible (the face-centered hexagonal lattice in **Fig. 5–4 a**, where every point is equidistant from its six nearest neighbors) to a highly clustered pattern composed of randomly located centers around which the probability of occurrence of a point decreases exponentially with distance (**Fig. 5–4 f**). Of course most maps will have patterns intermediate between these extremes, and the problem becomes one of determining where the observed pattern lies within the spectrum of possible distributions. For example, most people would intuitively regard the distribution of points in **Figure 5–4 c** as random. However, intuition is wrong, because the map was created by dividing the map area into a 4×4 array of regular cells and then placing four points at random within each cell (except in the shortened bottom row, which received only two points per cell). The distribution therefore has both random and regular aspects and is more uniform in density than a purely random arrangement such as **Figure 5–4 d**.

The pattern of points on a map is said to be *uniform* if the density of points in any subarea is equal to the density of points in all other subareas of the same size and shape. The pattern is *regular* if the spacings between points repeat, as on a grid. That is, the distance between a point i and another point j lying in some specified direction from i is the same for all pairs of points i and j on the map. Obviously, a regular pattern also will be uniform, but the converse is not necessarily true. A *random* pattern can be created if any subarea is as likely to contain a point as any other subarea of the same size, regardless of the subarea's location, and the placement of a point has no influence on the placement of any other point. In an aggregated or *clustered* pattern, the probability of occurrence of a point varies in some inverse manner with distances to preexisting points.

299

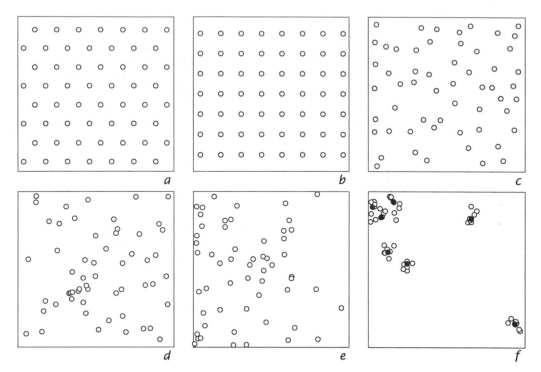

Figure 5–4. Some possible patterns of points on maps. Each map contains 56 points. (*a*) Points regularly spaced on a face-centered hexagonal grid or network. Every point is equidistant from six other points. (*b*) Points regularly arranged on a square grid. (*c*) Sets of four points placed randomly within each cell of a regular 4 × 4 grid. The bottom row contains only two points per cell. (*d*) Points located by a bivariate uniform random process. (*e*) Nonuniform pattern of points produced by logarithmic scaling of the *X*-axis. (*f*) Points located by randomly placing seven cluster centers (black points) and moving eight points a random direction and logarithmically scaled distance from each center.

A uniform density of data points is important in many types of analysis, including trend-surface methods which we will discuss later. The reliability of contour maps is directly dependent upon the total density of control points as well as their uniformity of distribution. However, most geologic researchers have been content with qualitative judgments of the adequacy and representativeness of the distribution of their data. Even though the desirability of a uniform density of observations is often cited, the degree of uniformity is seldom measured. The tests necessary to determine uniformity are very simple, and it is unfortunate that many geologists seem unaware of them. These tests are, however, extensively used by geographers. Haggett, Cliff, and Frey (1977); Getis and Boots (1978); Cliff and Ord (1981); and Bailey and Gatrell (1995) provide an introduction to this literature.

Uniform density

A map area may be divided into a number of equal-sized subareas (sometimes called *quadrats*) such that each subarea contains a number of points. If the data points are distributed uniformly, we expect each subarea to contain the same number of points. This hypothesis of no difference in the number of points per subarea can be tested using a χ^2 method, and is theoretically independent of the shape or

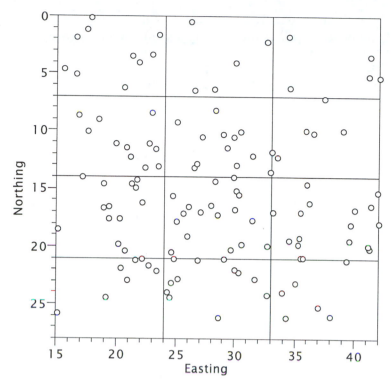

Figure 5–5. Locations of 123 exploratory holes drilled to top of Ordovician rocks (Arbuckle Group) in central Kansas. Map has been divided into 12 cells of equal size.

orientation of subareas. However, the test is most efficient if the number of subareas is a maximum (this increases the degrees of freedom), subject to the restriction that no subarea contain fewer than five points. The expected number of points in each subarea is

$$E = \frac{N}{k} \tag{5.8}$$

where N is the total number of data points and k is the number of subareas. A χ^2 test of goodness of fit of the observed distribution to the expected (uniform) distribution is

$$\chi^2 = \sum_{i=1}^{k} \frac{(O_i - E)^2}{E} \tag{5.9}$$

where O_i is the observed number of data points in subarea i and E is the expected number. The test has $\nu = k - 2$ degrees of freedom, where k is the number of subareas.

As an example of the application of this test, consider the data-point distribution shown in **Figure 5–5**. These are the locations of 123 holes drilled in the search for oil in the Ordovician Arbuckle stratigraphic succession in central Kansas. These data are listed in file ARBUCKLE.TXT. In **Figure 5–5**, the map area has been divided into 12 equal subareas, each of which we expect to contain about ten points, if the points are uniformly distributed. The observed number of points in each subarea and the computations necessary to find the test value are given in **Table 5–1**. This test has $\nu = 10$ degrees of freedom, so the critical value of χ^2 at the 5% ($\alpha = 0.05$) significance level is 18.3. The computed test value of $\chi^2 = 17.0$ does not exceed this, so we conclude that there is no evidence suggesting that the quadrats are unevenly populated. Note that the test applies only to the uniformity of point densities between areas of a specified size and shape. It is possible that we could select quadrats of different sizes or shapes that might not be uniformly populated, especially if they were smaller than those used in this test.

Table 5–1. Number of wells in 12 subareas of central Kansas.

Observed Number of Points	$\dfrac{(O - E)^2}{E}$
10	0.006
5	2.689
5	2.689
11	0.055
13	0.738
5	2.689
12	0.299
16	3.226
16	3.226
9	0.152
13	0.738
8	0.494
TOTAL = 123	$\chi^2 = 16.995$[a]

[a]Test value is not significant at the $\alpha = 0.05$ level.

Random patterns

Establishing that a pattern is uniform does not specify the nature of the uniformity, for both regular and random patterns are expected to be homogeneous. For many purposes, verifying uniformity is sufficient; but, if we desire more information about the pattern, we must turn to other tests. If points are distributed at random across a map area, even though the coverage is uniform, we do not expect exactly the same number of points to lie within each subarea. Rather, there will be some preferred number of points that occur in most subareas and there will be progressively fewer subareas that contain either more points or fewer. This is apparent in the example we just worked; although our hypothesis of uniformity specified that we expect about ten observations in each subarea, we actually found some areas that contained more than ten and some that contained fewer.

You will recall that the Poisson probability distribution is the limiting case of the binomial distribution when p, the probability of a success, is very small and $(1 - p)$ approaches 1.0. The Poisson distribution can be used to model the occurrence of rare, random occurrences in time, as it was used in Chapter 4, or it can be used to model the random placement of points in space. Although the Poisson distribution, like the binomial, uses the numbers of successes, failures, and trials in the calculation of probabilities, it can be rewritten so that neither the number of failures nor the total number of trials is required. Rather, it uses the number of points per quadrat and the density of points in the entire area to predict how many quadrats should contain specified numbers of points. These predicted or expected numbers of quadrats can be used in a χ^2 procedure to test whether the points are distributed at random within the area.

As an application, we can determine if oil discoveries in a basin occur at random or are distributed in some other fashion. It is not intuitively obvious that the Poisson distribution can be expressed in a form appropriate for this problem, so we will work through its development.

Assume a basin has an area, a, in which m discovery wells are randomly located. The **density** of discovery wells in the basin is designated λ, and is simply

$$\lambda = \frac{m}{a} \tag{5.10}$$

The basin may be divided into small lease tracts, each of area A (here the term "tract" is equivalent to "quadrat"). In turn, each tract may be divided into n extremely small, equal-sized subareas which we might regard as potential drilling sites. The probability that any one of these extremely small subareas contains a discovery well tends toward zero as n becomes infinitely large.

The area of each drilling site is A/n. The probability that a site contains a discovery well is

$$p = \lambda \frac{A}{n}$$

and the probability that it does not contain a discovery well is

$$1 - p = \left(1 - \lambda \frac{A}{n}\right)$$

We wish to investigate the probability that r of the n drilling sites within a tract contain discovery wells, and $n - r$ drilling sites do not. The probability of a specific combination of discovery and nondiscovery well sites within a tract is

$$P = \left(\lambda \frac{A}{n}\right)^r \left(1 - \lambda \frac{A}{n}\right)^{n-r}$$

However, within a tract, there are $\binom{n}{r}$ combinations of the n drilling sites, of which r contain discovery wells and all are equally probable. The probability that a tract will contain exactly r discovery wells is therefore

$$P(r) = \binom{n}{r} \left(\lambda \frac{A}{n}\right)^r \left(1 - \lambda \frac{A}{n}\right)^{n-r}$$

Note that this is simply the binomial probability of r discovery wells on n drilling sites.

The combinations can be expanded into factorials,

$$P(r) = \frac{n(n-1)(n-2)\cdots(n-r+1)}{r!} \frac{(\lambda A)^r}{n^r} \left(1 - \frac{\lambda A}{n}\right)^n \left(1 - \frac{\lambda A}{n}\right)^{-r}$$

Rearranging and canceling terms yields

$$P(r) = \left(1 - \frac{1}{n}\right)\left(1 - \frac{2}{n}\right)\cdots\left(1 - \frac{r-1}{n}\right)\left(1 - \frac{\lambda A}{n}\right)^{-r}\left[\left(1 - \frac{\lambda A}{n}\right)^n \frac{(\lambda A)^r}{r!}\right] \tag{5.11}$$

As n becomes infinitely large, all of the fractions that contain n in their denominator become infinitesimally small and vanish, so all terms inside parentheses simply become equal to 1. The terms inside the brackets simplify to

$$P(r) = e^{(-\lambda A)} \frac{(\lambda A)^r}{r!} \tag{5.12}$$

Note that n, the number of drilling sites, has vanished from the equation leaving only the discovery-well density, λ, the number of discovery wells, r, and the area, A, of the tracts. This is an expression of the Poisson distribution, as applied to the probability of rare, random events (discovery wells) occurring within geographic areas. Also note that λA is simply the mean number of wells per tract, because it is the product of the density of discovery wells times the area of a tract. In practice, we estimate λA from the total number of discovery wells, m, and the total number of tracts, T

$$\lambda A = \frac{m}{T} \tag{5.13}$$

We can now perform a χ^2 test to see if the number of wells per tract matches that expected if the wells are randomly located according to the Poisson model. The number of tracts that contain exactly r discovery wells can be found by

$$n_r = mP(r)$$

$$= me^{(-\lambda A)} \frac{(\lambda A)^r}{r!} \tag{5.14}$$

If λA is estimated by m/T, the equation becomes

$$n^r = me^{(-m/T)} \frac{(m/T)^r}{r!} \tag{5.15}$$

Figure 5–6 shows the locations of discovery wells in part of the Eastern Shelf area of the Permian Basin in Fisher and Noland counties of Texas. The area has been divided into a 10×16 grid of 160 tracts, or quadrats, each containing approximately 10 mi^2. Since there are 168 discovery wells in the area, the mean number of wells per tract is

$$\frac{m}{T} = \frac{168}{160} = 1.05$$

We can count the number of tracts in the map that contain no discovery wells, exactly one discovery, two discoveries, and so forth. Using Equation (5.15), we can also calculate the expected number of tracts that contain these same numbers of wells. The expected and observed numbers of tracts for the Permian Basin area are given in **Table 5–2**. This table contains all of the figures necessary to calculate a χ^2 test of goodness of fit, which is essentially a comparison of the two histograms shown in **Figure 5–7**. The last three categories must be combined so that the observed number of tracts is equal to or greater than five

$$\chi^2 = \frac{(70 - 56.0)^2}{56.0} + \frac{(42 - 58.8)^2}{58.8} + \frac{(26 - 30.9)^2}{30.9}$$

$$+ \frac{(17 - 10.8)^2}{10.8} + \frac{(5 - 3.5)^2}{3.5} = 13.28$$

The test statistic has $c - 2$ degrees of freedom, where c is the number of categories (one degree of freedom is lost because the expected frequencies are constrained to sum to 160, and a second degree of freedom is required for estimation of the parameter λ). For $c = 5$ categories, there are three degrees of freedom.

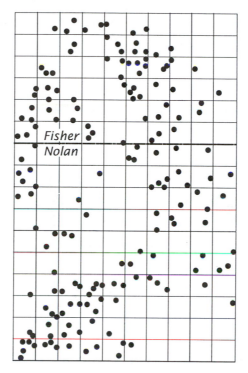

Figure 5–6. Locations of oil-field discovery wells in part of the Eastern Shelf area of the Permian Basin, Fisher and Noland counties, Texas. Quadrats are approximately 10 mi^2 in size.

Table 5–2. Calculation of expected numbers of tracts containing r discoveries in eastern part of Permian Basin, Texas, assuming a Poisson distribution.

Number of Discoveries Per Tract (r)	Poisson Equation	Probability Tract Contains r Discoveries	Number of Tracts Expected	Number of Tracts Observed
0	$P_{(0)} = e^{(-1.05)} \frac{1.05^0}{0!}$	0.3499	56.0	70
1	$P_{(1)} = e^{(-1.05)} \frac{1.05^1}{1!}$	0.3674	58.8	42
2	$P_{(2)} = e^{(-1.05)} \frac{1.05^2}{2!}$	0.1929	30.9	26
3	$P_{(3)} = e^{(-1.05)} \frac{1.05^3}{3!}$	0.0675	10.8	17
4	$P_{(4)} = e^{(-1.05)} \frac{1.05^4}{4!}$	0.0177	2.8	3
5	$P_{(5)} = e^{(-1.05)} \frac{1.05^5}{5!}$	0.0037	0.6	1
6	$P_{(6)} = e^{(-1.05)} \frac{1.05^6}{6!}$	0.0007	0.1	1
TOTALS		0.9998	160.0	160

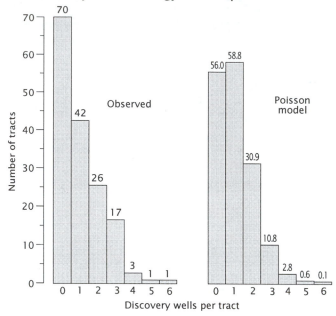

Figure 5–7. Histograms showing observed numbers of discovery wells per tract in an area of the Permian Basin, and the number expected if fields are distributed randomly according to a Poisson model.

The critical value of χ^2 for $\nu = 3$ and $\alpha = 0.05$ is 7.81. The test statistic far exceeds this value, so we must reject the hypothesis of equality between the observed and expected distributions and conclude that the Poisson model is not appropriate. Oil discoveries have not been made randomly within this area of the Permian Basin.

In the process of fitting the Poisson model to this data, we have generated some information that may provide additional insight into the nature of the spatial distribution. The mean number of discoveries per tract is estimated by Equation (5.13). The variance in number of discoveries per tract is

$$s^2 = \frac{\sum_{i=1}^{T} (r_i - m/T)^2}{T - 1} \tag{5.16}$$

where r_i is the number of discoveries in the ith tract. The summation extends over all T tracts. The alternative results of comparing the estimated mean and variance are

$m/T > s^2$ Pattern more uniform than random
$m/T = s^2$ Pattern random
$m/T < s^2$ Pattern more clustered than random

Of course, some difference between m/T and s^2 may arise due to random variation in the particular set of tracts chosen. The statistical significance of the observed difference may be tested by a t-test based on the standard error of the mean, which is the variance that would be expected in values of m/T if a basin were repeatedly sampled by different sets of tracts of the same size. The standard error in the mean number of discoveries per tract is

$$s_e = \sqrt{2/(T - 1)} \tag{5.17}$$

The t-test compares the ratio between m/T and s^2, which should be equal to 1.0 if the two statistics are the same

$$t = \frac{\left(\frac{m/T}{s^2}\right) - 1.0}{s_e} \tag{5.18}$$

The test has $T - 1$ degrees of freedom.

For the eastern Permian Basin area, the variance in number of wells per tract is

$$s^2 = \frac{231.6}{159} = 1.46$$

The standard error of the mean number of wells per tract can be estimated as

$$s_e = \sqrt{\frac{2}{159}} = 0.112$$

The t-statistic for the test of equivalence of the mean and variance is

$$t = \frac{(1.05/1.46) - 1.0}{0.112} = -8.86$$

At a significance level of $\alpha = 0.05$ and 159 degrees of freedom, the critical value of t for a two-tailed test is 1.96; the computed statistic far exceeds this and so we may conclude as we did in the χ^2 test that the spatial distribution is not random. Since the variance is significantly greater than the mean, we must also conclude that discovery wells are areally clustered.

Clustered patterns

Many naturally occurring spatial distributions show a pronounced tendency toward clustering. This is especially true of certain biological variables, such as presence of specific organisms or occurrences of an infectious disease. The descendants of a sedentary parent, perhaps a coral or a tree, tend to grow nearby, leading to development of densely populated areas surrounded by areas that are relatively barren. Clustered patterns of points can be modeled by many theoretical distributions, most of which can be regarded as combinations of two or more simpler distributions. One of the distributions describes the locations of the centers of clusters, while the other describes the pattern of individual points around the centers of the clusters.

The negative binomial distribution can be used to model the occurrence of clustered points in space in a manner equivalent to the use of the Poisson to model randomly arranged points. An extensive discussion with citations to studies in many fields is given by Ripley (1981). Griffiths (1962, 1966) advocated the use of the negative binomial as an appropriate model for the occurrence of oil fields and ore bodies.

One derivation of the negative binomial is as a compound Poisson and logarithmic distribution with clusters of points randomly located within a region; individual points within a cluster follow a logarithmic distribution. In the formulation appropriate for describing spatial patterns, the negative binomial is

$$P(r) = \binom{k + r - 1}{r} \left(\frac{p}{1 + p}\right)^r \left(\frac{1}{1 + p}\right)^k \tag{5.19}$$

In terms of the oil-field distribution problem we have just considered, r is the number of discovery wells in a tract, p is the probability that a given drilling site contains a discovery well, and k is a measure of the degree of clustering of the discoveries. If k is large, clustering is less pronounced and the spatial distribution approaches the Poisson, or randomness. As k approaches zero, the pattern of clustering becomes more pronounced. The density, λ, is equal to

$$\lambda = kp \qquad (5.20)$$

If k is not an integer (and in general it will not be), this combinatorial equation cannot be solved. Then, the following approximation must be used:

$$P(0) = \frac{1}{(1+p)^k}$$
$$P(r) = \frac{(k+r-1)(p/1+p)}{r} P(r-1) \qquad (5.21)$$

As with the Poisson distribution, λ is estimated by the average density of discoveries per tract, m/T. The clustering parameter, k, is estimated by

$$k = \frac{(m/T)^2}{s^2 - (m/T)} \qquad (5.22)$$

where s^2 is the variance in number of discovery wells per tract. Then, the probability p can be estimated as

$$p = \frac{\lambda}{k} = \frac{(m/T)}{k} \qquad (5.23)$$

We can apply the negative binomial model to the data on discovery wells in the eastern part of the Permian Basin (**Fig. 5–6**) to see if this distribution can adequately describe their spatial distribution. The mean and variance of the number of discovery wells per tract have already been found: $m/T = 1.05$ and $s^2 = 1.46$. The clustering effect can be estimated using Equation (5.22)

$$k = \frac{1.05^2}{1.46 - 1.05} = 2.69$$

In turn, the probability of a discovery well occurring in a tract is

$$p = \frac{1.05}{2.69} = 0.390$$

Using the approximation equations, the probability that a given tract will contain no discovery wells is

$$P(0) = \frac{1}{(1+0.390)^{2.69}} = 0.4124$$

The probability that a tract will contain exactly one discovery well is

$$P(1) = \frac{(2.69+1-1)(0.390/1.390)}{1} \times 0.4124 = 0.3112$$

Table 5–3. Expected numbers of tracts containing r discoveries in eastern part of Permian Basin, Texas, assuming a negative binomial distribution.

Number of Discoveries Per Tract (r)	Probability Tract Contains r Discoveries	Number of Tracts Expected	Number of Tracts Observed
0	0.4124	66.0	70
1	0.3112	49.8	42
2	0.1611	25.8	26
3	0.0706	11.3	17
4	0.0281	4.5	3
5	0.0106	1.7	1
6	0.0038	0.6	1
TOTALS	0.9988	159.7	160

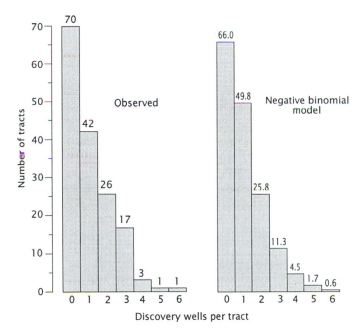

Figure 5–8. Histograms showing observed numbers of discovery wells per tract in an area of the Permian Basin, and the number expected in a clustered (negative binomial) model.

The probabilities that a tract will contain exactly two, three, or other number of discovery wells can be calculated in a similar fashion. Then, the expected number of tracts containing r discoveries can be determined simply by multiplying these probabilities by 160, the total number of tracts. **Table 5–3** gives the expected numbers of tracts for up to six discoveries per tract.

The numbers of tracts containing exactly r discoveries as predicted by the negative binomial model are compared to the corresponding observed numbers of tracts in **Figure 5–8**. The goodness of fit of the negative binomial can be tested by a χ^2 test exactly like that used to check the fit of the Poisson model. Again, it is necessary to combine the final three categories so a frequency of five or more is obtained. The test statistic is $\chi^2 = 4.82$, with $(5 - 2 = 3)$ degrees of freedom. This is less than the critical value of χ^2 for $\alpha = 0.05$ and $\nu = 3$, so we cannot reject the negative binomial as a model of the spatial distribution of discovery wells in the eastern part of the Permian Basin. Keep in mind that this is not equivalent to proof that the wells do follow a negative binomial model, because it is possible that some other clustered model might provide an even better fit. However, the negative binomial does generate a spatial distribution that is statistically indistinguishable from the one observed.

Nearest-neighbor analysis

An alternative to quadrat analysis is **nearest-neighbor analysis**. The data used are not the numbers of points within subareas, but the distances between closest pairs of points. Since it is not necessary to select a quadrat size, nearest-neighbor procedures avoid the possibility of finding that a pattern is random at one scale but not at another. Also, since there are usually many more pairs of nearest neighbors than quadrats, the analysis is more sensitive. A good introduction to nearest-neighbor techniques is given by Getis and Boots (1978). Ripley (1981) provides a review of theory and applications in several fields, as do Cliff and Ord (1981). Shaw and Wheeler (1994) and Bailey and Gatrell (1995) discuss computational aspects of nearest-neighbor analyses.

Nearest-neighbor analysis compares characteristics of the observed set of distances between pairs of nearest points with those that would be expected if the points were randomly placed. The characteristics of a theoretical random pattern can be derived from the Poisson distribution. If we ignore the effect of the edges of our map, the expected mean distance between nearest neighbors is

$$\bar{\delta} = \frac{1}{2}\sqrt{A/n} \tag{5.24}$$

where A is the area of the map and n is the number of points. You will recall that A/n is the point density, λ. The sampling variance of $\bar{\delta}$ is given by

$$\sigma_{\bar{\delta}}^2 = \frac{(4 - \pi)A}{4\pi n^2} \tag{5.25}$$

If we work out the constants,

$$\sigma_{\bar{\delta}}^2 = \frac{0.06831\,A}{n^2} \tag{5.26}$$

The standard error of the mean distance between nearest neighbors is the square root of $\sigma_{\bar{\delta}}^2$

$$s_e = \frac{0.26136}{\sqrt{A/n^2}} \tag{5.27}$$

The distribution of $\bar{\delta}$ is normal provided n is greater than 6, so we can use the simple z-test given in Chapter 2 to test the hypothesis that the observed mean distance between nearest neighbors, $\overline{d}$, is equal to the value of $\overline{\delta}$ from a random pattern of points of the same density. The test is

$$z = \frac{\overline{d} - \overline{\delta}}{s_e} \tag{5.28}$$

This is the form of the nearest-neighbor test that is commonly presented, but unfortunately it has a serious defect for most practical purposes. The expected value $\overline{\delta}$ assumes that edge effects are not present, which means that the observed pattern of points must extend to infinity in all directions if $\overline{d}$ and $\overline{\delta}$ are to be validly compared. Since the map does not extend indefinitely, the nearest neighbors of points near the edges must lie within the body of the map, and so $\overline{d}$ is biased toward a greater value (Upton and Fingleton, 1985). There are several corrections for this problem. If data are available beyond the limits of the area being analyzed, the map can be surrounded by a *guard region*. Then, nearest-neighbor distances between points inside the map and points in the guard region can be included in the calculation of $\overline{d}$. Alternatively, we can consider our map to be drawn not on a flat plane but on a torus. In this case, in the right map edge would be adjacent to the left edge and the top adjacent to the bottom. The nearest neighbor of a point along the right edge of the map might lie just inside the left edge (this concept should be familiar to anyone who has contoured point densities on stereonets). Another way of regarding this particular correction is to imagine that the pattern of points repeats in all directions, like floor tiles. Any point lying adjacent to an edge of the map has the opportunity to find a point across the edge that may be a closer neighbor than the nearest point within the map.

A third correction involves adjusting $\overline{d}$ so that the boundary effects are included in its expected value. Using numerical simulation, Donnelly (1978) found these alternative expressions for the theoretical mean nearest-neighbor distance and its sampling variance:

$$\overline{\delta} \approx \frac{1}{2}\sqrt{\frac{A}{n}} + \left(0.514 + \frac{0.412}{\sqrt{n}}\right)\frac{p}{n} \tag{5.29}$$

and

$$s_{\overline{\delta}}^2 \approx 0.070\frac{A}{n^2} + 0.035p\frac{\sqrt{A}}{n^{5/2}} \tag{5.30}$$

In these approximations, p is the perimeter of the rectangular map. Note that if the map has no edges, as when it is considered to be drawn on a torus, p is zero and these equations are identical to equations (5.24) and (5.26).

The expected and observed mean nearest-neighbor distances can be used to construct an index to the spatial pattern. The ratio

$$R = \frac{\overline{d}}{\overline{\delta}} \tag{5.31}$$

is the *nearest-neighbor statistic* and ranges from 0.0 for a distribution where all points coincide and are separated by distances of zero, to 1.0 for a random distribution of points, to a maximum value of 2.15. The latter value characterizes a distribution in which the mean distance to the nearest neighbor is maximized. The distribution has the form of a regular hexagonal pattern where every point

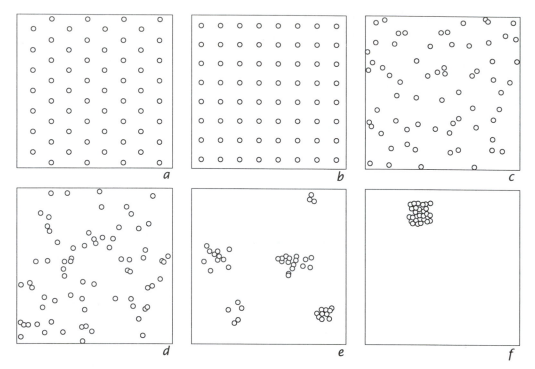

Figure 5–9. Nearest-neighbor statistics, R, for patterns of points on maps. (*a*) Points in a regular hexagonal network, $R = 2.15$. (*b*) Points in a regular square network, $R = 2.00$. (*c*) Points placed randomly within regular hexagonal cells, $R = 1.26$. (*d*) Points placed at random locations, $R = 0.91$. (*e*) Points placed randomly within five random clusters, $R = 0.34$. (*f*) Points placed randomly within a single cluster, $R = 0.13$. Point density, λ, is the same for all patterns. From Olea (1982).

is equidistant from six other points. **Figure 5–9** shows a series of patterns with different values of the nearest-neighbor statistic, all having the same point density.

We will illustrate the application of the nearest-neighbor method using the map shown in **Figure 5–10**. The "map" actually represents a polished facing stone on the front of a bank in a university town. It provides an interesting subject of study for an igneous petrology class. The stone is black anorthosite and contains small, scattered, euhedral crystals of magnetite. The instructor uses the slab to demonstrate a variety of topics, including examples of numerical techniques in petrography. For pedagogical purposes, it has been decreed that the slab is mounted in its original orientation. That is, it represents a vertical surface; "down" is toward the bottom of the slab. The map shows the location of all visible magnetite grains on the surface. Coordinates of each grain, in centimeters from the lower left corner of the slab, are listed in file BANK.TXT. Are magnetite grains uniformly distributed across the surface, or do they tend to be clustered? Is the density of crystals greater near the bottom of the slab than near the top? These and similar questions are of great importance in determining the petrogenesis of an igneous rock, and can be effectively investigated using the techniques we have discussed. Test the hypothesis of uniform, random distribution of crystals by both quadrat and nearest-neighbor analysis. This problem may be done by hand by measuring distances directly on **Figure 5–10**, or the distances may be computed using the

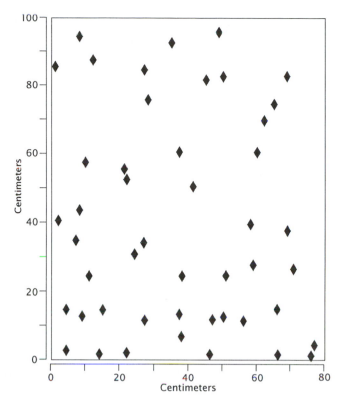

Figure 5–10. Representation of a polished slab of anorthosite facing stone showing locations of magnetite crystals listed in file BANK.TXT.

coordinates in file BANK.TXT. Ripley (1981, p. 175–181) gives an exhaustive analysis of these data, using a variety of techniques.

Distribution of Lines

Some naturally occurring patterns are composed of lines, such as lineaments seen on satellite images, the tracery of joints exposed on a weathered granite surface, or the microfractures seen in a thin section of a deformed rock. Just as a set of points can form a pattern that ranges from uniform to tightly clustered, so can sets of lines. Of course, lines are more complex than points because they possess length and orientation, as well as location. Their analysis is correspondingly more difficult, and statistical methods suitable for the study of patterns of lines seem less well developed than those applied to patterns of points. Few studies have examined the distribution of lengths of lines, except for some work on the lognormal distribution (Aitchison and Brown, 1969). A small number of workers have investigated the spacing between lines in a pattern, a problem analogous to nearest-neighbor analysis of points (Miles, 1964; Dacey, 1967). A much larger body of literature exists on the orientation of lines, a topic we will consider in the next section.

We can define a random pattern of lines as one in which any line is equally likely to cross any location, and any orientation of the crossing line is also equally likely. Such random patterns can be generated in many ways; one procedure consists of

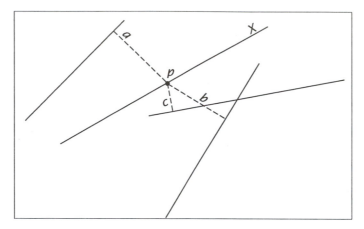

Figure 5–11. Calculation of nearest-neighbor distances between lines. Point p is chosen at random on a line X. Dashed lines a, b, and c are perpendiculars drawn from point p to nearby lines. The shortest of these, perpendicular line c, is the distance to the nearest neighbor of line X. The process is repeated to find the nearest-neighbor distances for all lines.

choosing two pairs of coordinates from a random number table, then drawing a line through them. Another consists of drawing a radius at a randomly chosen angle, measuring out along the radius a random distance from the center, then constructing a perpendicular to the radial line. Repeating either procedure will result in patterns of lines that are statistically indistinguishable.

We can define a measure of line density that is analogous to λ, the point density:

$$\lambda = L/A \tag{5.32}$$

The quantity L is simply the total length of lines on the map, which has an area A. λ is the parameter that determines the form of the Poisson distribution; as we would expect, the Poisson model describes the distribution of many properties of a pattern formed by random lines.

The distribution of distances between pairs of lines can be examined by calculating a nearest-neighbor measure. We must first randomly pick a point on each of the lines in the map. From each point, the distance is measured to the nearest line, in a direction perpendicular to that line. The mean nearest-neighbor distance $\overline{d}$ is the average of these measurements. The procedure is illustrated in **Figure 5–11**.

Dacey (1967) has determined that the expected nearest-neighbor distance $\overline{\delta}$ for a pattern of random lines is

$$\overline{\delta} = \frac{0.31831}{\lambda} \tag{5.33}$$

and that the expected variance is

$$\sigma_{\overline{\delta}}^2 = \frac{0.10132}{\lambda^2} \tag{5.34}$$

From the expected variance and the number of lines in the pattern, we can find the standard error of our estimate of the mean nearest-neighbor distance. The standard error is

$$s_e = \sqrt{\frac{\sigma_{\overline{\delta}}^2}{n}} \tag{5.35}$$

This allows us to calculate a simple z-statistic for testing the significance of the difference between the expected and observed mean nearest-neighbor distance:

$$z = \frac{\overline{d} - \overline{\delta}}{s_e} \tag{5.36}$$

The test is two-tailed; if the value of z is not significant, we conclude that the observed pattern of lines cannot be distinguished from a pattern generated by a random (Poisson) process. We can also create a nearest-neighbor index identical to that used for point patterns by taking the ratio of the observed and expected mean nearest-neighbor distances, or $\overline{d}/\overline{\delta}$. The index is interpreted exactly as is the index for point patterns.

This test will work for sets of lines that are straight or curved, provided the lines do not reverse direction frequently. Also, the lines should be at least one and one-half times longer than the average distance between the lines. If the number of lines on the map is small, the estimated density should be adjusted by the factor $(n-1)/n$, where n is the number of lines in the pattern. The estimate of the line density is, therefore

$$\lambda = \frac{(n-1)L}{nA} \tag{5.37}$$

A simple alternative way of investigating the nature of a set of lines on a map involves converting the two-dimensional pattern into a one-dimensional sequence. We can do this by drawing a **sampling line** at random across the map and noting where the line intersects the lines in the pattern. The distribution of intervals between the points of intersection along the sampling line will provide information about the spatial pattern. We can test this one-dimensional sequence using methods presented in Chapter 4. If a single sampling line does not provide enough intersections for a valid test, we can draw a randomly oriented continuation of the sampling line from the point where the sampling line intersects the last line on the map, and a second randomly oriented continuation from the last line on the map intersected by this continuation, and so on (**Fig. 5–12**). The zigzag path of the sampling line is a **random walk**, and the succession of intersections can be

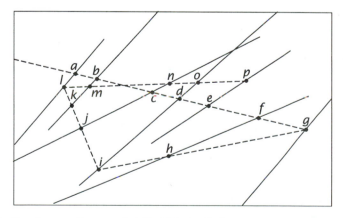

Figure 5–12. Random-walk sampling line (dashed) drawn across pattern of lines on a map. Intersections along sampling line form a sequence of intervals, $a-b$, $b-c$, ..., $o-p$, that can be tested for randomness.

treated as though they occurred along a single, straight sampling line. This and other methods for investigating the density of patterns of lines are reviewed by Getis and Boots (1978). A computer program for computing nearest-neighbor distances, orientation, and other statistical measures of patterns of lines is given by Clark and Wilson (1994).

Analysis of Directional Data

Directional data are an important category of geologic information. Bedding planes, fault surfaces, and joints are all characterized by their attitudes, expressed as strikes and dips. Glacial striations, sole marks, fossil shells, and water-laid pebbles may have preferred orientations. Aerial and satellite photographs may show oriented linear patterns. These features can be measured and treated quantitatively like measurements of other geologic properties, but it is necessary to use special statistics that reflect the circular (or spherical) nature of directional data.

Following the practice of geographers, we can distinguish between **directional** and **oriented** features. Suppose a car is traveling north along a highway; the car's motion has direction, while the highway itself has only a north–south orientation. Strikes of outcrops and the traces of faults are examples of geologic observations that are oriented, while drumlins and certain fossils such as high-spired gastropods have clear directional characteristics.

We may also distinguish observations that are distributed on a circle, such as paleocurrent measurements, and those that are distributed spherically, such as measurements of metamorphic fabric. The former data are conventionally shown as **rose diagrams**, a form of circular histogram, while the latter are plotted as points on a projection of a hemisphere. Although geologists have plotted directional measurements in these forms for many years, they have not used formal statistical techniques extensively to test the veracity of the conclusions they have drawn from their diagrams. This is doubly unfortunate; not only are these statistical tests useful, but the development of many of the procedures was originally inspired by problems in the Earth sciences.

Figure 5–13 is a map of glacial striations measured in a small area of southern Finland; the measurements are listed in **Table 5–4** and contained in file FINLAND.TXT. The directions indicated by the striations can be expressed by plotting them as unit vectors or on a circle of unit radius as in **Figure 5–14 a**. If the circle is subdivided into segments and the number of vectors within each segment counted, the results can be expressed as the rose diagram, or circular histogram, shown as **Figure 5–14 b**.

Nemec (1988) pointed out that many of the rose diagrams published by geologists violate the basic principal on which histograms are based and, as a consequence, the diagrams are visually misleading. Recall that areas of columns in a histogram are proportional to the number (or percentage) of observations occurring in the corresponding intervals. For a rose diagram to correctly represent a circular distribution, it must be constructed so that the areas of the wedges (or "petals") of the diagram are proportional to class frequencies. Unfortunately, most rose diagrams are drawn so that the radii of the wedges are proportional to frequency. The resulting distortion may suggest the presence of a strong directional trend where none exists (**Fig. 5–15**).

Figure 5–13. Map showing location and direction of 51 measurements of glacial striations in a 35-km^2 area of southern Finland.

Table 5–4. Vector directions of glacial striations measured in an area of southern Finland; measurements given in degrees clockwise from north.

23	105	127	144	171
27	113	127	145	172
53	113	128	145	179
58	114	128	146	181
64	117	129	153	186
83	121	132	155	190
85	123	132	155	212
88	125	132	155	
93	126	134	157	
99	126	135	163	
100	126	137	165	

If we define a radius for a sector of a rose diagram that represents either one observation, or 1%, we can easily calculate the appropriate radii that represent any number of observations or relative frequencies,

$$r_f = r_u\sqrt{f} \tag{5.38}$$

where r_u is the unit radius representing one observation or 1%, f is the frequency (in counts or percent) of observations within a class, and r_f is the radius of the class sector. In other words, the radius should be proportional to the square root of the frequency rather than to the frequency itself.

Rose diagrams, even if properly scaled, suffer from the same problems as ordinary histograms; their appearance is extremely sensitive to the choice of class widths and starting point and they exhibit variations similar to the histogram

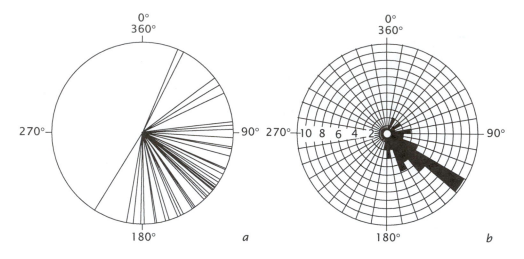

Figure 5–14. Directions of glacial striations shown on Figure 5–13. (*a*) Directions plotted as unit vectors. (*b*) Directions plotted as a rose diagram showing numbers of vectors within successive 10° segments.

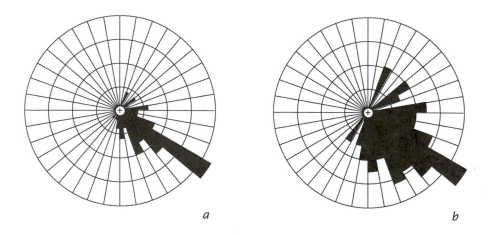

Figure 5–15. Rose diagram of glacial striations shown on Figure 5–13 plotted in 10° segments. (*a*) Length of petals proportional to frequency. (*b*) Area of petals proportional to frequency.

examples shown in **Figure 2–11** on p. 30. Wells (1999) provides a computer program that quickly constructs rose diagrams with different conventions and also includes an assortment of graphical alternatives that may be superior to conventional rose diagrams for some uses (**Fig. 5–16**).

To compute statistics that describe characteristics of an entire set of vectors, we must work directly with the individual directional measurements rather than with a graphical summary such as a rose diagram. (Note that the following discussion uses geological and geographic conventions in which angles are measured clockwise from north, or from the positive end of the Y-axis. Many papers on directional statistics follow a mathematical convention in which angles are measured counterclockwise from east, or from the positive end of the X-axis.)

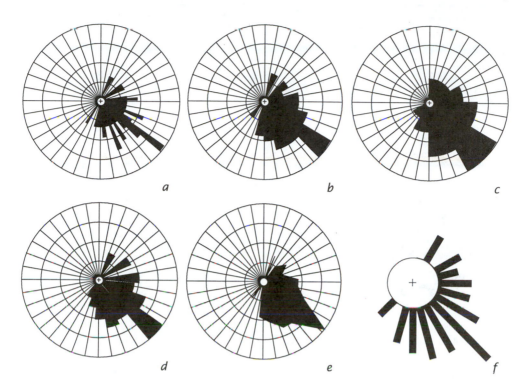

Figure 5–16. Effect of choice of segment size and origin on appearance of rose diagrams. Data are directions of glacial striations from file FINLAND.TXT: (*a*) 5° segments, 0° origin, outer ring 20%; (*b*) 15° segments, 0° origin, outer ring 30%; (*c*) 30° segments, 0° origin, outer ring 40%; (*d*) 15° segments, 10° origin—compare to (*b*). Alternative graphical forms include (*e*) kite diagram, 15° segments, 0° origin—sometimes used in statistical presentations; (*f*) circular histogram, 15° segments, 0° origin—widely used to plot wind directions.

The dominant direction in a set of vectors can be found by computing the **vector resultant**. The X- and Y-coordinates of the end point of a unit vector whose direction is given by the angle θ are

$$X_i = \cos \theta_i$$
$$Y_i = \sin \theta_i$$
(5.39)

Three such vectors are shown plotted in **Figure 5–17**. Also shown is the vector resultant, R, obtained by summing the sines and cosines of the individual vectors:

$$X_r = \sum_{i=1}^{n} \cos \theta_i$$
$$Y_r = \sum_{i=1}^{n} \sin \theta_i$$
(5.40)

From the resultant, we can obtain the **mean direction**, $\overline{\theta}$, which is the angular average of all of the vectors in a sample. It is directly analogous to the mean value of a set of scalar measurements

$$\overline{\theta} = \tan^{-1} (Y_r / X_r) = \tan^{-1} \left(\sum_{i=1}^{n} \sin \theta_i \Big/ \sum_{i=1}^{n} \cos \theta_i \right)$$
(5.41)

Obviously, the magnitude or length of the resultant depends in part on the amount of dispersion in the sample of vectors, but it also depends upon the number of

319

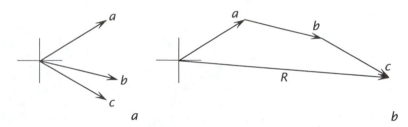

Figure 5–17. Determination of mean direction of a set of unit vectors. (*a*) Three vectors taken from Figure 5–16. (*b*) Vector resultant, *R*, obtained by combining the three unit vectors. Order of combination is immaterial.

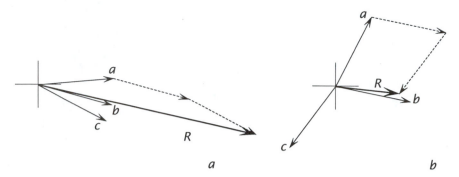

Figure 5–18. Use of length of resultant to express dispersion in a collection of unit vectors. (*a*) Three vectors tightly clustered around a common direction. Resultant *R* is relatively long, approaching the value of n. (*b*) Three widely dispersed vectors; resultant length is less than 1.0.

vectors. In order to compare resultants from samples of different sizes, they must be converted into a standardized form. This is done simply by dividing the coordinates of the resultant by the number of observations, n

$$\overline{C} = X_r/n = \tfrac{1}{n} \sum_{i=1}^{n} \cos \theta_i$$
$$\overline{S} = Y_r/n = \tfrac{1}{n} \sum_{i=1}^{n} \sin \theta_i$$

(5.42)

Note that these coordinates also define the centroid of the end points of the individual unit vectors.

The resultant provides information not only about the average direction of a set of vectors, but also on the spread of the vectors about this average. **Figure 5–18 a** shows three vectors that deviate only slightly from the mean direction. The resultant is almost equal in length to the sum of the lengths of the three vectors. In contrast, three vectors in **Figure 5–18 b** are widely dispersed; their resultant is very short. The length of the resultant, *R*, is given by the Pythagorean theorem:

$$R = \sqrt{X_r^2 + Y_r^2} = \sqrt{\left(\sum_{i=1}^{n} \cos \theta_i \right)^2 + \left(\sum_{i=1}^{n} \sin \theta_i \right)^2}$$

(5.43)

The length of the resultant can be standardized by dividing by the number of observations. The standardized resultant length can also be found from the standardized end points

$$\overline{R} = \frac{R}{n} = \sqrt{\overline{C}^2 + \overline{S}^2}$$

(5.44)

The quantity $\overline{R}$, called the ***mean resultant length,*** will range from zero to one. It is a measure of dispersion analogous to the variance, but expressed in the opposite sense. That is, large values of $\overline{R}$ indicate that the observations are tightly bunched together with a small dispersion, while values of $\overline{R}$ near zero indicate that the vectors are widely dispersed. **Figure 5–19** shows sets of vectors having different values of $\overline{R}$. In order to have a measure of dispersion that increases with increasing scatter, $\overline{R}$ is sometimes expressed as its complement, the ***circular variance***

$$s_o^2 = 1 - \overline{R} = (n - R)/n \tag{5.45}$$

Other directional statistics can be computed, including circular analogs of the standard deviation, mode, and median. Equations for these are given in convenient table form by Gaile and Burt (1980).

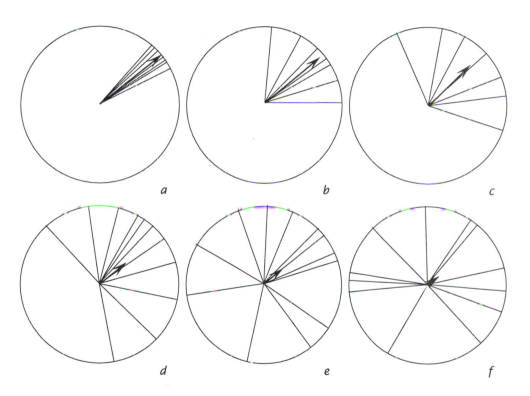

Figure 5–19. Sets of unit vectors illustrating the value of $\overline{R}$ produced by different dispersions of vectors. In all examples, the mean direction is 52°: (*a*) $\overline{R} = 0.997$, (*b*) $\overline{R} = 0.90$, (*c*) $\overline{R} = 0.75$, (*d*) $\overline{R} = 0.55$, (*e*) $\overline{R} = 0.40$, (*f*) $\overline{R} = 0.10$.

Orientation data must be modified before mean directions or measures of dispersion can be calculated. Since the orientation of any feature may be expressed as either of two opposite directions, some convention must be adopted to avoid inflating the dispersion of the measurements. Krumbein (1939) hit upon a novel solution to this problem while studying the orientations of stream pebbles. If all of the measured angles are doubled, the same angles will be recorded regardless of which directional sense of the oriented features is used. As an example, consider a fault trace that strikes northeast–southwest. Its orientation could equally well be recorded as 45° or as 225°. If we double the angles, we obtain $45° \times 2 = 90°$ and $225° \times 2 = 450°$, which becomes $(450° - 360°) = 90°$.

Mean direction, mean resultant length, and circular variance can be found in the usual manner after the orientation angles have been doubled. To recover the true mean orientation, simply divide the calculated mean direction by two. This may be illustrated using the data in file CAROLINA.TXT, which contains orientations of the major axes of 99 "Carolina bays"—ellipsoidal depressions on the southern part of the U.S. Atlantic Coastal Plain. The origin of these geomorphic features (**Fig. 5–20**) was at one time the subject of intense controversy; the depressions were attributed to causes as diverse as meteorite impact, karstic solution, or deflation (Prouty, 1952). Subsequent studies (Rasmussen, 1959) have favored a composite origin involving differential solution by groundwater and eolian removal of material. **Figure 5–21** shows rose diagrams of the axial orientations of the "bays" plotted (a) incorrectly as vectors (resulting in a bimodal distribution), (b) as vectors whose angles have been doubled, and then (c) as vectors after dividing the doubled angles of (b) by two, and also plotting the complements of the vectors. Although each measurement is plotted twice in this diagram, it yields the correct impression of a symmetrical distribution.

Testing hypotheses about circular directional data

In order to test statistical hypotheses about circularly distributed data, we must have a probability model of known characteristics against which we can test. There are circular analogs of the univariate distributions discussed in Chapter 2. The most useful of these is the ***von Mises distribution***. It is a circular equivalent of the normal distribution and similarly possesses only two parameters, a mean direction, $\overline{\theta}$, and a concentration parameter, κ. The von Mises distribution is unimodal and symmetric about the mean direction. As the concentration parameter increases, the likelihood of observing a directional measurement very close to the mean direction increases. If κ is equal to zero, all directions are equally probable, and a circular uniform distribution results. **Figure 5–22 a** shows the form of the von Mises distribution for several values of κ. The distribution can also be shown in conventional form as in **Figure 5–22 b**; note that the horizontal scale is given in degrees and corresponds to a complete circle.

It is difficult to determine κ directly, but the concentration parameter can be estimated from $\overline{R}$ if we assume that the data are a sample from a population having a von Mises distribution. Appendix Table **A.9** gives maximum likelihood estimates of κ for a calculated $\overline{R}$. We will use these estimated values of κ in some subsequent statistical tests.

Test for randomness.—The simplest hypothesis that can be statistically tested is that the directional observations are random. In other words, there is no preferred direction, or the probability of occurrence is the same for all directions. If we assume that the observations come from a von Mises distribution, the hypothesis is equivalent to stating that the concentration parameter, κ, is equal to zero, because then the distribution becomes a circular uniform. In formal terms, the null hypothesis and alternative are

$$H_0 : \kappa = 0$$
$$H_1 : \kappa > 0$$

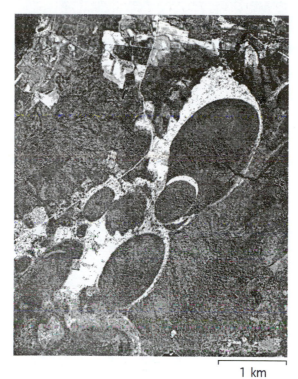

Figure 5–20. Aerial photo of "Carolina bays," subparallel ellipsoidal depressions on Atlantic Coastal Plain of southeastern U.S. in Bladen County, North Carolina (Prouty, 1952).

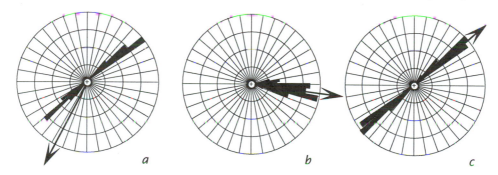

a   *b*   *c*

Figure 5–21. Effect of doubling angular direction in order to calculate mean orientation. (*a*) Orientations of major axes of 99 "Carolina bays" plotted as vector directions. Resultant mean direction is 207° and is near zero in length ($\overline{R} = 0.008$). (*b*) Orientation measurements plotted as vector directions after angles are doubled. Distribution is no longer bimodal. Resultant reflects correct trend of doubled angles and is near unity in length (mean direction is 97.4°; $\overline{R} = 0.98$). (*c*) Orientations replotted at original angles and their complement. True resultant direction (48.7°) is found by halving resultant direction in (*b*).

The test is extremely simple and involves only the calculation of $\overline{R}$ according to Equation (5.44). This statistic is compared to a critical value of $\overline{R}$ for the desired level of significance. If the observations do come from a circular uniform distribution, we would expect $\overline{R}$ to be small, as in **Figure 5–19 f**. However, if the computed statistic is so large that it exceeds the critical value, the null hypothesis must be rejected and the observations may be presumed to come from a population having

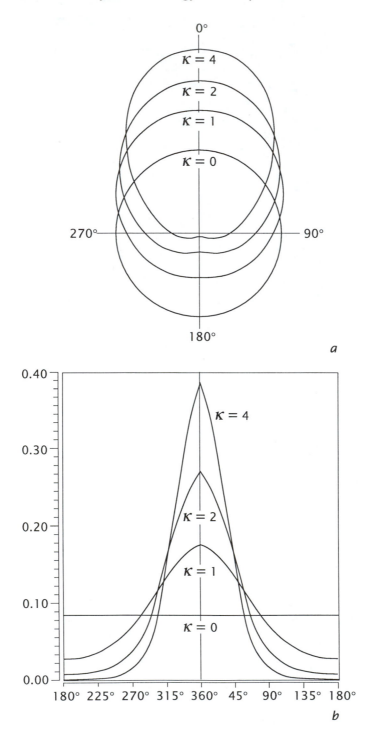

Figure 5–22. Von Mises distributions having different concentration parameters. (*a*) Distribution plotted in polar form. (*b*) Distribution plotted as conventional probability distribution. Note that horizontal axis is given in degrees. After Gumbel, Greenwood, and Durand (1953).

a preferred orientation. This test was originally developed by Lord Rayleigh at the turn of the nineteenth century; a modern derivation is given by Mardia (1972). Appendix Table **A.10** gives critical values of $\overline{R}$ for Rayleigh's test of a preferred trend for various levels of significance and numbers of observations.

Remember that Rayleigh's test presumes that the observed vectors are sampled from a von Mises distribution. That is, the population of vectors is either uniform (if $\kappa = 0$) or has a single mode or preferred direction. If the vectors are actually sampled from a bimodal distribution such as that shown in **Figure 5–21 a**, the test will give misleading results.

We will test the measurements of Finnish glacial striations at a 5% level of significance to determine if they have a preferred direction. Since there are 51 observations, Appendix Table **A.10** yields a critical value of $\overline{R}_{50,5\%} = 0.244$. The test statistic is simply the normalized resultant, $\overline{R}$. The sum of the cosines of the vectors is $X_r = -25.793$ and the sum of the sines is $Y_r = 31.637$. The resultant length is

$$R = \sqrt{(-25.793)^2 + (31.637)^2} = 40.819$$

which, when divided by the sample size, yields a mean resultant length of

$$\overline{R} = 40.819/51 = 0.800$$

Since the computed value of $\overline{R}$ far exceeds the critical value, we reject the null hypothesis that the concentration parameter is equal to zero. The striations must have a preferred trend.

Test for a specified trend.—On some occasions we may wish to test the hypothesis that the observations correspond to a specified trend. For example, the area of Finland where the measurements of glacial striations were taken is located within a broad topographic depression aligned northwest–southeast at approximately 105°. Does the mean direction of ice movement, as indicated by the striations, coincide with the axial direction of this depression?

Exact tests of the hypothesis that a sample of vectors has been taken from a population having a specified mean direction require the use of extensive charts in order to set the critical value (Stephens, 1969). A simpler alternative is to determine a **confidence angle** around the mean direction of the sample and see if this angle is sufficiently broad to encompass the hypothetical mean direction. This confidence angle is based on the standard error of the estimate of the mean direction, $\overline{\theta}$, and thus considers both the size of the sample and its dispersion.

Before computing the confidence angle, the Rayleigh test should be applied to confirm that a statistically significant mean direction does exist. Then the mean resultant length $\overline{R}$ must be computed and the concentration parameter κ estimated using Appendix Table **A.9**. The approximate standard error of the mean direction, given in radians, is

$$s_e = 1 \left/ \sqrt{n\overline{R}\kappa} \right. \tag{5.46}$$

Since the standard error is a measure of the chance variation expected from sample to sample in estimates of the mean direction, we can use it to define probabilistic limits on the location of the true or population mean direction. Assuming that estimation errors are normally distributed, the interval

$$\overline{\theta} \pm z_\alpha s_e \tag{5.47}$$

325

should capture (or include) the true population mean direction α% of the time. For example, if we collected 100 random samples of the same size from a population of vectors and computed the mean directions and 95% confidence intervals around each, we would expect that all but about five of those intervals would contain the true mean direction. Of course, we would not know which five of the intervals failed to capture the true direction, so we must assign a probabilistic caveat to all of them. We might, for example, make the statement that "the interval, plus and minus so many degrees around the mean direction of this particular sample, contains the true population mean direction. The probability that this statement is incorrect is 5%."

We have already applied Rayleigh's test and rejected the hypothesis of no trend in the observations of the striations. The approximate standard error of the mean direction can now be found:

$$s_e = \frac{1}{\sqrt{51 \cdot 0.8004 \cdot 2.87129}} = \frac{1}{10.826} = 0.0924 \text{ radians} = 5.29°$$

Therefore, the probability is 95% that the interval

$$129.2° \pm 1.96 \times 5.29°$$

contains the population mean direction. In other words,

$$118.8° \leq \overline{\theta} \leq 139.6°$$

Since this interval does not include the direction of alignment of the topographic depression, we must conclude that the axis of the depression does not coincide with the mean direction of the striations.

Test of goodness of fit.—A simple nonparametric alternative to the Rayleigh test of uniformity involves dividing the unit circle into a convenient number of angular segments. If these segments are equal in size and the observed vectors are distributed at random, we should observe approximately equal numbers of vectors in each segment. The number actually observed can be compared to those expected by a χ^2 test. The expected frequency in each segment must be at least 5, and there should be between $n/15$ and $n/5$ segments. The χ^2 statistic is computed in the usual manner (see Eq. 2.45) and has $k-1$ degrees of freedom, where k is the number of segments.

The same procedure can be used to test the goodness of fit of the observed vectors to other theoretical models, such as a von Mises distribution with a specified concentration parameter κ greater than zero and a specified mean direction $\overline{\theta}$. Computing the expected frequencies, however, can be complicated. Examples are given by Gumbel, Greenwood, and Durand (1953) and Batschelet (1965).

Testing the equality of two sets of directional vectors.—We may sometimes wish to test hypotheses about the equivalence of two samples or collections of directional measurements. For example, we may have paleocurrent measurements in two different stratigraphic units and want to determine if their mean directions are the same, or we may wish to see if the orientations of lineaments seen on a satellite image coincide with the orientations of faults known to exist in the photographed area. At a much smaller scale, we may want to compare the alignment of elongated pores in thin sections from two cored samples of sandstone from a petroleum reservoir.

The equality of two mean directions may be tested by comparing the vector resultants of the two groups to the vector resultant produced when the two sets of measurements are combined or pooled. If the two samples actually are drawn from the same population, the resultant of the pooled samples should be approximately equal to the sum of their two resultants. If the mean directions of the two samples are significantly different, the pooled resultant will be shorter than the sum of their resultants.

If κ is a large value (greater than 10), an F-test statistic can be computed by

$$F_{1,n-2} = \frac{(n-2)\left(R_1 + R_2 - R_p\right)}{(n - R_1 - R_2)} \tag{5.48}$$

where n is the total number of observations, R_1 and R_2 are the resultants of the two samples of vectors, and R_p is the resultant of the set of vectors after the two groups have been pooled.

Using Appendix Table **A.9**, we can estimate the value of κ from $\overline{R}_p$, the length of the mean resultant of the two pooled samples. If κ is smaller than 10 but greater than 2, then a more accurate F-test is

$$F_{1,n-2} = \left(1 + \frac{3}{8\kappa}\right)\frac{(n-2)\left(R_1 + R_2 - R_p\right)}{(n - R_1 - R_2)} \tag{5.49}$$

If κ is less than 2, special tables (such as those given in Mardia, 1972) are necessary. It is also possible to test the equality of the concentration parameters of two sets of vectors, but the computations are involved. Refer to Mardia (1972) for a detailed discussion, and to Gaile and Burt (1980) for a worked example from geomorphology.

A fold belt, expressed topographically as the Naga Hills and their extensions, occurs at the juncture between the Indian subcontinent and the Indochinese peninsula. Apparently related to compressive movements that created the Himalayas, the fold belt includes a series of subparallel anticlines along the eastern border of Bangladesh. Oil and gas have been found in structural traps in this region, so delineation of the folds is of economic as well as scientific interest. Presumably the folds occur, perhaps with reduced magnitude, to the west of the Naga Hills, but are concealed by modern sediments deposited by the Ganges River and its tributaries. Unfortunately, reflection seismic data that could reveal the buried structures are very sparse.

Interpretations of Landsat satellite images of this region indicate numerous lineations of unknown origin. It is possible that the lineations reflect subsurface folds, and if so, they may provide valuable clues to structural geology and possible petroleum deposits.

Figure 5–23 is a map of eastern Bangladesh showing the traces of axial planes of major exposed anticlines and the larger lineaments measured on Landsat images. The orientations of these two sets are shown on **Figure 5–24**. Because the lines have no sense of direction, the plots are bimodal, and we must double the observed angles to obtain the correct distribution of vectors. **Table 5–5** lists the orientations of both the axial planes and the lineaments, which also are contained in file BANGLA.TXT. There is an obvious difference between the two sets, but is this difference statistically significant or could it have arisen through the vagaries of sampling?

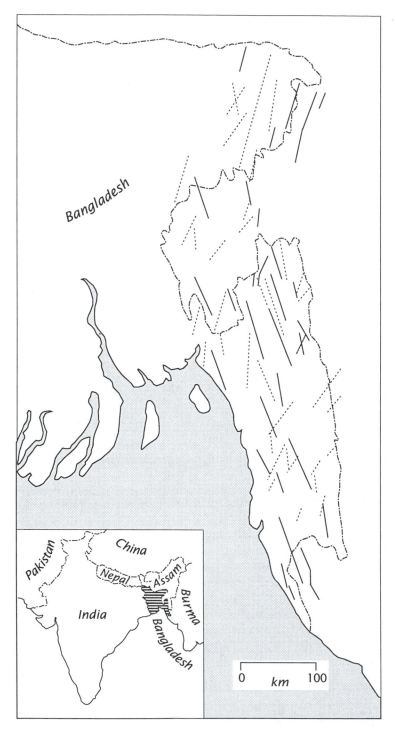

Figure 5–23. Map of eastern Bangladesh showing axial planes of major anticlines (solid lines) and large lineaments interpreted from Landsat images (dashed lines).

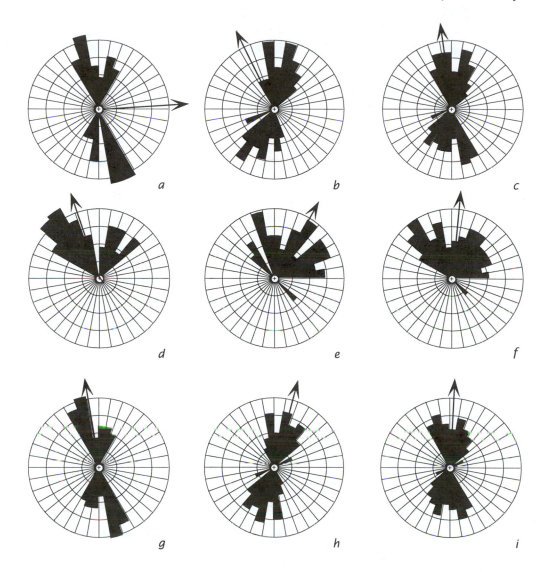

Figure 5–24. Rose diagrams of orientation data from eastern Bangladesh. Mean orientations indicated by arrows. Top row shows plots of vector directions from file BANGLA.TXT: (*a*) Anticlinal axes (mean direction is 86.2°; $\overline{R} = 0.05$). (*b*) Lineaments (mean direction is 334.6°; $\overline{R} = 0.15$). (*c*) Pooled vectors (mean direction is 352.5°; $\overline{R} = 0.70$). Middle row shows plots of doubled vector directions: (*d*) Anticlinal axes (mean direction is 341.5°; $\overline{R} = 0.85$). (*e*) Lineaments (mean direction is 30.1°; $\overline{R} = 0.77$). (*f*) Pooled vectors (mean direction is 5.3°; $\overline{R} = 0.74$). Bottom row shows orientations replotted at original angles and their complements. True resultant directions found by halving resultant directions shown in middle row: (*g*) Anticlinal axes (mean direction is 350.8°; $\overline{R} = 0.85$). (*h*) Lineaments (mean direction is 15.0°; $\overline{R} = 0.77$). (*i*) Pooled vectors (mean direction is 2.6°; $\overline{R} = 0.74$).

To test the hypothesis that the mean directions of the anticlinal axes and the Landsat lineaments are the same, we must first compute the resultants of each of the two groups and the resultant of the two groups combined. The resultant of the

Table 5–5. Orientation of axial planes of anticlines and Landsat lineations in eastern Bangladesh; measurements given in degrees clockwise from north.

Anticlinal Axes, $n = 34$				Landsat Lineaments, $n = 36$			
12	16	14	5	350	32	15	8
192	202	169	163	214	192	16	26
186	186	24	344	356	218	198	221
343	346	161	341	350	18	221	342
339	150	169	336	160	205	35	337
351	156	159	352	2	171	196	14
152	162	341	181	184	246	175	25
348	158	156		354	213	26	212
330	162	20		42	354	13	202

34 doubled measurements of the fold axial planes is $R_1 = 38.97$ and the resultant of the 36 doubled measurements of the Landsat lineament is $R_2 = 27.79$. The two groups can be combined into a pooled collection of 70 observations that has a resultant of $R_p = 51.73$. The mean resultant of the pooled group is

$$\overline{R}_p = \frac{51.73}{70} = 0.74$$

and by use of Appendix Table **A.9**, we can estimate the concentration factor as $\kappa = 2.2893$.

Since κ is greater than 2 but less than 10, the appropriate test statistic is given by Equation (5.49). Substituting values we have calculated into that equation gives

$$F = \left(1 + \frac{3}{8\,(2.2893)}\right) \left(\frac{(70 - 2)\,(38.97 + 27.79 - 51.73)}{(70 - 38.97 - 27.79)}\right) = 367.11$$

The test has $v_1 = 1$ and $v_2 = (70 - 2)$ degrees of freedom. From the values of F in Appendix Table **A.3**, we can interpolate to find the critical value for F at the 5% level of significance ($\alpha = 0.05$) with 1 and 68 degrees of freedom; the value is $F = 3.99$. Since the test value far exceeds the critical value, we must regretfully conclude that the Landsat lineaments and the fold axes are not drawn from a common population. Although Landsat lineaments may be useful guides for exploration, in this region they apparently do not reflect the trends of structural folds.

Spherical Distributions

Statistical tests of directional data distributed in three dimensions have been developed only in recent years, in part because the mathematics of the distributions are very complicated. However, geologic problems that involve three-dimensional vectors are exceedingly common, and we should not shy away from the use of the available statistical techniques for their interpretation. Some of these methods require matrix algebra, although the matrices are not large, and the extraction of eigenvalues and eigenvectors. The geometric interpretation of eigenvectors presented in Chapter 3 will be of direct application. The mathematics are closely related to multivariate procedures described in Chapter 6. Here we deal with three physical dimensions; later we will apply the same steps to the analysis of multidimensional data in which each "dimension" is a different geologic variable.

Examples of three-dimensional directional data in the Earth sciences include measurements of strike and dip taken for structural analyses, vectorial measurements of the geomagnetic field, directional permeabilities measured on cores from petroleum reservoirs, measurements of orientation and dip of crossbeds, and determinations of crystallographic axes for petrofabric studies.

As with two-dimensional data, we must first establish a standard method of notation. We can regard three-dimensional directional observations as consisting of vectors; since we are concerned primarily with their angular relationships, these can be considered to be of unit length. If all of the directional measurements from an area are collected together at a common origin, the tips of the unit vectors will lie on the surface of a sphere; hence the term *spherical distribution*.

Some oriented features do not have a sense of direction and can be referred to as *axes*. Examples include the lines of intersections between sets of dipping planes, axes of revolution, and perpendiculars to planes. In addition, it is sometimes advantageous to disregard the directional aspect of vectors and to treat them as axes.

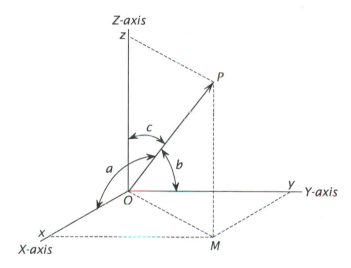

Figure 5–25. Notational system for three-dimensional vector OP in space defined by Cartesian axes X, Y, and Z. Angles between OP and the axes are a, b, and c.

Standard mathematical notation utilizes three Cartesian coordinates to describe a vector in space (**Fig. 5–25**). The direction of the vector OP is specified by the cosines of the angles between the vector and each of the coordinate axes. The coordinates of the point P are equal to

$$x = \cos a$$
$$y = \cos b \qquad (5.50)$$
$$z = \cos c$$

Since the vector is considered to have unit length,

$$x^2 + y^2 + z^2 = 1 \qquad (5.51)$$

An alternative notation uses polar coordinates. Using spherical angles, we can define the direction of vector OP by ϕ, the angle between the X-axis and the projection of the vector on the X-Y plane, and by θ, the inclination with respect to the Z-axis (**Fig. 5–26**). In effect, θ defines the "latitude" of the vector, while ϕ defines its "longitude." The relationship between these spherical polar coordinates and Cartesian coordinates is

$$x = \sin \theta \cos \phi$$
$$y = \sin \theta \sin \phi \qquad (5.52)$$
$$z = \cos \theta$$

Measurements of geologic properties are often given in terms of strike and dip, rather than as spherical angles or three-dimensional Cartesian coordinates. Also, the coordinate notation used by geologists differs from that conventionally used in mathematics. If we regard the positive end of the X-axis as corresponding to north, the positive end of the Y-axis as corresponding to east, and the positive end of the Z-axis as vertically downward, we have defined a Cartesian system in which dips are expressed as positive angles (Mardia, 1972).

The notation is illustrated in **Figure 5–27** for a vector OP defined by the strike and dip of its enclosing plane. The line OM is the **azimuth**, or projection of OP onto the horizontal X-Y plane; it is perpendicular to the line of strike. The angle S is the **angle of strike**, measured clockwise from $0°$ at the north. D is the **dip**, measured as a positive angle from OM downward. The x-, y-, and z-coordinates of point P are

$$x = -\sin S \cos D$$
$$y = \cos S \cos D \qquad (5.53)$$
$$z = \sin D$$

Middleton (2000) points out that the use of strike, although traditional, is potentially confusing because the strike is an orientation rather than a direction. It is less ambiguous to specify the dip and azimuth of an inclined geological feature. If A is the azimuth measured clockwise from $0°$ at the north, the x-, y-, and z-coordinates of point P can be found by

$$x = \cos D \cos A$$
$$y = \cos D \sin A \qquad (5.54)$$
$$z = \sin D$$

Conversion becomes more complicated if strike is measured in quadrant convention, which requires specifying the direction of dip. Refer to Watson (1970) for a more complete discussion.

Once we have the spherical measurements in terms of X-, Y-, and Z-coordinates of the vector end points, it is a simple matter to compute the mean direction and spherical variance. This is done in a manner analogous to the computation of the circular mean and variance. The mean direction is given by the resultant, R, of the unit vectors. Its length is

$$R = \sqrt{\left(\sum x_i\right)^2 + \left(\sum y_i\right)^2 + \left(\sum z_i\right)^2} \qquad (5.55)$$

In normalized form, this is

$$\overline{R} = R/n \qquad (5.56)$$

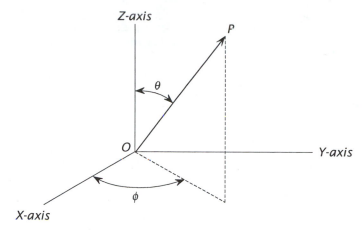

Figure 5–26. Notational system for three-dimensional vector OP in space defined by spherical angles ϕ and θ.

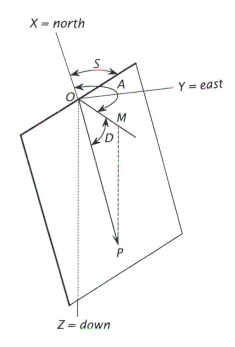

Figure 5–27. Strike and dip notation for three-dimensional vector OP. Angle S, measured clockwise from north, is the strike of the surface containing vector OP. Angle A, also measured clockwise from north, is the azimuth of vector OP. The plane OMP is perpendicular to the dipping surface. The angle D is the dip.

The direction of the resultant with respect to the three coordinate axes is given by the cosines of the angles between the resultant and these axes:

$$\begin{aligned} \cos \alpha &= \sum x_i/n \\ \cos \beta &= \sum y_i/n \\ \cos \gamma &= \sum z_i/n \end{aligned} \qquad (5.57)$$

If the observations are tightly clustered around a common direction, the resultant R will be a large number, approaching n. If the observations are scattered, R will be small. As in the case of circular distributions, R can be used as a measure of concentration and can be expressed as the **spherical variance**:

$$s_s^2 = (n - R)/n = \left(1 - \overline{R}\right) \qquad (5.58)$$

These methods for determining the mean direction and spherical variance work well if the vectors are not too widely dispersed. Under certain conditions, however, the mean direction may be misleading. Suppose the dips of nearly flat-lying beds are measured; some dip gently to the west, others a few degrees to the east. Since dip is taken as a vector direction pointing into the lower hemisphere, the vector resultant of the east and west dips will be vertically downward! Of course the length of the resultant will be near zero, so the spherical variance will be large, indicating an extremely high dispersion among the vectors.

If these dips are regarded as nondirectional axes rather than vectors, their two ends project into both the upper and lower hemispheres; it is apparent then that the lines representing the east and west dips are closely related. The mean axis, computed using the eigenvector technique described below, will be horizontal and will pass through the bundle of dip axes.

Matrix representation of vectors

You will recall from Chapter 3 that the rows of a matrix can be represented graphically by vectors. Conversely, measurements of vector directions can be expressed in matrix form. The eigenvalues and eigenvectors of such a matrix will provide information about the arrangement of the vectors in space. However, in order to express a set of vectors in the appropriate matrix form we must first review a few points of geometry, starting in two-dimensional space. Note that we now revert to conventional mathematical notation in which the positive end of the X-axis is east, the positive end of the Y-axis is north, and angles are measured counterclockwise.

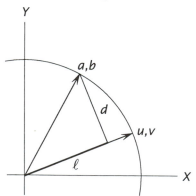

Figure 5–28. Projection of vector a, b onto vector u, v. The length of projection of a, b onto u, v is ℓ. The distance of vector end point a, b from the vector u, v is d.

The geometric relationship that gives the projection of one vector onto another is the scalar product of the two vectors (**Fig. 5–28**). If the two are assumed to be unit vectors, the Cartesian coordinates of their end points are the same as their directional cosines with respect to the X- and Y-axes. The projection is

$$\ell = au + bv \qquad (5.59)$$

where ℓ is the length of the projection of vector a, b onto vector u, v. (It is also the length of the projection of u, v onto a, b.)

As can be seen in **Figure 5–28**, the vector a, b is the hypotenuse of a right triangle whose sides are the projection ℓ on vector u, v and the perpendicular distance d. The Pythagorean theorem defines the relationship between these sides and can be rewritten for this instance as

$$d^2 = 1 - \ell^2 = 1 - (au + bv)^2 \tag{5.60}$$

Any number of vectors can be projected onto the line u, v by Equation (5.59), and their squared distances from the line u, v can be determined by Equation (5.60). The sum of these squared distances is

$$M = \sum_{i=1}^{n} d_i^2 = n - \sum_{i=1}^{n} (a_i u + b_i v)^2 \tag{5.61}$$

M can be regarded as a ***moment of inertia*** of the end points of the vectors around the line u, v. This equation can be further generalized to three dimensions by introducing the third spatial coordinate

$$M = \sum_{i=1}^{n} d_i^2 = n - \sum_{i=1}^{n} (a_i u + b_i v + c_i w)^2 \tag{5.62}$$

It is possible to express Equation (5.62) in matrix form. First, the coordinates of the line are given as a column vector $\mathbf{U}$

$$\mathbf{U} = \begin{bmatrix} u \\ v \\ w \end{bmatrix}$$

We also define the matrix $\mathbf{B}$

$$\mathbf{B} = n\mathbf{I} - \mathbf{T}$$

where $\mathbf{T}$ is a 3×3 matrix of the sums of squares and cross products of the direction cosines of the vectors:

$$\mathbf{T} = \begin{bmatrix} \sum a_i^2 & \sum a_i b_i & \sum a_i c_i \\ \sum b_i a_i & \sum b_i^2 & \sum b_i c_i \\ \sum c_i a_i & \sum c_i b_i & \sum c_i^2 \end{bmatrix}$$

The matrix $\mathbf{B}$ therefore has the form

$$\mathbf{B} = \begin{bmatrix} n - \sum a_i^2 & \sum a_i b_i & \sum a_i c_i \\ \sum b_i a_i & n - \sum b_i^2 & \sum b_i c_i \\ \sum c_i a_i & \sum c_i b_i & n - \sum c_i^2 \end{bmatrix}$$

The moment of inertia of the vectors about the direction $\mathbf{U}$ is simply

$$M = \mathbf{U}'\mathbf{B}\mathbf{U}$$

Rather than determining the moment around an arbitrary line $\mathbf{U}$, we may find a unique line around which the moment of inertia will be the maximum possible. The coordinates of this line are given by the first eigenvector of the matrix $\mathbf{B}$. If λ_1

335

is the first eigenvalue of $\mathbf{B}$ and $\mathbf{b}_i$ is its associated eigenvector, then you will recall from Chapter 3 that

$$\lambda_1 = \mathbf{b}'_1 \, \mathbf{B} \, \mathbf{b}_1$$

That is, λ_1 is the moment of inertia of the vectors around the first eigenvector. This means that the sum of the squared distances from the tips of the vectors to the first eigenvector is the maximum possible, or that the eigenvector is simultaneously as nearly perpendicular to all of the vectors as it is possible to be.

The moment of inertia around the second eigenvector is the greatest possible for any line that is orthogonal to the first eigenvector. The third eigenvector must be orthogonal to the other two, and must also account for all of the remaining squared distances to the vector tips. Since the three eigenvectors define an orthogonal framework fully equivalent to the original set of Cartesian axes, the third eigenvector must define a line along which the moment of inertia is a minimum. That is, the line will be oriented simultaneously as close as possible to all of the vectors.

If two vectors are diametrically opposite, as in **Figure 5–29**, they both will be the same perpendicular distance away from the eigenvector $\mathbf{b}_1$ and will have exactly the same influence on the location of the eigenvector. This means that the sense of direction of the vectors is lost; they are indistinguishable from axes. For this reason, eigenvector methods are preferable for the examination of spherically distributed data in those instances where ambiguity may result from an arbitrary distinction between vectors in the upper and lower hemispheres.

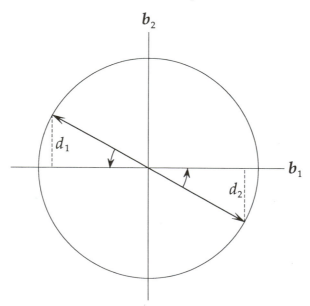

Figure 5–29. Projection of two diametrically opposite vectors onto eigenvector $\mathbf{b}_1$. The distances d_1 and d_2 are identical and act in the same rotational sense.

The eigenvalues provide direct information about the distribution of the vectors. Mardia (1972) distinguishes four cases.

1. λ_1 is large, while λ_2 and λ_3 are both small. This means that the sum of the squares of the perpendiculars between the vector end points and the axis corresponding to the first eigenvector is very large. Most of the observations must

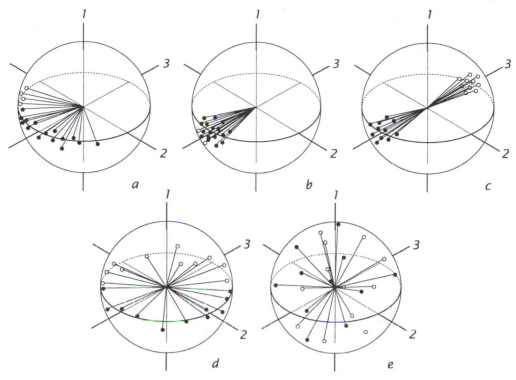

Figure 5–30. Patterns of vectors in the unit sphere. (*a*) Partial girdle pattern in the plane containing eigenvectors 2 and 3. (*b*) Unimodal distribution of vectors around eigenvector 3. (*c*) Bimodal distribution of vectors around eigenvector 3. (*d*) Complete girdle pattern in the plane containing eigenvectors 2 and 3. Their eigenvalues are identical or nearly so. (*e*) Uniform distribution. Eigenvalues are all approximately equal.

lie in the plane containing eigenvectors 2 and 3, forming a girdle distribution (**Fig. 5–30 a, d**).

2. λ_1 and λ_2 are both large, while λ_3 is small. The perpendicular distances from the end points to the first and second eigenvectors must be very large, but the distances to the third eigenvector must be small. The observations are clustered around the end of the third eigenvector (**Fig. 5–30 b, c**). Either a bimodal or unimodal distribution will yield the same result; they can be distinguished by the value of $\overline{R}$, which will be large for the unimodal case.

3. Two eigenvalues are identical. This is actually a variation of case 1. The observations form a symmetrical girdle around the axis corresponding to the unique eigenvalue (**Fig. 5–30 d**).

4. All three eigenvalues are identical. The distribution is uniform, as the perpendicular directions from the points are the same for all three orthogonal axes. There is no preferred arrangement of the points on the unit sphere (**Fig. 5–30 e**).

Woodcock (1977) has generalized this classification by graphing logs of ratios of the eigenvalues [ln (λ_1/λ_2) versus ln (λ_2/λ_3)]. On his diagrams, all possible patterns of points on a sphere fall into specific regions. This form of graphical analysis may be especially useful with petrofabric data. **Figure 5–31** shows one of Woodcock's diagrams.

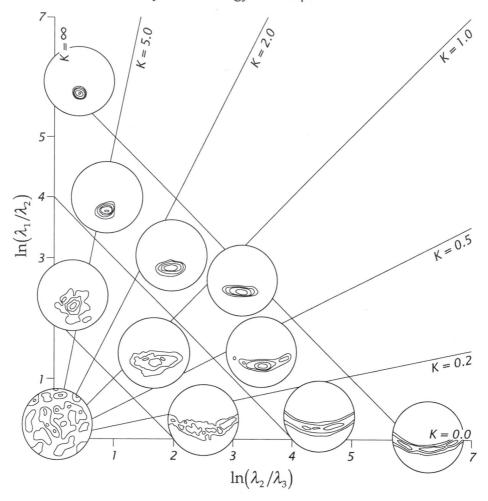

Figure 5–31. Classification of patterns of vectors on the unit sphere, according to the logarithms of the ratios of their eigenvalues. Typical petrofabric diagrams are shown for different ratios. K is the ratio $\ln(\lambda_1/\lambda_2)/\ln(\lambda_2/\lambda_3)$. Adapted from Woodcock (1977).

Displaying spherical data

Although perspective drawings of unit spheres such as those in **Figure 5–30** are useful for illustrative purposes, they cannot convey detailed information about the distribution of the vectors. Conventionally, three-dimensional vectors are shown by projecting their end points onto a plane. Since the points actually lie on the surface of a sphere, portraying them in two dimensions requires use of a projection equation. Geologists traditionally have used the equal-area polar Lambert projection, which is referred to as a "Schmidt net." Crystallographers have preferred the equal-angle polar stereographic projection, or "Wulff net."

The relevant projections can be explained with the aid of **Figure 5–32** for a Wulff net and **Figure 5–33** for a Schmidt net. In both illustrations, the tip of vector OP is projected onto the horizontal plane that passes through the center of the unit sphere. The horizontal plane forms the desired net or map. The vector OP is defined in terms of its azimuth, θ, measured clockwise from north, and its dip,

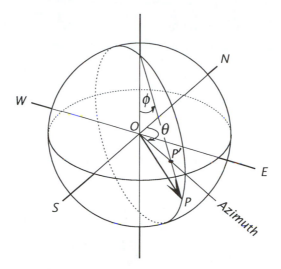

Figure 5–32. Projection of spherical vector OP defined by azimuth θ and dip angle d onto equal-angle Wulff net. Point P is projected as P' onto horizontal plane at center of unit sphere.

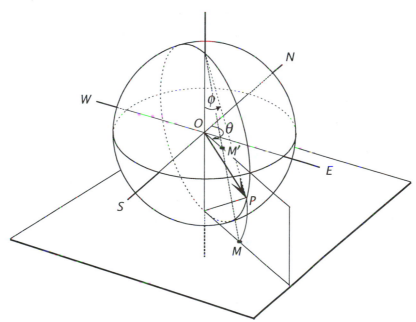

Figure 5–33. Projection of spherical vector OP defined by azimuth and dip angle d onto equal-area Schmidt net. Point P is first projected as M on tangent plane, then projected as M' onto horizontal plane at center of unit sphere.

d, measured downward from the horizontal in the plane of the azimuth. In the equal-angle Wulff net, point P is projected onto the middle plane along a line from P to the zenith of the unit sphere. The projected point P' is found by

$$OP' = \tan \phi \qquad (5.63)$$

where

$$\phi = 45° - d/2 \qquad (5.64)$$

Then, the X- and Y-coordinates of point P' on the Wulff net can be determined by

$$X = OP' \sin \theta$$
$$Y = OP' \cos \theta$$

(5.65)

Construction of an equal-area Schmidt net and the projection of points onto the net is slightly more complicated. First, the point P must be projected onto a tangent plane below the unit sphere by means of a circular arc. This yields point M on the tangent plane. This is then projected onto the Schmidt net to determine point M'. This requires that Equation (5.63) be modified to

$$OM' = \sqrt{2} \sin \phi$$

(5.66)

The X- and Y-coordinates of the projected point on the Schmidt net are found by Equation (5.65), with the substitution of M' for P'. Middleton (2000) provides a concise discussion of spherical projection, and MATLAB® code that creates and plots points on Wulff or Schmidt nets. Texts on structural geology, such as Ragan (1985) and Suppe (1985), also have more detailed discussions.

Figure 5–34 shows a set of vectors within the unit sphere and their projection onto an equal-area diagram. It is necessary to distinguish between vectors that go into the lower hemisphere and those that go into the upper. Because geologists often describe vectors in terms of their "plunge," a word with a "downward" connotation, they conventionally plot the lower hemisphere of the unit sphere.

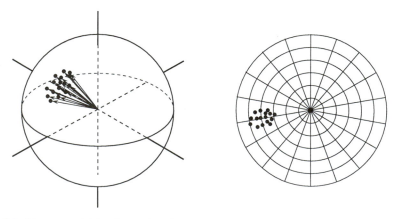

Figure 5–34. Vectors within the unit sphere and their projection onto an equal-area polar diagram.

In addition to vectors, it is sometimes necessary to plot the three-dimensional orientations of planes such as faults and fracture surfaces. If a plane is placed through the center of the unit sphere, its intersection with the sphere will form a great circle (**Fig. 5–35 a**). However, it is easier to represent the plane by an axis (called a "pole") that is perpendicular to the plane at the origin. Conventionally, geologists plot the intersection of a pole with the lower hemisphere, although it seems more logical to plot its intersection with the upper hemisphere. Then, the projection of the pole to a plane that dips to the west, for example, will plot on the left-hand or "western" side of a diagram (**Fig. 5–35 b**).

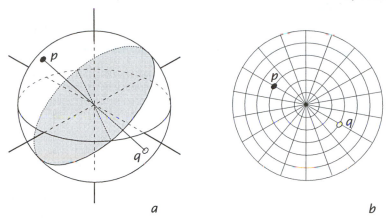

Figure 5–35. Plotting of poles to planes. (*a*) Plane and its poles in the unit sphere. (*b*) Poles to a plane projected onto an equal-area diagram. Point *p* is the projection onto the upper hemisphere, point *q* is the projection onto the lower hemisphere.

Very large sets of three-dimensional data can result in diagrams that contain so many points that their general pattern cannot be seen. In such instances the local density of points can be contoured by counting the number of points that lie within a small area of the diagram. This can be done conveniently only if an equal-area projection (Schmidt net) is used. The projection is covered with a regular array of grid nodes and the number of points within a fixed radius of each node is counted. Usually the radius is set so that it describes a circle containing 10% of the total area. Since the grid nodes are closer together than the radius, the successive areas overlap and there tends to be a gradual change in density from one part of the diagram to another. The small diagrams on **Figure 5–31** show typical contoured patterns obtained in petrofabric studies.

Testing hypotheses about spherical directional data

The simpler tests of three-dimensional orientation are extensions of those used with circular data. As in that instance, we need a probability model of known characteristics against which we can test. A widely used model is the ***Fisher distribution***, an extension of the von Mises distribution and a spherical equivalent of the normal curve. The Fisher distribution is characterized by two parameters: a mean vector direction $\bar{\theta}$ and a dispersion κ. Because we are dealing with three dimensions, the mean vector has three elements, each a direction cosine with respect to the three coordinate axes.

The mean vector is estimated by the direction cosines of the resultant (Eq. 5.57). The dispersion can be approximated by

$$\kappa = (n - 2)/(n - R) \tag{5.67}$$

which is sufficiently accurate if κ is "large," or greater than about 10. Mardia (1972) gives a table of more exact estimates based on the size of the normalized resultant, $\bar{R}$.

A test of randomness.—As in the circular case, the simplest hypothesis that can be tested is that the data are distributed uniformly in all directions, which is equivalent

341

to stating that the concentration parameter κ is zero. The null hypothesis and alternative are

$$H_0 : \kappa = 0$$

$$H_1 : \kappa > 0$$

The test statistic is calculated in the same manner as it is using circular data, and consists of the normalized resultant $\overline{R}$. This statistic is then compared to a critical value of $\overline{R}$ for the desired level of significance. A table of critical values is given in Appendix Table **A.11**. If the calculated value of $\overline{R}$ exceeds the table value, the hypothesis that the observations are uniformly distributed is rejected at the specified level of significance.

It is also possible to test hypotheses about the specific orientation of the mean vector, and to construct a "cone of confidence" around the vector. These tests, however, require extensive tables such as those published by Stephens (1967) and Mardia (1972). Two-sample tests for the equivalency of the mean directions of two sets of observations can also be performed. The necessary tables are also given in Stephens (1967) and Mardia (1972).

Fractal Analysis

The word "fractal" was coined by Benoit Mandelbrot (1982) from the Latin word *fractus*, meaning broken, which he applied to objects that were too irregular to be described by ordinary Euclidean geometry. The term encompasses a wide range of objects, from regularly repeating forms that have an exact mathematical definition to naturally occurring stochastic patterns that can only be described in a probabilistic way. All fractals possess fine structure; that is, the more closely they are examined, the more detail will be perceived. In addition, many fractals are self-similar, so that enlarged parts resemble the whole in some way. Other fractals are self-affine, which means that enlarged parts must be rescaled in some manner (usually by compression or expansion in one dimension) before they are self-similar. Some authors (*e.g.*, Turcotte, 1997) have emphasized the similarity between the words "fractal" and "fractional," in reference to the dimensional geometry of these objects. A simple curved line in Euclidean space can be described as one-dimensional, since only one value (the distance along the line) is required to locate any point with respect to another. However, the more closely we examine a fractal line, the more complexity we perceive and the greater distance between any two points. In effect, a fractal line has a higher dimensionality which lies somewhere between the one dimension of a simple line and the two dimensions of a surface. Similarly, we can locate points on a surface in Euclidean space using only two coordinates, hence the surface is two-dimensional. A fractal surface, however, possesses roughness at every level of magnification, so distances that we measure on the surface are a function of scale. The surface has a dimensionality that is fractionally greater than two.

The concept of dimensionality can be illustrated by noting how the "size" of an object behaves as its linear dimension increases (Bourke, 1991). If the linear dimension of a line segment is doubled, then obviously the length (the one-dimensional characteristic measure of size) of the line is increased by two times. If the linear dimensions of a rectangle are doubled, then the area (the two-dimensional characteristic measure of size) is increased by four times. If the linear dimensions of a box are doubled, the box's volume (the three-dimensional characteristic measure of size) increases by eight times. The relationship between dimension, D, linear scaling, L, and the resulting increase in size, S, can be generalized and written as

$$S = L^D \qquad (5.68)$$

This simply tells us what we intuitively know: if we scale a two-dimensional object, the area increases by the square of the scaling; and if we scale a three-dimensional object, the volume increases by the cube of the scaling. The equation can be rearranged to give an expression for dimension that depends on how size changes as a function of linear scaling:

$$D = \frac{\log S}{\log L} \qquad (5.69)$$

For ordinary Euclidean shapes, D is always an integer. However, if we linearly scale a fractal object and then measure its characteristic size, the resulting dimension will be a fractional value that exceeds the Euclidean dimension.

Like all other mathematical concepts, fractals are an abstraction. Many natural objects appear fractal-like over certain ranges of scale, and it may be very useful to regard them as fractals for analytical or modeling purposes. However, unlike mathematical abstractions, the fractal nature of real objects changes or disappears at some extreme scale. Of course the same thing is true of conventional geometric figures—there are no truly straight lines or perfect circles in nature!

Although the essential characteristic of a fractal object is its dimension, this attribute is difficult to define in a universally applicable manner. A formal definition of fractals requires use of the ***Hausdorff dimension***, a concept defined in set-theoretic terms that escapes intuitive description. Adler (1983) provides a brief definition; Falconer (1990) devotes an entire chapter to the topic. Most introductory books on fractals gloss over the Hausdorff dimension, as we do here. Instead, we will turn to the various operational definitions of fractal dimension. As Falconer (1990) strongly cautions, when discussing fractal dimensions it is necessary not only to cite numerical values but also to state how the values were determined.

One measure of dimensionality is applicable to a self-similar object made up of m copies of itself scaled by a factor r. This measure is called a ***similarity*** fractal dimension and is defined as $D_S = -\log m / \log r$. Unfortunately, this definition is useful only for mathematically constructed linear fractals such as those shown at several stages of iteration in **Figure 5–36**. The first example is a ***Gosper island*** (Gardner, 1967), which starts as a line and evolves into three copies of itself, each scaled by $1/\sqrt{7}$ at each iteration. The resulting object has a fractal dimension of $D_S = -\log 3 / \log \frac{1}{\sqrt{7}} = 1.13$. The ***Sierpinski gasket*** results from scaling three copies of the starting segment by $1/2$; it has a fractal dimension of $D_S = -\log 3 \big/ \log \frac{1}{2} = 1.59$. A ***dragon curve*** is formed by two copies of the starting segment, each scaled by $1 \big/ \sqrt{2}$. The fractal dimension of a dragon curve is $D_S = -\log 2 \big/ \log \frac{1}{\sqrt{2}} = 2.00$.

Ruler procedure

The most straightforward way of estimating the fractal dimension of an arbitrary line, such as the coastline of an island shown on a map or the margin of a pore in a carbonate rock as seen in a photomicrograph, is the ***ruler*** or ***compass*** method. The outline is approximated by a series of straight-line segments of constant length. We can imagine creating such a straight-line approximation by setting a drafting compass to a fixed width and "walking" the compass along the line of interest.

Figure 5–36. (*a*) Gosper island fractal after two, three, and seven iterations. (*b*) Sierpinski gasket after three, five, and six iterations. (*c*) Dragon curve after six, eight, and eleven iterations. Curves created using the *FractaSketch* program (Wahl, 1995).

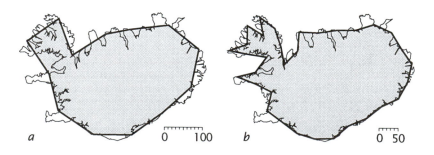

Figure 5–37. Ruler method of estimating fractal dimension. Light line is digitized coastline of Iceland; heavy lines are ruler approximations: (*a*) 100-km ruler lengths, (*b*) 50-km ruler lengths.

Figure 5–37 illustrates the method applied to a map of the coastline of Iceland. In **Figure 5–37 a**, the compass has been set to a span corresponding to a distance of 100 km at the scale of the map; about 14 steps of the compass are required to encircle Iceland. This means that at the 100-km resolution defined by the compass setting, the length of the Icelandic coastline is estimated to be 1400 km. If we set the compass to a different scale distance, perhaps 50 km as in **Figure 5–37 b**, we will find that many more steps are required to measure the coastline and the result will be a different total length, 1650 km in this instance. **Table 5–6** lists the number of steps required to circumscribe the coastline of Iceland at several settings of the compass and the resulting estimate of the total length of shoreline.

Table 5–6. Estimating the fractal dimension of the coast of Iceland by the ruler method.

Length of line segments, r	Number of segments, n	Estimate of length, nr	log $1/r$	log nr
200	6	1200	−5.298	7.09
100	14	1400	−4.605	7.24
80	20	1600	−4.382	7.38
50	33	1650	−3.912	7.41
40	46	1840	−3.689	7.52
20	112	2240	−2.996	7.71

Note that the length of the measuring step, r, seems inversely related to the estimated perimeter or total length of the coastline, nr. Equivalently, we can state that there is a direct relation between resolution (the reciprocal of the measuring step) and perimeter, as we can see in **Figure 5–38 a**. We can imagine that in the limit the measuring step will become infinitesimally small and the estimated length of the coastline will approach infinity. The relationship between the resolution and the length of the line being measured is more apparent in the log–log plot of **Figure 5–38 b**.

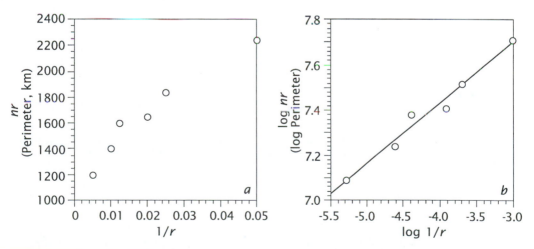

Figure 5–38. (*a*) Perimeter (km) of Iceland coastline versus resolution (inverse of length) of ruler. (*b*) Logarithms of perimeter versus logarithms of resolution.

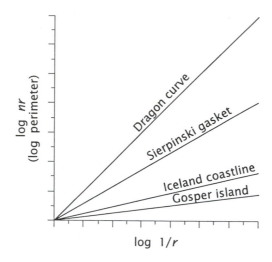

Figure 5–39. Log perimeters versus log resolutions (inverse of lengths) of rulers for self-similar fractal constructions shown in Figure 5–36, with relationship for coastline of Iceland. Fractal dimensions are equal to slope of fitted lines, plus one.

We can use this relationship to estimate the **compass** or **ruler** fractal dimension, D_C. The procedure was originally developed by L.F. Richardson, and is sometimes called **Richardson's dimension** (Wahl, 1995). It involves estimating the slope of a line fitted by regressing log nr (the logarithm of the perimeter of the fractal object) onto the logarithm of the reciprocal of the measuring step. That is, we fit the linearized equation

$$\log\, nr = \alpha + \beta\, \log\, \left(\frac{1}{r}\right) \qquad (5.70)$$

The slope of this line, β, is related to the fractal dimension by

$$D_C = 1 + \beta \qquad (5.71)$$

If we measure several total lengths of coastline for different measuring steps, we can use the least-squares regression methods of Chapter 2 to estimate β. For the Iceland data in **Table 5–6**, we find that $\beta = 0.27$, so the fractal dimension is $D_C = 1.27$. The goodness of fit of the regression is $R^2 = 0.98$.

The same procedure of comparing the measured perimeter of an object to the size of the measuring step can, of course, be applied to mathematically constructed fractals such as those in **Figure 5–36**. The estimated compass dimensions will not be exact but usually will be within a few percent of the true values. The results for these mathematical fractals are shown on the log–log plot of **Figure 5–39**; for comparison purposes, the line fitted to data for the Icelandic coastline is also shown.

Grid-cell procedure

A popular way to compute the fractal dimension of a line on a map is the **grid-cell** method, in which the map is covered with a succession of square grids, and we simply count the number of squares in which the line appears. The line will be seen in only a few squares of a coarse grid, but if the grid squares are small, we will count a large number of them that include parts of the line. We can obtain an estimate of the length of the line by multiplying the number of occupied squares by the size of a square. In effect, the size of the grid we have placed over the map determines the resolution at which the length of the line is measured. **Figure 5–40** illustrates the process on the map of Iceland we used to illustrate the ruler method. Two sets of grids are shown, consisting of squares that are 100 km and 50 km on a side. (Not shown are additional grids composed of 20-km, 11.5-km, 5.7-km and 2.85-km squares.)

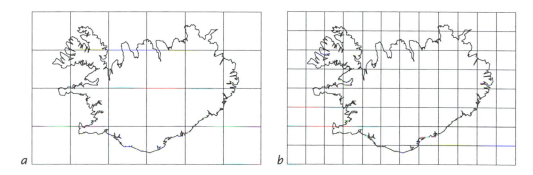

Figure 5–40. Grid-cell method of estimating fractal dimension: (*a*) 100-km square cells, (*b*) 50-km square cells.

From counts of the number of squares that contain part of the coastline of Iceland, we can construct **Table 5–7**. The column labeled *s* contains the dimensions of the grid cells, column $N(s)$ contains the number of cells counted, and column $s \times N(s)$ is the product of these two and is an estimate of the perimeter or length of the Icelandic coastline. The final two columns contain the logarithms of the reciprocal of the cell size ($\log 1/s$) and the number of cells ($\log N(s)$).

Table 5–7. Estimating the fractal dimension of the coast of Iceland by the grid-cell method.

Size of grid cells s	Number of square cells $N(s)$	Estimate of perimeter $s \times N(s)$	$\log 1/s$	$\log N(s)$
100	19	1900	−4.6052	2.9444
50	46	2300	−3.9120	3.8286
20	155	3100	−2.9957	5.0434
11.5	321	3692	−2.4423	5.7714
5.7	743	4235	−1.7405	6.6107
2.85	1645	4688	−1.0473	7.4055

Points defined by values in the final two columns of **Table 5–7** will tend to fall along a straight line when plotted (**Fig. 5–41 a**). The fractal dimension for the grid-cell procedure is given by the slope of the regression of $\log N(s)$ on $\log 1/s$. In this example, the regression line is $\log (N(s)) = 8.79 + 1.26 \log (1/s)$. The estimated fractal dimension of the coastline of Iceland is $D_G = 1.26$.

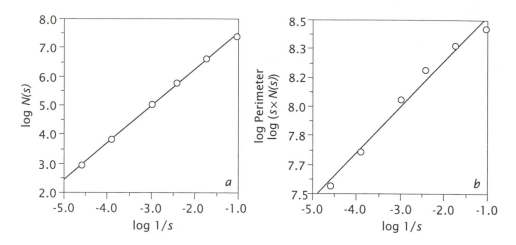

Figure 5–41. Grid cell count statistics for Iceland coastline. (*a*) Logarithm of number of cells occupied versus logarithm of reciprocal of cell size. Slope of fitted line is $\beta = 1.26$. (*b*) Logarithm of perimeter versus logarithm of reciprocal of cell size. Slope of fitted line is $\beta = 0.26$.

We also can use Richardson's method of calculating the fractal dimension, since the sum of the products of the cell sizes times the number of cells of each size containing part of the object's outline can be considered an estimate of the object's perimeter. If we regress $\log (s \times N(s))$, the logarithm of the perimeter, on $\log 1/s$, we obtain the plot shown in **Figure 5–41b**. The slope is $\beta = 0.26$, yielding the same estimated fractal dimension of $D_G = 1 + 0.26 = 1.26$. The small number of cell sizes used in this example limits the accuracy of the computation. The same digital coastline of Iceland was analyzed using Bourke's (1993) FDC program which computes sets of 12 successive sizes of grids, starting at a size of only one pixel on the image (roughly corresponding to a cell dimension of 2.8 km). The program returns an estimate of $D_G = 1.31$.

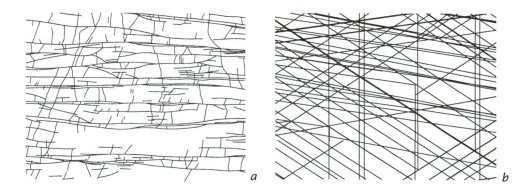

Figure 5–42. (*a*) Plan view of fracture pattern in sandstone, Wasatch Formation (Tertiary) in Piceance Basin, northwest Colorado, traced from photomosaic. Image is approximately 40 ft (12 m) across. Grid-cell fractal dimension is $D_G = 1.52$ (from Lorenz, Teufel, and Warpinski, 1991). (*b*) Line pattern from part of fractal dragon curve. Grid-cell fractal dimension is $D_G = 1.56$.

A fractal need not be a single, contiguous object. An image consisting of lines such as fracture traces can have a fractal dimension. **Figure 5–42 a** shows a plan view of fractures in sandstone of the Tertiary Wasatch Formation, traced from a photomosaic of an exposure in the Piceance Basin of northwestern Colorado (Lorenz, Teufel, and Warpinski, 1991). Using the grid-cell method, we can estimate the fractal dimension of this image as $D_G = 1.53$. The fractal nature comes from the sets of parallel lines. Within the sets, there are parallel subsets of more closely spaced lines, and within them, even more closely spaced subsets, and so on. A roughly similar pattern (**Fig. 5–42 b**) can be created from a linear fractal such as a dragon curve by scaling the initial copies so they are larger instead of smaller than the starting segment. The fractal dragon curve will grow endlessly upon iteration (Wahl, 1995). This example has an estimated grid-cell fractal dimension of $D_G = 1.56$.

One area in which the fractal concept has proved to be especially useful is in the description of fluid flow through porous materials, especially of immiscible fluids such as crude oil being displaced by water in a reservoir. The invading water forms a complex, branching pattern that can best be characterized by its fractal dimension. Then, simulations can be made based on a stochastic model called "diffusion-limited aggregation" (DLA) that has properties nearly identical to the observed natural process (Feder and Jøssang, 1995). **Figure 5–43 a** shows the result of

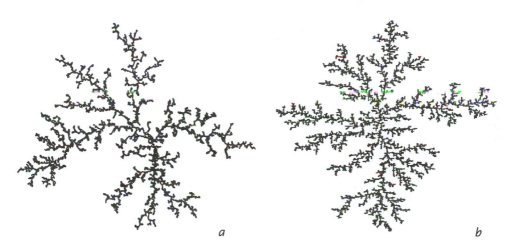

a   *b*

Figure 5–43. Viscous fingering of immiscible fluids in a porous medium. (*a*) Photograph of radial displacement of epoxy by air in packed glass beads. (*b*) Computer simulation based on diffusion-limited aggregation (DLA) stochastic model (Meakin and Fowler, 1995).

an experiment in which a viscous liquid (epoxy resin) filled the pores of an artificial reservoir rock composed of packed glass spheres. The epoxy was displaced by air injected at the center of the image, forming an intricate branching pattern. The fractal dimension of the object, as measured by the grid-cell procedure, is $D_G = 1.57$. **Figure 5–43 b** is a digital simulation based on a DLA model, which has a theoretical fractal dimension of $D_S = 1.71$. The fractal dimension estimated by the grid-cell method is $D_G = 1.64$. Similar radial flow experiments have been performed in a variety of artificial materials, yielding similar results that closely accord with DLA simulations (Meakin and Fowler, 1995).

In applying the grid-cell method of estimating fractal dimensions, the user must exercise judgment in the choice of measurement grids. Grids whose cells are smaller than the width of lines in the object simply measure the proportion of black in the image. Grid cells that are too large relative to the object will fail to detect fine structure.

We can see these effects in an analysis of the drainage pattern of the Wolf River in northeastern Kansas, taken from digital cartography captured at a scale of 1:24,000. The drainage network is shown in **Figure 5–44**, and **Figure 5–45** is a plot of $\log (s \times N(s))$ versus $\log 1/s$ over a wide range of cell sizes. Grid dimensions

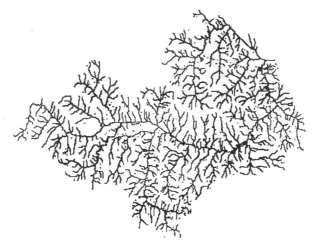

Figure 5–44. Drainage pattern of Wolf River in northeastern Kansas. East–west dimension is approximately 36 mi.

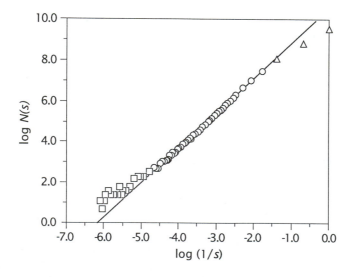

Figure 5–45. Grid-cell count statistics for Wolf River in northeastern Kansas. Triangles indicate cells so small that multiple cells fall within individual lines. Squares indicate cells so large that fine structure is not detected. Line used to estimate slope is fitted only to cell measurements indicated by circles. Slope of fitted line is $\beta = 1.71$.

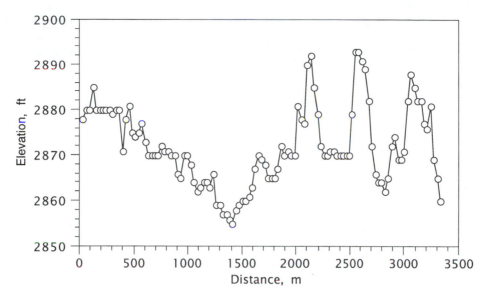

Figure 5–46. Topographic profile measured along north–south line over Pleistocene sand dune area in southwestern Kansas. Elevation measurements at 30-m spacings.

judged to be either too small or too large are indicated by symbols. The fractal dimension can be estimated correctly only from the slope of a line fitted to counts from cells in the middle of the range of grid sizes. If all the grid data applied to the Wolf River drainage pattern are used, the fractal dimension of the stream network is estimated to be $D_G = 1.49$. Using only the central portion of the data yields an estimate of $D_G = 1.71$, which is closer to the theoretical dimension of a space-filling process (Rodríguez–Iturbe and Rinaldo, 1997).

Spectral procedures

The Fourier transform methods we used for spectral analysis in Chapter 4 also can provide estimates of fractal dimensions, especially of lines defined by x–y coordinate pairs in which x increases monotonically. The spectral procedure has been widely used to estimate the fractal dimensions of topographic and bathymetric profiles, and profiles of the texture of soil surfaces and fracture planes (Gallant and others, 1994). Such profiles are ideal for spectral analysis since the records are almost never double valued (this would require an overhang along the profile).

The first step is to compute a periodogram or discrete power spectrum of the profile line. Then, the logarithm of the power, s_k^2, is plotted against the logarithm of the inverse of the harmonic number, k. The slope of a linear regression of $\log s_k^2$ on $\log 1/k$ yields an estimate of the fractal dimension, D_F:

$$D_F = \frac{5 - \beta}{2} \tag{5.72}$$

where β is the estimated slope coefficient found by the regression. Some authors (Turcotte, 1997) regress log power against log frequency rather than log harmonic number; this changes the intercept of the fitted regression but not the estimate of the slope.

Figure 5–46 is a profile taken from a digital topographic model produced by the U.S. Geological Survey for part of the Garden City East quadrangle in southwestern

Kansas. The north–south profile crosses an area of Pleistocene sand dunes. The digital record is given in file GARDENNS.TXT. Elevation measurements have been made at 30-m spacings. **Figure 5–47** is the periodogram of the profile. A cross plot and fitted linear regression of log s_k^2 versus log $1/k$ is shown in **Figure 5–48**.

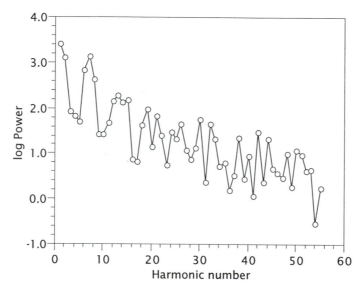

Figure 5–47. Periodogram of topographic profile in Figure 5–46.

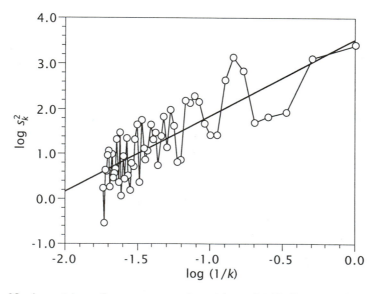

Figure 5–48. Logarithm of power versus logarithm of $1/k$ (inverse of wave number) for values in periodogram in Figure 5–47. Slope of fitted line is $\beta = 1.69$.

The coefficient of the fitted slope is $\beta = 1.69$, so the spectral estimate of fractal dimension is

$$D_F = \frac{5 - 1.69}{2} = 1.66$$

Higher dimensional fractals

Estimating the fractal dimension, D_2, of a convoluted surface is conceptually similar to estimating the fractal dimension of an object intermediate between a line and a surface. One approach is a three-dimensional extension of box counting in which the space enclosing the fractal object is filled with a meshwork of cubes. A count is made of the cubes which contain part of the object's surface. The cubes are iteratively replaced with increasingly larger cubes, and counted, until the entire surface is enclosed within a single cube. A plot is made of the number of cubes, $N(s)$, versus the resolution of the cubes, $1/s$. As before, the fractal dimension D_2 is given by the slope of the regression of $\log N(s)$ on $\log 1/s$, except that for fractal surfaces the slope will lie between 2 and 3.

A box-counting procedure is not appropriate for many surfaces that have limited relief. Examples include digital topography and bathymetry and the planar surfaces of faults, fractures, and cleavage planes. Turcotte (1997) discusses a two-dimensional Fourier transform procedure that can be used to estimate the fractal dimension D_2, which is a characterization of the roughness or complexity of the surface. The data consist of measurements of the height of the surface at points arranged in a regular array or square grid. A discrete Fourier transformation is made of the square array (after removing the mean and detrending, if necessary), yielding a square matrix of Fourier coefficients. These are combined into estimates of the mean power for each harmonic number, averaged across direction. The logarithms of the mean power values are plotted against logarithms of the inverses of the harmonic numbers, producing a graph very similar to a plot of $\log N(s)$ against $\log 1/s$. Regression is used to estimate the slope of this line, which in turn provides an estimate of D_{2F}, the fractal dimension of the surface:

$$D_{2F} = \frac{7 - \beta}{2} \tag{5.73}$$

Figure 5–49 is a topographic map of the dune area south of Garden City, Kansas. The profile in **Figure 5–46** was taken from this area. **Figure 5–50** is a graphical representation of the two-dimensional Fourier transform of the digital map data, and **Figure 5–51** is a plot of the logarithm of the distance-averaged power versus the logarithm of the inverse of the harmonic number (or period); the slope of the fitted linear regression provides an estimate of the fractal dimension of the topographic surface. For this example, the slope is $\beta = 2.53$ so the fractal dimension is estimated to be $D_{2F} = 2.23$.

Rather than estimating the fractal dimension directly, we could characterize the roughness of a surface by estimating the fractal dimensions of profiles across the surface using the spectral method. Turcotte (1997) notes that the two types of fractal dimension are approximately related by $D_{2F} = 1 + D_F$, but in our example, the difference between the two estimates seems greater.

Figure 5–49. Topographic map of dune area in southwestern Kansas near Garden City, Kansas. Contour interval is 10 ft. Map covers an area of approximately 4 mi².

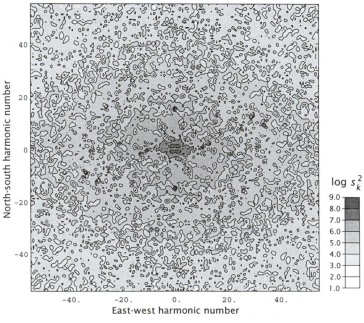

Figure 5–50. Two-dimensional power spectrum of map shown in Figure 5–49. Power is given in logarithms.

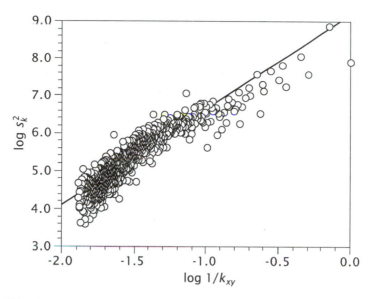

Figure 5–51. Log direction-averaged power versus log inverse wave number for power spectrum shown in Figure 5–50. Slope of fitted line is $\beta = 2.53$.

Shape

Shape is an extremely difficult property to measure, or even to define in a precise manner. Perhaps this is why there are so many proposed shape measures, none of which have been entirely satisfactory. A shape measure should possess several desirable properties. Obviously, objects with different shapes should yield different measures and similar shapes should yield similar values regardless of the size or orientation of the objects. Unfortunately, a shape measure possessing these properties may be a chimera; it has been proven mathematically that no single measure can be unique to only one shape (Lee and Sallee, 1970).

Earth scientists have attempted to characterize a broad spectrum of shapes ranging from relatively simple forms such as the projected outlines of sand grains to the more complex patterns presented by fossil organisms. Geomorphologists have been especially prolific in the creation of shape measures, applying them to the study of drainage basins, drumlins, and coral atolls, among other landforms. The shapes of oil fields, as defined by their axial ratios, have been investigated, as have the shapes of certain types of structural traps. Extensive literature reviews are given by Moellering and Rayner (1979) and by Clark (1981); both articles also discuss the theoretical aspects of shape measurement.

Table 5–8 contains a selection of single-value measures of shape, gleaned from the geologic and geographic literature. This list is by no means exhaustive. Most of these are calculated from a few basic measurements, such as axial lengths, perimeters, and areas. Some involve comparisons with standard forms, such as a circle. Any of these shape measurements can be used in the same manner as any other variable. Summary statistics, such as means and variances, can be calculated for the shape measurements from a collection of objects, although there is no guarantee that the measurements will follow a normal distribution.

Dryden and Mardia (1998) do not regard shape as a single measure, but rather as the total geometrical information remaining when location, scale, and orientation

Table 5–8. Measures of shape used in the geologic and geographic literature; only nondimensional measures are listed.

1. Measures based on axial ratios: Form $F = \frac{\ell}{w}$

   Elongation $E = \frac{w}{\ell}$

   Circularity $C_1 = \sqrt{\frac{\ell w}{\ell^2}}$

2. Measures based on perimeters: Grain Shape Index $GSI = \frac{p}{\ell}$

   Shape Factor $SF_1 = \frac{p_c}{p}$

   $SF_2 = \frac{p}{p_c} \times 100$

3. Measures including both perimeters and areas: Circularity $C_2 = \frac{4A}{p^2}$

   $C_3 = \frac{4A}{\ell p}$

   Compactness $K_1 = \frac{2\sqrt{\pi A}}{p}$

   $K_2 = \frac{p^2}{4\pi A}$

   Thinness Ratio $TR = 4\pi \left(\frac{A}{p^2} \right)$

4. Measures based on areas: Circularity $C_4 = \sqrt{\frac{A}{A_c}}$

   Shape Factor $SF_3 = \frac{A_i}{A}$

   $SF_4 = \frac{A_c - A_i}{A}$

   $SF_5 = \frac{A}{A_c} \times 100$

5. Measures based on areas and axis length: Form Ratio $FR = \frac{A}{\ell^2}$

   Ellipticity Index $EI = \frac{\pi(1/2\ell)\ell}{A}$

6. Other measures: Circularity $C_5 = \sqrt{\frac{D_i}{D_c}}$

   Mean Radius $\overline{R} = \frac{\sum R_j}{n}$

   Radial Variance $s_R^2 = \frac{\sum (R_j - \overline{R})^2}{n}$

   Mean Side $\overline{S} = \frac{\sum S_j}{n}$

   Variance of Side $s_S^2 = \frac{\sum (S_j - \overline{S})^2}{n}$

Original sources of measures given in Folk (1968) and Moellering and Rayner (1979).

Key:

A Area of object

A_c Area of smallest enclosing circle

A_i Area of largest inscribed circle

D_c Diameter of smallest enclosing circle

D_i Diameter of largest inscribed circle

ℓ Length of long axis

n Number of sides, considered as a polygon

p Perimeter of object

p_c Perimeter of a circle having the same area as object

R_j jth radius of object, measured from centroid to edge

S_j Length of jth side of object, considered as a polygon

w Width of object perpendicular to long axis

are removed from an object. Two objects are said to have the same shape if one of them can be moved, resized, and rotated until its features will exactly coincide with the features of the second object when the two are superimposed. The extent to which the two objects do not coincide is an expression of the difference in their shapes. Obviously, using this approach, more than single measurements are necessary to define objects. These authors distinguish three types of *landmarks*, or points of correspondence, that can be recognized on each object in a population. An *anatomical* landmark is a point, assigned by an expert, that has some biological or functional significance. Examples include the pedicle opening of a brachiopod, the nasion of a primate skull, and the stream exit point of a drainage basin. *Mathematical* landmarks are located according to some geometrical property of the objects, such as the points of greatest curvature or farthest distance from the centroids. *Pseudo-landmarks* are convenient points (often equally spaced) located either around the outline or between anatomical or mathematical landmarks on an object.

Once landmarks have been identified or assigned, measurement of distances or angles between them can take place and the measurements (or ratios of them) can be used as variables for comparing one object with another. Often this involves multivariate analyses such as those we will examine in the next chapter. This approach is sometimes called "multivariate morphometrics"; it is discussed in detail by Reyment, Blackith, and Campbell (1984). However, the process of converting landmark coordinates to distances, angles, and ratios sacrifices information about the geometry of the objects, so Dryden and Mardia (1998) advocate working directly with the landmarks themselves.

Recent advances in shape analysis have been spurred by development of automated image-analysis equipment and computer algorithms for various shape-recognition procedures collectively known as *Procrustes analysis*. Kendall (1989) reviews applications in archaeology, astronomy and geography. Applications in biology and medicine are summarized by Bookstein (1991).

The initial step in Procrustes analysis is to transform all objects so they have the same size and orientation within a suitable coordinate system. Dryden and Mardia (1998) describe and provide equations for several transformation procedures, the simplest of which converts objects to *Bookstein coordinates*. To perform the procedure, first select two *key landmarks* that can be identified on every object. Translate and rotate one of the objects until its key landmarks lie on a horizontal baseline; define the Y-coordinate of this baseline as 0.0. Move the object along the baseline until the two key landmarks are equidistant from a central vertical baseline whose X-coordinate is defined as 0.0 (the landmark to the left of the baseline will have a negative X-coordinate). Scale both the X- and Y-axes by multiplying by a constant such that the X-coordinates of the two landmarks are -0.5 and 0.5. Since the same scaling factor is used for both axes, the shape of the object is not changed. Repeat the process for each object. When the objects are drawn on the same set of Bookstein coordinates, the two key landmarks of all the objects will coincide at $(-0.5, 0)$ and $(0.5, 0)$. The coordinates of other landmarks will not coincide unless the objects are identical in shape.

Figure 5–52 shows interior and exterior views of the pedicle valve of the Ordovician orthid brachiopod, *Resserella* sp., with the positions of 11 landmarks indicated. A typical specimen measures about 2 cm between landmarks 9 and 10. The X, Y-coordinate pairs for ten specimens of *Resserella* sp., measured in centimeters from

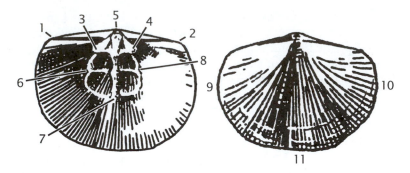

Figure 5–52. Interior and exterior views of the pedicle valve of the Ordovician orthid brachiopod *Resserella* sp., with landmarks indicated. Lateral terminations of the hinge line (landmarks 1 and 2) are designated key landmarks. From Moore, Lalicker, and Fischer (1952).

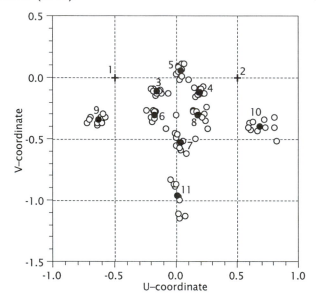

Figure 5–53. Cross plot of Bookstein U, V coordinates of landmarks measured on ten specimens of *Resserella* sp. Centroids of landmarks indicated by solid dots. Numbers correspond to landmarks in Figure 5–52.

an arbitrary origin in the lower left, are given in file RESSEREL.TXT. These Cartesian coordinates can be converted to Bookstein coordinates using the following equations, in which landmarks 1 and 2 (the cardinal extremities, or the outer ends of the hinge line on the pedicle valve) have been chosen as the key landmarks:

$$u_j = \left\{ (x_2 - x_1)\left(x_j - x_1\right) + (y_2 - y_1)\left(y_j - y_1\right) \right\} \Big/ D_{1,2}^2 - 0.5$$
$$v_j = \left\{ (x_2 - x_1)\left(y_j - y_1\right) - (y_2 - y_1)\left(x_j - x_1\right) \right\} \Big/ D_{1,2}^2 \tag{5.74}$$

where $j = 3, \ldots, k$ and $D_{1,2}^2 = (x_2 - x_1)^2 + (y_2 - y_1)^2$.

A scatter plot of the U, V coordinate pairs is shown in **Figure 5–53**. Each set of brachiopod measurements has been shifted, rotated, and scaled so that key landmarks 1 and 2 coincide for every specimen. The shapes of the brachiopods are not altered by this process, so the scatter in the plotted points for each landmark indicates the variability within the specimens, after adjustment for differences in overall size. We can find an average shape by taking the arithmetic means of the Bookstein coordinates. We also can test whether two or more sets of objects have the same shapes, but this requires multivariate statistics and will be deferred until the next chapter. There, we will compare measurements of *Resserella* sp. to measurements of a second set of seven brachiopod fossils (file ORTHID.TXT) to see if the shapes of specimens in the two sets are significantly different.

Fourier measurements of shape

The relative merits of alternative shape measurements have been argued with some heat, and perhaps still provide a fruitful area for a master's thesis or two. However, we will turn our attention to the more promising, multivalued shape descriptors. Among these are various modifications of the Fourier transformation used in Chapter 4 to analyze time series.

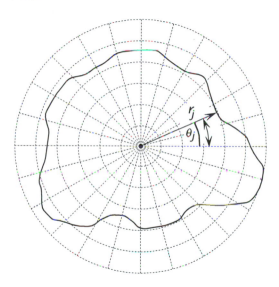

Figure 5–54. Projected outline of sand grain expressed in polar coordinates. Coordinate pairs are formed as the length, $r_j c$, and angle, θ_j, of radii drawn from the centroid to the edge.

The coordinates of a closed line, such as an outline of the projection of a sand grain or a fossil shell, can be expressed in polar form as in **Figure 5–54**. One of the two coordinates is the angular orientation of a radius extending from a point within the outline; the other is the distance along this radius from the central point to the outline.

This immediately introduces a problem because we must be able to specify the placement of the point within the object's outline that defines the center of the polar coordinate system. If the center is moved, the distances along all the radii will change, and the Fourier transform will be different as well. If we wish to compare several different shapes, we must identify equivalent points within each of them as the center of the coordinate system. If this is not done, we will not be able to tell if the differences we see between Fourier spectra result from differences in shape or differences in our choice of the origin.

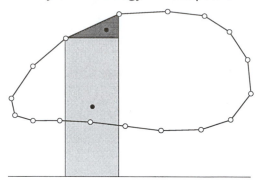

Figure 5–55. Determination of centroid by trapezoidal approximation. Digitized form is divided into trapezoids consisting of a rectangle and a right triangle. Points indicate their centroids.

For some applications, unique points or landmarks can be identified within each shape that can serve as the origin of the polar coordinate system. This was done, for example, by Kaesler and Waters (1972), who measured radii extending from a characteristic muscle scar in ostracodes to the outlines of their shells. Unfortunately, most shapes such as sand grains, pebbles, or the outlines of salt domes lack distinctive points that could serve as common centers of reference. We can, however, compute a unique point within any closed shape that can serve as the origin of the coordinate system.

The ***centroid*** represents the center of gravity of a form and is unique for each object. If the periphery of an object is represented by a series of Cartesian coordinates, such as those generated by a digitizer, the centroid can be found by integrating the X- and Y-coordinates. A simple procedure for integration uses trapezoidal approximation.

A series of points has been placed around the periphery of an object in **Figure 5–55.** Pairs of these points can be used in combination with the axes to define trapezoids. From an arbitrary starting point, we construct a series of trapezoids, moving counterclockwise around the figure. The center of gravity of each trapezoid can be found, and these combined to yield the center of gravity of the figure. The coordinates of the centroid of the figure are

$$\overline{X} = \frac{\sum \overline{X}_t A_t}{\sum A_t} \qquad (5.75)$$

and

$$\overline{Y} = \frac{\sum \overline{Y}_t A_t}{\sum A_t} \qquad (5.76)$$

where $\overline{X}_t$ is the X-coordinate of the centroid of the tth trapezoid, $\overline{Y}_t$ is its Y-coordinate, and A_t is the area of the trapezoid.

The centroids of the trapezoidal areas can be found in turn by simple geometry. Each trapezoid can be broken down into a rectangle and a right triangle. The rectangular portion has its centroid located at a point midway between each of the four sides. The centroid of the triangular portion is located at coordinates one-third of the way between the right angle and the two acute angles. The centroids of the two portions are combined by weighting each according to their respective areas. These operations can be combined and simplified into the following expressions, which provide the terms needed in Equations (5.75) and (5.76):

$$A_t = \tfrac{1}{2}\left(y_{i+1} + y_i\right)\left(x_i - x_{i+1}\right)$$

$$\overline{X}_t A_t = \tfrac{1}{6}\left(x_{i+1}^2 + x_{i+1}x_i + x_i^2\right)\left(y_{i+1} - y_i\right) \qquad \textbf{(5.77)}$$

$$\overline{Y}_t A_t = \tfrac{1}{6}\left(y_{i+1}^2 + y_{i+1}y_i + y_i^2\right)\left(x_{i+1} - x_i\right)$$

The procedure that yields the location of the centroid of a complicated figure can be performed rapidly by computer. The accuracy of the location depends upon the number of points placed around the perimeter.

We have now established the centroid of the figure and can proceed to draw radii from the centroid to the figure's perimeter. The lengths of these radii, together with their angular orientation, provide the polar coordinate pairs which can be analyzed by the Fourier techniques introduced in the preceding chapter. The analysis will yield the spectrum of the closed figure, and from the spectrum we may deduce many interesting things about the shape of the figure. The circular Fourier spectrum has all of the desirable properties of the ordinary Fourier spectrum: it contains all of the information contained in the original figure, the successive harmonics are independent of one another, and each spectral value is a measure of the "contribution" or amount of the total variance added by the corresponding harmonic wave form.

In order to make the circular Fourier transformation computationally tractable, the radii must be spaced at equal angular increments, just as in ordinary Fourier analysis samples must be equally spaced in time or distance. Unfortunately, it is unlikely that the original coordinates placed around the perimeter of the figure and used to determine the centroid will be in the proper locations to form equal angles when these points are connected to the centroid. We must find a new set of points along the perimeter that do define a set of equal angles with the centroid, either by remeasuring or by interpolation between the existing points.

Either procedure introduces a new complication, however, because a centroid calculated from the new set of points is unlikely to coincide exactly with the centroid determined from the original set of points. Unless the radii are measured from the true centroid, a distortion will be introduced into the Fourier spectrum. The effect is analogous to an off-center wheel that "wobbles" up and down once every revolution. This wobble contributes to the first harmonic, which otherwise will be zero. Since spectra usually are standardized for comparative purposes, the presence of a spurious first harmonic will reduce the relative magnitudes of successive harmonics. Boon, Evans, and Hennigar (1982) describe an iterative procedure that produces successively closer approximations to the true centroid of a set of coordinates spaced at equal angles around the perimeter of an object.

Converting the coordinate system from rectangular to polar form has the effect of "unrolling" the closed outline (**Fig. 5–56**). The techniques of ordinary Fourier analysis are obviously applicable to the unrolled figure.

You will recall from Chapter 4 that a continuous, single-valued time series can be represented as the sum of a series of sine and cosine terms, expressed in the form of the Fourier transform of Equation (4.105) on p. 268. The coefficients can be estimated using Equations (4.109) and (4.110). These equations can be modified to find the polar Fourier transform of a closed shape. If we place n radii at equal angles around a complete circle, the angle between each pair of radii is $2\pi/n$. Therefore, the angle from the origin to the jth radius is $2\pi j/n$. Designating the

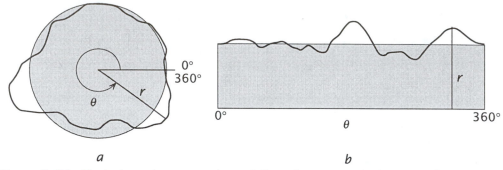

Figure 5–56. Equivalency between polar and Cartesian representations. (*a*) Grain outline shown in polar notation. (*b*) Polar coordinates plotted as r versus θ.

angular direction of the jth radius as θ_j, the length of the jth radius as r_j, and substituting into Equation (4.109):

$$\alpha_k = \frac{2}{n}\sum_{j=1}^{n} r_j \cos k\theta_j$$
$$\beta_k = \frac{2}{n}\sum_{j=1}^{n} r_j \sin k\theta_j \tag{5.78}$$

As in single Fourier analysis, some of the terms can be simplified. The coefficient β_0 is equal to zero, and α_0 is equal to the mean radius

$$\alpha_0 = \overline{r} = \frac{\sum_{j=1}^{n} r_j}{n} \tag{5.79}$$

The effect of size can be eliminated from an analysis by dividing all radii by the mean radius; then, α_0 will always be equal to one. For the reason discussed earlier, the coefficients of the first harmonic, α_1 and β_1, should be zero if the radial measurements are taken from the true centroid.

Once the α and β coefficients have been determined for a series of harmonics, the interpretation of the circular Fourier spectrum proceeds in a manner directly analogous to the interpretation of a conventional Fourier spectrum. The amplitudes and phase angles of the polar harmonics can be found from the Fourier coefficients using Equations (4.111) and (4.112) in exactly the same manner as they are used to find the spectrum of discrete time series.

It is usually most convenient to combine the α and β coefficients to form the power, or variance, spectrum which directly expresses the contribution made by each harmonic to the shape of the figure. Successive approximations of the analyzed form can be constructed by inserting the α and β coefficients into the Fourier equation. As harmonics are added, the reconstruction will become increasingly detailed until it exactly matches the original digitized outline. (This will occur at the Nyquist frequency, when $k = (n-1)/2$.) **Figure 5–57** shows the reconstruction of the shape of a sand grain. The zeroth harmonic produces a circle whose diameter is equal to the mean diameter of the original outline. There is no first harmonic because the grain has been properly centered around its centroid. The second harmonic changes the form of the reconstruction to an ellipse, the third harmonic adds a "triangular" component, the fourth a "square" component, and so on. Most simple closed figures can be closely replicated with fewer than ten harmonics (**Fig. 5–58**).

The proportion of the variation in shape accounted for by the successive harmonics is given by the power spectrum. Usually a small number of the lowest

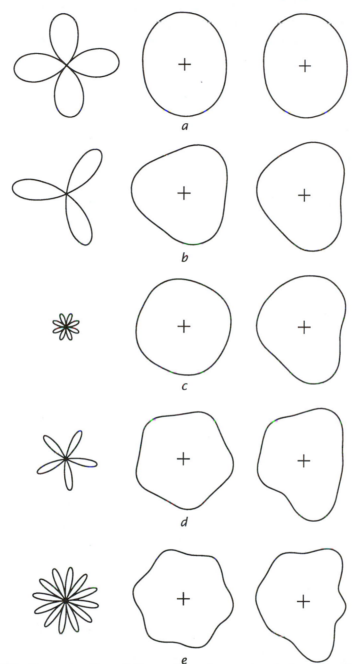

Figure 5–57. Reconstruction of the outline of a sand grain by polar Fourier series. (*a*) Plot of second harmonic (left), second harmonic plus circle corresponding to mean radius or zeroth harmonic (center), cumulative sum of harmonics (right). (*b*) Third harmonic. (*c*) Fourth harmonic. (*d*) Fifth harmonic. (*e*) Sixth harmonic.

harmonics will account for almost all of the variation in simple forms such as projections of sand grains (**Fig. 5–59**). The higher harmonics reflect smaller and smaller details of the outline and have been used by sedimentologists as measures of "surface roughness." The higher harmonics of the outlines of quartz sand grains are

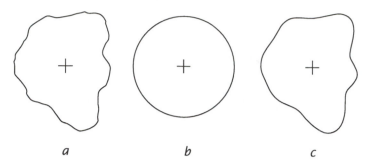

a   *b*   *c*

Figure 5–58. Comparison between (*a*) digitized grain outline, (*b*) zeroth harmonic or circle of mean radius, and (*c*) reconstruction based on six harmonics.

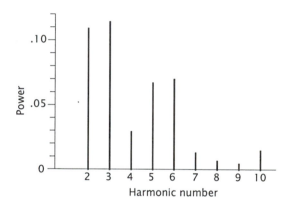

Figure 5–59. Power spectrum of digitized sand grain shown in Figure 5–56.

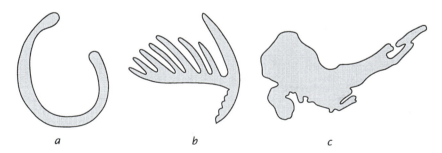

a   *b*   *c*

Figure 5–60. Examples of shapes which cannot be analyzed by polar Fourier analysis because of double-valued radii. (*a*) Shoreline of a mid-Pacific atoll. (*b*) Projection of the conodont *Ligonodina*. (*c*) Outline of a granite pluton in southern Ontario.

said to reflect provenance or transport history (many studies are cited in Ehrlich, Brown, and Yarus, 1980), but caution should be used in making such interpretations from parts of the Fourier spectrum. The standard errors of the estimates of Fourier coefficients are very large, being of the same order of magnitude as the estimates themselves. The higher harmonics of grain outlines typically have extremely low power, perhaps five or more orders of magnitude below the power contained in the low harmonics. Considering their standard errors, it seems difficult to attach much significance to such frequencies. In addition, the effects of aliasing are most pronounced at the high frequencies, making determination of true spectral values even more difficult.

Polar Fourier transformation has certain limitations, the most conspicuous of which is that only single-valued outlines can be analyzed. That is, a radius drawn from the centroid must intercept the perimeter only once. This means that the procedure cannot be used for extremely convoluted shapes, such as those shown in **Figure 5–60**.

Other Fourier procedures permit the analysis of more complicated forms, which may be double-valued in a polar representation. One method converts the original outline into a series of angular deviations. First, equally spaced points are placed around the perimeter of the object. The direction from the first to the second point is found, and then the angular difference between this direction and the direction from the second to the third point (**Fig. 5–61**). As this process is repeated, the original series of X- and Y-coordinates defining the form is replaced with a series of angular deviations between the successive points. This new series can be analyzed by conventional Fourier procedures, although the spectrum may be difficult to interpret because the Y-variable is not a distance but an angular change. This and similar transforms are discussed by Clark (1981).

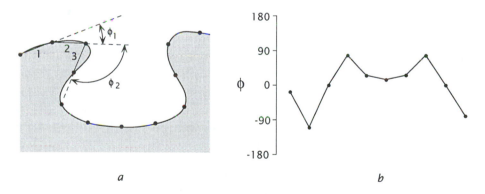

a   *b*

Figure 5–61. Conversion of digitized coordinates into angular deviations. (*a*) The straight lines are the approximation formed by digitizing the original shape at equally spaced points. Angular change between segments 1 and 2 is given by ϕ_1; change between segments 2 and 3 is given by ϕ_2. (*b*) Plot of angular change ϕ between successive segments. It is this record that is analyzed by Fourier methods.

Some closed forms are bilaterally symmetrical, or possess a distinctive line or fold that allows them to be oriented along an axis. The form can then be "split" and one-half "folded over" by reversing the signs of one coordinate along the symmetry axis for half of the form (**Fig. 5–62**). The object will then appear as a sinusoidal line that can be analyzed by ordinary Fourier analysis. This approach may be useful for the study of the shapes of certain invertebrate fossils and also has been used to characterize the shapes of arrowheads.

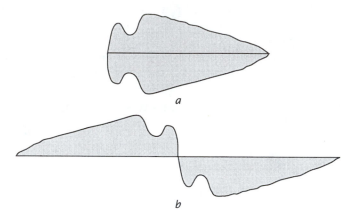

Figure 5–62. Outline of Late Paleolithic flint arrowhead. (*a*) Shape shown in conventional representation. (*b*) Transformation of outline by reversing one-half of figure along axis of symmetry.

Spatial Analysis by ANOVA

Regional studies, particularly of geochemical and environmental properties, require knowledge of the spatial scale at which variation occurs in the property of interest. If we know that a property changes most pronouncedly over a specific distance, it would be inefficient to collect observations at shorter intervals. Conversely, we would run the risk of missing important changes if we sampled at greater distances. Agricultural statisticians have developed special analyses of variance and combined them with field sampling patterns to partition the variance of a spatially distributed property into components that are attributable to different spatial scales. These ANOVA's are hierarchical or nested designs, in which the nested factors correspond to a progression of smaller geographical sampling units. The statistical design is an extension of the two-level nested experimental design discussed in Chapter 2. Sampling designs based on nested ANOVA's have been widely used in regional geochemical studies by the U.S. Geological Survey (Miesch, 1976; Severson and Tidball, 1979) and the Geological Survey of Canada (Garrett, 1983, 1994), and underlie the sampling methodology used in the IUGS project to produce the Geochemical Map of the World (Darnley and others, 1995).

Krumbein and Slack (1956) were among the first to apply a nested ANOVA to geological data. They examined the regional pattern of radioactivity in the Pennsylvanian Brereton shale, a black, fissile shale that overlies the No. 6 Coal in southwestern Illinois. The shale can be sampled in strip mines and in numerous boreholes that have been drilled to assess coal reserves. The region of Illinois where the Brereton shale has been sampled is shown in **Figure 5–63**. The study area, about 200 mi long from north to south and about 20 mi wide, has been divided into 11 square cells, each 18 × 18 mi. In turn, each cell is subdivided into 6 × 6-mi square areas corresponding to political townships. The hierarchical sampling scheme consists of the random selection of two townships within each cell, then the random selection of two strip-mine pits from within each selected township. At the lowest level, four measurements of shale radioactivity are made within each

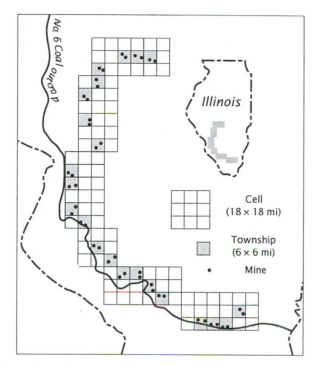

Figure 5–63. Map of southwestern Illinois showing sampling locations of the Brereton shale and outcrop pattern of No. 6 Coal. Grid for balanced nested sampling plan is superimposed. After Krumbein and Slack (1956).

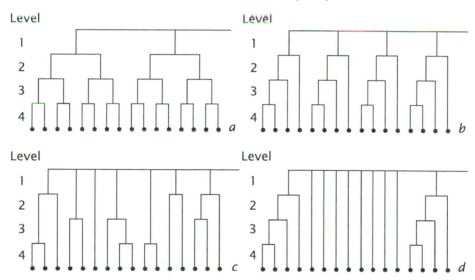

Figure 5–64. Schematics for 4-level nested designs. (*a*) Balanced design. (*b*) Staggered unbalanced design. (*c*) Highly fractionated unbalanced design. (*d*) Inverted unbalanced design. After Garrett and Goss (1980).

selected mine. The sampling pattern is superimposed on **Figure 5–63** and a schematic diagram of the nested ANOVA is shown in **Figure 5–64a**. The data are contained in file BRERETON.TXT. Like other measures of radioactivity, the observations are counts produced by a Poisson process. The raw data are transformed in subsequent calculations by taking their square roots, which stabilizes their variance.

Because Krumbein and Slack's design is balanced (*i.e.*, equal numbers of subdivisions within each nested branch), conventional analysis of variance methods can be applied to data. This results in the ANOVA shown below. The model used by Krumbein and Slack (1956) tests for homogeneity, that is, whether subdivisions nested within a level are more variable than the levels themselves. The mean square for one level is compared to the mean square for the succeeding level; for example, $MS_{townships}$ is compared to $MS_{mines\ within\ townships}$. The only significant differences are between samples within mines, that is, at a level corresponding to the smallest spatial separation between observations. Almost 85% of the variation in radioactivity of the Brereton shale occurs at this level of sampling, over distances of a half-mile or less. The variation at all other spatial scales is not statistically significant, which implies that a very widely spaced sampling pattern would be as effective in capturing any spatial variability in radiation as a more expensive, dense, regional sample-collection pattern. Conversely, to map the spatial variation of shale radioactivity in detail would require taking measurements of the Brereton shale at a very small scale; a second ANOVA described by Krumbein and Slack (1956) shows that most of the differences between observations occur over a spacing of 2 ft or less.

Source of Variance	Sum of Squares	Degr. of Freedom	Mean Squares	Variance	Percent Variance	*F*-tests
Cells	3.7556	10	0.3756	0.0035	1.82	1.1783[a]
Townships	3.5227	11	0.3202	0.0138	7.25	1.528[b]
Mines	4.6103	22	0.2096	0.0120	6.29	1.297[c]
Samples (Error)	21.3272	132	0.1616	0.1616	84.65	
Total	33.2158	1785	0.1898	0.1909	100.00	

[a] $p = 0.397$ (not significant). Tests significance of $MS_{cells}/MS_{townships\ within\ cells}$.

[b] $p = 0.17$ (not significant). Tests significance of $MS_{townships}/MS_{mines\ within\ townships}$.

[c] $p = 0.1848$ (not significant). Tests significance of $MS_{mines}/MS_{samples\ within\ mines}$.

A major drawback to a classical balanced nested sampling design is the large numbers of observations required if the design has many levels of spatial scale. This has led to development of unbalanced designs which are much more parsimonious in their use of observations. The contributions by Garrett and his associates have been especially noteworthy in the design of regional geochemical sampling schemes; these include special computer programs for computing unbalanced nested ANOVA's which require unorthodox procedures to estimate statistics such as the appropriate error mean squares and degrees of freedom for error (Garrett and Goss, 1980).

Schematics of several possible unbalanced nested ANOVA designs are shown in **Figure 5–64 b–d**; compare these to the balanced design shown in **Figure 5–64 a**.

Although conventional analysis of variance programs may not be able to correctly analyze data collected by such unbalanced designs, the data can be analyzed using special-purpose programs (Garrett and Goss, 1980) or by powerful general-purpose programs (usually identified as "General Linear Model" or GLM routines).

With over 107,000 tons of mercury extracted since it opened in 1490, the Idria mine in Slovenia was the second largest producer of mercury in the world. Although the mine ceased commercial operation in 1990, the deleterious effects of mining continue; it is estimated that 27% of the mercury produced was lost into the surficial environment. As part of an environmental study, 123 sediment samples for chemical analysis were collected in Slovenia using a staggered nested sampling pattern. A series of grids representing spacings of 25 km, 5 km, 1 km, and 0.2 km were superimposed on a map of the country. A sample was collected at the center of each 25×25-km cell. Then, three more sampling sites were located within each 25×25-km cell, one at the center of a randomly chosen 5×5-km cell, another at the center of a randomly chosen 1×1-km cell within the 5×5-km cell, and the third at the center of a randomly chosen 0.2×0.2-km cell within the 1×1-km cell. Finally, some samples were split and analyzed twice to provide information on laboratory variability. The data are given in file SLOVENIA.TXT; sample locations are given in kilometers on a Gauss–Krueger coordinate system (similar to UTM coordinates). Because the distribution of Hg values is highly skewed, values were transformed by taking their logarithms prior to further statistical analysis.

A nested ANOVA similar to that described in Chapter 2 (p. 88) can be used to test the relative variability in mercury concentration associated with successive levels in the hierarchy. The following ANOVA shows that there are statistically significant large regional trends in Hg concentration (detected by the 25×25-km grid), no significant variation at the smaller scale of the 5×5-km grid, and significant variation at a local scale of about 1×1 km or less (Pirc, Bidovec, and Gosar, 1994). Note that the mean square for each level is compared to the mean square for error. The result has been interpreted as reflecting a regional trend of mercury contamination across Slovenia, with levels declining to the west away from the site of the

Source of Variance	Sum of Squares	Degr. of Freedom	Mean Squares	Variance	Percent Variance	F-Tests
25 × 25 Cells	6.993	27	0.259	0.031	28.89	14.94[**]
5 × 5 Cells	2.175	22	0.099	0.002	2.18	5.7[*]
1 × 1 Cells	2.660	27	0.099	0.043	39.54	5.68[*]
0.2 × 0.2 Cells	0.933	28	0.033	0.015	13.42	1.92[NS]
Error	0.312	18	0.259	0.017	15.98	
Total	13.073	122	0.107			

[*] = Significant.   [**] = Highly Significant.   [NS] = Not significant.

Idria mine. Superimposed on this regional trend are scattered local anomalies of higher mercury contamination.

Nested analyses of variance serve a dual purpose; they provide an accepted methodology for designing efficient regional sampling plans and checking the efficacy of the sampling, and, at the same time, provide insight into the scale of spatial variability of the geological property. The current tendency is to use geostatistical procedures for spatial analysis, partly because of the greater apparent resolution available from a semivariogram, which is continuous rather than discrete. Nevertheless, ANOVA's are a valuable tool for the design of regional sampling programs.

Computer Contouring

The objective of contour mapping is to portray the form of a surface. A contour map is a type of three-dimensional graph or diagram, compressed onto a flat, two-dimensional piece of paper. The axis running across the page is variable X_1, and the axis running up and down the page is X_2; these axes correspond to east–west and north–south geographic coordinates. The mapped property is shown by means of isolines, or lines of equal value, which represent an offset in the vertical dimension, Y. The isolines may represent height above sea level, thickness of a stratigraphic unit, or some other quantity. (Many authors denote the three axes as X, Y, and Z, corresponding to the two geographic coordinates and the mapped variable. To maintain consistency with later topics, we will continue to use Y as a dependent variable that is expressed as a function of independent variables X_i.) Contour lines, strictly speaking, are isolines of elevation, whether they represent a topographic surface or the top of a subsurface formation. Geologists are casual in their use of terminology, however, and usually call any isoline a contour, whether it depicts elevation, thickness, porosity, composition, or some other property.

The contours on a map connect points of equal value, and the space between two successive contour lines contains only points whose values fall within the interval defined by the contour lines. In most circumstances the value of the surface cannot be determined at every possible location, nor can its value be measured at any specific point we might choose. Usually only scattered measurements of the surface are made at relatively few control points, such as well locations, seismic shot points, or sites where assay samples have been taken.

A contour map may be drawn by hand, as has been done for hundreds of years, or it may be drawn by computer. Drawing a map by computer usually involves an intermediate step—the construction of a mathematical model of the surface—that must be performed before the contour lines themselves can be constructed.

A computer contouring program traces out contour lines by a precise mathematical relationship based on the geometry of the control points. A geologist, however, not only contours the control points but also incorporates his concepts and ideas about what the surface should look like. If these preconceived ideas are indeed correct, a competent geologist may be able to create a map superior to a computer-made product. On the other hand, if the geologist's preconceptions are erroneous, his finished map is likely to be wrong.

Experiments have offered some reassuring insight into the nature of computer-produced maps (Dahlberg, 1975). In one experiment, experienced petroleum geologists were pitted against a widely used computer contouring program. The test data consisted of structural elevations from a collection of wells drilled into and around a Devonian reef in Alberta (the data are given in file LEDUC.TXT and will be used later). Information from only a small number of the wells actually available was presented to the participants. The objective was to assess the relative capabilities of man and machine in creating a realistic structural contour map.

All the maps tended to be very much alike at and near control points, but differed radically in uncontrolled areas. Some geologists produced "better" structural maps than the computer program, in the sense that their representations were closer to the structural configuration revealed by the complete data set. Other geologists, however, were seriously in error. Most interestingly, the computer-contoured map coincided almost exactly with the average of the manually produced maps. That is, between control points some geologists tended to bend their contour lines in one direction while others bent them in the opposite direction. Most drew their lines through a common middle ground, and only a few deflected their contour lines significantly one way or another. The computer-drawn contours passed almost precisely through the middle of this bundle of lines. In this experiment, the computer program behaved like an "average" geologist.

Dahlberg (1975) also provided the comparison shown in **Figure 5–65** of the thickness of sand measured in 13 wells in a hypothetical area. The depositional

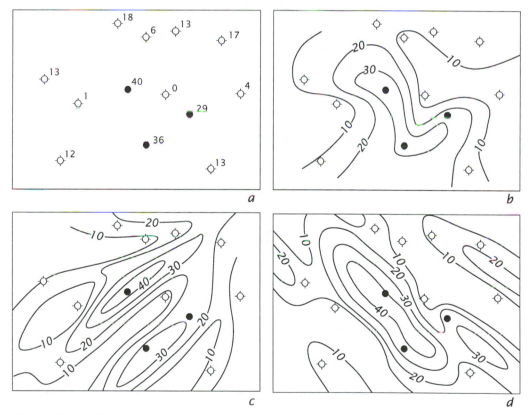

Figure 5–65. Thickness of sand unit encountered in 13 hypothetical wells. Depositional environment of sand is transitional marine, with sea to southwest, land to northeast, and shoreline trending northwest–southeast. (*a*) Posting of data with sand thicknesses in feet. (*b*) Computer-produced contour map of sand thickness. (*c*) Contour map by geologist who interpreted sand as distributary channel fill. (*d*) Contour map by geologist who interpreted sand as offshore bars. After Dahlberg (1975).

environment of the sand was transitional marine with a seaward direction to the southwest and a shoreline that trended northwest-southeast. If the sands are interpreted as distributary channel sands, the thickness of the sand might be contoured as shown in **Figure 5–65 c**. However, if the sands are considered to be offshore bars, the resulting contoured isopach map might resemble **Figure 5–65 d**. Both of these interpretations honor the data equally well. In comparison, **Figure 5–65 b** is an isopach map produced by a computer contouring routine. The computer map is based solely on the known thicknesses of sand in the boreholes and the geometrical arrangement of the wells; the computer knows nothing about paleogeography. This mechanical interpretation is almost surely incorrect in detail and inferior to one of the two manually drawn maps. Unfortunately, one of the manual maps is completely wrong, and we cannot tell which one it is! Because of the strict adherence to geometrical rules built into its contouring program, the computer yields a "neutral" interpretation that may serve as a valuable constraint on the geologist's imagination (for a contrary view, see discussions on interpretative contouring in Hamilton and Jones, 1992).

One of the strongest arguments for computer contouring is that it creates a mathematical model of the mapped surface that can also be used for other purposes. Among the operations that can be performed are the mapping of derivatives of the surface, calculating volumes beneath the surface, and various surface-to-surface operations such as isopaching (subtracting one surface from another).

It should be emphasized that many computer contouring procedures are *ad hoc*; there is little theoretical basis for the various methodologies that have been employed. Rather, they are based on familiar premises about the way surfaces behave and results achieved through practical experience.

Several assumptions must be embodied in any computer algorithm used to create the mathematical model from which a contour map is constructed. The completed map reflects these assumptions; the model is reasonable and the map is a realistic representation of the surface only if these assumptions are valid. In general, a contouring program is designed to map a surface that is (a) single-valued at every point, (b) continuous everywhere within the map area, and (c) autocorrelated over a distance greater than the typical spacing between control points.

If there is only one possible value that a mapped property can have at a specific geographic location, that property is ***single-valued***. An example is the elevation of a stratigraphic horizon, as measured in a well, to be used in constructing a structural contour map. Only in very unusual geological circumstances, such as an overturned fold, could a stratigraphic unit occur at more than one elevation at a single location.

Some important geological variables are not obviously single-valued. Measurements of porosity or chemical composition, for example, are statistical in nature and repeated sampling and analysis at a single location may result in a suite of values. This results both from errors in measurement and from random variations in the pieces of rock that are analyzed. Most automatic contouring programs cannot accommodate multiple values at a location, nor a range or interval of values. However, it is possible to reduce multiple observations or distributions to a single, representative value such as the average or mean which can then be mapped.

Automatic contouring procedures involve interpolation between control points and extrapolation beyond the control points. Because of the mathematical methods involved, all values obtained by interpolation lie on a continuous, sloping surface between the control points. If the real surface contains discontinuities such as faults, these will not be recognized by the contouring program but will be mapped simply as areas of very steep slope. Faults or other discontinuities that are known in advance of mapping can be accommodated by procedures that in effect insert boundaries into the map. The mapping program will draw the surfaces on opposite sides of boundaries as though they were entirely separate maps. The corresponding mathematical model will have an abrupt change in numerical values along the boundaries. However, it has not yet proved possible to create a contouring program that automatically recognizes unidentified faults or breaks in a surface.

Contouring programs incorporate the commonsense assumption that the value of a surface at one point is closely related to values at nearby points and less closely related to values at more distant locations. This assumption that the variable being mapped is positively autocorrelated over at least short distances is expressed in some procedures by the selection of all the nearest control points around a location to be evaluated, which is then estimated at that location as some type of average value. If a surface is highly autocorrelated, all of the neighboring control points will have approximately the same value, and their average will be a reasonable estimate of the intermediate location. In contrast, if the surface is poorly autocorrelated, neighboring control points will have little relation to one another, and will be unrelated to the value at the location to be estimated. Under such conditions it may be impossible to make reasonable interpretations about the nature of the surface between control points.

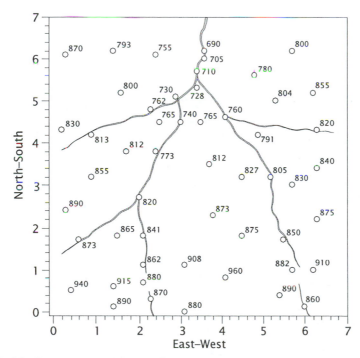

Figure 5–66. Survey control points for topographic mapping problem. Geographic coordinates in 50-ft units from origin in southwest corner of map. Elevations in feet above sea level. Data listed in file NOTREDAM.TXT.

Figure 5–66 is a map used in an introductory geology course to teach the concepts of contouring. Values at the points are topographic elevations measured in

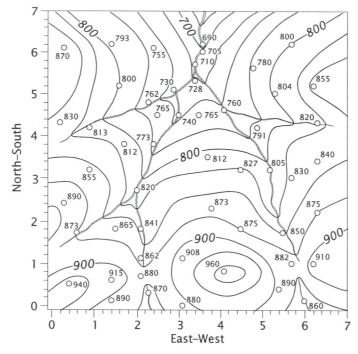

Figure 5–67. Contour map of topographic data produced by manual contouring. Contour interval is 25 ft. Note inferred effect of streams on form of contour lines.

a small area by a surveying class; the X_1- and X_2-coordinates are given in arbitrary units (1 map unit = 50 ft) from an origin in the southwest corner of the map, but this is done only for convenience. The control locations could equally well be expressed by any numerical Cartesian coordinate system. The data for each control point, in the form of an east–west coordinate, a north–south coordinate, and the elevation of the ground surface relative to sea level, are contained in file NOTREDAM.TXT and will be used for several examples in this and the following section. **Figure 5–67** shows a typical manually contoured map of these data. Note that the stream traces on the map have been used to guide the paths of contour lines.

Contouring by triangulation

The first computer programs for contouring were direct implementations of methods used by surveyors for the hand mapping of topography (IBM, 1965). Almost every undergraduate geologist and engineer has experienced a manual equivalent of this process as an exercise in plane-table mapping or surveying. In such programs, control points, assumed to be located without any particular regularity, are first connected by straight lines. This forms a mesh of triangles that covers the map (**Fig. 5–68**). By interpolating down the sides of the triangles, locations are found where the ground elevation is a constant, specified value. Connecting these points of equal elevation produces a contour line. In effect, the surface is modeled as a series of flat, triangular plates, each held at its corners by a control point. Contouring consists of drawing horizontal lines across these tilted plates.

It is obvious that if the control points were connected with a different set of straight lines, a different set of triangular plates would be defined and a different

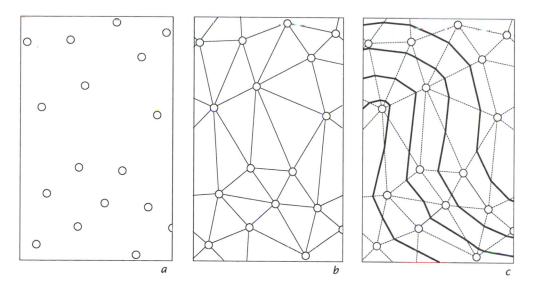

Figure 5–68. Triangulation method of estimating positions of contour lines. (*a*) Irregularly spaced data points. (*b*) Triangles formed across map area with data points as vertices. (*c*) Contour lines drawn through sides of triangles where points of specified elevation are found by linear interpolation (after IBM, 1965).

set of contour lines would result. In some early contouring programs, simply entering the data points in a different sequence could result in contour lines with a conspicuously different appearance. To avoid this problem, attempts were made to select a unique, "optimal" set of triangles for mapping. Usually this meant that individual triangles should be as near to equilateral as possible, or that the triangles should have the minimum possible height, or that the longest leg of each triangle should be the shortest possible (Gold, Charters, and Ramsden, 1977). Unfortunately no algorithm was available that could ensure the construction of a triangular mesh having any of these properties. This was a serious impediment in early contouring programs, because it meant starting with an arbitrary arrangement of triangles and, through iteration, trying to adjust the triangular mesh until an optimum configuration was obtained. This often resulted in exorbitant run times and led to the almost complete abandonment of triangulation algorithms. Their place was taken by procedures that used the control points only to make estimates of the surface at the nodes of a regular grid, and then contoured this grid rather than the control points themselves.

Triangulation procedures have now resumed their former importance, especially in geographic information systems (GIS) where a meshwork of triangles across a surface, augmented with topologic information, is referred to as a TIN (triangulated irregular network) data model (ESRI, 1992). Algorithms that produce almost optimal triangular networks on the first pass have been written by Gold, Charters, and Ramsden (1977) and by McCullagh and Ross (1980); a review is given by Watson (1992). These networks, referred to as **_Delaunay triangulations_**, are uniquely defined for a given set of points. In addition, the triangles formed are as nearly equiangular as possible, and the longest sides of the triangles are as short as possible. This means that the greatest distances over which interpolations must be made to find contour levels are smaller than in other triangular nets.

375

We can imagine that each point in a field of scattered points is surrounded by an irregular polygon so that any point location within the polygon will be closer to the enclosed point than the location will be to any other point (**Fig. 5–69 a**). Conversely, every location outside a particular polygon is closer to some other point than it is to the point within the polygon. This is the most compact division of space possible. Sets of closed figures that have these properties are called Thiessen, Dirichlet, or Voronoi polygons. They arise in many diverse fields. Geographers use Thiessen polygons to model zones of influence around competing cities. Metallurgists model the growth of crystals in a cooling melt by Voronoi polyhedra—the three-dimensional equivalent of polygons. A mass of soap bubbles forms an easily observed network of Voronoi polyhedra.

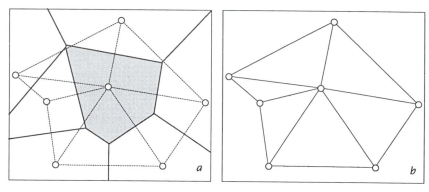

Figure 5–69. (*a*) Thiessen polygon (shaded) around central point. All locations within polygon are closer to central point than to any other point. (*b*) Delaunay triangle network formed by connecting central points of Thiessen neighbors.

The Thiessen polygon that encloses a specific point is immediately surrounded by other Thiessen polygons, each of which also encloses a single point. These points are called "Thiessen neighbors." If these neighboring points are connected by straight lines, the result is a Delaunay triangular network (**Fig. 5–69 b**). The sets of Thiessen polygons and the Delaunay triangles are both unique for any arrangement of points.

The process of triangulation consists of determining the Thiessen neighbors of successive points on a map. In **Figure 5–70 a** the neighbors of point *a* are to be found. We first assume that a nearby point *b* is a neighbor and construct a circle whose diameter is defined by the line *a b*. If no other points occur within the circle, *b* is indeed a neighbor of *a*. If a point is found in the circle, it replaces point *b*.

The search for the next neighbor proceeds in a clockwise manner around point *a*. The circle is expanded as shown in **Figure 5–70 b** so that points *a* and *b* lie on its perimeter. The interior of the circle is checked to see if any points are enclosed. If one point is found, it is the second Thiessen neighbor. If two or more points are found, the correct second neighbor must be determined. This is done by redrawing the circle so that points *a* and *b*, and the candidate point *c*, all lie on its perimeter. If a check of the interior of this circle discloses an enclosed point *d*, this is the true third neighbor. Next, a circle is constructed that includes point *a* and point *d* on its perimeter, and a search is made for the fourth Thiessen neighbor *e* in its interior (**Fig. 5–70 c**). The process repeats (**Fig. 5–70 d**) until eventually point *b* will be rediscovered as a Thiessen neighbor; then, all of the neighbors of *a* have

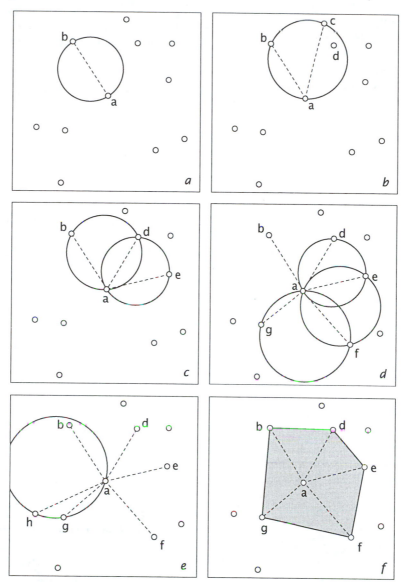

Figure 5–70. Determining Thiessen or natural neighbors of point a. (*a*) Nearby point b is selected as possible neighbor and circle with diameter $a\,b$ is constructed. If circle contains no other points, b is a neighbor. (*b*) Next point in clockwise direction is selected as possible neighbor and circle with points a, b, and c on perimeter constructed. Circle encloses point d, so c is not a natural neighbor. (*c*) Circle with points a, d, and candidate point e is constructed. Circle contains no other points so e is a natural neighbor. (*d*) Circle construction proceeds in clockwise direction and finds point f, then point g. (*e*) Final circle through points a, g, and candidate point h encloses starting point b, so h is not a natural neighbor. (*f*) All natural neighbors of central point a are connected, forming a Thiessen polygon (shaded) around a.

been identified (**Fig. 5–70 e**). Connecting these neighbors will form the triangular network around a (**Fig. 5–70 f**).

One of the Thiessen neighbors is now designated as a new point a around which a search will be conducted, and the entire process begins again. The network grows, spreading like a wave across the map until every point is included. **Figure 5–71** shows a completed triangulation of the student map data in file NOTREDAM.TXT. To achieve computational efficiency, elaborate prior sorting of the point coordinates is done so that the searches first consider only the most likely candidates as neighbors. Although the process is complicated to describe, McCullagh and Ross (1980) state that the number of necessary steps to triangulate a set of n points is proportional to $n \log n$. In contrast, the number of operations required by algorithms that adjust the triangular net by trial and error may be proportional to n^3 and the number of steps in a gridding procedure is proportional to n^2.

The assumption that the triangles represent tilted flat plates obviously results in a very crude approximation of a surface. A better approximation can be achieved using curved or bent triangular plates, particularly if these are made to join smoothly across the edges of the triangles. Several procedures have been used for this purpose. One of the earliest involved finding the three neighbors closest to the faces of a triangle, then fitting a second-degree polynomial trend surface to these and to the points at the vertices of the triangle. A second-degree trend surface is dome- or basin-shaped, and is defined by six coefficients. This means that the fitted surface will pass exactly through all six points. The equation can then be evaluated to find a series of locations having a specified elevation. These are connected to form the contours, which will be curved rather than straight lines.

Even though adjacent plates are fitted using common points, their trend surfaces will not coincide exactly along the line of overlap. This means there may be abrupt changes in direction where contour lines cross from one triangular plate to another. One way of correcting this problem is to "blend" the contour lines from the two surfaces by averaging them.

A more elegant procedure uses three-dimensional equivalents of the spline functions introduced in Chapter 4. These are described in detail by Tipper (1979). Surface interpolation equations used in finite-element analysis can also be adapted to contouring (Gold, Charters, and Ramsden, 1977; McCullagh, 1981). Watson (1992) discusses several alternative spline functions that can be fitted to an irregular network and provides an extensive guide to the literature. In most of these procedures, the initial triangles are subdivided into smaller triangles and surface interpolations are made using Bézier coefficients, bicubic polynomials, tensors or other similar functions. Most of these interpolants require that derivatives of the surface be calculated from slopes or gradients estimated from values in the adjacent sets of triangles. These procedures usually result in the coefficients for a linear equation that estimates values at specified locations. All estimates for locations within a triangle lie on a smoothly curving surface that merges continuously with the curved surfaces in adjacent triangles. Continuity across the sides of adjoining triangles is achieved because the triangles share vertices and have common slopes at their edges.

In contour mapping, the equation used to estimate values of the surface is reversed. The value of $\hat{Y}$ is set at the desired contour level, and for several selected X_1-coordinates the corresponding X_2-coordinates are found (or vice-versa). The result is a series of points having constant values of $\hat{Y}$. A contour line can be drawn merely by connecting the points.

Figure 5–71 shows the set of Delaunay triangles for the NOTREDAM.TXT data, constructed using the freeware program Triangulate developed by Bourke (1993). The shaded area includes the triangular net and is defined by the outermost control

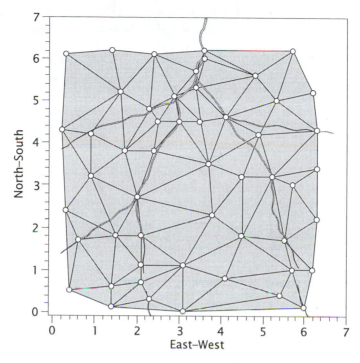

Figure 5-71. Delaunay triangular network for survey control points of Figure 5-66.

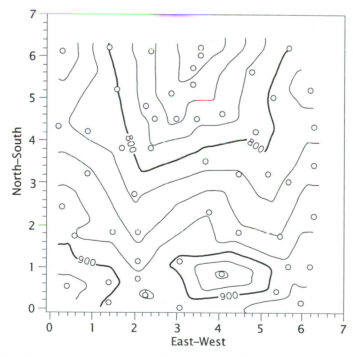

Figure 5-72. Contour map of topographic data, given in file NOTREDAM.TXT, produced by contouring program that uses triangular mesh for mathematical model of surface. Contour interval is 25 ft.

points. In order to draw contours beyond this enclosing polygon, "pseudopoints" would have to be placed along the boundary of the map.

The finished map, as contoured by a triangulation procedure, is given in **Figure 5–72**. Although very similar to maps produced using gridding algorithms, there are differences in detail. The most obvious of these are abrupt changes in direction of contour lines, as in the southwest corner and in the east-central part of the map. These are areas where the Delaunay triangles are extremely acute and there are pronounced differences in slope of the triangular facets. Along the margins of a map, such features could be corrected by judicious insertion of pseudopoints, but they cannot be altered within the interior of a map unless additional data points are provided.

Contouring by gridding

Gridding is the process of determining values of a surface at a set of locations that are arranged in a regular pattern, usually square, which completely covers the mapped area. In general, values of the surface are not known at these uniformly spaced points and so they must be estimated from the irregularly located control points where values of the surface are known. The locations where estimates are made are referred to as *grid points* or *grid nodes*.

In this procedure, we first construct a mathematical model of the surface which has the form of a pattern of tilted square plates. In simple algorithms, these plates are flat. In more complex procedures they are curved, and each blends smoothly into the adjacent plates. The mathematical model is constructed purely for practical reasons. It is much easier to draw contour lines through an array of regularly spaced grid nodes than it is to draw them through the irregular pattern of the original points. All the possible ways for a contour line to enter and leave the square defined by four uniformly spaced grid nodes are known, and a line-drawing algorithm can easily be written to handle all of these possibilities. A contour line can be drawn simply by tracing the path of the line from one square of the grid to the next. Determining the path of a contour line through the irregular pattern of control points, as is done in triangulation algorithms, is much more difficult. Since the individual points cannot be connected to form a regular array, the possible paths that a contour line might follow cannot be known in advance. Also, the explicit X_1- and X_2-coordinates of all intermediate points involved in the calculation of the path of the contour lines must be retained in computer memory. In a grid or regular array of estimated values, the X_1- and X_2-coordinates are implied by their position in the array. This saves memory, but more importantly, it speeds computation. **Figure 5–73** illustrates the process. The original data (Jones, Hamilton, and Johnson, 1986) consist of elevations of an unconformity at the top of the Mississippian Lamont sandstone in an area in Ohio, measured in wells and estimated from return times along seismic traverses; coordinates and subsea elevations are listed in file LAMONT.TXT. The regular grid to which values will be interpolated is represented by the intersections of sets of horizontal and vertical dotted lines on **Figure 5–73 a**, along with the well locations and seismic shot points. The completed contour map, based on the grid of interpolated values, is shown in **Figure 5–73 b**.

The grid nodes, or intermediate locations where values of the surface must be estimated, usually are arranged in a square pattern so the distance between nodes in one direction is the same as the distance between them in the perpendicular direction. In most contouring programs, the spacing is under user control, and

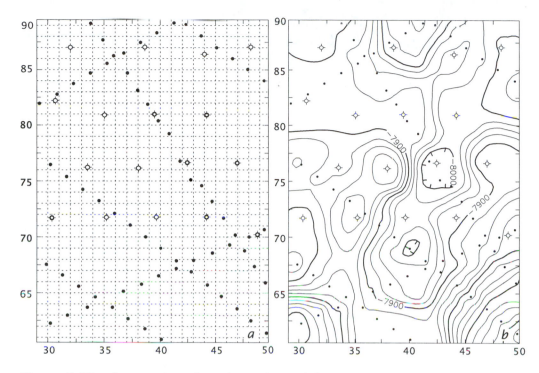

Figure 5–73. Construction of regular grid model for contour mapping. (*a*) Posting of drill-hole control points (dry hole symbols) and seismic control points (solid symbols) in an area of Ohio. Regular grid of interpolated values indicated by dashed lines. (*b*) Contour map of unconformity at top of Mississippian Lamont sandstone, in feet subsea elevation. Contour interval 25 ft. Data from Jones, Hamilton, and Johnson (1986) in file LAMONT.TXT.

is one of many parameters that must be chosen before a surface can be gridded and mapped. The area enclosed by four grid nodes is called a **grid cell**; if a large size is chosen for the grid cells, the resulting map will have low resolution and a coarse appearance, but can be computed quickly. Conversely, if the grid cells are small in size, the contour map will have a finer appearance, but will be more computationally expensive to produce.

Since gridding algorithms estimate only a single location from a collection of nearby control points, the estimation procedure is repeatedly applied across the map area until the entire map is covered by a regular grid or mesh of estimated values. Once the regular grid of estimates has been constructed, contour lines can be laced rapidly through this numerical array.

In some computer contouring packages, the initial grid estimation step may be followed by one or more additional steps in which the grid estimates are "refined." Typically, the grid nodes in the immediate vicinity of each control point are recalculated, using both original control points and the initial estimates at the nearby grid nodes. This may result in a map surface that comes nearer to passing exactly through the control points than would be possible otherwise.

Gridding, or the calculation of the regular array of estimated values, involves three essential steps. First, the control points must be sorted according to their geographic coordinates. Second, from the sorted files, the control points surrounding a grid node to be estimated must be searched out. Third, the program must estimate the value of that grid node by some mathematical function of the values in these neighboring points. Sorting greatly affects the speed of operation, and hence the cost of using a contouring program. However, it has no effect on the accuracy of the estimates, so we will not consider it further. Watson (1992) provides a detailed discussion of sorting algorithms in his Section 2.1. Both the search procedure and the mathematical function do have significant effects on the form of the final map.

The most obvious function that could be used to estimate the value of a surface at a specific location on a map is the simple calculation of an average of the known values of the surface at the closest control points. In effect, all of these surrounding known values are projected horizontally to the location to be estimated (**Fig. 5–74 a**). Then, a composite estimate can be made by averaging the values, usually weighting the closest points more heavily than distant points. If this is done on a regular grid over the entire map area, the resulting map has certain characteristics. The highest and lowest areas on the surface contain control points, and most interpolated grid nodes lie at intervening values, since an average cannot be outside the range of the values from which it was calculated. At grid nodes beyond the outermost control points, the estimates are extrapolations and will be close to the average value of the nearest control points.

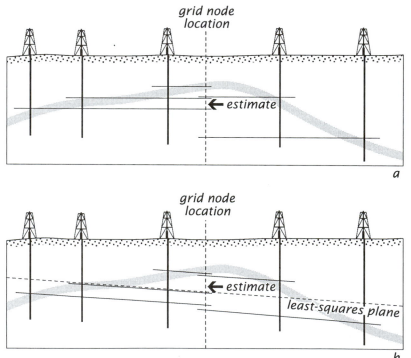

Figure 5–74. (*a*) Cross-section showing control points (wells) where value of surface elevation of formation top (indicated by shading) is known. Estimate at grid node (dashed line) is made by projecting values horizontally, then averaging. (*b*) Cross-section in which estimate at grid node is made by fitting least-squares plane to known values, then evaluating equation of plane at grid-node location.

An original data point, i (where i is an arbitrary integer from 1 to n, perhaps representing position in a data file or order of measurement), has coordinates X_{1i} in the east-west direction, X_{2i} in the north-south direction, and elevation Y_i. In

Figure 5 73 a, we have superimposed a regular grid of nodes on the posting of the sample points. We can imagine that the grid nodes also are numbered sequentially, from 1 to k. Grid node k has coordinates X_{1k} and X_{2k} and an estimated elevation of $\hat{Y}_k$. We are going to estimate $\hat{Y}_k$ from the nearest m data points; therefore, we must be able to search out the points nearest to each grid node and calculate the distances from these points to the nodes. The search procedure may be simple or elaborate; we will consider different alternatives later. Now we will assume we have by some method located the m nearest data points to grid node k. The distance, $D_{i,k}$, from observation point i to grid node k is found by the Pythagorean equation:

$$D_{ik} = \sqrt{(X_{1k} - X_{1i})^2 + (X_{2k} - X_{2i})^2} \tag{5.80}$$

Having found the distances D_{ik} to the m nearest data points, we now can estimate the grid node elevation, $\hat{Y}_k$, from these. The estimate is

$$\hat{Y}_k = \sum_{i=1}^{m} (Y_i/D_{ik}) \Big/ \sum_{i=1}^{m} (1/D_{ik}) \tag{5.81}$$

This type of algorithm is sometimes called a "moving average," because each node in the grid is estimated as the average of values at control points within a neighborhood that is "moved" from grid node to grid node. **Figure 5–75** is a contour map of the survey data in file NOTREDAM.TXT produced by a moving-average procedure.

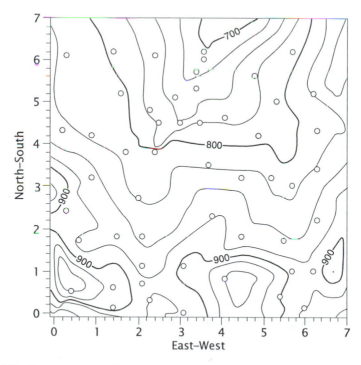

Figure 5–75. Contour map of topographic data produced by contouring program that uses a 61 × 61 regular grid estimated by a moving-average procedure for mathematical model of surface. Contour interval is 25 ft.

Such algorithms can be considered to be special cases of a more general set of procedures that involve the fitting of planes or curved surfaces to control points

383

within a neighborhood, as illustrated in **Figure 5–74 b**. First, all control points within a specified neighborhood around a grid node to be estimated are found. We can then imagine that the values at these points approximately define a sloping plane. This plane can be expressed as a first-degree trend surface, whose coefficients can be found by least squares. The coefficients of the plane are calculated in exactly the same manner as the coefficients of a trend surface, except, of course, only points within the neighborhood are used.

Once the equation of the plane has been found, values for X_{1k} and X_{2k} that correspond to the location of the grid node can be inserted and the equation evaluated. This will yield $\hat{Y}_k$, which is the estimate of the surface for that grid node. The process of fitting a plane and evaluating it to estimate the surface is repeated for every node in the grid. The plane represents a "general slope" of the surface around the grid node. In effect, the values of the surface at control points within the neighborhood are projected parallel to this sloping plane, and then averaged at the grid node (**Fig. 5–74 b**). If the fitted plane does not slope but rather is completely flat, an estimate made by this method will be the same as that found by the moving-average method (**Fig. 5–74 a**).

Gridding procedures that fit planes in this manner are sometimes called "piecewise linear least squares." A variant called "piecewise quadratic least squares" differs only by the fitting of a curved surface rather than a plane. A quadratic or second-degree trend surface is dome-, bowl-, or saddle-shaped and is defined by an equation that contains squares and cross products of the X_1- and X_2-coordinates.

Since these algorithms consider the slope of the surface in a neighborhood, they may perform better than simple moving-average methods when used for interpolation between control points. Grid values can be estimated that are either higher or lower than the range of values at the control points, so the uppermost and lowermost areas of the surface need not correspond exactly with the known values. However, beyond the limits of control, extrapolation may create extreme values that are completely unwarranted. This occurs because any slopes that exist near the margins of the controlled part of the map are continued without constraint beyond the data. Use of a fitted quadratic surface may make this problem even worse.

A somewhat more complicated type of algorithm computes the dip or slope (sometimes called the "local gradient") of the mapped surface within a neighborhood around each control point. This local slope is estimated using the same linear least-squares algorithm that is used to fit a plane to control points around a grid node. That is, it uses a least-squares method, and determines the coefficients of a plane that comes as close as possible to the values of the surface at the surrounding control points.

This algorithm must proceed in two phases. First, a neighborhood is defined around a control point and all the other control points within the neighborhood are found. Then a plane is fitted to the known values of the surface at these points using least squares. However, the plane is constrained so that it must pass exactly through the value at the central control point. The coefficients of this plane, which define the slope of the surface at the central point, are stored along with the Y-value for that point. In the second phase, a neighborhood is defined around each grid node to be estimated. The control points within this neighborhood are found, and the equations for the planes at each of the control points are evaluated for the location of the grid node. The different estimates from the planes are then

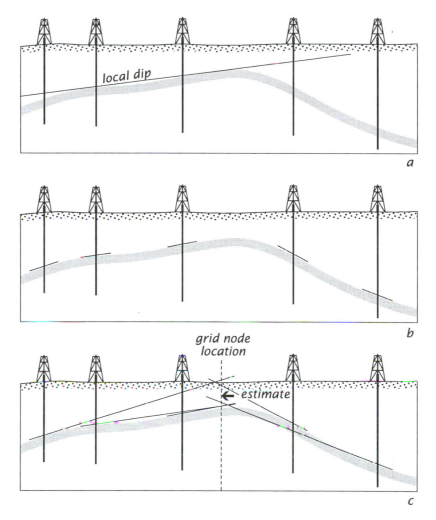

Figure 5–76. (*a*) Cross-section showing control points (wells) where value of surface elevation of formation top (indicated by shading) is known. Dip of surface at well is estimated by fitting least-squares plane to values in nearby wells. (*b*) Surface dips estimated at all nearby wells. (*c*) Dips projected to grid-node location (dashed line) and averaged to form estimate.

weighted and combined. In effect, the slopes of the surface at the control points are projected to the grid node where they are averaged (**Fig. 5–76**).

Variants of this algorithm, sometimes referred to as "linear projection," are among the most popular of those used in commercial contouring packages. Some programs incorporate modifications of this procedure in which quadratic surfaces rather than planes are fitted to the control points. These algorithms are especially good within areas that are densely controlled by uniformly spaced data points. Like the piecewise linear least-squares methods, they have the distressing habit of creating extreme projections when used to estimate grid nodes beyond the geographic limits of the data. An example of a map produced by linear projection based on the survey data in file NOTREDAM.TXT is shown in **Figure 5–77**.

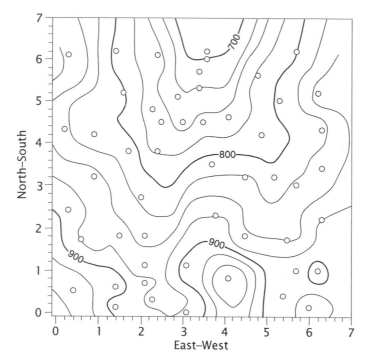

Figure 5–77. Contour map of topographic data produced by contouring program that uses a 61 × 61 regular grid estimated by linear projection of dips for mathematical model of surface. Contour interval is 25 ft.

The control points used to estimate a grid node, whether they are projected or not, ordinarily are weighted. The weightings vary according to the distances between the grid node being estimated and the control points. **Figure 5–78** shows graphs for a number of commonly used weighting functions. Most contouring programs allow the user to select from among a variety of such functions.

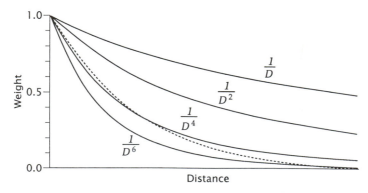

Figure 5–78. Distance weighting functions used in contouring programs. Dotted line indicates scaled inverse distance-squared weighting.

The weights that are assigned to the control points according to the weighting function are adjusted to sum to 1.0. Therefore, the weighting function actually assigns proportional weights and expresses the relative influence of each control point. A widely used version of the weighting process assigns a function whose

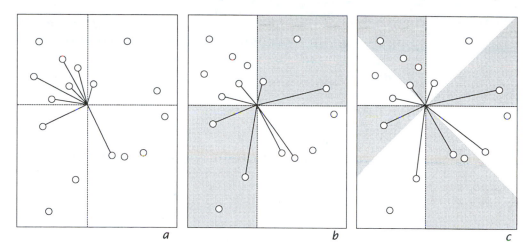

Figure 5–79. (*a*) Search technique that locates n nearest neighbors around grid node being estimated. No constraints are placed on radial distribution of control points. (*b*) Quadrant search pattern used to insure equitable distribution of control points around grid node being estimated. (*c*) Octant constraint on search pattern around grid node being estimated.

exact form depends upon the distance from the location being estimated and the most distant point used in the estimation, or in one variant, over the distance to the outer limits of the neighborhood. The inverse distance-squared weighting function is then scaled so that it extends from one to zero over this distance. The process can be expressed in a single equation,

$$w_D = (1 - D/D_{\max})^2 \Big/ (D/D_{\max}) \tag{5.82}$$

Like the other weighting functions, the sum of the weights is set equal to 1.0. A graph of this function is shown as the dotted line in **Figure 5–78**.

The most obvious differences between various contouring programs are in the search methods employed. These are the algorithms used to select the data points within the local neighborhood around the grid location to be estimated. The simplest selection technique is called a nearest-neighbor search (**Fig. 5–79 a**), which locates some specified number of control points or well locations that are closest to the grid node being estimated. A set of possible nearest-neighbor control points are selected from the complete data collection by sorting on the X_1- and X_2-coordinates of the points. The Euclidean distances from the grid node to each of these points are then calculated, and a specified number of the closest points are found.

An objection to a simple nearest-neighbor search is that it may find that all nearby points lie in a narrow wedge on one side of the grid node that is to be estimated. The resulting estimate is essentially unconstrained, except in one direction. This may be avoided by restricting the search in some way that ensures that the control points are equitably distributed about the location to be estimated. **Figure 5–79 b** illustrates one mode of radial constraint called a *quadrant search*. Some minimum number of control points must be taken from each of the four quadrants around the grid node being calculated. An elaboration on the quadrant search is an *octant search* (**Fig. 5–79 c**), which introduces a further constraint on the radial distribution of the points used in the estimating equation. A specified number of control points must be found in each of the octants surrounding the grid node being estimated. This search method is one of the more elegant procedures currently employed and is widely used in commercial programs.

The unconformity at the top of the Mississippian Lamont sandstone in an area of Ohio is shown in **Figure 5–80 a** as a contour map based only on elevations estimated from seismic return times. The map has been constructed using a subset of the data from Jones, Hamilton, and Johnson (1986) given in file LAMONT.TXT. Because seismic shot points are closely spaced along widely separated traverses, this map, made using a nearest-neighbor search, shows characteristic sudden changes in slope (shaded area) wherever the search algorithm switches from points found primarily on one traverse to points found on an adjacent traverse. Abrupt transitions are not apparent in **Figure 5–80 b**, which shows the same data mapped using an octant search pattern.

Any constraints on the search for nearest control points, such as a quadrant or octant requirement, will obviously expand the size of the neighborhood around the grid node being estimated. This is because some nearby control points are likely to be passed over in favor of more distant points in order to satisfy the requirement that only a few points may be taken from a single sector. Unfortunately the autocorrelation of a typical geological surface decreases with increasing distance, so these more remote control points are less closely related to the location being estimated. This means the estimate may be poorer than if a simple nearest-neighbor search procedure were used.

A matter of major concern to those who use and evaluate contouring programs is how well the resulting contours "honor the data points." That is, how closely does the mathematical model accord with the control points that were used in its construction? Since the contour lines are drawn based on relationships between the grid nodes rather than between the control points themselves, it is possible for a contour line to be drawn correctly with respect to the grid values but incorrectly with respect to the original data points. This is especially apt to happen if the grid is relatively coarse. Usually the errors are very small and are easily overlooked in areas where the surface is complex in shape or where slopes on the surface are steep so that the contour lines are closely spaced. In areas of very gentle slope, however, a small discrepancy between the grid node values and a control point may be sufficient to displace a contour line some distance from the point that it is supposed to honor. This may result in a conspicuous instance of a contour line passing by a control point on the wrong side. This problem, of course, does not occur with triangulation procedures because the model of the surface is formed by the control points themselves.

Appearance of a contour map is not a reliable guide to how well the underlying mathematical model represents the original control points, and certainly is not a guide to how well the model might predict values of the surface at unsampled locations. There is no formal statistical theory that allows us to predict, on theoretical grounds alone, which contouring procedure might be superior. In any given situation, the performance of a particular algorithm is determined by the complexity of the surface being mapped, the density and arrangement of the control points, the size of the grid and, of course, the algorithm itself. Empirical tests of how well various gridding algorithms perform using typical subsurface data have been published by Davis (1976).

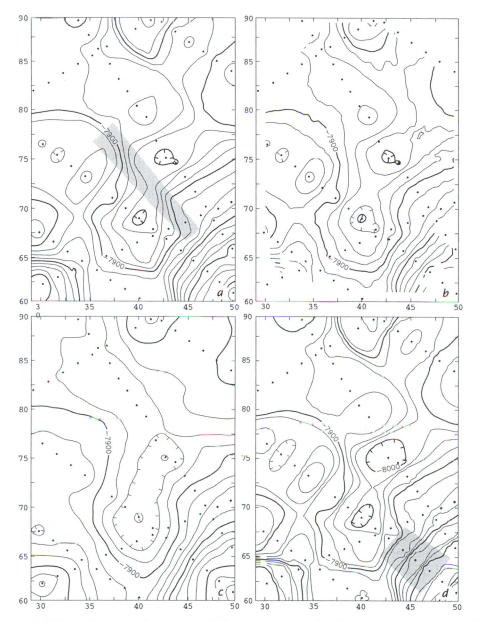

Figure 5–80. Influence of search technique and weighting function on contours when control points are poorly located. Data (solid dots) are elevations (ft subsea) of top of Mississippian Lamont sandstone in an area of Ohio, estimated from seismic return times. Contour interval is 250 ft. (*a*) Grid estimates based on nearest-neighbor search with scaled $1/D^2$ weighting function. Compare contours in shaded area with (*b*) grid estimates based on octant search and same weighting function. (*c*) Grid estimates based on $1/D$ weighting function and nearest neighbor search. (*d*) Grid estimates based on $1/D^6$ weighting function and nearest neighbor search. Compare contours in shaded area with equivalent area in (*c*).

Representing known data as precisely as possible is only one objective of contour mapping. Computer scientists concerned with the display of images tend to emphasize this aspect of contouring and the construction of the underlying mathematical model (Watson, 1992). Geologists, particularly those doing petroleum exploration, tend to see computer contouring as a graphical display tool for geological modeling. They are especially concerned with techniques for manipulating the data and the data model to support their interpretations. The books by Jones, Hamilton, and Johnson (1986) and Hamilton and Jones (1992) reflect this viewpoint. A more statistical approach is to regard contour mapping as spatial prediction or a way of estimating, with the smallest possible error, values of the surface at locations where measurements have not yet been made (Cressie, 1993). Later, we will examine kriging, a geostatistical estimation procedure whose performance can be evaluated on theoretical grounds. The ability of the many empirical contouring algorithms to produce accurate estimates at locations where no control exists can only be checked by cross validation in which a small proportion of the available control points are removed from the data set prior to mapping. Comparisons are made between the true values at these "blind" locations and the estimates made by the mapping program. Then, the omitted points can be returned to the data set, another random set selected for removal, and the process repeated over and over (Davis, 1976).

The distressing (although not surprising) conclusion from cross-validation studies of empirical contouring techniques is that the various objectives we might wish to fulfill with a contouring procedure may not be mutually attainable. In order to faithfully reproduce or honor the original control points with a gridding-type algorithm, it is necessary to utilize a weighting function that drops off extremely rapidly with distance and uses only a few nearest neighbors. Such an algorithm produces poor predictions or estimates in areas where control is limited. To make the best prediction of values of a surface at unsampled locations, we must use many control points in the calculation of each grid node, and weight distant points relatively heavily. Unfortunately, the smooth, generalized surface created in this manner performs poorly in terms of honoring the original control points. The effect of different weighting functions is shown in **Figure 5–80 c**, in which inverse distance ($1/D$) weighting is used, and in **Figure 5–80 d**, which is made using $1/D^6$ weighting. Compare the contour lines in the shaded area, where low weighting produces smooth gradients but pronounced local weighting results in an unrealistic "stair-step" pattern that closely conforms to values at the seismic shot points. Some commercial contouring programs attempt to circumvent this impasse by creating an initial coarse grid with an algorithm designed to achieve good predictions in areas of low control, then regrid on a finer mesh in the immediate vicinity of control points (Watson, 1992). The resulting "blended" surface does at least partially satisfy some objectives of contouring. However, the surface may be marked by peculiar small features that surround each control point and occur nowhere else on the map.

We have presented several examples of contour maps produced by different algorithms applied to the same topographic data (file NOTREDAM.TXT). **Figure 5–67** shows a manually drawn contour map, **Figure 5–72** shows a contour map produced from an irregular triangular network, and **Figures 5–75** and **5–77** show maps in which an intermediate grid of 61 rows and 61 columns has been produced. Simple weighted projection of nearby control points is used for the first of these; the second uses weighted projections of the slopes calculated around each data point to estimate the grid nodes. All of these maps are roughly similar, except that the manually contoured map incorporates information about the effects of streams on topography, additional information not available to the contouring programs. Ripley (1981) provides several additional examples of these same data mapped by other procedures.

Problems in contour mapping

Since constructing a contour map by computer is an *ad hoc* process, it is not surprising that it may not be entirely successful in all circumstances. Problems commonly arise when the assumptions about the surface being mapped are violated, particularly the assumption that the surface is continuous everywhere without breaks or discontinuities.

Edge effects are unrealistic gradients projected at the edges of a map. A mapped surface is unconstrained beyond the outermost control points, so gradients near the limits of control may be extrapolated in an unrealistic manner. Edge effects are most pronounced for global functions such as trend surfaces, may be severe for gridding algorithms that use dip projection, and are less troublesome for algorithms based on weighted averaging. However, all contour maps are subject to edge effects to one degree or another.

Figure 5–81 shows a subsurface structural contour map of the Wilburton gas field of southeastern Oklahoma. Contours, in feet subsea, are drawn on the top of the Spiro Sand (basal Atokan) of Pennsylvanian age, which is the producing interval. The field is structurally complex, formed of a succession of folded thrust sheets related to the Ouachita Mountains. The map in **Figure 5–81 a** is based on data from 86 gas-producing wells within the Wilburton field. Although the map is shown as a rectangular area, the contoured surface actually extends only a short distance beyond the convex hull defined by the outermost wells, as indicated by the shaded band. Edge effects exist within this band, whose width depends upon the grid spacing, the variability of the surface and density of control points near the edges, and the contouring algorithm used.

Edge effects cannot be eliminated but can be pushed beyond the area of interest by inserting additional control points in a "guard region" outside the map proper. This has been done in **Figure 5–81 b** in which the 86 producing wells have been supplemented with structural information from 16 holes drilled beyond the margin of the field. Edge effects have now been moved out to the lighter shaded band so that most areas of the field are no longer subject to uncontrolled extrapolations. It should be noted that edge effects are not confined to the outer edges of a map, but occur around the margins of any large map area that is devoid of control points.

The *zero isopach* problem arises when the quantity zero (0.0) is used to indicate the absence of a property at a location, and not as a value on a continuous scale that extends to negative values. On an isopach map, zero means that the stratigraphic unit whose thickness is being mapped is missing, because of either erosion or nondeposition. It is not logically possible for a stratigraphic unit to have "negative thickness," so zero values actually indicate that the surface being mapped is discontinuous and not present everywhere within the map area. Unfortunately, unless a contouring algorithm has been specifically written to avoid this problem, a program will assume that all values at control points are valid measurements and will attempt to draw zero contour lines through every control point whose value is zero. If the program uses distance-weighted averaging, the surface will be 0.0 only exactly at the control points and will have small values elsewhere, as in the map of

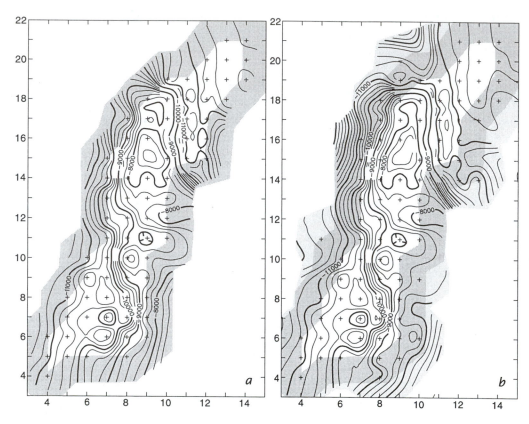

Figure 5–81. Edge effects in contour maps of top of Pennsylvanian Spiro Sand in Wilburton gas field, Oklahoma (ft subsea). Contour interval is 250 ft. (*a*) Map based on 86 producing wells in field. Extent of edge effects indicated by shading. (*b*) Map after addition of 16 control points in "guard region" around field. Edge effects have been moved out to area of lighter shading beyond field margin.

initial saturated thickness in a small oil reservoir in western Kansas shown in **Figure 5–82 a**. This is because the grid estimates are averages of control points whose values are zero (they are outside the reservoir and hence have no oil saturation) and other control points within the reservoir which have some actual oil-saturated thickness.

A more dramatic manifestation of the zero isopach problem appears when using a dip-projection gridding algorithm (**Fig. 5–82 b**). The contouring program will calculate the dips of the surface between interior points in the reservoir and wells outside the field margin. The projected dips will reach a value of 0.0 exactly at the wells immediately outside the field and will continue downward, creating increasing estimates of "negative saturated thicknesses." If additional control points having zero values exist beyond the points that define the field's margin, the projected surface will be forced to bend upward in order to pass through these points and outlying "bumps" of positive oil saturated thickness will be created. An example can be seen along the northwest edge of the field shown in **Figure 5–82 b**.

One *ad hoc* procedure that can be used to produce maps having a more acceptable appearance is to estimate dips near the edges of the field, extrapolate these slopes to the locations of the innermost control points that have been assigned

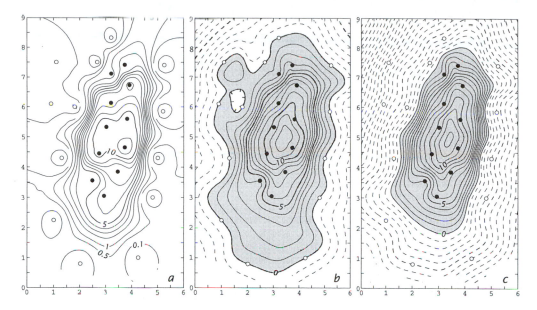

Figure 5–82. Zero isopach problem shown in map of initial saturated thickness in western Kansas oil field. Contour interval is 1 ft. (*a*) Contour map made using weighted-averaging algorithm with contours added at 0.5 and 0.1 ft. Zero contour line does not appear on map. (*b*) Contour map made using dip-projection algorithm. Zero contour line goes through dry holes on field margin and contours extend to negative values (dashed). (*c*) Zero values in dry holes replaced with negative numbers representing extrapolated slopes. Zero contour line lies between dry holes and producing wells.

zero values, and replace the zeros with "pseudovalues" of negative thicknesses estimated by the slope projections. The zero isopach will then fall inside the innermost wells where the saturated thickness is known to be zero (**Fig. 5–82 c**). A less arbitrary approach locates the zero isopach by a probabilistic assessment of whether a location is inside or outside the field boundary (Pawlowsky, Olea, and Davis, 1993).

Claims are sometimes made that TIN contouring algorithms are not affected by the zero isopach problem, but this is not true unless the program distinguishes between interval or ratio measurements which actually are 0.0 and nominal zeros that indicate a property is missing.

A third common problem contouring programs may encounter is proper mapping of *faulted* surfaces, that is, surfaces containing discontinuities. Without additional information, it is impossible for a contouring algorithm to distinguish between a steep slope and an abrupt discontinuity between two planes of a surface. **Figure 5–83** shows a contour map of subsurface structural configuration in a small, structurally controlled oil prospect. An abrupt change in elevation of about 80 to 100 ft is aligned along a trace S 20° W from a point whose coordinates are E 17.0, N 11.5 on the northern edge of the map. In **Figure 5–83 a** this structural change is interpreted as an area of unusually steep slope. In **Figure 5–83 b** it is interpreted as a fault whose western side is downthrown relative to the eastern side. Data for this illustration were adapted from Banks and Sukkar (1992) and are listed in file BANKSAND.TXT.

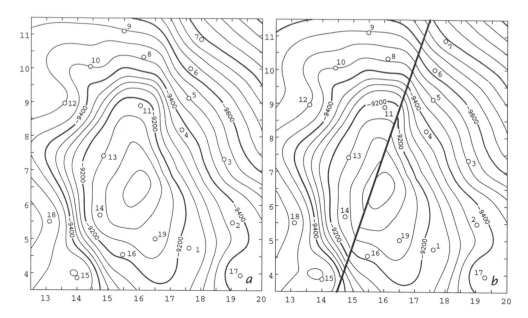

Figure 5–83. Effect of fault on structure contour map of small oil prospect. Data in feet subsea, contour interval 50 ft. (*a*) Map produced without assuming fault. (*b*) Map produced when fault with approximately 80-ft throw is assumed.

If a fault is inserted into a map, a contouring algorithm simply treats the fault line as a boundary, and does not search for control points across the fault. Therefore, contour lines on one side of the fault are drawn independently of the contour lines on the other side. If the fault dies out within the area of the map, the contouring algorithm must search for control points around the end of the map, and the discontinuous surface should blend into a single surface around the end of the fault. Special provisions must be made to draw the resulting contour lines correctly.

TIN algorithms have advantages for the contouring of faulted surfaces because a fault trace defines the edges of triangles that bound the fault on both sides. Special provisions must be made so the triangular plates do not join smoothly across the fault line, and contour lines are not drawn from one plate across the fault to the adjacent plate.

More elaborate approaches first produce a model of the throw of faults within a map area and use this to restore observations of the surface to their original, unfaulted values. These restored values are contoured, then vertical offsets representing the faults are reintroduced. In this way, trends in one area of the map are continued across faults. If the data represent sedimentary properties or stratigraphic thicknesses, this complicated procedure may result in more realistic maps because the trends existed before faulting. However, in the mapping of subsurface structure, this approach presumes that structural deformation preceded and was independent of faulting.

Extensions of contour mapping

One reason gridding-type contour algorithms have proved very popular is because the regular mathematical models or grids they create can be easily transformed in useful ways. Two or more different surfaces can be combined or compared even though they do not have a common set of control points, provided the surfaces are represented by grids having the same numbers of rows and columns. *Grid-to-grid operations* include addition (sometimes called "stacking"), subtraction (sometimes called "isopaching"), and multiplication or division.

A more elaborate form of surface modification is *filtering*. A small numerical matrix having an odd number of rows and columns (the "filter") is progressively centered over nodes of a map grid. The corresponding elements in the grid and filter are multiplied and their products summed. The sum becomes the value in a new grid at a location that corresponds to the center of the filter. The filter is then moved by one row or one column to a new position in the grid and the process repeated. By moving the filter back and forth until it has covered the entire grid matrix, a new grid is created that is a weighted average of the values in the original grid. Typically, the elements of the small matrix (often referred to as "filter weights") are chosen so they sum to 1.0. Then, the average value of the filtered map grid will be the same as the average of elements in the original grid.

The most common types of filters applied to map grids accomplish *smoothing* by suppressing small-scale variation in the surface, or *sharpening*, by accentuating small-scale variation. A smoothing filter has weights that are nearly equal, so the central value in the filtered grid is an average of the values in several surrounding rows and columns of the grid matrix. **Figure 5–84 a** is a structural contour map of the top of the Pennsylvanian Lansing–Kansas City Limestone in Graham County, Kansas. A 5×5 smoothing filter has been applied to this map, producing **Figure 5–84 b**. A sharpening filter consists of a large positive central weight surrounded by small negative weights; it has the effect of accentuating contrasts between adjacent elements in a grid. The map in **Figure 5–84 c** was processed using a sharpening filter.

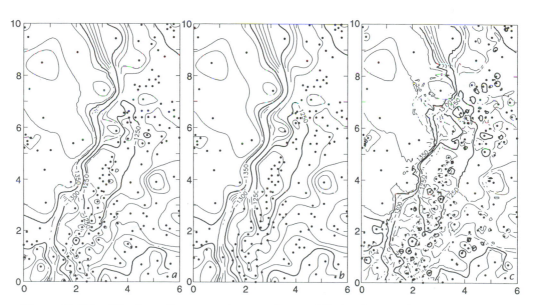

Figure 5–84. Filtering of structural map of top of Pennsylvanian Lansing–Kansas City limestone in area of northwestern Kansas. Data in feet subsea, contour interval 10 ft. (*a*) Unfiltered contour map. (*b*) Map after application of 5×5 smoothing filter. (*c*) Map after application of 5×5 sharpening filter.

Usually, filter weights are symmetrical around the central value, so the filtering is nondirectional. By increasing or decreasing the filter weights along a selected orientation, features in the map that share this orientation can be emphasized or suppressed. Other adjustments in the size of a filter and the relative magnitudes of the filter weights can emphasize features of specified size. A simple introduction to such spatial-domain filtering is given by Zurflueh (1967).

The *derivative* of a surface gives the rate of change in value of the surface, or its slope. The derivative of a surface can be shown by contour lines, but these lines connect locations having equal slopes or changes in value rather than equal values of the surface. The derivative can be taken with respect to an arbitrary direction as $\tan \phi = \partial Y / \partial d$, where Y is the mapped variable and d is distance in the arbitrary direction θ. A directional derivative is approximated at a grid node by extending imaginary opposite lines parallel to direction θ for distances $d = D/4$, where D is the spacing between rows and columns in the grid. Values of the surface at the endpoints of the lines, $\hat{Y}_1$ and $\hat{Y}_2$, are estimated by double linear interpolation from the nodes which define the grid cells that contain the imaginary lines. The derivative in the direction θ for grid element $\hat{Y}_{r,c}$ is computed by

$$\hat{Y}_{r,c} = \tan \phi \approx \left(\hat{Y}_2 - \hat{Y}_1 \right) \big/ 2d \qquad (5.83)$$

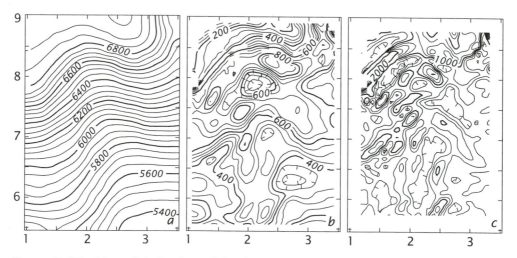

Figure 5–85. Map of derivatives of formline structural map of area in Raton Basin, Colorado. Contours in arbitrary units of relative elevation; contour interval 50 units. (*a*) Structure formline map. (*b*) Generalized first-derivative map showing maximum slopes. (*c*) Generalized second-derivative map showing areas of maximum curvature.

A *generalized derivative* can be calculated which is a measure of the maximum rate of change at a point on a surface, regardless of the direction of slope of the surface. Double linear interpolation is used to find the highest and lowest points a distance d (determined in the same manner as above) from the grid node being evaluated. The difference between these two points is then used to calculate the maximum slope at the grid node. The resulting contour map expresses the maximum rate of change in the surface. A second-derivative map can be calculated by taking the derivative of a derivative map. This will indicate areas where the change in the curvature of the surface is greater. An application is shown in **Figure 5–85 a,** which is a formline structural map based on a photogeologic interpretation of an assumed geologic horizon in the Raton Basin of southern Colorado. Formlines indicate the general shapes of structures but do not express true vertical elevations. Although this area contains potential hydrocarbon source beds of Cretaceous age, there are few suitable reservoir rocks in the overlying strata. However, thick silicified sandstones and conglomerates of Tertiary age might be prospective if the rocks are highly fractured. **Figure 5–85 b** is a generalized derivative map made from the formline structural map and shows areas of maximum structural slope. **Figure 5–85 c** is a second-derivative map, calculated by taking the derivative of **Figure 5–85 b**, which indicates areas of maximum change in slope, corresponding to areas of maximum structural folding. Under the theory that maximum deformation corresponds to maximum development of fractures in brittle beds, this map can be used to define potential prospects.

Trend Surfaces

Trend analysis is the geology profession's name for a mathematical method of separating map data into two components—that of a regional nature, and that attributable to local fluctuations. This separation has been done intuitively or graphically by geologists for years. Petroleum geologists, for example, refer to "regional dip" or "basinal configuration" as opposed to "local structure." Petrologists may speak of the "regional grain" of a metamorphic terrain. Geophysicists have long been accustomed to the concept of "regional trends" and "local anomalies." All of these expressions imply a belief that any given observation is the outcome of two interacting geologic forces or sets of forces—that which shaped the region or general geologic setting, and that which caused small areas to deviate from the regional pattern. Obvious examples may be drawn from structural geology. The Tertiary basins of Wyoming are the result of major movement along faults that extend deep into the crust. Within basins, folded structures have developed as the result of such diverse local agents as gravity sliding, minor antithetic faulting, failure of incompetent beds in areas of high dip, and so forth. If we regard the geometric shape of a basin as a regional structure, these smaller structures represent local deviations.

What we consider to be either "regional" or "local" is largely subjective. It depends in part upon the size of the region being examined. If the buried Precambrian surface of the United States is under consideration, the basins and intervening mountain ranges of Wyoming will appear as anomalies or local deviations, as will the Black Hills, the Ozark Dome, the Michigan Basin, and other major features of the basement. Within an individual Wyoming basin, "regional" and "local" take on quite different meanings.

The availability of data exerts a very real influence on determining the nature of regional trends and local deviations. It is fruitless, for example, to search for meaningful local features whose suspected size approaches the spacing between sample points. Such features, whether they exist or not, simply cannot be detected. The relationship between the size of detectable features and the spacing of control points can be calculated for regular nets of points (Singer and Wickman, 1969; McCammon, 1977), but not for the less tractable situation of irregularly distributed points.

Even the purpose of the geologic examination exerts an influence on our concept of "regional" and "local" spatial components. In a South African gold mine, for example, the only "deviations" of interest may be those which exceed a certain arbitrary value determined by the economics of the operation. On the other hand, a petroleum exploration group reexamining an area may be searching for small structural anomalies, knowing that the larger structures in the region have been adequately tested. The larger features are then considered to be part of the "regional trend."

For purposes of illustration, we can consider the set of contoured map observations shown in **Figure 5–86 a**. The map represents the subsurface structural configuration of the top of the Lansing–Kansas City Group of Pennsylvanian age in an area of northwestern Kansas. The X_1 (easting) and X_2 (northing) map coordinates are given in miles from an arbitrary origin; Y is in feet subsea. Data are given in file GRAHAM.TXT at the Web sites. The subsurface structure may be separated into "regional" and "local" components in a variety of ways. Suppose we decide that the regional trend has the form of a dipping plane placed through the observations in some manner (**Fig. 5–86 b**). The data may then be separated into this major linear trend and three large anomalies, one positive (or above the trend) and two negative (**Fig. 5–86 c**). However, we may decide that a parabolic function would better represent the regional trend than a dipping plane. In that case, we might construct the trend shown in **Figure 5–86 d**. The parabolic trend differs greatly from the linear trend shown in **Figure 5–86 b**, and the distribution of deviations is consequently different (**Fig. 5–86 e**). We could postulate still more complex forms for the regional trend, such as the fifth-order polynomial function fitted to the data in **Figure 5–86 f**, resulting in smaller and smaller deviations (**Fig. 5–86 g**). Eventually, our trend and samples would coincide, at which point there would be no residuals. Then, of course, there would be no separation of the data into components, and the purpose of the exercise would be defeated.

An obvious question at this point is how can the data be objectively separated into two components if the definition of the components is entirely subjective? This may be done if, instead of a geologic definition of trend and deviation, we use an operational definition that specifies the way in which our data are to be treated. A trend may then be defined as a linear function of the geographic coordinates of a set of observations so constructed that the squared deviations of the observations from the trend are minimized. Let us examine the three parts of this definition carefully.

1. It is based on the ***geographic coordinates***. This means that an observation, whether it is the elevation of a horizon or the amount of gold in a vein, is considered to be in part a function of the geographic location of the observation.

2. The trend is a ***linear function***. That is, it has the form $\hat{y}_i = b_0 + b_1 x_i + b_2(f x_i)\ldots$, where the b's are coefficients and the x_i's are some function or combination of the geographic coordinates of the locations where observations have been made. The equation will yield values of $\hat{y}$ that are the trend components of the observations. Because the equation is linear, the terms of the equation are added together.

3. The specific linear function chosen for the trend must ***minimize the squared deviations*** from the trend. The appearance of squared deviations at this point may bring to mind some material from Chapter 2. The sum of the squared deviations from the mean defines the variance of the sample. If we substitute "line" or "plane" for mean in this sentence, we see that the trend can be regarded as a function having the smallest variance about it. You may also recall

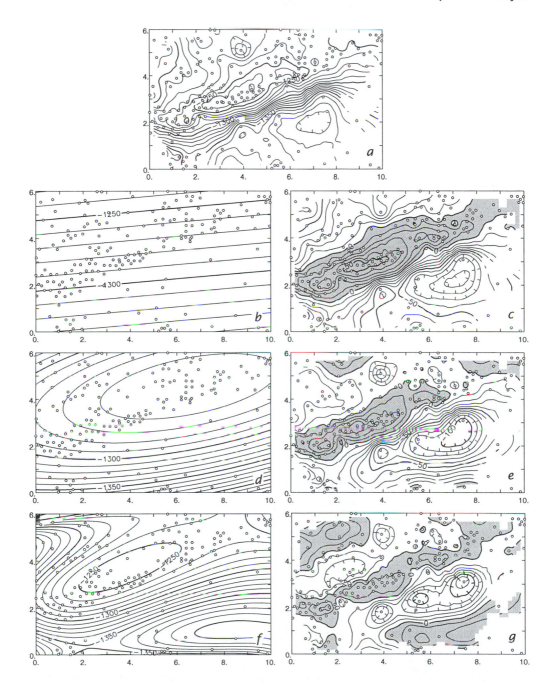

Figure 5–86. Concept of the trend illustrated by fitting trend surfaces to subsurface structural elevations of top of Pennsylvanian Lansing–Kansas City Group in part of Graham County, Kansas. Open dots are well locations. Contour interval is 10 ft. Horizontal scale in miles. (*a*) Subsurface structure contour map. (*b*) First-order trend. (*c*) Deviations from first-order trend. Positive residuals are shaded. (*d*) Second-order trend. (*e*) Deviations from second-order trend. (*f*) Fifth-order trend. (*g*) Deviations from fifth-order trend.

that we defined a line of regression in a similar fashion. Indeed, trend analysis is an adaptation of the statistical field of multiple regression, and the computational procedures of trend-surface analysis have been borrowed directly from statistical regression. In some instances, we can even use the powerful tests of hypotheses of multiple regression on geologic problems.

In Chapter 4 we computed a line of regression of Y on X, which was the line of best estimate of values of $\hat{y}_i$ for any specified value of x_i. The equation of the line, $\hat{y}_i = b_0 + b_1 x_i$, was found by solving a set of **normal equations**

$$\sum_{i=1}^n y_i = b_0 + b_1 \sum_{i=1}^n x_i$$
$$\sum_{i=1}^n x_i y_i = b_0 \sum_{i=1}^n x_i + b_1 \sum_{i=1}^n x_i^2$$

for the unknown coefficients b_0 and b_1. We can easily expand this set of equations to the case where there are two independent variables, such as mutually perpendicular geographic coordinates. A linear trend surface is an equation of this type:

$$\hat{y}_i = b_0 + b_1 x_{1i} + b_2 X_{2i} \tag{5.84}$$

That is, a geologic observation, y_i, may be approximated as a linear function consisting of some constant value b_0 related to the mean of the observations, plus a slope b_1 of the east–west coordinate x_1 and a slope b_2 of the north–south coordinate x_2. Because the equation contains three unknowns, we need three normal equations to find the solution:

$$\sum_{i=1}^n y_i = b_0 n + b_1 \sum_{i=1}^n x_{1i} + b_2 \sum_{i=1}^n x_{2i}$$
$$\sum_{i=1}^n x_{1i} y_i = b_0 \sum_{i=1}^n x_{1i} + b_1 \sum_{i=1}^n x_{1i}^2 + b_2 \sum_{i=1}^n x_{1i} x_{2i} \tag{5.85}$$
$$\sum_{i=1}^n x_{2i} y_i = b_0 \sum_{i=1}^n x_{2i} + b_1 \sum_{i=1}^n x_{1i} x_{2i} + b_2 \sum_{i=1}^n x_{2i}^2$$

Solving this series of simultaneous equations will give the coefficients of the best-fitting linear trend surface—"best-fitting" as defined by the least-squares criterion. The equations can be rewritten into matrix form as

$$\begin{bmatrix} n & \sum x_1 & \sum x_2 \\ \sum x_1 & \sum x_1^2 & \sum x_1 x_2 \\ \sum x_2 & \sum x_1 x_2 & \sum x_2^2 \end{bmatrix} \cdot \begin{bmatrix} b_0 \\ b_1 \\ b_2 \end{bmatrix} = \begin{bmatrix} \sum y \\ \sum x_1 y \\ \sum x_2 y \end{bmatrix} \tag{5.86}$$

The similarity between this matrix equation and Equation (4.43), on p. 209, should be obvious. Both may be regarded as curve fitting using two independent variables. Here, our two variables are X_1 and X_2, mutually perpendicular geographic coordinates. In fitting a quadratic curve to data along a line, the two variables were X and X^2. However, there is basically no difference between the two procedures. As an example of the fitting of a linear trend surface, we will consider the following problem.

The Anglo–Barren Oil Company (ABOC) has been given a concession in a remote part of northeastern Africa. The area is extremely inhospitable, nearly inaccessible, and almost completely unknown geologically. Under the terms of the concession, ABOC must drill ten wells during the year or forfeit their rights. The company management has decided to drill a series of widely spaced exploratory holes to develop the geologic background necessary for continued prospecting. Well locations

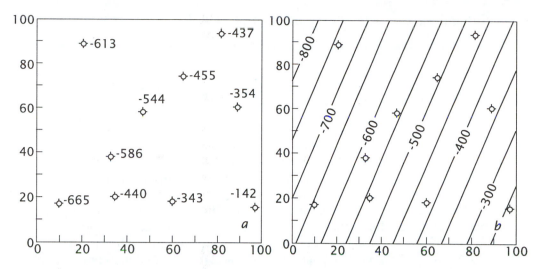

Figure 5–87. Anglo–Barren Oil Company concession in northeastern Africa. (*a*) Posting of elevations of basal Cretaceous contact in concession wells. (*b*) Linear trend fitted to elevations of basal Cretaceous contact. Contour interval is 50 m.

within the 100 km × 100 km concession are shown in **Figure 5–87 a** and are listed in **Table 5–9** and in file ABOC.TXT. Geographic coordinates are given in kilometers from the southwest corner of the concession, along with elevations (in meters subsea) of the bottom of the Cretaceous. We wish to determine the linear trend and, from an examination of residuals, pick areas that seem the most promising for additional exploration.

Table 5–9. Geographic coordinates, elevation of base of Cretaceous, trend-surface estimate of Cretaceous base, and residual $(Y - \hat{Y})$.

X_1 (km)	X_2 (km)	Y (m)	$\hat{Y}$ (m)	$(Y - \hat{Y})$ (m)
10.0	17.0	−665.0	−606.6	−58.3
21.0	89.0	−613.0	−695.7	82.7
33.0	38.0	−586.0	−537.8	−48.1
35.0	20.0	−440.0	−492.8	52.8
47.0	58.0	−544.0	−510.2	−33.7
60.0	18.0	−343.0	−369.2	26.2
65.0	74.0	−455.0	−455.5	0.5
82.0	93.0	−437.0	−411.5	−25.4
89.0	60.0	−345.0	−313.0	−40.9
97.0	15.0	−142.0	−186.1	44.1

We must first accumulate the sums, sums of powers, and sums of cross products required in Equation (5.86). The necessary entries are

$$\sum X_1 = 539 \qquad \sum x_1^2 = 36,934 \qquad \sum Y = -4,579$$
$$\sum X_2 = 482 \qquad \sum x_2^2 = 31,692 \qquad \sum X_1 Y = -211,098$$
$$\sum X_1 X_2 = 27,030 \qquad \sum X_2 Y = -232,337$$

401

Substituting these values into Equation (5.86) gives

$$
\begin{bmatrix}
10 & 539 & 482 \\
539 & 36{,}943 & 27{,}030 \\
482 & 27{,}030 & 31{,}692
\end{bmatrix}
\cdot
\begin{bmatrix}
b_0 \\
b_1 \\
b_2
\end{bmatrix}
=
\begin{bmatrix}
-4{,}579 \\
-211{,}098 \\
-232{,}337
\end{bmatrix}
$$

This matrix equation can be solved by methods described in Chapter 3. The solutions are

$$
b_0 = -621.0 \qquad b_1 = 4.8 \qquad b_2 = -2.0
$$

Having obtained the coefficients of the linear equation we can calculate the expected or trend values, $\hat{y}_i$, of the base of the Cretaceous at each of the ten wells. These, along with the deviations $y_i - \hat{y}_i$, are listed in **Table 5–9**. We can also calculate measures of how well the trend surface corresponds to the observations by using Equations (4.18)–(4.24), which we developed for the fitting of a line. In particular, we can compute the total variation as the sum of squares of the dependent variable, depth:

$$
SS_T = \sum_{i=1}^{n} y_i^2 - \frac{\left(\sum_{i=1}^{n} y\right)^2}{n} = 215{,}324.9
$$

For the estimated values of $\hat{y}_i$ in **Table 5–9**, we can calculate the sum of squares due to the trend or regression:

$$
SS_R = n \sum_{i=1}^{n} \hat{y}_i^2 - \frac{\left(\sum_{i=1}^{n} \hat{y}\right)^2}{n} = 193{,}861.4
$$

The difference between these gives the sum of squares due to residuals or deviations from the trend. (The change in notation from SS_E used in Chapter 4 to SS_D reflects the fact that deviations from trend surfaces may not be "errors" in the statistical sense.) The numerical value is

$$
SS_D = SS_T - SS_R = 21{,}463.5
$$

Now, the percent goodness of fit of the trend surface is

$$
100\% \cdot R^2 = SS_R / SS_T = 90.0\%
$$

The coefficient of multiple correlation is

$$
R = \sqrt{R^2} = 0.95
$$

From these extremely high values, we would conclude that the base of the Cretaceous in this area is an almost smooth, uniformly dipping plane. Deviations from the surface are relatively small, as a glance at **Table 5–9** will confirm. Apparently, the basal Cretaceous contact is a gently dipping surface of low relief, as shown in **Figure 5–87 b**. Although this simple analysis seems sufficient for this example, we have already discussed the possibility that a geologic trend may not be a plane, and may be extremely complex. Furthermore, we very rarely have any prior knowledge about what the functional form of the trend should be. Physicists can state that the path of a falling projectile should be a parabola because they know something of the controlling forces—that is, the acceleration of gravity and the conservation of momentum. Geologists can seldom speak with any authority about what form a geologic surface or distribution "should" take. Instead, they do the next best thing; they approximate the unknown function with one of arbitrary nature. In particular, they use a polynomial expansion of the linear trend surface, introducing powers and cross products of the geographic coordinates. Polynomials are extremely flexible and, if expanded to sufficiently high orders, can conform to very complex surfaces.

A note of caution should be injected at this point. Polynomial functions are used for geologic trend analysis merely as a matter of convenience. The equations that are necessary to find the coefficients of the trend may be established easily and solved by computer. Use of polynomials in no way intimates a belief that geologic processes are polynomial functions or even that they are linear. The nature of geologic processes, unknown and perhaps ultimately unknowable, can only be approximated by a polynomial expansion. Other approximations (or in rare cases, model equations) may be more appropriate in specific instances.

As we saw in Chapter 4, a least-squares line may be expanded to a second-order curve (parabola) by adding a squared term to the linear equation:

$$y_i = b_0 + b_1 x_i + b_2 x_i^2$$

An equivalent expansion gives a second-degree trend surface:

$$y_i = b_0 + b_1 x_{1i} + b_2 x_{2i} + b_3 x_{1i}^2 + b_4 x_{2i}^2 + b_5 x_{1i} x_{2i} \qquad (5.87)$$

Note that the equation contains terms which are the squares of the two geographic coordinates and a cross-product term, $x_1 x_2$. The expansion of this trend-surface equation to higher powers should be obvious. Each pair of geographic coordinates is simply raised to a higher power, creating two new variables, then the appropriate cross products of the two coordinates are calculated to give other new variables. For example,

$$y_i = b_0 + \overbrace{b_1 + b_2 x_{2i}}^{1} + \overbrace{b_3 x_{1i}^2 + b_4 x_{2i}^2 + b_5 x_{1i} x_{2i}}^{2}$$
$$\underbrace{+\ b_6 x_{1i}^3 + b_7 x_{2i}^3 + b_8 x_{1i}^2 x_{2i} + b_9 x_{1i} x_{2i}^2}_{3} \qquad (5.88)$$

is a third-degree trend surface. The first-degree coefficients are b_1 and b_2. The coefficients b_3, b_4, and b_5 are second-degree, because the variables in these terms are of the form $x_3 = x_1 \cdot x_1$, $x_4 = x_2 \cdot x_2$, and $x_5 = x_1 \cdot x_2$. That is, the variables are the products of multiplying the original variables together. Similarly, b_6, b_7, b_8, and b_9 are third-degree coefficients, as the variables in these terms result from the multiplication of the original variables together twice; that is, $x_6 = x_1 \cdot x_1 \cdot x_1$, $x_7 = x_2 \cdot x_2 \cdot x_2$, $x_8 = x_1 \cdot x_1 \cdot x_2$, and $x_9 = x_1 \cdot x_2 \cdot x_2$.

Petroleum explorationists have made good use of trend-surface analysis in the search for oil and gas in central Alberta, Canada. The Alberta Basin is extremely petroliferous, with production coming primarily from Lower Cretaceous sands and from much deeper carbonate reefs of Upper Devonian age. Many more wells penetrate the shallower Cretaceous horizons than the Devonian, especially near the Rocky Mountain front where reef prospects may be 15,000 ft or more deep. The Devonian reefs are thick accumulations of carbonates in the Leduc Formation, overlain and surrounded by shales that represent lagoonal and open marine clastic sediments. The carbonate reefs were rigid and incompressible, while the enclosing

403

fine-grained clastic sediments of the Ireton shale were compacted to a fraction of their original thickness as the basin subsided and deposition continued. This differential compaction created drape structures over the buried reefs; these structures persist in the overlying rocks, although their magnitudes become less on shallower horizons.

Deep-seated compaction features are not readily apparent on the Cretaceous horizons because of the overriding effect of the strong regional dip (**Fig. 5–88**).

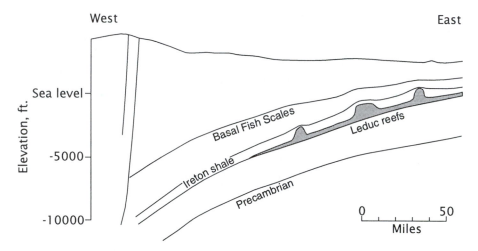

Figure 5–88. Diagrammatic cross-section across western margin of Alberta Basin, Canada, showing Cretaceous Basal Fish Scales marker unit and Upper Devonian Ireton shale and Leduc reefs.

Closure at depth is expressed only as slight changes in local gradient that appear on structural maps of Cretaceous horizons as subtle variations in the spacing of contour lines. However, if the strong regional dip component could be removed from these maps, presumably the underlying drape features would appear as closed structures. Since the density of wells that bottom in the Cretaceous is relatively great, their analysis could provide valuable information about possible prospects in the Devonian, even though few wells penetrate the deeper horizon.

The Basal Fish Scales is a black shale occurring near the boundary between the Lower and Upper Cretaceous. The unit derives its name from the abundant fish scales it contains. It is also characterized by numerous bentonite beds which produce conspicuous spikes on gamma-ray logs because of their high radioactivity. Bentonites are synchronous and so form excellent markers for regional correlation. The top of the Basal Fish Scales can be picked with exceptional consistency on log traces because of the pronounced response from a thick, persistent bentonite. The map shown in **Figure 5–89a** includes an area of about 3500 mi^2 in west-central Alberta, and was constructed using picks of this bentonite at the top of the Basal Fish Scales in 360 exploratory holes. The map area lies on the western margin of the Alberta Basin, immediately in front of the overthrust zone marking the Rocky Mountain front. Beds dip downward to the southwest, with increasing dip as they approach the western edge of the basin. Within the map area, depths to the Basal Fish Scales range from slightly in excess of 1000 ft below sea level in the northeast, to almost 5000 ft below sea level in the southwest.

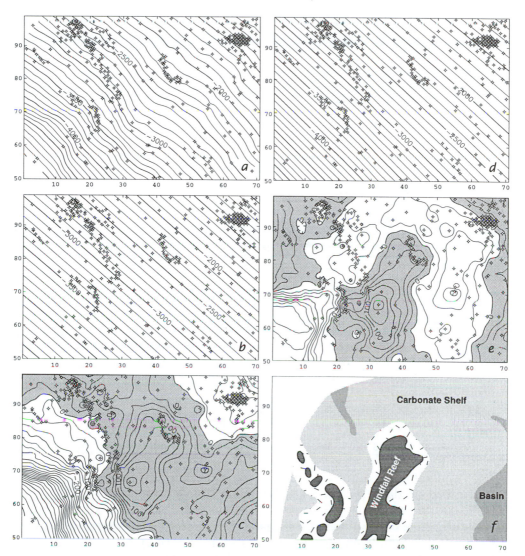

Figure 5–89. Subsurface maps of Cretaceous Basal Fish Scales in part of Alberta Basin, Canada. Contours in feet below sea level. Contour interval is 100 ft. Horizontal scale in miles. (*a*) Structure contour map. (*b*) First-order trend surface. (*c*) Residuals from first-order trend. Positive residuals are shaded. (*d*) Second-order trend surface. (*e*) Residuals from second-order trend. (*f*) Paleogeographic interpretation of Upper Devonian, showing reefs, carbonate platform, and deep basin areas.

Figures **5–89 b** and **d** are first- and second-degree polynomial trend surfaces of the Basal Fish Scales; statistics relating to the fit of the trend surfaces are given in **Table 5–10**. Both trend surfaces provide an extremely good fit to the observations, although the second degree is significantly better than the first. In this application, statistical tests of significance are not an appropriate guide for the selection of degree of trend surface because the problem is not one of statistical estimation. Rather, the objective is to simulate closely the regional features of the structure contour map so subtraction of the trend will remove the regional component of

Table 5–10. Statistics of first- and second-degree polynomial trend surfaces fitted to elevation of Basal Fish Scales Alberta.

First-Degree Trend Surface			Second-Degree Trend Surface		
% Goodness of Fit (R^2)	Correlation Coefficient (R)	Trend-Surface Equation	% Goodness of Fit (R^2)	Correlation Coefficient (R)	Trend-Surface Equation
98.6%	0.993	$Y = -6351.4$	99.7%	0.999	$Y = -7993.3$
		$+29.3\,X_1 + 33.8\,X_2$			$+63.4\,X_1 + 59.2 X_2$
					$-0.1\,X_1^2 - 0.1\,X_2^2$
					$-0.3\,X_1\,X_2$

structure. In effect the trend surface is used as a high-pass filter, removing the large-scale structural variation from the map and leaving small-scale features behind.

Figure 5–89 c shows residuals from the first-degree trend surface and **Figure)** 5–89 e shows residuals from the second-degree trend surface; positive deviations are shaded. **Figure 5–89 f** is a paleogeographic map representing the Upper Devonian, reconstructed from well and seismic information. Note the strong coincidence between positive deviations from the second-order trend fitted to the Basal Fish Scales and the locations of major Devonian reefs, particularly the Windfall Reef. Removal of the regional trend has successfully isolated minor components of structure representing drape over these reefs. Several major oil fields were discovered in this area using this exploration technique. Data used to create this example are given in file LEDUC.TXT.

Still unanswered is the question, "What is a high (or low) goodness of fit or correlation in a trend surface?" In geologic studies, we must often rely on experience and intuition to provide an answer. For example, in structural analyses performed on data from Kansas, Oklahoma, Texas, Wyoming, and California in the U.S., and from England, Canada, and other localities, we have judged first- and second-degree correlations of less than 0.3 as "poor fits." Correlations of 0.4–0.6 yield interpretable trend and residual maps, and those higher than 0.7 are regarded as conforming closely to the original data points. Again, the purpose of the investigation must be kept firmly in mind while interpreting goodness of fit. In all of these structural studies, we examined basins with relatively simple shapes and searched for relatively small residuals compared to the total size of the basins. Third- and fourth-degree polynomials usually provided very high fits (over 0.8). As a reference value, randomly generated data sets having the same range of values as actual data yielded correlations of up to 0.3 for fourth-degree polynomials. Trends that appear to be realistic can be extracted from random data, so geologists should proceed cautiously in the interpretation of trend and residual maps that have low goodness of fit.

Subsurface structure in part of Graham County, Kansas, is shown in the form of a contour map in **Figure 5–86 a** and as a block diagram in **Figure 5–90 a**. **Figure 5–90 b** through **Figure 5–90 h** show block diagrams of trend surfaces of successively greater polynomial orders that have been fitted to these same data.

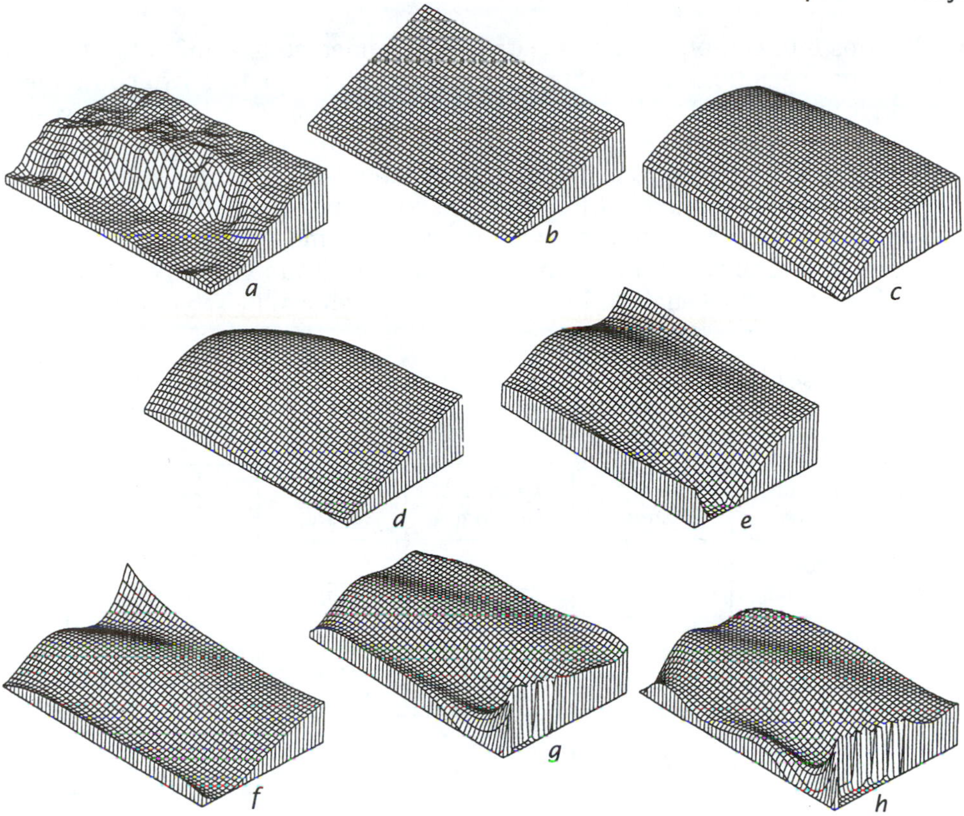

Figure 5–90. Block diagrams of trends fitted to subsurface structural elevations of top of Pennsylvanian Lansing–Kansas City Group in part of Graham County, Kansas. (*a*) Subsurface structure. (*b*) First-order trend. (*c*) Second-order trend. (*d*) Third-order trend. (*e*) Fourth-order trend. (*f*) Fifth-order trend. (*g*) Sixth-order trend. (*h*) Seventh-order trend.

Statistical tests of trends

The goodness of fit of a trend surface may be tested statistically by comparing the variance due to regression or trend to the variance due to deviations from the trend. You will recall from Chapter 2 that tests of equality of variances involve the F-distribution and are valid only if the data satisfy certain conditions. If these assumptions are justifiable, we may regard the b_k coefficients found by least squares to be estimates of the true population regression coefficients, β_k, and we may test hypotheses about their nature. We must assume that the population of the dependent variable is normally distributed about the regression, and that its variance does not change with changes in the independent variables. In addition, the samples must be drawn without bias from this population. The last condition is often the most difficult to satisfy, especially in a structural analysis based on oil wells, as the locations of the wells presumably were not chosen by random selection! Testing of statistical hypotheses about trend surfaces is perhaps most easily defended when dealing with data such as geochemical assays of samples collected according to a planned experimental design.

The significance of a trend or regression may be tested by performing an analysis of variance, which is the process of separating the total variation of a set of observations into components associated with defined sources of variation. This, of course, has been done by dividing the total variation of Y into two components, the trend (or regression) and the residuals (or deviations). The degrees of freedom associated with total variation in a trend analysis are $(n-1)$, where n is the number of observations. The degrees of freedom associated with the trend (or regression) are determined by the number of coefficients in the polynomial equation fitted to the data. Degrees of freedom for deviation are the number of degrees of freedom associated with total variation minus those that are accounted for by the trend, or $\nu_D = \nu_T - \nu_R$. A formal analysis of variance table is shown in **Table 5–11**.

Table 5–11. General ANOVA for significance of regression of kth-degree polynomial trend surface. Number of coefficients in trend-surface equation is m, number of data points is n.

Source of Variation	Sum of Squares	Degrees of Freedom	Mean Squares	F-Test
Polynomial Regression	SS_R	$m-1$	MS_R	MS_R / MS_D
Deviation from Polynomial	SS_D	$n-m$	MS_D	
Total Variation	SS_T	$n-1$		

The mean squares are found by dividing the various sums of squares by the appropriate degrees of freedom. Reducing sums of squares to mean squares converts them to estimates of variance that may be compared using an F-distribution. The MS_D is the variance about the trend surface; MS_R is the variance of the trend surface about its mean. If the regression is significant, the deviation about the trend will be small compared to the variance of the trend itself.

In a general test of a trend-surface equation, the ratio of interest is that between variance due to the trend and variance due to deviation from the trend. The F-test gives a probabilistic answer to the question of whether the variances being examined have been obtained by random sampling from the same population. Or, is the trend not significantly different from a random scatter of observations about their mean? An affirmative answer may be interpreted as meaning that (*a*) the distribution of Y is random and independent of values of $X_1, \ldots, X_m$ or (*b*) the distribution of Y may be in part a function of $X_1, \ldots, X_m$ but the wrong functional model has been fitted to the data.

In more formal terms, the F-test for significance of fit is a test of the hypothesis and alternative

$$H_0 : \beta_1 = \beta_2 = \ldots = \beta_m = 0$$
$$H_1 : \text{not all } \beta_k = 0$$

(5.89)

The hypothesis to be tested is that the partial regression (trend) coefficients are equal to zero, or in other words, there is no trend. If the computed value of F exceeds the table value of F, this hypothesis is rejected and the alternative, H_1, is accepted.

In polynomial trend-surface analysis, it is customary for some investigators to fit a series of equations of successively higher degrees to the data. In such an analysis, a succession of trend sums of squares will be produced, each larger than the preceding sum. The analysis of variance table may be expanded to analyze the contribution of the additional trend or partial regression coefficients and give a measure of the appropriateness of increasing the order of the equations. The test is developed by finding the difference in sums of squares due to the higher polynomial trend minus the sums of squares due to fitting the lower order trend. This difference is divided by the difference in degrees of freedom for the two trends, giving a mean square due to increasing the degree of the polynomial trend surface. This mean square is then divided by the mean square due to deviation from the higher order polynomial trend surface. If the resulting F-value is significant, the deleted order was contributing to the trend and should be retained. If the value is not significant, nothing has been gained by fitting the higher order polynomial. An ANOVA table for testing the significance of a higher order polynomial trend surface is given in **Table 5–12**.

Table 5–12. General ANOVA for the significance of increasing the degree of a polynomial trend surface from degree p to degree $(p + 1)$; polynomial equation of degree p has k coefficients; equation of degree $(p + 1)$ has m coefficients; number of observations is n.

Source of Variation	Sum of Squares	Degrees of Freedom	Mean Squares	F-Test
Regression of Degree $(p + 1)$	$SS_{R(p+1)}$	m	$MS_{R(p+1)}$	$\dfrac{MS_{R(p+1)}}{MS_{D(p+1)}}$ [a]
Deviation from Degree $(p + 1)$	$SS_{D(p+1)}$	$n - m$	$MS_{D(p+1)}$	
Regression of Degree p	$SS_{R(p)}$	k	$MS_{R(p)}$	$\dfrac{MS_{R(p)}}{MS_{D(p)}}$ [b]
Deviation from Degree p	$SS_{D(p)}$	$n - k$	$MS_{D(p)}$	
Regression due to increase from p to $(p + 1)$	$SS_{R(I)} =$ $SS_{R(p+1)} - SS_{R(p)}$	$m - k$	$MS_{R(I)}$	$\dfrac{MS_{R(I)}}{MS_{D(p+1)}}$ [c]
Total Variation	SS_T	$n - 1$		

[a] Test of significance of the $(p+1)$-degree trend surface.
[b] Test of significance of the p-degree trend surface.
[c] Test of significance of increase in fit of the $(p+1)$-degree trend over the p-degree trend.

The F-test for significance of added terms is a test of the hypothesis and alternative

$$H_0 : \beta_{k+1} = \beta_{k+2} = \ldots = \beta_m = 0$$
$$H_1 : \text{not all } \beta_{k+1} \ldots \beta_m = 0$$

(5.90)

The null hypothesis states that trend coefficients after the kth term are all equal to zero, or they do not contribute to the trend represented by the first through

kth coefficients. (Remember that the polynomial trend surface of degree p contains k coefficients, whereas the polynomial equation of the $(p + 1)$ trend contains m coefficients.) Again, if the computed value of F exceeds the table value at the specified level of significance, the hypothesis is rejected. The test procedure is given for an equivalent curvilinear model by Li (1964, p. 176–181).

In some instances we may be interested in assessing the effect of a single trend coefficient. This may be done simply by omitting that term from the polynomial equation and recomputing the sums of squares due to the regression or trend and that due to deviation from the trend. The contribution of the deleted term is the difference in the two sums of squares due to the trend. Significance of this term may be tested by computing the ratio between the mean square due to the deleted term and the mean square due to the complete regression. The F-ratio has one and $(n - 1)$ degrees of freedom. The ANOVA for the significance of a single deleted coefficient is given in **Table 5–13**.

Table 5–13. ANOVA for testing the significance of a single deleted coefficient; the complete polynomial regression equation contains m coefficients, not counting the b_0 term; the regression equation contains only $(m - 1)$ coefficients after deleting the kth coefficient; number of observations is n.

Source of Variation	Sum of Squares	Degrees of Freedom	Mean Squares	F-Test
Regression of All Terms	SS_R	m	MS_R	MS_R / MS_D [a]
Deviation	SS_D	$n - m - 1$	MS_D	
Regression omitting kth Term	SS_{R-1}	$m - 1$	MS_{R-1}	MS_{R-1} / MS_{D-1} [b]
Deviation	SS_{D-1}	$n - m - 2$	MS_{D-1}	
Regression of kth Term Only	$SS_{RK} = SS_R - SS_{R-1}$	1	MS_{RK}	MS_{RK} / MS_D [c]
Total Variation	SS_T	$n - 1$		

[a] Test of significance of the p-degree trend surface.
[b] Test of significance of the p-degree trend surface fitted without the kth coefficient.
[c] Test of significance of the kth coefficient alone.

Individual terms in the regression equation may also be tested by calculating the increase in SS_R when a new variable is added. However, this is not an advisable practice, because the tendency may be to regard all further terms as nonsignificant after encountering several successive nonsignificant coefficients. This may not be the case. In trend-surface analysis, the complete set of terms for the next higher order should be added, and then the added individual terms tested one by one by elimination. Incomplete sets of higher degree terms should not be added blindly, unless there are compelling reasons to do so. In one early example, because of computer limitations, third-order "hypersurfaces" were fitted to oil-gravity data with an equation that contained no cubic cross-product terms. The resulting regression was considerably different than that produced from the same data when they were rerun on a larger computer using a program that fitted a complete cubic equation. Additional terms are especially apt to be significant if the correlation coefficients of lower orders are small. The two data sets presented previously from files GRAHAM.TXT and LEDUC.TXT are representative of problems in structural trend-surface analysis. The purpose of both previous investigations was to seek areas where the structural surface departed from a polynomial trend. In those problems, the distribution of deviations was such that the appropriateness of statistical tests of the significance of the trend would be suspect. However, in the following example, experimental and sampling conditions seem adequate to justify application of regression tests.

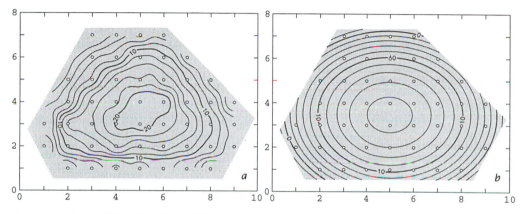

Figure 5–91. Polished section of Mexican sphalerite crystal showing locations of microprobe analyses for iron content. Contour interval is 2% Fe. (*a*) Contoured data. (*b*) Second-order polynomial trend surface.

Figure 5–91 shows a planar view of a single crystal of sphalerite collected from a mine in northern Mexico. Investigators are interested in the distribution of iron throughout the crystal, so it has been carefully cleaved through its center and the surface has been polished. The iron content at 10-Å spots spaced 1 mm apart has been determined by electron microprobe. The sample points are shown as dots on **Figure 5–91 a**, which also shows contours of the iron concentration in 2% intervals; the data are given in file SPHALRT.TXT. Although use of iron content in sphalerite as a temperature indicator has been criticized because of the possibility of nonequilibrium conditions at the time of crystallization, the researchers postulate that the growing shell of the crystal was in equilibrium with the ore solution at all times. Therefore, the average composition of the crystal may be inadequate as a temperature indicator, but the composition of successive thin shells of the crystal may define a temperature-change curve. Excluding the possibility of zoning, the simplest model for the distribution of iron in the crystal would be a gradual increase or decrease symmetrically outward from the center. The fitting of a second-degree polynomial trend surface of iron content on the coordinates of the sample points seems an appropriate way of testing this hypothesis; the resulting trend surface is shown as a contour map in **Figure 5–91 b**. The various sums of squares necessary for the ANOVA table for testing the significance of the polynomial regression are provided in **Table 5–14**. The fitted regression is highly significant.

Table 5–14. ANOVA for significance of regression of second-degree polynomial trend surface fitted to iron concentration in sphalerite crystal. Trend surface has five coefficients. There are 49 microprobe measurements in percent Fe.

Source of Variation	Sum of Squares	Degrees of Freedom	Mean Squares	*F*-Test
2nd-Degree Trend	1464.80	5	292.96	52.34[**]
Deviation from 2nd-Degree Trend	240.68	43	5.60	
Total Variation	1705.48	48		

[**] $F_{5,43;\,.99} = 3.50$.

Two trend-surface models

You may have noted that the preceding discussions imply that two basically different types of geologic problems are analyzed using trend-surface methods. On one hand, trend surfaces are fitted to structural data in an attempt to isolate "local structures." It has been demonstrated empirically that in a sedimentary basin these residuals may be associated with structurally or hydrodynamically trapped oil. Alternatively, trend surfaces have been used to define regional trends in petrographic and geochemical data. These two applications differ in objectives and in underlying assumptions, but their common methodology has obscured these differences.

Trend surfaces fitted to structural data can be represented by the model equation

$$Y_i = \beta_0 + \beta_1 X_{1i} + \beta_2 X_{2i} + \ldots + \beta_m X_{2i}^m + (\gamma_i + \varepsilon_i) \tag{5.91}$$

The equation states that a given observation (elevation of a formation top) is equal to the sum of a constant term related to the means of the geographic coordinates, plus a polynomial expansion of degree p of the geographic coordinates, plus a local component, γ, plus a randomly distributed measurement error, ε. The latter two terms are considered to be confounded, and are the parameters of interest. The trend surface itself is merely a means to the end of delineating areas of interesting deviations.

In contrast, a trend surface fitted to petrographic or similar data usually is represented by an ordinary response surface model:

$$Y_i = \beta_0 + \beta_1 X_{1i} + \beta_2 X_{2i} + \ldots + \beta_m X_{2i}^m + \varepsilon_i \tag{5.92}$$

The response surface model equation is similar in all respects to the trend-surface equation except that the local component γ is not considered to be present. Interest is centered upon the nature of the trend itself, as expressed in the estimates of the β's, or polynomial coefficients.

Petrographic and geochemical variates usually are characterized by high variance among replicates. This variance arises because of inhomogeneities within the analyzed material, local or small-scale variation in composition (on a scale larger than an individual sample but smaller than the distance between samples), and errors in instruments or analytical procedures. In a typical study these are confounded and produce a normal distribution of error about each observation. Although each source of variance could be isolated and measured by replication at several levels within the experiment, this may not be done because of economic or other factors.

A trend surface or regression on geographic variables can be fitted appropriately to data containing random errors if certain basic assumptions seem reasonable. These assumptions require that the deviations, ε_i, be randomly and normally distributed about the trend, that the deviations have zero mean, and that they have constant variance. This in turn means that the deviations are independent of one another. If these conditions are fulfilled, the regression may be tested for significance and inferences can be drawn about the trend. The appropriate statistical tests are presented in **Tables 5–11** to **5–13**. Many other special statistical designs have been created and are widely used in such fields as agriculture and chemical engineering; an introduction to these is given in Mendenhall (1968, chapter 10). Koch and Link (1980, p. 217-222) discuss a geologic application of one such design. The important aspect of such inferences is that they pertain to the trend. This is emphasized in **Figure 5–92a**, where the observed values of y_i can be seen to fall within the distribution of errors around the regression.

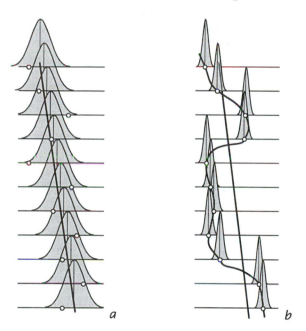

Figure 5–92. Distribution of error variance in two different trend-surface models. (*a*) Polynomial regression model in which observations are assumed to occur randomly within the error distribution about the model regression line. (*b*) "Structural residual" model in which deviations from trend are local features superimposed on a simpler regional shape modeled by a trend surface. Deviations in replicates of observations are centered about some mean value for each point rather than around the polynomial model.

In the fitting of trend surfaces to structural data, observations (typically elevations of a geologic horizon) are not replicated. Although it is possible to re-log a well, it usually is not possible to drill a second, nearby well which would constitute a replicate sample. Repeated downhole measurements may vary by a few feet depending upon depth and amount of stretch in the cable, but this source of experimental error is always one or more orders of magnitude smaller than deviations in trend-surface analysis. Lack of replicates means that local variation cannot be assessed. However, it may be rationalized that this source of error is also inconsequential, as the drill hole is not "sampling" from a population of surfaces. Only one formation top exists and the only variance is due to the negligible measurement error. Thus, all residual variance in trend-surface analysis can be ascribed to lack of fit.

In terms of the model equation, this is equivalent to saying that ε_i is negligibly small compared to y_i. Although the random error, ε_i, has zero mean and is independent for all values of y_i, it cannot be isolated because replicates have not been taken. The confounded term $(\varepsilon_i + y_i)$ also has a mean of zero, but it is not, in general, independent for values of y_i. In fact, the purpose of the analysis is to define regions of specified size in X_1 and X_2 over which $(\varepsilon_i + y_i)$ are correlated. **Figure 5–92 b** shows a theoretical distribution of ε_i about the structural surface. For the most part, deviations of the surface from the polynomial model do not reflect the magnitude of the error term but rather the local component, y_i.

The difference between these two model equations for trend surfaces is reflected in the way in which autocorrelation among residuals is regarded. In polynomial regression analysis, autocorrelation is considered to be a violation of model assumptions and to invalidate (or seriously weaken) inferences drawn from the analysis. Such an attitude may be appropriate for petrographic and geochemical data, because replication tends to reveal a normal distribution of error around relatively simple regressions. Although analyses for lack of fit may suggest more complex regressions, in general the error term is sufficiently large to account for all deviations.

In contrast, geologists applying trend-surface analysis to structural data are seeking areas of autocorrelated residuals. As stated before, almost all structural deviations are ascribable to lack of fit, and the presence of autocorrelated residuals indicates a region larger than the sampling interval where the surface deviates from the polynomial model in a consistent direction. Either large areas of autocorrelated residuals or single points of large deviations are of interest in petroleum exploration because they may indicate regions where local structures (y_i) are strongly developed. Since the deviations are not randomly distributed, standard tests of significance of the regression cannot be used. This generally is not regarded as a drawback, as the regression itself is only of incidental interest.

Pitfalls

At this point it may be appropriate to emphasize some factors which can adversely affect trend-surface analyses, or indeed, any type of map analysis. These cautionary statements have been repeated many times in the literature, but they seem to be ignored more times than not (Chayes, 1970). The effect of these factors can range from a slight distortion of the trend (a form of bias) to total invalidation of the results.

The first and most obvious point is that adequate control must be present. At an absolute minimum, the number of data points must exceed the number of coefficients in the polynomial equation or the results of the regression are invalid. If statistical tests are to be run, the number of control points determines the degrees of freedom and these must be sufficiently large so a meaningful F-test can be performed. If the degrees of freedom for deviation are small (a consequence of the number of polynomial coefficients approaching the number of data points), only extremely high correlations will be found significant. Furthermore, the power of the test (the probability of not committing a type II error) decreases drastically with small sample size. Of course, the number and spacing of control points has a direct influence on the size of local deviations that can be detected in trend-surface analysis of structural data, and relates to the resolution at which local structures can be defined.

Ordinarily, we do not consider control points beyond the boundaries of our map. Often the map area may even extend slightly beyond the actual lateral extent of the data points, there being few (or no) control points on the map boundaries. In such circumstances, there are almost no constraints on the form of the trend surface near the edges of the map. Whatever slope exists in the region of control is extrapolated without limit along the map boundaries. This creates the "edge effects" that we considered briefly in the section on contour mapping. If a high-order trend is being fitted to data, extrapolated values near the edges of the map may reach astronomical proportions. Minor edge effects will exist even if the entire map area is uniformly covered with control points up to the boundary. Therefore, it is good practice to have data over an area in excess of the size of the area to be mapped. This forms a "buffer region" around the map in which edge effects are concentrated; the control points in this region constrain the form of the trend surface within the map area proper. The necessary width of the buffer depends primarily on the density of available control. If the map contains many control points, a narrow bordering strip will suffice. If control density is low, a much wider belt around the map will be necessary to absorb edge effects.

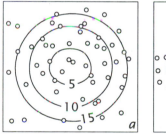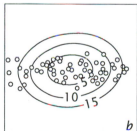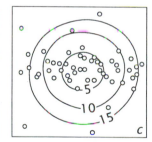

Figure 5–93. Influence of control-point distribution on trend surfaces. (*a*) Original test surface with randomly placed control points. (*b*) Distorted trend produced by sampling original test surface along a narrow band. (*c*) Nearly correct trend produced by adding a few outlying points to the narrow band of observations.

The arrangement of the data points within the map also has a pronounced effect on the form of the regression. The illustrations in **Figure 5–93** are taken from an evaluation of the effect of data distributions on polynomial trend surfaces (Doveton and Parsley, 1970, p. B198–B202). A series of points was placed at random locations on a basin-shaped surface, and a second-degree trend was calculated. The regression equation was then used to calculate values of the dependent variable at points placed according to various sampling patterns. Ideally, surfaces fitted to these points would be identical to the original form from which the data were derived. **Figure 5–93a** shows the surface produced by randomly distributed points; the fit is over 95% and the trend is essentially identical to the original model. In **Figure 5–93b**, however, the sample points are distributed along a narrow band. The fit of the surface is still good (93%), but the form of the regression has become badly

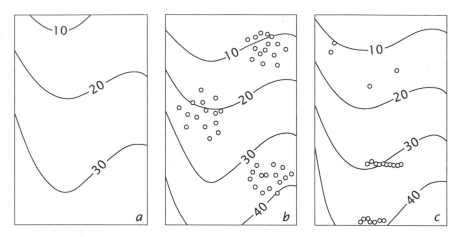

Figure 5–94. Influence of clustering of control points on trend surfaces. (*a*) Original third-order test surface. (*b*) Trend surface obtained from control points placed in clusters on the test surface. (*c*) Trend surface obtained from control points located along traverses across parts of the test surface.

distorted parallel to the sample pattern. Only a few control points outside the band are necessary to correct this bias, however, as can be seen in **Figure 5–93 c** (also 93% fit). These tests demonstrate that the shape of the area defined by control points can seriously affect the form of a polynomial surface fitted to the data. If the control points do not occupy an approximately equidimensional area, the trend surface will be elongated parallel to the pattern of points. Remember that these examples are for idealized models having no local or random component; distortions may be more severe if small-scale "noise" is present.

Cautionary remarks about the deleterious effect of clustered data points on trend surfaces have also been published. Clustering or bunching of control points is especially apt to be troublesome in petroleum exploration using trend-surface methods, because wells are most abundant on and around known oil fields. These areas may exert an undue influence on the regional trend, although there is evidence that this effect is not as severe as sometimes feared (Doveton and Parsley, 1970, p. B200–B201). **Figure 5–94 a** shows a cubic trend-surface model used to generate data points for analysis of the clustering effect. Points were selected in various clustered patterns and used in an attempt to recreate the original surface. **Figure 5–94 b** shows the trend surface created using a strongly clustered data point distribution. The surface accounts for 99% of the original variance. An even more radically clustered distribution is shown in **Figure 5–94 c**, but the cubic surface accounts for 100% of the total variance. Both **Figure 5–94 b** and **5.94 c** are essentially identical to the original surface. These experiments suggest that trend-surface methods may be more robust against effects of clustering than is commonly supposed. Again, it should be remembered that these experimental tests are essentially noise free; more severe distortions may be expected in the presence of local variation.

Kriging

The concept of a regionalized variable introduced in Chapter 4 describes a naturally occurring property that has characteristics intermediate between a truly random variable and a variable that is completely deterministic. This concept of regionalized variables is central to geostatistics, a branch of applied statistics that deals with spatially distributed properties. Many geological surfaces, both real and conceptual, can be regarded as regionalized variables. These surfaces are continuous from place to place and hence must be spatially correlated over short distances. However, widely separated points on an irregular surface tend to be statistically independent. The degree of spatial continuity of a regionalized variable can be expressed by a semivariogram or spatial covariance function, as discussed in Chapter 4. If measurements have been made at scattered sampling points and the form of the semivariogram or covariogram is estimated by an appropriate model, it is possible to estimate the value of the surface at any unsampled location. The estimation procedure is called *kriging*, named after D.G. Krige, a South African mining engineer and pioneer in the application of statistical techniques to mine evaluation.

Kriging is a family of generalized linear regression techniques in which the value of a property at an unsampled location is estimated from values at neighboring locations. Kriging estimates require prior knowledge in the form of a model of the semivariogram (Eq. 4.96) or the spatial covariance (Eq. 4.102). Kriging differs from classical linear regression in that it does not assume that variates are independent, nor does it assume that observations are a random sample.

Geostatistics models a regionalized variable as a *random function*, which is analogous to the concept of an ensemble in time-series analysis except that a random function is defined over a one-, two-, or three-dimensional space rather than over time. At every location within the area or volume of interest there exists a random variable. The set of actual observations, taken over the entire area or volume, constitutes a single random realization just as an observed time series constitutes a single, partial realization of an ensemble. This theoretical construct allows us to treat regionalized variables in a probabilistic manner, even though only one observation can exist at a location.

For historical reasons, the nomenclature and symbolism of geostatistics differ from that developed in other areas of applied statistics and used elsewhere in this book. To provide some compatibility with current geostatistical notation, we will use a symbology that generally corresponds to the usage of Deutsch and Journel (1998), Hohn (1999), and Olea (1999), which in turn reflect the *Geostatistical Glossary and Multilingual Dictionary* (Olea, 1991).

An observation of a regionalized variable is denoted $Z(\mathbf{x}_i)$, indicating that property Z has been measured at point location $\mathbf{x}_i$; that is, $\mathbf{x}_i$ represents a one-, two-, or three-column vector of geographic coordinates. The term $Z(\mathbf{x})$ simply refers to all locations $\mathbf{x}$ in general, and not to a specific observation. $Z(\mathbf{x}_0)$ designates a location where an estimate of the regionalized variable is to be made. Other subscripts (such as $\mathbf{x}_j$, $\mathbf{x}_k$, or $\mathbf{x}_2$) denote unique spatial locations, and the index has no particular significance. The expected or mean value of $Z(\mathbf{x})$ is m, the spatial covariance is $\text{cov}(\mathbf{h})$, and the semivariance is $\gamma(\mathbf{h})$, where $\mathbf{h}$ is the vector distance between $Z(\mathbf{x})$ and $Z(\mathbf{x} + \mathbf{h})$. Although $\mathbf{h}$ was introduced in Chapter 4 as the integer *lag*, here it is generalized to a continuous measure of distance. We also will use the notation $\text{cov}(\mathbf{x}_i, \mathbf{x}_k)$, denoting the spatial covariance over a distance $h = |\mathbf{x}_i - \mathbf{x}_k|$. If $i = k$, then $h = 0$ and the spatial covariance is simply the variance of $Z(\mathbf{x})$. Similarly, $\gamma(\mathbf{x}_i, \mathbf{x}_k)$ is the value of the semivariogram over a distance $h = |\mathbf{x}_i - \mathbf{x}_k|$. Recall that for a distance $h = 0$, the value of the semivariogram is zero.

The specific kriging application of interest in this section is making regular grids of estimates from which contour maps can be drawn. Unlike conventional contouring algorithms, kriging produces map grids that have statistically optimal properties. Kriging is an exact interpolator; that is, values estimated by kriging will be exactly the same as the values of the observations at the same locations. Kriging estimates are unbiased, so the expected value of the estimates is the same as the expected value of the observations. Perhaps most importantly, the error variances of kriging estimates are the minimum possible of any linear estimation method, and they can be estimated at every location where a kriging estimate is made. This provides a way of expressing the uncertainty of a contoured surface.

Simple kriging

As the name suggests, *simple kriging* is the mathematically least complicated form of kriging. It is based on three assumptions: (1) The observations are a partial realization of a random function $Z(\mathbf{x})$, where $\mathbf{x}$ denotes spatial location; (2) The random function is second-order stationary, so the mean, spatial covariance, and semivariance do not depend upon $\mathbf{x}$; (3) The mean is known.

The third assumption would seem to severely limit the applicability of simple kriging, but surprisingly, there are a number of circumstances where the mean is known in advance. For example, if the variable being mapped has been standardized, the mean of a standardized variable is zero. If the variable consists of residuals from a function fitted by least squares (such as trend-surface residuals or residuals from a regression model), the mean is also zero.

The kriging estimator is a weighted average of values at control points $Z(\mathbf{x})$ inside a neighborhood around the location $Z(\mathbf{x}_0)$ where the estimate is to be made. Every point inside such a neighborhood is related in some degree to the central location, $Z(\mathbf{x}_0)$, so theoretically all control points in the neighborhood could provide information about its value. As a practical matter, only the closest points are significantly important, so we can limit consideration to a subset of the k nearest observations. Each observation is weighted in a manner that reflects the spatial covariances among the observations and between the observation and the central location. If k observations are used in the estimate, we must simultaneously consider the k^2 covariances among all pairs of observations $Z(\mathbf{x}_i)$ and $Z(\mathbf{x}_k)$, as well as the k covariances between the observations and location $Z(\mathbf{x}_0)$. The definition for the simple-kriging estimate is:

$$\hat{Z}(\mathbf{x}_0) = m + \sum_{i=1}^{k} \lambda_i \left[Z(\mathbf{x}_i) - m \right] \qquad (5.93)$$

which has k kriging weights, λ_i, that must be estimated. We introduce the requirement that the weights must result in estimates, $\hat{Z}(\mathbf{x}_0)$, that have the minimum variance for the errors, $\hat{Z}(\mathbf{x}_0) - Z(\mathbf{x}_0)$. This is analogous to a problem that was discussed in Chapter 4, that of fitting a regression line so that the sum of the squared deviations from the fitted line was the smallest value possible. As in regression, we must solve a set of simultaneous normal equations to find the unknown weights (for a discussion of the optimization process, see Olea, 1999).

$$\left.\begin{array}{l} \sum_{i=1}^{k} \lambda_i \text{cov}(\mathbf{x}_i, \mathbf{x}_1) = \text{cov}(\mathbf{x}_0, \mathbf{x}_1) \\[2ex] \sum_{i=1}^{k} \lambda_i \text{cov}(\mathbf{x}_i, \mathbf{x}_2) = \text{cov}(\mathbf{x}_0, \mathbf{x}_2) \\[1ex] \cdots\cdots\cdots\cdots\cdots\cdots\cdots\cdots\cdots \\[1ex] \sum_{i=1}^{k} \lambda_i \text{cov}(\mathbf{x}_i, \mathbf{x}_k) = \text{cov}(\mathbf{x}_0, \mathbf{x}_k) \end{array}\right\} \quad (5.94)$$

The necessary numerical values for the covariances are supplied by the model of the spatial covariance. We can also use the semivariogram model to formulate an equivalent set of simultaneous normal equations because, as illustrated in **Figure 4–66**, the semivariogram and spatial covariance function have a reciprocal relationship.

$$\left.\begin{array}{l} \sum_{i=1}^{k} \lambda_i (\text{cov}(\mathbf{x}_i, \mathbf{x}_i) - y(\mathbf{x}_i, \mathbf{x}_1)) = \text{cov}(\mathbf{x}_i, \mathbf{x}_i) - y(\mathbf{x}_0, \mathbf{x}_1) \\[2ex] \sum_{i=1}^{k} \lambda_i (\text{cov}(\mathbf{x}_i, \mathbf{x}_i) - y(\mathbf{x}_i, \mathbf{x}_2)) = \text{cov}(\mathbf{x}_i, \mathbf{x}_i) - y(\mathbf{x}_0, \mathbf{x}_2) \\[1ex] \cdots\cdots\cdots\cdots\cdots\cdots\cdots\cdots\cdots\cdots\cdots\cdots\cdots\cdots \\[1ex] \sum_{i=1}^{k} \lambda_i (\text{cov}(\mathbf{x}_i, \mathbf{x}_i) - y(\mathbf{x}_i, \mathbf{x}_k)) = \text{cov}(\mathbf{x}_i, \mathbf{x}_i) - y(\mathbf{x}_0, \mathbf{x}_k) \end{array}\right\} \quad (5.95)$$

Either set of normal equations will return the same set of kriging weights, provided the spatial covariance model and semivariogram model are equivalent.

We can recast the kriging normal equations into matrix form, in which the matrix equation is simpler to solve. First, we define three matrices:

$$\Lambda = \begin{bmatrix} \lambda_1 \\ \lambda_2 \\ \vdots \\ \lambda_i \end{bmatrix}$$

$$W = \begin{bmatrix} \text{cov}(\mathbf{x}_1, \mathbf{x}_1) & \text{cov}(\mathbf{x}_1, \mathbf{x}_2) & \cdots & \text{cov}(\mathbf{x}_1, \mathbf{x}_k) \\ \text{cov}(\mathbf{x}_2, \mathbf{x}_1) & \text{cov}(\mathbf{x}_2, \mathbf{x}_2) & \cdots & \text{cov}(\mathbf{x}_2, \mathbf{x}_k) \\ \vdots & \vdots & \ddots & \vdots \\ \text{cov}(\mathbf{x}_k, \mathbf{x}_1) & \text{cov}(\mathbf{x}_k, \mathbf{x}_2) & \cdots & \text{cov}(\mathbf{x}_k, \mathbf{x}_k) \end{bmatrix}$$

or

$$\begin{bmatrix} \text{cov}(\mathbf{x}_i, \mathbf{x}_i) - y(\mathbf{x}_1, \mathbf{x}_1) & \text{cov}(\mathbf{x}_i, \mathbf{x}_i) - y(\mathbf{x}_1, \mathbf{x}_2) & \cdots & \text{cov}(\mathbf{x}_i, \mathbf{x}_i) - y(\mathbf{x}_1, \mathbf{x}_k) \\ \text{cov}(\mathbf{x}_i, \mathbf{x}_i) - y(\mathbf{x}_2, \mathbf{x}_1) & \text{cov}(\mathbf{x}_i, \mathbf{x}_i) - y(\mathbf{x}_2, \mathbf{x}_2) & \cdots & \text{cov}(\mathbf{x}_i, \mathbf{x}_i) - y(\mathbf{x}_2, \mathbf{x}_k) \\ \vdots & \vdots & \ddots & \vdots \\ \text{cov}(\mathbf{x}_i, \mathbf{x}_i) - y(\mathbf{x}_k, \mathbf{x}_1) & \text{cov}(\mathbf{x}_i, \mathbf{x}_i) - y(\mathbf{x}_k, \mathbf{x}_2) & \cdots & \text{cov}(\mathbf{x}_i, \mathbf{x}_i) - y(\mathbf{x}_k, \mathbf{x}_k) \end{bmatrix}$$

and

$$B = \begin{bmatrix} \text{cov}(\mathbf{x}_0, \mathbf{x}_1) \\ \text{cov}(\mathbf{x}_0, \mathbf{x}_2) \\ \vdots \\ \text{cov}(\mathbf{x}_0, \mathbf{x}_k) \end{bmatrix} \quad \text{or} \quad \begin{bmatrix} \text{cov}(\mathbf{x}_i, \mathbf{x}_i) - y(\mathbf{x}_0, \mathbf{x}_1) \\ \text{cov}(\mathbf{x}_i, \mathbf{x}_i) - y(\mathbf{x}_0, \mathbf{x}_2) \\ \vdots \\ \text{cov}(\mathbf{x}_i, \mathbf{x}_i) - y(\mathbf{x}_0, \mathbf{x}_k) \end{bmatrix} \quad (5.96)$$

Then, the simple-kriging coefficients are found by

$$\Lambda = W^{-1}B \tag{5.97}$$

The k observations of the regionalized variable within the neighborhood around the location where an estimate is desired are used to form the matrix

$$Y = \begin{bmatrix} Z(\mathbf{x}_1) - m \\ Z(\mathbf{x}_2) - m \\ \vdots \\ Z(\mathbf{x}_k) - m \end{bmatrix} \tag{5.98}$$

The simple-kriging estimate of the regionalized variable at location $\mathbf{x}_0$ is

$$\hat{Z}(\mathbf{x}_0) = m + Y'\Lambda = m + Y'W^{-1}B \tag{5.99}$$

and the variance of the estimate is

$$\sigma^2(\mathbf{x}_0) = \text{cov}\,(\mathbf{x}_i, \mathbf{x}_i) - B'\Lambda = \text{cov}\,(\mathbf{x}_i, \mathbf{x}_i) - B'W^{-1}B \tag{5.100}$$

If the matrices W and B are composed of semivariances rather than spatial covariances, Equation (5.100) becomes

$$\sigma^2(\mathbf{x}_0) = B'\Lambda \tag{5.101}$$

since the semivariance for $h = 0$ is zero.

Rather than demonstrating simple kriging with a numerical example, we will proceed directly to the somewhat more complicated procedure of ordinary kriging. The numerical examples for ordinary kriging will also illustrate the steps of simple kriging.

Ordinary kriging

The assumptions of *ordinary kriging* are the same as those for simple kriging, with one exception. The mean of the regionalized variable is assumed to be constant throughout the area of interest, but the requirement that the value of the mean be known in advance is dropped. Removing the requirement for prior knowledge of the mean greatly extends the applicability of kriging.

The simple-kriging estimator (Eq. 5.93) can be rewritten as

$$\hat{Z}(\mathbf{x}_0) = m\left(1 - \sum_{i=1}^{k} \lambda_i\right) + \sum_{i=1}^{k} \lambda_i\, Z(\mathbf{x}_i)$$

If we can force the sum of the weights to equal 1.0, the quantity inside the parentheses will be 0.0 and the first term, which includes the mean, will vanish. The kriging estimate then will be independent of the value of the mean. By using the Lagrange method of multipliers (James and James, 1992), we can incorporate into the problem of minimizing the error variance the constraint that the sum of the λ's must equal 1.0. This involves the insertion of a Lagrange multiplier, μ, into the set of kriging normal equations, which increases the number of unknown coefficients that must be estimated and expands the matrices defined in Equation (5.96) to

$$
\Lambda = \begin{bmatrix} \lambda_1 \\ \lambda_2 \\ \vdots \\ \lambda_k \\ \mu \end{bmatrix} \quad \text{or} \quad \begin{bmatrix} \lambda_1 \\ \lambda_2 \\ \vdots \\ \lambda_k \\ -\mu \end{bmatrix}
$$

$$
W = \begin{bmatrix}
\text{cov}(x_1, x_1) & \text{cov}(x_1, x_2) & \cdots & \text{cov}(x_1, x_k) & 1 \\
\text{cov}(x_2, x_1) & \text{cov}(x_2, x_2) & \cdots & \text{cov}(x_2, x_k) & 1 \\
\vdots & \vdots & \ddots & \vdots & \vdots \\
\text{cov}(x_k, x_1) & \text{cov}(x_k, x_2) & \cdots & \text{cov}(x_k, x_k) & 1 \\
1 & 1 & \cdots & 1 & 0
\end{bmatrix}
$$

$$
\text{or} \quad \begin{bmatrix}
\gamma(x_1, x_1) & \gamma(x_1, x_2) & \cdots & \gamma(x_1, x_k) & 1 \\
\gamma(x_2, x_1) & \lambda(x_2, x_2) & \cdots & \gamma(x_2, x_k) & 1 \\
\vdots & \vdots & \ddots & \vdots & \vdots \\
\gamma(x_k, x_1) & \gamma(x_k, x_2) & \cdots & \gamma(x_k, x_k) & 1 \\
1 & 1 & \cdots & 1 & 0
\end{bmatrix}
$$

$$
B = \begin{bmatrix} \text{cov}(x_0, x_1) \\ \text{cov}(x_0, x_2) \\ \vdots \\ \text{cov}(x_0, x_k) \\ 1 \end{bmatrix} \quad \text{or} \quad \begin{bmatrix} \gamma(x_0, x_1) \\ \gamma(x_0, x_2) \\ \vdots \\ \gamma(x_0, x_k) \\ 1 \end{bmatrix} \quad \text{(5.102)}
$$

As before, the elements of W and B are derived from the spatial covariance function or the semivariogram model.

We also require a vector of the k observations around location x_0 where the estimate $\hat{Z}(x_0)$ of the regionalized variable is desired. This vector is simply

$$
Y = \begin{bmatrix} Z(x_1) \\ Z(x_2) \\ \vdots \\ Z(x_k) \\ 0 \end{bmatrix} \quad \text{(5.103)}
$$

We first estimate the ordinary-kriging weights by

$$
\Lambda = W^{-1}B \quad \text{(5.104)}
$$

using either the covariance or semivariance versions of the matrices. The ordinary-kriging estimate of the regionalized variable at location x_0 is

$$
\hat{Z}(x_0) = Y'\Lambda = Y'W^{-1}B \quad \text{(5.105)}
$$

The ordinary-kriging estimation variance is

$$
\sigma^2(x_0) = B'\Lambda = B'W^{-1}B \quad \text{(5.106)}
$$

for the semivariance versions of the matrices, or equivalently

$$\sigma^2(\mathbf{x}_0) = \text{cov}\,(\mathbf{x}_1, \mathbf{x}_1) - \mathbf{B}'\mathbf{\Lambda} = \text{cov}\,(\mathbf{x}_1, \mathbf{x}_1) - \mathbf{B}'\mathbf{W}^{-1}\mathbf{B}$$

if the spatial covariance versions of the matrices are used.

Many practitioners prefer to formulate ordinary kriging in terms of spatial covariances because then the diagonal elements of matrix $\mathbf{W}$ will be the largest elements in the matrix. If semivariances are used, the diagonal elements of $\mathbf{W}$ are all zero—in the past, there have been concerns that such a condition could lead to an unstable solution. However, modern computer algorithms have no difficulty in solving the matrix equation using entries from either a spatial covariance function or a semivariogram, so there is no computational reason to prefer one form of the equation over the other.

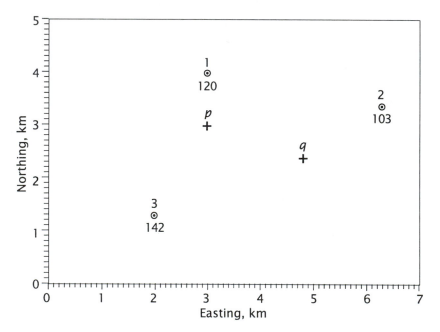

Figure 5–95. Map showing water-table elevations (in meters) at three observation wells. Estimates of the water-table elevation will be made at locations p and q. Coordinates given in kilometers from an arbitrary origin.

To demonstrate ordinary kriging, we will estimate the elevation of the water table at point $\mathbf{x}_p$ on the map shown in **Figure 5–95**. The estimate will be made from known elevations measured in three observation wells. The map coordinates of the wells and the distances between them are given in **Table 5–15**. We will assume that a prior structural analysis has produced the experimental semivariogram and model shown in **Figure 5–96**; the model is linear with a slope of 4.0 m^2/km within a neighborhood of 20 km. Values of the semivariance corresponding to distances between the wells are also given in **Table 5–15**; these may be read directly off the semivariogram or calculated from the slope.

Table 5–15. Coordinates of wells and locations to be estimated, distances, and semivariances. Coordinates measured from an arbitrary origin in southwest corner of map.

Well or Location	Easting km	Northing km	Elevation m
x_1	3.0	4.0	120
x_2	6.3	3.4	103
x_3	2.0	1.3	142
x_4	3.8	2.4	115
x_5	1.0	3.0	148
x_p	3.0	3.0	
x_q	4.9	2.2	

Distances in km above diagonal, semivariances in m^2 below diagonal.

	x_1	x_2	x_3	x_4	x_5	x_p	x_q
x_1	0	3.354	2.879	1.789	2.236	1.000	2.617
x_2	13.416	0	4.785	2.693	5.315	3.324	1.844
x_3	11.517	19.142	0	2.110	1.972	1.972	3.036
x_4	7.155	10.770	8.438	0	2.864	1.000	1.118
x_5	8.944	21.260	7.889	11.454	0	2.000	3.981
x_p	4.000	13.297	7.889	4.000	8.000	0	
x_q	10.469	7.376	12.146	4.472	15.925		0

The equations that must be solved to find the weights, λ_i, in this example are

$$\lambda_1 0.0 \ + \ \lambda_2 13.42 \ + \ \lambda_3 11.52 \ + \ \mu \ = \ 4.00$$
$$\lambda_1 13.42 \ + \ \lambda_2 0.0 \ + \ \lambda_3 19.14 \ + \ \mu \ = 13.30$$
$$\lambda_1 11.52 \ + \ \lambda_2 19.14 \ + \ \lambda_3 0.0 \ + \ \mu \ = \ 7.89$$
$$\lambda_1 \ + \ \lambda_2 \ + \ \lambda_3 \ + \ 0 \ = \ 1.00$$

Set in matrix form, this is

$$
\begin{bmatrix}
0 & 13.42 & 11.52 & 1.00 \\
13.42 & 0 & 19.14 & 1.00 \\
11.52 & 19.14 & 0 & 1.00 \\
1.00 & 1.00 & 1.00 & 0
\end{bmatrix}
\times
\begin{bmatrix}
\lambda_1 \\
\lambda_2 \\
\lambda_3 \\
-\mu
\end{bmatrix}
=
\begin{bmatrix}
4.00 \\
13.30 \\
7.89 \\
1.00
\end{bmatrix}
$$

The inverse of the left-hand matrix can be found using the procedure described in Chapter 3, although it may be necessary to rearrange the order of the equations to avoid having zeros along the main diagonal. The inverse is

$$
\begin{bmatrix}
-0.0655 & 0.0295 & 0.0360 & 0.1897 \\
0.0295 & -0.0394 & 0.0099 & 0.4146 \\
0.0360 & 0.0099 & -0.0459 & 0.3958 \\
0.1897 & 0.4146 & 0.3958 & -10.1201
\end{bmatrix}
$$

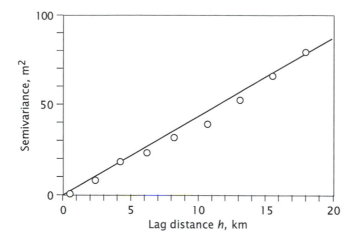

Figure 5–96. Linear semivariogram of water-table elevations in an area which includes the map in Figure 5–93. Semivariogram has a slope of 4.0 m²/km within a 20-km neighborhood.

The unknown weights can now be found by post-multiplying the transpose by the right-hand vector of the semivariances, yielding

$$\Lambda = \begin{bmatrix} \lambda_1 \\ \lambda_2 \\ \lambda_3 \\ -\mu \end{bmatrix} = \begin{bmatrix} 0.6039 \\ 0.0867 \\ 0.3094 \\ -0.7266 \end{bmatrix}$$

The estimate of the elevation of the water table at location $\mathbf{x}_p$ is found by inserting the appropriate weights in the linear equation (Eq. 5.105)

$$\hat{Z}(\mathbf{x}_p) = 0.6039(120) + 0.0867(103) + 0.3094(142)$$
$$= 125.3 \text{ m}$$

Similarly, the kriging estimation variance is the weighted sum of the semivariances for the distances from the control points to the location of the estimate

$$\sigma^2(\mathbf{x}_p) = 0.6039(4) + 0.0867(12.1) + 0.3094(7.9) - 0.7266(1.0)$$
$$= 5.28 \text{ m}^2$$

The standard error of the kriging estimate is simply the square root of the kriging estimation variance, or

$$\sigma(\mathbf{x}_p) = \sqrt{5.28} = 2.3 \text{ m}$$

If we assume the errors of estimation are normally distributed about a true value, we can use the standard error as a confidence band around the kriging estimates. The probability that the true elevation of the water table at point $\mathbf{x}_p$ is within one standard error above or below the value estimated is 68%, and the probability is 95% that the true elevation lies within 1.96 standard errors. That is, the water-table elevation at this location must be within the interval

$$Z(\mathbf{x}_p) = 125.3 \pm 4.5 \text{ m, with 95\% probability}$$

At every point on this map we can use ordinary kriging to estimate the elevation of the water table and can also determine the standard errors of these estimates. From these we can construct two maps; the first is based on the kriging estimates themselves and is a "best guess" of the configuration of the mapped variable. The second is an error map showing the confidence envelope that surrounds this estimated surface; it expresses one particular aspect of the reliability of the kriged surface. It is based entirely on the geometric arrangement and distances between observations used in the estimation process and on the degree of spatial continuity of the regionalized variable as expressed by the spatial covariance or semivariogram model. It does not consider any other sources of variation, such as sampling variance that might be revealed by replication of the observations. In areas of poor control, the error map will show large values, indicating that the estimates are subject to high variability. In areas of dense control the error map will show low values, and at the control points themselves, the estimation error will be zero.

We must be careful that the data do not include duplicate points, even if these represent valid replicate measurements made at a common location. Including the same location more than once results in identical rows and columns in matrix $\mathbf{W}$, with the consequence that $\mathbf{W}$ becomes singular and impossible to invert.

The system of equations used to find the kriging weights must be solved for every estimated location (unless the samples are arranged in an absolutely regular pattern so the distances between points remain the same and we neglect locations near the edges of the map). If we wish to estimate the elevation of the water table at point $\mathbf{x}_q$ on **Figure 5–95**, the distances between $\mathbf{x}_q$ and the three observation wells must be considered. The distance from $\mathbf{x}_q$ to well 1 is 2.62 km, from $\mathbf{x}_q$ to well 2 is 1.84 km, and from $\mathbf{x}_q$ to well 3 is 3.04 km. The corresponding semivariances (plus 1, which represents the sum of the weights) were taken from **Figure 5–96** and are

$$\mathbf{B} = \begin{bmatrix} 10.47 \\ 7.38 \\ 12.15 \\ 1 \end{bmatrix}$$

Since the arrangement of the observation wells remains the same, all distances between the wells are unchanged and the left-hand side of the set of kriging simultaneous equations is unchanged. The inverse is likewise unchanged, so multiplying the inverse by the new vector of semivariances will yield weights for estimating the elevation of the water table at point $\mathbf{x}_q$. This new set of weights is

$$\Lambda = \begin{bmatrix} 0.1588 \\ 0.5531 \\ 0.2881 \\ -0.2697 \end{bmatrix}$$

The estimate of the water-table elevation is

$$\hat{Z}(\mathbf{x}_q) = 0.1588(120) + 0.5531(103) + 0.2881(142)$$
$$= 116.9 \text{ m}$$

and the kriging estimation variance is

$$\sigma^2(\mathbf{x}_q) = 0.1588(9.6) + 0.5531(6.3) + 0.2881(12.0) - 0.2697(1)$$
$$= 8.97 \text{ m}^2$$

The standard error of the kriging estimate at x_q is

$$\sigma(x_q) = \sqrt{8.97} = 3.0 \text{ m}$$

so the elevation of the water table at point x_q can be expressed as

$$\hat{Z}(x_q) = 116.9 \pm 5.9 \text{ m, with 95\% probability}$$

if we assume the estimation error is normally distributed.

The groundwater surface is lower at point x_q than at x_p, and the standard error is greater, reflecting the greater total distance to the observation wells. If one of the control points is changed, some of the distances between wells are also changed and the system of equations must be solved anew. In **Figure 5–97**, observation well 4 has been drilled at a location nearer to site x_p, and a water-table elevation of 115 m measured for the regionalized variable. This well is a replacement for well 2. The interpoint distances and corresponding semivariances for well 4 are included in **Table 5–15**. The set of kriging simultaneous normal equations is now

$$\begin{bmatrix} 0 & 7.16 & 11.52 & 1 \\ 7.16 & 0 & 8.44 & 1 \\ 11.52 & 8.44 & 0 & 1 \\ 1 & 1 & 1 & 0 \end{bmatrix} \times \begin{bmatrix} \lambda_1 \\ \lambda_4 \\ \lambda_3 \\ -\mu \end{bmatrix} = \begin{bmatrix} 4.00 \\ 4.00 \\ 7.89 \\ 1.0 \end{bmatrix}$$

whose solution is

$$\begin{bmatrix} \lambda_1 \\ \lambda_4 \\ \lambda_3 \\ -\mu \end{bmatrix} = \begin{bmatrix} 0.4545 \\ 0.3858 \\ 0.1598 \\ -0.6001 \end{bmatrix}$$

The new estimate of the water-table elevation at point x_p, using information from well 4 instead of well 2, is

$$\hat{Z}(x_p) = 0.4545(120) + 0.3858(115) + 0.1598(142)$$
$$= 121.6 \text{ m}$$

The kriging estimation variance of this new estimate is

$$\sigma^2(x_p) = 0.4545(4.0) + 0.2858(4.0) + 0.1595(7.9) - 0.6001(1)$$
$$= 4.0 \text{ m}^2$$

The standard error of the kriging estimate at point x_p is now

$$\sigma(x_p) = \sqrt{4.00} = 2.0 \text{ m}$$

which is a somewhat lower value than was found when using observation well 2 rather than well 4. This illustrates the fact that the estimation errors are reduced if the control points are closer to the location where the estimate is to be made.

Suppose one of the control points coincides with the location to be estimated. Then, one of the values on the right-hand side of the matrix equation becomes zero, and the remaining values become equal to some of the values in the left-hand matrix. We can determine the effect of this change in our example in **Figure 5–97** by assuming that an observation well 5 is drilled at location x_p, and the water level

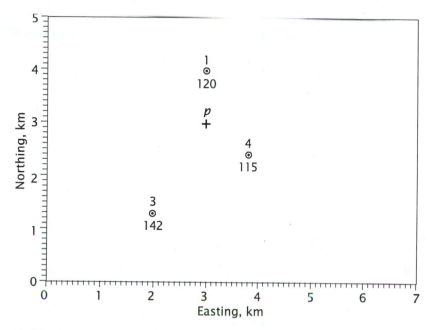

Figure 5–97. Map showing water-table elevations (in meters) at three observation wells. Well 4 is closer to location p being estimated than is well 2 in Figure 5–95.

is measured at 125 m. The distance between any point x_i and well 5 is now the same as the distance between any point x_i and location x_p. The semivariances are likewise the same, so the set of simultaneous equations becomes

$$
\begin{bmatrix}
0 & 4.00 & 11.52 & 1 \\
4.00 & 0 & 7.89 & 1 \\
11.52 & 7.89 & 0 & 1 \\
1 & 1 & 1 & 0
\end{bmatrix}
\times
\begin{bmatrix}
\lambda_1 \\
\lambda_5 \\
\lambda_3 \\
-\mu
\end{bmatrix}
=
\begin{bmatrix}
4.00 \\
0 \\
7.89 \\
1.00
\end{bmatrix}
$$

The vector of weights can be calculated and is, as we should expect,

$$
\begin{bmatrix}
\lambda_1 \\
\lambda_5 \\
\lambda_3 \\
-\mu
\end{bmatrix}
=
\begin{bmatrix}
0.00 \\
1.00 \\
0.00 \\
0.00
\end{bmatrix}
$$

If these weights are used to estimate $\hat{Z}(x_p)$, we see that the estimated elevation is exactly equal to the measured value of the water level in well 5.

$$
\begin{aligned}
\hat{Z}(x_p) &= 0.00(120) + 1.00(125) + 0.00(142) \\
&= 125.0 \text{ m}
\end{aligned}
$$

Also, as we should expect, the kriging estimation variance is

$$
\begin{aligned}
\sigma^2(x_p) &= 0.00(4.0) + 1.00(0) + 0.00(7.9) + 0.00(1) \\
&= 0.00 \text{ m}^2
\end{aligned}
$$

This demonstrates what is meant by the oft-heard statement that kriging is an "exact interpolator"; it predicts the actual values measured at the known points, and does this with zero error. Of course, we do not ordinarily produce estimates for locations already known, but this does occasionally occur when using ordinary kriging for contouring. If any of the control points happen to coincide with grid nodes, kriging will produce the correct, error-free values. We also can be assured that the estimated surface must pass exactly through all control points (if the contouring grid is sufficiently fine so that all observation locations coincide with grid nodes), and that confidence bands around the estimated surface go to zero at the control points.

In these examples we have assumed that each estimate is made using only three control points in order to simplify the mathematics as much as possible. In actual practice we would expect to use more observations, perhaps many more, in making each estimate. Every observation used in an estimate must be weighted, and finding each weight requires another equation. Most contouring routines use 16 or more control points to estimate every grid intersection, which means a set of at least 17 simultaneous equations must be solved for every grid node location when using ordinary kriging.

In theory, the number of points needed to estimate a location varies with the local density and arrangement of control and the continuity of the regionalized variable. All control points within the neighborhood around the location to be estimated provide information and should be considered. In practice, many of these points may be redundant, and their use will improve the estimate only slightly. Practical rules-of-thumb have been developed for contour mapping by kriging, which limit the number of control points actually needed to a subset of the points within the zone of influence or neighborhood. The optimum number of control points is determined by the semivariogram and the spatial pattern of the points (Myers, 1991; Olea, 1999). The structural analysis thus plays a doubly critical role in kriging; it provides the semivariogram necessary to solve the kriging equations, and also determines the neighborhood size within which the control points are selected for each estimate.

Universal kriging

A significant shortcoming of ordinary kriging is that the procedure is not valid unless the regionalized variable being mapped has a constant mean. In the presence of a trend, or systematic change in average value, the ordinary-kriging estimator is not optimal. Computed estimates will be shifted upwards or downwards from their true values, so the estimation error variance will be inflated and no longer the minimum possible.

Universal kriging is a further generalization of the kriging procedure that removes the restriction that the regionalized variable must have a constant mean. In effect, universal kriging treats a first-order nonstationary regionalized variable as though it consisted of two components. The *drift* is the average or expected value of a regionalized variable within a neighborhood, and is the slowly varying, nonstationary component of the surface. The *residuals* are the differences between the actual observations and the drift. Obviously, if the drift is removed from a regionalized variable, the residuals must be stationary and ordinary kriging can be applied to them. Universal kriging performs in one simultaneous operation what otherwise would require three steps: The drift is estimated and removed to form stationary residuals at the control points; the stationary residuals are kriged to obtain estimated residuals at unsampled locations; and, the estimated residuals are combined with the drift to obtain estimates of the actual surface. Instead of performing this complicated series of operations, we can expand the kriging system of equations to incorporate additional Lagrangian multipliers that represent the constraints imposed by the addition of a drift. Then the kriging weights will be estimated as though the drift effect had been removed from the regionalized variable, although the original observations (which include a trend) are used directly.

The drift is somewhat analogous to a trend surface, except that the drift is estimated locally rather than globally. The drift may be modeled as a linear combination of the geographic coordinates of the observations within a local neighborhood. The usual models used in geostatistics are a first- or second-degree polynomial:

$$m(\mathbf{x}_0) = \alpha_0 + \sum_{i=1}^{k}(\alpha_1 x_{1,i} + \alpha_2 x_{2,i})$$

$$m(\mathbf{x}_0) = \alpha_0 + \sum_{i=1}^{k}(\alpha_1 x_{1,i} + \alpha_2 x_{2,i} + \alpha_3 x_{1,i}^2 + \alpha_4 x_{2,i}^2 + \alpha_5 x_{1,i} x_{2,i})$$

(5.107)

where in two dimensions, $x_{1,i}$ represents the easting coordinate of observation i and $x_{2,i}$ represents the northing coordinate at the same location (actually, they may represent coordinate pairs at any orientation provided the two coordinate axes are orthogonal). The α_j are the unknown drift coefficients.

In ordinary kriging, we must have prior knowledge of the spatial continuity of the regionalized variable in the form of a model of the semivariance or spatial covariance. We face a similar requirement for universal kriging, with the additional complication that the semivariance or spatial covariance model must be of the residuals from the regionalized variable. As discussed in Chapter 4, this means that a structural analysis must be performed to determine the best combination of neighborhood size and expression for the drift. This is not a trivial undertaking, because the drift model and neighborhood size are interdependent.

The expressions for the drift may be incorporated into the system of normal equations used to find the kriging weights as additional constraints. Solving this expanded set of simultaneous equations will produce a set of weights for the kriged estimate which incorporate the effect of the specified drift within the local neighborhood. The drift expressions relate the geographic coordinates of each control point to the geographic coordinates of the location being kriged.

The complexity of the drift model, the size of the neighborhood, and the model of the semivariogram or covariogram of the residuals from the drift are interrelated. This means that the variance of the residuals depends in part on the somewhat arbitrary specification of the drift. To perform universal kriging, we must estimate the kriging weights and a set of Lagrange multipliers that insure unbiasedness and account for the local drift. Since more coefficients are being estimated, additional simultaneous equations, incorporating more terms, must solved. This in turn means a larger sample of control points within the neighborhood must be used in order to have sufficient degrees of freedom to solve the equations. The simplest example of universal kriging consists of estimating the regionalized variable, $\hat{Z}(\mathbf{x}_0)$, when we assume that the drift is linear and that we have determined an appropriate model of the semivariogram of the residuals from the drift. The linear drift model given in Equation (5.107) has two coefficients and ordinary kriging requires at least three coefficients, so we will need at least five observations for each estimate. In practical applications we typically use many more observations for each estimated location, as many as 16 to 32 control points.

The matrices defined in Equation (5.102) must be expanded to include the Lagrangian multipliers that represent the additional constraints imposed by the presence of a drift. Only the semivariance forms of the matrices are shown; the equivalent spatial covariance matrices can be deduced by comparing Equation (5.108) to Equation (5.102).

$$
\Lambda = \begin{bmatrix} \lambda_1 \\ \lambda_2 \\ \vdots \\ \lambda_k \\ \mu_0 \\ \mu_1 \\ \mu_2 \end{bmatrix}
$$

$$
W = \begin{bmatrix}
\gamma(\mathbf{x}_1,\mathbf{x}_1) & \gamma(\mathbf{x}_1,\mathbf{x}_2) & \cdots & \gamma(\mathbf{x}_1,\mathbf{x}_k) & 1 & x_{1,1} & x_{2,1} \\
\gamma(\mathbf{x}_2,\mathbf{x}_1) & \gamma(\mathbf{x}_2,\mathbf{x}_2) & \cdots & \gamma(\mathbf{x}_2,\mathbf{x}_k) & 1 & x_{1,2} & x_{2,2} \\
\vdots & \vdots & \ddots & \vdots & \vdots & \vdots & \vdots \\
\gamma(\mathbf{x}_k,\mathbf{x}_1) & \gamma(\mathbf{x}_k,\mathbf{x}_2) & \cdots & \gamma(\mathbf{x}_k,\mathbf{x}_k) & 1 & x_{1,k} & x_{2,k} \\
1 & 1 & \cdots & 1 & 0 & 0 & 0 \\
x_{1,1} & x_{1,2} & \cdots & x_{1,k} & 0 & 0 & 0 \\
x_{2,1} & x_{2,2} & \cdots & x_{2,k} & 0 & 0 & 0
\end{bmatrix}
\tag{5.108}
$$

$$
B = \begin{bmatrix} \gamma(\mathbf{x}_0,\mathbf{x}_1) \\ \gamma(\mathbf{x}_0,\mathbf{x}_2) \\ \vdots \\ \gamma(\mathbf{x}_0,\mathbf{x}_k) \\ 1 \\ x_{1,p} \\ x_{2,p} \end{bmatrix}
$$

Note that in the notation used in these matrices, $\mathbf{x}_i$ represents a vector of the coordinates of point i, while $x_{1,i}$ is a scalar value representing the location of point i along coordinate axis 1 (perhaps east–west) and is the first element of the coordinate vector of point i.

The vector of universal-kriging weights Λ is found by Equation (5.104), except that W and B have the definitions given above in Equation (5.108). As in simple kriging and ordinary kriging, we require a vector of the observations within the neighborhood that will be used to estimate $\hat{Z}(\mathbf{x}_0)$. This vector Y will have $k + d + 1$ elements, of which the final $d + 1$ elements (corresponding to the Lagrange multipliers) are zeros.

$$
Y = \begin{bmatrix} Z(\mathbf{x}_1) \\ Z(\mathbf{x}_2) \\ \vdots \\ Z(\mathbf{x}_k) \\ 0 \\ 0 \\ 0 \end{bmatrix}
\tag{5.109}
$$

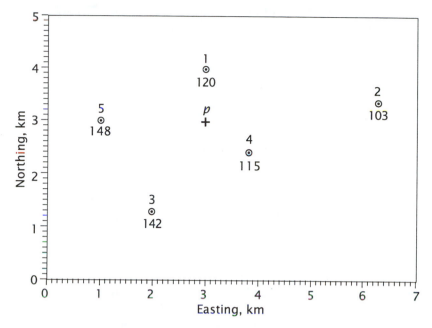

Figure 5–98. Map showing water-table elevations (in meters) at five observation wells. Estimates of the water-table elevation will be made by universal kriging at location p and at the southwest corner of the map.

The estimate $\hat{Z}(\mathbf{x}_0)$ is found by Equation (5.105). The universal-kriging variance is found by Equation (5.106). In both equations, the definitions of $\mathbf{W}$ and $\mathbf{B}$ are those given in Equation (5.108) above.

We may extend our example problem, which is based on data from a western Kansas aquifer, to demonstrate the steps in universal kriging. **Figure 5–98** shows the locations of five observation wells that will be used to estimate the drift and universal-kriging estimate of the water-table elevation at location $\mathbf{x}_p$. We will assume that **Figure 5–96** now represents the estimated semivariogram for the residuals from a first-degree drift, and that it is linear in form with a slope of 4.0 m^2/km. All of the basic information required is given in **Table 5–15**, which also includes the necessary semivariances. The equation that must be solved to estimate the water-table elevation at location $\mathbf{x}_p$ is:

$$
\begin{bmatrix}
0 & 13.4 & 11.5 & 7.2 & 8.9 & 1 & 3.0 & 4.0 \\
13.4 & 0 & 19.1 & 10.8 & 21.3 & 1 & 6.3 & 3.4 \\
11.5 & 19.1 & 0 & 8.4 & 7.9 & 1 & 2.0 & 1.3 \\
7.2 & 10.8 & 8.4 & 0 & 11.5 & 1 & 3.8 & 2.4 \\
8.9 & 21.3 & 7.9 & 11.5 & 0 & 1 & 1.0 & 3.0 \\
1 & 1 & 1 & 1 & 1 & 0 & 0 & 0 \\
3.0 & 6.3 & 2.0 & 3.8 & 1.0 & 0 & 0 & 0 \\
4.0 & 3.4 & 1.3 & 2.4 & 3.0 & 0 & 0 & 0
\end{bmatrix}
\times
\begin{bmatrix}
\lambda_1 \\
\lambda_2 \\
\lambda_3 \\
\lambda_4 \\
\lambda_5 \\
\mu_0 \\
\mu_1 \\
\mu_2
\end{bmatrix}
=
\begin{bmatrix}
4.0 \\
13.3 \\
7.9 \\
4.0 \\
8.0 \\
1 \\
3.0 \\
3.0
\end{bmatrix}
$$

Solving the equation gives a set of eight coefficients, of which the first five are the universal-kriging weights and the final three are Lagrange multipliers.

$$
\begin{bmatrix} \lambda_1 \\ \lambda_2 \\ \lambda_3 \\ \lambda_4 \\ \lambda_5 \\ \mu_0 \\ \mu_1 \\ \mu_2 \end{bmatrix} = \begin{bmatrix} 0.4119 \\ -0.0137 \\ 0.0934 \\ 0.4126 \\ 0.0957 \\ -0.7245 \\ 0.0660 \\ 0.0229 \end{bmatrix}
$$

The estimate of the water-table elevation at location $\mathbf{x}_p$ is

$$
\hat{Z}(\mathbf{x}_p) = 0.4119(120) - 0.0137(103) + 0.0934(142) + 0.4126(115) + \\ 0.0957(148) = 122.9 \text{ m}
$$

which is only slightly different than the results obtained from three observations without assuming a drift. The kriging estimation error variance can be calculated by premultiplying the vector of right-hand terms, $\mathbf{B}$, by the transpose of the solution vector, $\boldsymbol{\lambda}$. The estimation error variance is 3.9 m^2.

This example does not illustrate a major distinction between ordinary kriging and universal kriging with drift because, in this instance, the two procedures yield almost identical estimates. Ordinary kriging, however, in common with other weighted-averaging methods, does not extrapolate well beyond the convex hull of the control points. That is, most estimated values will lie on the slopes of the surface and the highest and lowest points on the surface usually will be defined by control points. Suppose we estimate the water-table elevation at a location where it seems obvious that the surface should be outside the interval defined by the observation wells within the neighborhood. The water table appears to dip from west to east, dropping almost 40 m between the observation well at location $\mathbf{x}_2$ and the well at location $\mathbf{x}_3$. If this dip continues, we would expect water levels higher than 142 m at locations on the western side of the map area, and levels below 103 m at locations on the eastern side.

We will estimate the water-table elevation in the extreme southwest corner of the map, at coordinates $\mathbf{x} = 0, 0$. We will first use ordinary kriging and our five observation wells. This will yield the following set of ordinary-kriging weights:

$$
\begin{bmatrix} \lambda_1 \\ \lambda_2 \\ \lambda_3 \\ \lambda_4 \\ \lambda_5 \\ \mu \end{bmatrix} = \begin{bmatrix} -0.1221 \\ 0.0110 \\ 0.7523 \\ -0.0307 \\ 0.3895 \\ 7.9235 \end{bmatrix}
$$

The estimate of the water level is

$$
\hat{Z}(\mathbf{x}_0) = -0.1221(120) + 0.0110(103) + 0.7523(142) - \\ 0.0307(115) + 0.3895(148) = 147.4 \text{ m}
$$

As we expect, the estimate is based almost entirely on the closest observation wells and is within the interval defined by the highest and lowest values used in the estimate. The ordinary-kriging estimation error variance is $\sigma^2(\mathbf{x}_0) = 17.3$ m^2.

If a first-degree drift is assumed, the universal-kriging coefficients are

$$
\begin{bmatrix} \lambda_1 \\ \lambda_2 \\ \lambda_3 \\ \lambda_4 \\ \lambda_5 \\ \mu_0 \\ \mu_1 \\ \mu_2 \end{bmatrix} = \begin{bmatrix} -0.5594 \\ -0.3020 \\ 1.3133 \\ 0.1451 \\ 0.4030 \\ 26.3832 \\ -1.7940 \\ -4.1795 \end{bmatrix}
$$

Using these weights yields an estimate of the water table of

$$
\hat{Z}(\mathbf{x}_0) = -0.5594(120) - 0.3020(103) + 1.3133(142) + \\
0.1451(115) + 0.4030(148) = 164.6 \text{ m}
$$

which is much higher than the highest observed elevation in the neighborhood. Universal kriging has considered the dip of the water table, or drift, within the local neighborhood and has projected this to the location being kriged. The estimation error variance is $\sigma^2(\mathbf{x}_0) = 26.8 \text{ m}^2$, a much larger value that includes both the uncertainty in the estimate of the drift and the uncertainty in the estimate of the regionalized variable, which is the residual from the drift.

The universal-kriging equations and the numerical examples are given here in terms of semivariances; but, as in simple and ordinary kriging, an equivalent formulation can be made in terms of spatial covariances. Provided the models for the semivariogram of the residuals and the covariogram of the residuals are equivalent, the two approaches will yield identical results. In practice, the universal-kriging equations usually are set up and solved using spatial covariances in order to avoid possible numerical instabilities when inverting the W matrix.

Calculating the drift.—Kriging finds a set of weights that relates the semivariances between the observations to the semivariances between the observations and location $\mathbf{x}_0$. Suppose, however, that all of the available observations were beyond the range. The semivariance, $\gamma(\mathbf{x}_0, \mathbf{x}_i)$, between location $\mathbf{x}_0$ and a distant observation, $\mathbf{x}_i$, is identical for all distant observations i, and equal to the variance σ_0^2 of the regionalized variable (or if a drift is present, equal to the variance of the residuals from the drift). For these same distant points, the spatial covariance, $\text{cov}(\mathbf{x}_0, \mathbf{x}_i)$, equals zero for all points i. That is, the first k elements of the right-hand vector of Equation (5.108) used to estimate the universal-kriging weights are identical and equal to the variance (if the semivariogram version of the equation is used) or equal to 0 (if the spatial covariance version is used). Either version of the universal-kriging equation implies that the regionalized variable $Z(\mathbf{x}_0)$ is statistically independent of the values of $Z(\mathbf{x})$ at each of the distant observations. We cannot use these distant points to estimate a value $\hat{Z}(\mathbf{x}_0)$, which is more precise than an estimate based on the global properties of the regionalized variable; i.e., an estimate based exclusively on these distant points is simply an estimate of the drift, $\hat{m}(\mathbf{x}_0)$, itself.

We can estimate the drift $\hat{m}(\mathbf{x}_0)$ at a location $\mathbf{x}_0$ even if the observations used are within a neighborhood smaller than the range around $\mathbf{x}_0$. All that is necessary is to replace the semivariances (or spatial covariances) on the right-hand side of the universal-kriging equations with the semivariances (or spatial covariances) that would be observed if location $\mathbf{x}_0$ were so far from the observations that it was independent of their values. Since the semivariance for all distances beyond the range is equal to the variance of the residuals, the terms for the right-hand side can be set equal to the variance of the residuals, s_0^2. Unfortunately we again arrive at a circular impasse, because we cannot know s_0^2 until the drift has been calculated. Fortunately, the structural analysis allows us to make an *a priori* estimate of s_0^2, because it is equal to the semivariance at the sill, or beyond the range. The situation is somewhat less complicated if we express the universal-kriging equations in spatial covariance form, since we know that the spatial covariance for residuals between points separated by distances greater than the range must be 0.0.

It should be emphasized that the drift is an arbitrary but convenient construct that is necessary to remove nonstationarity in the regionalized variable. It has no physical significance. There may be many alternative combinations of drift model, neighborhood size, and estimated semivariogram that will satisfactorily model the structure of a regionalized variable. The choice of a specific combination depends upon the availability of data, computational convenience, and other considerations. We will continue our simple numerical example, assuming that the drift in water-table elevations is linear and that the semivariogram for the residuals is also linear. From a structural analysis, we may conclude that the range of the regionalized variable extends up to 30 km. Since the slope of the semivariogram is 4 m²/km, the variance beyond the range should be about 4×30, or $120\,\mathrm{m}^2$ (if we formulate the drift estimation problem in terms of spatial covariances, the equivalent values are $0\,\mathrm{m}^2$). The right-hand part of the kriging matrix will have the following appearance when using semivariances for calculating the drift, $\hat{m}(\mathbf{x}_0)$:

$$\mathbf{M} = \begin{bmatrix} 120 \\ 120 \\ 120 \\ 120 \\ 120 \\ 1 \\ 0 \\ 0 \end{bmatrix}$$

The left-hand side, $\mathbf{W}$, is unchanged from the universal-kriging calculation, so all that is necessary to estimate the drift is to solve the equation,

$$\mathbf{\Delta} = \mathbf{W}^{-1}\,\mathbf{M} \tag{5.110}$$

This yields a set of five universal-kriging drift coefficients, δ_k, which are used to compute the drift, plus three constant terms that contribute to the estimation error variance for the drift. The solution vector, $\mathbf{\Delta}$ is

$$\mathbf{\Delta} = \begin{bmatrix} 0.1311 \\ 0.3702 \\ 0.2202 \\ -0.1587 \\ 0.4372 \\ 109.2283 \\ -0.9048 \\ 0.4935 \end{bmatrix}$$

The drift $\hat{m}(\mathbf{x}_0)$ is estimated by multiplying the elevations at the observation wells by the appropriate drift coefficients and summing.

$$\hat{m}(\mathbf{x}_0) = 0.1311(120) + 0.3702(103) + 0.2202(142) - 0.1587(115)$$
$$+ 0.4372(148) = 131.6 \text{ m}$$

The drift estimation error variance is found by

$$\sigma_m^2(\mathbf{x}_0) = \mathbf{M}'\mathbf{\Delta} = 229.3 \text{ m}^2$$

As usual, the standard error of the estimate is the square root of the estimation variance, or

$$\sigma_m(\mathbf{x}_0) = \sqrt{229.35} = 15.1 \text{ m}$$

Again, by assuming normality, we can make probabilistic statements in the form of a confidence interval about the true value of the drift. For example, the probability is 95% that the true linear drift lies within the interval 131.6 ± 29.6 m, or between 102.0 and 161.2 m. (We must keep in mind that there is nothing "real" about the drift in a physical sense. The confidence interval merely indicates the uncertainty due to distance between observations and their spatial arrangement with respect to the point where the drift is being estimated.)

In order to keep the numerical examples in this section computationally tract-able, many simplifications have been made. For example, the number of obser-vations used is the smallest possible. In actual applications, many more control points should be considered because this will improve the accuracy of the kriged estimate and reduce the estimation error. Also, the semivariogram is assumed to be linear because, having only one parameter, this is the simplest model possible. A strictly linear semivariogram would not have a sill, but rather would continue on to an infinite variance. Here we have in effect assumed that the semivariogram is linear up to some value, then breaks and becomes a constant. Armstrong and Jabin (1981) have pointed out that semivariograms possessing sudden changes in slope may lead to unstable solutions and the calculation of negative variances. It would be better to use a continuous function such as the spherical model to represent the semivariogram, although it would complicate the calculations somewhat. The un-certainty in specifying a correct *a priori* estimate of the variance of the residuals is not too troublesome, because only the estimation error for the drift depends upon this parameter. The drift itself will be calculated correctly regardless of the value chosen as σ_0^2.

An example.—You may now appreciate that kriging, even the highly simplified vari-ations we have considered, can be arithmetically tedious. The practical application of kriging to a real problem is only possible with the use of a computer because the estimations must be made repeatedly for a large number of locations to character-ize the changes in a regionalized variable throughout an area. As an example we will consider a subset of data representing the elevation of the water table of the High Plains aquifer in a six-county area of central Kansas (Olea and Davis, 2000). The locations of the 161 observation wells, listed in file AQUIFER.TXT, are given as UTM coordinates in meters. Water-table elevations are given in feet above sea level.[†] A structural analysis shows that the water table can be regarded as a nonstationary regionalized variable having a first-order drift. A series of semivariograms were run to determine a direction perpendicular to the slope of the drift. A semivariogram (**Fig. 5–99**) calculated along direction N 2° W is drift-free and has a range of 63, 400 m and a sill of 1546 ft^2. A spherical semivariogram model was fitted to the experimental semivariances by weighted regression as discussed in Chapter 4.

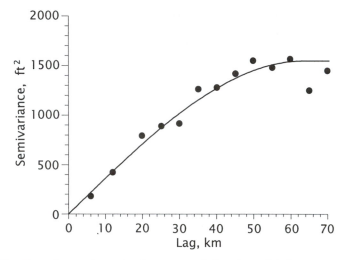

Figure 5–99. Experimental semivariogram and fitted model for water-table elevations in High Plains aquifer in an area of central Kansas along drift-free orientation of N 2° W. Lag distance in meters, semivariance in square feet. Model is spherical with range = 63,388.8 and sill = 1546.4.

Universal kriging was used to generate estimates of the water table at 7575 locations at the intersections of 75 columns and 101 rows equally spaced across the map area. The individual universal-kriging estimates were based on a maximum of 24 control wells, selected by an octant search around the location being estimated. Each estimate required the solution of as many as 27 simultaneous equations. The completed map of water-table elevation is shown in **Figure 5–100 a**.

In addition to the map of the water table itself, kriging was also used to produce the map of the standard error of the estimates shown in **Figure 5–100 b**. The standard error is zero at each of the 161 observation wells, but increases with distance away from the known control points. If we assume the kriging errors are normally distributed, we can assert that there is a 95% probability the true surface of the groundwater table lies within an interval defined by plus or minus 1.96 times the indicated value; that is, within $\pm 1.96\,\sigma$. For example, at point A the water table is estimated to be at about 1750 ft. Because there are relatively few observation wells near this location, the map of standard error indicates a value of over 15 ft.

[†] Use of the "footric" system of mixed English and metric measurements is not recommended practice but sometimes is unavoidable. Well locations, given in latitude and longitude on the geoid, must be projected into a flat Cartesian coordinate system. This can be done most conveniently by a Universal Transverse Mercator (UTM) projection, which yields coordinates in meters from a zone origin. Local usage requires that depths to the water table be expressed in feet. Similar mixed-unit usage is common in oil-well logs, where depths may be given in meters but measurements are recorded at 6-in. intervals.

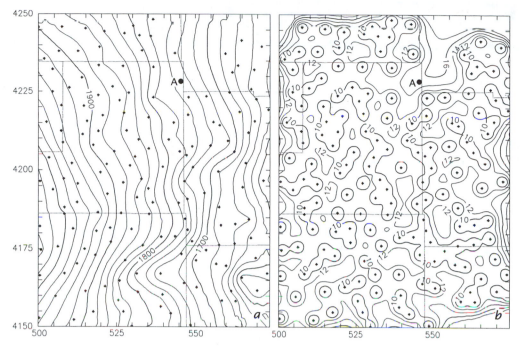

Figure 5–100. (*a*) Map showing elevation of water table in High Plains aquifer in an area in central Kansas. Map produced by universal kriging, assuming a first-order drift. Contours in feet above sea level. Crosses indicate observation wells. Geographic coordinates in kilometers from origin of UTM Zone 14. (*b*) Map of kriging standard error of estimates of water-table elevations in High Plains aquifer. Contour interval is 2 ft.

Therefore the true elevation at this point, with 95% confidence, must be 1750 ± 30 ft, or between 1720 and 1780 ft.

In this application there is little geologic significance to the drift itself, but it can be mapped if desired. **Figure 5–101a** shows the first-order drift of the water-table elevation; **Figure 5–101b** is the standard error map for the drift. The residuals from the drift, found by subtracting the drift map from the kriged map, are shown in **Figure 5–102**. Areas of negative residuals where the water table is lower than the drift are shown by shading.

Block kriging

Simple, ordinary, and universal kriging can be collectively considered as **punctual kriging**, in which the support consists of dimensionless points only. We will now briefly consider a further generalization of kriging in which the support for estimates of the regionalized variable consists of relatively large areas or volumes. **Block kriging** is the general name for a collection of kriging procedures that include a change in support for the regionalized variable, most typically from point observations to estimates that represent the average of areas or volumes. Olea (1999) provides a succinct development of block kriging. A less formal discussion is given by Isaaks and Srivastava (1989). Many of the earlier references on geostatistics emphasize block kriging because the original applications of geostatistics involved estimating the values of mining units such as stope blocks from drill-core assays—this requires a change of support from points to volumes.

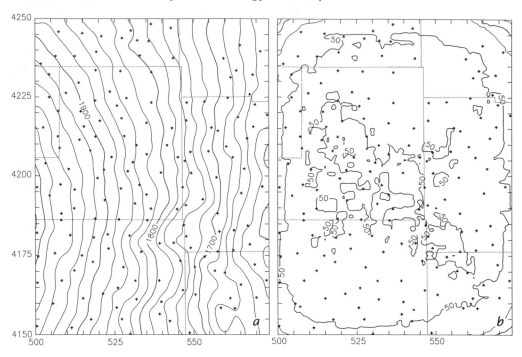

Figure 5–101. (*a*) Map showing first-order drift of water-table elevation in the High Plains aquifer of an area in central Kansas. Contours are in feet above sea level. (*b*) Map of standard error of first-order drift of the water table in the High Plains aquifer. Contour interval is 50 ft.

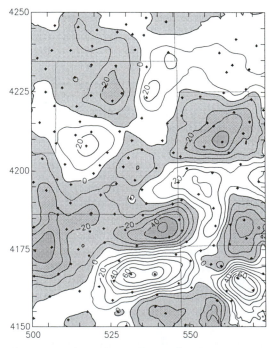

Figure 5–102. Map of residuals from first-order drift of water-table elevation in the High Plains aquifer of an area in central Kansas. Contour interval is 10 ft. Shaded areas indicate negative residuals where water table is lower than drift.

The most commonly used version of block kriging is the counterpart of ordinary kriging and involves solving a set of normal equations which are equivalent to Equation (5.104). However, the spatial covariances or semivariances in the right-hand vector B of Equation (5.104) no longer represent relationships between point observations, x_i, and a point location, x_0. Instead they are averages of the spatial covariances or semivariances between the point observations x_i and all possible points within A, the area or volume of interest. Actually, the elements of the right-hand vector B represent the covariances or semivariances between the observations and A, integrated over the area or volume of A. As a practical matter, the integration must be approximated by averages of spatial covariances or semivariances between the observations and point locations within A.

Examples of block kriging include the estimation of average percent copper in mining blocks measuring $10 \times 10 \times 3$ m in an open-pit copper mine in Utah, using assays of diamond-drill cores as data. The assay values can be regarded as point observations in three-dimensional space. A similar problem is the estimation of average porosities within $200 \times 200 \times 20$-ft compartments in a western Kansas oil field. The compartments correspond to the cells in a black-oil reservoir simulation model. Observations are apparent porosities calculated at selected depths from well logs. A two-dimensional example of block kriging, based on measurements from auger samples, is the estimation of chromite content (in kg/m^2) within uniform, 50×100-m rectangular areas of a black-sand beach deposit in the Philippines.

One of the first steps in block kriging is deciding how the blocks are to be discretized or subdivided into subareas or subvolumes in order to approximate integration of the spatial covariance or semivariance over the blocks. Each smaller subarea or subvolume is represented by the geographic coordinates of its center point and the spatial covariances or semivariances are determined between an observation x_i and each of the center points within a block. All of these covariances or semivariances are then averaged to determine the point-to-block spatial covariance, $cov(A, x_i)$, or semivariance, $y(A, x_i)$, between block A and observation x_i. The subdivision of a block should always be regular and should be the same for all blocks if possible. In principle, the greater the number of subdivisions of a block, the more accurate will be the approximation of the integrated spatial covariance or semivariance over the block. As a practical matter, Isaaks and Srivastava (1989) found that $4 \times 4 = 16$ subdivisions of areas or $4 \times 4 \times 4 = 64$ subdivisions of volumes are adequate.

To provide a simple demonstration of block kriging, we will adapt the problem shown in **Figure 5–95** to the estimation of the average water-table elevation within the 1-km square area A shown in **Figure 5–103**. Although this is not a particularly realistic application for block kriging, it allows us to compare our results to those obtained by punctual kriging. The left-hand matrix, W, of semivariances between observations remains unchanged, as does its inverse, W^{-1}.

To determine the vector of semivariances between the observations and block A, we must subdivide the block into an appropriate number of subareas and determine the distances to the observations. Distances to observation x_1 are shown in **Figure 5–104** and are listed, with the corresponding semivariances, in **Table 5–16**. Distances and semivariances to other observations are found in the same manner.

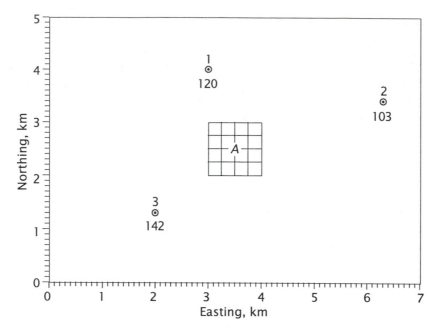

Figure 5–103. Map showing water-table elevations (in meters) at three observation wells. Estimates of the average water-table elevation will be made for block A. Coordinates given in kilometers from an arbitrary origin.

The average point-to-block semivariances form the elements of vector **B**:

$$\mathbf{B} = \begin{bmatrix} \gamma(A, x_1) \\ \gamma(A, x_2) \\ \gamma(A, x_3) \\ 1 \end{bmatrix} = \begin{bmatrix} 6.42 \\ 11.82 \\ 7.76 \\ 1.0 \end{bmatrix}$$

The vector of block-kriging coefficients resulting from the operation $\mathbf{W}^{-1}\mathbf{B}$ is

$$\Lambda = \begin{bmatrix} 0.416 \\ 0.188 \\ 0.396 \\ -0.425 \end{bmatrix}$$

Inserting these block-kriging weights into the kriging equation results in an estimate of the average water table within the 1-km^2 area A. The estimated block average elevation is 125.5 m, nearly the same as the punctual estimate of 125.1 m at location $\mathbf{x}_p$ produced by ordinary kriging. However, the block-kriging estimation variance is 7.54 m^2, which is larger than the punctual-kriging variance of 5.25 m^2.

Block kriging is most often used in mining problems, either in two or three dimensions. As a simple two-dimensional example, we will consider the 164 observations in file RANIGANJ.TXT, which represent the coordinates of drill holes in the Raniganj coal field near Calcutta, India. The variable is "Useful Heat Value" (UHV), expressed in kilocalories/kilogram and derived from measurements of ash and moisture content of the coal. UHV is used to set the market price of coal in India.

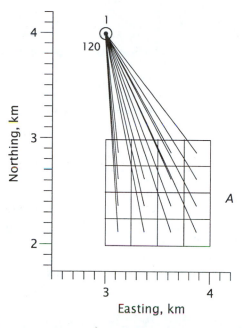

Figure 5–104. Lines connecting observation well 1 (Fig. 5–103) to centers of subareas of block A. Corresponding distances and semivariances given in Table 5–16.

Table 5–16. Distances and semivariances from subcell centroids of block A to observation well 1.

Easting km	Northing km	Distance km	Semivariance m^2
3.125	2.875	1.132	4.528
3.125	2.625	1.381	5.523
3.125	2.375	1.630	6.519
3.125	2.125	1.879	7.517
3.375	2.875	1.186	4.743
3.375	2.625	1.425	5.701
3.375	2.375	1.668	6.671
3.375	2.125	1.912	7.649
3.625	2.875	1.287	5.148
3.625	2.625	1.510	6.042
3.625	2.375	1.741	6.964
3.625	2.125	1.976	7.906
3.875	2.875	1.425	5.701
3.875	2.625	1.630	6.519
3.875	2.375	1.846	7.382
3.875	2.125	2.069	8.276

Drill holes have been placed on a rough grid spacing of about 500 m × 500 m across the mine, although the grid has some gaps in places and in other places additional drillings have been made. Reserve estimates are based on 31 blocks that cover the minable extent of the Raniganj coal field; each block is 1 km² in area.

The spatial continuity of UHV can be modeled by a Gaussian semivariogram with a nugget effect. An appropriate model has a magnitude of $240,000$ $(kcal/kg)^2$ for the nugget, coefficients of $631,000$ $(kcal/kg)^2$ for the sill minus the nugget, and 3607.76 m for the apparent range. (It may be convenient to scale the data by dividing by 1000 to convert the UHV's to kcal/g. This will avoid extremely large covariances in the kriging matrices.) **Figure 5–105** shows average block values of UHV in the Raniganj coal field estimated by block kriging. Because the average value changes abruptly between one block and adjacent blocks, the results cannot be represented as a contoured map. Instead, the block average values are shown as shades of gray applied uniformly across each block. Each block has an associated standard error of the block-kriging estimate, but these also cannot be shown in contoured form. Because most of the 31 blocks in the Raniganj field have about the same standard error (approximately 1400 kcal/kg), a shaded map would be relatively featureless and is not shown.

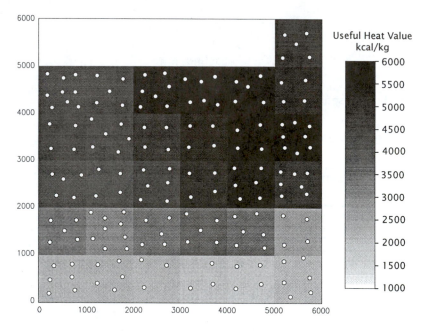

Figure 5–105. Map of average Useful Heat Value (UHV) in 1-km^2 blocks of the Raniganj coal field in India.

For comparison, a punctual-kriging map computed using universal kriging and assuming a first-order drift is shown in **Figure 5–106**. The contour interval is 500 kcal/kg and the range of contours extends from below 2000 kcal/kg to over 6000 kcal/kg. The nugget effect is expressed by the "bulls-eye" pattern of contours that surround some individual drill holes.

We have only touched on some of the topics that constitute the field of geostatistics and the family of kriging estimators. Other kriging techniques include *indicator kriging*, in which the regionalized variable is coded into dichotomous classes (1 and 0) and the predicted value can be interpreted as a probability of the occurrence of one class; *disjunctive kriging*, which requires assumptions about the form of the distribution of the regionalized variable; *multigaussian kriging*, which involves a normal score (Z) transformation; and numerous other, less widely used

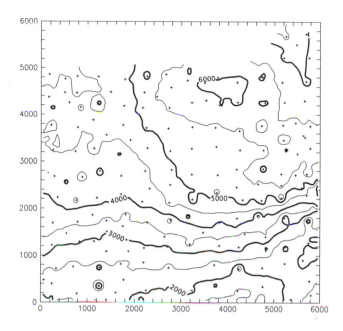

Figure 5–106. Contour map of Useful Heat Value (UHV) in the Raniganj coal field, India, estimated by universal kriging.

variants of kriging estimators. *Cokriging* is a family of procedures that includes equivalents to the standard kriging techniques, but is used to estimate values of several regionalized variables simultaneously when they have been measured at the same locations within the same sampling area. The spatial covariances of kriging are supplemented with spatial cross-covariances between the regionalized variables. *Cross validation* is a method for checking the performance of a kriging procedure by iteratively removing observations from the data set, estimating values at the missing locations, and then comparing the estimates to the true values. Finally, *simulation* is playing an increasingly important role in geostatistical studies. This involves creating a randomized realization having the same spatial statistics as the observed regionalized variable. In *conditional simulation*, the random realizations are constrained so they have the same values as the real observations at the control points.

EXERCISES

Exercise 5.1

Determine the probability of detecting a magnetic anomaly in the form of an ellipse 2 mi long × 1 mi wide (approximately 1000 acres in area), using a search consisting of parallel survey lines spaced 1 mi apart. Assume that the anomaly may occur anywhere within the search area and may be oriented in any manner.

In the interests of economy, the distance between the magnetic survey lines may be doubled. What effect will this have on the probability of detecting a magnetic anomaly?

443

Exercise 5.2

An area in central Alabama is being evaluated for the location of an oil-fired power plant. Because the power plant must be built on extremely solid underpinnings, a seismic survey will be run to determine if the bedrock is suitable. The near-surface geology consists of a sequence of nearly flat-lying limestones and shales, overlain by a thin veneer of unconsolidated sands and clays. The uppermost bedrock unit is the Oligocene Glendon Limestone; severe weathering has led to development of karstic features on its upper surface. Drilling in the proposed power plant site has confirmed the presence of rubble-filled collapse features in the limestone. In the nearby national forest, the Glendon Limestone is exposed along river bluffs where the collapse features typically are about 500 ft in diameter. If a high-resolution seismic survey consisting of a square grid of lines spaced 700 ft apart is run in the area, what is the probability that a sinkhole of typical size will be missed?

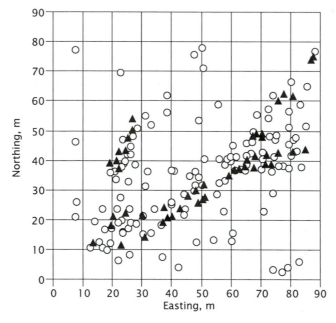

Figure 5–107. Spatial distribution of vegetation across test area in southern Arizona. Circles are creosote bush and triangles are brittlebush.

Exercise 5.3

Calibrating airborne or satellite systems that remotely sense ground conditions requires the extensive collection of "ground-truth" information about details of the Earth's surface that might have an effect on the electromagnetic spectrum being measured. Thermal infrared radiation from the ground is influenced primarily by topography, soil type, soil moisture, and vegetation. As part of a study to calibrate remote sensing instruments, a high-angle photograph was taken of a test area in southern Arizona and the location of each plant in the scene was noted; the coordinates of individual plants are given in file THERMAL.TXT. **Figure 5–107** is a map of the test area with a superimposed 10-m grid and locations of individual plants indicated by symbols. Plants were classified into two types; creosote bush (*Larrea tridentata*) and similar forms, and brittlebush (*Encelia farinose*) and other plants of similar habit. The scene is to be generalized for use in a mathematical model of thermal radiation; this requires a statistical description of the distribution of plants across the area. Using quadrat and nearest-neighbor procedures, produce appropriate locational statistics for the vegetation. Do the two plant types differ in their spatial distributions?

Exercise 5.4

To determine the directions of principal fracturing in oil fields in the Midland Basin of Texas, lineaments believed to represent surface traces of fractures within the areas of the fields were interpreted from aerial photomosaics (Alpay, 1973). The resulting patterns for three fields near Odessa, Texas, are shown in **Figure 5–108**. Borehole televiewers run in the Grayburg Dolomite (Permian) reservoir interval of the Odessa North field (**Fig. 5–108 a**) indicate that fractures within the reservoir

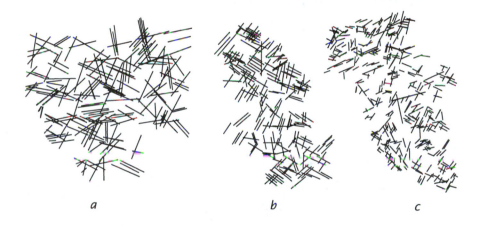

a   *b*   *c*

Figure 5–108. Fracture patterns over oil fields near Odessa, Texas, interpreted from lineaments on aerial photomosaics. (*a*) Odessa North field, producing from Grayburg Dolomite. (*b*) Odessa Northwest field, producing from Grayburg Dolomite. (*c*) Odessa West field, producing from San Andres Limestone.

trend approximately N 75° E (data are given in file ODESSAN.TXT). Dye injection tests also suggest channeling in approximately the same direction. Does this correspond with the mean vector direction of fracturing?

The Odessa Northwest field (**Fig. 5–108 b**) also produces from the Grayburg Dolomite (file ODESSANW.TXT). Gas-injection pressure histories suggest fracturing along N 35° E and N 55° W directions. Do either of these orientations seem to reflect the mean vector direction of surface fractures?

The Odessa West field (**Fig. 5–108 c**) produces from the Permian San Andres Limestone (file ODESSAW.TXT). Water-flood breakthrough problems indicate channeling through the saturated interval in directions that are approximately N 30° E and N 75° E. Does the mean vector direction of the fractures correspond to either of these orientations?

Compare the mean directions of fractures from the three fields. Are they significantly different?

Exercise 5.5

The relationships between quartz and feldspar grains in a granite reflect in part the history of crystallization of the granitic melt (Rosenberg and Riller, 2000). If the quartz grains are euhedral, the apparent dihedral angles between faces of adjacent quartz grains where they are in contact with feldspar can be measured on electron photomicrographs. (The angles are apparent because they are the intersections of three-dimensional surfaces with the plane of the polished section.) The dihedral angles formed in quartz-granitic melt systems are 22°–23°, whereas the angles formed during natural equilibration in the solid state at high temperatures are about 110°. Quartz-quartz-feldspar apparent dihedral angles were measured on 110 images of granite from the Murray pluton of central Ontario, Canada, and are listed in file DIHEDRAL.TXT. Plot the apparent angles on a circular histogram and determine their statistics, including the median apparent angle which approximates the true dihedral angle between grain faces. Do the results suggest that the grains originated from a melt or in a solid state? (Keep in mind that dihedral angles range only from 0° to 180°, so your histogram should have the form of a half-rose.)

Exercise 5.6

During drilling by the R.V. Glomar Challenger of Site 146 (15.13° N, 69.38° W) in the Caribbean Sea part of Leg 15 of the Deep Sea Drilling Project, several diabase sills of Late Cretaceous age were cored. The natural remanent magnetism of the oriented cores was measured on 12 plugs cut from the core; the data are listed in file GLOMAR.TXT. The measurements are given in terms of declination (degrees clockwise from north) and inclination (degrees downward from the horizontal plane). According to the National Geophysical Data Center, the magnetic declination and inclination at this site at the time the cores were taken were 354.3° and 46.4°. Plot the remanent magnetic measurements on a Schmidt or Wulff net and calculate the mean direction and spherical variance. Do the paleomagnetic readings differ significantly from the modern magnetic field? What does this suggest about the movement of tectonic plates in this region since the Cretaceous?

Exercise 5.7

File WASATCH.TXT contains 20 observations of azimuth and dip of cross-bedding measured on the Eocene Cathedral Bluffs Member of the Wasatch Formation in Carbon County, Wyoming (from Steinmetz, 1962). Plot the observations on a Schmidt net and calculate their mean azimuth and dip, as well as the mean vector resultant and the spherical variance. Does the plot suggest that the data are drawn from a single population of crossbeds? Calculate the matrix of directional covariances for the 20 observations and find its eigenvalues. Does the appearance of the plot accord with expectations based on the eigenvalues?

Exercise 5.8

Surficial material in piedmont areas of the Laguna Mountains near Yuma, Arizona, are composed of coarse, unsorted gravels underlain by sand. These alluvial pediments may become deeply dissected by numerous gullies and washes, creating an extremely rugged badlands topography. Aerial photographs taken at a low sun angle reveal a complicated, interfingering, digitate pattern (**Fig. 5–109**). Examine the photograph using the grid cell measurement method and determine the fractal dimension of the pattern. (Files YUMA.PIC and YUMA.TIF contain bitmap images of the aerial photo in **Figure 5–109**, in formats suitable for use on Macintosh or PC operating systems.)

Figure 5–109. High-contrast aerial photograph of piedmont area of Laguna Mountains, Arizona, taken at low sun angle. Area is 1 mi^2.

Exercise 5.9

A topographic map of an area of Pleistocene sand dunes near Garden City, Kansas, was shown in **Figure 5–49**. The spectral method for estimating fractal dimension was demonstrated using elevations measured along a north–south profile across the center of this area (**Fig. 5–46**; file GARDENNS.TXT). A second, perpendicular profile has been measured in an east–west direction and is given in file GARDENEW.TXT. Using this second profile, estimate the fractal dimension of the dune topography by the spectral method. Remember that before estimating the power spectrum of a trace you must confirm that the data are stationary, and if not, you must level the data prior to calculating the Fourier transform.

Exercise 5.10

File RESSEREL.TXT contains coordinates of 11 landmarks measured on ten specimens of the Ordovician orthid brachiopod, *Resserella* sp.; a typical specimen is shown in **Figure 5–52**, on p. 358. The same landmark points have been measured on seven specimens of a similar, but somewhat larger brachiopod; the measurements are contained in file ORTHID.TXT. The coordinates are given in centimeters from an arbitrary origin in the lower left. Convert the Cartesian coordinates to Bookstein coordinates so the shapes of the two collections of brachiopods can be compared. Do this by plotting the landmarks for all the brachiopods on one graph, using different symbols for the two sets. Plot bivariate means and 90% bivariate probability ellipses for the landmarks in each set. Do the ellipses drawn on one set enclose the means of the landmarks for the other set? What does this suggest about the similarity in shape of the two collections of brachiopods?

Exercise 5.11

Figure 5–110 is a photomicrograph of ferruginous ooids in an arenite of the Triassic Nubia Formation of Egypt. Iron-rich ooids may have formed by secondary growth within sediments, or by replacement of carbonate ooliths, in shallow water near the mouths of rivers draining tropical lateritic soils. The shape of the ooids may shed some light on their origin. File FEOOID.TXT contains polar coordinates of the outlines of seven of the ooids in the photomicrograph, measured at 5° increments. The first column contains angles (in degrees counterclockwise from the positive end of the X-axis) of radii extending from the centroid to the perimeter of the

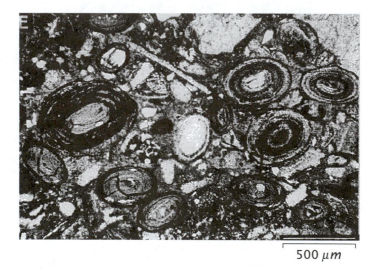

500 μm

Figure 5–110. Thin section of arenite from the Triassic Nubia Formation, Egypt, showing ferruginous ooids associated with bioclasts and siliciclasts. Scale is 500 μm. From Garzanti (1991).

grain. The second column contains lengths of the radii in microns (0.001 mm). Characterize the shapes of the grains by their power spectra and comment on the degree of similarity in their shapes.

Exercise 5.12

File SLOFEPB.TXT contains measurements of iron and lead made at 129 locations in Slovenia as part of a study of soil geochemistry by Pirc, Bidovec, and Gosar (1994), as described on p. 369–370. The data were collected following the same nested design used in the mercury contamination study, and are coded in the file in the same manner as data in file SLOVENIA.TXT. Perform a nested ANOVA on log Pb, which is mostly of anthropogenic origin, and determine the proportion of total variance attributable to the various spatial scales. Repeat the analysis on log Fe, whose concentration in soil primarily reflects geologic factors. Are there differences in the spatial scale of variability between the two elements?

Exercise 5.13

Some of the most productive oil reservoirs in North America occur in reefs of Permian age that have been discovered in west Texas. The dolomitic reef facies is enclosed in impermeable shales, usually with abrupt lateral changes in lithology. Isopach maps of the thickness of the dolomitic interval are useful for predicting where production wells might best be located, but contouring the thickness of a unit that abruptly thins and is absent over part of the mapped area is challenging. In this circumstance, a thickness of 0.0 indicates that the unit is absent. File REEF.TXT gives thicknesses in feet of the dolomitic carbonate facies encountered in wells whose coordinates (eastings and northings) are given in feet from an arbitrary origin located southwest of the map area. The symbol code is 5 = dry hole, reef facies absent; 3 = dry hole, reef facies present; 4 = producing well. Produce an isopach map of the reef interval, taking into consideration the discontinuous nature of the mapped property. Explain the possible cause of any features that appear to be unrealistic along the reef margin (*i.e.*, at the zero isopach) and describe how such features might be controlled. Data were kindly provided by R.B. Banks, Scientific Computer Applications, Inc.

Exercise 5.14

Reflections interpreted on seismic profiles often are contoured to develop an understanding of the subsurface structural configuration of the reflecting horizons. The data are given as coordinates locating the common midpoints along the seismic profiles and the two-way travel times to the reflecting event of interest. File SEISMIC.TXT gives information collected from five intersecting seismic lines over a structure in southern California. Coordinates are given in feet from an arbitrary origin to the southwest of the mapped area; reflection times are given in milliseconds. Contour the data using two alternative gridding procedures, one that uses a nearest-neighbor search criterion, and one that uses an octant search constraint. Which approach produces the contour map that is most acceptable in appearance? Contour the data using an algorithm that projects slopes and compare the resulting map with one created using a weighted-averaging algorithm. What combination of contouring options seems to be best for these data? Generate a triangular network among the seismic data points and contour the resulting irregular mesh. What steps could you take to improve the contour map? Data were kindly provided by R.B. Banks, Scientific Computer Applications, Inc.

Exercise 5.15

The Tonga–Kermadec region north of New Zealand (**Fig. 5–111**) is marked by deep oceanic trenches, volcanic islands, and a pattern of earthquake hypocenters located at increasingly greater depths. These define a Benioff subduction zone which can be statistically modeled as a dipping plane. Data in file TONGA.TXT consist of latitudes, longitudes, and depths in kilometers of hypocenters of earthquakes in the Tonga–Kermadec region [taken from the report of the National Earthquake Information Center (NEIC), "Preliminary Determination of Epicenters for the Period June–August, 1972"]. Since longitudes and latitudes are nearly equal at these low latitudes, it is not necessary to convert degrees latitude and longitude to Cartesian coordinates. Fit a first-order trend surface to the data. Because the 3-mo record can be considered a random sample from the continuous record of earthquakes in the region, analysis of variance can be used to test the statistical significance of the fitted trend. The trend coefficients give the apparent slope of the fitted plane in the east-west and north-south directions and can be converted into true dip by finding their vector resultant.

Figure 5–111. Location of earthquake epicenters (dots) in Tonga–Kermadec region of New Zealand for the period June–August, 1972 (from NEIC).

Figure 5–112. Location of earthquake epicenters (dots) in a region along the west coast of South America for the period June–August, 1972 (from NEIC).

Exercise 5.16

The Pacific Plate underthrusts the South American Plate on the eastern side of the Pacific Ocean, leading to frequent earthquakes in Chile, Bolivia, and Peru (**Fig. 5–112**). File ANDES.TXT contains earthquake hypocenters for this region during the period June–August, 1972. The hypocenters define the subduction surface, but unlike that of the Tonga Trench (Exercise 5.15), the South American subduction surface is curved. A higher order polynomial trend may be a better model, but the statistical significance of a higher order trend over a simpler linear model should be tested. Latitudes and longitudes can be treated as Cartesian coordinates in this exercise. Fit a series of polynomial trend surfaces up to the fourth order to this data and check the significance of each successive increase in complexity. What polynomial seems the most appropriate model for the subduction surface in this region?

Exercise 5.17

The file ARBUCKLE.TXT contains well locations and subsea elevations of the top of the Ordovician Arbuckle Group in an area of central Kansas. A posting of the well locations is shown in **Figure 5–5**. Well coordinates are given in arbitrary units (one unit equals approximately 6 mi) measured from an origin to the northwest of the map area; elevations are in feet below sea level. The distribution of oil and gas fields seems to be related to broad regional structure patterns which can be investigated by trend-surface analysis. Produce a structure contour map of the top of the Arbuckle Group, then calculate and map the first-, second-, and third-degree trend surfaces and residuals. How do the patterns of residuals change? Compare the trend residual maps to **Figure 5–113**, which is a generalized map of the locations of major oil and gas fields in central Kansas (both within the Arbuckle and in overlying units). Do trend-surface residuals and the distribution of fields seem to have some correspondence? Do you think trend-surface analysis could be used for petroleum exploration on such a regional scale?

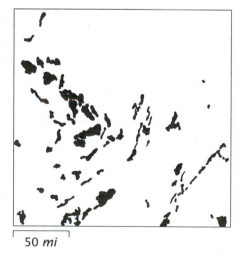

50 *mi*

Figure 5–113. Locations of major oil and gas fields in central Kansas.

Exercise 5.18

Both files SLOVENIA.TXT and SLOFEPB.TXT give the locations where soil samples were collected (in Gauss–Krueger coordinates, which are equivalent to UTM coordinates, but with a different origin). Calculate semivariograms for log Fe, log Pb, and log Hg using these coordinates. How does your interpretation of the scale of spatial variability based on semivariograms compare with your interpretation based on nested analysis of variance?

Exercise 5.19

Geophysical data such as that described in Exercise 5.13 and given in file SEISMIC.TXT have characteristics that make them simultaneously both well and poorly suited for geostatistical analysis. The two-way travel times are measured at uniform, closely spaced intervals along straight profiles, which are ideal for estimating semivariograms or spatial covariance functions. On the other hand, the seismic lines are widely spaced, so kriging estimates at locations between lines have large standard errors. Sort the data in file SEISMIC.TXT into individual seismic lines and calculate experimental semivariograms (or spatial covariances) for each line. If the experimental semivariograms for lines that are nearly parallel are similar, the lines can be combined to improve modeling of the semivariogram. Is there an indication of spatial anisotropy? Determine an appropriate model for the spatial structure of the reflection-time data and use this model to krige the reflecting surface, producing both a map of the estimated surface and a map of the standard error of the kriged estimate. Compare your ordinary-kriging contour map with the maps you made using arbitrary contouring procedures in Exercise 5.13. Describe the similarities and differences between the maps with respect to the uncertainties expressed in the kriging standard error map.

SELECTED READINGS

Adler, R.J., 1983, Hausdorff dimension, *in* Kotz, S., N.L. Johnson, and C.B. Read [Eds.], *Encyclopedia of Statistical Sciences,* Vol. 3: John Wiley & Sons, Inc., New York, p. 580–584.

Aitchison, J., and J.A.C. Brown, 1969, *The Lognormal Distribution: With Special Reference to Its Uses in Economics*: Cambridge Univ. Press, Cambridge, U.K., 176 pp.

Alpay, O.A., 1973, Application of aerial photographic interpretation to the study of reservoir natural fracture systems: *Jour. Petroleum Technology,* v. 25, no. 1, p. 37–45.

Anderson, R.Y., 1961, Solar-terrestrial climatic patterns in varved sediments: *Annals New York Academy of Science,* v. 95, p. 424–439.

Anderson, R.Y., and D.W. Kirkland, 1960, Origin, varves, and cycles of Jurassic Todilto Formation, New Mexico: *Bull. Am. Assoc. Petroleum Geologists,* v. 44, p. 37–52.

Antevs, E., 1925, *Retreat of the Last Ice Sheet in Eastern Canada*: Geol. Survey Canada, Memoir 146, 142 pp.

Armstrong, M., 1998, *Basic Linear Geostatistics*: Springer-Verlag, Berlin, 153 pp.

Armstrong, M., and R. Jabin, 1981, Variogram models must be positive-definite: *Mathematical Geology,* v. 13, no. 5, p. 455–459.

Bailey, T.C., and A.C. Gatrell, 1995, *Interactive Spatial Data Analysis*: Longman Scientific & Technical, Harlow, Essex, U.K., 413 pp., diskette. *Includes a DOS program for spatial analysis.*

Banks, R.B., 1991, Contouring algorithms: *Geobyte,* v. 6, no. 5, p. 15–23. *A good, nontechnical summary of contouring algorithms.*

Banks, R.B., and J.K. Sukkar, 1992, Computer processing of multiple 3-D fault blocks containing multiple surfaces: *Geobyte,* v. 7, no. 4, p. 58–62.

Barton, C.C., and P.R. LaPointe [Eds.], 1995, *Fractals in Petroleum Geology and Earth Processes*: Plenum Press, New York, 317 pp.

Batschelet, E., 1965, *Statistical Methods for the Analysis of Problems in Animal Orientation and Certain Biological Rhythms*: American Institute of Biological Sciences Monograph, Washington, D.C., 57 pp.

Bookstein, F.L., 1991, *Morphometric Tools for Landmark Data: Geometry and Biology*: Cambridge Univ. Press, Cambridge, U.K., 435 pp. *A thorough treatment of shape analysis based on landmarks, with biological and medical examples.*

Boon, III, J.D., D.A. Evans, and H.F. Hennigar, 1982, Spectral information from Fourier analysis of digitized quartz grain profiles: *Mathematical Geology,* v. 14, no. 6, p. 589–605. *Points out the importance of determining the true centroid in polar Fourier analysis, especially as applied to the study of grain shape.*

Bourke, P., 1989, *Triangulate User Notes,* Version 1.7: Downloadable from Internet website www. swin.edu.au/astronomy/pbourke/software/, 5 pp. Select Triangulate. *See also* www. swin.edu.au/astronomy/pbourke/modelling/ triangulate/.

Bourke, P., 1991, *An Introduction to Fractals*: Internet website www.swin.edu.au/astronomy/pbourke/fractals/fracintro/, 26 pp.

Bourke, P., 1993, *Fractal Dimension Calculator* (FDC) *User Manual*: Internet website www.swin.edu.au/astronomy/pbourke/fractals/fracdim/, 10 pp.

Caldenius, C.D., 1938, Carboniferous varves measured at Peterson, New South Wales: *Geol. Foren Stockholm Forh.,* v. 60, p. 349–364.

Carr, J.R., and W.B. Benzer, 1991, On the practice of estimating fractal dimension: *Mathematical Geology,* v. 23, p. 945–958.

Chayes, F., 1970, On deciding whether trend surfaces of progressively higher order are meaningful: *Bull. Geol. Soc. America,* v. 81, p. 1273–1278.

Chilès, J.-P., and P. Delfiner, 1999, *Geostatistics: Modeling Spatial Uncertainty*: John Wiley & Sons, Inc., New York, 695 pp. *A detailed treatment of geostatistics at an advanced level, from the viewpoint of the "French school."*

Clark, C.D., and C. Wilson, 1994, Spatial analysis of lineaments: *Computers & Geosciences,* v. 20, no. 7/8, p. 1237–1258.

Clark, M.W., 1981, Quantitative shape analysis: A review: *Mathematical Geology,* v. 13, no. 4, p. 303–320. *An extensive review of particle shape analysis, with emphasis on Fourier techniques.*

Clark, W.A.V., and P.L. Hosking, 1986, *Statistical Methods for Geographers*: John Wiley & Sons, Inc., New York, 518 pp.

Cliff, A.D., and J.K. Ord, 1981, *Spatial Processes: Models and Applications*: Pion Ltd., London, 266 pp. *This advanced text emphasizes spatial autocorrelation and analysis of the pattern of points. Most of the applications are to discontinuous geographic variables.*

Cramér, H., 1946, *Mathematical Methods of Statistics*: Princeton Univ. Press, Princeton, N.J., 575 pp.

Cressie, N.A.C., 1993, *Statistics for Spatial Data,* Rev. ed.: John Wiley & Sons, Inc., New York, 900 pp. *A statistician's view of geostatistics, in a formal treatment.*

Dacey, M.F., 1967, Description of line patterns: *Northwestern Studies in Geography,* v. 13, p. 277–287. *One of the few discussions of patterns formed by lines in a plane.*

Dahlberg, E.C., 1975, Relative effectiveness of geologists and computers in mapping potential hydrocarbon exploration targets: *Mathematical Geology,* v. 7, no. 5/6, p. 373–394.

Darnley, A.G., and others, 1995, *A Global Geochemical Database for Environmental and Resource Management*: UNESCO Publ., New York, 122 pp.

Davis, J.C., 1976, Contouring algorithms, *in* AUTOCARTO II, *Proc. International Symposium on Computer-Assisted Cartography*: U.S. Bureau of the Census, Washington, D.C., p. 352–359.

Davis, J.C., and R.J. Sampson, 1992, Trend surface analysis: *Geobyte,* v. 7, no. 4, p. 38–43.

Deutsch, C.V., and A.G. Journel, 1998, *GSLIB—Geostatistical Software Library and User's Guide,* 2nd ed.: Oxford Univ. Press, New York, 369 pp., CD-ROM. *Although primarily a software user's guide, this provides a succinct introduction to the "Stanford school" of geostatistics.*

Donnelly, K.P., 1978, Simulations to determine the variance and edge effect of total nearest neighbor distance, *in* Hodder, I. [Ed.], *Simulation Studies in Archaeology*: Cambridge Univ. Press, Cambridge, U.K., p. 91–95.

Doveton, J.H., and A.J. Parsley, 1970, Experimental evaluation of trend surface distortions induced by inadequate data-point distributions: *Inst. Mining and Metallurgy Trans.,* Sec. B, p. B197–B208. *The discussion of the influence of data-point distribution on trend surfaces in this chapter is based in large part on these experiments. Figures 5.93 and 5.94 are adapted from this article.*

Dryden, I.L., and K.V. Mardia, 1998, *Statistical Shape Analysis*: John Wiley & Sons, Inc., New York, 347 pp. *Emphasizes Procrustes procedures and the application of multivariate statistical methods.*

Ehrlich, R., J.P. Brown, and J.M. Yarus, 1980, The origin of shape frequency distributions and the relationship between size and shape: *Jour. Sedimentary*

Petrology, v. 50, no. 2, p. 475–484. *One of a series of articles by these authors on the use of circular Fourier analysis in the study of grain shape.*

ESRI, 1992, *Surface Modeling with TIN,* 2nd ed.: Environmental Systems Research Institute, Inc., Redlands, Calif., 258 pp.

Falconer, K.J., 1990, *Fractal Geometry: Mathematical Foundations and Applications*: John Wiley & Sons, Ltd., Chichester, U.K., 288 pp. *Although written at a relatively high mathematical level, this text is very readable and complete.*

Feder, J., and T. Jøssang, 1995, Fractal patterns in porous media flow, *in* Barton, C.C., and P.R. LaPointe [Eds.], *Fractals in Petroleum Geology and Earth Processes*: Plenum Press, New York, p. 179–226.

Fisher, N.I., T. Lewis, and B.J.J. Embleton, 1987, *Statistical Analysis of Spherical Data*: Cambridge Univ. Press, Cambridge, U.K., 329 pp. *The definitive treatment of spherical data analysis.*

Folk, R.L., 1968, *Petrology of Sedimentary Rocks*: Hemphill's, Austin, Texas, 184 pp. *Discusses traditional measures of grain shape.*

Gaile, G.L., and J.E. Burt, 1980, Directional statistics: Concepts and techniques in modern geography: *GeoAbstracts,* No. 25, Univ. East Anglia, Norwich, U.K., 39 pp.

Gallant, J.C., and others, 1994, Estimating fractal dimension of profiles: A comparison of methods: *Mathematical Geology,* v. 26, p. 455–481.

Gardner, M., 1967, Mathematical recreations—Dragon curves: *Scientific American,* v. 226, no. 4 (April), p. 117–120.

Garrett, R.G., 1983, Sampling methodology, *in* Howarth, R.J. [Ed.], *Statistics and Data Analysis in Geochemical Prospecting,* Vol. 2: Elsevier Scientific Publ. Co., Amsterdam, p. 83–110. *An introduction to nested ANOVA sampling designs.*

Garrett, R.G., 1994, The distribution of cadmium in A horizon soils in the prairies of Canada and adjoining United States: Current Research 1994-B, Geol. Survey of Canada, p. 73–82.

Garrett, R.G., and T.I. Goss, 1980, UANOVA: A FORTRAN IV program for unbalanced nested analysis of variance: *Computers & Geosciences,* v. 6, p. 61–68.

Garzanti, E., 1991, Non-carbonate intrabasinal grains in arenites: Their recognition, significance and relationship to eustatic cycles and tectonic setting: *Jour. Sedimentary Petrology,* v. 61, no. 6, p. 959–975.

Getis, A., and B. Boots, 1978, *Models of Spatial Processes: An Approach to the Study of Point, Line, and Area Patterns*: Cambridge Univ. Press, Cambridge, U.K., 198 pp. *Discusses numerous stochastic models that generate point and line patterns.*

Gold, C.M., T.D. Charters, and J. Ramsden, 1977, Automated contour mapping using triangular element data structures and an interpolant over each irregular triangular domain: *Computer Graphics,* v. 11, no. 2, p. 170–175.

Griffith, D.A., and C.G. Amrhein, 1991, *Statistical Analysis for Geographers*: Prentice Hall, Inc., Englewood Cliffs, N.J., 478 pp.

Griffith, D.A., and C.G. Amrhein, 1997, *Multivariate Statistical Analysis for Geographers*: Prentice Hall, Inc., Upper Saddle River, N.J., 345 pp.

Griffiths, J.C., 1962, Frequency distributions of some natural resource materials: *Mineral Industries Experiment Station Circ.,* v. 63, no. 7, Pennsylvania State Univ., p. 174–198. *This and the following reference discuss the negative binomial distribution as a model for the occurrence of mineral deposits and oil fields.*

Griffiths, J.C., 1966, Exploration for natural resources: *Jour. Operations Research Soc. America,* v. 14, no. 2, p. 189–209.

Gumbel, E.J., J.A. Greenwood, and D. Durand, 1953, The circular normal distribution: Tables and theory: *Jour. American Statistical Assoc.,* v. 48, p. 131–152.

Hadikusumo, D., 1961, Volcanological survey of Indonesia for the period 1950–1957: *Bull. Republik Indonesia,* Dasar/Pertambangan, Djawatan Geologi-Bandung, Dept. Perindustrian, 122 p.

Haggett, P., A.D. Cliff, and A. Frey, 1977, *Locational Analysis in Human Geography,* 2nd ed.: John Wiley & Sons, Inc., New York, 605 pp. *An advanced book containing techniques which may be unfamiliar to many geologists. Part 2, "Methods in locational analysis," discusses sampling, classification of regions, and testing of statistical hypotheses about spatial relationships.*

Hamilton, D.E., and T.A. Jones [Eds.], 1992, *Computer Modeling of Geologic Surfaces and Volumes,* Computer Applications in Geology Series, No. 1: Am. Assoc. Petroleum Geologists, Tulsa, Okla., 297 pp. *A collection of articles that emphasize the use of contouring programs for modeling and display of subsurface structure.*

Harbaugh, J.W., J.C. Davis, and J. Wendebourg, 1995, *Computing Risk for Oil Prospects: Principles and Programs*: Pergamon Press, Oxford, U.K., 452 pp., 2 diskettes.

Hohn, M.E., 1999, *Geostatistics and Petroleum Geology,* 2nd ed.: Kluwer Academic Publishers, Dordrecht, The Netherlands, 235 pp.

IBM, 1965, *Numerical Surface Techniques and Contour Map Plotting*: Data Processing Applications, International Business Machines, White Plains, N.Y., 35 pp.

Isaaks, E.H., and R.M. Srivastava, 1989, *Applied Geostatistics*: Oxford Univ. Press, New York, 561 pp. *A very readable introduction to the "Stanford school" of geostatistics, using a simulated regionalized variable throughout.*

James, R.C., and G. James, 1992, *Mathematics Dictionary,* 5th ed.: Van Nostrand Reinhold, New York, 548 pp.

Jones, N.L., and J. Nelson, 1992, Geoscientific modeling with TINs: *Geobyte,* v. 7, no. 4, p. 44–49.

Jones, T.A., D.E. Hamilton, and C.R. Johnson, 1986, *Contouring Geologic Surfaces with the Computer*: Van Nostrand Reinhold, New York, 314 pp. *Subsurface modeling using contouring programs, as developed by EXXON.*

Jones, T.A., and G.L. Krum, 1992, Pitfalls in computer contouring, Part II: *Geobyte,* v. 7, no. 4, p. 31–37.

Kaesler, R.L., and J.A. Waters, 1972, Fourier analysis of the ostracode margin: *Bull. Geol. Soc. America,* v. 83, no. 4, p. 1169–1178.

Kaye, B.H., 1994, *A Random Walk Through Fractal Dimensions,* 2nd ed.: VCH Verlag GmbH, Weinheim, Germany, 427 pp.

Kendall, D.G., 1989, A survey of the statistical theory of shape: *Statistical Science,* v. 4, no. 1, p. 87–120.

Kendall, M.G., and P.A.P. Moran, 1963, *Geometrical Probability*: Harner Publ. Co., New York, 125 pp. *Chapter 2 is concerned with the distribution of points on a plane.*

Koch, G.S., Jr., and R.F. Link, 1980, *Statistical Analysis of Geological Data*: Dover Publications, Inc., New York, 850 pp. *Trend-surface analysis is discussed in Chapter 9. Problems of systematic searching and the applications of response surfaces are given in Chapter 12.*

Korn, H., and H. Martin, 1951, Cyclic sedimentation in varved sediments of the Nama system in South-West Africa: *Trans. Geol. Soc. South Africa,* v. 54, p. 65–67.

Krum, G.L., and T.A. Jones, 1992, Pitfalls in computer contouring, Part I: *Geobyte,* v. 7, no. 3, p. 30–35.

Krumbein, W.C., 1939, Preferred orientation of pebbles in sedimentary deposits: *Jour. Geology,* v. 47, p. 673–706.

Krumbein, W.C., and H.A. Slack, 1956, Statistical analysis of low-level radioactivity of Pennsylvanian black fissile shale in Illinois: *Bull. Geol. Soc. America,* v. 67, p. 739–762. *Data in file BRERETON.TXT are taken from Table 2 of this article.*

Lambie, Fred, 1981, An analysis of the probability of hitting an arbitrary elliptical target with sets of parallel search lines: Unpub. Report, Terrasciences Inc., Lakewood, Colo., 17 pp.

Lee, D.R., and G.T. Sallee, 1970, A method of measuring shape: *Geographical Review,* v. 60, no. 4, p. 555–563.

Li, J.C.R., 1964, *Statistical Inference,* Vol. 2: Edwards Bros. Inc., Ann Arbor, Mich., 575 pp. *Chapter 30 discusses tests of curvilinear regressions, which are directly extendible to trend surfaces.*

Lorenz, J.C., L.W. Teufel, and N.R. Warpinski, 1991, Regional fractures, I: A mechanism for the formation of regional fractures at depth in flat-lying reservoirs: *Bull. Am. Assoc. Petroleum Geologists,* v. 75, p. 1714–1737.

Mandelbrot, B.B., 1982, *The Fractal Geometry of Nature*: W.H. Freeman & Co., San Francisco, 460 pp. *The seminal book on fractals, by the author who coined the term.*

Mandelbrot, B.B., D.E. Passoja, and A.J. Paullay, 1984, Fractal character of fracture surfaces of metals: *Nature,* No. 308, p. 721–722.

Mardia, K.V., 1972, *Statistics of Directional Data*: Academic Press, London, 357 pp. *A complete treatment of statisical methods appropriate for both two- and three-dimensional directional data. Many of the examples are taken from geology.*

McCammon, R.B., 1977, Target intersection probabilities for parallel-line and continuous-grid types of search: *Mathematical Geology*, v. 9, no. 4, p. 369–383. *Contains equations and graphs for determining the probabilities of hitting an elliptical target with a search pattern consisting of regularly spaced lines.*

McCullagh, M.J., 1981, Creation of smooth contours over irregularly distributed data using local surface patches: *Geographical Analysis*, v. 13, no. 1, p. 51–63. *This and the following article discuss calculation of a triangular mesh connecting irregularly spaced points, and the contouring of a surface using this mesh.*

McCullagh, M.J., and C.R. Ross, 1980, Delaunay triangulation of a random data set for isarithmic mapping: *Cartographic Journal*, v. 17, no. 2, p. 93–99.

Meakin, P., and A.D. Fowler, 1995, Diffusion-limited aggregation in the earth sciences, *in* Barton, C.C., and P.R. LaPointe [Eds.], *Fractals in Petroleum Geology and Earth Processes*: Plenum Press, New York, p. 227–261.

Mendenhall, W., 1968, *Introduction to Linear Models and the Design and Analysis of Experiments*: Wadsworth Publ. Co., Inc., Belmont, Calif., 465 pp. *Chapter 10 describes the fitting of response surfaces to experimental designs.*

Middleton, G.V., 2000, *Data Analysis in the Earth Sciences Using* MATLAB®: Prentice Hall, Inc., Upper Saddle River, N.J., 260 pp., diskette. *Excellent, brief introductions to contouring, trend analysis, and kriging with MATLAB routines. Also discusses plotting and analysis of directional data.*

Miesch, A.T., 1976, Geochemical survey of Missouri—Methods of sampling, laboratory analysis, and statistical reduction of data: U.S. Geol. Survey, Prof. Paper 954-A, 39 pp.

Miles, R.E., 1964, Random polygons determined by lines in a plane, I and II: *Proc. National Academy of Sciences*, v. 52, p. 901–907 and 1157–1160.

Moellering, H., and J.N. Rayner, 1979, Measurement of shape in geography and cartography: *Numerical Cartography Laboratory Report*, No. SOC77-11318, Ohio State Univ., 109 pp. *An extensive review of shape analysis, including polar Fourier methods.*

Moore, R.C., C.G. Lalicker, and A.G. Fischer, 1952, *Invertebrate Fossils*: McGraw-Hill, Inc., New York, 766 pp.

Myers, D.E., 1991, Interpolation and estimation with spatially located data: *Chemometrics and Intelligent Laboratory Systems*, v. 11, no. 3, p. 209–228.

Nemec, W., 1988, The shape of the rose: *Sedimentary Geology*, v. 59, p. 149–152.

Neumann van Padang, M., 1951, *Catalogue of the Active Volcanoes of the World Including Solfatara Fields*, Part I—*Indonesia*: International Volcanological Assoc., Naples, Italy, 271 pp.

Olea, R.A., 1982, Systematic approach to sampling spatial functions, Vol. 1: Ph.D. dissertation, Univ. Kansas, Lawrence, Kansas, 277 pp.

Olea, R.A. [Ed.], 1991, *Geostatistical Glossary and Multilingual Dictionary*: Oxford Univ. Press, New York, 177 p. *The definitive guide to geostatistical terms and usage.*

Olea, R.A., 1999, *Geostatistics for Engineers and Earth Scientists*: Kluwer Academic Publishers, Boston, Mass., 303 pp. *A very detailed and systematic discussion of geostatistics, written at a mathematically accessible level.*

Olea, R.A., and J.C. Davis, 2000, Year 2000 sampling analysis and mapping of water levels in the High Plains aquifer of Kansas: Kansas Geol. Survey, Open-File Report No. 2000–13, Lawrence, Kansas, 34 p., 5 plates.

Pawlowsky, V., R.A. Olea, and J.C. Davis, 1993, Boundary assessment under uncertainty: A case study: *Mathematical Geology,* v. 25, p. 125–144.

Pirc, S., M. Bidovec, and M. Gosar, 1994, Distribution pattern of mercury in Slovenia, *in* Roark, A. [Ed.], *Proc. 2nd International Symposium on Environmental Contamination in Central and Eastern Europe,* Budapest, Hungary, 20–23 Sept.: Tech. Univ. Budapest/Florida State Univ., Tallahassee, Fla., p. 857–861.

Prouty, W.F., 1952, Carolina bays and their origin: *Bull. Geol. Soc. America,* v. 63, p. 167–224.

Ragan, D.M., 1985, *Structural Geology: An Introduction to Geometrical Techniques,* 3rd ed.: John Wiley & Sons, Inc., New York, 393 pp.

Rasmussen, W.C., 1959, Origin of the "bays" and basins of the Atlantic coastal plain (Abs.): *Bull. Geol. Soc. America,* v. 70, p. 1660.

Reyment, R.A., R.E. Blackith, and N.A. Campbell, 1984, *Multivariate Morphometrics,* 2nd ed.: Academic Press, London, 233 pp. *Although concentrating on classification procedures, this text also discusses characterization and measurement.*

Ripley, B.D., 1981, *Spatial Statistics*: John Wiley & Sons, Inc., New York, 252 pp.

Rodríguez-Iturbe, I., and A. Rinaldo, 1997, *Fractal River Basins: Chance and Self-Organization*: Cambridge Univ. Press, Cambridge, U.K., 547 pp.

Rosenberg, C.L., and U. Riller, 2000, Partial-melt topology in statically and dynamically recrystallized granite: *Geology,* v. 28, no. 1, p. 7–10.

Sen, N., and P.K. Pal, 1978, A study on optimisation of sample size for proximate and ultimate analyses of coal core samples: *Minetech,* v. 3, no. 2, Central Mine Planning & Design Institute, Ranchi, India, p. 11–25.

Severson, R.C., and R.R. Tidball, 1979, Spatial variation in total element concentration in soil within the Northern Great Plains coal region: U.S. Geol. Survey, Prof. Paper 1134–A, p. A1–A18.

Shaw, G., and D. Wheeler, 1994, *Statistical Techniques in Geographical Analysis,* 2nd ed.: David Fulton Publ. Ltd., London, 359 pp. *Discusses both quadrat and nearest-neighbor techniques, with worked numerical examples.*

Singer, D.A., and F.E. Wickman, 1969, Probability tables for locating elliptical targets with square, rectangular, and hexagonal point-nets: *Mineral Sci. Experiment Station Spec. Publ. 1-69,* Pennsylvania State Univ., 100 pp.

Snedecor, G.W., and W.G. Cochran, 1989, *Statistical Methods,* 8th ed.: Iowa State Univ. Press, Ames, Iowa, 503 pp. *This classical text on agricultural and industrial statistics discusses nested ANOVA sampling designs under the heading, "split plots."*

Steinmetz, R., 1962, Analysis of vectorial data: *Jour. Sedimentary Petrology,* v. 32, p. 801–812.

Stephens, M.A., 1967, Tests for the dispersion and for the modal vector of a distibution on a sphere: *Biometrika,* v. 54, p. 211–223.

Stephens, M.A., 1969, Tests for randomness of directions against two circular alternatives: *Jour. American Statistical Assoc.,* v. 64, no. 325, p. 280–289.

Suppe, J., 1985, *Principles of Structural Geology*: Prentice Hall, Inc., Englewood Cliffs, N.J., 537 pp.

Tipper, J.C., 1979, Surface modelling techniques: Series on Spatial Analysis No. 4, Kansas Geol. Survey, Lawrence, Kansas, 108 pp.

Turcotte, D.L., 1997, *Fractals and Chaos in Geology and Geophysics, 2nd ed.*: Cambridge Univ. Press, Cambridge, U.K., 398 pp.

Upton, G.J.G., and B. Fingleton, 1985, *Spatial Data Analysis by Example,* Vol. 1, *Point Pattern and Quantitative Data*: John Wiley & Sons Ltd., Chichester, U.K., 410 pp. *Most applications are to geography and biology, but of interest to Earth scientists as well.*

Uspensky, J.V., 1937, *Introduction to Mathematical Probability*: McGraw–Hill, Inc., New York, 411 pp.

Vail, O.E., 1917, Lithologic evidence of climatic pulsations: *Science,* v. 46, p. 90–93.

Wahl, B., 1995, *Exploring Fractals on the Macintosh*: Addison–Wesley Publ. Co., Reading, MA, 352 pp., diskette. *A popular introduction to fractals.*

Watson, D.F., 1992, *Contouring: A Guide to the Analysis and Display of Spatial Data*: Pergamon Press, Oxford, U.K., 321 pp., diskette. *Contouring from the computer scientist's perspective.*

Watson, G.S., 1970, Orientation statistics in the earth sciences: *Bull. Geol. Inst. Uppsala,* v. 2, no. 9, p. 73–89.

Watson, G.S., 1983, *Statistics on Spheres*: John Wiley & Sons, Inc., New York, 238 pp.

Wells, N.A., 1999, ASTRA.BAS: A program in QuickBasic 4.5 for exploring rose diagrams, circular histograms and some alternatives: *Computers & Geosciences,* v. 25, p. 641–654. *The computer program can be downloaded from the International Association for Mathematical Geology website at www.iamg.org/.*

Woodcock, N.H., 1977, Specification of fabric shapes using an eigenvalue method: *Bull. Geol. Soc. America,* v. 88, p. 1231–1236.

Zurflueh, E.G., 1967, Applications of two-dimensional linear wavelength filtering: *Geophysics,* v. 32, p. 1015–1035.

Chapter 6
Analysis of Multivariate Data

In previous chapters we have considered the analysis of data consisting of only a single variable measured on each specimen or observational unit. In Chapters 4 and 5 we also considered the influence of the temporal or geographic coordinates of the sample points. We will now examine techniques for the analysis of multivariate data, in which each observational unit is characterized by several variables. Multivariate methods allow us to consider changes in several properties simultaneously. Examples of data appropriate for multivariate analysis abound in geology. They include chemical analyses, where the variables may be percentage compositions or parts per million of trace elements; measures on streams, such as discharge, suspended sediment load, depth, dissolved solids, pH, and oxygen content; and paleontologic variables, perhaps a large number of measurements made on specimens of an organism. Dozens of other examples quickly spring to mind. Some are simple extensions of problems we have considered previously; others are entirely new classes of problems.

Multivariate methods are extremely powerful, for they allow the researcher to manipulate more variables than can otherwise be assimilated. They are complicated, however, both in their theoretical structure and in their operational methodology. For some of the procedures, statistical theory and tests have been worked out only for the most restrictive set of assumptions. The nature and behavior of the tests under more relaxed, general assumptions (such as those necessary for most real-world problems) are inadequately known. In fact, some of the procedures we will consider have no theoretical statistical basis at all, and tests of significance have yet to be devised. Nevertheless, these methods seem to hold the most promise for fruitful returns in geological investigations. Most of the problems in geology involve complex and interacting forces which are impossible to isolate and study individually. Often a meaningful decision as to the relative worth of one of a number of possible variables cannot be made. The best course of action frequently is to examine as many facets of a problem as possible, and sort out, *a posteriori*, the major factors. The methods discussed in this chapter can be a significant help.

Multiple Regression

The first topic we will consider in our final chapter is actually a familiar subject under a new and more general guise. This is multiple regression, which includes polynomial curve fitting (discussed in Chapter 4) and trend-surface analysis (discussed in Chapter 5). However, we will now remove the restrictions that limited us to considerations of change as a function of temporal or spatial coordinates. Any observed variable can be considered to be a function of any other variable measured on the same samples. In Chapter 4 we considered changes in moisture content that occurred with changes in depth in the sediment. We could equally well have measured the montmorillonite content of the sediment in the core and examined the changes in water content that may accompany changes in montmorillonite percentage. In fact we could have measured several variables, perhaps organic content, mean grain size, and bulk density, and we could have examined the differences in water content associated with changes in each or all of these variables. In a sense, variables may be considered as dimensions, and their values as coordinates, so we can envision changes occurring "along" a dimension defined by a variable such as mineral content. Casting variables as dimensions is nothing new; we perform this every time we plot two variables against one another, because we are substituting spatial scales in the plot for the original scales on which the variables were measured. Such interchangeability is explicit in the references to "p-dimensional space" which abound in the literature of multivariate analysis. Just as trend surfaces are a generalization of curve-fitting procedures to two-dimensional space, multiple regression is a further generalization to "many-dimensional" space.

We will not consider multiple regression in great detail because the theoretical and computational essentials have been presented in earlier chapters. You will recall from Chapter 4 that polynomial regressions (having one independent variable) can be represented in a model equation of the general form

$$y_i = \beta_0 + \beta_1 x_{1i} + \beta_2 x_{1i}^2 \cdots + \beta_m x_{1i}^m + \varepsilon_i \tag{6.1}$$

The model states that the value of a dependent variable, y_i, at a location i is equal to a constant term plus the sum of a series of powers of an independent variable, x_{1i}, also observed at location i, plus a random error that is unique for location i. A least-squares solution to a linear equation of this type can be found by solving a set of normal equations for the β coefficients. These can be expressed in matrix form as

$$S_{XY} = S_{XX} \, b \tag{6.2}$$

with a solution

$$b = S_{XX}^{-1} \, S_{XY} \tag{6.3}$$

where S_{XY} is a column matrix of the sums of cross products of y, with $x_1, x_1^2, \ldots,$ x_1^m; S_{XX} is a matrix of sums of squares and cross products of the $x_1, x_1^2, \ldots, x_1^m$ powers; and b estimates β, the column matrix of unknown regression coefficients. In Chapter 4, we found the entries in the various matrices by labeling rows and columns and cross multiplying.

Although we regarded this problem as involving only one independent variable (or two, in the case of trend-surface analysis as discussed in Chapter 5), it can be regarded as containing m independent variables. This can readily be seen if we rewrite the model equation as

$$y_i = \beta_0 + \beta_1 x_{1i} + \beta_2 x_{2i} + \cdots + \beta_m x_{mi} + \varepsilon_i \tag{6.4}$$

and define the variables as $x_1 = x_1$, $x_2 = x_1^2$, $x_3 = x_1^3$, and so forth. Thus, the regression procedures we have considered up to this point have simply involved the definition of the independent variables in a specific manner.

A regression of any m independent variables upon a dependent variable can be expressed as in Equation (6.4). The normal equations that will yield a least-squares solution can be found by appropriate labeling of the rows and columns of the matrix equation and cross multiplying to find the entries in the body of the matrix. For three independent variables, we obtain

$$
\begin{array}{c}
\begin{array}{cccc} X_0 & X_1 & X_2 & X_3 \end{array} \\
\begin{array}{c} X_0 \\ X_1 \\ X_2 \\ X_3 \end{array}
\left[\quad\quad\quad\quad \right]
\end{array}
\begin{bmatrix} b_0 \\ b_1 \\ b_2 \\ b_3 \end{bmatrix}
=
\begin{array}{c} Y \\ \left[\quad \right] \end{array}
$$

where, again, x_0 is a dummy variable equal to 1 for every observation. The matrix equation, after cross multiplication, is

$$
\begin{bmatrix}
n & \sum x_1 & \sum x_2 & \sum x_3 \\
\sum x_1 & \sum x_1^2 & \sum x_1 x_2 & \sum x_1 x_3 \\
\sum x_2 & \sum x_1 x_2 & \sum x_2^2 & \sum x_2 x_3 \\
\sum x_3 & \sum x_1 x_3 & \sum x_2 x_3 & \sum x_3^2
\end{bmatrix}
\begin{bmatrix} b_0 \\ b_1 \\ b_2 \\ b_3 \end{bmatrix}
=
\begin{bmatrix}
\sum y \\ \sum x_1 y \\ \sum x_2 y \\ \sum x_3 y
\end{bmatrix}
\tag{6.5}
$$

The β's in the regression model are estimated by the b's, the sample **partial regression coefficients**. They are called partial regression coefficients because each gives the rate of change (or slope) in the dependent variable for a unit change in that particular independent variable, *provided* all other independent variables are held constant. Some statistics books emphasize this point by using the notation

$$y_i = b_0 + b_{1.23} x_{1i} + b_{2.13} x_{2i} + b_{3.12} x_{3i} + \varepsilon_i$$

The coefficient $b_{1.23}$, for example, is read "the regression coefficient of variable x_1 on y as variables x_2 and x_3 remain constant." In general, these coefficients will differ from the **total regression coefficients**, which are the simple regressions of each individual x variable on the dependent y variable. We ordinarily expect multiple regression coefficients to account for more of the total variation in y than will any of the total regression coefficients. This is because multiple regression considers all possible interactions within combinations of variables as well as the variables themselves.

We will consider a problem in geomorphology to illustrate a typical application of multiple regression. For this study, a well-dissected area of relatively homogeneous geology was selected in eastern Kentucky. The study region contains many drainage basins of differing sizes; from these, all third-order basins were chosen, and several variables were measured on each. The order of a drainage basin is defined by the number of successive levels of junctions on its stream from the stream's sources to the point where it joins another stream of equal or higher order. Thus, a third-order basin has two levels of junctions within its boundaries. Basin size, however, may be defined by many alternative methods. One of these is basin magnitude, which essentially is a count of the number of sources in the basin. A collection of basins of specified order may contain many different magnitudes. The relationship between magnitude and order of streams in drainage basins is shown in **Figure 6–1**. Seven variables were measured on the collection of third-order basins:

Y — Basin magnitude, defined by the number of sources.
X_1 — Elevation of the basin outlet, in feet.
X_2 — Relief of the basin, in feet.
X_3 — Basin area, in square miles.
X_4 — Total length of the stream in the basin, in miles.
X_5 — Drainage density, defined as total length of stream in basin/basin area.
X_6 — Basin shape, measured as the ratio of inscribed to circumscribed circles.

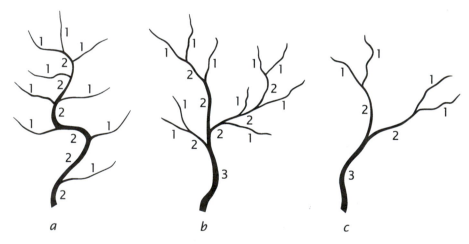

Figure 6–1. Contrast between stream magnitude and stream order. (*a*) Tenth-magnitude stream of second order. (*b*) Tenth-magnitude stream of third order. (*c*) Fourth-magnitude stream of third order. Magnitude is based on number of joining streams; order is based on succession of joining.

Our problem is to determine the influence of the six independent X variables on variable Y. Multiple regression, using basin magnitude as the dependent variable, is an appropriate technique. From the regression, the influence that all the variables have on basin magnitude can be assessed. File KENTUCKY.TXT contains measurements on these variables for 50 third-order basins in eastern Kentucky, taken from Krumbein and Shreve (1970). The significance of the linear relationship can be tested by analysis-of-variance methods presented in Chapter 4. **Table 4–9** (p. 197), for example, outlines the ANOVA for simple linear regression which may be expanded to multiple regression by changing the various degrees of freedom to account for additional variables. The modified ANOVA is shown in **Table 6–1**. The

Table 6–1. ANOVA for multiple regression with m independent variables.

Source of Variation	Sum of Squares	Degrees of Freedom	Mean Squares	F-Test
Linear Regression	SS_R	m	MS_R	MS_R / MS_D
Deviation	SS_D	$n - m - 1$	MS_D	
Total Variation	SS_T	$n - 1$		

Table 6–2. Completed ANOVA for the significance of regression of six geomorphic variables on basin magnitude.[1]

Source of Variation	Sum of Squares	Degrees of Freedom	Mean Squares	F-Test
Linear Regression	1800.70	6	300.12	11.38[**]
Deviation	1134.12	43	26.38	
Total Variation	2934.82	49		

[1] Regression equation: $y_i = -2.24 + 0.01\, X_{1i} + 0.02\, X_{2i} - 23.28\, X_{3i} + 6.26\, X_{4i} - 0.20\, X_{5i} - 11.66\, X_{6i}.$ $R = 0.78.$

[**] $p < 0.0001$ (highly significant).

completed ANOVA for multiple regression on basin magnitude is shown in **Table 6–2**. The regression coefficients are also shown.

In multiple-regression problems, we usually are interested in the relative effectiveness of the independent variables as predictors of the dependent variable. We cannot determine this from a direct examination of the regression coefficients, however, because their magnitudes are dependent upon the magnitudes of the variables themselves, which in part reflect the units of measurement. This is apparent in trend-surface analysis, where coefficients of higher orders almost invariably decrease in absolute size, even though higher orders may make greater contributions to the trend than lower orders. This results from the fact that a geographic coordinate, raised to a power as it is in high orders, is much larger in magnitude than the original coordinate. The higher order regression coefficients become correspondingly smaller.

Fortunately, it is easy to standardize the partial regression coefficients by converting them to units of standard deviation. The standard partial regression coefficients, B_k, are found by

$$B_k = b_k \frac{s_k}{s_y} \tag{6.6}$$

where s_k is the standard deviation of variable x_k and s_y is the standard deviation of y. Because the standard partial regression coefficients are all expressed in units of standard deviation, they may be compared directly with each other to determine the most effective variables.

To compute the matrix of sums of squares and products necessary in the normal equation set, we found the diagonal entries, $\sum x_k^2$. It is a simple matter to convert these sums of squares to corrected sums of squares, SS_k, and then to the standard deviations necessary to compute the partial correlation coefficients. However, it is possible to solve the normal equations in a manner that will yield the standardized partial regression coefficients directly, and gain an important computational advantage in the process.

The major sources of error in multiple regression occur in the creation of the entries in the S_{XX} matrix and during the inversion process. The sums of squares of the variables may become so large that significant digits are lost by truncation. If the entries in the S_{XX} matrix differ greatly in their magnitudes, an additional loss of digits may occur during inversion, especially if high correlations exist among the variables. Some computer programs may be capable of retaining only one or two significant digits in the coefficients, and with certain data sets retention may even be worse. Studies have shown that calculations using double-precision arithmetic may not be sufficient to overcome this problem. However, a few simple modifications in our computational procedure will gain us two to six significant digits during computation and greatly increase the accuracy of the computed regression (Longley, 1967, p. 821–827).

The most obvious step that can be taken is to convert all observations to deviations from the mean. This reduces the absolute magnitude of variables and centers them about a common mean of zero. As an inevitable consequence, the coefficient b_0 will become zero, so the matrix equation can be reduced by one row and one column. This simple step may gain several significant digits. However, we also may reduce the size of entries in the matrix still further by converting them all to correlations. This is equivalent to expressing the original variables in the standard normal form of zero mean and unit standard deviation. The matrix equation for regression then has the form

$$\mathbf{R_{XX}\, B = R_{XY}} \tag{6.7}$$

which can be solved by the operation

$$\mathbf{B = R_{XX}^{-1}\, R_{XY}} \tag{6.8}$$

where $\mathbf{R_{XY}}$ represents the column vector of correlations between y and the x_k independent variables. The $m \times m$ matrix of correlations between the x_k variables is represented by $\mathbf{R_{XX}}$. For example, the normal equation for three independent variables has the form

$$
\begin{bmatrix} 1 & r_{12} & r_{13} \\ r_{21} & 1 & r_{23} \\ r_{31} & r_{32} & 1 \end{bmatrix}
\begin{bmatrix} B_1 \\ B_2 \\ B_3 \end{bmatrix}
=
\begin{bmatrix} r_{x_1 y} \\ r_{x_2 y} \\ r_{x_3 y} \end{bmatrix}
\tag{6.9}
$$

Note that the equation has one less row and column than the equivalent equation using the original variables (Eq. 6.5).

Computing the regression equation in standardized form has the disadvantage that the correlation matrix must be created first, increasing the computational effort. In order to preserve accuracy, the correlations must be calculated using the definitional equation for the sums of products (Eq. 2.23; p. 40) rather than with the computational form for correlation given in Equation (2.28). This is because Equation (2.28) involves squaring the quantities $\sum x_j^2$ and $\sum x_k^2$. If these sums are large, the squares may be inaccurate because of truncation. This problem is avoided if the means are subtracted from each observation prior to calculation of the sums of squares. The sums of squares are then found by Equations (2.19) and (2.23). This process requires that the data be handled twice—first to calculate the means, and then to subtract out this quantity during calculations. Although this involves a significant increase in labor if computations are performed by hand, the additional effort is trivial on a digital computer. Also, the resulting coefficients must be "unstandardized" if they are to be used in a predictive equation with raw data. However, these disadvantages are more than offset by the increased stability and accuracy of the matrix solution, and the standardized coefficients provide a way of assessing the importance of individual variables in the regression. Partial regression coefficients can be derived from the standardized partial regression coefficients by the transformation

$$b_k = B_k \frac{s_y}{s_k} \tag{6.10}$$

The constant term, b_0, can be found by

$$b_0 = \overline{Y} - b_1\overline{X}_1 - b_2\overline{X}_2 - \ldots - b_m\overline{X}_m \tag{6.11}$$

Although the various sums of squares change if the data are standardized (*i.e.*, the correlation form of the matrix equation is used), the ratios of the sums of squares remain the same. Therefore, tests of significance based on standardized regression are identical to those based on an unstandardized regression. Quantities such as the coefficient of multiple correlation (R) and percentage of goodness of fit ($100\% \ R^2$) also remain unchanged.

We can compare the partial regression coefficients between basin magnitude and the other six basin properties in both raw and standardized form:

$$\mathbf{b}' = \begin{bmatrix} -2.244 & 0.005 & 0.226 & -0.233 & 0.063 & -0.002 & -0.117 \end{bmatrix}$$

$$\mathbf{B}' = \begin{bmatrix} 0.000 & 0.049 & 0.284 & -0.458 & 0.975 & -0.120 & -0.163 \end{bmatrix}$$

Although the standardized partial regression coefficients suggest that the basin properties having the most pronounced relationship with basin magnitude are x_2 (relief), x_3 (area), and x_4 (stream length), these values do not take into account the uncertainty associated with each estimated parameter. The easiest way to consider this aspect is by expanding the analysis of variance to test the significance of each independent variable.

The sum of squares attributable to a single variable, x_j, can be determined by calculating $SS_{R(m)}$ for the regression with all m variables, calculating $SS_{R(m-1)}$, which is the sum of squares for regression using all variables except the jth variable, then finding the difference. This process can be repeated for each independent variable in turn, in order to assess the contribution that each makes to the total regression. Fortunately, there is an easier way to calculate the individual regression sums of squares, which simply requires dividing the square of each partial regression coefficient by the diagonal elements of $\mathbf{S_{XX}^{-1}}$ that correspond to each of the variables. If we designate $\mathbf{C_{XX}} = \mathbf{S_{XX}^{-1}}$, then

$$SS_{R(x_j)} = b_j^2 \big/ c_{jj} \tag{6.12}$$

Once the regression sums of squares of the individual variables have been calculated, they can be entered into an expanded ANOVA table such as that shown in **Table 6–3** and tested for significance. The F-test ratios are formed from the mean squares due to partial regression with each of the individual variables in the numerators, and the mean square due to deviation from the regression model as the denominator. Each F-test has 1 and $(n - m - 1)$ associated degrees of freedom. The F-tests will not change if the calculations are based on standardized partial regression coefficients.

Table 6–3. ANOVA for testing the significance of partial regression of individual variables.

Source of Variation	Sum of Squares	Degrees of Freedom	Mean Squares	F-Test
Regression	SS_R	m	MS_R	MS_R/MS_D
Addition due to x_1	SS_{R1}	1	MS_{R1}	MS_{R1}/MS_D
...	...	...	...	...
Addition due to x_m	SS_{Rm}	1	MS_{Rm}	MS_{Rm}/MS_D
Deviation from regression	SS_D	$n - m - 1$	MS_D	
Total Variation	SS_T	$n - 1$		

A complete ANOVA for testing the significance of the partial regression of each geomorphic variable on basin magnitude is given in **Table 6–4**. Although basin relief, basin area, and stream length all have the largest standardized partial regression coefficients, the contribution to the total regression made by basin area is not statistically significant. This is because the partial regression coefficient for basin area has an associated high standard error.

Although the standardized partial regression coefficients provide a guide to the most effective variables in the regression, they are not an infallible index to the "best possible" regression equation. Suppose you examine the regression equation and decide two variables are contributing a negligible amount to the regression and can be discarded. When one of the variables is omitted and the regression is recalculated, the goodness of fit and the regression equation, of course, change. Now suppose you decide to discard the second variable; again the regression changes. But the change might be quite different from the change that would occur if the first discarded variable were still in the regression. This occurs because the interaction effects of the two discarded variables with other variables cannot be assessed without recomputing the regression. If we want to search through a large set of variables and "weed out" those which are not helpful in the problem, we must do more than simply examine the partial regression coefficients.

Table 6–4. Completed ANOVA for testing the significance of regression of individual geomorphic variables on basin magnitude.

Source of Variation	Sum of Squares	Degrees of Freedom	Mean Squares	F-Test
Regression	1800.70	6	300.12	11.38^a
Outlet Elevation	5.69	1	5.69	0.22^b
Basin Relief	201.43	1	201.43	7.64^c
Basin Area	46.17	1	46.17	1.75^d
Stream Length	243.95	1	243.95	9.25^e
Drainage Density	22.91	1	22.91	0.87^f
Basin Circularity	67.99	1	67.99	2.58^g
Deviation	1134.12	43	26.38	
Total Variation	2934.82	49		

[a] $p < 0.0001$ (highly significant).
[b] $p = 0.645$ (not significant).
[c] $p = 0.008$ (highly significant).
[d] $p = 0.193$ (not significant).
[e] $p = 0.004$ (highly significant).
[f] $p = 0.357$ (not significant).
[g] $p = 0.116$ (not significant).

Increasing the number of independent variables in the regression equation will always increase the SS_R (except in the situation where a new variable is perfectly correlated with a previous variable). However, the increase may not be significant. The loss of degrees of freedom for deviations may offset the reduction in SS_D, and actually increase the mean squares due to deviation. If this happens, the F-ratio for the significance of the regression will decrease, and the addition of another variable has actually detracted from the regression. To determine the very best possible regression (in the sense of having the most significant F-ratio), all possible combinations of the variables would have to be examined. This is possible if we are dealing with few variables, but the number of possible variable combinations is equal to $2^m - 1$, and the computational effort is formidable if m is large. Other procedures are available which yield a nearly optimal regression with much less effort. These include schemes such as the backward elimination procedure, the forward selection procedure, stepwise regression, and stagewise regression. These methods may not find identical regression equations in a large selection of possible variables, but all will produce approximately equivalent results. A consideration of each is beyond the scope of this book; we will be content with a brief description of one of the techniques. These methods are well described in some of the texts listed in the Selected Readings at the end of the chapter, especially in Marascuilo and Levin (1983) and in Draper and Smith (1998).

The backward elimination procedure consists of computing a regression including all possible variables and selecting the least significant variable. The selection proceeds by examining the standardized partial regression coefficients for the smallest value and then recomputing the regression, omitting that variable. The significance of the deleted variable is tested by the analysis of variance shown in **Table 6–3**. If the variable is not making a significant contribution to the regression, it is permanently discarded. The reduced regression model is then fitted to the data, a new set of standardized partial regression coefficients for the reduced equation is calculated, and the process is repeated. At each step, the regression equation is reduced by one variable, until all remaining variables are significant.

It is instructive to examine the collection of six independent variables measured on river basins (file KENTUCKY.TXT) and see if any can be discarded without significantly affecting the multiple regression on basin magnitude. We can find a minimal set of regressions by examining the standardized partial regression coefficients, deleting the smallest of these, and recomputing the regression. Repeatedly running a multiple-regression program obviously is less efficient than using a stepwise computer program, but it has the advantage that every step in the process can be examined closely. When you are confident that you understand the elimination process and the changes that occur in the regression coefficients, you may turn to a more automated procedure.

Although multiple regression is "multivariate" in the sense that more than one variable is measured on each observational unit, it really is a univariate technique because we are concerned only with the variance of one variable, y. Behavior of the independent variables, the x's, is not subject to analysis.

The next topic we will consider is discriminant function analysis, which involves identification or the placing of objects into predefined groups. The discrimination between two alternative groups is a process that is computationally intermediate between univariate procedures and true multivariate methods in which many variables are considered simultaneously. Two groups, each characterized by a set of multiple variables, can be discriminated by solving a set of simultaneous equations almost identical to those involved in multiple regression. The right-hand vector of the matrix equation, however, does not contain cross products between independent variables and a single dependent variable, but rather differences between the multivariate means of the two groups that are to be discriminated.

Tests of discriminant functions involve multivariate extensions of simple univariate statistical tests of equality. These will be considered next, followed by a discussion of multivariate classification, or the sorting of objects into homogeneous groups. We will then consider eigenvector techniques, including principal component and factor analysis. The final topics will include multivariate extensions of discriminant analysis and multiple regression.

This list of topics is certainly not all-inclusive. However, the subjects have been chosen because they have found special utility in the Earth sciences. They include a wide variety of computational techniques and encompass many fundamental concepts. An understanding of the theory and operational procedures involved in these methods should provide you with a sufficient background to evaluate other multivariate techniques as well.

Discriminant Functions

One of the most widely used multivariate procedures in Earth science is the discriminant function. We will consider it at length for two reasons: discrimination is a powerful statistical tool and it can be regarded as either a way to treat univariate problems related to multiple regression, or multivariate problems related to the statistical tests we will discuss later. Discriminant functions therefore provide an additional link between univariate and multivariate statistics.

First, however, we must define the process of **discrimination**, and carefully distinguish it from the related process of classification. Suppose we have assembled two collections of shale samples of known freshwater and saltwater origin. We may have determined their origin from an examination of their fossil content. A number of geochemical variables have been measured on each specimen, including the content of vanadium, boron, iron, and so forth. The problem is to find the linear combination of these variables that produces the maximum difference between the two previously defined groups. If we find a function that produces a significant difference, we can use it to allocate new specimens of shale of unknown origin to one of the two original groups. In other words, new shale samples, not containing diagnostic fossils, can then be categorized as marine or freshwater on the basis of the linear discriminant function of their geochemical components. [This problem was considered by Potter, Shimp, and Witters (1963).]

Classification can be illustrated with a similar example. Suppose we have obtained a large, heterogeneous collection of shale specimens, each of which has been geochemically analyzed. On the basis of the measured variables, can the shales be separated into groups (or **clusters**, as they are commonly called) that are both relatively homogeneous and distinct from other groups? The process by which this can be done has been highly developed by numerical taxonomists, and will be considered in a later section. There are several obvious differences between these procedures and those of discriminant function analysis. A classification is internally based; that is, it does not depend on *a priori* knowledge about relations between observations as does a discriminant function. The number of groups in a discriminant function is set prior to the analysis, while in contrast the number of clusters that will emerge from a classification scheme cannot ordinarily be predetermined. Similarly, each original observation is defined as belonging to a specific group in a discriminant analysis. In most classification procedures, an observation is free to enter any cluster that emerges. Other differences will become apparent as we examine these two procedures. The result of a cluster analysis of shales would be a classification of the observations into several groups. It would then be up to us to interpret the geological meaning (if any) of the groups so found.

A simple linear discriminant function transforms an original set of measurements on a specimen into a single **discriminant score**. That score, or transformed variable, represents the specimen's position along a line defined by the linear discriminant function. We can therefore think of the discriminant function as a way of collapsing a multivariate problem down into a problem which involves only one variable.

Discriminant function analysis consists of finding a transform which gives the maximum ratio of the difference between two group multivariate means to the multivariate variance within the two groups. If we regard our two groups as forming clusters of points in multivariate space, we must search for the one orientation along which the two clusters have the greatest separation while each cluster simultaneously has the least inflation. This can be graphically shown for two-dimensional cases, as in **Figure 6–2**, which is a scatter plot of the two groups of data listed in file SANDS.TXT. One group contains grain-size statistics of modern beach sands collected along the Gulf Coast in Texas; the second group contains grain-size statistics for sands collected offshore in the Gulf of Mexico. Both data sets consist of two variables, the median grain size and the grain-size sorting coefficient. Although the two clusters of points overlap, it is apparent that a line of division could be placed between the two clusters such that most of the beach sands would be on one side and most offshore sands would be on the other. An adequate separation between the sands of the two groups cannot be made using either median grain size or sorting coefficient alone. However, it is possible to find the orientation of an axis along which the two sets of sands are separated the most and inflated the least. The coordinates of this axis are the coefficients of the linear discriminant function.

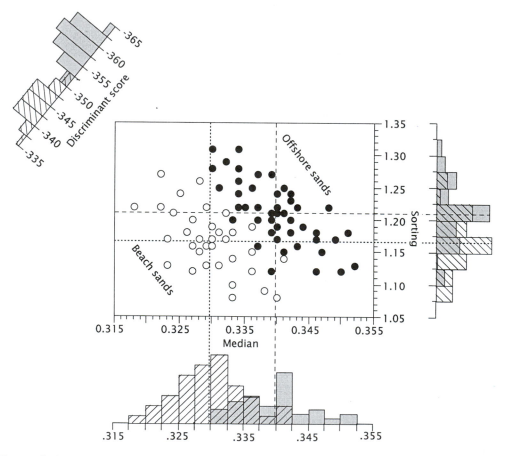

Figure 6–2. Plot of distributions of median grain size and sorting coefficient for samples of modern sands, with scatter plot of both variables. Samples indicated by open circles are beach sands, those indicated by solid dots are offshore sands. Dashed lines indicate bivariate means of the two groups. Distribution of discriminant scores also is shown along line parallel to discriminant axis.

One method that can be used to find the discriminant function is regression; however, the dependent variable consists of the differences between the multivariate means of the two groups. In matrix notation, we must solve an equation of the form

$$S\lambda = D \qquad (6.13)$$

where S is an $m \times m$ matrix of pooled variances and covariances of the m variables. The coefficients of the discriminant equation are represented by a column vector of the unknown lambdas. Lowercase lambdas (λ) are used by convention to represent the coefficients of the discriminant function. These are exactly the same as the betas (β) used (also by convention) in regression equations. They should not be confused with lambdas used to represent eigenvalues in principal component or factor analyses.

The right-hand side of the equation consists of the column vector of m differences between the means of the two groups, which we will refer to as A and B. You will recall from Chapter 3 that such an equation can be solved by inversion and multiplication, as

$$\lambda = S^{-1}D \qquad (6.14)$$

where S^{-1} is the inverse of the variance–covariance matrix formed by pooling the matrices of the sums of squares and cross products of the two groups, A and B. To compute the discriminant function, we must determine the various entries in the matrix equation. The mean differences are found simply by

$$d_j = \overline{A}_j - \overline{B}_j = \frac{\sum_{i=1}^{n_a} a_{ij}}{n_a} - \frac{\sum_{i=1}^{n_b} b_{ij}}{n_b} \qquad (6.15)$$

In this notation, a_{ij} is the ith observation on variable j in group A and $\overline{A}_j$ is the mean of variable j in group A, which is the arithmetic average of the n_a observations of variable j in group A. The same conventions apply to group B. The multivariate means of groups A and B can be regarded as forming two vectors. The difference between these multivariate means therefore also forms a vector

$$D = \overline{A} - \overline{B}$$

or, in expanded form,

$$\begin{bmatrix} d_1 \\ d_2 \\ \vdots \\ d_m \end{bmatrix} = \begin{bmatrix} \overline{A}_1 \\ \overline{A}_2 \\ \vdots \\ \overline{A}_m \end{bmatrix} - \begin{bmatrix} \overline{B}_1 \\ \overline{B}_2 \\ \vdots \\ \overline{B}_m \end{bmatrix}$$

To construct the matrix of pooled variances and covariances, we must compute a matrix of sums of squares and cross products of all variables in group A and a similar matrix for group B. For example, considering only group A,

$$SP_{Ajk} = \sum_{i=1}^{n_a} a_{ij}a_{ik} - \frac{\sum_{i=1}^{n_a} a_{ij} \sum_{i=1}^{n_a} a_{ik}}{n_a}$$

Here, a_{ij} denotes the ith observation of variable j in group A as before, and a_{ik} denotes the ith observation of variable k in the same group. Of course, this quantity will be the sum of squares of variable k whenever $j = k$. Similarly, a matrix of sums of squares and cross products can be found for group B:

Table 6–5. Matrices necessary to compute discriminant function between beach sands and offshore sands listed in file SANDS.TXT.

Vector mean of beach sands:	$\begin{bmatrix} 0.3297 & 1.1674 \end{bmatrix}$
Vector mean of offshore sands:	$\begin{bmatrix} 0.3399 & 1.2100 \end{bmatrix}$
Vector of mean differences:	$\begin{bmatrix} -0.0101 & -0.0426 \end{bmatrix}$
Corrected sums of squares for beach sands:	$\begin{bmatrix} 0.000925 & -0.004886 \\ -0.004886 & 0.075662 \end{bmatrix}$
Corrected sums of squares for offshore sands:	$\begin{bmatrix} 0.001384 & -0.008440 \\ -0.008440 & 0.107000 \end{bmatrix}$
Pooled variance–covariance matrix:	$\begin{bmatrix} 0.000029 & -0.000687 \\ -0.000687 & 0.002312 \end{bmatrix}$
Inverse of pooled variance–covariance matrix:	$\begin{bmatrix} 59{,}098.3047 & 4311.6403 \\ 4311.6403 & 747.0581 \end{bmatrix}$

$$SP_{Bjk} = \sum_{i=1}^{n_b} b_{ij} b_{ik} - \frac{\sum_{i=1}^{n_b} b_{ij} \sum_{i=1}^{n_b} b_{ik}}{n_b}$$

We will denote the sums of products matrix from group A as $\mathbf{S}_A$ and that from group B as $\mathbf{S}_B$. The matrix of pooled variance can now be found as

$$\mathbf{S} = \frac{\mathbf{S}_A + \mathbf{S}_B}{n_a + n_b - 2} \tag{6.16}$$

Remember this equation for the pooled variance; we will use it later in a T^2 test of the equality of the multivariate means of the two groups. Although the amount of mathematical manipulation that must be performed to calculate the coefficients of a discriminant function appears large, it actually is less formidable than it seems at first glance. To demonstrate, we can calculate a discriminant function between the two groups of observations in file SANDS.TXT. Group A consists of the beach sands and Group B consists of the offshore sands.

Table 6–5 contains the calculations necessary to find the two vectors of multivariate means and the two matrices of sums of squares and products. From these, the matrix of pooled variances is calculated. We now have all of the entries necessary to estimate the discriminant function coefficients:

$$
\begin{array}{ccc}
\mathbf{S} & \mathbf{D} & \boldsymbol{\lambda} \\
\begin{bmatrix} 59,098.305 & 4311.640 \\ 4311.640 & 747.058 \end{bmatrix} \cdot \begin{bmatrix} -0.010 \\ -0.043 \end{bmatrix} = \begin{bmatrix} -783.442 \\ -75.602 \end{bmatrix}
\end{array}
$$

The set of λ coefficients we have found are entries in the discriminant function equation which has the form

$$
\begin{aligned}
R_i &= \lambda_1 x_{1i} + \lambda_2 x_{2i} \\
&= -783.442 x_{1i} - 75.602 x_{2i}
\end{aligned} \tag{6.17}
$$

Equation (6.17) is a linear function; that is, all the terms are added together to yield a single number, the discriminant score, R_i. In a two-dimensional example, we can plot the discriminant function as a line on the scatter diagram of the two original variables. It is a line through the plot whose slope, α, is

$$
\alpha = \lambda_2 / \lambda_1 \tag{6.18}
$$

Substitution of the midpoint between the two group means into the discriminant function equation yields the discriminant index, R_0. That is, for each value of x_{ji} in Equation (6.17), we insert the terms

$$
x_{j\cdot} = \frac{\overline{A}_j + \overline{B}_j}{2} \tag{6.19}
$$

In our example,

$$
\begin{aligned}
R_0 &= (-783.442 \cdot 0.335) + (-75.602 \cdot 1.189) \\
&= -352.146
\end{aligned}
$$

The discriminant index, R_0, is the point along the discriminant function line that is exactly halfway between the center of group A and the center of group B. Next, we may substitute the multivariate mean of group A into the equation (that is, we set $x_j = \overline{A}_j$) to obtain R_A and substitute the multivariate mean of group B (setting $x_j = \overline{B}_j$) to obtain R_B. The centers of the two original groups projected onto the axis defined by the discriminant function are R_A and R_B.

For group A,

$$
\begin{aligned}
R_A &= (-783.442 \cdot 0.330) + (-75.602 \cdot 1.167) \\
&= -346.560
\end{aligned}
$$

and for group B,

$$
\begin{aligned}
R_B &= (-783.442 \cdot 0.340) + (-75.602 \cdot 1.210) \\
&= -357.732
\end{aligned}
$$

The three points may be plotted as in **Figure 6–3**. In fact, every observation in the analysis can be entered into the equation and its position along the discriminant function located. These values are the *raw discriminant scores*. This has been done on **Figure 6–3**; note that a few members of group A are located on the

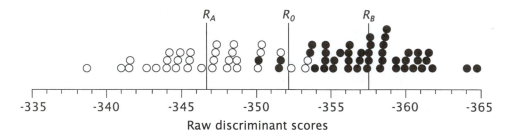

Figure 6–3. Projection of beach and offshore sands onto discriminant function line shown in Figure 6–2. R_A is projection of bivariate mean of beach sands, R_B is projection of bivariate mean of offshore sands, and R_0 is discriminant index.

group B side of R_0 and a few members of group B are located on the group A side. These are observations that have been misclassified by the discriminant function. The ***misclassification ratio***, or percent of observations that the discriminant function places into the wrong group, is sometimes taken as an indication of the function's discriminatory power. However, the misclassification ratio is biased and can be misleading because it is calculated by reusing the observations that were used to estimate the coefficients of the discriminant function in the first place. It seems likely that the function may be less successful in correctly classifying new observations. Reyment and Savazzi (1999) discuss alternative ways of evaluating the goodness of a discriminant function.

We have calculated the ***raw discriminant function*** which yields raw scores whose units are products of the units of measurement attached to the original variables. There actually are an infinity of discriminant functions that will maximize the difference between the two groups, but all of these alternatives are proportional to the classical, or raw, solution. If $\boldsymbol{\lambda}$ is the vector of coefficients determined by Equation (6.14), then all sets $c\,\boldsymbol{\lambda}$ (where c is an arbitrary constant), will serve equally well. Although different computer programs may yield sets of coefficients that seem to be different, all of them are proportional to each other. Alternative choices include:

1. The raw coefficients are divided by the pooled mean squares within groups, or

$$c = MS_W^{-1}$$

   where

$$MS_W = \boldsymbol{\lambda}'\,\mathbf{S}\,\boldsymbol{\lambda}$$

   This standardizes the coefficients to dimensionless z-scores.

2. The raw coefficients are first divided by MS_W, then rescaled by dividing every coefficient by the first coefficient, which becomes equal to 1.

3. Each raw coefficient is divided by the square root of the sum of the squared raw coefficients, or

$$c = \left(\sum\nolimits_{j=1}^{m} \lambda_j^2\right)^{-1/2}$$

   The sum of the squares of the transformed coefficients will then be equal to 1.

Tests of significance

If we are willing to make some assumptions about the nature of the data used in the discriminant function, we can test the significance of the separation between the two groups. Five basic assumptions about the data are necessary: (a) the observations in each group are randomly chosen, (b) the probability of an unknown observation belonging to either group is equal, (c) variables are normally distributed within each group, (d) the variance–covariance matrices of the groups are equal in size, and (e) none of the observations used to calculate the function were misclassified. Of these, the most difficult to justify are (b), (c), and (d). Fortunately, the discriminant function is not seriously affected by limited departures from normality or by limited inequality of variances. Justification of (b) must depend upon *a priori* assessment of the relative abundance of the groups under examination. If the assumption of equal abundance seems unjustified, a different assumption may be made, which will shift the position of R_0. [See Anderson (1984, chapter 6) for an extensive discussion of alternative decision rules for discrimination.]

The first step in a test of the significance of a discriminant function is to measure the separation or distinctness of the two groups. This can be done by computing the distance between the centroids, or multivariate means, of the groups. The measure of distance is derived directly from univariate statistics. We can obtain a measure of the difference between the means of two univariate samples, $\overline{X}_1$ and $\overline{X}_2$, by simply subtracting one from the other. However, this difference is expressed in the same units as the original observations. If the difference is divided by the pooled standard deviation, we obtain a **standardized difference** in which the difference between the means of the two groups is expressed in dimensionless units of standard deviation, or z-scores:

$$d = \frac{\overline{X}_1 - \overline{X}_2}{s_p} \tag{6.20}$$

When both sides of Equation (6.20) are squared, the denominator is the pooled variance of the two samples, s_p^2:

$$d^2 = \frac{\left(\overline{X}_1 - \overline{X}_2\right)^2}{s_p^2} \tag{6.21}$$

Suppose that instead of a single variable, two variables are measured on each observation in the two groups. The difference between the bivariate means of the two groups can be expressed as the ordinary Euclidean, or straight-line, distance between them. Again denoting the two groups as A and B,

$$\text{Euclidean distance} = \sqrt{\left(\overline{A}_1 - \overline{B}_1\right)^2 + \left(\overline{A}_2 - \overline{B}_2\right)^2} \tag{6.22}$$

In general, if m variables are measured on each observation, the straight-line distance between the multivariate means of the two groups is

$$\text{Euclidean distance} = \sqrt{\sum_{j=1}^{m} \left(\overline{A}_j - \overline{B}_j\right)^2} \tag{6.23}$$

The square of the Euclidean distance is $\sum_{j=1}^{m} \left(\overline{A}_j - \overline{B}_j\right)^2$; you can verify that this is the same as the matrix product,

$$\text{Euclidean distance}^2 = \mathbf{D}'\mathbf{D} \tag{6.24}$$

The Euclidean distance and its square, unfortunately, are expressed as hodge-podges of the original units of measurement. To be interpretable, they must be standardized. Comparison with Equation (6.20) suggests that standardization must involve division by the multivariate equivalent of the variance, which is the variance–covariance matrix **S**. Of course, division is not a defined operation in matrix algebra, but we can accomplish the same end by multiplying by the inverse. Multiplying Equation (6.24) by the inverse of the variance–covariance matrix yields the standardized squared distance,

$$D^2 = \mathbf{D}' \, \mathbf{S}^{-1} \, \mathbf{D} \tag{6.25}$$

This standardized measure of difference between the means of two multivariate groups is called *Mahalanobis' distance*. Substituting quantities from **Table 6–5** into Equation (6.25), we obtain

$$D^2 = \begin{bmatrix} -0.010 & -0.043 \end{bmatrix} \begin{bmatrix} 59,098.305 & 4311.640 \\ 4311.640 & 747.058 \end{bmatrix} \begin{bmatrix} -0.010 \\ -0.043 \end{bmatrix}$$
$$= 11.172$$

Interestingly, we can obtain exactly the same distance measure by substituting the vector of mean differences into the discriminant function equation itself:

$$D^2 = \begin{bmatrix} -0.010 & -0.043 \end{bmatrix} \begin{bmatrix} -783.442 \\ -75.602 \end{bmatrix}$$
$$= 11.172$$

Mahalanobis' distance can be visualized on **Figure 6–3**, where it is equal to the distance between R_A and R_B.

The significance of Mahalanobis' distance can be tested using a multivariate equivalent of the t-test of the equality of two means, called Hotelling's T^2 test. We will discuss this test more extensively in the next section. Here, we simply note that it has the form

$$T^2 = \frac{n_a n_b}{n_a + n_b} D^2 \tag{6.26}$$

and can be transformed to an F-test. The test of multivariate equality, using this more familiar statistic, is

$$F = \left(\frac{n_a + n_b - m - 1}{(n_a + n_b - 2)\,m} \right) \left(\frac{n_a n_b}{n_a + n_b} \right) D^2 \tag{6.27}$$

with m and $(n_a + n_b - m - 1)$ degrees of freedom. The null hypothesis tested by this statistic is that the two multivariate means are equal, or that the distance between them is zero. That is,

$$H_0 : \quad D = 0$$

against

$$H_1 : \quad D > 0$$

The appropriateness of this as a test of a discriminant function should be apparent. If the means of the two groups are very close together, it will be difficult to tell them apart, especially if both groups have large variances. In contrast, if the two means are well separated and scatter about the means is small, discrimination will be relatively easy. As an exercise, it may be instructive to calculate the significance of the discriminant function for the example we have just worked.

Not all of the variables we have included in the discriminant function will be equally useful in distinguishing one group from another. We may wish to isolate those variables that are not especially helpful and eliminate them from future analyses. Selecting the most effective set of discriminators for discriminant function analysis would seem to be analogous to selecting the most efficient predictors in multiple regression. The problem, however, is more complicated because the "dependent" or predicted variable in a discriminant function is composed of differences between two sets of the same variables that are used as "independent" predictors of the discrimination. Unlike regression, where the sums of squares of y do not change as different variables x_j are added to the equation, the sums of squares of the differences between groups A and B do change as variables are added or deleted.

Some idea of the effectiveness of the variables as discriminators can be gained by computing the **standardized differences**,

$$D_j = \frac{\overline{A}_j - \overline{B}_j}{s_{pj}} \tag{6.28}$$

This is simply the difference between the means of the two groups A and B for variable j, divided by the pooled standard deviation of variable j. Since the measure does not consider interactions between variables, it is useful only as a general guide to discriminating power. Stepwise discriminant analysis programs may use standardized differences in choosing the order in which variables are added to the discriminant function. Marascuilo and Levin (1983) discuss "after-the-fact" contrast procedures that can be used to select the most important variables. However, the significance of different combinations of variables can be tested only by computing the various functions and determining the relative amounts of separation the different equations produce between the two groups. To avoid bias, such tests should be run on independent random samples.

Discriminant function analysis provides a natural transition between two major classes of multivariate statistical techniques. On one hand, it is closely related to multiple regression and trend-surface analysis. On the other, it can be expressed as an eigenvalue problem, related to principal component analysis, factor analysis, and similar multivariate methods. There are advantages to the use of eigenvectors in calculating the discriminant function, because they allow us to simultaneously discriminate between more than two groups. However, we will delay a consideration of this topic until we examine the basic elements of eigenvector analysis and some of the simpler eigenvector techniques.

Multivariate Extensions of Elementary Statistics

In Chapter 2, we considered some simple geologic problems that could be examined by elementary statistical methods. We will begin our consideration of multivariate methods in geology with some direct extensions of these simple tests. You will recall that the variation measured in most naturally occurring phenomena could be described by the normal distribution. This is a reflection of the central limit theorem, which states that observations which are the sums of many independently operating processes tend to be normally distributed as the number of effects becomes large. It is this tendency that allows us to use the normal probability distribution as a basis for statistical tests and provides the starting point for the development of the t-, F-, and χ^2 distributions and others. The concept of the normal distribution can be extended to include situations in which observational units consist of many variables.

Suppose we collect rocks from an area and measure a set of properties on each specimen. The measurements may include determinations of chemical or mineralogical constituents, specific gravity, magnetic susceptibility, radioactivity, or any of an almost endless list of possible variables. We can regard the set of measurements made on an individual rock as defining a vector $\mathbf{X}_i = [\, x_{1i} \quad x_{2i} \quad \cdots \quad x_{mi} \,]$, where there are m measured characteristics or variables. If a sample of observations, each represented by vectors X_i, is randomly selected from a population that is the result of many independently acting processes, the observed vectors will tend to be multivariate normally distributed. Considered individually, each variate is normally distributed and characterized by a mean, μ_j, and a variance, σ_j^2. The *joint probability distribution* is a p-dimensional equivalent of the normal distribution, having a vector mean $\mu = [\, \mu_1 \quad \mu_2 \quad \cdots \quad \mu_m \,]$ and a variance generalized into the form of a diagonal matrix:

$$\Sigma = \begin{bmatrix} \sigma_1^2 & 0 & \cdots & 0 \\ 0 & \sigma_2^2 & \cdots & 0 \\ \vdots & \vdots & \ddots & \vdots \\ 0 & 0 & \cdots & \sigma_m^2 \end{bmatrix}$$

In addition to these obvious extensions of the normal distribution to the multivariate case, the multivariate normal distribution has an important additional characteristic. This is the covariance, cov_{jk}, which occupies all of the off-diagonal positions of the matrix Σ. Thus, in the multivariate normal distribution, the mean is generalized into a vector and the variance into a matrix of variances and covariances. In the simple case of $m = 2$, the probability distribution forms a three-dimensional bell curve such as that in **Figure 2–19**, shown as a contour map in **Figure 6–4**. Although the distributions of variables x_1 and x_2 are shown along their respective axes, the essential characteristics of the joint probability distribution are better shown by the major and minor axes of the probability density ellipsoid. Many of the multivariate procedures we will discuss are concerned with the relative orientations of these major and minor axes.

One of the simplest tests we considered in Chapter 2 was a t-test of the probability that a random sample of n observations had been drawn from a normal population with a specified mean, μ, and an unknown variance, σ^2. The test, given in Equation (2.45) on p. 70, can be rewritten in the form

$$t = \frac{(\overline{X} - \mu)\sqrt{n}}{\sqrt{s^2}} \tag{6.29}$$

An obvious generalization of this test to the multivariate case is the substitution of a vector of sample means for $\overline{X}$, a vector of population means for μ, and a variance–covariance matrix for s^2. We have defined the vector of population means as μ, so a vector of sample means can be designated $\overline{\mathbf{x}}$. Similarly, Σ is the matrix of population variances and covariances, so S represents the matrix of sample variances and covariances. Both $\overline{\mathbf{x}}$ and μ are taken to be column vectors, although equivalent equations may be written in which they are assumed to be row vectors. A column vector of differences between the sample means and the population means

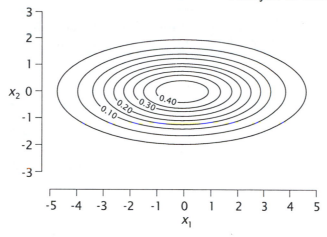

Figure 6–4. Contour map of bivariate normal probability distribution. See Figure 2.19 on p. 40 for perspective diagram of same distribution.

is obtained by subtracting these two vectors. Substituting these quantities directly into Equation (6.29) gives

$$t = \frac{(\overline{\mathbf{x}} - \boldsymbol{\mu})\sqrt{n}}{\sqrt{\mathbf{S}}}$$

Unfortunately, there is no equally obvious way of solving this equation so that it yields a single value of t. We must reduce the vectors and the matrix to single numbers if we wish to apply this test. If we were to multiply the column vector $(\overline{\mathbf{x}} - \boldsymbol{\mu})$ by a row vector having the same number of elements, the result would be a single number. We will therefore define an arbitrary row vector, $\mathbf{A}$, whose transpose is a column vector, $\mathbf{A}'$. Multiplication of the column vector of differences $(\overline{\mathbf{x}} - \boldsymbol{\mu})$ by the row vector $\mathbf{A}$ gives a single number, and premultiplication of $\mathbf{S}$ by $\mathbf{A}$ and postmultiplication by $\mathbf{A}'$ also yields a single number. That is, our test has become

$$t = \frac{\mathbf{A}(\overline{\mathbf{x}} - \boldsymbol{\mu})\sqrt{n}}{\mathbf{A}\sqrt{\mathbf{S}}\,\mathbf{A}'}$$

However, we have also changed what we are testing, from a null hypothesis of

$$H_0 : \boldsymbol{\mu}_1 = \boldsymbol{\mu}_0$$

to

$$H_0^* : \mathbf{A}\boldsymbol{\mu}_1 = \mathbf{A}\boldsymbol{\mu}_0$$

The original hypothesis, H_0, is true only if the new hypothesis, H_0^*, holds for all possible values of $\mathbf{A}$. It is sufficient, however, to test only the maximum possible value of the test statistic, because if H_0^* is rejected for any value of $\mathbf{A}$, the hypothesis H_0 is also rejected. With a bit of mathematical manipulation, we can determine the conditions under which a maximum test statistic will result for any arbitrary vector $\mathbf{A}$. This involves introducing the constraint $\mathbf{A}\mathbf{S}\mathbf{A}' = 1$ and expressing the equation in a form that incorporates a determinant. In the process, we can eliminate the troublesome square roots by squaring the equation. This also squares the test value, which is referred to as **Hotelling's** T^2, in honor of Harold Hotelling, the

American statistician who formulated this generalization of Student's t. When all operations are complete, we find that the test statistic can be expressed as

$$T^2 = n\,(\overline{\mathbf{x}} - \boldsymbol{\mu})'\,\mathbf{S}^{-1}\,(\overline{x} - \boldsymbol{\mu}) \qquad (6.30)$$

That is, the arbitrary vector $\mathbf{A}$ is equal to the vector of differences between the means, $(\overline{\mathbf{x}} - \boldsymbol{\mu})$. We must find the inverse of the variance–covariance matrix, premultiply this inverse by a row vector of differences, $(\overline{\mathbf{x}} - \boldsymbol{\mu})'$, and then postmultiply by a column vector of these same differences. The test statistic is a multivariate extension of the t-statistic, Hotelling's T^2. Critical values of T^2 can be determined by the relation

$$F = \frac{n - m}{m(n - 1)} T^2 \qquad (6.31)$$

where n is the number of observations and m is the number of variables, allowing us to use conventional F-tables rather than special tables of the T^2 distribution. More complete discussions of this and related tests are given in texts on multivariate statistics such as Overall and Klett (1983), Harris (1985), Krzanowski (1988), and Morrison (1990).

Although the expression of this test in a form such as Equation (6.30) is easy, computation of a test value for an actual data set may be very laborious. For example, suppose we have measured the content of four elements in seven lunar samples. We wish to test the hypothesis that these samples have been drawn from a population having the same mean as terrestrial basalts. Assume we take our values for the populations' means from the *Handbook of Physical Constants* (Clark, 1966, p. 4). Hotelling's T^2 seems appropriate to test the hypothesis that the vector of lunar sample means is no different than the vector of basalt means given in this reference.

We must first compute the vector of four sample means and the 4×4 matrix of variances and covariances. The vector of differences between sample and population means, $(\overline{\mathbf{x}} - \boldsymbol{\mu})$, must also be computed. Next, we must find the inverse of the variance–covariance matrix, or $\mathbf{S}^{-1}$. We then must perform two matrix multiplications, $(\overline{\mathbf{x}} - \boldsymbol{\mu})'\mathbf{S}^{-1}(\overline{\mathbf{x}} - \boldsymbol{\mu})$, and multiply by n to produce T^2. From this description, you can appreciate that the computational effort becomes increasingly greater as the number of variables grows larger.

The data for the seven lunar samples are listed in **Table 6–6**, with the "population" means from Clark. Intermediate values in the computation of T^2 are also given, with the final test value of T^2 and the equivalent F-statistic, which has m and $(n - m)$ degrees of freedom. The test statistic of $F = 73.11$ far exceeds the critical value of $F_{4,3,0.01} = 28.71$, so we conclude that the mean composition of the sample of lunar basalts is significantly different than the mean composition of the population of terrestrial basalts.

We have dwelled on the T^2 test against a known mean not because this specific test has greater utility in geology than other multivariate tests, but to illustrate the close relationship between conventional statistics and multivariate statistics. Multivariate equivalents can be formulated directly from most univariate tests with the proper expansion of the basic assumptions. However, the transition from ordinary algebra to matrix algebra often obscures the underlying similarity between the two applications. Although we usually regard multivariate methods as an extension of univariate statistics, univariate, or ordinary, statistical analysis should be considered as a special subset of the general area of multivariate analysis.

Table 6–6. Abundances of four elements in seven lunar samples and mean abundances of same elements in terrestrial basalts (after Wanke and others, 1970).

Lunar Samples	Si	Al	Fe	Mg
1	19.4	5.9	14.7	5.0
2	21.5	4.0	15.7	3.7
3	19.2	4.0	15.4	4.3
4	18.4	5.4	15.2	3.4
5	20.6	6.2	13.2	5.5
6	19.8	5.7	14.8	2.8
7	18.7	6.0	13.8	4.6
MEANS	19.66	5.31	14.69	4.19
"Population" Means	22.10	7.40	10.10	4.00
Differences	−2.44	−2.09	4.59	0.19

Variance–covariance matrix:

$$\begin{bmatrix} 1.179524 & -0.307619 & 0.059286 & 0.079286 \\ -0.307619 & 0.868095 & -0.683095 & 0.301905 \\ 0.059286 & -0.683095 & 0.801429 & -0.546905 \\ 0.079286 & 0.301905 & -0.546905 & 0.891429 \end{bmatrix}$$

Inverse of variance–covariance matrix:

$$\begin{bmatrix} 1.061478 & 0.994883 & 0.817269 & 0.070054 \\ 0.994883 & 5.209577 & 5.336676 & 1.421289 \\ 0.817269 & 5.336676 & 7.660054 & 2.819468 \\ 0.070054 & 1.421289 & 2.819468 & 2.363995 \end{bmatrix}$$

$T^2 = 584.78$
$F = 73.10$

In the remaining discussion in this section, we will consider multivariate tests that are the m-dimensional equivalent of some of the tests we considered in Chapter 2. However, we will not point out the details of the extrapolation from the univariate to the general case as we have done with the T^2 test. These derivations can be found in many texts on multivariate statistics, some of which are listed in the Selected Readings at the end of this chapter.

Equality of two vector means

The test we have just considered is a one-sample test against a specified population mean vector. Suppose instead we have collected two independent random samples and we wish to test the equivalency of their mean vectors. We assume that the two samples are drawn from multivariate normal populations, both having the same unknown variance–covariance matrix Σ. We wish to test the null hypothesis

$$H_0 : \boldsymbol{\mu}_1 = \boldsymbol{\mu}_0$$

against
$$H_1 \ : \ \boldsymbol{\mu}_1 \neq \boldsymbol{\mu}_0$$

The null hypothesis states that the mean vector of the parent population of the first sample is the same as the mean vector of the parent population from which the second sample was drawn.

The test we must use is a multivariate equivalent of Equation (2.48) on p. 73. In that two-sample t-test, we used a pooled estimate of the population variance based on both samples. Accordingly, we must compute a pooled estimate, S_p, of the common variance–covariance matrix from our two multivariate samples. This is done by calculating a matrix of sums of squares and products for each sample. We can use the terminology of discriminant functions and denote the matrix of sums of squares and cross products of sample A as S_A; similarly, the matrix from sample B is S_B. The pooled estimate of the variance–covariance matrix is

$$S_p = (n_A + n_B - 2)^{-1} (S_A + S_B) \tag{6.32}$$

We must next find the difference between the two mean vectors, $\mathbf{D} = \bar{\mathbf{x}}_A - \bar{\mathbf{x}}_B$. Our T^2 test has the form

$$T^2 = \frac{n_A n_B}{n_A + n_B} \mathbf{D}' S_p^{-1} \mathbf{D} \tag{6.33}$$

The significance of the T^2 test statistic can be determined by the F-transformation:

$$F = \frac{n_A + n_B - m - 1}{(n_A + n_B - 2)m} T^2 \tag{6.34}$$

which has m and $(n_A + n_B - m - 1)$ degrees of freedom (Morrison, 1990).

Equality of variance–covariance matrices

An underlying assumption in the two preceding tests is that the samples are drawn from populations having the same variance–covariance matrix. This is the multivariate equivalent of the assumption of equal population variances necessary to perform t-tests of means. In practice, an assumption of equality may be unwarranted, because samples which exhibit a high mean often will also have a large variance. You will recall from Chapter 4 that such behavior is characteristic of many geologic variables such as mine-assay values and trace-element concentrations. Equality of variance–covariance matrices may be checked by the following "test of generalized variances" which is a multivariate equivalent of the F-test (Morrison, 1990).

Suppose we have k samples of observations, and have measured m variables on each observation. For each sample a variance–covariance matrix, S_k, may be computed. We wish to test the null hypothesis

$$H_0 \ : \ \boldsymbol{\Sigma}_1 = \boldsymbol{\Sigma}_2 = \cdots = \boldsymbol{\Sigma}_k$$

against the alternative

$$H_1 \ : \ \boldsymbol{\Sigma}_i \neq \boldsymbol{\Sigma}_j$$

The null hypothesis states that all k population variance–covariance matrices are the same. The alternative is that at least two of the matrices are different. Each variance–covariance matrix S_i is an estimate of a population matrix $\boldsymbol{\Sigma}_i$. If the parent populations of the k samples are identical, the sample estimates may be combined to form a pooled estimate of the population variance–covariance matrix. The pooled estimate is created by

$$\mathbf{S}_p = \left(\left(\sum_{i=1}^{k} n_i\right) - k\right)^{-1} \sum_{i=1}^{k} (n_i - 1)\, \mathbf{S}_i \tag{6.35}$$

where n_i is the number of observations in the ith group and the summation over n_i gives the total number of all observations in all k samples. This equation is algebraically equivalent to Equation (6.32) when $k = 2$.

From the pooled estimate of the population variance-covariance matrix, a test statistic, M, can be computed:

$$M = \left\{\left(\sum_{i=1}^{k} n_i\right) - k\right\} \ln \left|s_p^2\right| - \sum_{i=1}^{k} \left\{(n_i - 1) \ln \left|s_i^2\right|\right\} \tag{6.36}$$

The test is based on the difference between the logarithm of the determinant of the pooled variance–covariance matrix and the average of the logarithms of the determinants of the sample variance–covariance matrices. If all the sample matrices are the same, this difference will be very small. As the variances and covariances of the samples deviate more and more from one another, the test statistic will increase. Tables of critical values of M are not widely available, so the transformation

$$C^{-1} = 1 - \frac{2m^2 + 3m - 1}{6\,(m + 1)\,(k - 1)} \left(\sum_{i-1}^{k} \frac{1}{n_i - 1} - \frac{1}{\left(\sum_{i=1}^{k} n_i - k\right)}\right) \tag{6.37}$$

can be used to convert M to an approximate χ^2 statistic:

$$\chi^2 \approx MC^{-1} \tag{6.38}$$

The approximate χ^2 value has degrees of freedom equal to $v = (1/2)(k - 1)$. If all the samples contain the same number of observations, n, Equation (6.37) can be simplified to

$$C^{-1} = 1 - \frac{(2m^2 + 3m - 1)(k + 1)}{6(m + 1)k(n - 1)} \tag{6.39}$$

The χ^2 approximation is good if the number of k samples and m variables do not exceed about 5 and each variance–covariance estimate is based on at least 20 observations.

To illustrate the process of hypothesis testing using multivariate statistics, we will work through the following problem. Note that the number of observations is just sufficient for some of the approximations to be strictly valid; we will consider them to be adequate for the purposes of this demonstration.

In a local area in eastern Kansas, all potable water is obtained from wells. Some of these wells draw from the alluvial fill in stream valleys, while others tap a limestone aquifer that also is the source of numerous springs in the region. Residents prefer to obtain water from the alluvium, as they feel it is of better quality. However, the water resources of the alluvium are limited, and it would be desirable for some users to obtain their supplies from the limestone aquifer.

In an attempt to demonstrate that the two sources are equivalent in quality, a state agency sampled wells that tapped each source. The water samples were analyzed for chemical compounds that affect the quality of water. Some of the data

Table 6–7. Multivariate statistics for cation composition of water samples collected from wells in an area of eastern Kansas: x_1 = silica, x_2 = iron, x_3 = magnesium, x_4 = sodium + potassium, x_5 = calcium. Data given in file WELLWATR.TXT.

Vector mean of water from wells in limestone

$$\bar{\mathbf{x}}_L = [\,9.760 \quad 13.955 \quad 30.935 \quad 25.930 \quad 33.270\,]$$

Vector mean of water from wells in alluvium

$$\bar{\mathbf{x}}_A = [\,12.055 \quad 16.080 \quad 34.465 \quad 29.910 \quad 25.055\,]$$

Variance-covariance matrix of water from wells in limestone, $|\mathbf{S}_L| = 1.8838 \cdot 10^8$

$$\mathbf{S}_L = \begin{bmatrix} 5.1615 & 0.5134 & 7.3683 & -1.4103 & -3.4402 \\ 0.5134 & 21.0247 & 10.6948 & -4.0896 & -25.3972 \\ 7.3683 & 10.6948 & 102.8045 & -38.5269 & -58.1689 \\ -1.4103 & -4.0896 & -38.5269 & 98.8654 & 7.2520 \\ -3.4402 & -25.3972 & -58.1689 & 7.2520 & 290.8706 \end{bmatrix}$$

Variance-covariance matrix of water from wells in alluvium, $|\mathbf{S}_A| = 2.1777 \cdot 10^8$

$$\mathbf{S}_A = \begin{bmatrix} 5.6394 & 0.7333 & 8.6868 & -2.9822 & -4.7095 \\ 0.7333 & 23.1733 & 12.7656 & -4.5593 & -26.9878 \\ 8.6868 & 12.7656 & 103.3982 & -42.3949 & -58.1232 \\ -2.9822 & -4.5593 & -42.3949 & 106.9525 & 9.2199 \\ -4.7095 & -26.9878 & -58.1232 & 9.2199 & 275.1616 \end{bmatrix}$$

Pooled variance-covariance matrix, $|\mathbf{S}_p| = 2.0351 \cdot 10^8$

$$\mathbf{S}_p = \begin{bmatrix} 5.4005 & 0.6233 & 8.0275 & -2.1962 & -4.0749 \\ 0.6233 & 22.0990 & 11.7302 & -4.3244 & -26.1925 \\ 8.0275 & 11.7302 & 103.1013 & -40.4609 & -58.1461 \\ -2.1962 & -4.3244 & -40.4609 & 102.9089 & 8.2360 \\ -4.0749 & -26.1925 & -58.1461 & 8.2360 & 283.0661 \end{bmatrix}$$

Inverse of pooled variance-covariance matrix

$$\mathbf{S}_p^{-1} = \begin{bmatrix} 0.2101 & 0.0027 & -0.0178 & -0.0024 & -3.0820 \cdot 10^{-4} \\ 0.0027 & 0.0521 & -0.0036 & 4.9006 \cdot 10^{-4} & 0.0041 \\ -0.0178 & -0.0036 & 0.0148 & 0.0051 & 0.0023 \\ -0.0024 & 4.9006 \cdot 10^{-4} & 0.0051 & 0.0116 & 7.2056 \cdot 10^{-4} \\ -3.0820 \cdot 10^{-4} & 0.0041 & 0.0023 & 7.2056 \cdot 10^{-4} & 0.0044 \end{bmatrix}$$

from these analyses are given in the file WELLWATR.TXT. The variance-covariance matrices, inverses, and determinants for the two data sets and for the pooled data are given in **Table 6–7**. From these we can test the equivalence of the two vector means. We will assume that the samples have been drawn randomly from multivariate normal populations.

We must first test the assumption that the variance-covariance matrices for the two samples are equivalent using the test statistic M given in Equation (6.36):

$$M = (20 + 20 - 2) \ln 2.0351 \cdot 10^8 - (19 \ln 1.8838 \cdot 10^8 + 19 \ln 2.1777 \cdot 10^8)$$
$$= 0.1804$$

The transformation factor, C^{-1}, must also be calculated to allow use of the χ^2 approximation:

$$C^{-1} = 1 - \frac{2 \cdot 5^2 + 3 \cdot 5 - 1}{6(5+1)(2-1)} \left(\frac{1}{19} + \frac{1}{19} - \frac{1}{40-2} \right)$$

$$= 0.8637$$

The χ^2 statistic is approximately $0.1804 \cdot 0.8637 = 0.1558$, with degrees of freedom equal to $\nu = 1/2(2-1)(5)(5+1) = 15$.

The critical value of χ^2 for $\nu = 15$ with a 5% level of significance is 25.00. The computed statistic is less than this value and does not fall into the critical region, so we may conclude that there is nothing in our samples which suggests that the variance–covariance structures of the parent populations are different. We may pool the two sample variance–covariance matrices and test the equality of the multivariate means using the T^2 test of Equation (6.33):

$$T^2 = \frac{20 \cdot 20}{20+20} 1.4847 = 14.847$$

The value 1.4847 is the product of the matrix multiplications $\mathbf{D}' \mathbf{S}_p^{-1} \mathbf{D}$ specified in Equation (6.33). The T^2 statistic may be converted to an F-statistic by Equation (6.34):

$$F = \frac{(20+20-5-1)}{(20+20-2)5} 14.847 = 2.657$$

Degrees of freedom are $\nu_1 = 5$ and $\nu_2 = (20+20-5-1) = 34$. The critical value for F with 5 and 34 degrees of freedom at the 5% ($\alpha = 0.05$) level of significance is 2.49. Our computed test statistic just exceeds this critical value, so we conclude that our samples do, indeed, indicate a difference in the means of the two populations. In other words, there is a statistically significant difference in composition of water from the two aquifers. This simple test will not pinpoint the chemical variables responsible for this difference, but it does substantiate the natives' contention that they can tell a difference in the water!

Multivariate techniques equivalent to the analysis-of-variance procedures discussed in Chapter 2 are available. In general, these involve a comparison of two $m \times m$ matrices that are the multivariate equivalents of the among-group and within-group sums of squares tested in ordinary analysis of variance. The test statistic consists of the largest eigenvalue of the matrix resulting from the comparison. We will not consider these tests here because their formulation is complicated and their applications to geologic problems have been, so far, minimal. This is not a reflection on their potential utility, however. Interested readers are referred to chapter 5 of Griffith and Amrhein (1997), which presents worked examples of MANOVA's applied to problems in geography. Koch and Link (1980) include a brief illustration of the application of multivariate analysis of variance to geochemical data. Statistical details are discussed by Morrison (1990).

Cluster Analysis

Cluster analysis is the name given to a bewildering assortment of techniques designed to perform classification by assigning observations to groups so each group is more or less homogeneous and distinct from other groups. This is the special forte of taxonomists, who attempt to deduce the lineage of living creatures from their characteristics and similarities. Taxonomy is highly subjective and dependent upon the individual taxonomist's skills, developed through years of experience. In this respect, the field is analogous in many ways to geology. As in geology, researchers dissatisfied with the subjectivity and capriciousness of traditional methods have sought new techniques of classification which incorporate the massive data-handling capabilities of the computer. These workers, responsible for many of the advances made in numerical classification, call themselves numerical taxonomists.

Numerical taxonomy has been a center of controversy in biology, much like the suspicion that swirled around factor analysis in the 1930's and 1940's and provoked acrimonious debates among psychologists. As in that dispute, the techniques of numerical taxonomy were overzealously promoted by some practitioners. In addition, it was claimed that a numerically derived taxonomy better represented the phylogeny of a group of organisms than could any other type of classification. Although this has yet to be demonstrated, rapid progress in genotyping suggests that an objective phylogeny may someday be possible. The conceptual underpinnings of taxonomic methods such as cluster analysis are incomplete; the various clustering methods lie outside the body of multivariate statistical theory, and only limited tests of significance are available (Hartigan, 1975; Milligan and Cooper, 1986; Bock, 1996). Although cluster analysis has become an accepted tool for researchers and there are an increasing number of books on the subject, a more complete statistical basis for classification has yet to be fashioned. In spite of this, many of the methods of numerical taxonomy are important in geologic research, especially in the classification of fossil invertebrates and the study of paleoenvironments.

The purpose of cluster analysis is to assemble observations into relatively homogeneous groups or "clusters," the members of which are at once alike and at the same time unlike members of other groups. There is no analytical solution to this problem, which is common to all areas of classification, not just numerical taxonomy. Although there are alternative classifications of classification procedures (Sneath and Sokal, 1973; Gordon, 1999), most may be grouped into four general types.

1. *Partitioning methods* operate on the multivariate observations themselves, or on projections of these observations onto planes of lower dimension. Basically, these methods cluster by finding regions in the space defined by the m variables that are poorly populated with observations, and that separate densely populated regions. Mathematical "partitions" are placed in the sparse regions, subdividing the variable space into discrete classes. Although the analysis is done in the m-dimensional space defined by the variables rather than the n-dimensional space defined by the observations, it proceeds iteratively and may be extremely time-consuming (Aldenderfer and Blashfield, 1984; Gordon, 1999).

2. *Arbitrary origin methods* operate on the similarity between the observations and a set of arbitrary starting points. If n observations are to be classified into k groups, it is necessary to compute an asymmetric $n \times k$ matrix of similarities between the n samples and the k arbitrary points that serve as initial group centroids. The observation closest or most similar to a starting point is combined with it to form a cluster. Observations are iteratively added to the nearest cluster, whose centroid is then recalculated for the expanded cluster.

3. **Mutual similarity procedures** group together observations that have a common similarity to other observations. First an $n \times n$ matrix of similarities between all pairs of observations is calculated. Then the similarity between columns of this matrix is iteratively recomputed. Columns representing members of a single cluster will tend to have intercorrelations near $+1$, while having much lower correlations with nonmembers.

4. **Hierarchical clustering** joins the most similar observations, then successively connects the next most similar observations to these. First an $n \times n$ matrix of similarities between all pairs of observations is calculated. Those pairs having the highest similarities are then merged, and the matrix is recomputed. This is done by averaging the similarities that the combined observations have with other observations. The process iterates until the similarity matrix is reduced to 2×2. The progression of levels of similarity at which observations merge is displayed as a dendrogram.

Hierarchical clustering techniques are most widely applied in the Earth sciences, probably because their development has been closely linked with the numerical taxonomy of fossil organisms. Because of the widespread use of heirarchical techniques, we will consider them in some detail.

Suppose we have a collection of objects we wish to arrange into a hierarchical classification. In biology, these objects are referred to as "operational taxonomic units" or OTU's (Sneath and Sokal, 1973). We can make a series of measurements on each object which constitutes our data set. If we have n objects and measure m characteristics, the observations form an $n \times m$ data matrix, **X**. Next, some measure of resemblance or similarity must be computed between every pair of objects; that is, between the rows of the data matrix. Several coefficients of resemblance have been used, including a variation of the correlation coefficient $\hat{r}_{ij}$ in which the roles of objects and variables are interchanged. This can be done by transposing **X** so rows become columns and *vice versa*, then calculating $\hat{r}_{ij}$ in the conventional manner (Eq. 2.28; p. 43), following the matrix algorithm given in Chapter 3. Although called "correlation," this measure is not really a correlation coefficient in the conventional sense because it involves "means" and "variances" calculated across all the variables measured on two objects, rather than the means and variances of two variables.

Another commonly used measure of similarity between objects is a standardized m-space Euclidean distance, d_{ij}. The distance coefficient is computed by

$$d_{ij} = \sqrt{\frac{\sum_{k=1}^{m} (x_{ik} - x_{jk})^2}{m}} \tag{6.40}$$

where x_{ik} denotes the kth variable measured on object i and x_{jk} is the kth variable measured on object j. In all, m variables are measured on each object, and d_{ij} is the distance between object i and object j. As you would expect, a small distance indicates the two objects are similar or "close together," whereas a large distance indicates dissimilarity. Commonly, each element in the $n \times m$ raw data matrix **X** is standardized by subtracting the column means and dividing by the column standard deviations prior to computing distance measurements. This ensures that each variable is weighted equally. Otherwise, the distance will be influenced most strongly by the variable which has the greatest magnitude. In some instances this may be desirable, but unwanted effects can creep in through injudicious choice of measurement units. As an extreme example, we might measure three perpendicular axes on a collection of pebbles. If we measure two of the axes in centimeters and the third in millimeters, the third axis will have proportionally ten times more influence on the distance coefficient than either of the other two variables.

Other measures of similarity that are less commonly used in the Earth sciences include a wide variety of *association coefficients* which are based on binary (presence–absence) variables or a combination of binary and continuous variables. The most popular of these are the *simple matching coefficient, Jaccard's coefficient*, and *Gower's coefficient*—all ratios of the presence–absence of properties. They differ primarily in the way that mutual absences (called "negative matches") are considered. Sneath and Sokal (1973) discuss the relative merits of these and other coefficients of association. *Probabilistic similarity coefficients* are used with binary data and consider the gain or loss of information when objects are combined into clusters. Again, Sneath and Sokal (1973) provide a comprehensive summary.

Computation of a similarity measurement between all possible pairs of objects will result in an $n \times n$ symmetrical matrix, **C**. Any coefficient c_{ij} in the matrix gives the resemblance between objects i and j. The next step is to arrange the objects into a hierarchy so objects with the highest mutual similarity are placed together. Then groups or clusters of objects are associated with other groups which they most closely resemble, and so on until all of the objects have been placed into a complete classification scheme. Many variants of clustering have been developed; a consideration of all of the possible alternative procedures and their relative merits is beyond the scope of this book. Rather, we will discuss one simple clustering technique called the *weighted pair-group method* with *arithmetic averaging*, and then point out some useful modifications to this scheme.

Extensive discussions of hierarchical and other classification techniques are contained in books by Jardine and Sibson (1971), Sneath and Sokal (1973), Hartigan (1975), Aldenderfer and Blashfield (1984), Romesburg (1984), Kaufman and Rousseeuw (1990), Backer (1995), and Gordon (1999). Diskettes containing clustering programs are included in some of the these books or are available separately at modest cost. In addition, most personal computer programs for statistical analysis contain modules for hierarchical clustering.

Table 6–8 contains measurements made on six greywacke thin sections, identified as $A, B, \ldots, F$. The values represent the average of the apparent maximum diameters of ten randomly chosen grains of quartz, rock fragment, and feldspar and the average of the apparent maximum diameters of ten intergranular pores in each thin section. The table also gives a symmetric matrix of similarities, in the form of "correlation" coefficients calculated between the six thin sections.

The first step in clustering by a pair-group method is to find the mutually highest correlations in the matrix to form the centers of clusters. The highest correlation (disregarding the diagonal element) in each column of the matrix in **Table 6–8** is shown in boldface type. Specimens A and B form mutually high pairs, because A most closely resembles B, and B most closely resembles A. C and D also form mutually high pairs. E most closely resembles D, but these two do not form a mutually high pair because D resembles C more than it does E. To qualify as a mutually high pair, coefficients c_{ij} and c_{ji} must be the highest coefficients in their respective columns.

We can indicate the resemblance between our mutually high pairs in a diagram such as **Figure 6–5 a**. Object C is connected to D at a level of $\hat{r} = 0.99$, indicating

Table 6–8. Average apparent grain diameters measured on thin sections of six greywackes and matrix of "correlations" between thin sections. Highest "correlation" in each column is indicated in boldface type.

Specimen	Average diameters in mm			
	Pore	Quartz	Rock frag-ment	Feldspar
A	0.24	1.78	0.69	3.32
B	0.48	2.07	2.41	4.78
C	0.76	4.05	1.2	3.21
D	0.23	2.98	0.85	2.06
E	0.04	3.33	3.39	2.63
F	1.98	0.98	2.01	2.02

"Correlations" on initial iteration

	A	B	C	D	E	F
A	1	**0.9110**	0.7671	0.7041	0.4401	−0.1067
B	**0.9110**	1	0.5393	0.4996	0.5704	**0.1680**
C	0.7671	0.5393	1	**0.9910**	0.5873	−0.7187
D	0.7041	0.4996	**0.9910**	1	**0.6647**	−0.7675
E	0.4401	0.5704	0.5873	0.6647	1	−0.3883
F	−0.1067	0.168	−0.7187	−0.7675	−0.3883	1

"Correlations" on second iteration

	AB	CD	E	F
AB	1	0.394	0.505	**0.031**
CD	0.394	1	**0.626**	−0.744
E	**0.505**	**0.626**	1	−0.388
F	0.031	−0.744	−0.388	1

"Correlations" on third iteration

	AB	CDE	F
AB	1	**0.450**	**0.031**
CDE	**0.450**	1	−0.566
F	0.031	−0.566	1

"Correlations" on fourth iteration

	ABCDE	F
ABCDE	1	−0.268
F	−0.268	1

the degree of their mutual similarity. In the same manner, A and B are connected at a level of $\hat{r} = 0.91$. This is the first step in the construction of a ***dendrogram***, or tree diagram, which is the most common way of displaying the results of clustering.

Next, the similarity matrix must be recomputed, treating grouped or clustered elements as a single element. There are several methods for doing this. In the simple technique we are considering, new correlations between all clusters and unclustered objects are recalculated by simple arithmetic averaging. For example, the new correlation between cluster CD and object E is equal to the sum of the correlations of the elements common to both CD and E, divided by 2 (that is, $\hat{r} = (0.5873 + 0.6647)/2 = 0.626$). **Table 6–8** contains the results of these

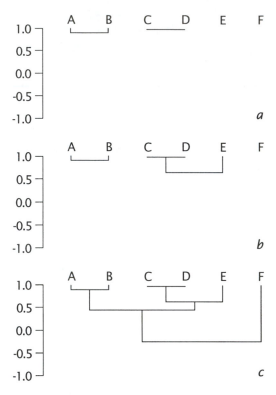

Figure 6–5. (*a*) Dendrogram with initial clusters, *CD* and *AB*. (*b*) Connection of object *E* to initial cluster *CD*. (*c*) Final connection of two clusters *AB* and *CDE*, and connection of isolated object *F* to *CDE*, completing dendrogram.

recalculations. Again, the highest correlations in each column are shown in bold-face type.

The clustering procedure is now repeated; mutually high pairs are sought out and clustered. In this cycle, object *E* joins cluster *CD* (**Fig. 6–5 b**) to form cluster CDE. The correlations between cluster CDE and other clusters or individual objects such as *F* are again found by adding together the common elements and dividing by 2. This process is repeated again and again until all objects and clusters are joined together. The final matrix of similarities will be a 2×2 matrix between the last remaining object and everything else collected into a single cluster, as shown in **Table 6–8**. This indicates that cluster *ABCDE* has a resemblance of $\hat{r} = -0.27$ with object *F*. Our dendrogram can then be completed (**Fig. 6–5 c**).

Clustering is an efficient way of displaying complex relationships among many objects. However, the process of averaging together members of a cluster and treating them as a single new object introduces distortions into the dendrogram. This distortion becomes increasingly apparent as successive levels of clusters are averaged together. We can evaluate the severity of this distortion by examining what numerical taxonomists call the ***matrix of cophenetic values***. This is nothing more than a matrix of apparent correlations contained within the dendrogram. For example, the dendrogram in **Figure 6–5** implies that the correlations between *C*, *D*, and *E*, on one hand, with *A* and *B*, on the other, are all $\hat{r} = 0.45$. Similarly, the correlation between *F* and *E* is the same as the correlation between *F* and *D*, or between *F* and any of the other objects. Only the correlations between *A* and *B* and between

C and *D* are correct. The lower half of the matrix in **Table 6–9** contains cophenetic values extracted from the dendrogram; actual correlations are shown in the upper diagonal half. We can obtain a visual impression of the degree of distortion in the dendrogram by plotting elements in the cophenetic value matrix against elements in the original correlation matrix (**Fig. 6–6**). If the two matrices were identical, the plot would form a straight line. Deviations indicate distortions in the dendrogram; if a point falls above the line, the correlation expressed in the dendrogram is too high. Conversely, if a point falls below the line, averaging has resulted in a correlation which is lower than the true correlation. A numerical measure of the similarity between the two matrices can be found simply by computing the ***cophenetic correlation*** between equivalent elements. Only one-half of the matrices, either above or below the diagonal, need be used because both matrices are symmetrical about the diagonal. In our example, the cophenetic correlation is $r = 0.90$.

Table 6–9. Actual correlations (above diagonal) and apparent or cophenetic correlations (below diagonal) extracted from dendrogram in Figure 6–5.

	A	B	C	D	E	F
A	1	0.911	0.767	0.704	0.440	−0.107
B	0.911	1	0.539	0.500	0.570	0.168
C	0.450	0.450	1	0.991	0.587	−0.719
D	0.450	0.450	0.991	1	0.665	−0.768
E	0.450	0.450	0.020	0.020	1	−0.388
F	−0.268	−0.268	−0.268	−0.268	−0.268	1

The essential features of the weighted pair-group method of cluster analysis with arithmetic averaging can be summarized in list form:

1. The correlation coefficient is used as a similarity measure.
2. Pairs of objects or clusters with the highest similarities are clustered or linked first.
3. Two objects can be connected only if they have mutually highest similarities with each other.
4. After two objects or clusters are combined, similarity coefficients with all other objects are found by averaging.

An obvious modification to this scheme is to incorporate some other similarity measure. Although many measures have been proposed, only two are widely used: the correlation coefficient and the distance coefficient. If the raw data are standardized prior to computing the similarity coefficient, correlations and distance coefficients can be directly transformed from one to the other. Dendrograms constructed from the two measures generally are similar. However, the distance coefficient is not constrained within the range ±1.0 as is the correlation coefficient, so it may produce more effective dendrograms if a few of the objects are very dissimilar to the others.

Suppose we measure seven variables on each of three objects. These might be, for example, size measurements on three fossils, or chemical analyses of three rocks. If we plot each measurement as shown in **Figure 6–7**, we find that two of the

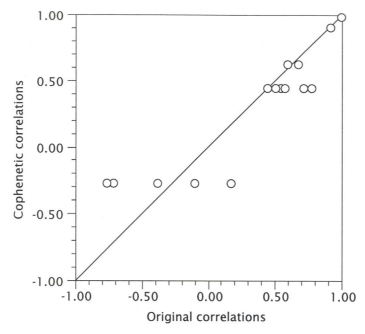

Figure 6–6. Plot of cophenetic values derived from dendrogram in Figure 6–5 against equivalent original correlations contained in Table 6–9. If the dendrogram represented the structure of the correlation matrix exactly, all points would fall on the diagonal line. Deviations from this line represent distortions in the dendrogram.

objects are similar in their relationships among the variables. These two will form a more or less parallel pattern, as do *a* and *b* in the diagram. The third object, *c*, may form a divergent pattern, but be much "closer" to the set of measurements of one of the other objects. In the illustration, *a* and *b* are highly correlated, or have a high linear relationship, but *b* and *c* have the smallest distance between them. If the variables were measurements of lengths on fossils, perhaps brachiopod shells, we would conclude that *a* and *b* are alike with regard to shape, but *b* and *c* are nearly the same size. If the variables were the percentage of heavy elements in ore samples, we might conclude that samples *a* and *b* are compositionally alike, but *a* is dilute with respect to *b*. The element concentrations of *b* and *c* are similar, but the chemical ratios are different.

The second attribute in our list needs little explanation. The correlation coefficient indicates greatest similarity at high positive values, while the distance coefficient indicates greatest similarity with the smallest distance. Therefore, correlations must be linked or interconnected at high values, and distance coefficients must be linked at low values.

Table 6–10 contains both distance and correlation coefficients for 11 carbonate minerals. Variables are five physical and optical properties measured on the minerals (data are listed in file CARBONAT.TXT). The resulting dendrograms from each similarity matrix are shown in **Figure 6–8**. Although there are some similarities (the high association of strontianite and witherite, for example), most of the hierarchical pattern is different.

Our criterion for linking two objects to form a cluster requires that both have mutually the highest correlation with each other. Other criteria are possible. A

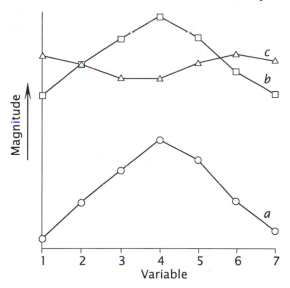

Figure 6–7. Plot of variables measured on three objects. Profiles a and b are highly correlated but are separated by a large distance. Profiles b and c are negatively correlated but are "close" in distance.

Table 6–10. Similarity measurements between 11 carbonate minerals*; upper diagonal half contains distance coefficients, d_{ij}; lower diagonal half contains correlation coefficients, r_{ij}. Raw data given in file CARBONAT.TXT have been standardized prior to clustering.

	ca	ma	rh	si	do	ar	st	wi	ce	az	sm	
	0.00	2.37	2.64	4.14	1.65	1.74	1.58	1.87	6.05	3.57	3.93	ca
		0.00	1.85	2.56	0.75	0.80	1.43	2.29	6.10	2.88	2.43	ma
ca	1.00		0.00	1.59	1.80	1.89	2.01	1.85	4.34	2.58	1.40	rh
ma	−0.59	1.00		0.00	2.91	2.98	3.26	3.33	4.52	3.15	0.60	si
rh	0.70	0.06	1.00		0.00	0.57	1.10	1.84	5.99	3.03	2.75	do
si	−0.26	0.90	0.48	1.00		0.00	0.88	1.74	5.97	2.57	2.79	ar
do	−0.21	0.91	0.41	0.94	1.00		0.00	0.80	5.66	2.62	2.92	st
ar	−0.74	0.94	−0.11	0.80	0.77	1.00		0.00	5.16	2.84	2.96	wi
st	−0.57	0.21	−0.80	−0.23	−0.01	0.21	1.00		0.00	4.94	4.31	ce
wi	0.21	−0.58	−0.56	−0.85	−0.56	−0.66	0.76	1.00		0.00	2.91	az
ce	0.64	−0.99	0.01	−0.86	−0.88	−0.98	−0.24	0.58	1.00		0.00	sm
az	−0.59	−0.15	−0.63	−0.30	−0.50	0.15	0.07	−0.23	0.06	1.00		
sm	−0.49	0.97	0.11	0.86	0.93	0.84	0.28	−0.41	−0.93	−0.34	1.00	
	ca	ma	rh	si	do	ar	st	wi	ce	az	sm	

*ca = calcite; ma = magnesite; rh = rhodochrosite; si = siderite; do = dolomite; ar = aragonite; st = strontianite; wi = witherite; ce = cerussite; az = azurite; sm = smithsonite.

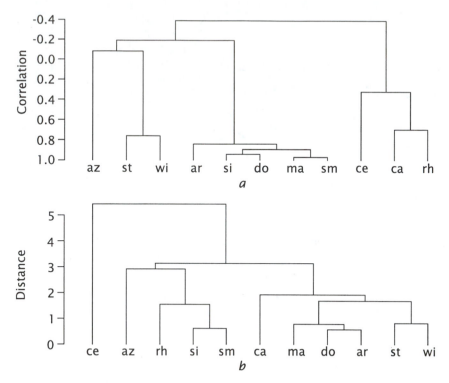

Figure 6–8. (*a*) Dendrogram of 11 carbonate minerals constructed by weighted pair–group method using arithmetic averaging of correlation coefficients. Original matrix given in Table 6–10. Cophenetic correlation coefficient = 0.92. (*b*) Dendrogram constructed by same method but using distance coefficients. Cophenetic correlation coefficient = 0.80.

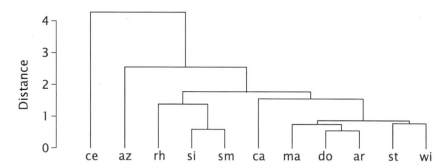

Figure 6–9. Dendrogram of distance matrix in Table 6–10 clustered by single linkage method. Cophenetic correlation coefficient = 0.90.

simple method is known as ***single linkage*** clustering, which connects objects to clusters on the basis of the highest similarity between the object and any object in the cluster. Results of clustering the distance matrix in **Table 6–10** by single linkage are shown in **Figure 6–9**. Because objects are allowed to enter a cluster on the basis of their distance to the closest of any object already in a cluster, linkage tends to occur at lower distances than in pair-group methods. In addition to the compression of the dendrogram, some connections may be different, especially in later iterations of clustering. In this example, the only difference is in greater apparent similarities within the early formed clusters.

This leads directly to the final feature in our list, arithmetic averaging of the similarity measures of objects that have been clustered. In a single linkage method, no averaging is done at all. The methods illustrated in **Figures 6–8 a** and **6–8 b** and in the initial example (**Fig. 6–5**) are called *averaged* or *weighted* techniques, although the name perhaps should be "equally weighted." In **Figure 6–5 a**, C and D are joined at the onset of clustering, followed by A and B. The correlations of the new cluster CD are found by combining the C and D rows and columns and dividing each entry by two. Next, object E enters the CD cluster and the correlations of the new cluster CDE are found by combining the CD cluster row and column with the E row and column and dividing by 2. That is, CD is treated as though it were a single object, when in fact it is two. The new object E has twice the influence on the correlations of the cluster CDE as do either C or D. Objects added late in the clustering procedure have a greater influence on the similarity matrix than do those linked earlier.

Unweighted average or *centroid* methods attempt to avoid this by weighting each cluster proportionally to the number of objects within it during the clustering process. For example, having formed cluster CD, we could then link object E to form the new cluster CDE. However, the similarity measures for this new cluster would be found by summing the correlations of C with all elements except D and E, the correlations of D with all elements except C and E, and the correlations of E with all elements except C and D. That is, we would sum the correlations of all of the original elements in the cluster. Then, each sum would be divided by three. This, in principle, gives each object in the cluster equal influence on the similarity characteristics of the entire cluster. The drawback of this technique is opposite to that of weighted methods: late entries into a large cluster have almost no influence on the similarity measures of the cluster. **Figure 6–10** shows a dendrogram of the distance data in **Table 6–10** clustered by unweighted averaging.

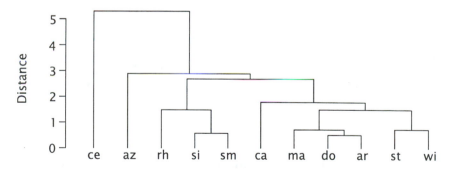

Figure 6–10. Dendrogram of distance matrix in Table 6–10, clustered by unweighted average method. Cophenetic correlation coefficient = 0.93.

One of the most widely used measures for determining the order of clustering is ***Ward's criterion***, which is based on concepts related to analysis of variance (Ward, 1963). Basically, clusters are iteratively combined in a manner that minimizes the sums of squares within the clusters at each step. Define the sum of squares within a cluster, k, as $SS_k = \sum (x_{ik} - \overline{x}_k)^2$, where $\overline{x}_k$ is the centroid of cluster k and x_{ik} is an individual object within the cluster. The sum of squares between clusters A and B is $SS_{A,B} = \sum (\overline{x}_A - \overline{x}_B)^2$. Ward's method uses the change in sums of squares between clusters as two clusters are merged to guide the hierarchical clustering process. The original procedure was modified by Wishart (1969) to use a measure of **dissimilarity**, which made the methodology compatible with other clustering algorithms. Define the dissimilarity between cluster A and cluster B as

$$d_{A,B}^2 = \frac{2n_A n_B}{n_A + n_B} \sum (\overline{x}_A - \overline{x}_B)^2 \qquad (6.41)$$

We want to merge two clusters if the operation will result in the smallest change in dissimilarity of all possible mergers of two clusters. Because of Wishart's modification, we can determine how much the dissimilarity will change without actually calculating the within-cluster similarities of the possible merged clusters.

Suppose A and B are two preexisting clusters that might be merged to form a new cluster, C. We want to know the effect this will have on a fourth cluster, D. We can find the dissimilarity between clusters C and D without actually forming cluster C and calculating its centroid:

$$d_{C,D}^2 = \frac{n_A + n_D}{n_C + n_D} d_{A,D}^2 + \frac{n_B + n_D}{n_C + n_D} d_{B,D}^2 + \frac{n_D}{n_C + n_D} d_{A,B}^2 \qquad (6.42)$$

All of the necessary dissimilarities have already been calculated on a previous iteration, and the size of cluster C will simply be the sum of the sizes of clusters A and B, so the dissimilarity defined by Equation (6.42) can easily be determined. Kaufman and Rousseeuw (1990) provided code for adapting a conventional clustering algorithm to utilize Ward's criterion, which now is incorporated in many statistical packages. Since the procedure results in clusters that have the minimum within-cluster variance, it often is considered more statistically appealing than other cluster procedures. However, there is no evidence that it is superior to alternative methods in practice.

We can illustrate the effect of the four different linkage strategies by considering a very simple clustering problem where only two variables are measured on each object. Then, all relationships between the objects can be shown in two dimensions, as in **Figure 6–11**. The distances between objects on the diagram are directly proportional to the degree of dissimilarity between them. Four of the objects, A through D, form a tight cluster. The dotted lines indicate the order in which the four have been joined together. A somewhat less similar object, E, also has been joined into the cluster. A sixth object, labeled F, is now being considered for possible inclusion in the growing cluster. The point M_1 is the centroid of points A through E, and M_2 is the average of object E and cluster $ABCD$.

Using a single linkage criterion, object F will be joined to the cluster if the distance CF is smaller than the distance to any other object in any other cluster. In an unweighted average or centroid linkage, object F will join the cluster if the distance $M_1 F$ is smaller than the distance to the centroid of any other group. In weighted pair–group or average linkage, the candidate object F will join if the distance $M_2 F$ is smaller than the distance to the average of any other cluster. (Note that the point M_2 is halfway between the average of the cluster $ABCD$ and object E, which was admitted on the previous cycle.) Finally, with complete linkage, object F will join the cluster if the distance EF is less than the distance to the most distant point in any other cluster.

Faced with such a welter of alternative methods, all yielding slightly different results, a researcher may justifiably ask which is best. Unfortunately, there is

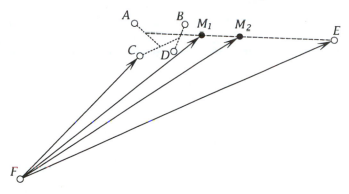

Figure 6–11. Diagram representing manner in which objects characterized by two variables enter a cluster. Objects A, B, C, and D form a cluster. Object E has joined this cluster, and object F is a candidate to join on the next iteration. M_1 is the centroid of objects A through E. M_2 is the average of (or midpoint between) object E and the last average of objects A through D.

no clear answer to this crucial question. Experience suggests that weighted pair-group methods tend to be superior to either single linkage or unweighted average methods. Relative superiority is defined by the tendency to produce the highest cophenetic correlation coefficient, which is interpreted as indicating low distortion in the dendrogram. Cophenetic correlations below about 0.8 may indicate such severe distortion in the dendrogram at lower linkages that the diagram is misleading. Jardine and Sibson (1971), however, argued for single linkage as the only procedure yielding entirely reproducible results, although Gordon (1999) points out that much of the argument is circular. Distance matrices usually cluster more successfully than do correlation matrices, in the sense that they yield higher cophenetic correlations. Distance matrices also seem to be less susceptible to drastic changes among different clustering methods. However, only limited statistical tests are available for hierarchical clustering, and little statistical theory has been developed or applied. [For some methods of clustering, a certain amount of theoretical justification has been developed. For examples, see Switzer (1970), Hartigan (1975), Everitt (1993), and Gordon (1999).] Most researchers who use clustering methods experiment with a variety of similarity measures and clustering techniques and choose the combination that yields the most satisfactory results with their data.

A pragmatic consideration may dictate the choice of a clustering procedure. Most hierarchical techniques may require creation and manipulation of an inordinately large matrix of similarities if the number of objects is large. (In the fields of ecology and archaeology, studies involving thousands of objects are not unusual.) Clustering procedures using a limited number of arbitrary cluster centers were devised to offset this computational difficulty. Probably the most widely used of these is the **k-*means*** procedure of McQueen (1967). Here, k points characterized by m variables are designated (either by the user or arbitrarily by the program) as initial seeds of clusters. A matrix of similarities between the k seeds and the n observations is calculated, and the closest or most similar observations are clustered with the nearest seeds. Centroids are then calculated for each of these k initial clusters and the process iterates exactly like a hierarchical procedure. In principle, the initial centroid of each cluster will rapidly shift toward the true center of a growing cluster as the influence of the real observations overwhelms that of the arbitrary starting points. The advantage of the k-means procedure is that only a $k \times n$

rectangular matrix of similarities is necessary, rather than an $n \times n$ square matrix. If k is small (5 to 10) and n is large (1000 or more), the process may be faster than a hierarchical method by more than two orders of magnitude. The disadvantage of the k-means method is that a suboptimal clustering may result if the arbitrary starting points do not fall within divergent clusters. This may lead to premature merger of the centroids and failure to detect outlying clusters.

The many considerations that enter into the choice of a clustering procedure introduce a certain subjectivity into a process designed to promote objectivity, but the cophenetic correlation provides a measure of guidance. The great benefits of cluster analysis are that it provides a relatively simple and direct way to classify objects, and it presents results in a manner that is both familiar and easy to understand.

Introduction to Eigenvector Methods, Including Factor Analysis

The term "factor analysis" is applied loosely to a number of related computational procedures. These procedures share the common objective of attempting to reveal a simple underlying structure that is presumed to exist within a set of multivariate observations. This structure is expressed in the pattern of variances and covariances between variables and the similarities between observations. Factor methods all operate by extracting the eigenvalues and eigenvectors from a square matrix produced by multiplying a data matrix (or some transformation of a data matrix) by its transpose. The basic mathematical operation is exactly the same as that demonstrated in Chapter 3. Upon this fundamental objective and methodology there is superimposed an array of embellishments and variations, often of an arbitrary nature, that has resulted in a bewildering assortment of computational techniques. Often the similarities among these methods are obscured by the complicated mathematical notation and terminology used by the various practitioners.

Factor analysis was originally developed by experimental psychologists in the 1930's, and much factor terminology has meaning only within the context of this field. Indeed, the very name "factor" alludes to hypothetical mental attributes, referred to as "factors of the mind." Sociologists and biometricians also have contributed to the richness of the jargon of factor analysis, helping to create a controversial and poorly understood methodology that extends the beguiling promise of instant insight to the researcher faced with more data than discernment.

Dozens of methodological variants of factor analysis have been developed. We will examine some of the most widely used of these, but will not dwell deeply on the philosophical or mathematical arguments that accompany them. Rather, we will explore some of the mathematical relationships that exist between a data matrix, its matrices of cross products, and their eigenvalues and eigenvectors.

Factor methods can be divided into two broad classes, called R-mode and Q-mode techniques. The first is concerned with interrelations between variables, and operates by extracting eigenvalues and eigenvectors from a covariance or correlation matrix. The second is concerned with the relationships between objects, often as an attempt to discern patterns or groupings within their arrangement in multivariate space. Most *Q-mode* analyses proceed by extracting the eigenvalues and eigenvectors from a matrix of similarities between all possible pairs of objects. *R-mode* techniques are statistical procedures, in the sense that the data are regarded as samples taken from a much larger population and the results pertain to the general properties or behavior of the variables. Because Q-mode methods focus on the similarities between individuals in the data set, they are not usually amenable to statistical analysis.

The first step in either R-mode or Q-mode analysis is to convert the data matrix into a square, symmetric matrix that expresses either the degrees of interrelationships between the variables, or between the objects on which these variables are measured. This is done by pre- or postmultiplying the data matrix by its transpose. In the simplest case, when a matrix of raw data, $\mathbf{X}$, consisting of n rows of observations and m columns of variables is premultiplied by its transpose, $\mathbf{X}'$, the result is a square matrix, $\mathbf{R}$, which is $m \times m$.

$$\mathbf{R} = \mathbf{X}' \, \mathbf{X}$$

The elements of $\mathbf{R}$ will consist of the raw sums of squares and cross products of the m variables. That is,

$$r_{jk} = \sum_{i=1}^{n} x_{ij} x_{ik}$$

where j and k are two columns of the data matrix. If the means of each variable are subtracted from the observations so that the variables' means equal zero, the matrix $\mathbf{R}$ will contain the covariances between the m variables. If the data are standardized so that each variable has a mean of zero and a standard deviation of one, the matrix $\mathbf{R}$ will contain the correlation coefficients between the m variables. Alternatively, the data matrix $\mathbf{X}$ may be postmultiplied by its transpose $\mathbf{X}'$ to yield the square symmetric matrix $\mathbf{Q}$, which has n rows and n columns.

$$\mathbf{Q} = \mathbf{X} \mathbf{X}'$$

If $\mathbf{X}$ contains raw observations, $\mathbf{Q}$ contains the squares and cross products for all pairs of objects, summed across variables. That is, the elements of $\mathbf{Q}$ are

$$q_{il} = \sum_{j=1}^{m} x_{ij} x_{lj}$$

where i and l are two rows of the data matrix. In most investigations, we have many more objects than variables, so $\mathbf{Q}$ may be much larger than $\mathbf{R}$, even though both are created from the same original data matrix, $\mathbf{X}$.

Most geological applications of factor analysis have then proceeded to extract eigenvalues and eigenvectors from either $\mathbf{R}$ or $\mathbf{Q}$. However, it is obvious that there must be a close link between the two, since both are generated from the same data set. This link was established in the early days of factor analysis, but its implications were overlooked until very recent times. In part, this is because the psychologists and sociologists who were responsible for most of the pioneering work in factor analysis concerned themselves exclusively with R-mode studies. Not until biologists and geologists became interested in factor analysis were Q-mode techniques widely used, and most of these applications were direct adaptations of R-mode methods in which $\mathbf{Q}$ was simply substituted for $\mathbf{R}$. Since $\mathbf{Q}$ often may be an enormous matrix, these direct approaches were not feasible until large, general-purpose computers became available in the late 1950's. Unfortunately, the basic relationship between $\mathbf{R}$ and $\mathbf{Q}$ was overlooked, and the early methods for Q-mode factor analysis often were unduly complex (for example, see Imbrie and Purdy,

1962, and Ondrick and Srivastava, 1970). Recent authors have exploited the duality between the *R*- and *Q*-modes, achieving a great simplification in the computations of *Q*-mode factor analysis (Jöreskog, Klovan, and Reyment, 1976; David, Dagbert, and Beauchemin, 1977; Zhou, Chang, and Davis, 1983; Jobson, 1991; Reyment and Savazzi, 1999).

Eckart–Young theorem

The critical interrelationships between a data matrix and the eigenvalues and eigenvectors of its two cross-product matrices are expressed in the Eckart-Young theorem, first given by these two authors in their classic article in the first volume of *Psychometrika*, which appeared in 1936. The **Eckart-Young theorem** is the cornerstone of several multivariate techniques, including factor analysis. It states that for any real matrix, **X**, two orthogonal matrices, **V** and **U**, can be found for which the product is a real diagonal matrix with no negative elements. A proof of the theorem is provided by Johnson (1963); we will examine the consequences of this theorem as it pertains to factor analysis, adopting a numerical example originally devised by Burt (1937).

Probably the most significant implementation of the Eckhart-Young theorem is in **singular value decomposition**, or SVD, a set of techniques that are widely used in numerical analysis. The original algorithm for singular value decomposition was devised by Golub and Reinsch in 1971 and has been modified and extended by many authors. SVD algorithms are now incorporated into most computer programs for the solution of simultaneous equations (especially singular or ill-conditioned sets of equations) and for the extraction of eigenvalues and eigenvectors from large data sets (Press and others, 1992).

Table 6–11. Measurements made on four specimens of goniatite ammonoids, representing species of the genus *Manticoceras*.

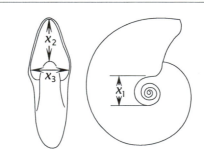

	Umbilical diameter (mm)	Height of whorl (mm)	Width of whorl (mm)
	X_1	X_2	X_3
Species A	4	27	18
Species B	12	25	12
Species C	10	23	16
Species D	14	21	14
Means	10	24	15

The measurements in **Table 6–11** are typical of those that might be obtained in a simple geological investigation. We will pretend they are measurements made by a paleontologist on the shells of four goniatite ammonoid specimens, representing species of the genus *Manticoceras*. The variables include diameter of the umbilicus or exposed part of the inner whorls, height of the outer whorl at the peristome or shell opening, and width of the outer whorl at the peristome. The mean of each variable may be subtracted from each observation to simplify calculations. The resulting data matrix is

$$\mathbf{X} = \begin{bmatrix} -6 & 3 & 3 \\ 2 & 1 & -3 \\ 0 & -1 & 1 \\ 4 & -3 & -1 \end{bmatrix}$$

Rearranging the matrices in the statement of the Eckart-Young theorem shows that the data matrix can be viewed as the product of three other matrices

$$\mathbf{X} = \mathbf{V}\mathbf{\Lambda}\mathbf{U}' \tag{6.43}$$

where $\mathbf{V}$ is an $n \times r$ matrix whose columns are **orthonormal**. This means that $\mathbf{V}'\mathbf{V} = \mathbf{I}$, where $\mathbf{I}$ is $r \times r$. Likewise, $\mathbf{U}$ is an $m \times r$ matrix whose columns are orthonormal, so $\mathbf{U}'\mathbf{U} = \mathbf{I}$, where $\mathbf{I}$ is also $r \times r$. $\mathbf{\Lambda}$ is an $r \times r$ square matrix containing r positive elements along the diagonal. These are called the **singular values** of $\mathbf{X}$; all off-diagonal elements of $\mathbf{\Lambda}$ are zero.

The minor product matrix $\mathbf{R} = \mathbf{X}'\mathbf{X}$ is of size $m \times m$ and has r nonzero eigenvalues and $m - r$ eigenvalues that are equal to zero. The nonzero eigenvalues are equal to the square of the singular values in matrix $\mathbf{\Lambda}$. That is,

$$\mathbf{\Lambda}^2 = \mathbf{I}\boldsymbol{\lambda}'$$

or equivalently,

$$\mathbf{\Lambda} = \mathbf{I}\sqrt{\boldsymbol{\lambda}}' \tag{6.44}$$

where $\boldsymbol{\lambda}$ is a vector containing the r nonzero eigenvalues of $\mathbf{R}$. The major product matrix $\mathbf{Q} = \mathbf{X}\mathbf{X}'$ is of size $n \times n$ but also has only r nonzero eigenvalues. These are identical to the eigenvalues extracted from $\mathbf{R}$, except there are additional eigenvalues that are all equal to zero if n, the number of objects, is larger than m, the number of variables.

Furthermore, the columns of the matrix $\mathbf{U}$ contain the eigenvectors of $\mathbf{R}$ that are associated with each eigenvalue λ. The columns of $\mathbf{V}$ contain the eigenvectors from $\mathbf{Q}$. Since the eigenvalues from both $\mathbf{R}$ and $\mathbf{Q}$ are identical, there must be a relationship between the two sets of eigenvectors $\mathbf{U}$ and $\mathbf{V}$. This relationship is

$$\mathbf{V} = \mathbf{X}\mathbf{U}\mathbf{\Lambda}^{-1}$$

or

$$\mathbf{U} = \mathbf{X}'\mathbf{V}\mathbf{\Lambda}^{-1} \tag{6.45}$$

Recall that eigenvectors usually are calculated so that the sum of their squared elements is equal to 1.0. That is, the eigenvectors are **normalized** to be of unit length. If the normalized eigenvectors are multiplied by their corresponding singular values (or square roots of their eigenvalues), the vectors are scaled so their lengths are proportional to the magnitudes of their singular values. In the context of factor analysis, the vector formed by multiplying an eigenvector by its corresponding singular value is referred to as a **factor**, but for the sake of generality we will use the term **principal vector**. The individual elements of a principal vector are called *loadings*; they relate the vector to the original variables. In matrix notation, the R-mode principal vectors are

$$\mathbf{A}^R = \mathbf{U}\mathbf{\Lambda} \qquad (6.46)$$

The loadings represent the proportion or weighting that must be assigned to each variable in order to project the objects onto the principal vectors as *scores*. They also represent the correlations of the individual variables with the principal vectors. The corresponding equation for the principal vectors in Q-mode analysis is

$$\mathbf{A}^Q = \mathbf{V}\mathbf{\Lambda} \qquad (6.47)$$

and the loadings are the proportions of each individual object necessary to project the variables onto the principal vectors.

R-mode scores are found by multiplying the data by the loadings, or

$$\mathbf{S}^R = \mathbf{X}\mathbf{A}^R \qquad (6.48)$$

which project the n individual objects onto the principal vectors. For a specific observation, i,

$$s_{ik} = \sum_{j=1}^{m} a_{jk} x_{ji}$$

or

$$s_{ik} = a_{1k} x_{i1} + a_{2k} x_{i2} + \ldots + a_{mk} x_{im}$$

where s_{ik} is the score of the ith observation on the kth principal vector, x_{ij} is the value of variable j measured on object i, and a_{jk} is the loading of variable j on principal vector k. In turn, a_{jk} is the product of element j of the kth eigenvector, times the square root of the kth eigenvalue.

In a similar manner, Q-mode scores are found by multiplying the transpose of the data matrix by the Q-mode loadings:

$$\mathbf{S}^Q = \mathbf{X}'\mathbf{A}^Q \qquad (6.49)$$

This equation will project the m variables onto the principal vectors.

Some algebraic rearrangements will demonstrate the relationship between R- and Q-mode loadings and scores. Equation (6.46) defines R-mode loadings as $\mathbf{A}^R = \mathbf{U}\mathbf{\Lambda}$, and the Eckart–Young theorem defines $\mathbf{U}$ as $\mathbf{U} = \mathbf{X}'\mathbf{V}\mathbf{\Lambda}^{-1}$. Multiplying both sides by $\mathbf{\Lambda}$,

$$\mathbf{U}\mathbf{\Lambda} = \mathbf{X}'\mathbf{V}\mathbf{\Lambda}^{-1}\mathbf{\Lambda}$$
$$\mathbf{A}^R = \mathbf{X}'\mathbf{V}$$

Q-mode scores are defined by Equation (6.49) as $\mathbf{S}^Q = \mathbf{X}'\mathbf{A}^Q$ and Q-mode loadings, $\mathbf{A}^Q$, are defined as

$$\mathbf{A}^Q = \mathbf{V}\mathbf{\Lambda}$$

Substituting, $\mathbf{S}^Q = \mathbf{X}'\mathbf{V}\mathbf{\Lambda}$. But, $\mathbf{X}'\mathbf{V} = \mathbf{A}^R$, so

$$\mathbf{S}^Q = \mathbf{A}^R\mathbf{\Lambda} \qquad (6.50)$$

Similar manipulations will show that

$$\mathbf{S}^R = \mathbf{A}^Q\mathbf{\Lambda} \qquad (6.51)$$

So, Q-mode scores are proportional to R-mode loadings, and *vice versa*. The constant of proportionality is equal to $\mathbf{\Lambda}$, the singular values. Equivalent expressions are

$$\mathbf{A}^R = \mathbf{S}^Q \mathbf{\Lambda}^{-1} \tag{6.52}$$

and

$$\mathbf{A}^Q = \mathbf{S}^R \mathbf{\Lambda}^{-1} \tag{6.53}$$

This means that if we perform an R-mode analysis, we can also automatically perform a Q-mode analysis, since both the Q-mode loadings and scores can be obtained from the R-mode solution.

We can illustrate these relationships using measurements on ammonoid shells introduced earlier. The R-mode or minor product matrix is obtained by premultiplying the data matrix by its transpose:

$$\mathbf{X}'\mathbf{X} = \mathbf{R}$$

$$\begin{bmatrix} -6 & 2 & 0 & 4 \\ 3 & 1 & -1 & -3 \\ 3 & -3 & 1 & -1 \end{bmatrix} \cdot \begin{bmatrix} -6 & 3 & 3 \\ 2 & 1 & -3 \\ 0 & -1 & 1 \\ 4 & -3 & -1 \end{bmatrix} = \begin{bmatrix} 56 & -28 & -28 \\ -28 & 20 & 8 \\ -28 & 8 & 20 \end{bmatrix}$$

The eigenvalues of $\mathbf{R}$ are $\lambda_1 = 84$, $\lambda_2 = 12$, $\lambda_3 = 0$. Since the final eigenvalue is zero, the matrix $\mathbf{\Lambda}^2$ has a rank of only two rather than three. That is,

$$\mathbf{\Lambda}^2 = \begin{bmatrix} 84 & 0 \\ 0 & 12 \end{bmatrix}$$

so

$$\mathbf{\Lambda} = \begin{bmatrix} 9.165 & 0.0 \\ 0.0 & 3.464 \end{bmatrix}$$

The eigenvectors of $\mathbf{R}$ are

$$\mathbf{U} = \begin{bmatrix} 0.8165 & 0.0 & 0.5774 \\ -0.4082 & 0.7071 & 0.5774 \\ -0.4082 & -0.7071 & 0.5774 \end{bmatrix}$$

Because the final eigenvalue is zero, the last column of $\mathbf{U}$ disappears, leaving the 3×2 matrix

$$\mathbf{U} = \begin{bmatrix} 0.8165 & 0.0 \\ -0.4082 & 0.7071 \\ -0.4082 & -0.7071 \end{bmatrix}$$

The matrix of R-mode loadings, $\mathbf{A}^R$, is given by Equation (6.46):

$$\begin{bmatrix} 0.8165 & 0.0 \\ -0.4082 & 0.7071 \\ -0.4082 & -0.7071 \end{bmatrix} \cdot \begin{bmatrix} 9.165 & 0.0 \\ 0.0 & 3.464 \end{bmatrix} = \begin{bmatrix} 7.4832 & 0.0 \\ -3.7412 & 2.4494 \\ -3.7412 & -2.4494 \end{bmatrix}$$

We can now project the four specimens onto the R-mode principal vectors by computing their scores by Equation (6.48):

$$\begin{bmatrix} -6 & 3 & 3 \\ 2 & 1 & -3 \\ 0 & -1 & 1 \\ 4 & -3 & -1 \end{bmatrix} \cdot \begin{bmatrix} 7.4832 & 0.0 \\ -3.7412 & 2.4494 \\ -3.7412 & -2.4494 \end{bmatrix} = \begin{bmatrix} -67.3 & 0.00 \\ 22.4 & 9.8 \\ 0.0 & -4.9 \\ 44.9 & -4.9 \end{bmatrix}$$

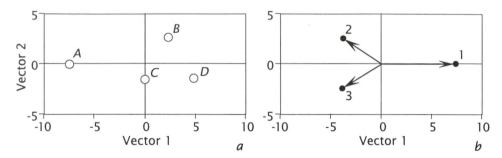

Figure 6–12. Scores of ammonoid species plotted on principal vectors 1 (horizontal) and 2 (vertical). (*a*) R-mode scores of four ammonoid species. (*b*) Q-mode scores of three variables measured on ammonoid species.

The scores can be graphically displayed by plotting them in the space defined by the orthogonal principal vectors. **Figure 6–12 a** shows the four ammonoid specimens plotted on the first and second vectors.

A Q-mode analysis begins by postmultiplying the data matrix by its transpose:

$$\mathbf{XX'} = \mathbf{Q}$$

$$\begin{bmatrix} -6 & 3 & 3 \\ 2 & 1 & -3 \\ 0 & -1 & 1 \\ 4 & -3 & -1 \end{bmatrix} \cdot \begin{bmatrix} -6 & 2 & 0 & 4 \\ 3 & 1 & -1 & -3 \\ 3 & -3 & 1 & -1 \end{bmatrix} = \begin{bmatrix} 54 & -18 & 0 & -36 \\ -18 & 14 & -4 & 8 \\ 0 & -4 & 2 & 2 \\ -36 & 8 & 2 & 26 \end{bmatrix}$$

The matrix **V** can be converted into the matrix of Q-mode loadings by Equation (6.47):

$$\mathbf{V\Lambda} = \mathbf{A}^Q$$

$$\begin{bmatrix} -0.8018 & 0.0 \\ 0.2673 & 0.8165 \\ 0.0 & -0.4082 \\ 0.5345 & -0.4082 \end{bmatrix} \cdot \begin{bmatrix} 9.165 & 0.0 \\ 0.0 & 3.464 \end{bmatrix} = \begin{bmatrix} -7.3485 & 0.0 \\ 2.4498 & 2.8284 \\ 0.0 & -1.4140 \\ 4.8987 & -1.4140 \end{bmatrix}$$

Q-mode scores are the projections of variables onto the principal vectors and are found by multiplying the transpose of the data matrix by the Q-mode loadings.

$$\mathbf{X'} \mathbf{A}^Q = \mathbf{S}^Q$$

$$\begin{bmatrix} -6 & 2 & 0 & 4 \\ 3 & 1 & -1 & -3 \\ 3 & -3 & 1 & -1 \end{bmatrix} \cdot \begin{bmatrix} -7.3485 & 0.0 \\ 2.4498 & 2.8284 \\ 0.0 & -1.4140 \\ 4.8987 & -1.4140 \end{bmatrix} = \begin{bmatrix} 68.6 & 0.0 \\ -34.3 & 8.5 \\ -34.3 & -8.5 \end{bmatrix}$$

We can plot the three variables measured on the ammonoid specimens projected onto the plane defined by the first two Q-mode principal axes, as in **Figure 6–12 b**.

We may now confirm that Q-mode scores are proportional to R-mode loadings using Equation (6.50):

$$\mathbf{A}^R \mathbf{\Lambda} = \mathbf{S}^Q$$

$$\begin{bmatrix} 7.4832 & 0.0 \\ -3.7412 & 2.4494 \\ -3.7412 & -2.4494 \end{bmatrix} \cdot \begin{bmatrix} 9.165 & 0.0 \\ 0.0 & 3.464 \end{bmatrix} = \begin{bmatrix} 68.6 & 0.0 \\ -34.3 & 8.5 \\ -34.3 & -8.5 \end{bmatrix}$$

Finally, we may demonstrate the Eckart-Young theorem by re-creating the data matrix $\mathbf{X}$ from its orthonormal parts:

$$\mathbf{X} = \mathbf{V}\mathbf{\Lambda}\mathbf{U}'$$

$$= \begin{bmatrix} -0.8018 & 0.0 \\ 0.2673 & 0.8165 \\ 0.0 & -0.4082 \\ 0.5345 & -0.4082 \end{bmatrix} \cdot \begin{bmatrix} 9.165 & 0.0 \\ 0.0 & 3.464 \end{bmatrix} \cdot \begin{bmatrix} 0.8165 & -0.4082 & -0.4082 \\ 0.0 & 0.7071 & -0.7071 \end{bmatrix}$$

$$= \begin{bmatrix} -6 & 3 & 3 \\ 2 & 1 & -3 \\ 0 & -1 & 1 \\ 4 & -3 & -1 \end{bmatrix}$$

In this simple numerical example we have taken a multivariate set of observations and resolved it into a smaller number of principal vectors. We have also shown that R-mode solutions are equivalent to those obtained by Q-mode analysis. Both of these are critical points that will be referred to repeatedly in the following discussions in which we will investigate some of the elaborations that are placed upon the relatively simple structure we have just examined.

As noted at the beginning of this section, "factor analysis" is a catchall term including a multitude of techniques that involve the extraction of eigenvalues and eigenvectors from a matrix of the cross products of a data set. Factor analysis is also used (more correctly) in a strict sense to mean a statistical procedure by which a data matrix is decomposed into a prescribed number of uncorrelated factors and a residual set of "unique" random variations. Other important eigenvalue methods include principal component analysis (PCA), correspondence analysis, and in the Q-mode, principal vector and principal coordinates analysis.

The characteristics of the various eigenvalue procedures will be illustrated with an artificial data set similar to the one published by Cooley and Lohnes (1971, p. 133–136). We are all familiar with the problem of specifying the "size" of something. Should we measure a length, a width, an area, a volume, or some ratio of these? How is the concept of "size" to be separated from that of "shape"? To investigate these questions, we have created a set of 25 objects in the form of rectangular blocks (**Fig. 6–13**). The three dimensions of the blocks were chosen randomly and range up to ten units. All shapes and sizes are equally probable in the resulting collection, from a cube less than one unit on a side, through rod-like prisms and flat plates, to a $10 \times 10 \times 10$ cube. A variety of measurements have been made on each of the blocks, and these constitute our variables, which are:

x_1 = long axis,

x_2 = intermediate axis,

x_3 = short axis,

x_4 = longest diagonal,

x_5 = ratio $\dfrac{\text{radius of smallest circumscribed sphere}}{\text{radius of largest inscribed sphere}}$,

x_6 = ratio $\dfrac{\text{long axis + intermediate axis}}{\text{short axis}}$,

x_7 = ratio $\dfrac{\text{surface area}}{\text{volume}}$.

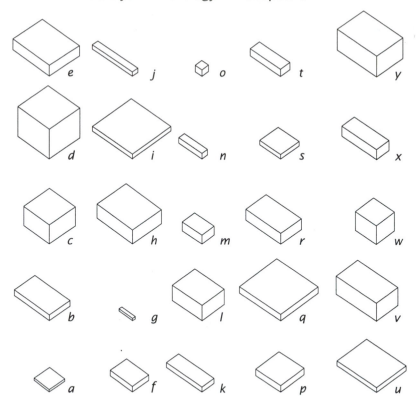

Figure 6–13. Twenty-five blocks with random lengths, widths, and heights. Data are given in file BOXES.TXT.

File BOXES.TXT contains the 25 observations that are illustrated by the collection of blocks shown in **Figure 6–13**. Note that this data set has some interesting properties; it should possess at most only three independent dimensions, as variables x_4 through x_7 have been created by various combinations of the length, width, and height measurements. Also, the data contain a certain amount of induced correlation caused by the inherent nature of the variables. The long axis of each block, by definition, must be longer than the intermediate axis, which in turn must be longer than the short axis. This means that if, for example, length and width are plotted against each other, the scatter of points is confined to the lower diagonal half of the diagram (**Fig. 6–14**). This induces a positive correlation that is significantly greater ($r = 0.58$) than that expected from two independent variables.

Of course, we ordinarily would not collect measurements that we know to be dependent upon one another. Unfortunately, it happens more often than not that geologic variables are interrelated; compositional variables contain induced correlations because they are parts of a whole, and taxonomic measurements may be interrelated because of the effect of size. In this artificial example, we know what interdependencies exist, and we can expect to see certain consequences of these interdependencies in the outcomes of our analyses. This may help us understand the results that we obtain from analyses of real data, where the possibilities of similar interdependencies cannot be established before the fact. Illenberger (1991) has examined the mathematical characteristics of this data set.

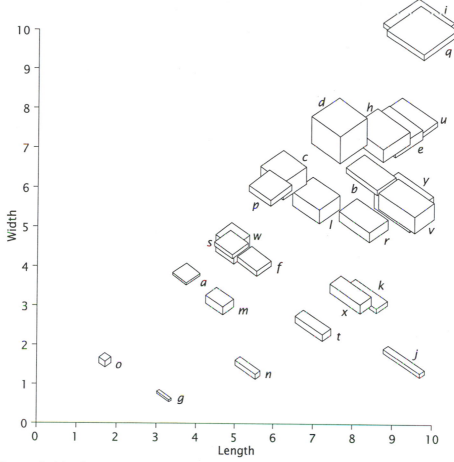

Figure 6–14. Cross plot of lengths versus widths of random blocks. Because width must be less than length, all blocks plot in lower diagonal half of diagram. This induces a correlation of $r = 0.58$ between length and width. Length and width are given in arbitrary units.

Principal Component Analysis

The first major procedure that will be considered in this section is called principal component analysis, or PCA. **Principal components** are nothing more than the eigenvectors of a variance–covariance matrix or a correlation matrix. By themselves they may provide significant insight into the structure of the matrix, and they often may be interpreted in much the same manner as factors. Many factor analysis schemes employ principal components as starting points for the analysis. For this reason, and because their derivation and interpretation are more straightforward, we will begin with a discussion of principal components. Also, as we have noted, geologists have been rather confused in their use of terminology; most of the published studies that geologists have called "factor analyses" actually are principal component analyses. The "factors" cited in these articles properly should be called components. Some authors have found the confusion so great that they have resorted to labeling factor analysis as "true factor analysis" to distinguish it from component analysis (Jöreskog, Klovan, and Reyment, 1976).

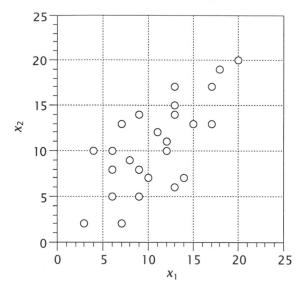

Figure 6–15. Scatter diagram of bivariate data from Table 6–12.

Table 6–12. Twenty-five bivariate observations with variance of $x_1 = 20.277$, variance of $x_2 = 24.060$, and covariance, $cov_{1,2} = 15.585$.

x_1	x_2	x_1	x_2
3	2	12	10
4	10	12	11
6	5	13	6
6	8	13	14
6	10	13	15
7	2	13	17
7	13	14	7
8	9	15	13
9	5	17	13
9	8	17	17
9	14	18	19
10	7	20	20
11	12		

Suppose we measure two variables on a collection of objects, perhaps lengths and widths of brachiopod shells, and obtain the data shown plotted in **Figure 6–15** and listed in **Table 6–12**. (The data are also included as the first two columns in file TABLE612.TXT.) The variance of x_1 is 20.277, the variance of x_2 is 24.060, and the covariance between the two is 15.585. (Note that we are carrying more digits than the data warrant; we will round to the appropriate level when calculations are complete.) These constitute the elements of the variance–covariance matrix **S**, which is

$$S = \begin{bmatrix} 20.277 & 15.585 \\ 15.585 & 24.060 \end{bmatrix}$$

You will recall from Chapter 3 that a matrix can be expressed in geometric form as a series of vectors in multidimensional space. We regard each row of the matrix as giving the coordinates of the end-point of the vector which represents that row. A 2×2 matrix can be plotted on a flat diagram, as in **Figure 6–16**. Furthermore, these vectors can be considered to define axes within an m-dimensional ellipsoid. The eigenvectors of the matrix yield the orientations of the **principal axes** of the ellipsoid, and the eigenvalues represent half the lengths of each successive principal axis. (Another way of stating this is to say that the eigenvalues represent the lengths of the **semiaxes**, which are the equivalents in an ellipse of the radii of a circle.) In Chapter 3, we interpreted arbitrary matrices in this manner, but it is obvious that variance–covariance matrices can be interpreted geometrically with equal facility. The PCA method is concerned with finding these axes and measuring their magnitudes.

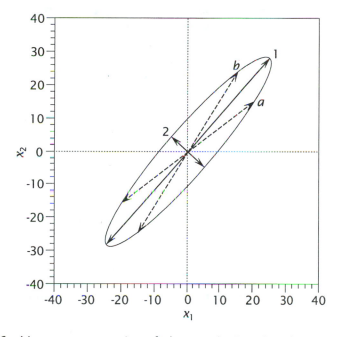

Figure 6–16. Vector representation of elements in 2×2 variance–covariance matrix **S** calculated from data in Table 6–12. Dashed arrow a represents first column of matrix; dashed arrow b represents second column. Solid arrow 1 is the first principal axis of the enclosing ellipse and represents the first column of eigenvector matrix **U**; solid arrow 2 is the second principal axis and represents the second column of **U**. Lengths of axes are equal to the first and second eigenvalues. The first principal component accounts for 86% of total variance and the second accounts for 14%.

If we measure m variables on a collection of objects, we can compute an $m \times m$ matrix, **S**, of variances and covariances. From **S**, we can extract m eigenvalues and m eigenvectors. Eigenvectors are always mutually orthogonal, and in addition the m eigenvalues will be nonnegative because the variance–covariance matrix is symmetrical.

We can compute the eigenvalues and eigenvectors of our 2×2 matrix **S** and plot the vectors. The first eigenvector, expressed in standardized form is

$$\mathbf{U}_1 = \begin{bmatrix} 0.663 \\ 0.748 \end{bmatrix}$$

which means that the vector corresponding to the first principal axis of the ellipsoid slopes 0.663 units in x_1 for every 0.748 units in x_2. The first eigenvalue is 37.868, which we can plot as the length of the principal semiaxis. The second eigenvector is

$$\mathbf{U}_2 = \begin{bmatrix} -0.748 \\ 0.663 \end{bmatrix}$$

which you will recognize as being at right angles to the first. (The sign of an element in an eigenvector is arbitrary, provided all signs are changed if one is changed.) The second eigenvalue is 6.469, which we can use as the length of the second principal semiaxis. These geometric relationships are shown in **Figure 6–16**. Keep in mind that we have plotted vectors from a variance–covariance matrix, so the measurements on the diagram are given in the same units as the variances, or in this instance, the square of units of length.

We can define the total variance in a data set as the sum of the individual variances. Because the variances are located along the diagonal of the variance–covariance matrix, this is equivalent to finding the trace of the matrix. In our example, the total variance is $20.277 + 24.060 = 44.337$. The first variable contributes 20.3/44.3 or about 46% of the total variance, and the second contributes the remainder, about 54%. You will recall from Chapter 3 that the sum of the eigenvalues of a matrix is equal to the trace of the matrix, so the total of our two eigenvalues is $37.868 + 6.469 = 44.337$. Since these eigenvalues represent the lengths of the two principal semiaxes, the axes also represent the total variance of the data set, and each accounts for an amount of the total variance equal to the eigenvalue divided by the trace. The first principal axis contains 37.9/44.4 or about 86% of the total variance, whereas the second axis represents only 14%. In other words, if we measure the variation in our data set along the first principal axis, we can represent four-fifths of the total variation in the observations. It inevitably happens that at least one of the principal axes will be more efficient (in terms of accounting for total variance) than any of the original variables. On the other hand, at least one of the axes must be less efficient than any of the original variables.

If we perform the matrix multiplication $\mathbf{S}^R = \mathbf{XU}$, where $\mathbf{X}$ is the 25×2 matrix of original observations listed in **Table 6–12** and $\mathbf{U}$ is the 2×2 matrix whose columns contain the eigenvectors of $\mathbf{S}$, we create a new 25×2 matrix of values, or *principal component scores*. The first column of scores will have a variance exactly equal to the first eigenvalue, or 37.868, and the second column of scores will have a variance equal to the second eigenvalue, or only 6.469. Because the scores are measured along axes at right angles to each other, the covariance (and the correlation) between the two columns will be zero. **Table 6–13** contains the data from **Table 6–12** transformed in this manner; the scores are plotted in **Figure 6–17**.

The matrix multiplication in effect performs the pair of calculations,

$$s_{1i} = 0.663x_{1i} + 0.748x_{2i}$$
$$s_{2i} = 0.663x_{2i} - 0.748x_{1i}$$

The first equation projects the ith observation onto the first principal axis by multiplying the observed values of x_1 and x_2 by the corresponding elements of the first

Table 6–13. Principal component scores of data from Table 6–12 calculated by projecting original data onto principal axes. Variance of 1st principal component = 37.868. Variance of 2nd principal component = 6.469.

PCA 1	PCA 2	PCA 1	PCA 2
3.485	−0.918	15.44	−2.346
10.130	3.638	16.18	−1.683
7.718	−1.173	13.11	−5.746
9.962	0.816	19.09	−0.442
11.460	2.142	19.84	0.221
6.137	−3.910	21.34	1.547
14.370	3.383	14.52	−5.831
12.040	−0.017	19.67	−2.601
9.707	−3.417	21.00	−4.097
11.950	−1.428	23.99	−1.445
16.440	2.550	26.15	−0.867
11.870	−2.839	28.22	−1.700
16.270	−0.272		

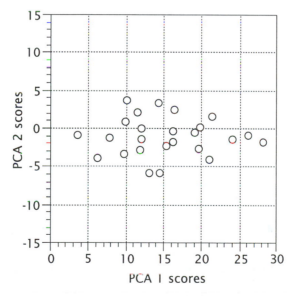

Figure 6–17. Plot of principal component scores–projections of observations in Table 6–12 onto their principal components. Variance of component 1 scores is 37.868, variance of component 2 scores is only 6.469.

eigenvector and summing. The same observation is projected onto the second principal axis by the second equation. The elements of the eigenvectors that are used to compute the scores of observations are called the ***principal component loadings*** and are simply coefficients of the linear equation defined by the eigenvector. In articles describing principal component analyses (often erroneously referred to as "factor" analyses), statements are sometimes made about the "loading of variable A onto factor 1." This is nothing more than a way of referring to the coefficient in eigenvector i which corresponds to variable a.

Let us look again at our original data set. We have found the eigenvectors of the covariance matrix and determined that the first eigenvector accounts for about 86% of the total variance. Suppose we decide that it is imperative that we reduce our system to only one variable. This could be done by discarding either variable x_1 or x_2, but this would cause a loss of 46% or 54% of the total variance, depending upon which variable we retained. If, however, we convert our observations to scores on the first principal axis, we sacrifice only 14% of the total variation in our data set. It should be emphasized that the variance along the first principal axis will be greater than the variance along any other possible axis or straight line drawn through the data points. However, the variance along this principal axis is not as great as the sum of the variances along both original axes x_1 and x_2, and if the second eigenvector is eliminated, an unavoidable loss of variance must result.

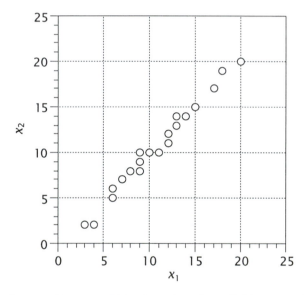

Figure 6–18. Scatter diagram of bivariate data from Table 6–12 after placing variable x_2 in rank order.

Suppose we take the data in **Table 6–12** and rank each observation. (Variable x_1 is already in rank order; variable x_2 is placed in rank order in column 3 of file TABLE612.TXT.) The data will plot as shown in **Figure 6–18**. Ranking causes the variables to become highly correlated, and this is reflected in the covariance, which is now 21.918. Because we are using the same values and merely changing their order, the variances of x_1 and x_2 remain unchanged. If we extract the eigenvectors and eigenvalues of this new variance-covariance matrix, we discover that the eigenvectors are almost the same:

$$\mathbf{U} = \begin{bmatrix} 0.676 & 0.737 \\ 0.737 & -0.676 \end{bmatrix}$$

However, the two new eigenvalues are radically different. Eigenvalue $\lambda_1 = 44.168$, so the first principal axis now accounts for 44.2/44.4 or over 99% of the total variance in the data. The second principal axis is so short ($\lambda_2 = 0.169$) that it is almost

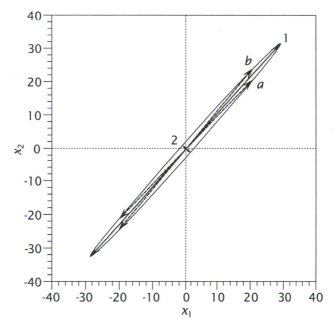

Figure 6–19. Vector representation of variance–covariance matrix **S** and its eigenvectors computed from ranked data shown in Figure 6–18.

impossible to plot on our diagram (**Fig. 6–19**). Obviously, we could discard the second principal component and lose very little of the variance in the data set. If we do this, we are representing our original bivariate data by a single new variable (defined by the first principal component), and we have reduced the dimensionality of our data from two to one.

Rather than ranking the observations, we might randomize them as in **Figure 6–20**. (Column 4 of file TABLE612.TXT contains values of x_2 in random order; randomizing one variable is sufficient to achieve a random relationship between both variables.) Randomization destroys all correlation between the two variables, resulting in a covariance that is essentially zero ($cov_{1,2} = 0.460$). The variances, of course, remain the same because we are using the same data and have simply rearranged their order. If we extract the eigenvalues and eigenvectors of this variance–covariance matrix, we will obtain the two vectors

$$U = \begin{bmatrix} 0.119 & 0.993 \\ 0.993 & -0.119 \end{bmatrix}$$

The two eigenvalues are almost identical, $\lambda_1 = 24.115$ and $\lambda_2 = 20.222$. What we have found are the two principal axes of an ellipse which is almost circular (**Fig. 6–21**). This accords with what we expect, because the correlation between the two original variables is nearly zero; hence the two original axes are nearly orthogonal. Because the two original variable axes are almost equal in magnitude, the axes define a near-circular ellipse. No other set of axes, even those found by PCA, will be significantly better than the two original variables. In this situation, there is no transformation of the original data which will allow us to reduce the number of variables without a significant loss of information. Of course, it is very unlikely that genuine data sets would exhibit zero correlation between all variables.

515

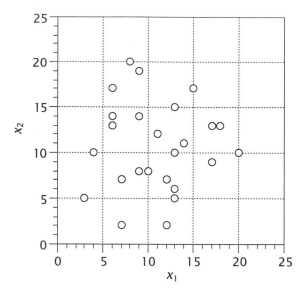

Figure 6–20. Scatter diagram of bivariate data from Table 6–12 after randomizing order of variable x_2.

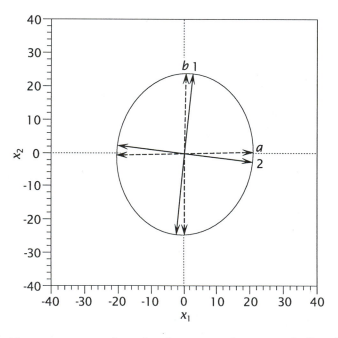

Figure 6–21. Vector representation of variance–covariance matrix **S** and its eigenvectors computed from randomized data shown in Figure 6–20.

In this simple example we first computed a variance-covariance matrix whose elements reflected the original units of measurement. Provided all variables are expressed in the same or commensurate units, the principal components will properly reflect the relative importance of the different variables. However, principal component analysis is sensitive to the magnitudes of the measurements, so if the lengths of the brachiopod shells were given in centimeters while their widths had been given in millimeters, the width variable would have exerted ten times more influence than the length variable on the outcome.

An obvious way around this difficulty is to standardize all variables so they have means of 0.0 and variances of 1.0. Then, the elements of the variance–covariance matrix will consist of correlations, and the principal components will be produced in dimensionless form. Standardization tends to inflate variables whose variance is small and reduce the influence of variables whose variance is large. This may be undesirable but unavoidable if the original variables are expressed in different, incompatible units. In most geologic studies the relative magnitudes of the variables are important, so we should strive to work with the original variables and the variance–covariance matrix. In circumstances where we must standardize the variables in order to make them compatible, we must remember that a property that seems relatively insignificant may exert a strong influence on the analysis. Also, if the components are calculated from the correlation matrix, scores must be calculated from standardized, and not raw, variables.

We can examine the effects of principal component analysis on a larger data set using the measurements on 25 randomly generated blocks given in file BOXES.TXT. The matrix of variances and covariances between the seven variables, representing measurements of different aspects of the geometry of the boxes, is given along with eigenvalues and eigenvectors of the matrix in **Table 6–14**. The eigenvectors are the coordinates of the principal component axes of the block data.

The first two eigenvectors account for 93% of the variance in the data set. The first, accounting for 60% of the total variance, heavily weights the contributions of variables x_5 and x_6. Both of these are ratios which contain the short axis measurement (x_3) in the denominator. Therefore, we infer that the first principal component measures differences in the relative thickness of blocks. An examination of **Figure 6–22**, which is a plot of the blocks at positions corresponding to their scores on the first two principal component axes, shows that this is the case; very flat blocks are placed far to the right, whereas equidimensional blocks are placed on the extreme left. A separation of rod- and plate-shaped blocks is not apparent along the first principal component, as either form may have very abbreviated short axes. We may conclude that the first component reflects the height of a block relative to its overall size.

The second eigenvector is heavily weighted on all three axes (x_1, x_2, x_3) and on the length of the major diagonal (x_4). We may interpret this as a general reflection of size, an interpretation borne out by **Figure 6–22**. Along the second principal component, the blocks are sorted according to their size, with the smallest at the bottom and the largest at the top.

Interpreting the meanings (if any) of the principal component loadings is sometimes called "reification." Possibly some analysts feel the use of this term makes this subjective process more respectable. Plotting the principal component loadings as bar graphs may aid interpretation. Remember from Chapter 3 that eigenvectors are standardized so that the squares of the elements sum to one. Therefore, a loading reflects only the relative importance of a variable within a principal component, and does not reflect the importance of the component itself. However, if the components are transformed by multiplying each element of the eigenvectors by the square roots of their associated eigenvalues, the resulting bar graphs will be comparable (Reyment and Savazzi, 1999).

Table 6–14. Variance–covariance matrix of seven variables measured on 25 blocks. Only the lower half of the symmetric matrix is shown.

	x_1	x_2	x_3	x_4	x_5	x_6	x_7
x_1	5.400						
x_2	3.260	5.846					
x_3	0.779	1.465	2.774				
x_4	6.391	6.083	2.204	9.107			
x_5	2.155	1.312	−3.839	1.611	10.714		
x_6	3.035	2.877	−5.167	2.783	14.774	20.776	
x_7	−1.996	−2.370	−1.740	−3.283	2.252	2.622	2.594

Eigenvalues

λ	34.491	18.999	2.539	0.806	0.341	0.033	0.003
% trace	60.3	33.2	4.4	1.4	0.6	0.0	0.0
cum. %	60.3	93.5	97.9	99.3	99.9	99.9	100.0

Four largest eigenvectors

Eigenvector	1	2	3	4
x_1	0.164	0.422	0.645	0.090
x_2	0.142	0.447	−0.713	0.050
x_3	−0.173	0.257	−0.130	−0.629
x_4	0.170	0.650	0.146	−0.212
x_5	0.546	−0.135	0.105	−0.165
x_6	0.768	−0.133	−0.149	0.062
x_7	0.073	−0.313	0.065	−0.720

Although PCA produces a result in general agreement with our expectations, the separation of shapes is not definitive in this experiment. The third principal component accounts for only 4% of the variance of the data set and is essentially comprised of the longest (x_1) versus the intermediate axis (x_2). However, it is adequate, when used in conjunction with the first two components, to completely separate platelike blocks from rod-shaped blocks. This result is not unexpected, because all of the variables in the experiment were created from three independent variables, the lengths of the long, intermediate, and short axes of the boxes. Although two components are sufficient to express most of the variation in the data, a third independent component is necessary to recapture all of the essential details. In this example, we can see that PCA may be a powerful tool to determine the true number of linearly independent vectors that exist in a matrix. Therefore, it can measure the redundancy in the original set of variables.

As an example of the application of PCA to a geologic problem, we will consider the data given in file BARATARA.TXT, modified from Krumbein and Aberdeen (1937, table 1). These are 50 grain-size analyses of bottom samples collected in Barataria Bay, Louisiana. The bay, a large embayment on the west side of the Mississippi Delta, includes a variety of depositional environments characterized by a

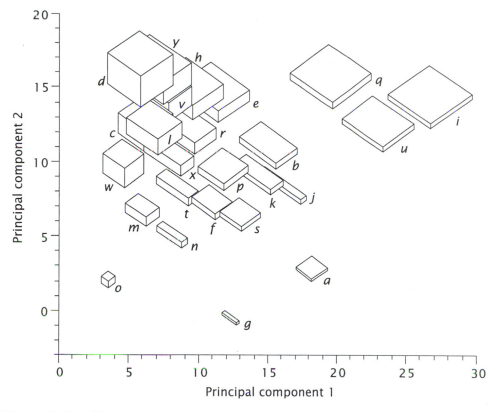

Figure 6–22. Principal component scores of block data plotted on first two principal components. Horizontal axis is component 1, vertical axis is component 2. Blocks shown plotted at the respective positions of their scores on the two components.

range of sediment types. The analyses are sieve and pipette grain-size separations made at 1ϕ intervals. Raw data consist of the weight percent of sediment in each size fraction. For pedagogical purposes, we have combined the "pan fraction" into the smallest grain-size class, $(7 - 8\phi)$, then recalculated the percentages so each analysis sums to 100%. This forms a closed data set subject to the constant-sum constraint; subsequently, we will investigate the effect of closure on our initial results.

Variables are the proportions of each sample which fall within specified size fractions. These are the same variables used by sedimentary petrologists to compute either graphic or moment statistics such as the mean, sorting, and skewness of sediment grain size. By PCA, we can examine the interactions between the various size fractions and find the most efficient linear combination of them, the term "most efficient" meaning the one that accounts for the greatest amount of total variance. We would expect the loadings on the first principal component to approximate in some way the mean, as this statistic traditionally is regarded as being the most efficient of all possible statistics.

The first step in our analysis is to compute the matrix of variances and covariances (**Table 6–15**). Standardization is not necessary in this problem, because all variables of the raw data are measured in the same units. As noted, the data form a closed array because the variables sum to 100% for every observation, raising

Table 6–15. Variance–covariance matrix of grain-size measurements made on sediments from Barataria Bay, Louisiana. Only the lower half of the symmetric matrix is shown.

	$1-2\phi$	$2-3\phi$	$3-4\phi$	$4-5\phi$	$5-6\phi$	$6-7\phi$	$7-8\phi$
$1-2\phi$	5.9541						
$2-3\phi$	−4.9738	460.4453					
$3-4\phi$	−3.0179	51.6221	325.5359				
$4-5\phi$	−0.6235	−242.0367	−117.2327	185.3487			
$5-6\phi$	0.5543	−104.0509	−96.2505	71.2172	51.0437		
$6-7\phi$	0.7240	−88.3899	−86.3467	59.4483	41.8634	39.5347	
$7-8\phi$	1.3828	−72.6162	−74.3104	43.8788	35.6228	33.1663	32.8759

Table 6–16. Eigenvalues and eigenvectors (principal components) of covariance matrix in Table 6–15.

Eigenvalues

λ	707.2973	320.2525	38.4045	6.6661	4.0921	2.0111	0.0000
% trace	65.6	29.7	3.4	0.6	0.4	0.2	0.0
cum. %	65.6	95.3	98.8	99.4	99.8	100.0	100.0

Four largest eigenvectors

Eigenvector	1	2	3	4
$1-2\phi$	0.0068	0.0010	0.0800	−0.8989
$2-3\phi$	−0.7193	0.5080	−0.2553	0.1253
$3-4\phi$	−0.3666	−0.8360	0.0054	0.1543
$4-5\phi$	0.4723	−0.0286	−0.7828	0.0640
$5-6\phi$	0.2360	0.1252	0.2560	0.3352
$6-7\phi$	0.2034	0.1201	0.2881	0.1869
$7-8\phi$	0.1675	0.1103	0.4087	0.0331

troublesome questions about induced negative correlations. The covariance matrix in **Table 6–15** is "overdetermined," that is, it has more rows and columns than necessary. Obviously, if we know A, B, and C, and the total $(A + B + C)$, we have more information than we really need, and one of the three variables is superfluous. Of necessity, one of the eigenvalues of a matrix resulting from such data must be zero, as is the seventh eigenvalue in this example.

The principal components or eigenvectors of the Barataria Bay data are given in **Table 6–16**. Note that the first two components alone account for over 95% of the variance in the data set. **Figure 6–23** is a plot of variable loadings on the two components. It is obvious from the plots that the first principal component essentially represents the relative proportion of fine and very fine sand in the sediment, or the sand/(silt + clay) ratio. The second component represents the ratio between fine and very fine sand, all other variables being weighted near zero. These two components alone are sufficient to account for almost all of the variance in the original data set, which suggests that very little of importance is contained within

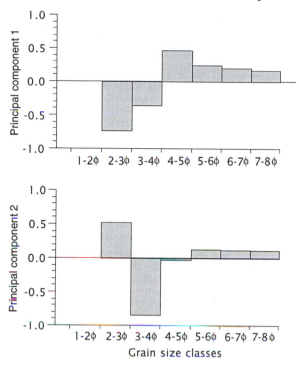

Figure 6–23. Loadings of variables on first two principal components of Barataria Bay data.

the subdivisions of the silt and clay-sized fractions. The differences between the sediments can be almost completely described by only two variables.

We can verify our analysis by computing the scores of the observations on the first two principal components. **Figure 6–24** is a scatter diagram of scores for the first two principal components; the five different sediment categories are shown by symbols. Compare the separation between sediment types in this diagram with that shown in **Figure 6–25**, which is a plot of the median grain size versus sorting (quartile deviation). Perhaps even more interesting is **Figure 6–26**, which is a plot of percentage of sand versus the ratio of fine to very fine sand; these are variables suggested by the nature of the loadings on the first two principal components. Each of these diagrams is approximately equally efficient in terms of separating the five sediment types. However, the amount of experimental effort necessary to find the data plotted in **Figure 6–26** is a fraction of that required to plot **Figure 6–25**. Rather than laborious sieve and pipette separation of samples into seven size intervals, only two simple sieving operations are necessary. In addition, results of the PCA suggest that sediments in the bay may profitably be regarded as mixtures of two populations, sand-sized material and silt–clay material. In this example, principal components not only suggest a new way of looking at the composition of these sediments, but also indicate a possible modification of experimental techniques that will result in considerable savings in effort with negligible loss of information. This experiment, with minor modifications, was given in Davis (1970). It is instructive to compare these results with those obtained by Klovan (1966) from a Q-mode factor analysis of the same data.

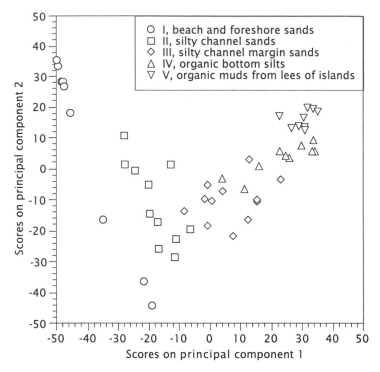

Figure 6–24. Principal component scores of Barataria Bay data plotted on first two principal component axes. Symbols correspond to five sediment types recognized by Krumbein and Aberdeen (1937).

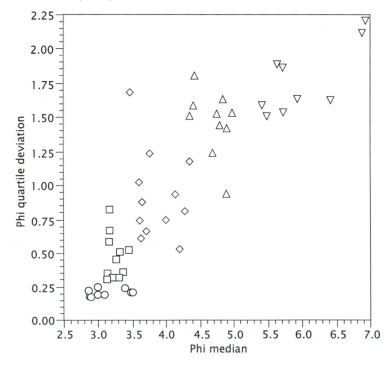

Figure 6–25. Plot of median grain size vs. sorting (quartile deviation) of Barataria Bay sediments. Symbols correspond to those in Figure 6–24.

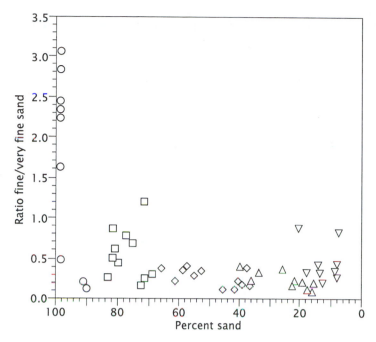

Figure 6–26. Plot of percentage of sand vs. ratio of fine to very fine sand in Barataria Bay sediments. Symbols correspond to those in Figure 6–24.

Closure effects on principal components

As noted, the data in file BARATARA.TXT form a closed array, as every observation sums to 100%. The resulting covariance matrix (**Table 6–15**) inevitably must contain negative covariances that are induced solely by the constant-sum constraint; other covariances also may be distorted to a lesser degree so that they do not reflect the true strengths of relationships among the variables. Since the principal components are calculated from the covariance matrix, the component loadings and scores must reflect the closure constraint. The most obvious consequence of closure is that the final eigenvalue of the covariance matrix must be identically equal to zero (**Table 6–16**), confirming that the true dimensionality of the data is one less than the number of variables.

We can examine the effect that closure has on the Barataria Bay data by making a centered logratio transformation of the original data and repeating our principal component analysis. The centered logratio transformation of an observation consists of taking the logarithm of each variable after the variables have been divided by the geometric mean calculated across all variables. Equivalently, an observation can be transformed by first taking the logarithms of each variable, calculating the average across the logarithms of all variables, then subtracting this average from each variable's logarithm. The resulting data set will have the same number of rows and columns as the original data set, and the covariance matrix of the transformed data will also be of the same dimensions as the original covariance matrix. The centered logratio covariance matrix will be singular, just as the original raw covariance matrix is singular because it is calculated from a closed data set. The difference is that the original closed data are constrained to the simplex, a bounded $m - 1$

523

Table 6–17. Variance–covariance matrix of centered logratio-transformed grain-size measurements made on sediments from Barataria Bay, Louisiana. Only the lower half of the symmetric matrix is shown.

	$1-2\phi$	$2-3\phi$	$3-4\phi$	$4-5\phi$	$5-6\phi$	$6-7\phi$	$7-8\phi$
$1-2\phi$	2.9217						
$2-3\phi$	1.6460	4.0679					
$3-4\phi$	1.0190	2.6724	2.1019				
$4-5\phi$	−1.1340	−2.0267	−1.0035	1.7643			
$5-6\phi$	−1.4736	−2.0763	−1.5277	1.0164	1.5750		
$6-7\phi$	−1.4348	−2.1466	−1.6296	0.7921	1.3371	1.6637	
$7-8\phi$	−1.5442	−2.1366	−1.6324	0.5914	1.1492	1.4182	2.1544

dimensional space in which the sum of all the variables is a constant. The logratio-transformed data also occupy an $m - 1$ dimensional space, but the values are no longer bounded.

Table 6–17 contains the variance–covariance matrix of the Barataria Bay grain-size data after they have been transformed by Aitchison's centered logratio procedure. The original data set includes numerous zero values; a trace amount equal to 0.1 has been added to each of these in order to determine their logarithms. (The magnitude of the added trace amount affects the transformed values and the elements of the covariance matrix, but in this example there is only a minor effect on the principal components.) The most obvious differences between the two covariance matrices in **Table 6–15** and **Table 6–17** are the changes in magnitude of the elements caused by conversion to logarithms. Next, we note that the signs of the covariances have been reversed in the coarsest grain-size category.

Table 6–18. Eigenvalues and eigenvectors (principal components) of centered logratio covariance matrix in Table 6–17.

Eigenvalues

λ	11.6017	1.9768	1.4614	0.5788	0.2279	0.0774	0.0000
% trace	72.9	12.4	9.2	3.6	1.4	0.5	0.0
cum. %	72.9	85.3	94.4	98.1	99.5	100.0	100.0

Four largest eigenvectors

Eigenvector	1	2	3	4
$1-2\phi$	−0.3508	−0.8424	−0.1489	0.0010
$2-3\phi$	−0.5555	0.4074	−0.1428	−0.1901
$3-4\phi$	−0.3845	0.3093	0.2388	0.2207
$4-5\phi$	0.2702	−0.0908	0.7430	0.2874
$5-6\phi$	0.3303	0.0540	0.0688	−0.5507
$6-7\phi$	0.3414	0.0343	−0.2125	−0.3853
$7-8\phi$	0.3490	0.1282	−0.5464	0.6171

The eigenvalues and eigenvectors from the variance-covariance matrix of the centered logratio transformed data are given in **Table 6–18**, which can be compared to **Table 6–16**. The first two eigenvectors are shown as a principal component loading diagram in **Figure 6–27**, which can be compared to **Figure 6–23**. The most conspicuous change is the increase in importance of the coarsest size fraction, medium sand, which is insignificant in the original analysis. The interpretation of the first principal component is unchanged; it is the proportion of sand in the samples. After logratio transformation, the second component represents the ratio between medium sand and fine plus very fine sand, whereas before transformation it represented the ratio between fine and very fine sand.

Logratio transformation has a more conspicuous effect on principal component scores of the Barataria Bay sediments, as can be seen by comparing cross plots of the first and second components before and after logratio transformation. **Figure 6–28** shows a plot of the component scores calculated from logratio transformed data. Both axes in the figure have been plotted on the same scale to facilitate comparison with the plot of scores from untransformed data in **Figure 6–24**. The untransformed scores appear to form a systematic U-shaped pattern of points. Such a pattern is a characteristic feature of principal component plots of closed data, and represents a quadratic distortion caused by the closure constraint. Note that after logratio transformation, this pattern is no longer apparent.

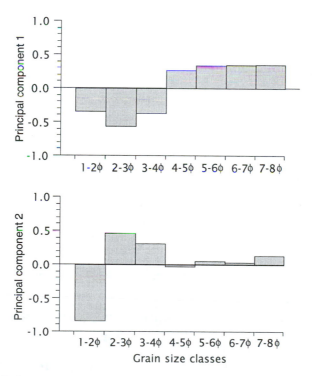

Figure 6–27. Loadings of variables on first two principal components of centered logratio-transformed Barataria Bay data.

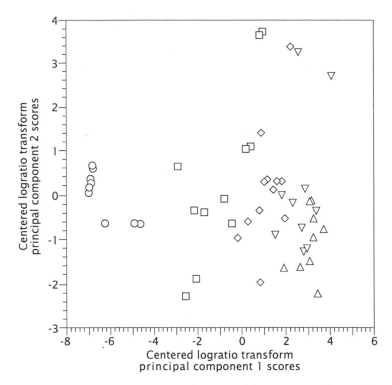

Figure 6–28. Principal component scores of centered logratio-transformed Barataria Bay data plotted on first two principal component axes. Symbols correspond to those in Figure 6–24.

R-Mode Factor Analysis

Principal component analysis consists of a linear transformation of m original variables to m new variables, where each new variable is a linear combination of the old. The process is performed in a fashion that requires that each new variable accounts for, successively, as much of the total variance as possible. When m new variables have been computed, all of the original variance will be accounted for. Nothing is said about probability, hypotheses, or testing, because PCA is not, strictly speaking, a statistical procedure. Rather, it is a mathematical manipulation. However, it assumes some of the characteristics of statistical procedures when decisions are made to discard some new variables or components as being inconsequentially small. Some statistical tests are available for checking the significance of discarded variables, but these are based on highly restrictive assumptions and are seldom applicable. A review of these is given by Morrison (1990), and Jackson (1991) provides a set of semi-empirical procedures for estimating the number of significant components. Principal component analysis belongs to that category of techniques, including cluster analysis, in which appropriateness is judged more by performance and utility than by theoretical considerations.

Factor analysis is somewhat different, for it is commonly regarded as a statistical technique. It relies on a set of assumptions about the nature of the parent population from which the samples were drawn. These assumptions provide the rationale for the operations that are performed, and the manner in which results are interpreted. Some factor procedures even provide tests of significance (Lawley and Maxwell, 1971; Krzanowski, 1988), although these are not widely used.

In factor analysis, the relationship within a set of m variables is regarded as reflecting the partial correlations of each of the variables with p mutually uncorrelated underlying, or latent, factors. The usual assumption is that $p < m$. Variance in the m variables is therefore derived from variance in the p factors, but in addition a contribution is made by unique sources which independently affect the m original variables. The p underlying factors are referred to as **common factors** and the independent contribution is summarized as a **unique factor**. The factor model may be expressed as

$$x_j = \sum_{r=1}^{p} a_{j.r} f_r + \varepsilon_j \tag{6.54}$$

where x_j is the jth observed variable, f_r is the rth common factor, p is the specified number of factors, and ε_j is random variation unique to the original variable x_j. Because there are m original variables x_j, there are m random variables ε_j; taken together, these constitute the unique factor. The coefficient $a_{j.r}$ is the loading of the jth variate on the rth factor. It corresponds to a loading or weight in principal components.

We may assume that the x_j are multivariate normally distributed. The variances and covariances form an $m \times m$ matrix, S. From Equation (6.54) we can determine that the diagonal elements of S, the variances of the m variables, should be

$$s_{jj}^2 = \sum_{r=1}^{p} a_{j.r}^2 + s_{\varepsilon_{jj}}^2 \tag{6.55}$$

and the off-diagonal elements, or covariances, should be

$$\text{cov}_{jk} = \sum_{r=1}^{p} a_{j.r} a_{k.r} \tag{6.56}$$

The hypothesis underlying factor analysis may be expressed in matrix notation in the following way. The observed matrix of variances and covariances, S, is the product of an $m \times p$ matrix of factor loadings, A, multiplied by its transpose, plus an $m \times m$ diagonal matrix of unique variances, $\boldsymbol{\varepsilon}$

$$S = AA' + \boldsymbol{\varepsilon} \tag{6.57}$$

Multiplying an $m \times p$ matrix by its transpose will create an $m \times m$ matrix, however, it will have only p positive eigenvalues and associated eigenvectors. If $p = m$, the matrix $\boldsymbol{\varepsilon}$ will vanish and our problem is equivalent to principal component analysis. In cases where $p < m$, we must estimate the matrix of parameters, A, which are the loadings on the factors, and the unique variance, $\boldsymbol{\varepsilon}$. Note that the factor model requires that p, the number of factors, be known prior to analysis. This implies that the investigator has some insight into the probable nature of the factors, and can predict a suitable number of factors to be extracted. If p cannot be specified, the partition of variances between the common factors and the unique factor becomes indeterminate. This is a point sometimes overlooked by experimenters who wish to use factor analysis for "fishing expeditions." Stated in another way, the number of factors, p, the matrix of factor loadings, A, and the unique variances, $\boldsymbol{\varepsilon}$, are all interrelated. They cannot all be estimated simultaneously, so it is necessary to introduce various constraints in order to find a unique solution. The simplest constraint is to assume some prior value for p, the number of factors. Unfortunately, in most geological problems the number of possible factors is not knowable in advance, and may even be an important objective in a study. Another approach is to assume some prior limit for either $\mathbf{AA'}$ or $\boldsymbol{\varepsilon}$ and to extract factors until this limit is reached. However this makes the definition of latent factors an arbitrary matter, which is contrary to the factor model.

We will examine two of the many factor analysis schemes that have been suggested. The principal components approach to factor analysis starts by extracting the eigenvalues and eigenvectors of the correlation matrix, and then discarding the less important of these. This does not lead to a "true" factor solution, but the mathematics are relatively straightforward and this is the approach used almost universally in the Earth sciences when factor analysis is employed. We will also take a brief look at the method of maximum likelihood, which does yield "true" factors. Unfortunately the underlying mathematics are so involved that most authors dismiss them, considering them to be too complicated to describe.

Although the first method of factor analysis we will consider utilizes principal components, the computation of eigenvalues and eigenvectors is different in two respects. First, the eigenvalue operation is always performed on a standardized variance–covariance (or correlation) matrix. This assures that all variables are weighted equally, and also allows us to convert the principal component vectors into factors. Second, the eigenvectors, which are calculated in normalized form (the squared elements of each eigenvector sum to 1.0), are scaled so that they define vectors whose lengths are proportional to the variation they represent. These scaled eigenvectors are the factors of the data set.

The conversion of normalized eigenvectors into factors does not affect the directions of the vectors, only their lengths. It is done by multiplying every element in a normalized eigenvector by the corresponding singular value, or square root of the corresponding eigenvalue. The resultant factor is a vector that is weighted proportionally to the amount of total variance it represents.

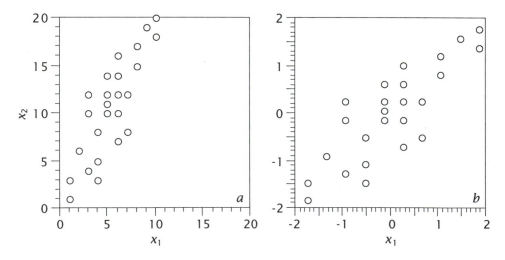

Figure 6–29. Data set to show effect of standardization. (*a*) Raw data have means of $\overline{X}_1 = 5$ and $\overline{X}_2 = 10$. (*b*) Data standardized to have zero means and unit standard deviations. Note that both variables now cover same range.

We can demonstrate the effect of standardization using the artificial data set plotted in **Figure 6–29 a** and listed in file FACTOR.TXT. The raw data have a variance–covariance matrix,

$$S = \begin{bmatrix} 6.240 & 10.953 \\ 10.953 & 27.727 \end{bmatrix}$$

whose eigenvalues are $\lambda_1 = 32.326$ and $\lambda_2 = 1.641$, accounting for 95.2% and 4.8% of the trace of S, respectively. The eigenvalues can be set as the diagonal elements of a matrix, Λ^2:

$$\Lambda^2 = \begin{bmatrix} 32.326 & 0 \\ 0 & 1.641 \end{bmatrix}$$

The eigenvectors form the columns of a matrix, U:

$$U = \begin{bmatrix} 0.387 & -0.922 \\ 0.922 & 0.387 \end{bmatrix}$$

If the data are standardized by subtracting their means and dividing by their standard deviations (**Fig. 6–29 b**), the standardized variance-covariance (or correlation) matrix is

$$R = \begin{bmatrix} 1.000 & 0.833 \\ 0.833 & 1.000 \end{bmatrix}$$

the eigenvalue matrix Λ^2 becomes

$$\Lambda^2 = \begin{bmatrix} 1.833 & 0 \\ 0 & 0.167 \end{bmatrix}$$

and the successive eigenvalues account for 91.6% and 8.4% of the trace of the correlation matrix. The eigenvector matrix U becomes

$$U = \begin{bmatrix} 0.707 & 0.707 \\ 0.707 & -0.707 \end{bmatrix}$$

The eigenvectors can be converted into factors by Equation (6.46):

$$A^R = U\Lambda$$

$$= \begin{bmatrix} 0.707 & 0.707 \\ 0.707 & -0.707 \end{bmatrix} \begin{bmatrix} \sqrt{1.833} & 0 \\ 0 & \sqrt{0.167} \end{bmatrix}$$

$$= \begin{bmatrix} 0.957 & 0.289 \\ 0.957 & -0.289 \end{bmatrix}$$

The elements of the factor are the **factor loadings**. If we have performed the conversion correctly, the sum of the squares of the factor loadings should equal the eigenvalues:

$$0.957^2 + 0.957^2 = 1.833$$
$$0.289^2 + (-0.289)^2 = 0.167$$

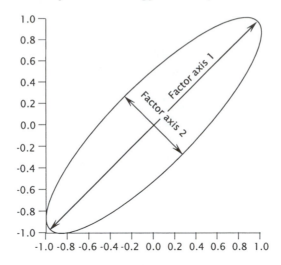

Figure 6–30. Plot of two factors extracted from standardized bivariate data shown in Figure 6–29 b.

The two factor axes are shown plotted in **Figure 6–30**. The orientations of the factors are the same as the original eigenvectors, but their lengths are now equal to the square roots of the eigenvalues. Since the eigenvalues represent the proportion of the total variance accounted for by the eigenvectors and the lengths of the factors are proportional to the eigenvalues (or rather, to the singular values or square roots of the eigenvalues), the factors also represent the variances (or more correctly, the standard deviations). Because the normalized eigenvectors are multiplied by the corresponding singular values, each factor loading is proportional to the square root of the amount of variance contributed by that variable to the factor. In our example, the first factor accounts for $1.833/2.000 = 91.6$ of the total variance of our data. Of this, $0.957^2/1.833 = 50.0\%$ is derived from variable x_1 and $0.957^2/1.833 = 50.0\%$ is derived from variable x_2. Similarly, factor 2 accounts for $0.167/2.00 = 8.4\%$ of the total variance, deriving $0.289^2/0.167 = 50.0\%$ from variable x_1 and $-0.289^2/0.167 = 50.0\%$ from variable x_2. One hundred percent of the variance of variable x_1 is accounted for in the two factors, as is 100% of the variance of variable x_2. (The reciprocal nature of the influence of these variances on the two factors is an inevitable consequence of working with a 2×2 correlation matrix. This relationship is not generally found in larger matrices.)

If we square the elements in the factor loading matrix, $\mathbf{A}^R$, and sum within each variable, the totals are the *communalities* of each variable retained in the factors. That is,

$$\begin{array}{cc} & \text{Factors} \qquad\qquad \text{Communalities} \\ & \quad 1 \qquad\qquad 2 \\ \text{Variables} \begin{array}{c} x_1 \\ x_2 \end{array} & \left[\begin{array}{cc} 0.957^2 & 0.289^2 \\ 0.957^2 & -0.289^2 \end{array} \right] = \left[\begin{array}{c} 1.00 \\ 1.00 \end{array} \right] \end{array}$$

The communalities are symbolically represented as h_j^2, where the subscript refers to the jth variable. If we extract m factors from an $m \times m$ matrix of variances and covariances, the communalities are equal to the original variances. Because our variables are standardized with zero mean and unit standard deviation, these communalities will be 1.00. However, if we extract fewer than m factors, the communalities will be less than the original variances, and provide an index to the efficiency of our reduced set of factors. For example, if we were to retain only one factor from our 2×2 factor matrix, the communalities would be

$$h_1^2 = 0.957^2 = 0.916$$

$$h_2^2 = 0.957^2 = 0.916$$

That is, retaining only a single factor still accounts for 92% of the variance of variable x_1 and 92% of the variance of variable x_2.

Of course, the magnitude of the communalities is dependent upon the number of factors that are retained, bringing us up against one of the major problems of this approach to factor analysis. How many factors should be retained? There is, unfortunately, no simple answer and this question is one of the major sources of argument among factor analysts. Early experimental psychologists solved the problem in a very straightforward way; they extracted as many factors as their current ruling theory demanded. Another equally pragmatic approach is to extract only two or three factors, because this is the maximum number that can conveniently be displayed as scatter diagrams, and any number larger than this increases the dimensionality of the problem to the point where it again becomes difficult to grasp.

Some analysts recommend retaining all factors that have eigenvalues greater than one. That is, retain all factors that contain greater variance than the original standardized variables. In most instances, only a few factors will contain most of the variance in the data set, and this recommendation is useful. However, if the original variables are only weakly correlated or uncorrelated, half or more of their factors may have eigenvalues greater than one. You will be left not only with an inordinate number of factors, but may discover that none of them are interpretable anyway. If the factor theory is applicable to a given data set (*i.e.*, the variances observed are the result of partial correlations between variables and the underlying factors), a few factors should account for a very high percentage of the variance, and communalities will be high. If a large number of factors must be retained to account for much of the original variance, or if the communalities of the first few factors are low, the factor model probably is not appropriate.

Before continuing to the next step in factor analysis, which is rotation of the factor axes to "simple structure," let us apply what we have covered so far to an example. We will use the data on the blocks shown in **Figure 6–13** and contained in file BOXES.TXT. We will retain two factors, because intuition tells us two factors should be involved, one a size factor and the other an expression of shape. The matrix of standardized variances and covariances is given in **Table 6–19**, which also contains the eigenvalues and the eigenvectors corresponding to the four largest eigenvalues. These account for almost 100% of the standardized variance in the seven variables. We will retain the first two of these eigenvectors and convert them to factors. This is done by multiplying the normalized eigenvectors by the corresponding singular values (square roots of the eigenvalues) to yield the factor loadings. The $m \times p$ factor loading matrix, $\mathbf{A}^R$, with the communalities of the variables and the unique component, are shown in **Table 6–20**.

If m factors were retained from a set of m variables, the original standardized covariance matrix, $\mathbf{R}$, could be re-created by multiplying together all possible pairs of factor loadings and summing across the factors. Of course, when only $p < m$ factors are retained, the original covariance matrix cannot be reproduced exactly.

Table 6–19. Correlation matrix of seven variables measured on 25 blocks. Only the lower half of the symmetric matrix is shown.

	x_1	x_2	x_3	x_4	x_5	x_6	x_7
x_1	1.000						
x_2	0.580	1.000					
x_3	0.201	0.364	1.000				
x_4	0.911	0.834	0.439	1.000			
x_5	0.283	0.166	−0.704	0.163	1.000		
x_6	0.287	0.261	−0.681	0.202	0.990	1.000	
x_7	−0.533	−0.609	−0.649	0.676	0.427	0.357	1.000

Eigenvalues

λ	3.395	2.805	0.437	0.278	0.081	0.003	0.000
% trace	48.5	40.1	6.2	4.0	1.2	0.0	0.0
cum. %	48.5	88.6	94.8	98.8	100.0	100.0	100.0

Four largest eigenvectors

Eigenvector	1	2	3	4
x_1	0.405	−0.293	−0.667	−0.089
x_2	0.432	−0.222	0.698	0.034
x_3	0.385	0.356	0.148	−0.628
x_4	0.494	−0.232	−0.119	−0.210
x_5	−0.128	−0.575	0.029	−0.111
x_6	−0.097	−0.580	0.174	0.006
x_7	−0.481	−0.130	0.018	−0.735

Table 6–20. Factor matrix and communalities for two factors and unique component from block data.

	Factor 1	Factor 2	Communality	Unique
x_1	0.747	−0.491	0.798	0.202
x_2	0.795	−0.373	0.771	0.229
x_3	0.710	0.596	0.860	0.140
x_4	0.910	−0.389	0.979	0.021
x_5	−0.2353	−0.963	0.983	0.017
x_6	−0.178	−0.971	0.976	0.024
x_7	−0.886	−0.218	0.833	0.167

For variables j and k, the ***reproduced covariance***, $\hat{r}_{jk}$, is

$$\hat{r}_{jk} = a_{j1}a_{k1} + a_{j2}a_{k2} + \ldots + a_{jp}a_{kp} \tag{6.58}$$

where a_{j1}, for example, is the loading of the jth variable on factor 1. If we denote the factor loadings as the matrix $\mathbf{A}^R$, the matrix equivalent of Equation (6.58) is

$$\hat{\mathbf{R}} = \mathbf{A}^R \mathbf{A}^{R'} \tag{6.59}$$

If the factor loadings are considered to constitute column vectors. The **residual standardized variance-covariance matrix** (or residual correlation matrix, as it also is called) can be found by subtraction:

$$R_{res} = R - \hat{R} \qquad\qquad (6.60)$$

The reproduced and residual covariance matrices for our block example are given in **Table 6-21**. The residual matrix is a measure of the inability of two factors to account for all of the variability in the original data set.

Table 6-21. Reproduced correlation matrix for two factors extracted from random block data. Residual correlation matrix contains correlations between variables unaccounted for by the two factors. Only the lower half of each symmetric matrix is shown.

Reproduced correlation matrix

	x_1	x_2	x_3	x_4	x_5	x_6	x_7
x_1	0.798						
x_2	0.777	0.771					
x_3	0.238	0.343	0.860				
x_4	0.870	0.869	0.414	0.979			
x_5	0.297	0.172	−0.741	0.161	0.983		
x_6	0.343	0.220	−0.706	0.216	0.978	0.975	
x_7	−0.555	−0.623	−0.759	−0.721	0.419	0.370	0.833

Residual correlation matrix

	x_1	x_2	x_3	x_4	x_5	x_6	x_7
x_1	0.202						
x_2	−0.196	0.229					
x_3	−0.037	0.021	0.140				
x_4	0.041	−0.035	0.024	0.021			
x_5	−0.014	−0.006	0.037	0.002	0.017		
x_6	−0.057	0.041	0.025	−0.013	0.012	0.024	
x_7	0.021	0.015	0.110	0.046	0.008	−0.013	0.167

Factor rotation

Although factor analysis may reduce the dimensionality of a problem to manageable size, the meaning of the factors may be difficult to deduce. Under factor theory, this may be the result of the fact that positions of the p orthogonal factor axes in m space are constrained by $m - p$ unnecessary axes, which also must be placed orthogonally through the sample space. However, we need only p factor axes to explain our data. If we "chop off" the extraneous orthogonal axes, it seems possible to further rotate the factors and perhaps find a better position for them. This we can do by a variety of rotational procedures. The particular technique we will employ is called **Kaiser's varimax** scheme, which has as its objective the moving of each factor axis to positions such that projections from each variable onto the factor axes are either near the extremities or near the origin. The method operates by adjusting the factor loadings so they are either near ±1 or near zero. For each factor, there may be a few significantly high loadings and many insignificant loadings. Interpretation, in terms of original variables, is thus made easier. However, in certain instances rigid rotation of the factor axes will not improve the analysis, and may even confuse the results further. This may indicate that the factors are oblique, or intercorrelated, or it may imply that the factor model is inappropriate.

The varimax criterion involves maximization of the variance of the loadings on the factors. We may define the variance, s_k^2, of the loadings on the kth factor as

$$s_k^2 = \frac{p \sum_{j=1}^{m} \left(a_{jp}^2 / h_j^2 \right)^2 - \left(\sum_{j=1}^{m} a_{jp}^2 / h_j^2 \right)^2}{p^2}$$

(6.61)

where, as before, p is the number of factors, m is the number of original variables, a_{jp} is the loading of variable j on factor p, and h_j^2 is the communality of the jth variable. The quantity we wish to maximize is

$$V = \sum_{k=1}^{p} s_k^2$$

(6.62)

The variance is calculated from the factor loadings, a_{jp}, which are corrected by dividing each by its communality, h_j^2. In other words, only the common part of the variance of each variable is considered, removing the constraint imposed by the $m - p$ additional components necessary to account for all of the variance of each variable. Maximizing the variance implies maximizing the range of the loadings, which tends to produce either extreme (positive or negative) or near-zero loadings, satisfying the purpose of factor rotation.

No simple analytical scheme exists whereby the varimax criterion may be maximized. Rather, rotation of the factor axes must be performed iteratively, a sort of trial-and-error process. The factor axes are adjusted two at a time, holding the other axes stationary. After all axes have been adjusted, the process is repeated until the increase in variance of the loadings with each iteration drops below some specified cutoff level.

The varimax rotation process can best be illustrated with an example. We will attempt to "clean up" the factors produced from our artificial random block data set by rotating the two factors we retained. In **Figure 6–31 a**, the loadings of the seven original variables on factor 1 are plotted against the loadings on factor 2. Connecting the plotted points with the origin yields a diagram in which the factor loadings are shown as vectors. The orientation of the vectors with respect to the factor axes reflects their degree of correlation with the factors. The length of a vector is proportional to the communality of the variable represented by the vector. If two factors completely account for all of the variation in an original variable, the communality is one and will lie somewhere along a circle of unit radius in the diagram. In the example, all communalities are high, so the vectors representing the seven original variables all extend to near the unit circle.

Varimax rotation changes the factor loadings so that the original variables have either a high positive or high negative correlation (near ±1) with a factor, or a correlation near zero. **Figure 6–31 b** shows the positions of the variables on the factor axes after rotation. Note that the positions of the variables with respect to each other are not altered by rotation; only their relation to the factor axes is changed. Also note that the lengths of the vectors remain unchanged.

Similarly, relationships among the observations themselves are not changed by rotation, although the position of individual objects within the space defined by the

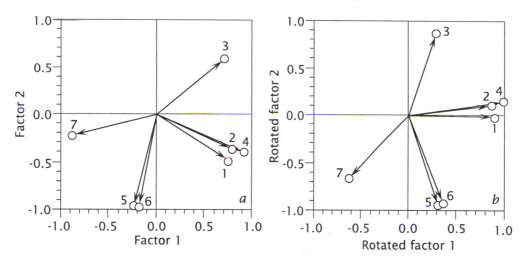

Figure 6–31. (*a*) Plot of loadings on two raw factors extracted from measurements on 25 random blocks. (*b*) Plot of loadings on two factors rotated by the varimax criterion. Variables are those measured on 25 random blocks listed in file BOXES.TXT.

factor axes is altered. **Figure 6–32** is a plot of factor scores, similar to the plot of principal component scores given in **Figure 6–22**. Note that two sets of factor axes are shown on the diagram, one for unrotated factor scores and the other for scores after varimax rotation. The first factor does indeed seem to reflect overall size of the blocks, as smaller blocks are located on the left and larger blocks are placed on the right. The second factor separates equidimensional shapes at the top, with plates and rods lower on the second factor. In this instance, varimax rotation does not seem to contribute to our ability to interpret the factors. Nor is the pattern of factor scores much different from that obtained by principal component analysis, although the relative importance of the first and second factor axes are reversed compared to the principal component axes.

Plotting of scores on factors (either rotated or unrotated) is more complex than the plotting of principal component scores. Principal components are linear transformations, so we were able to plot PCA scores simply by projecting our original observations onto the principal axes. In factor analysis, however, the scores represent estimates of the contributions of various factors to each original observation. Because the factors themselves are estimated from these same data, the computation of factor scores is a somewhat circular process and the factor scores have a high uncertainty. Also, the scores are not unique unless additional constraints are introduced. Some of the clearer discussions of these problems are given by Krzanowski (1988) and Morrison (1990); also see the procedure given by Harman (1967, p. 348–350). In psychometrics, the factors themselves are usually the items of interest, and factor scores are not often needed. Consequently, the calculation of factor scores has received comparatively little attention. The factor scores may be an important part of a geological factor study, however, and it is essential that we be able to compute them.

We will refer to our original data as the $n \times m$ matrix $\mathbf{X}$ in which the n rows are observations and the m columns are variables. Following the method used for principal component analysis, it seems as though we could compute a matrix of

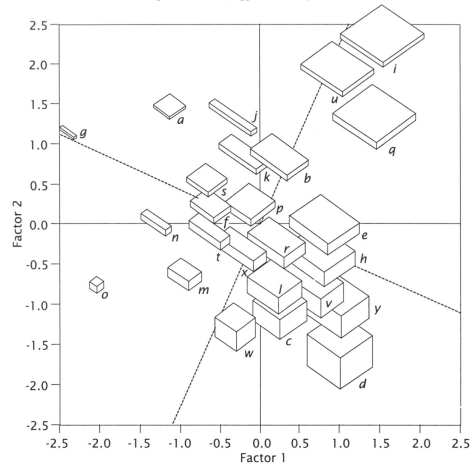

Figure 6–32. Plot of factor scores on first two factors of random block data. Horizontal axis is factor 1, vertical axis is factor 2. Blocks shown plotted at the respective positions of their scores on the two factors. Varimax rotated factor axes shown as dashed lines.

factor scores, S^R, by multiplying this data matrix by the matrix of factor loadings, A^R; that is, by performing the operation $XA^R = S^R$.

If we retain p factors, the loading matrix, A^R, will be $m \times p$ and the matrix of scores, S^R, will be $n \times p$. However, you will recall that the original data represent not only the factors, but also unique variation (Eq. 6.54). Therefore, the matrix of scores computed in this manner will reflect in part the covariance structure of the original m variables as well as the structure of the p factors. The influence of the unique part of the original variables must, in effect, be divided out of these scores to obtain true factor scores. This may be done by multiplying the equation by the inverse of the covariance matrix, S:

$$XS^{-1}A^R = \hat{S}^R \tag{6.63}$$

The inverse of the covariance matrix is $m \times m$ and the matrix of factor loadings is $m \times p$, so the matrix of "true" factor scores, $\hat{S}^R$, is $n \times p$, which is dimensionally correct. This operation will return factor scores which are free of the unique component present in each of the original observations.

Although computationally direct, this method of finding the factor scores is not implemented in some programs, especially earlier software written when personal computers could not accommodate large arrays. The covariance matrix S may be very large, especially in the Q-mode analyses which we will consider later, and inversion may be difficult because of its size. However, by using an algebraic identity, we can invert a $p \times p$ matrix of covariances derived from the factors and obtain the same result. Ordinarily p is much smaller than m, simplifying the calculations, although the number of matrix operations is increased.

We first premultiply the matrix of factor loadings by its transpose to create a square $p \times p$ matrix. This smaller matrix is then inverted and the inverse premultiplied by the factor loading matrix to yield a matrix of score coefficients, **B**:

$$\mathbf{B} = \mathbf{A}^R (\mathbf{A}^{R'} \mathbf{A}^R)^{-1}$$

These score coefficients can then be used to compute the true factor scores by matrix multiplication:

$$\mathbf{XB} = \hat{\mathbf{S}}^R$$

This sequence of matrix operations can be expanded and expressed entirely in terms of the factor loading matrix, $\mathbf{A}^R$:

$$\mathbf{XA}^R (\mathbf{A}^{R'} \mathbf{A}^R)^{-1} = \hat{\mathbf{S}}^R \tag{6.64}$$

The same procedure is used to produce factor scores on unrotated or varimax rotated factor axes. Note that the data matrix, **X**, contains standardized variables, and not raw variables.

The problem of specifying p, the number of factors to be retained, becomes critical at this point. The number of factors affects the magnitudes of the reproduced and residual correlation matrices, the communalities, and the loadings on the unique component. The factor loadings themselves are not affected. That is, if $p = 2$ factors are extracted from a data set, the loadings on factor 1 and factor 2 are not altered by the extraction of a third factor. However, if we extract and rotate two factors, the loadings may be radically different from those obtained if we extract and rotate three factors from the same data. The factors obtained when $p = 2$ are unconstrained during rotation. The same two factors are not as free to rotate if $p = 3$ because of the constraint of the third orthogonal axis that must also be accommodated in the m-dimensional space defined by the variables.

The varimax rotational scheme preserves the orthogonality of the factor axes. Even though the factors no longer coincide with the principal axes of the variance–covariance ellipsoid, they are at right angles to one another, thus uncorrelated. A host of rotational schemes exists in which the requirement of orthogonality is relaxed and factor axes may be oblique to one another. In some instances the resulting oblique factors are more readily interpretable because more extreme loadings may be obtained on the factors. However, some philosophical difficulties exist with oblique rotation schemes. For one, the factor model assumes that the observed matrix of variances and covariances results from correlations between the m variables and p mutually uncorrelated factors. Relaxing the restriction on orthogonality introduces intercorrelations between factors which would seem to be a violation of the original set of assumptions. If the factors themselves are correlated, relationships between factors and original variables are much more complex than the model assumes because interactions exist between pairs of variables and pairs of factors. The presence of intercorrelation also brings up the disquieting suspicion that perhaps the oblique factors are themselves nothing more than the result of correlations with some "superfactor" hidden still farther from direct observation.

Factor analysis was first devised to explain the interrelationships in large numbers of variables by the presence of a few factors. Original applications were accompanied by theory which specified the expected nature of the factors, thus allowing their interpretation. However, when factor analysis is applied to problems in areas where no theory of structure exists, it is necessary to deduce the meaning of the factors. This is not always possible, because either no pattern emerges in the factor loadings or the theoretical framework of the problem is too poorly developed for adequate understanding. Rather than admit defeat, factor analysts have devised nonorthogonal rotation schemes that allow them to express the factors in terms of the original variables. Thus the process has come full circle from variables to factors for reduction in the size of the problem, back to variables for interpretation of the factors. This should not be taken to imply that oblique methods are useless; in certain problems significant results have been obtained using these techniques. However, if ordinary orthogonal factor methods fail to yield interpretable results, most novice practitioners should resign themselves to the admission that the problem is intractable using the factor approach, or that too little is known of causal relationships in the problem to allow interpretation. Many oblique factor studies of geologic data have led to trivial results, recapturing original variables relabeled as "factors" and yielding no more insight than could be gained from a careful inspection of the original correlation matrix. Oblique solutions introduce one more subjective decision into an already arbitrary process and probably should be avoided by all but the expert. Those interested should refer to Harman (1967, chapter 15) and to the earlier work by Thurstone (1947), especially his chapter 15.

Maximum likelihood factor analysis

We will now briefly consider an alternative method for R-mode factor analysis, the **maximum likelihood** procedure developed by Lawley (1940) and subsequently modified by many workers. It avoids some, but not all, of the problems that beset other factor techniques (Krzanowski, 1988). The maximum likelihood method does this by making certain initial assumptions about the nature of the factors and the unique variance. The factors are assumed to be normally distributed with means of zero and variances of one. The elements of the matrix of unique variances are also assumed to be normally distributed with a mean of zero and a variance ε_{jj}. All of the factors and the elements of the unique variance are further assumed to be independent. Finally, the observed matrix of variances and covariances is assumed to be adequate for estimation of Σ, the unobservable matrix of variances and covariances between the factors.

Deriving the maximum likelihood estimators of the factor loadings requires mathematical gyrations of exceeding complexity; in common with most authors, we will forego tracing their development. Interested readers are referred to Lawley and Maxwell (1971) and Jöreskog (1977); an especially readable, short description is contained in Morrison (1990) and a somewhat more complete explanation has been given by Krzanowski (1988). We will content ourselves with examining the computational steps involved, as these are implemented in several commonly available libraries of computer programs.

The maximum likelihood procedure begins with the same model equation as other forms of factor analysis,

$$S = A^R A^{R'} + \varepsilon_{jj} \tag{6.65}$$

However, the maximum likelihood estimates of the factor loadings must be developed iteratively. To clarify the steps, we now introduce the notation ${}_i a_{jr}$ which means the ith iteration in the estimation of the loading of variable j on factor r. Similarly, ${}_i \varepsilon_{jj,r}$ is the ith iteration in the approximation of the unique variance of variable j left after extraction of the rth factor.

The starting estimates of the loadings on the first factor, ${}_0 a_{j1}$, are based on the elements of the first eigenvector extracted from the observed matrix of variances and covariances, S. The elements of the eigenvector are scaled so that the sum of their squares is equal to the first eigenvalue. A starting approximation of the specific variance is made as

$$_0\varepsilon_{jj,1} = \mathrm{diag}\left(S - {}_0 a_{j1}\, {}_0 a'_{j1}\right) \tag{6.66}$$

(The operator **diag** means that only the diagonal elements of the matrix are retained; off-diagonal elements are zero.)

Next, we form the matrix

$$_0\varepsilon_{jj,1}^{-1/2} \left(S - {}_0\varepsilon_{jj,1}\right) {}_0\varepsilon_{jj,1}^{-1/2} \tag{6.67}$$

and extract its first eigenvalue and eigenvector. The eigenvector is again scaled so that the sum of its squared elements is equal to the eigenvalue; this we designate ${}_1 a_{j1}$. The estimate of the factor loading matrix at the end of the first iteration is

$$_1 A_1^R = {}_0\varepsilon_{jj,1}^{-1/2}\, {}_1 a_{j1} \tag{6.68}$$

The unique variance as estimated at the end of the first iteration is

$$_1\varepsilon_{jj,1} = \mathrm{diag}\left(S - {}_1 A_1^R\, {}_1 A^{R'}\right) \tag{6.69}$$

This is analogous to the initial estimate of the unique variance given in Equation (6.66). The process is repeated again from that point, using the new estimate of $\varepsilon_{jj,1}$. The iterations continue until ${}_i A_1^R$ and ${}_{i+1} A_1^R$ differ by no more than a trivial amount. The column vector, ${}_i A_1^R$, is the maximum likelihood estimate of the loadings on the first factor. The hypothesis is that the data contain only a single factor, so that

$$S = A_1^R A_1^{R'} + \varepsilon_{jj,1}$$

can be tested by a χ^2 procedure, described by Morrison (1990). If the hypothesis is rejected, additional factors must be estimated. These are found in an iterative process similar to that used to find the first factor, except that the process begins with the residual matrix, S_{res}

$$S_{res} = S - {}_i A_1^R\, {}_i A_1^{R'}$$

The algorithm used to find the initial factor is then repeated to find the second and subsequent factors, except that each iterative cycle begins with the residual matrix from the previous cycle of iterations. Unfortunately, the iterative process converges slowly and in some instances may not converge at all. Mostly through the efforts of Jöreskog (summarized in Jöreskog, 1977), a two-stage numerical optimization procedure has been developed that now is utilized in almost all modern programs for factor analysis. Krzanowski (1988) gives a compact description of the algorithm, and Jöreskog (1977) provides details.

The extensions of the maximum-likelihood factor process are similar to those of factor analysis based on principal components. The factors may be rotated to simple structures or even to oblique positions in the search for meaning. The problem of specifying p in advance of analysis has been avoided, and the factors are free of the bias inherent in factors extracted by simpler procedures. Unfortunately, the fundamental criticisms of factor analysis remain. In those fields with well-developed theories of causality, factor analysis may be especially useful. In geology, today's strongly held truths tend to be tomorrow's discredited conjectures, and factor interpretations probably will fare no better. The skeptical-minded are invited to read the critique of factor analysis in geology by Temple (1978) and the more developed criticism of the use of factor analysis in hydrology by Matalas and Reiher (1967).

Q-Mode Factor Analysis

We now turn to Q-mode factor analysis, where attention is devoted exclusively to interpretation of the interobject relationships in a data set, rather than to the intervariable (or covariance) relationships explored with R-mode factor analysis. The fact that the two are really equivalent has escaped most investigators, and has led to the creation of Q-mode procedures that are extremely cumbersome and computationally extravagant.

The first step in Q-mode analysis is to create an $n \times n$ matrix of similarities between samples. The correlation coefficient, however, may be considered inappropriate as a measure of similarity between samples because it requires calculation of "variances" across variables. The reasoning behind such a measure is at best obscure.

The most widely used measure of similarity in Q-mode factor analysis is the cosine θ coefficient of proportional similarity,

$$\cos \theta_{ij} = \frac{\sum_{k=1}^{m} x_{ik} x_{jk}}{\sqrt{\sum_{k=1}^{m} x_{ik}^2 \sum_{k=1}^{m} x_{jk}^2}} \qquad (6.70)$$

This expresses the similarity between object i and object j by regarding each as a vector defined in m-dimensional space. The cosine θ coefficient is the cosine of the angle between the two vectors. Note that the equation is very similar in form to the correlation coefficient (Eq. 2.28); if the variables are standardized, with means of zero and standard deviations of one, the two measures are numerically identical.

Cosine θ ranges from 1.0 for two objects whose vector representations coincide, to 0.0 for objects whose vectors are at 90°. Since cosine θ measures only angular similarity, it is sensitive only to the relative proportions of the variables and not to their absolute magnitudes. If, for example, measurements were made on two brachiopods which were identical in shape but not in size, the cosine θ similarity measure between them would be 1.0.

The $n \times n$ matrix of similarities may be generated most conveniently if cosine θ is calculated in two steps (Jöreskog, Klovan, and Reyment, 1976). First, every element in a row of the data matrix is divided by the square root of the sum of squares of the elements in that row,

$$w_{ik} = \frac{x_{ik}}{\sqrt{\sum_{k=1}^{m} x_{ik}^2}} \tag{6.71}$$

This standardizes the objects so that the squares of the variables measured on each object sum to one. Then, cosine θ is given by

$$\cos \theta_{ij} = \sum_{k=1}^{m} w_{ik} w_{jk} \tag{6.72}$$

In matrix notation, we first define an $n \times n$ diagonal matrix, $\mathbf{D}$, which contains the sums of squares of each row along the diagonal and zeros elsewhere. The standardization step is

$$\mathbf{W} = \mathbf{D}^{-1/2} \mathbf{X} \tag{6.73}$$

and the similarity matrix, $\mathbf{Q}$, is

$$\begin{aligned} \mathbf{Q} &= \mathbf{W}\mathbf{W}' \\ &= \mathbf{D}^{-1} \mathbf{X}\mathbf{X}' \mathbf{D}^{-1} \end{aligned} \tag{6.74}$$

The fact that the $n \times n$ matrix $\mathbf{Q}$ can have at most only m eigenvalues suggests that it should be possible to take advantage of the Eckart–Young theorem and to extract the eigenvectors from the $m \times m$ matrix $\mathbf{W}'\mathbf{W}$ rather than from the larger matrix $\mathbf{Q}$, and then transform the R-mode scores into Q-mode loadings and vice versa. This topic is considered in greater detail in the next section. Under certain circumstances the reciprocal relationship between R- and Q-mode factors and scores holds exactly, but this depends upon the nature of any scaling that is performed on the data matrix, $\mathbf{X}$.

File QMODE.TXT contains the Q-mode cosine θ similarity matrix for the random block data, and **Table 6–22** gives the eigenvalues and first three eigenvectors. As we would expect, the first few eigenvalues account for almost all of the variation among the blocks. Their eigenvectors can be converted into factor loadings by multiplying each element in a vector by the corresponding singular value (or square root of the corresponding eigenvalue). This is exactly the same procedure used in R-mode factor analysis. It scales the Q-mode factor axes so their lengths are proportional to the amount of variation between the objects that they contain. **Table 6–22** also lists the Q-mode factors that correspond to the first three eigenvectors.

In Q-mode analysis, we plot the loadings rather than the factor scores if we wish to see relationships between the objects in our sample. **Figure 6–33** is a plot of the first two Q-mode factors; the blocks are shown in positions representing their loadings on the factors. The arc on the diagram is part of a circle representing a communality of 1.00; if an object falls on the circle, the two factors account for all of its variability. Blocks that plot inside the circle are characterized by variability that is not represented by the two factors.

Figure 6–34 is a plot of the second and third Q-mode factors. This plot and that shown in **Figure 6–33** together represent 99% of the variation in the blocks, which is exactly what we expect from our knowledge of how the blocks were originally created. Although **Figure 6–33** shows a general progression from larger to smaller blocks, the distinction by shape is not as clear-cut as in R-mode factor analysis (**Fig. 6–32**). You will note that the second and third factors shown in **Figure 6–34**

Table 6–22. Eigenvalues, first three eigenvectors, and factor loading vectors of cosine θ similarity matrix (file QMODE.TXT) computed from BOXES.TXT data. Eigenvalues higher than 7 are identically zero.

Vector	Eigenvalue	Variance	Cumulative Variance
1	22.3024	89.21	89.21
2	1.9882	7.95	97.16
3	0.4545	1.82	98.98
4	0.2193	0.88	99.86
5	0.0323	0.13	99.99
6	0.0029	0.01	100.00
7	0.0004	0.00	100.00

	Eigenvector			Factor		
Block	1	2	3	F1	F2	F3
a	0.1780	−0.3710	−0.0416	0.8408	−0.5232	−0.0281
b	0.2085	−0.0969	−0.1592	0.9845	−0.1366	−0.1073
c	0.1961	0.2577	0.0628	0.9260	0.3633	0.0423
d	0.1859	0.3264	0.1051	0.8779	0.4602	0.0708
e	0.2079	0.1157	−0.1099	0.9816	0.1631	−0.0741
f	0.2087	−0.1122	−0.0613	0.9858	−0.1583	−0.0413
g	0.1607	−0.4139	0.4271	0.7587	−0.5836	0.2879
h	0.2042	0.1806	−0.0632	0.9644	0.2547	−0.0426
i	0.2004	−0.1783	−0.2521	0.9463	−0.2513	−0.1699
j	0.1997	−0.1894	−0.0659	0.9431	−0.2670	−0.0444
k	0.2055	−0.1417	−0.1168	0.9706	−0.1998	−0.0788
l	0.2022	0.2099	0.0158	0.9549	0.2960	0.0107
m	0.2103	0.0605	0.1124	0.9933	0.0853	0.0758
n	0.2045	−0.1301	0.1498	0.9658	−0.1834	0.1010
o	0.1833	−0.0913	0.7009	0.8655	−0.1288	0.4725
p	0.2095	−0.0458	−0.1089	0.9893	−0.0646	−0.0734
q	0.2078	−0.0534	−0.2167	0.9814	−0.0752	−0.1461
r	0.2085	0.1097	−0.0635	0.9847	0.1547	−0.0428
s	0.2026	−0.1917	−0.0826	0.9567	−0.2703	−0.0557
t	0.2090	−0.0068	0.0095	0.9870	−0.0096	0.0064
u	0.2018	−0.1791	−0.2295	0.9532	−0.2525	−0.1547
v	0.1993	0.2329	0.0014	0.9410	0.3284	0.0009
w	0.1943	0.2606	0.1569	0.9176	0.3674	0.1058
x	0.2073	0.0927	−0.0101	0.9791	0.1307	−0.0068
y	0.1957	0.2699	0.0199	0.9243	0.3806	0.0135

produce patterns of loadings that are superior for classification purposes to the first factor, even through the first eigenvector is an order of magnitude larger than the second.

In *R*-mode factor analysis, the data are centered about zero by subtracting the means from each observation prior to factoring. This is not done in *Q*-mode factor analysis, with the result that the first *Q*-mode factor is simply a vector from the origin to the centroid of the set of objects. In effect, this first factor expresses "size," or the combined magnitudes of the variables measured on each object. Typically, all of the loadings on the first factor will be positive in sign and about equal in magnitude. Since the first factor reveals very little about the structure of the data,

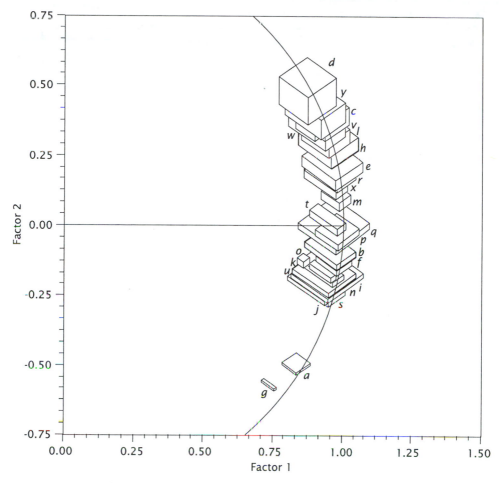

Figure 6–33. Plot of Q-mode factor loadings on the first two factor axes of block data, calculated from cosine θ similarity matrix (file QMODE.TXT). Horizontal axis is factor 1, vertical axis is factor 2. Circular arc represents communality of 1.00.

it is usually regarded as a "nuisance factor" and discarded. The second and third factors, as in **Figure 6–34**, are considered to be much more revealing.

The geologic example we will consider is taken from igneous petrology. File IGNEOUS.TXT contains data on the major chemical constituents of 20 rock specimens taken from a complex and apparently differentiated igneous body. By Q-mode analysis, we hope to place each rock in its proper place within the differentiation series.

The succession of specimens within the sequence between end members can be shown by plotting the rock analyses as loadings on pairs of factors. In this example, the communalities of all specimens on the first two factors is almost 1.00, so the specimens define vectors that plot as radii of a circle. The angles between different radii are a measure of the similarity between rocks, as expressed by the cosine θ coefficient. The cosine θ similarity matrix between observations is given in **Table 6–23**, and the factor matrix is given in **Table 6–24**. The first and second factor loadings are shown plotted in **Figure 6–35**. Note that the first factor is essentially a "nuisance factor," expressing only the information that the major constituents

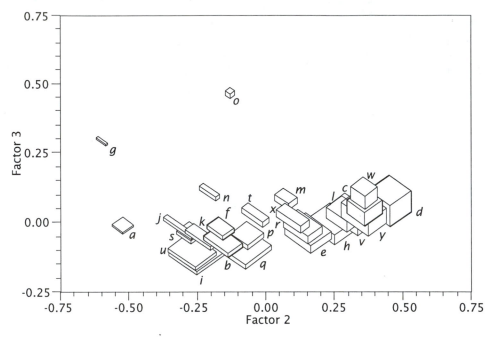

Figure 6–34. Plot of Q-mode factor loadings on second and third factor axes of block data, calculated from cosine θ similarity matrix.

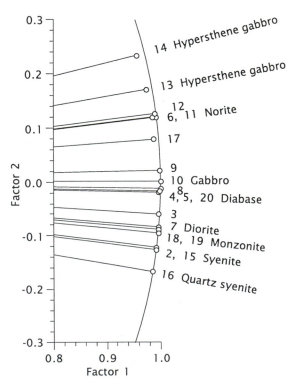

Figure 6–35. Plot of factor loadings from Q-mode analysis of cosine θ similarity matrix of 20 igneous rock analyses.

Table 6–23. Cosine θ coefficients between 20 samples of igneous rock from a differentiated pluton.

Sample	1	2	3	4	5	6	7	8	9	10	11	12	13	14	15	16	17	18	19	20
1	1.000																			
2	0.997	1.000																		
3	0.994	0.997	1.000																	
4	0.996	0.991	0.994	1.000																
5	0.993	0.988	0.989	0.997	1.000															
6	0.972	0.968	0.981	0.987	0.984	1.000														
7	0.998	0.999	0.998	0.995	0.992	0.977	1.000													
8	0.995	0.991	0.995	0.998	0.996	0.989	0.995	1.000												
9	0.991	0.986	0.990	0.998	0.998	0.992	0.991	0.998	1.000											
10	0.992	0.985	0.988	0.998	0.996	0.991	0.991	0.997	0.999	1.000										
11	0.966	0.961	0.975	0.978	0.974	0.996	0.972	0.981	0.984	0.984	1.000									
12	0.971	0.966	0.978	0.985	0.985	0.993	0.974	0.990	0.990	0.988	0.981	1.000								
13	0.948	0.943	0.958	0.970	0.973	0.984	0.951	0.973	0.978	0.973	0.965	0.993	1.000							
14	0.934	0.922	0.940	0.950	0.937	0.970	0.936	0.957	0.952	0.956	0.972	0.969	0.945	1.000						
15	0.998	1.000	0.996	0.992	0.989	0.967	0.999	0.991	0.987	0.986	0.960	0.965	0.941	0.921	1.000					
16	0.997	0.997	0.989	0.985	0.982	0.953	0.995	0.985	0.978	0.978	0.948	0.951	0.922	0.911	0.997	1.000				
17	0.979	0.967	0.973	0.990	0.984	0.980	0.973	0.987	0.987	0.990	0.970	0.982	0.969	0.965	0.968	0.961	1.000			
18	0.996	0.998	0.997	0.994	0.994	0.977	0.998	0.994	0.992	0.990	0.968	0.975	0.958	0.925	0.998	0.992	0.971	1.000		
19	0.999	0.998	0.995	0.995	0.991	0.973	0.999	0.994	0.990	0.991	0.968	0.970	0.946	0.934	0.999	0.996	0.975	0.997	1.000	
20	0.992	0.993	0.998	0.996	0.993	0.989	0.996	0.998	0.996	0.993	0.980	0.988	0.972	0.947	0.992	0.984	0.977	0.997	0.993	1.000

of the rocks sum to nearly 100%. Variation along the second factor is much more informative, as it ranks the rocks in the order expected in a differentiation series. The first two factors were retained for rotation; rotated factor loadings are given in **Table 6–25**. A plot of these loadings will reveal a pattern that is essentially identical to that shown in **Figure 6–35**, except that the loading vectors will be rotated by approximately 45°. In this instance, factor rotation provides no additional insight into the relationships among the observations. **Table 6–25** also contains the scores of the original variables (chemical constituents) on the two factors. The scores indicate that both rotated factors are heavily influenced by the relative abundance of silica plus alumina, conforming to traditional interpretations of igneous rock sequences. However, anyone tempted to draw inferences based on Q-mode loadings are strongly cautioned to ponder Reyment and Savazzi's (1999) acerbic criticism of this practice—in addition to theoretical objections, Q-mode loadings are notoriously unstable. Q-mode factor analysis should be used as a graphical tool, and not for inference.

A few final remarks should be made about Q-mode factor analysis. The technique often is used in an attempt to identify end members—real or hypothetical objects having extreme properties. Intermediate objects may be regarded as mixtures of the two end members, just as the igneous rocks shown in **Figure 6–35** could be interpreted as being from a gradational sequence between hypothetical sialic and femic extremes.

Q-mode factor analysis is also used for classification, an objective that is essentially the same as cluster analysis, yet is much more costly in terms of computing time. If the object of an analysis is to search through samples for groups or clusters,

Table 6–24. First five factors extracted from cosine θ matrix in Table 6–23. Eigenvalues 8 through 20 are all zero.

Sample	1	2	3	4	5	Communality
1	0.9948	−0.0910	0.0242	0.0324	0.0069	0.9996
2	0.9918	−0.1223	0.0081	−0.0177	−0.0268	0.9997
3	0.9958	−0.0587	0.0085	−0.0457	−0.0344	0.9983
4	0.9989	−0.0126	−0.0070	0.0357	0.0178	0.9997
5	0.9963	−0.0191	−0.0596	0.0297	0.0353	0.9986
6	0.9904	0.1188	−0.0133	−0.0594	0.0309	0.9997
7	0.9959	−0.0838	0.0191	−0.0235	−0.0086	0.9998
8	0.9996	0.0010	−0.0017	0.0112	−0.0132	0.9996
9	0.9983	0.0204	−0.0336	0.0055	0.0391	0.9997
10	0.9978	0.0223	−0.0049	0.0291	0.0498	0.9994
11	0.9833	0.1202	0.0550	−0.0988	0.0746	0.9997
12	0.9890	0.1259	−0.0512	0.0008	−0.0538	0.9995
13	0.9721	0.1719	−0.1552	0.0066	−0.0365	0.9999
14	0.9561	0.2323	0.1691	0.0146	−0.0527	0.9997
15	0.9918	−0.1257	0.0102	−0.0084	−0.0137	0.9998
16	0.9844	−0.1665	0.0458	0.0203	−0.0113	0.9994
17	0.9866	0.0783	0.0214	0.1316	0.0259	0.9980
18	0.9950	−0.0870	−0.0367	−0.0275	−0.0089	0.9998
19	0.9945	−0.0946	0.0296	0.0035	0.0066	0.9989
20	0.9981	−0.0161	−0.0236	−0.0395	−0.0295	0.9995

	1	2	3	4	5	6	7
Eigenvalues	19.6249	0.2298	0.0699	0.0408	0.0233	0.0074	0.0026
Variance	98.12	1.15	0.35	0.20	0.12	0.04	0.01
Cumulative variance	98.12	99.27	99.62	99.83	99.94	99.98	100.00

either cluster analysis or *R*-mode factor analysis may provide more efficient means for doing so. If significant factors can be extracted by *R*-mode methods, scatter diagrams of factor scores or cluster diagrams usually will reveal the relationships between samples. As an example, **Figure 6–36** is a dendrogram constructed by weighted pair–group averaging of the matrix of correlations between the igneous rock data from file IGNEOUS.TXT. The relative arrangement of samples is almost exactly the same as that obtained by *Q*-mode analysis.

A word about closure

Investigators sometimes resort to *Q*-mode analyses because they believe this will free their data from the troublesome constraints imposed by closure. They reason that a *Q*-mode analysis, in effect, represents a transposition of the data matrix, **X**, because the roles of variables and observations are exchanged. Closure arises in compositional data because the rows of **X** sum to a constant such as 100%. If **X** is transposed, the rows will no longer sum to a fixed value (in fact, the sums will change with the number of observations), so the closure effect should vanish. Unfortunately, the Eckhart–Young theorem demonstrates that such reasoning is fallacious; there is an inescapable link between *R*- and *Q*-mode analyses, and a

Table 6–25. Rotated factor loadings and varimax factor scores.

Sample	1	2	Communality
1	0.7851	0.6177	0.9980
2	0.8044	0.5929	0.9986
3	0.7636	0.6418	0.9950
4	0.7342	0.6774	0.9980
5	0.7368	0.6709	0.9929
6	0.6377	0.7671	0.9950
7	0.7809	0.6236	0.9988
8	0.7254	0.6878	0.9993
9	0.7111	0.7009	0.9970
10	0.7094	0.7020	0.9960
11	0.6316	0.7632	0.9814
12	0.6319	0.7712	0.9940
13	0.5879	0.7930	0.9745
14	0.5348	0.8259	0.9681
15	0.8068	0.5904	0.9995
16	0.8295	0.5556	0.9968
17	0.6628	0.7350	0.9796
18	0.7825	0.6207	0.9976
19	0.7873	0.6148	0.9979
20	0.7360	0.6744	0.9965
Variance	52.311	46.962	
Cumulative variance	52.311	99.272	

Varimax Factor Score Matrix

Variable	Factor 1	Factor 2
x_1	70.2648	5.6766
x_2	14.5830	8.1431
x_3	4.5006	−1.0267
x_4	−16.2185	24.5371
x_5	−19.4934	28.8944
x_6	−6.5178	16.8625
x_7	11.4400	−6.9660
x_8	11.1204	−7.2130

row-wise constant-sum constraint must also manifest itself in a Q-mode analysis. This equivalence is briefly discussed by Aitchison (1986). Applying a centered logratio transformation to compositional data will "open up" the measurements and remove the closure constraint. If this transform is applied to the major oxide data in file IGNEOUS.TXT, however, we will see that the resulting cosine θ similarity matrix of the transformed data is very similar to the cosine θ matrix calculated from the raw data. As a consequence, the Q-mode factor solution after transformation is almost identical to the original solution shown in **Figure 6–35**.

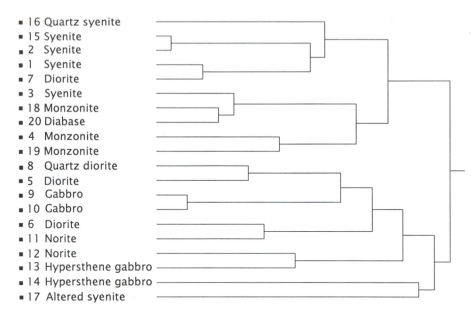

- 16 Quartz syenite
- 15 Syenite
- 2 Syenite
- 1 Syenite
- 7 Diorite
- 3 Syenite
- 18 Monzonite
- 20 Diabase
- 4 Monzonite
- 19 Monzonite
- 8 Quartz diorite
- 5 Diorite
- 9 Gabbro
- 10 Gabbro
- 6 Diorite
- 11 Norite
- 12 Norite
- 13 Hypersthene gabbro
- 14 Hypersthene gabbro
- 17 Altered syenite

Figure 6–36. Cluster analysis of chemical analyses of igneous rocks. The clustering shows essentially the same arrangement as Q-mode factor analysis.

Principal Coordinates Analysis

Principal coordinates analysis is a widely used Q-mode technique. Most applications have been in quantitative biology and paleontology, although the method has also been used in petrology. Principal coordinates analysis was popularized by Gower (1966), and is extensively discussed in Jöreskog, Klovan, and Reyment (1976) and in Reyment, Blackith, and Campbell (1984). A succinct discussion, with computer program, is given by Reyment and Savazzi (1999); Jackson (1991) describes the methodology in the broader context of multidimensional scaling (MDS). The objective of principal coordinates analysis is similar to that of Q-mode factor analysis. We can imagine that our observations define locations in multidimensional space. Unfortunately, the "true" orthogonal axes of this space are not given by the measurement scales along which the observed variables are defined, as these may be partially redundant or correlated (nonorthogonal). From the distances between the observations, principal coordinates analysis attempts to determine the number and orientation of the true axes of the space occupied by the observations.

Reyment and Savazzi (1999) provide a step-by-step guide to principal coordinates, which we will follow here. Our data consist of an $n \times p$ array of observations, **X**, which we may assume have been corrected for the variable means (that is, $x_{ik} = X_{ik} - \overline{X}_{.k}$, where X_{ik} is an original observation and $\overline{X}_{.k}$ is the mean of variable k). The first step is to calculate an $n \times n$ matrix of dissimilarities in the form of distances between the n objects. Any of a number of distance measures can be used, including the squared Euclidean distance:

$$d_{ij} = \sum_{k=1}^{p} \left(x_{ik} - x_{jk} \right)^2 \tag{6.75}$$

Another widely used distance measure is the Gower distance,

$$g_{ij} = \frac{\sum_{k=1}^{p} |x_{ik} - x_{jk}|/\text{range}_k}{p} \tag{6.76}$$

To compute the Gower distance between object i and object j, the absolute difference between them for variable k is divided by the range of variable k. This gives a number in the interval 0.0 to 1.0, with a small number representing close similarity and a number approaching 1.0 representing maximum dissimilarity among the objects in the set. The calculations are repeated for all p variables measured on objects i and j, then summed and divided by the number of variables to give the Gower distance, g_{ij}. In computing the Gower distance, no assumptions are made about the nature of the data; observations may be nominal or ordinal, or of higher rank. In fact, the data matrix may consist of mixtures of different types of numbers, such as counts of the number of plates in the calyxes of crinoids, the lengths of their arms, and the ratios of calyx heights to diameters.

The measures of distance between all possible pairs of objects are assembled in an $n \times n$ dissimilarity matrix, $\mathbf{D}$. This matrix will be symmetrical and have 0's down the diagonal and positive values elsewhere.

Each row of the matrix $\mathbf{D}$ is summed and divided by n to give the row averages; each column of $\mathbf{D}$ is also summed and divided by n to give the column averages. These are designated $\overline{d}_i.$ and $\overline{d}._k$, respectively. The grand mean, $\overline{\overline{d}}_{..}$, of all elements of $\mathbf{D}$ is also found. The distances d_{ik} are now transformed to give a new matrix $\mathbf{A}$ by the operation

$$a_{ij} = \frac{1}{2}\left[\overline{d}_i. + \overline{d}._j - \overline{\overline{d}}_{..} - d_{ij}\right] \tag{6.77}$$

When using squared Euclidean distances, the matrix $\mathbf{A}$ can be found directly from $\mathbf{X}$ (which you will recall contains the observations corrected for their variable means) by

$$\mathbf{A} = \mathbf{X}\,\mathbf{X}' \tag{6.78}$$

We may regard the n objects as being located in the p-dimensional space defined by the variables. The transformation in Equation (6.77) moves the origin of the coordinate system describing the p dimensions so that it coincides with the centroid of the cloud of points. The operation also closes the data set, as all rows and columns now sum to zero, so one eigenvalue of $\mathbf{A}$ is forced to become zero. This tends to increase the relative magnitudes of the first few eigenvalues.

Next, the eigenvalues and eigenvectors are extracted from $\mathbf{A}$. At most, only the first $p - 1$ eigenvalues will be greater than zero. We can discard all eigenvectors that correspond to zero-valued eigenvalues, leaving us with $p - 1$ eigenvectors $\mathbf{v}_1, \mathbf{v}_2, \ldots, \mathbf{v}_{p-1}$ to constitute the columns of a matrix $\mathbf{V}$. The $p - 1$ nonzero eigenvalues of $\mathbf{A}$ can be placed on the diagonal elements of a square matrix $\mathbf{\Lambda}$ whose off-diagonal elements are zero. The matrix of coordinates of the observations along the principal coordinate axes is determined by

$$\mathbf{S} = \mathbf{V}\mathbf{\Lambda}^{1/2} \tag{6.79}$$

That is, the principal coordinates consist of the eigenvectors multiplied by their corresponding singular values. The relative importance of each coordinate may be

Table 6–26. Eigenvalues and eigenvectors extracted from principal coordinates matrix **A** of box data. Eigenvalues 8 through 25 are zero.

Eigenvalues:	489.4904	87.8171	10.1569	4.5053	1.3734	0.1354	0.0104
Percent:	82.4767	14.7968	1.7114	0.7591	0.2314	0.0228	0.0017
Cumulative percent:	82.4767	97.2735	98.9849	99.7440	99.9754	99.9983	100.0000

Eigenvectors

	1	2	3	4	5	6	7
a	0.08030	−0.39286	0.26417	0.09806	0.29129	−0.37424	−0.06185
b	0.12287	−0.04275	−0.10690	−0.14287	−0.00127	−0.16318	0.07504
c	−0.11634	0.19535	0.19145	0.07817	0.09323	0.21303	0.20551
d	−0.01922	0.37387	0.17148	0.46398	0.34721	0.08599	−0.31579
e	0.06595	0.18098	0.00815	−0.09669	−0.35062	0.05102	0.02179
f	−0.08789	−0.11676	0.08561	−0.20222	0.01683	−0.04036	0.15220
g	−0.09787	−0.42428	−0.02806	0.67017	−0.45114	−0.05463	0.12512
h	0.01850	0.22475	0.02615	−0.03276	−0.29024	0.03121	0.17013
i	0.53772	−0.07650	0.22684	−0.00538	0.07577	0.09712	−0.11190
j	0.15702	−0.15801	−0.56410	0.13850	0.23085	0.54275	0.17619
k	0.08184	−0.10766	−0.29042	−0.09083	0.15260	−0.03745	0.08411
l	−0.10149	0.16134	0.05149	−0.02619	−0.01490	−0.01511	0.32072
m	−0.24712	−0.05585	0.06382	−0.19249	0.07526	0.00033	0.25749
n	−0.18762	−0.15892	−0.16851	−0.05314	−0.00081	0.03606	−0.25683
o	−0.40385	−0.19910	0.20864	−0.06930	−0.09469	0.17566	−0.27999
p	−0.02583	−0.03732	0.19875	−0.20128	−0.11395	0.22591	0.04525
q	0.34607	0.08975	0.14661	−0.07872	−0.34104	0.25873	−0.34616
r	−0.03630	0.11021	−0.11705	−0.12665	−0.13251	−0.14513	0.04097
s	−0.03710	−0.17825	0.18382	−0.16654	0.06292	−0.01310	0.08198
t	−0.12324	−0.04223	−0.24096	−0.13769	0.04122	−0.07066	−0.32513
u	0.40009	−0.09904	0.04823	−0.04997	0.13664	−0.19769	0.14037
v	−0.02111	0.25878	−0.18907	0.08219	−0.04293	−0.33616	−0.03047
w	−0.20979	0.11366	0.17383	0.06190	0.32283	0.18146	0.05995
x	−0.09769	0.05783	−0.26476	−0.11763	0.00815	−0.18457	−0.36914
y	0.00209	0.32301	−0.07923	0.19738	−0.02070	−0.26698	0.14042

Table 6–27. First two principal coordinates of box data.

Box	Principal Coordinate 1	Principal Coordinate 2	Box	Principal Coordinate 1	Principal Coordinate 2
a	1.7766	−3.6815	n	−4.1510	−1.4893
b	2.7184	−0.4006	o	−8.9350	−1.8658
c	−2.5740	1.8306	p	−0.5715	−0.3497
d	−0.4252	3.5036	q	7.6566	0.8411
e	1.4591	1.6960	r	−0.8031	1.0328
f	−1.9445	−1.0942	s	−0.8208	−1.6704
g	−2.1653	−3.9756	t	−2.7266	−0.3957
h	0.4093	2.1062	u	8.8518	−0.9281
i	11.8968	−0.7169	v	−0.4671	2.4250
j	3.4740	−1.4807	w	−4.6415	1.0651
k	1.8107	−1.0089	x	−2.1613	0.5419
l	−2.2454	1.5119	y	0.0462	3.0270
m	−5.4674	−0.5234			

assessed simply by calculating the percentage of the trace of **A** contained in each successive eigenvalue. Usually, only the first few coordinates are of interest, as these hopefully will account for most of the differences between the observations.

The final step is to plot the individual loadings on the principal coordinates; this is done by cross plotting the n elements (each corresponding to an object) of one principal coordinate against the elements of another for the first two (and possibly succeeding) principal coordinates.

We will use the artificial block data to illustrate principal coordinates analysis, so that the results may be compared to those obtained from other eigenvector techniques. File DISSIM.TXT contains the 25×25 matrix **D** of similarities between the individual blocks, calculated using the Euclidean distance measure as defined by Equation (6.75). File PCOORD.TXT contains the 25×25 matrix **A** of similarity measures after transformation by subtracting the row and column mean from every element, then adding the grand mean, as defined in Equation (6.77). It is from this matrix **A** that the eigenvalues and eigenvectors are extracted.

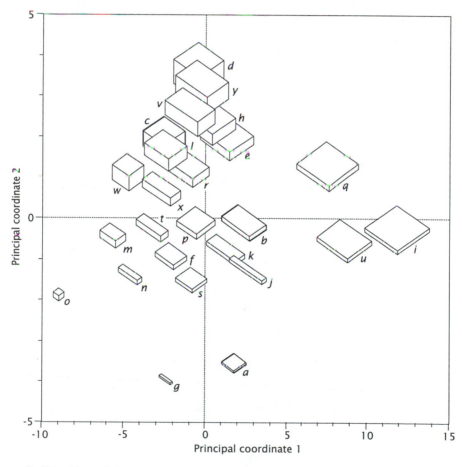

Figure 6–37. Plot of first two principal coordinates of random block data. Blocks are shown at positions corresponding to their loadings on the principal coordinates.

Table 6–26 gives the first seven eigenvalues extracted from matrix A; the eighth and all succeeding eigenvalues are zero. The first two eigenvalues account for 97.3%

of the total variation in the block data, and the third eigenvalue accounts for an additional 1.7%, or essentially all of the remaining variation. (Recall that the data were generated from only three independent variables. The small amount of variation not accounted for by the first, second, and third principal coordinates can be ascribed to various rounding errors in the calculations.) The first two principal coordinates are listed in **Table 6–27**. Each vector element corresponds to an individual observation, and is shown plotted in **Figure 6–37**. Compare the pattern of the principal coordinates solution to that obtained by *Q*-mode factor analysis (**Fig. 6–33**).

Correspondence Analysis

Factor analysis is designed for interval or ratio data—that is, measurements made on a continuous numerical scale. It is not appropriate for enumerative data, such as counts of the number of fossils of different types that are present in samples, or the number of fractures of different orientations that occur within mining blocks. Such nominal or ordinal observations may be all that are available, and in some instances it seems desirable to process them using an eigenvalue technique similar to factor analysis.

Problems in which only count data are available are very common in the social sciences. Surveys by questionnaire, for example, result in responses that can only be tallied in categories. Consequently, most of the research on the use of eigenvalue methods for analyzing such data has been done by sociologists and statisticians working on sociological problems. These data may be summarized conveniently in the form of contingency tables; the first work of the type we are considering was applied to such summaries and was known as "contingency table analysis" (Hirschfeld, 1935; Fisher, 1940). More recently, Benzécri and others (1980) have written extensively on the subject, and Benzécri's term "correspondence analysis" has come to be widely used, although as Reyment and Savazzi (1999) point out, this is a somewhat inaccurate translation of a French term which could be better (and more appropriately) rendered as "analysis of associations." Benzécri's work has provided the basis for many applications in geology, such as those of Teil (1975), Teil and Cheminée (1975), and David, Dagbert, and Beauchemin (1977). In these geological applications, however, the methods of Benzécri and his predecessors have been extensively modified. Hill (1974) describes the early history of correspondence analysis, and the interrelationships between the work of various authors. Detailed discussions of correspondence analysis and its extensions are contained in monographs by Lebart, Morineau, and Warwick (1984) and Greenacre (1984). Jackson (1991) fits correspondence analysis into the broader context of multidimensional scaling, and Reyment and Savazzi (1999) provide an extensive discussion accompanied by a computer program.

Correspondence analysis proceeds by operating on a matrix derived from a contingency table which has been transformed so that the elements of the table can be regarded as conditional probabilities. Because of the nature of the transformation (actually a form of scaling), relationships between rows and columns of the transformed table are the same as those within the original data matrix. This means that the Eckart–Young theorem holds exactly, and *R*- and *Q*-mode solutions are equivalent.

The raw data matrix, $\mathbf{X}$, has n rows that represent observations and m columns of variables. The elements themselves are tallies. In the usual situation, $n > m$, which is assumed in the following discussion but is not a required condition. In a problem from paleoecology, for example, the columns might consist of different microfossil varieties, the rows could be cuttings taken from different stratigraphic intervals within a well, and the entries in the table would be counts of the number of individual specimens of each variety of microfossil recovered from the cuttings. The total number of individuals is simply the sum of all of the elements in the data matrix, or

$$N = \sum_{i=1}^{n} \sum_{j=1}^{m} x_{ij}$$

The sum of the ith row

$$r_i = \sum_{j=1}^{m} x_{ij}$$

is the number of microfossils of all types that have been found in the cuttings from interval i. The sum of the jth column,

$$c_j = \sum_{i=1}^{n} x_{ij}$$

is the number of microfossils of variety j that have been recovered from the cuttings taken from all intervals. The tallies can be converted to percentages of the total, which may be regarded as probabilities:

$$p_{ij} = \frac{x_{ij}}{N} = \frac{x_{ij}}{\sum_{i=1}^{n} \sum_{j=1}^{m} x_{ij}} \tag{6.80}$$

These p_{ij} values are the **joint probabilities** that specific microfossils will be found in the cuttings from specific intervals in the well. The row totals divided by the grand total give the **marginal probabilities**

$$p_{i\cdot} = \frac{r_i}{N} = \frac{\sum_{j=1}^{m} x_{ij}}{\sum_{i=1}^{n} \sum_{j=1}^{m} x_{ij}} \tag{6.81}$$

which are the probabilities that specific intervals of cuttings will contain microfossils, without considering what the fossils might be. The column totals, treated in the same manner, give the marginal probabilities

$$p_{\cdot j} = \frac{c_j}{N} = \frac{\sum_{i=1}^{n} x_{ij}}{\sum_{i=1}^{n} \sum_{j=1}^{m} x_{ij}} \tag{6.82}$$

which are the probabilities that specific microfossils will occur, regardless of which interval of cuttings they are in. If the joint probabilities are divided by their corresponding marginal probabilities, the results are the **conditional probabilities**

$$p_{(i|j)} = \frac{p_{ij}}{p_{\cdot j}}$$

and

$$p_{(j|i)} = \frac{p_{ij}}{p_{i\cdot}} \tag{6.83}$$

The first of these conditional probabilities describes the situation in which it is given that a specific microfossil, j, will occur, and we want the probability that it will be found in the cuttings from interval i. The second conditional probability, based on the row totals, gives the probability that the microfossil we find will be of variety j, when it is given that we will find it in well cuttings from the ith interval.

In Chapter 2 we noted that in a contingency table the observations could be expressed as proportions of the total number of observations. Then, if the rows and columns of the table are independent, these observations should be approximately equal to the products of the marginal probabilities of their respective rows and columns. If two variables j and k are closely related, however, all the expected values which occur in the jth and kth columns should be very similar. This suggests that it should be possible to express the degree of similarity by computing a cross product that would involve the observed and expected probabilities for all rows in the two columns being compared. Such a measure is used in correspondence analysis and is a form of the correlation coefficient between two variables (Kendall and Stuart, 1967). It is given by

$$\rho_{jk} = \sum_{i=1}^{n} \left(\frac{p_{ij} - p_{i\cdot} p_{\cdot j}}{\sqrt{p_{i\cdot} p_{\cdot j}}} \right) \left(\frac{p_{ik} - p_{i\cdot} p_{\cdot k}}{\sqrt{p_{i\cdot} p_{\cdot k}}} \right) \tag{6.84}$$

Here, p_{ij} is the "observed" probability in row i and column j of the body of the contingency table, and $p_{i\cdot} p_{\cdot j}$ is the "expected" probability computed as the product of the marginal probabilities. Expressed in the terms used in Chapter 2, this becomes

$$\rho_{jk} = \sum_{i=1}^{n} \left(\frac{O_{ij} - E_{ij}}{\sqrt{E_{ij}}} \right) \left(\frac{O_{ik} - E_{ik}}{\sqrt{E_{ik}}} \right) \tag{6.85}$$

The relationship between this expression and the χ^2 statistic applied to contingency tables becomes clearer if we square one of the terms

$$\left(\frac{O_{ij} - E_{ij}}{\sqrt{E_{ij}}} \right)^2 = \frac{(O_{ij} - E_{ij})^2}{E_{ij}}$$

We then see that the similarity measure used in correspondence analysis can be regarded as the product of two χ^2 values. This leads to the expression "χ^2 distance," which is sometimes applied to this measure (Lebart, Morineau, and Warwick, 1984). If the similarity measure ρ_{jk} is computed between all pairs of columns j and k, it will form a square $m \times m$ matrix, $\mathbf{D}$. The eigenvalues and eigenvectors can be extracted from this matrix; these are the principal axes of correspondence analysis. Note that expressing all of the elements in the contingency table as a proportion of the total ensures that the sum of the column totals (and the sum of the row totals) of $\mathbf{D}$ is equal to 1.00. Therefore, we have closed the data set, and one eigenvalue must be zero. This means the dimensionality of our problem is guaranteed to be reduced from m to $m - 1$, and hopefully to much less.

We can express the preceding operations entirely in matrix notation if we define the $n \times n$ diagonal matrix $\mathbf{R}$, whose diagonal elements are the row totals r_i and whose off-diagonal elements are all zero. Similarly, the $m \times m$ diagonal matrix, $\mathbf{C}$, contains the column totals c_j and zero off-diagonal elements. Then, we can transform the data matrix $\mathbf{X}$ into a matrix of conditional probabilities, $\mathbf{H}$, that has the same dimensions:

$$H = R^{-1/2} X C^{-1/2} \qquad (6.86)$$

(Since we are dealing with diagonal matrices, the operations $R^{-1/2}$ and $C^{-1/2}$ consist of replacing each diagonal element r_{ii} and c_{jj} with $1/\sqrt{r_{ii}}$ and $1/\sqrt{c_{jj}}$, respectively. The off-diagonal zero elements in each matrix, of course, remain zero.)

We now can determine D simply by premultiplying H by its transpose. That is,

$$D = H' H \qquad (6.87)$$

An alternative formulation replaces X with B, in which every element of X has been divided by the grand total, so that

$$B = \left(\sum \sum x_{ij} \right)^{-1} X \qquad (6.88)$$

That is, we convert the tallies in matrix X into joint probabilities, p_{ij}. Then, we again define diagonal matrices R and C as above, except that the row and column totals along the diagonals are based on the transformed data matrix B rather than the raw data matrix X. Now, R will contain the marginal row probabilities $p_{i\cdot}$ and C will contain the marginal column probabilities $p_{\cdot j}$. We now continue as in Equations (6.86) and (6.87). The new matrix D that results from Equation (6.88) will contain elements equal to

$$\rho_{jk} = \sum_{i=1}^{n} \frac{p_{ij} \, p_{ik}}{p_{i\cdot} \sqrt{p_{\cdot j} \, p_{\cdot k}}} \qquad (6.89)$$

The last eigenvalue we extract from D when calculated by Equation (6.85) will be trivial and exactly equal to 0.0. Because the data are not centered around zero prior to factoring when using Equation (6.89), the factor solution will contain an initial trivial eigenvalue that is identically equal to 1.0.

We may also calculate a matrix of cross products of the rows by

$$Q = HH' \qquad (6.90)$$

The eigenvalues of D and Q will be identical, except that Q will have $(n - m)$ additional eigenvalues, each of which will be zero. Each eigenvector of D can be converted to correspondence loadings by multiplying the eigenvector by its singular value, which is the square root of its eigenvalue. That is,

$$R\text{-mode loadings} = \sqrt{\lambda} \cdot R\text{-mode eigenvector}$$

In the matrix notation we have used earlier, the singular values of D can be thought of as occurring along the diagonal of an $m \times m$ matrix, Λ, whose off-diagonal elements are all zero. The eigenvectors of D form the columns of an $m \times m$ matrix, U. The matrix equation used to determine the R-mode loadings is then

$$A^R = U\Lambda \qquad (6.91)$$

The scores of each of the n observations on the m correspondence axes are simply

$$S^R = HA^R \qquad (6.92)$$

555

These scores can be plotted along axes defined by the R-mode correspondence axes in the same way as principal component or factor scores.

If, instead of extracting eigenvalues from $\mathbf{D}$, we extract them from $\mathbf{Q}$, we can calculate Q-mode correspondence loadings and scores. Again, loadings are found by multiplying the elements of the eigenvectors by the square roots of the associated eigenvalues,

$$\mathbf{A}^Q = \mathbf{V}\mathbf{\Lambda} \tag{6.93}$$

where $\mathbf{V}$ is the $n \times n$ matrix whose columns contain the n eigenvectors of $\mathbf{Q}$. The Q-mode scores are

$$\mathbf{S}^Q = \mathbf{H}'\mathbf{A}^Q \tag{6.94}$$

As in principal components, if we perform these operations keeping all m eigenvalues and their associated eigenvectors, the procedure is reversible and the original data could be recovered unchanged. However, the usual objective of correspondence analysis is to reduce the dimensionality of the problem so relationships between observations and variables can be shown in only two dimensions. Therefore, it is necessary to retain only the $p < m$ largest eigenvalues and their associated eigenvectors. That is, matrix $\mathbf{\Lambda}$ is $p \times p$, matrices $\mathbf{U}$ and A^R are $m \times p$, and the matrix of R-mode scores, $\mathbf{S}^R$, is $n \times p$. In Q-mode, the matrices $\mathbf{V}$ and $\mathbf{A}^Q$ are $n \times p$ and the matrix of Q-mode scores, $\mathbf{S}^Q$, is $m \times p$. Because of the Eckart–Young theorem and the fact that the scaling of the original data matrix affected both rows and columns of the original data matrix $\mathbf{X}$ in the same manner, there is a direct relationship between the R- and Q-mode solutions. This relationship is

$$\begin{aligned}\mathbf{A}^Q &= \mathbf{H}\mathbf{A}^R\mathbf{\Lambda}^{-1} \\ &= \mathbf{S}^R\mathbf{\Lambda}^{-1}\end{aligned} \tag{6.95}$$

In other words, the Q-mode correspondence loadings are equal to the R-mode correspondence scores, divided by the appropriate singular values. Thus, we can obtain a Q-mode solution by solving an R-mode problem, which is ordinarily a great computational advantage, as the R-mode similarity matrix $\mathbf{D}$ is usually much smaller than the Q-mode similarity matrix $\mathbf{Q}$. [In fact, in many commercial software packages, correspondence analysis is performed using a singular value decomposition (SVD) algorithm, whereby the data matrix is decomposed directly into its singular values, vectors, and scores.] Furthermore, we can plot both our samples and our variables in the same space, using the same axes (Gabriel, 1971). This can be done by converting the R-mode loadings and the Q-mode loadings so they both have the same metric. Scaling of the loadings is performed by

$$\hat{A}^R = C^{1/2}A^R$$

and

$$\hat{A}^Q = R^{1/2}A^Q \tag{6.96}$$

We will now use a geological data set to examine the "classical" application of correspondence analysis, which is the interpretation of enumerated data. **Table 6–28** contains counts of the number of conodonts extracted from 10-kg composite samples of rock collected from each stratigraphic unit in a sequence in eastern Kansas. The counts also are given in file CONO.TXT. The rocks are Missourian in age and represent four megacyclothems, or repetitions of lithologies, that may reflect cyclic changes in the depth of seawater. Each unit has been classified as part

Table 6-28. Counts of conodont tests recovered from 10-kg samples of rock. Columns are conodont varieties; rows are stratigraphic units that are members in a section of Missourian age in eastern Kansas. Megacyclothem classifications: paralic shale (O), shoal limestone (S), upper limestone (U), middle limestone (M), "phantom black" shale (P), black shale (B).

N	Class	Rock Unit	a	b	c	d	e	f	g	h	i	j	Total
1	M	South Bend Ls	13	10	0	0	37	0	0	0	0	0	60
2	O	Rock Lake Sh	0	0	0	0	11	0	0	0	0	0	11
3	U	Stoner Ls	4	2	1	51	26	1	0	0	0	0	85
4	B	Eudora Sh	0	7	1	207	350	0	0	34	14	3	606
5	M	Captain Creek Ls	8	28	6	0	60	0	0	0	0	0	102
6	O	Vilas Sh	145	20	5	0	10	0	0	0	0	0	180
7	U	Spring Hill Ls	5	134	8	0	353	1	0	4	0	0	505
8	P	Hickory Creek Sh	20	60	0	0	920	0	0	0	0	0	100
9	M	Merriam Ls	115	255	10	0	1140	0	0	0	0	0	1520
10	M	Bonner Springs Sh	1	0	0	0	3	0	0	0	0	0	4
11	S	Farley Ls	31	21	7	0	4	1	0	0	0	0	61
12	O	Island Creek Sh	100	5	0	0	5	0	0	0	0	0	110
13	U	Argentine Ls	0	39	1	0	80	0	1	0	0	0	121
14	P	Quindaro Sh	10	70	0	0	538	0	0	5	0	0	623
15	M	Frisbee Ls	3	78	5	0	450	0	0	3	0	0	539
16	O	Lane Sh	0	0	0	0	28	0	0	0	0	0	28
17	U	Raytown Ls	38	20	3	100	267	3	0	25	0	0	456
18	B	Muncie Creek Sh	15	8	0	243	515	0	10	85	55	13	946
19	M	Paola Sh	10	130	10	200	900	0	0	50	0	0	1300
20	O	Chanute Sh	117	20	0	63	57	0	0	7	0	0	264

*a – *Adetognathus*; b – *Ozarkodina*; c – *Aethotaxis*;
d – *Idiognathodus delicatus*; e – *I. elegantulus*; f – *Magnilaterella*;
g – *Hindeodella*; h – *Idioprioniodus*; i – *Gondolella*; j – Others.

of an idealized megacyclothem; the classifications are indicated in the table. Paleoecologists have suggested that conodonts were associated with specific depth zones, as are some modern pelagic organisms. If both lithologies and conodont assemblages were responses to changes in sea level, correspondence analysis should reveal patterns in conodont abundances that would be similar to the changes in lithology.

Because there are ten species of conodonts and 20 stratigraphic units, it is most convenient to construct our matrix of similarities between variables. **Table 6-29** gives the matrix of χ^2 similarity measures, its eigenvalues, and eigenvectors 2 and 3; eigenvector 1 is a "nuisance" or trivial factor that merely reflects magnitude of the total frequency in **Table 6-28**. Also given are the R- and Q-mode loadings on correspondence axes 2 and 3. These loadings are used to calculate the R- and Q-mode scores plotted in **Figure 6-38**. The megacyclothem categories for each stratigraphic unit are indicated. Both the samples (rock units) and variables (conodont species) can be plotted in the same space.

Figure 6-39 shows the relative depth ranges of the conodonts collected from the Missourian stratigraphic sequence. Note that the R-mode loadings plotted on

Table 6–29. χ^2 similarity matrix, eigenvalues, second and third R-mode eigenvectors and correspondence loadings, and second and third Q-mode eigenvectors and correspondence loadings for conodont abundance data.

	a	b	c	d	e	f	g	h	i	j
				χ^2 distances between conodont species						
a	0.4542	0.0907	0.0468	0.0494	0.1000	0.0092	0.0019	0.0141	0.0042	0.0020
b	0.0907	0.1622	0.0457	0.0348	0.2719	0.0085	0.0041	0.0200	0.0025	0.0012
c	0.0468	0.0457	0.0280	0.0115	0.0526	0.0075	0.0003	0.0050	0.0004	0.0002
d	0.0494	0.0348	0.0115	0.2439	0.1900	0.0092	0.0264	0.1007	0.0772	0.0370
e	0.1000	0.2719	0.0526	0.1900	0.7092	0.0067	0.0243	0.1020	0.0602	0.0209
f	0.0092	0.0085	0.0075	0.0092	0.0067	0.0052	0.0000	0.0007	0.0000	0.0000
g	0.0019	0.0041	0.0003	0.0264	0.0243	0.0000	0.0104	0.0186	0.0211	0.0104
h	0.0141	0.0200	0.0050	0.1007	0.1020	0.0007	0.0186	0.0557	0.0472	0.0229
i	0.0042	0.0025	0.0004	0.0772	0.0602	0.0000	0.0211	0.0472	0.0510	0.0248
j	0.0020	0.0012	0.0002	0.0370	0.0290	0.0000	0.0104	0.0229	0.0248	0.0121

Eigenvalues

λ^2	1.0000	0.4262	0.2549	0.0461	0.0362	0.0094	0.0042	0.0015	0.0007	0.0000
%	55.0	24.6	14.7	2.7	2.1	0.5	0.2	0.0	0.0	0.0
$\sum$%	55.0	79.6	94.3	97.0	99.1	99.6	99.8	99.9	100.0	100.0

R-mode eigenvectors

2	0.9470	0.0370	0.0758	−0.0957	−0.2743	0.01617	−0.0248	−0.0740	−0.0657	−0.0316
3	0.0757	−0.3357	−0.0648	0.7573	−0.2552	0.0115	0.1095	0.3244	0.3138	0.1512

R-mode correspondence loadings on conodont species a through j

2	0.6182	0.0241	0.0495	−0.0625	−0.1791	0.0106	−0.0162	−0.0483	−0.0429	−0.0206
3	0.0382	−0.1695	−0.0327	0.3824	−0.1288	0.0058	0.0553	0.1638	0.1584	0.0763

Q-mode eigenvectors

2	−0.0726	0.0183	0.0150	0.1375	−0.0270	−0.6265	0.0585	0.1212	−0.0241	−0.0205
	−0.2400	−0.5472	0.0333	0.0926	0.0913	0.0293	−0.0019	0.1557	0.1497	−0.3747
3	0.0503	0.0221	−0.2605	−0.4363	0.1062	−0.0201	0.2328	0.2320	0.3260	0.0070
	0.0519	−0.0431	0.1223	0.1943	0.2005	0.0353	−0.0620	−0.5978	−0.0952	−0.2092

Q-mode correspondence loadings on stratigraphic units 1 through 20

2	−0.0474	0.0120	0.0098	0.0898	−0.0177	−0.4090	0.0382	0.0791	−0.0157	−0.0134
	−0.1567	−0.3572	0.0217	0.0605	0.0596	0.0191	−0.0013	0.1017	0.0977	−0.2446
3	0.0254	0.0112	−0.1315	−0.2203	0.0536	−0.0102	0.1175	0.1171	0.1646	0.0035
	0.0262	−0.0217	0.0618	0.0981	0.1012	0.0178	−0.0313	−0.3018	−0.0481	−0.1056

Figure 6–38, which represent the conodont types, do seem to reflect the water depths at which these organisms lived. Conodonts from the shallowest marine environments appear on the positive end of correspondence axis 2, those from the deepest water appear on the positive end of correspondence axis 3, and those from environments of intermediate depth are found near the origin. By examining the positions of the Q-mode loadings on the same diagram, we can infer the water depths at which the various rock units were deposited. The classification of the various stratigraphic units into the megacyclothem pattern, indicated in **Table 6–28**, follows the use of Heckel and Baesemann (1975). We can note that paralic "outside shales" such as the Vilas, Bonner Springs, and Chanute shales do fall in

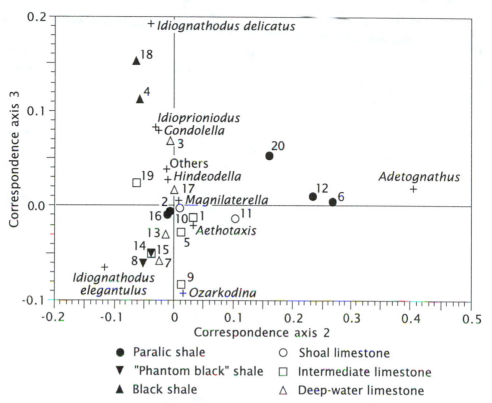

Figure 6–38. Plot of correspondence factor loadings for conodont abundance data given in Table 6–28 and file CONO.TXT. Megacyclothem classifications of stratigraphic units are shown by symbols on the figure.

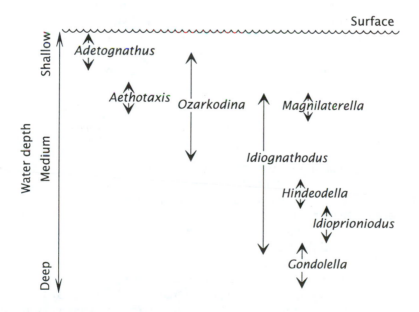

Figure 6–39. Relative depth ranges for conodonts collected from Missourian stratigraphic sequence.

the shallow-water part of the factor diagram, as does the Farley Limestone, which is classified as a shoal-water deposit. Phosphatic black shales such as the Eudora and Muncie Creek shales fall in the deep-water part of the diagram, with such fine-grained "upper limestones" as the Stoner and Raytown. Most rock units, however, plot near the origin, including the so-called "phantom black shales," which seem to be no different than most other rock types.

Correspondence analysis of these conodont data gives some insight into the nature of cyclic sedimentation in this particular sequence. The extremes of the marine environment do seem to yield distinctive lithologies that occur at characteristic positions within megacyclothems. Most lithologies, however, do not fit into a definitive pattern. In particular, the "phantom black shales" that presumably should have deep-water characteristics seem indistinguishable from other rock types that originated at intermediate depths.

Multidimensional Scaling

In geology, as well as in other areas, correspondence analysis has been applied to continuous data rather than to enumerated counts (see Teil, 1975, and David, Dagbert, and Beauchemin, 1977, for applications). This poses some conceptual problems, because the transformed observations cannot be regarded as probabilities, although some authors refer to them in this way (David, Dagbert, and Beauchemin, 1977). Since in general the grand total (as well as the row totals) will consist of a mixture of various types of measurements, the transformation process is strongly dependent upon units of measurement. In spite of this philosophical objection, correspondence analysis is routinely applied to sets of interval and ratio data. In these applications, the transformation process can be regarded as no more than an arbitrary procedure designed to close the data set, and to insure that rows and columns of the data matrix are scaled in equivalent manners when computing either $\mathbf{R}$ or $\mathbf{Q}$. The similarity measure often is referred to as the "profile distance"; it reflects the relative magnitude of the variables rather than their absolute values (Zhou, Chang, and Davis, 1983).

When applied in this manner, correspondence analysis is one variant in a class of procedures called ***multidimensional scaling***, whose objective is to represent multidimensional data as points along one or more scales. Depending upon the nature of the data, the scales may be ordinal, interval, or ratio. Green, Carmone, and Smith (1989) provide a complete guide to the various manifestations of multidimensional scaling. A more compact discussion is given by Jackson (1991).

We can use the artificial data on 25 randomly created blocks to study the behavior of correspondence analysis when applied to measured variables as a form of multidimensional scaling. The original data are listed in file BOXES.TXT; **Table 6–30** is the matrix of similarities or profile distances. The eigenvalues of the matrix are also given, as are the eigenvectors, the R-mode loadings, and the Q-mode loadings on the second and third multidimensional axes. **Figure 6–40** shows the second and third sets of R- and Q-mode scores plotted on the same axes; this diagram expresses 87.4% of the residual variation in the data after the trivial vector is discarded. The two sets of scores are obtained by multiplying the transformed data, $\mathbf{H}$, by the loadings matrix $\mathbf{A}^R$ and by $\mathbf{A}^Q$. From this presentation it is comparatively easy to see not only the similarities between the individual observations, but also the relative contributions of each of the original variables to the correspondence axes.

Table 6–30. Profile distances between seven variables measured on random block data. Also listed are the eigenvalues, second and third R-mode eigenvectors and MDS loadings, and second and third Q-mode eigenvectors and MDS loadings.

	Profile distances between block variables						
	1	2	3	4	5	6	7
1	0.1947	0.1542	0.1155	0.2189	0.1531	0.1743	0.0955
2	0.1542	0.1413	0.1015	0.1818	0.1197	0.1399	0.0690
3	0.1155	0.1015	0.0988	0.1383	0.0667	0.0753	0.0495
4	0.2189	0.1818	0.1383	0.2506	0.1689	0.1933	0.1047
5	0.1531	0.1197	0.0670	0.1689	0.1642	0.1885	0.1082
6	0.1743	0.1399	0.0753	0.1933	0.1885	0.2177	0.1204
7	0.0955	0.0690	0.0495	0.1047	0.1082	0.1204	0.1040

	Eigenvalues						
λ^2	1.0000	0.1214	0.0359	0.0110	0.0029	0.0001	0.0000
%	85.4	10.4	3.1	0.9	0.2	0.0	0.0
$\sum$%	85.4	95.8	98.9	99.8	100.0	100.0	

	R-mode eigenvectors						
2	0.1921	0.2782	0.4958	0.3112	−0.4126	−0.4712	−0.3883
3	−0.0538	−0.2149	0.4084	−0.0445	−0.1459	−0.2966	0.8203

	R-mode MDS loading vectors						
2	0.0669	0.0969	0.1728	0.1084	−0.1437	−0.1642	−0.1353
3	−0.0102	−0.0407	0.0774	−0.0084	−0.0276	−0.0562	0.1554

	Q-mode eigenvectors						
2	0.3475	0.0800	−0.2574	−0.3818	−0.1217	0.0800	0.4253
	−0.1858	0.1773	0.2035	0.1300	−0.2022	−0.0486	0.1149
	0.0741	0.0236	0.0367	−0.1050	0.1465	0.0112	0.1720
	−0.2396	−0.2425	−0.0786	−0.2933			
3	0.0096	−0.1434	0.0877	0.1557	−0.1054	−0.0209	0.5512
	−0.0505	−0.3419	−0.0101	−0.0775	0.0450	0.1313	0.1948
	0.4960	−0.0887	−0.2699	−0.0323	−0.0462	0.0630	−0.2771
	0.0416	0.1788	0.0391	0.0590			

	Q-mode MDS loading vectors						
2	0.1211	0.0279	−0.0897	−0.1330	−0.0424	0.0279	0.1482
	−0.0647	0.0618	0.0709	0.0453	−0.0705	−0.0169	0.0400
	0.0258	0.0082	0.0128	−0.0366	0.0510	0.0039	0.0599
	−0.0835	−0.0845	−0.0274	−0.1022			
3	0.0018	−0.0272	0.0166	0.0295	−0.0200	−0.0040	0.1044
	−0.0096	−0.0648	−0.0019	−0.0147	0.0085	0.0249	0.0369
	0.0939	−0.0168	−0.0511	−0.0061	−0.0088	0.0119	−0.0525
	0.0079	0.0339	0.0074	0.0110			

To demonstrate that multidimensional scaling can be applied to observations of any rank, and not just to interval- or ratio-scale measurements, we can convert the box data to ordinal rank by dividing each variable into discrete categories (small, medium, and large) and counting the number of measurements falling into each category. In effect, each continuous variable is replaced by three discrete variables. The transformed data consist of an array of ones and zeros and are given in file

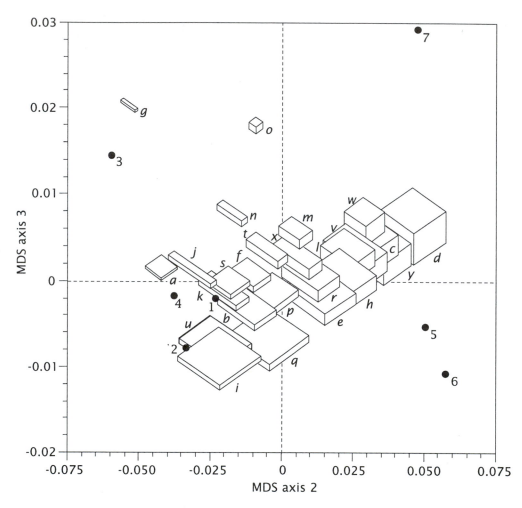

Figure 6–40. Plot of *R*- and *Q*-mode loadings on correspondence axis 2 (horizontal) and correspondence axis 3 (vertical) from block data. Blocks and variables shown plotted at their respective loadings on the two axes.

ORDNALBX.TXT. This discretization decreases the information in the data set, but considering the inexactitude of many geologic observations, the loss may not be important.

The ordinal data in the file are first converted into joint probabilities of occurrence, transformed, and the similarity matrices **R** or **Q** calculated. Note that **Q** will remain 25×25, but **R** will be expanded from 7×7 to 21×21. The successive eigenvalues, given in **Table 6–31**, are smaller and drop off more slowly than do the eigenvalues from the similarity matrix calculated from metric data (**Table 6–30**). Again, the first eigenvector is a trivial vector associated with an eigenvalue of 1 and should be discarded. The second and third eigenvalues account for 44.6% of the remaining trace of the similarity matrix. All eigenvalues after the fifteenth are identically zero. Even though the first few multidimensional axes do not seem to be as efficient as those calculated from **Table 6–30**, a plot of the loadings on the MDS axes shows patterns that are at least as meaningful as those obtained from metric data. **Figure 6–41** shows the *R*- and *Q*-mode loadings for the second and third MDS axes. On the diagram, the low, medium, and high classes for each of the seven variables are distinguished by symbols. Note that the "low" category of most variables lies on the left of the diagram, while the "high" categories plot on the right side. This demonstrates that the second MDS axis is essentially reflecting size. Compare this result with that obtained from metric data (**Fig. 6–40**), keeping in mind that the information content in the ordinal data is much lower.

Table 6–31. Eigenvalues from profile distance matrix based on ordinal data in file ORDNALBX.TXT. Matrix is 21 × 21 and represents three classes (low, medium, and high) of seven original variables.

λ^2	1.0000	0.4588	0.3791	0.3047	0.1889	0.1800	0.0817	0.0681
%	34.5	16.0	13.2	10.6	6.6	6.3	4.1	2.8
$\sum\%$	34.5	50.5	63.7	74.3	80.9	87.2	91.3	94.1

	0.0464	0.0250	0.0125	0.0086	0.0058	0.0003	(6 eigenvalues $=0$)
	2.4	1.6	0.9	0.4	0.3	0.2	0.1
	96.5	98.1	99.0	99.4	99.7	99.9	100.0

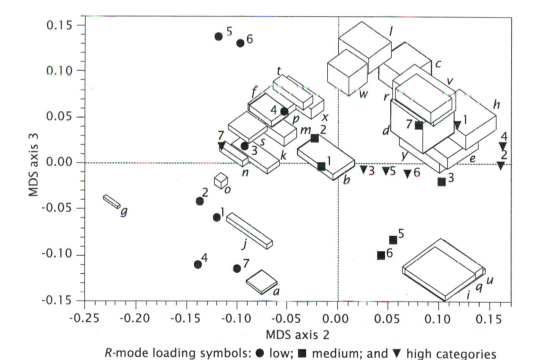

R-mode loading symbols: ● low; ■ medium; and ▼ high categories

Figure 6–41. Plot of *R*- and *Q*-mode loadings on correspondence axes 2 and 3 from block data expressed in ordinal classes (file ORDNALBX.TXT). Horizontal axis is correspondence axis 2; vertical axis is correspondence axis 3. Blocks are plotted at locations corresponding to *Q*-mode loadings.

File COOPERBA.TXT contains 16 properties measured on shales recovered from 19 exploratory holes drilled in Cooper Basin, Australia. The variables are used to assess petroleum source-rock potential and maturity and include an assortment of different types of data. Kerogen composition is based on visually estimated proportions of exinite, vitrinite, and inertinite seen in polished sections. Total organic carbon (TOC) is determined by pyrolysis. Solvent extraction, followed by chromatography, is used to estimate extracted organic material (EOM) which is composed of saturates, aromatics, asphaltics, and resins. Saturates and aromatics together make up the petroleum-like hydrocarbon fraction (HC). Various ratios of these measures (EOM/TOC, HC/TOC, and HC/EOM) are used in different source-rock evaluation schemes. Peaks on chromatograms are measured and the ratios of the peak heights of certain distinctive fractions (pristane/phytane and pristane/n-C_{17} ratios) are used to distinguish marine from nonmarine organic materials. Vitrinite reflectance is measured optically and indicates thermal maturity of kerogen in sediments. The variables differ in information content and include redundancies and closure effects; multidimensional scaling may help in their interpretation.

The 16×16 matrix of profile distances between variables is given in file COOP-ERD.TXT. **Table 6–32** lists the eigenvalues, first three *R*-mode eigenvectors and MDS loading vectors, and first three *Q*-mode eigenvectors and MDS loading vectors. *R*- and *Q*-mode scores have been calculated and MDS axes 2 and 3 are cross plotted in **Figure 6–42**. As expected, the first eigenvalue is equal to one and its associated *R*- and *Q*-mode eigenvectors are trivial and are discarded.

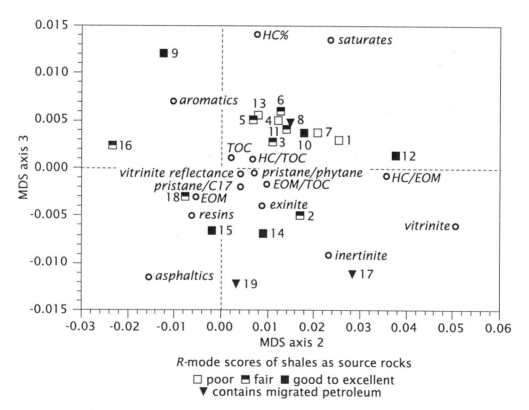

Figure 6–42. Plot of *R*- and *Q*-mode scores on multidimensional scaling axes 2 and 3 on source-rock maturity data from Cooper Basin, Australia. *Q*-mode scores are labeled with variable names.

Table 6–32. Eigenvalues, eigenvectors, and MDS loading vectors for R- and Q-mode multidimensional scaling of hydrocarbon source-rock potential measurements of shales in Cooper Basin, Australia.

Eigenvalues

1.0000	0.0753	0.0272	0.0152	0.0081	0.0032
0.0015	0.0013	0.0008	0.0002	0.0001	0.0001
0.0000	0.0000	0.0000	0.0000		

First three R-mode eigenvectors and MDS loading vectors

	Eigenvector			R-loading	
1	2	3	1	2	3
0.0364	0.1147	0.1238	0.0364	0.0315	0.0204
0.0713	0.6688	0.2199	0.0713	0.1835	0.0362
0.0518	0.3070	0.3347	0.0518	0.0843	0.0552
0.0293	0.0282	−0.0413	0.0293	0.0077	−0.0068
0.6750	−0.0720	0.1088	0.6750	−0.0198	0.0179
0.2228	0.3113	−0.5001	0.2228	0.0854	−0.0824
0.2572	−0.1385	−0.2584	0.2572	−0.0380	−0.0426
0.4291	−0.2075	0.4207	0.4291	−0.0570	0.0693
0.3393	−0.0842	0.1844	0.3393	−0.0231	0.0304
0.3402	0.0991	−0.5228	0.3402	0.0272	−0.0862
0.0232	0.1309	0.0591	0.0232	0.0359	0.0097
0.0122	0.0867	−0.0379	0.0122	0.0238	−0.0063
0.0503	0.4768	0.0209	0.0503	0.1308	0.0034
0.0160	0.0764	0.0135	0.0160	0.0210	0.0022
0.0092	0.0520	0.0597	0.0092	0.0143	0.0098
0.0071	0.0520	0.0194	0.0071	0.0143	0.0032

First three Q-mode eigenvectors and MDS loading vectors

	Eigenvector			Q-loading	
1	2	3	1	2	3
0.0627	0.3384	−0.1091	0.0627	0.0929	−0.0180
0.1093	0.2241	0.1511	0.1093	0.0615	0.0249
0.1061	0.1449	−0.1058	0.1061	0.0398	−0.0174
0.0888	0.1636	−0.1888	0.0888	0.0449	−0.0311
0.1187	0.0911	−0.1873	0.1187	0.0250	−0.0309
0.0801	0.1681	−0.2194	0.0801	0.0461	−0.0362
0.0661	0.2765	−0.1419	0.0661	0.0759	−0.0234
0.0800	0.2000	−0.1841	0.0800	0.0549	−0.0303
0.3185	−0.1632	−0.4465	0.3185	−0.0448	−0.0736
0.0763	0.2363	−0.1395	0.0763	0.0649	−0.0230
0.0918	0.1856	−0.1508	0.0918	0.0509	−0.0248
0.0533	0.5000	−0.0536	0.0533	0.1372	−0.0088
0.1019	0.1075	−0.2113	0.1019	0.0295	−0.0348
0.1789	0.1196	0.2518	0.1789	0.0328	0.0415
0.3577	0.0000	0.2413	0.3577	−0.0070	0.0398
0.6655	−0.3084	−0.0901	0.6655	−0.0846	−0.0148
0.1082	0.3768	0.3997	0.1082	0.1034	0.0659
0.3082	0.0000	0.0694	0.3082	−0.0135	0.0114
0.3087	0.0406	0.4446	0.3087	0.0111	0.0733

From the relationships between the hydrocarbon variables, we can deduce that changes along the second MDS axis reflect in part a terrestrial versus marine origin of organic material. Along the third MDS axis, more petroleum-like organic precursors lie toward the top of the diagram, and nonpetroleum organics lie toward the bottom. In combination, we would expect better source rocks to lie toward the upper left of the diagram and poorer source rocks to lie toward the lower right. We can see if our expectations are met by examining the pattern of symbols assigned to the observations, which have been subjectively classed by the geochemical laboratory as poor (open squares), fair (half-filled squares), or good to excellent (solid squares) source rocks. A complicating factor is that several specimens are believed to contain migrated petroleum (solid triangles).

Simultaneous *R*- and *Q*-Mode Analysis

Although the Eckart–Young theorem states that equivalent solutions can be obtained in either *R*- or *Q*-mode, in practice this may not be quite true. A plot of *R*-mode factor scores looks different than a plot of *Q*-mode factor loadings, as a comparison of **Figures 6–32** and **6–33** will attest. You will recall that an *R*-mode solution is derived from the symmetric minor product matrix $\mathbf{W'W}$, while a *Q*-mode solution is derived from the major product $\mathbf{WW'}$. Unfortunately, the scaling procedures that are used to create $\mathbf{W}$ from the original raw data, $\mathbf{X}$, are not the same in the two modes. For example, principal component analysis involves transformation of each element of $\mathbf{X}$ by dividing by the standard deviation of the columns, producing a scaled data matrix, $\mathbf{W}$. *Q*-mode factor analysis uses a standardization that includes dividing each element of $\mathbf{X}$ by the square root of the sum of the squares of the rows, also producing a scaled data matrix $\mathbf{W}$. However, the matrix $\mathbf{W}$ in principal component analysis is not identical to the matrix $\mathbf{W}$ produced as the first step in *Q*-mode factor analysis. The difference in scaling distorts the solution in one mode with respect to the other mode.

There are several ways around this problem. Obviously, if no scaling is done, the eigenvalues and eigenvectors of $\mathbf{X'X}$ are the same as those from $\mathbf{XX'}$, except that one or the other may have additional zero eigenvalues. *R*-mode scores will be proportional to *Q*-mode loadings, and *vice versa*. In addition, the *R*-mode and *Q*-mode loadings both occur in the space defined by the same eigenvectors, so both can be plotted on the same diagram, as in **Figure 6–43**.

Unfortunately, using the raw cross-product matrix has distinct disadvantages. Since no scaling has been done, the analysis is very sensitive to the choice of measurement units, and results may simply reflect the average magnitudes of the variables rather than their variances and covariances. Although such a method is a mathematically simple way to simultaneously construct combined *R*- and *Q*-mode plots, it is almost never used in practice.

A second solution is to scale $\mathbf{X}$ in a manner that treats rows and columns identically. This is done in correspondence analysis and in some variants of multidimensional scaling, where each element is divided by the product of the square roots of the row and column totals. Because the row and column transformations are symmetrical, *R*- and *Q*-mode scores can be plotted on the same scales.

A third alternative is to seek a way of scaling by rows that produces a meaningful measure of interrelation between the rows in the matrix $\mathbf{W}$, and at the same time results in a meaningful measure of interrelation between the columns. This proves

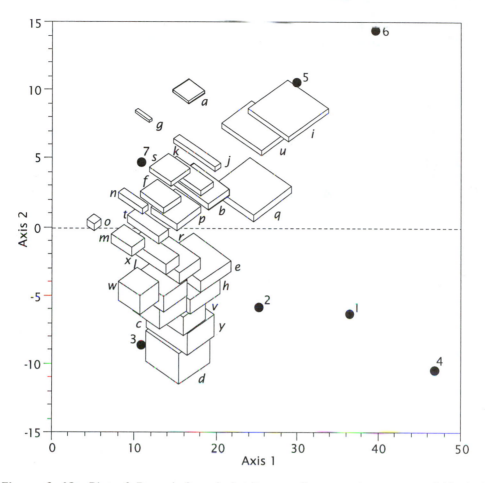

Figure 6–43. Plot of R- and Q-mode loadings on first two eigenvectors of block data, calculated from raw cross-product matrix $\mathbf{X'X}$. Blocks are plotted at locations corresponding to Q-mode loadings. Variables are plotted as points defined by R-mode loadings.

to be easier than might be supposed, and is the basis for at least two practical methods of simultaneous extraction of R- and Q-mode factors.

The elements of $\mathbf{X}$ can be standardized by subtracting the column (variable) means and dividing by the square root of n, the number of observations. That is,

$$w_{ij} = \frac{x_{ij} - \overline{X}_{\cdot j}}{\sqrt{n}} \tag{6.97}$$

Then, the minor product matrix $\mathbf{W'W}$ will contain the variances and covariances of the variables. At the same time, the major product matrix, $\mathbf{WW'}$, is equivalent to the principal coordinates matrix $\mathbf{Q}$ when the similarity between objects is defined by the Euclidean distance. That is,

$$q_{ij} = d_{ij} + \overline{\overline{d}}_{..} - (\overline{d}_{i\cdot} + \overline{d}_{\cdot j}) \tag{6.98}$$

and d_{ij} is an element of the distance matrix, $\mathbf{D}$:

$$d_{ij} = \frac{\sum_{k=1}^{m} (x_{ik} - x_{jk})}{\sqrt{n}} \tag{6.99}$$

Alternatively, we can standardize the elements of **X** by subtracting the column (variable) means and dividing by the product of the column (variable) standard deviations and the square root of n:

$$w_{ik} = \frac{x_{ik} - \overline{x}_{\cdot k}}{s_k \sqrt{n}} \tag{6.100}$$

The minor product matrix **W'W** will now contain the variances and covariances of the variables in standardized form, which of course are the correlations between the variables. Again, the major product matrix **WW'** is equivalent to one version of the principal coordinates matrix **Q**. Now, however, the distance matrix **D** contains Euclidean distances between the observations as defined by standardized variables, or

$$d_{ij} = \frac{\sum_{k=1}^{m} (x_{ik} - x_{jk}) \big/ s_k}{\sqrt{n}} \tag{6.101}$$

To perform simultaneous *R*- and *Q*-mode analysis, we first compute the minor product matrix **W'W** after scaling by either Equation (6.97) or (6.100). We then extract the eigenvalues and eigenvectors and compute *R*-mode loadings by multiplying each element of an eigenvector by the corresponding singular value or square root of its eigenvalue. If the eigenvectors form the columns of a matrix denoted **U**, this is

$$\mathbf{A}^R = \mathbf{U}\boldsymbol{\Lambda} \tag{6.102}$$

where again, $\boldsymbol{\Lambda}$ is a diagonal matrix of the singular values of **W'W**.

Next, we must find the *Q*-mode loadings, which can be found as the product of the scaled data matrix and the matrix of eigenvectors:

$$\mathbf{A}^Q = \mathbf{W}\mathbf{U} \tag{6.103}$$

Of course, we can also compute scores as well. *R*-mode scores are found by multiplying the scaled data matrix by the *R*-mode loadings matrix.

$$\mathbf{S}^R = \mathbf{W}\mathbf{A}^R \tag{6.104}$$

The *Q*-mode scores are found by

$$\begin{aligned}\mathbf{S}^Q &= \mathbf{W}'\mathbf{A}^Q \\ &= \mathbf{W}'\mathbf{W}\mathbf{U}\end{aligned} \tag{6.105}$$

Since the *R*-mode loadings, $\mathbf{A}^R$, give the coordinates of the variables as points in "factor space," and the *Q*-mode loadings, $\mathbf{A}^Q$, give the coordinates of objects in the same space, both sets of loadings can be plotted on the same diagram. Variables that plot close together are very similar. The Eckhart–Young theorem gives the relationship between the variables and the objects. Equation (6.43) can be rewritten as

$$\mathbf{W} = \mathbf{A}^Q \boldsymbol{\Lambda}^{-1/2} \mathbf{A}^{R'} \tag{6.106}$$

An element of **W** is thus equal to

$$w_{ij} = \frac{1}{\sqrt{\lambda_k}} \sum_{k=1}^{m} a_{ik}^Q a_{jk}^R \tag{6.107}$$

Observation w_{ij}, which is the scaled value of variable j observed on object i, can be regarded as the product of an object loading vector, a_i^Q, and a variable loading vector, a_j^R, multiplied by $1/\sqrt{\lambda_k}$. The magnitude of a vector product, you will recall from Chapter 5, is inversely related to the distance between the ends of the two vectors. Thus, the strength of the relationship between an object and a variable in the diagram is directly expressed by the distance between the object point and the variable point.

Although equivalent relationships exist between scores, they are not so neatly expressed in terms of similarities. The best way of displaying the results of simultaneous R- and Q-mode analysis is to plot the two sets of loadings along the same axes. This is done for the random block data in **Figure 6–44**, using the form of standardization given in Equation (6.97). If you compare **Figure 6–44** to **Figure 6–22**, you will note that the arrangement of the boxes differs only by scale constants and by inversion of the second axis. An eigenvector and corresponding loading vector and their inverses are equivalent and simply indicate the opposite sense of direction of the same vectors in space. An eigenvector can easily be inverted by multiplying by −1. From an R-mode perspective, the simultaneous R- and Q-mode procedure is identical to principal component analysis, so the critical matrices are the same as those given in **Table 6–14**.

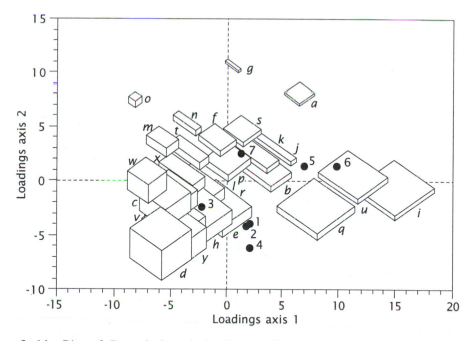

Figure 6–44. Plot of R- and Q-mode loadings on first two eigenvectors of block data, calculated from the variance–covariance matrix **W′W**. Blocks are plotted at locations corresponding to Q-mode loadings. Variables are plotted as points defined by R-mode loadings.

The duality between principal component analysis and principal coordinates analysis using the Euclidean distance was first pointed out by Gower (1966). However, the duality has not been widely exploited in geoscience applications, even though most multivariate analysis programs employ singular value decomposition, which is an implementation of the Eckart–Young theorem. A mathematical derivation of simultaneous *R*- and *Q*-mode factor analysis is given by Zhou, Chang, and Davis (1983), who also provide a number of geological examples of its application.

Table 6–33 and file RADIO.TXT contain measurements made along a profile across a small, highly radioactive quartz monzonite pluton that has intruded a chlorite–actinolite schist near Berea, Virginia. The 22 auger specimens collected along the profile were analyzed for the radioactive elements uranium, thorium, and potassium. Airborne radiometric measurements were also made at the same locations along the profile. The purpose of the study, originally conducted by Sherman, Bunker, and Bush (1971), was to relate the concentrations of the radioactive elements to the radiometric measurements.

The data were analyzed by Zhou, Chang, and Davis (1983), using the scaling given in Equation (6.100). **Table 6–34** contains the resulting 4×4 matrix of correlations, the eigenvalues, and *R*- and *Q*-mode loadings. **Figure 6–45** shows the first two *R*-mode and *Q*-mode loadings plotted in the same space. There is a clear distinction between samples from the pluton and those collected in the schist host rock and alluvial cover.

Table 6–33. Content of uranium, thorium, and potassium and airborne radiometric intensity along a traverse across quartz monzonite intrusive near Berea, Virgina.

No.	AERO[1]	U (ppm)	Th (ppm)	K (%)	Rock type
1	240	0.63	2.05	0.13	Chlorite–actinolite schist
2	360	2.18	5.31	0.31	Chlorite–actinolite schist
3	420	2.26	5.61	0.34	Chlorite–actinolite schist
4	500	1.71	6.44	0.7	Chlorite–actinolite schist
5	580	2.38	7.99	1.73	Quartz monzonite
6	700	3.83	8.32	4.26	Quartz monzonite
7	600	3.79	9.46	1.53	Quartz monzonite
8	650	4.09	14.71	3.11	Quartz monzonite
9	770	4.21	12	1.9	Quartz monzonite
10	930	4.72	12.78	2.92	Quartz monzonite
11	1020	6.24	16.31	2.29	Quartz monzonite
12	1000	5.24	14.51	1.88	Quartz monzonite
13	1000	4.73	15.79	4.64	Quartz monzonite
14	1040	4.67	10.3	4.17	Quartz monzonite
15	1150	5.08	13.11	3.97	Quartz monzonite
16	1000	5.27	13.4	4.36	Quartz monzonite
17	960	5.61	10.31	2.05	Quartz monzonite
18	420	2.33	6.83	0.47	Sand and gravel
19	370	2.64	9.88	0.58	Sand and gravel
20	400	2.29	6.02	0.34	Sand and gravel
21	480	2.32	6.14	0.32	Sand and gravel
22	730	5.94	12.86	1.35	Quartz monzonite

[1]Airborne ratiometric measurements in counts per second.

Table 6–34. Correlation matrix, eigenvalues, and R- and Q-mode loadings of composition and radiometric data from Berea, Virginia.

Correlation matrix

	AERO	U	Th	K
AERO	1.0000	0.8932	0.8241	0.8164
U	0.8932	1.0000	0.8852	0.6683
TH	0.8241	0.8852	1.0000	0.6859
K	0.8164	0.6683	0.6859	1.0000

Eigenvalues

λ	3.3923	0.3906	0.1549	0.0621
%	84.8	9.8	3.9	1.6
$\sum$ %	84.8	94.4	98.4	100.0

R-mode loading vectors 1 and 2

1	0.9611	0.9397	0.9240	0.8554
2	0.0496	−0.2673	−0.2476	0.5054

Q-mode loading vectors 1 and 2

1	−0.7511	−0.4889	−0.4482	−0.4075	−0.2160	0.1117
	−0.0799	0.2126	0.1151	0.3058	0.5078	0.3514
	0.5325	0.3604	0.5000	0.4856	0.2551	−0.4006
	−0.3067	−0.4428	−0.4058	0.2092		

2	0.1209	−0.0151	−0.0193	0.0413	0.09092	0.2940
	−0.0477	0.0027	−0.0758	0.0035	−0.2313	−0.1803
	0.1378	0.2112	0.1088	0.1264	−0.0927	−0.0355
	−0.1117	−0.0314	−0.0331	−0.2631		

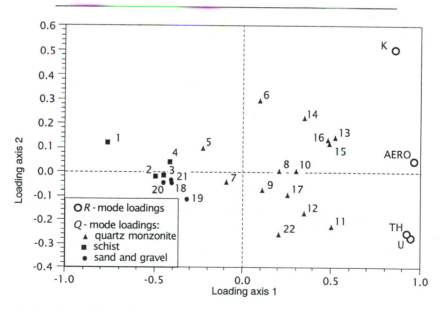

Figure 6–45. Plot of R- and Q-mode loadings on first two eigenvectors of radiometric and element data from Berea, Virginia, contained in file RADIO.TXT. Calculations are based on correlation matrix $\mathbf{W}'\mathbf{W}$ given in Table 6–34. From Zhou, Chang, and Davis (1983).

Multigroup Discriminant Functions

Multigroup discriminant analysis combines a rationale similar to that of analysis of variance with computational procedures based on eigenvector calculations. The problem is an extension of two-group discrimination, already discussed. As an example, suppose a paleontologist interested in sexual dimorphism of gastropods has collected specimens of modern whelks from several different seashore localities. By examining the soft parts of their bodies, the investigator can distinguish males from females. Individuals can then be categorized as males from locality A, females from locality A, males from locality B, and so forth. Multivariate measurements made on the shells alone can be used in a discriminant analysis to find combinations of measurements that allow the various categories of sex and location to be distinguished. Hopefully, the distinctions between males and females will be greater than the distinctions between localities. The project might provide insight into the characteristics of gastropod shells that would permit classification of fossil shells according to their sex.

The analogy with analysis of variance comes from the way in which variances and covariances can be partitioned among categories or groups. You will recall from Chapter 2 that in one-way analysis of variance, the total sum of squares (SS_T) is equal to the sums of squares within the groups (SS_W), plus the sums of squares between the groups (SS_B). Exactly the same structure is invoked in discriminant analysis.

The notation of multigroup discriminant analysis is complicated because we must consider not only objects and variables, but also the groups in which the objects occur. Therefore, we must initially use a notation with three subscripts in which x_{ijk} denotes the jth variable measured on object i which is a member of group k. A complicating factor is that all groups do not necessarily contain the same number of observations, so we must denote the number of observations in the kth group as n_k. We will assume the observations are classed into g distinct groups. If we add all of the groups together, we find that there is a grand total of $N = \sum_{k=1}^{g} n_k$ objects in the entire data collection, all characterized by a set of m variables.

The mean of the jth variable in the kth group is

$$\overline{X}_{\cdot jk} = \frac{\sum_{j=1}^{n_k} x_{ijk}}{n_k} \tag{6.108}$$

The grand mean of the jth variable is the average of all observations of variable j regardless of the group in which the observations are placed. The grand mean is equal to

$$\overline{\overline{X}}_{\cdot j\cdot} = \frac{\sum_{k=1}^{g} \sum_{i=1}^{n_k} x_{ijk}}{N} \tag{6.109}$$

The covariation between variable p and variable q for all observations, without regard to group, is

$$s_{pq} = \sum_{k=1}^{g} \sum_{i=1}^{n_k} \left(x_{ijk} - \overline{\overline{X}}_{\cdot p\cdot} \right) \left(x_{ijk} - \overline{\overline{X}}_{\cdot q\cdot} \right) \tag{6.110}$$

If we compute this measure for all possible pairs of variables, they will form an $m \times m$ symmetric matrix, **S**, that is referred to as the "matrix of total sums of products." We also can compute a measure of covariation between variable p and variable q within the g groups by

$$w_{pq} = \sum_{k=1}^{g} \sum_{i=1}^{n_k} \left(x_{ijk} - \overline{X}_{\cdot pk} \right) \left(x_{ijk} - \overline{X}_{\cdot qk} \right) \tag{6.111}$$

Again, for all possible pairs of variables, this will form an $m \times m$ matrix, W, that is the within-group sum of products. This quantity is equivalent to the sum of the matrices S_A and S_B used in simple two-group discriminant analysis. The final way in which we can express the variation is between groups:

$$b_{pq} = \sum_{k=1}^{g} n_k \left(\overline{X}_{\cdot pk} - \overline{\overline{X}}_{\cdot p\cdot} \right) \left(\overline{X}_{\cdot qk} - \overline{\overline{X}}_{\cdot q\cdot} \right) \tag{6.112}$$

This also forms an $m \times m$ matrix, B, which contains the between-group sums of products.

As in conventional analysis of variance, the combination of the within- and between-groups sums of products matrices are equal to the total sums of products matrix, S

$$S = B + W \tag{6.113}$$

We would like for the ratio B/W to be as large as possible. You will recognize that this ratio is a multivariate analogue of the F-ratio given by $F = MS_B/MS_W$, used to test the distinction between groups in an analysis of variance. If this F-ratio is large, the means of the groups are widely spread, while observations within groups are tightly clustered around their multivariate means. The problem of discriminant analysis is one of finding a set of linear weights for the variables that causes this ratio to be a maximum. If we refer to this set of weights as the vector A_1, discriminant analysis can be expressed as the finding of values for the elements of A_1 that cause the ratio

$$\frac{A_1' B A_1}{A_1' W A_1}$$

to be a maximum. Of course, we must place some constraints on A_1. In discriminant analysis, we usually specify that the denominator of this equation must be equal to one. That is,

$$A_1' W A_1 = 1$$

Under this constraint, the ratio will be a maximum when A_1 is the eigenvector corresponding to the largest eigenvalue of $W^{-1}B$. We can find a second set of linear weights, A_2, which are the elements of the eigenvector corresponding to the second largest eigenvalue. A third set of weights also can be found, as well as a fourth set, and so on. In this manner we can calculate a succession of discriminant functions along which the predefined groups are as distinct as possible. Because of the nature of eigenvectors, each eigenvector is orthogonal to the others, and each is successively the most efficient discriminator possible. A discriminant function can be computed for each positive eigenvalue. In general, the number of positive eigenvalues will be equal to the smaller of either $(g - 1)$ or m. Unfortunately, the matrix created by the operation $W^{-1}B$ is not symmetric, so its eigenvectors are not easily found. Most discriminant function programs compute eigenvectors using singular value decomposition (Jackson, 1991). Older programs first transform the matrix to symmetrical form, and then find a set of eigenvectors that in turn can be transformed into the required discriminant axes. The technique is described by Gnanadesikan (1977); critical steps are outlined by Maron (1982).

Essentially the same methodology is employed in multivariate analysis of variance (MANOVA) and in canonical variate analysis, although the motivations are different. Extensive descriptions are given by Marascuilo and Levin (1983) and Harris (1985); Reyment and Savazzi (1999) provide a short discussion with geological examples.

The observations used in the calculation of the discriminant function can be projected into the space defined by the discriminant axes. This is done by the matrix multiplication

$$\mathbf{S}^D = \mathbf{A}'\mathbf{X} \tag{6.114}$$

where $\mathbf{X}$ is the original $N \times m$ data matrix and $\mathbf{A}$ is an $m \times t$ matrix whose columns consist of the t largest eigenvectors to be used as discriminant functions. $\mathbf{S}^D$ is the matrix of **discriminant scores**. The centroid of each of the g groups can be projected into the discriminant space by

$$\overline{\mathbf{S}}^D = \mathbf{A}'\overline{\mathbf{X}}_k \tag{6.115}$$

where the matrix $\overline{\mathbf{X}}_k$ is $g \times m$ and contains the means of all variables for each group. If we consider the discriminant functions two at a time, we can plot the observations and the group centroids as scatter diagrams. The data usually are scaled in some way prior to plotting. Some programs standardize by subtracting the grand mean from each observation and dividing by the standard deviation calculated over the entire data set. Others form the divisor by pooling the within-group standard deviations. Marascuilo and Levin (1983) provide an instructive comparison of the different approaches.

Obviously, an observation of unknown origin can be projected into the discriminant space simply by premultiplying the data vector by the transpose of $\mathbf{A}$. The group affinity of the new observation may be apparent from its position on the scatter diagram, but it also is possible to compute a measure of its distance to the centroid of each group. The new observation is classified as belonging to the closest group.

To compute the generalized distances from a new observation, x_{ij0}, to each of the g group centroids, we must first determine all the differences, $(x_{ij0} - \overline{X}_{.jk})$, which can be arranged conveniently in a $g \times m$ matrix, $\mathbf{U}$. Then,

$$\mathbf{D}^2 = \mathbf{U}'\mathbf{A}\mathbf{A}'\mathbf{U} \tag{6.116}$$

This will provide the generalized distances from the new observation to each of the g groups, measured in the t-dimensional discriminant space. Alternatively, we can compute

$$\mathbf{D}^2 = \mathbf{U}'\mathbf{W}^{-1}\mathbf{U} \tag{6.117}$$

which yields the generalized distances from the new observation to the centroids of each group measured in the original, m-dimensional space. The implications of these and other alternative definitions of similarity between an observation and the different group centroids are discussed at length by Gnanadesikan (1977), who also provides a method for drawing confidence regions around the group centroids.

Discriminant functions are useful for determining if several groups, presumed to be different, are in fact distinct. We will examine an application of this type.

Saltwater is trapped in sedimentary rocks at the time they are formed in the marine environment. The chemical composition of the connate water is subsequently modified by ion exchange and other reactions, by mixing with other brines, and by dilution by infiltrating surface waters. Nonetheless, brines recovered during drillstem tests of wells may have relict compositional characteristics that provide clues to the origin or depositional environment of their source rocks.

Table 6–35. Chemical analyses of brines (in ppm) recovered from drillstem tests of three carbonate rock units in Texas and Oklahoma. Adapted from (Ostroff (1967).

No.	HCO$_3$	SO$_4$	Cl	Ca	Mg	Na	Group
			Ellenburger Dolomite				
1	10.4	30	967.1	95.9	53.7	857.7	1
2	6.2	29.6	1174.9	111.7	43.9	1054.7	1
3	2.1	11.4	2387.1	348.3	119.3	1932.4	1
4	8.5	22.5	2186.1	339.6	73.6	1803.4	1
5	6.7	32.8	2015.5	287.6	75.1	1691.8	1
6	3.8	18.9	2175.8	340.4	63.8	1793.9	1
7	1.5	16.5	2367	412	95.8	1872.5	1
Mean	5.6	23.1	1896.2	276.5	75	1572.3	–
			Grayburg Dolomite				
8	25.6	0	134.7	12.7	7.1	134.7	2
9	12	104.6	3163.8	95.6	90.1	3093.9	2
10	9	104	1342.6	104.9	160.2	1190.1	2
11	13.7	103.3	2151.6	103.7	70	2054.6	2
12	16.6	92.3	905.1	91.5	50.9	871.4	2
13	14.1	80.1	554.8	118.9	62.3	472.4	2
Mean	15.2	80.7	1375.4	87.9	73.4	1302.9	–
			Viola Limestone				
14	1.3	10.4	3399.5	532.3	235.6	2642.5	3
15	3.6	5.2	974.5	147.5	69	768.1	3
16	0.8	9.8	1430.2	295.7	118.4	1027.1	3
17	1.8	25.6	183.2	35.4	13.5	161.5	3
18	8.8	3.4	289.9	32.8	22.4	225.2	3
19	6.3	16.7	360.9	41.9	24	318.1	3
Mean	3.8	11.9	1106.4	180.9	80.5	857.1	–

Table 6–35 contains brine analyses, reported in equivalent parts per million (ppm), for oil-field waters from three carbonate units in Texas and Oklahoma. The data are also contained in file BRINE.TXT. Each brine sample has been collected from a different oil pool. Discriminant function analysis can be applied to these data to determine if they are distinctive. If they are, this suggests that the brine analyses might provide information about the nature of their original environment, since all three source rocks have approximately the same lithology and have undergone similar histories of burial.

Since there are three groups, only two discriminant functions can be calculated and only two positive eigenvalues will be found. The first of these, $\lambda_1 = 16.79$, accounts for 93.6% of the between-group variance and the second, $\lambda_2 = 1.14$, accounts

Table 6–36. Within- and between-groups covariance matrices for brine data from carbonate rock units in Texas and Oklahoma, with eigenvalues and eigenvectors for discriminant analysis.

	HCO_3	SO_4	Cl	Ca	Mg	Na
			Within-groups covariance matrix			
HCO_3	17.38	−62.64	−2387.04	−291.91	−152.33	−1993.74
SO_4	−62.64	561.82	7626.12	52.04	347.16	7709.21
Cl	−2387.04	7626.19	982107.42	107567.62	44839.24	833965.97
Ca	−291.91	52.04	107567.62	18681.70	6607.70	82178.23
Mg	−152.33	347.16	44839.24	6607.70	3342.17	35176.09
Na	−1993.74	7709.21	833965.97	82178.23	35176.09	721210.79
			Between-groups covariance matrix			
HCO_3	36.00	217.23	−157.92	−443.00	−16.43	539.07
SO_4	217.23	1310.84	−913.98	−2667.46	−99.37	3286.53
Cl	−157.92	−913.98	167033.16	26570.13	−890.65	141025.64
Ca	−443.00	−2667.46	26570.13	9097.59	59.66	14595.19
Mg	−16.43	−99.37	−890.65	59.66	13.07	−1075.95
Na	539.07	3286.53	141025.64	14595.19	−1075.95	131679.36

Eigenvalues: 16.7899   1.1394

Eigenvectors:

	HCO_3	SO_4	Cl	Ca	Mg	Na
1	−0.3765	−0.0468	0.0112	−0.0148	−0.0174	−0.0110
2	−0.0112	0.0248	0.0372	−0.0485	−0.0067	−0.0379

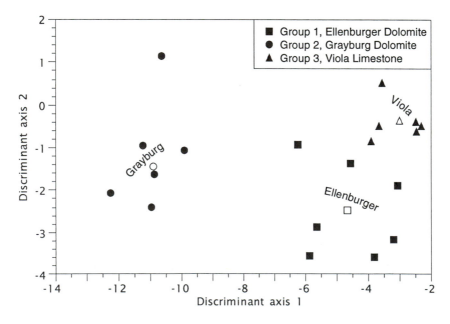

Figure 6–46. Plot of scores on first and second discriminant functions of oil-field brine compositions from Texas and Oklahoma.

for the remaining 6.4%. **Table 6–36** contains the within-groups covariance matrix and the between-groups covariance matrix, as well as the eigenvalues and eigenvectors extracted from the ratio matrix, $\mathbf{W}^{-1}\mathbf{B}$. (Note that we have previously used the matrix of the sums of squares and cross products. The results are identical.) **Figure 6–46** shows the two discriminant function axes, with the discriminant scores and the centroids for the three groups plotted.

The first discriminant function clearly separates brines from the Grayburg Dolomite (group 2) from those collected from the Ellenburger Dolomite (group 1) and Viola Limestone (group 3). Distinctions along the second discriminant function are less clear, with overlap between groups 1 and 2, and between 2 and 3. However, when viewed together, the two functions are adequate to completely separate the three groups. This encouraging result suggests that brines collected from formations having similar lithologies may retain unique, relatively homogeneous characteristics.

Canonical Correlation

We now turn to a multivariate technique that has the same computational basis as principal component and other eigenvector analyses, but which in its concept and objectives is closely related to multiple regression. You will recall that multiple regression is concerned with the relationship between a single dependent variable, Y, and a set of predictor variables, $X_1, X_2, \ldots, X_m$. An extension of this concern is the relationship(s) between a set of Y variables and a second set of X variables measured on the same objects. These relationships may be investigated by finding the linear combination of the X variables that gives the highest correlation with a linear combination of the Y variables.

Such correlations are called **canonical correlations** and the linear combinations are called **canonical variables**. In effect, we convert the set of X's into a single, new variable and the set of Y's into another single, new variable; we then determine the correlation between these two new variables. The conversion process is linear; that is, the original variables are each weighted and added together to yield the canonical variable. Applications of canonical correlation might include determining the relationship between a set of geochemical variables and a set of petrographic variables or the relationship between the petrophysical responses on logs from wells and formation properties measured on core samples from the same wells.

Because all the variables are measured on the same entities, the observations form a data matrix whose dimensions are $n \times (p + q)$, where p represents the number of Y variables and q the number of X variables. (For computational convenience, the smaller of the two sets of variables is called Y, so $p \leq q$.) The matrix of variances and covariances, $\mathbf{S}$, is $(p+q) \times (p+q)$ and can be thought of as composed of four parts: a $p \times p$ matrix, $\mathbf{S}_{yy}$, containing the variances and covariances of the Y variables; a $q \times q$ matrix, $\mathbf{S}_{xx}$, containing variances and covariances of the X variables; and the $p \times q$ matrix $\mathbf{S}_{xy}$ (and its transpose $\mathbf{S}'_{xy}$) which contain the covariances between the X's and Y's. That is,

$$\mathbf{S} = \begin{bmatrix} \mathbf{S}_{yy} & \mathbf{S}_{xy} \\ \mathbf{S}'_{xy} & \mathbf{S}_{xx} \end{bmatrix}$$

Although the matrix S may be thought of as being partitioned, it has the form of any other variance–covariance matrix. It is symmetrical around the diagonal, whose elements are the variances; the off-diagonal elements are the covariances.

We can denote the $n \times p$ part of the data matrix that contains the Y variables as Y and the $n \times q$ part of the data matrix that contains the X variables as X. The Y and X matrices can be transformed by multiplication by arbitrary vectors, which results in new variables that are linear combinations of the old:

$$YA$$
$$XB$$

where A is a $p \times 1$ vector and B is a $q \times 1$ vector. The variances of the two transformed variables will be

$$A'S_{yy}A$$
$$B'S_{xx}B$$

The covariance between the transformed X and Y variables will be

$$A'S_{xy}B$$

The objective of canonical correlation is to select elements of the two vectors A and B so that the covariance is maximized, subject to the constraint that the variances are equal to one. If the variances are initially standardized to equal one, the covariances are simultaneously standardized and become the correlations between the variables. By using eigenvalue techniques, values of the vectors A and B can be found that have the desired properties. We are guaranteed that the canonical correlation will be greater than the largest correlation between any original X variable and any original Y variable—that is, greater than any element in the matrix S_{xy}. This is true because we could immediately create linear combinations that would have a correlation this high by setting all elements of A and B to zero except for those that correspond to the two highest correlated variables, which would be set to one.

The equation that must be solved is very similar to the basic eigenvalue equation that occurs in principal component analysis:

$$|\Lambda - \lambda \cdot I| = 0 \qquad (6.118)$$

Here, Λ is a matrix that results from the multiplication of the various parts of the partitioned variance–covariance matrix, S. The matrix multiplication yields a product matrix, Λ, that is $q \times q$ and represents a pooling of the variances in the two sets of variables. That is,

$$\Lambda = S_{xx}^{-1} S_{xy}' S_{yy}^{-1} S_{xy} \qquad (6.119)$$

The matrix Λ is asymmetric, so we must either resort to heavy-handed methods to find the determinant and solve the equation or, as is more commonly done, use singular value decomposition.

The eigenvalue λ is numerically equal to the square of the correlation between the two canonical variables. Since the matrix Λ is $q \times q$, it will have q distinct eigenvalues, each of which represents the correlation between a different pair of canonical variables. Successive eigenvalues will be of decreasing magnitude, and each pair of canonical variables will be uncorrelated with all other pairs of canonical variables.

The vectors **B**, used to transform **X** into canonical variables, are found by determining the eigenvectors that correspond to the eigenvalues just found:

$$(\Lambda - \lambda \cdot \mathbf{I}) \, \mathbf{B} = 0$$

or

$$\left(\mathbf{S}_{xx}^{-1} \mathbf{S}_{xy}' \, \mathbf{S}_{yy}^{-1} \mathbf{S}_{xy} - \lambda \cdot \mathbf{I} \right) \mathbf{B} = 0 \tag{6.120}$$

Recall that an eigenvector is calculated simply by substituting an eigenvalue into the set of q simultaneous equations, and then finding the solution. Once the transformation vector **B** is found, the equivalent canonical transform for the Y's is given by

$$\mathbf{A} = \lambda^{-1/2} \mathbf{S}_{yy}^{-1} \mathbf{S}_{xy} \mathbf{B} \tag{6.121}$$

Of course, there will be a vector **A** and a vector **B** corresponding to each λ. Each vector pair will transform **X** and **Y** into canonical variates; the correlation between these new variables will be $R = \sqrt{\lambda}$.

We can illustrate canonical correlation by turning once again to our artificial data on random-sized blocks. The data fall naturally into two classes because variables X_1, X_2, and X_3 are the fundamental dimensions of the blocks, while variables X_4 through X_7 are derived from these. We can therefore define the first three variables as forming the set of Y variables, and the next four variables as forming the set of X variables.

The standardized variance–covariance (or correlation) matrix of the block data, partitioned for canonical correlation, is given in **Table 6–37**. The matrix Λ from which the eigenvalues must be extracted is also given in **Table 6–37**, as are its eigenvalues, the canonical correlations which they represent, and the **A** and **B** vectors corresponding to the largest canonical correlation. We can use the vector **A** to transform the Y variables into scores or canonical variates, and the **B** vector to transform the X variables into another set of scores or canonical variates. Because the canonical loadings are based on a standardized variance–covariance matrix, the canonical scores should be calculated using standardized variables. (If all variables are expressed in the same units of measurement, a canonical correlation can be based on the raw variance–covariance matrix and canonical scores calculated from the raw variables.) **Figure 6–47** shows a cross plot of the two sets of canonical variates from the block data. It is obvious that there is, indeed, a perfect linear relationship between the two sets when placed in canonical form. The relationship is primarily an expression of size. **Figure 6–48** is a similar cross plot of the second pair of canonical variates. This diagram also exhibits a very strong correlation even though the transformation is quite different, as it essentially orders the blocks by shape.

Comments made in the section on principal components about "reading" component loadings pertain equally to canonical transformations. The vectors **A** and **B** are weights used to transform the original variables **Y** and **X** into canonical variables. Under certain circumstances it may be possible to ascribe a physical meaning to a particular combination of weights; canonical variables can then be discussed in the same manner as true factors. However, there is no assurance that the pattern of weights will have any interpretable meaning, and the canonical correlations may simply reflect arbitrary mathematical combinations of the variables in the two sets. This is especially apt to be true if the canonical correlations between **X** and **Y** are weak. Marascuilo and Levin (1983), who have an exceptionally lucid chapter

Table 6–37. Partitioned matrix of standardized variances and covariances (correlations) for block data, with eigenvalues and corresponding canonical correlations, and vectors **A** and **B** for transforming original variables into canonical variables; **A** converts variables X_1, X_2, and X_3 to the left-hand canonical variable, and **B** converts variables X_4 through X_7 to the right-hand canonical variable.

Covariance matrix

$$
\begin{bmatrix}
1.0000 & 0.5803 & 0.2011 & 0.9113 & 0.2833 & 0.2865 & -0.5332 \\
0.5803 & 1.0000 & 0.3638 & 0.8337 & 0.1658 & 0.2611 & -0.6087 \\
0.2011 & 0.3638 & 1.0000 & 0.4386 & -0.7042 & -0.6805 & -0.6488 \\
0.9113 & 0.8337 & 0.4386 & 1.0000 & 0.1630 & 0.2023 & -0.6755 \\
0.2833 & 0.1658 & -0.7042 & 0.1630 & 1.0000 & 0.9902 & 0.4272 \\
0.2865 & 0.2611 & -0.6805 & 0.2023 & 0.9902 & 1.0000 & 0.3571 \\
-0.5332 & -0.6087 & -0.6488 & -0.6755 & 0.4272 & 0.3571 & 1.0000
\end{bmatrix}
$$

Pooled variance matrix Λ

$$
\begin{bmatrix}
0.9936 & -0.2173 & -0.1558 & -0.8650 \\
-0.2974 & 0.4103 & -0.3960 & 1.0900 \\
0.2688 & 0.4958 & 1.2826 & -0.6503 \\
0.0085 & -0.2732 & 0.2062 & -0.1851
\end{bmatrix}
$$

Eigenvalues and canonical correlations

λ	0.9991	0.8484	0.6538	0.0000
R	0.9996	0.9211	0.8086	0.0000

Canonical vectors **A** and **B**

First canonical correlation		Second canonical correlation		Third canonical correlation	
$\begin{bmatrix} 0.3460 \\ 0.2948 \\ 0.1363 \end{bmatrix}$	$\begin{bmatrix} 0.6462 \\ -0.5726 \\ 0.5021 \\ 0.0494 \end{bmatrix}$	$\begin{bmatrix} -0.0594 \\ 0.1489 \\ -0.1560 \end{bmatrix}$	$\begin{bmatrix} -0.0415 \\ -0.6350 \\ 0.7713 \\ 0.0136 \end{bmatrix}$	$\begin{bmatrix} 0.0956 \\ -0.0723 \\ -0.0403 \end{bmatrix}$	$\begin{bmatrix} -0.0321 \\ 0.7467 \\ -0.6594 \\ -0.0815 \end{bmatrix}$

on canonical correlation, discuss alternative methods for interpreting canonical weights.

There are two commonly used statistical tests for canonical correlations. One checks for the presence of any significant canonical correlation among the q canonical relationships. This test will be significant if one or more of the pairs of canonical variables are correlated. The second test procedure checks only the significance of the largest canonical correlation.

The ***test of overall significance*** was defined by Bartlett (1938) as

$$L = (1 - \lambda_1)(1 - \lambda_2)\ldots(1 - \lambda_q) \tag{6.122}$$

The quantity

$$\chi^2 = \left(\frac{p + q + 1}{2} - n + 1\right) \ln L \tag{6.123}$$

is distributed as χ^2 with pq degrees of freedom. The null hypothesis is that all of the canonical correlations are equal to zero. If the test statistic falls in the critical region, at least one of the canonical correlations is significantly greater than zero.

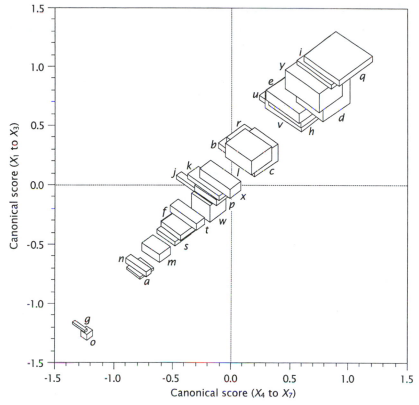

Figure 6–47. Cross plot of scores on the first pair of canonical variates of the random block data. Canonical scores of variables $X_1 - X_3$ are measured along horizontal axis and canonical scores of variables $X_4 - X_7$ are measured along vertical axis. Canonical correlation between first pair of canonical variables is $R = 1.00$.

A test of only the largest canonical correlation can be derived from Bartlett's test as

$$\chi^2 = \left(\frac{p + q + 1}{2} - n + 1\right) \ln\ (1 - \lambda_1) \tag{6.124}$$

where λ_1 is the largest eigenvalue. The test has approximately the degrees of freedom given by

$$df \approx p + q + 1 + \frac{[(p + 1)(q + 1)]^{2/3}}{2} \tag{6.125}$$

The number of degrees of freedom calculated by Equation (6.124) should be rounded down to the nearest whole number.

As a geological example, we will consider the data given in **Table 6–38** and in file LOGCORE.TXT which list well-log measurements made in a series of limestones of Pennsylvanian age encountered during the drilling of a well in northwestern Kansas. Logging tools measure gamma-ray intensity, sonic transmissivity, and electrical resistance of the interval of rock spanned by the tool. These measured properties reflect both the characteristics of the rock and those of the fluids in the pore spaces. The table also contains laboratory measurements made on cores taken from the same interval. These include permeability, porosity, and oil and water saturations.

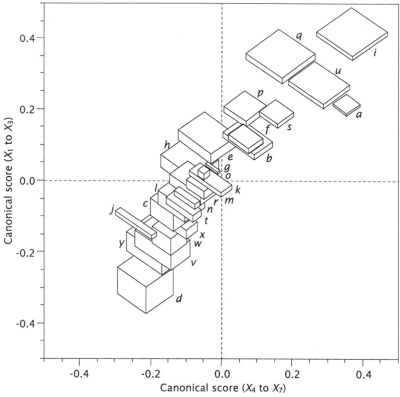

Figure 6–48. Cross plot of scores on the second pair of canonical variates of random block data. Canonical scores of variables $X_1 - X_3$ are measured along horizontal axis and canonical scores of variables $X_1 - X_7$ are measured along vertical axis. Canonical correlation between second pair of canonical variables is $R = 0.92$.

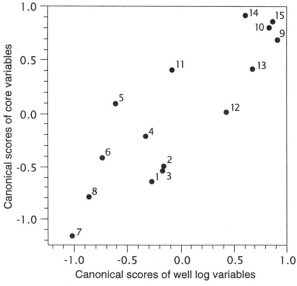

Figure 6–49. Cross plot of scores on canonical variates calculated from well-log responses (horizontal axis) against scores on canonical variates calculated from core measurements (vertical axis) from an exploratory well in northwestern Kansas. Canonical correlation between first pair of canonical variates is $R = 0.86$.

Table 6–38. Log and core measurements made on Pennsylvanian limestones encountered in well drilled in northwestern Kansas. y = gamma-ray intensity in gamma-ray units. Δt = sonic transit time in μsec/ft. R_t = microlaterolog resistivity in ohm-meters. K = permeability in millidarcies. ϕ = porosity in percent. S_o = oil saturation in percent. S_w = water saturation in percent.

No.		Δt	R_t	K	ϕ	S_o	S_w
1	3.1	64.0	28.8	0.1	3.9	28.2	53.8
2	3.4	69.0	25.1	0.4	7.0	17.2	55.6
3	3.4	65.0	38.0	0.1	6.1	24.6	54.2
4	2.8	62.0	15.1	0.4	6.2	19.3	63.0
5	2.5	56.0	58.9	0.1	5.9	15.3	73.0
6	2.3	56.0	61.7	0.3	4.7	14.9	61.6
7	2.3	54.0	129.0	0.2	6.2	29.0	37.1
8	2.6	60.0	110.0	1.6	12.7	26.7	34.6
9	6.0	97.0	5.2	0.0	3.0	0.0	96.6
10	5.2	67.0	18.2	18.0	18.9	26.4	32.3
11	3.9	82.0	26.9	6.5	18.4	19.0	48.3
12	4.7	80.0	12.9	2.5	17.9	16.8	48.0
13	5.1	77.0	12.0	0.1	12.3	11.7	72.6
14	5.0	79.0	11.0	0.0	10.4	0.0	91.4
15	6.1	81.0	70.8	0.0	5.2	0.0	97.8

In well-log analysis, logging tool measurements are transformed to yield estimates of porosity, oil saturation, and water saturation. These estimates are based on various combinations of log responses, calibrated against known standards. This suggests that we should find a significant canonical correlation between logging tool responses and core measurements.

Table 6–39 gives the standardized variance–covariance matrix of the seven variables in **Table 6–38**. The partitioning of the matrix into the submatrices S_{yy}, S_{xx}, and S_{xy} and the transpose of the matrix are indicated. The matrix Λ that results from the pooling of the variances and covariances in the two sets according to Equation (6.116) is also shown, as are the eigenvalues of Λ and the corresponding canonical correlations. The canonical scores produced by multiplying the standardized original variables by the canonical loading vectors **A** and **B** are shown as a cross plot in **Figure 6–49**. Because we usually are interested in estimating characteristics that would be observed in cores from well-log responses, the canonical scores from vector **A** have been placed on the X-axis. The correlation between the two canonical variables is 0.86, as expected.

The log intervals and cores are identified on the plot; an examination of the original data shows that log intervals having high canonical scores yielded cores that had a high fluid content, either because of high porosity, high water saturation, or high oil saturation. Intervals that scored low were associated with tight cores or those with low fluid saturations.

Table 6–39. Partitioned matrix of standardized variances and covariances (correlations) for well data in Table 6–38. Pooled variance matrix Λ is shown, along with eigenvalues of Λ and corresponding canonical correlations, transformation vector **A** that converts log measurements to canonical form, and transformation vector **B** that converts core measurements to canonical form.

Correlation matrix

Δt	1.0000	0.8675	−0.5236	0.2457	0.2417	−0.6728	0.5469
R_t	0.8675	1.0000	−0.5922	0.0273	0.1959	−0.6953	0.5624
	−0.5236	−0.5922	1.0000	−0.1906	−0.2098	0.3705	−0.3373
K	0.2457	0.0273	−0.1906	1.0000	0.6993	0.3190	−0.4865
ϕ	0.2417	0.1959	−0.2098	0.6993	1.0000	0.2160	−0.4878
S_o	−0.6728	−0.6953	0.3705	0.3190	0.2160	1.0000	−0.9285
S_w	0.5469	0.5624	−0.3373	−0.4865	−0.4878	−0.9285	1.0000

Pooled variance matrix Λ

$$\begin{bmatrix} 0.4017 & 0.1495 & -0.1248 & 0.1170 \\ 0.0611 & 0.1257 & -0.2832 & 0.2487 \\ 0.4458 & 0.1293 & 0.3188 & -0.1907 \\ 0.7079 & 0.3817 & -0.3103 & 0.3315 \end{bmatrix}$$

Eigenvalues and canonical correlations

λ	0.7521	0.3543	0.0713	0.0000
R	0.8672	0.5953	0.2670	0.0000

Canonical vectors **A** and **B**

First canonical correlation	Second canonical correlation	Third canonical correlation
$\begin{bmatrix} 0.6970 \\ -0.1610 \\ -0.1775 \end{bmatrix} \begin{bmatrix} 0.3838 \\ 0.3254 \\ 0.1151 \\ 0.8565 \end{bmatrix}$	$\begin{bmatrix} -0.4250 \\ 0.6805 \\ 0.2172 \end{bmatrix} \begin{bmatrix} -0.3442 \\ 0.0471 \\ -0.6796 \\ -0.6461 \end{bmatrix}$	$\begin{bmatrix} -0.2196 \\ 0.1010 \\ -0.1953 \end{bmatrix} \begin{bmatrix} -0.1549 \\ 0.3342 \\ 0.6306 \\ 0.6831 \end{bmatrix}$

EXERCISES

Exercise 6.1

Table 6–40 contains seven analyses for four elements in oceanic basalts from the Pacific Ocean; these are the same four elements analyzed on lunar basalts and listed in **Table 6–6** (p. 483). We will test the hypothesis that the mean vectors of lunar samples and Pacific samples are the same, assuming both samples are randomly drawn from multivariate normal populations with the same variance-covariance matrix. Is the null hypothesis of equality of multivariate means accepted at the 5% ($\alpha = 0.05$) significance level?

Table 6–40. Abundance of four elements in seven samples of basalts from the Pacific region. Quantities in percent.

	Si	Al	Fe	Mg
1	22.5	9.6	6.6	3.4
2	22.1	8.4	7.8	3.6
3	25.9	8.7	4.8	4.0
4	23.5	8.1	5.0	5.2
5	21.7	10.0	8.2	4.9
6	21.9	8.2	9.3	4.9
7	23.7	7.2	9.5	3.3

Exercise 6.2

The lunar basalts given in **Table 6–6** and the Pacific Ocean basalts in **Table 6–40** are both examples of compositional data, as are all chemical analyses of rocks. The elements listed account for slightly less than half the total constituents of each individual specimen, but are nonetheless subject to the closure constraint imposed by the constant sum condition. We can determine if closure has adversely affected our statistical tests by applying one of Aitchison's (1986) logratio transformations to the data and repeating the tests of equality of means. A simple form of logratio transform appropriate in this instance is

$$\hat{x}_{ij} = \log \frac{x_{ij}}{\left(1 - \sum_j^m x_{ij}\right)}$$

where x_{ij} is the percent of element j in basalt specimen i. We have simply set the divisor in the logratio to be "everything else" in the rock except the four measured elements. Compare the mean composition of the lunar basalts to Clark's terrestrial average, and to the mean of the sample of Pacific Ocean basalts after logratio transformation. Are your conclusions altered? It also is instructive to compare the correlations between elements in each set of data, both before and after logratio transformation. Discuss the differences you see.

Note that the logratio transformation and the test procedure described above are appropriate if we are interested in the differences between two sets of analyses when the constituents are considered as parts of a whole rock. A different approach would be appropriate if we were interested only in distinguishing the subcompositional components. Consider two analyses in which the ratios between elements are the same but values of one analysis are twice as large as values in the second analysis. Using the procedure outlined above, the denominator of the logratio will be much smaller for the rock whose analytical values are large, and the difference between the two specimens would be emphasized. In contrast, if we are interested only in differences within the subcomposition, we would recalculate the element values so each subcomposition sums to 100%, then perform either a centered logratio transformation or select one of the constituents to be the denominator. Under this procedure, we would determine that the two specimens are identical in their subcomposition.

Repeat the analysis under the assumption that we are interested in the mean subcompositional difference between the two samples of basalt. Do you reach the same conclusion as you did in the previous test?

Exercise 6.3

Using the approximate χ^2 test for the equivalency of covariance matrices (Eqs. 6.35–6.38), determine the equivalency of the variance-covariance matrices of the two basalt samples given in **Table 6–6** and **Table 6–40**. To compute the test statistic, you must find the determinants of the three 4×4 matrices, S_1, S_2, and S_p. Once the necessary determinants have been found, the test statistic can easily be calculated. For our purposes, we will assume that the sample sizes of the two data collections are sufficiently large so that the χ^2 approximation is accurate.

Exercise 6.4

Table 6–41 contains measurements made on a collection of Cambrian trilobites from the Great Basin region of the western United States; the data also are contained in file TRILOBIT.TXT. The specimens have been assigned to three genera by

Table 6–41. Ten variables measured as ratios on ten species of Cambrian trilobites collected in Utah.

Species	x_1	x_2	x_3	x_4	x_5	x_6	x_7	x_8	x_9	x_{10}
Aphelaspis brachyphasis	0.208	0.250	0.542	0.237	0.875	0.292	0.284	0.925	0.343	0.373
A. haguei	0.318	0.318	0.545	0.428	1.000	0.318	0.296	0.796	0.444	0.537
A. subditus	0.174	0.304	0.391	0.375	0.913	0.304	0.297	0.946	0.405	0.486
Dicanthopyge convergens	0.259	0.370	0.370	0.859	0.852	0.333	0.500	0.591	0.591	0.818
D. quadrata	0.250	0.350	0.500	0.615	0.900	0.351	0.434	0.783	0.478	0.652
D. reductus	0.316	0.421	0.474	0.736	1.158	0.421	0.500	0.675	0.500	0.775
Prehousia alata	0.136	0.409	0.273	0.469	1.000	0.136	0.269	0.769	0.327	0.423
P. indenta	0.192	0.308	0.269	0.628	0.923	0.154	0.308	0.795	0.308	0.436
P. prima	0.261	0.261	0.261	0.545	0.956	0.261	0.296	0.833	0.333	0.407
A. longispina	0.259	0.370	0.556	0.444	0.852	0.296	0.372	0.824	0.431	0.706

Key: x_1 = length of border/C
x_2 = length of brim/C
x_3 = length of palpebral lobes/C
x_4 = width of glabella/C
x_5 = width of fixed cheek/C
x_6 = length of genal spine/length of free cheek
　　　where C = correction for size of glabella = length of glabella.
x_7 = width of pygidial axis/D
x_8 = width of pleural axes/D
x_9 = length of pygidial axis/D
x_{10} = length of pygidium/D
　　　where D = correction for size of pygidium = width of pygidium.

conventional taxonomic procedures. Ten characters or variables have been measured on ten trilobites, each of which represents a separate species. Commonly, trilobites are found disaggregated. To avoid the confusion that would result if a pygidium from a large individual were accidentally associated with the cephelon of a small individual, all of the measurements have been converted to ratios. All measurements made on the glabella, or head, have been divided by the length of the glabella. Similarly, all measurements made on the pygidium, or tail, are divided by the width of the pygidium. The anatomical parts used in the variables are illustrated in **Figure 6–50**. Perform a cluster analysis on the trilobite data, after appropriate standardization, and see if numerical methods recover the same classification as that found by conventional taxonomy. Compute and utilize both correlation and distance as similarity measures. Which performs "best" in the sense of conforming to conventional taxonomy?

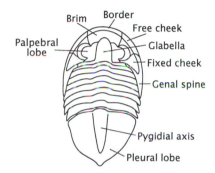

Figure 6–50. An opisthoparian trilobite, showing anatomical parts measured to create variables listed in Table 6–41.

Exercise 6.5

In Chapter 5 we discussed how shapes could be defined by the Cartesian coordinates of landmark points, k, placed around the outline of a form. Applying statistical tests directly to sets of such measurements is complicated, in part because each landmark consists of two (or three) coordinate values, X and Y (or X, Y, and Z for a three-dimensional object). A traditional approach to this problem has been to convert the landmark coordinate pairs into new, single-valued variables that are the distances between two landmarks. More recent developments consist of treating each coordinate pair as a single complex number, then projecting the complex coordinates as points in a k-dimensional "shape space."

We will consider a simpler, approximate procedure that uses Bookstein coordinates directly, and leave the more exact methods of morphometrics to specialized discussions such as Dryden and Mardia (1998). File ORTHID.TXT contains measurements of landmarks on shells of the Cambrian brachiopod *Resserella* sp., taken at the same positions as measurements on specimens of a smaller species of the same genera given in file RESSEREL.TXT and used in Chapter 5. There are measurements of ten individuals in each file.

The first step is to convert the Cartesian coordinates of the landmarks into Bookstein coordinates. Next we must define a mean shape for each sample; this is simply the arithmetic average of the Bookstein coordinates of the individual objects, each of which is a $(2k-4)$ vector. The mean shapes also are in the form of $(2k-4)$

vectors. We can define within-sample variances of the Bookstein coordinates, which form square, symmetric, $(2k-4) \times (2k-4)$ matrices of variances and covariances. Note that each value of u_i and v_i has been treated as though it were an independent variable. This is, of course, not true but holds approximately if the baseline for the Bookstein coordinates has been well-chosen and the variation between individual objects is "small"—meaning that the standard deviations of all landmarks (except landmarks 1 and 2) are less than about one-tenth the baseline distance after Bookstein transformation.

The statistical question is a classical one. Are the mean shapes of the two groups the same, considering the variation in individual shapes within the two groups? Solving it requires a multivariate test of the equivalence of two samples having n_1 objects in the first sample and n_2 objects in the second. The test statistic consists of Mahalanobis' squared distance between the means of the two samples

$$D^2 = (\overline{V} - \overline{W})' S_p^{-1} (\overline{V} - \overline{W})$$

where the pooled covariance matrix, S_p, is found in the usual manner by

$$S_p = (n_1 + n_2 - 2)^{-1} (n_1 S_v + n_2 S_w)$$

To test the hypothesis that the two mean vectors are the same, or equivalently, that the distance between the means is zero, we compute the test statistic

$$F = \frac{n_1 n_2 (n_1 + n_2 - (2k-4) - 1)}{(n_1 + n_2)(n_1 + n_2 - 2)(2k - 4)} D^2$$

which is distributed as F with a first degree of freedom equal to $2k - 4$ and a second degree of freedom equal to $n_1 + n_2 - (2k-4) - 1$. Transform the landmark coordinates for the brachiopod shapes in files RESSEREL.TXT and ORTHID.TXT to Bookstein coordinate form, determine their centroids and covariance matrices, and compute and test the significance of the Mahalanobis' distance between the two groups.

Exercise 6.6

The U.S. government has exploded a large number of nuclear devices underground at the Nevada Test Site, many at or below the water table. The fate of trace elements, including radionuclides, that may have been created in these tests and incorporated into the groundwater is poorly understood. In part, this is because there are few wells or natural springs whose waters can be sampled and analyzed. Farnham and others (2000) provide a table of 51 analyses for 19 trace elements in water collected from 22 wells and springs in the Nevada Test Site and adjacent Oasis Valley; these data are given in file OASISVAL.TXT. The file also includes the locations of the wells and springs in arbitrary Cartesian coordinates from an origin in the southwest corner of the study area.

We can determine if there are similarities among the water samples by performing a principal component analysis and examining a scatter plot of the scores of the first two components; waters having similar trace-element compositions will plot as closely spaced clusters of points. A similar result should be obtained by cluster analysis of the original data. By making a cumulative plot of the scores on the first principal component axis, we can also examine the question of whether there are natural groupings in the data. Such a plot may show "breaks" or abrupt increases in scores that represent boundaries between the groups. If we detect groups, we can determine if they are spatially contiguous by simply plotting each observation as a distinctive symbol representing its group membership. Finally, we can use techniques discussed in Chapter 5 to display the variation in trace-element concentrations over the study area by trend-surface analysis of principal component scores. Comment on the possible effects of the geographic distribution of wells and springs on your results.

Exercise 6.7

Krumbein and Aberdeen (1937) distinguished five classes of sediments in Barataria Bay, Louisiana, based on grain-size distributions. It may be interesting to examine the relative effectiveness of mean grain size, the first principal component of the grain-size distributions, and the percentage of sand in each sediment sample to distinguish between these five sediment types. The raw data and Krumbein and Aberdeen's (1937) classification are given in file BARATARA.TXT. We can test the relative efficacy of the three alternative measures by running a series of one-way analyses of variance using the five sediment types as groups. The ratio between the sums of squares among groups to total sum of squares is a measure of how tightly groups are clustered and separated from other groups. The variable which produces the highest ratio SS_A/SS_T is the most effective for characterizing the sediment types. Run the appropriate ANOVA's as outlined in Chapter 2, and determine which of the three variables is most effective.

Exercise 6.8

A study that may shed light on the utility of the multiplicity of statistics and quasi-statistics used to characterize sediments is to compute a variety of these measures and enter them as variables in a principal component analysis. The analysis may pick out interpretable combinations of statistical measures that are effective for characterizing sediments. Computational equations for numerous grain-size statistics are given in various reference books on sedimentary petrology such as Folk (1980). These measures can be computed for the raw grain-size distribution data given in file BARATARA.TXT. A principal component analysis of these new variables may be quite instructive. A similar study was done by Griffiths and Ondrick (1969, p. 86–88).

Exercise 6.9

Bahia de Guasimas is a coastal lagoon in the Sonora part of the Gulf of California. It is separated from the saltwater of the Gulf by a sand bar approximately 7 km in length. The lagoon has an area of 38 km², and receives almost no freshwater. It represents a relatively pristine lagoonal environment in this area of Mexico. A bottom core was taken in the center of the lagoon and the recovered sediments were analyzed for heavy metals by *CICTUS*, a research center at the University of Sonora, in Hermosillo, Mexico. The data are listed in file SONORA.TXT. We want to produce a multivariate characterization of the sediment composition, at least in terms of heavy metals, to serve as a baseline for comparison to other lagoons. Two other nearby lagoons are known to be receiving relatively large quantities of pollution, one in the form of agricultural runoff and the other in the form of sewage and industrial effluent; these were also sampled and are listed in Murillo (1991). Principal component analysis should be useful in condensing the multivariate data into more manageable dimensions. We also are interested to see if there are any trends in heavy-metal concentrations in the sediment with depth. This can be investigated by regression, plotting PCA scores as a function of depth.

Exercise 6.10

In Chapter 4 we compared the autocorrelation function and semivariogram of Fe% in the output stream from the beneficiation plant at the Eisenerz mine in Austria. In Exercise 4.15, we calculated spectra based on the hourly values of six compositional variables (Fe%, CaO%, SiO_2%, MgO%, MnO%, and Al_2O_3%) collected for 21 weeks. These data are given in file EISENERZ.TXT and are described in Exercise 4.15. In addition to periodicities and other features that may exist in the records of individual variables, more subtle expressions of changes in composition with time may be expressed in the multivariate record.

A simple way to examine a multivariate time series composed of several variables measured at common times is to compute the principal components of the variables and perform spectral analyses of the component scores. The correlation between principal component scores and the original variables will indicate the relative influence of each variable on each of the components, and the relative magnitudes of the successive eigenvalues will indicate how much of the original total variance is expressed in each set of scores. The spectral characteristics of the time records of the component scores can be determined by Fourier analysis, treating each set of scores as a univariate variable. Examine the multivariate time series in EISENERZ.TXT and determine the spectra of the principal component scores. Are there indications of trends in the scores? If so, the scores should be leveled and spectra computed from the residuals. Are there indications of periodicities in the scores? Can you suggest possible causes for any such periodicities by considering the correlations between the components and the original variables?

Exercise 6.11

A government survey group in northern Sweden is prospecting for heavy-metal deposits in densely forested mountains. Airborne magnetometer surveys have proved to be of limited value, so a geochemical prospecting approach based on stream-sediment analyses is being evaluated. Thirteen geochemical variables have been selected and two suites of measurements performed; the data are reported in parts per million (ppm). Group 1 consists of measurements of material from streams draining areas containing active mines or proven mineral deposits. Group 2 consists of similar measurements from streams draining areas that have been heavily prospected without results. Data are listed in file SWEDEN.TXT. From these, calculate the discriminant function between productive and nonproductive regions. Determine if the difference between the two groups is significant, and investigate the relative importance of the variables utilized. For the purposes of this exercise, we will assume that the parent populations of the two groups are multivariate normal. (Does this assumption seem realistic?) File SWEDEN.TXT also contains a set of measurements made on sediments from streams draining areas not known to have been prospected (Group 3). On the basis of the discriminant function, can any of these areas be selected as likely areas for prospecting?

Exercise 6.12

The data in file SWEDEN.TXT are compositional data; that is, the variables are parts of a whole and if all chemical elements had been measured, their sum would be a constant. Although the cations that actually were analyzed account for only 2% or less of the total composition, this subcomposition is subject to the same constant-sum constraint that affects all compositional data sets. However, the constraint is much less obvious when a subcomposition is a minor part of the total composition. Examine this problem by converting the data in file SWEDEN.TXT to logratios and recompute the discriminant function between productive and nonproductive regions. What do you conclude about the necessity for transformation in these circumstances?

Exercise 6.13

Most petroleum originated either where marine carbonates or marine shales have been deposited, or in terrestrial deltaic environments. The organic facies and depositional environments of source rocks tend to be reflected in the carbon isotope ratios of the volatile hydrocarbon fractions of the resulting oils. Oils that originated in marine carbonate environments have a greater proportion of ^{12}C compared to ^{13}C (expressed as *carbon isotope ratio*, $\delta^{13}C$) in the lighter volatile fraction than in the heavier volatile fraction. Oils from deltaic environments exhibit the opposite pattern, while oils that originated in marine-shale environments have intermediate characteristics. In addition, more traditional properties of crude oils such as API gravity, sulfur content, pristane to phytane ratio (Pr/Ph), and ratio of saturated to aromatic nonvolatile hydrocarbons (SAT/ARO) may be indicative of source environment, especially in combination with carbon isotope data. Chung and others (1994) have compiled a table of crude-oil properties of oils whose source environments are known; these are listed in file MARINEOL.TXT. Can the three environments be distinguished on the basis of these variables? Which variables seem most diagnostic and potentially useful for classifying oils of unknown origin?

Exercise 6.14

The Upper Jurassic Smackover Formation in the eastern Gulf Coast of the USA is extensively dolomitized. Many mechanisms seem to have been involved in the dolomitization process, which has created a regional dolostone body whose characteristics differ from place to place. Among the mechanisms were (1) dolomitization by seepage of seawater through an oolite sill which separated an intra-platform evaporite basin from the open sea; (2) reflux of hypersaline brines as coastal sabkhas prograded over the oolites; (3) intrusion of freshwater into buried sediments at shallow depths following sea-level lowering; (4) dolomitization in the contemporaneous downdip, mixed connate/freshwater zone; and (5) secondary dolomitization and alteration at depth. By means of petrographic fabric analysis and stratigraphic relationships, Prather (1992) determined the mechanisms of dolomitization in a large number of well cores from the Smackover Formation. Permeabilities and porosities of these cores were measured, as were carbon ($\delta^{13}C$) and oxygen ($\delta^{18}O$) isotope ratios; microprobe analyses were used to determine Fe, Mg, Na, Mn, Sr, and Ca contents. A selection of these data is contained in file DOLOMIT.TXT. Can the five mechanisms of dolomitization be recognized from these data alone? Which variables are most effective for distinguishing the different classes?

Exercise 6.15

File ZEOLITES.TXT contains data on the composition of 299 zeolites, culled from the literature by Alberti and Brigatti (1985). The original chemical formulae have been recalculated on the basis of 200 oxygen atoms and checked for charge balance, assuring compatibility among the different analyses. The zeolites have been classified into five families: heulandites, chabazites, erionites, phillipsites, and analcimes. In addition, each specimen has been categorized as either sedimentary or hydrothermal in origin. A zeolite is considered to be "sedimentary" when it is a significant constituent of a sedimentary rock, whether it formed during diagenesis or as a result of low-grade metamorphism. A "hydrothermal" zeolite occurs in veinlets, dikes, geodes, or fissures, with no evident reaction with the host rock.

The question to be examined is whether there are significant chemical differences between zeolites as a consequence of their origin. Such differences would be obscured by the inherent compositional distinctions between zeolite species, so it is necessary to consider both categorizations simultaneously to determine if the mode of origin is reflected in composition. This can be done by multivariate analysis of variance (MANOVA) and by multigroup discriminant function analysis. A study can be structured in different ways: as a series of two-group analyses, in which each zeolite family is considered separately; or as a multigroup analysis in which each combination of family and origin is considered a separate group. Other analysis schemes are possible. Can the origin of zeolites be deduced from their chemical differences? Which constituents are most effective in distinguishing between origins? Does this differ from zeolite family to family?

Exercise 6.16

File PROFILE.TXT gives "elevations" at equally spaced points along profiles across 14 hypothetical mountains that were used by Menke (1989) to demonstrate an application of principal component analysis that is sometimes called "empirical orthogonal function analysis." The points are numbered in order from left to right, so x_1 is the leftmost point, x_6 is the center point, and x_{11} is the rightmost point on each profile. There is a geometric relationship between the variables because they are ordered measurements taken along a continuous profile. This geometric relationship is expressed as autocorrelation among the variables, and emerges in the principal components. If the eigenvector elements are scaled so that they are proportional to the corresponding singular values (that is, they are converted into "factors") and then plotted, the geometric nature of the orthogonal transformation will be apparent; the plotted lines are approximations of successive moments of the profiles. Compute the first four factors of the profile data, plot the factor loadings, and determine "names" for each of the factors. It may help to plot factor scores and examine the positions of different profiles in factor space, much as scores of the random block data have been plotted.

Similar analyses and interpretations can be developed for many kinds of profile data, including stream profiles, surface-wave dispersion curves (plots of seismic velocity versus frequency), power spectra from random signals, oil production decline curves, and grain-size distributions. The key point of similarity is that the data consist of successive points measured along some type of continuous function.

Exercise 6.17

One of the objectives of remote sensing by satellite is to detect ground conditions that may indicate natural resources. The most conventional approach is to use satellite images in the same manner as aerial photographs to produce photogeologic maps and structural interpretations. However, it may be possible to use spectral information from satellite-borne scanners to directly detect favorable conditions on the ground. File THEMATIC.TXT contains readings from three Landsat Thematic Mapper visible and near-infrared bands (TM2, TM5, and the ratio of TM4 to TM5) made at locations in Nevada and Utah where the Mississippian Chainman Shale is exposed. Specimens of the shale have been collected from the same localities and analyzed by Rock-Eval pyrolysis and vitrinite reflectance to determine the thermal maturity of included hydrocarbons. If the laboratory measurements can be related to spectral signatures, it suggests that source-rock maturity could be mapped from satellite images in areas where the Chainman Shale is exposed; this could be a valuable exploration tool in the search for oil and gas. Canonical correlation seems the appropriate tool for relating Thematic Mapper responses to thermal maturity measurements. The data in file THEMATIC.TXT are adapted from a study by Rowan, Pawlewicz, and Jones (1992).

Exercise 6.18

The Oligocene 64-Zone sandstone is an important reservoir in the North Belridge field in California. The unit is a submarine fan deposit whose porosity has been modified by diagenesis, including compaction (which has reduced intergranular porosity), quartz cementation, carbonate cementation, feldspar dissolution, and growth of authigenic kaolinite. Taylor and Soule (1993) made point counts of thin sections from 63 intervals in the sandstone, yielding estimates of the percentages of four grain constituents, three cements, and two clay fractions. In addition, they counted the relative proportion of intergranular porosity and porosity due to feldspar dissolution, and independently measured porosity by a laboratory porosimeter. The data are given in file BELRIDGE.TXT.

Compute the canonical correlation between the three porosity measures as the Y variables, and the ten constituent variables as the X's. From the loadings, can you determine the main influences on porosity? Plot the canonical X and Y scores versus depth. Do the two canonical variables maintain a consistent relationship with depth?

Exercise 6.19

The Deep Sea Drilling Project (DSDP) has produced an enormous amount of information about the composition of the upper part of the oceanic crust. Most of the nonsedimentary rocks recovered are basalts, but the molten rocks often have reacted with seawater as they were extruded, producing serpentinites and greenstones. In order to compare the parent mineralogies of different rocks even though the original minerals may be highly altered, their chemical analyses have been recast into normative minerals.

Normative minerals are determined by a complicated algorithm that allocates oxides to silicate minerals in the order that they might crystallize from a melt under ideal conditions, starting with apatite and ilmenite and progressing through the feldspar series to quartz. If there is insufficient SiO_2 at any step, alternative minerals such as hypersthene and olivine are formed. Rollinson (1993) gives a succinct discussion of the calculations. Although there is a deterministic connection between the chemical analysis of a rock and its normative mineralogy, the relationship is nonlinear. We can examine the nature of this relationship by canonical correlation, using oxide composition as one set of variables and normative mineralogy as the other. File DEEPSEA.TXT contains eight major oxides and eight normative minerals for 24 drill-core specimens of oceanic basalt recovered during DSDP Leg 17 in the central Pacific Ocean (Bass and others, 1971). Compute the canonical correlations between the two sets of variables and assess the nature of the combinations. Note that both sets of variables are subcompositions; if complete, they would sum to 100%. Can you comment on the effect closure may have on your conclusions? How can you check for its effect?

SELECTED READINGS

Aitchison, J., 1986, *The Statistical Analysis of Compositional Data*: Chapman & Hall, London, 416 pp.

Alberti, A., and M.F. Brigatti, 1985, Dependence of chemistry on genesis in zeolites: Multivariate analysis of variance and discriminant analysis: *American Mineralogist,* v. 70, p. 805–813.

Aldenderfer, M.S., and R.K. Blashfield, 1984, *Cluster Analysis*: Series on Quantitative Applications in the Social Sciences, no. 07–044, Sage Publications, Inc., Beverly Hills, Calif., 87 pp. *This compact monograph is one of the best and most concise introductions to cluster analysis.*

Anderson, T.W., 1984, *An Introduction to Multivariate Statistical Analyses,* 2nd ed.: John Wiley & Sons, Inc., New York, 704 pp. *An advanced treatment of multivariate statistics, with discussions of T^2, discriminant functions, and principal components.*

Backer, E., 1995, *Computer-assisted Reasoning in Cluster Analysis*: Prentice Hall International Ltd., Hemel Hempstead, U.K., 367 pp., diskette.

Bartlett, M.S., 1938, Further aspects of the theory of multiple regression: *Proc. Cambridge Philosophical Soc.,* v. 34, p. 33–40.

Bass, M.N., and others, 1971, Volcanic rocks cored in the central Pacific, Leg 17, Deep Sea Drilling Project, *in* Roth, P.H., and J.R. Herring [Eds.], *Initial Reports of the Deep Sea Drilling Project*: National Science Foundation, v. 17, no. 14, p. 429–503.

Benzécri, J.-P., and others, 1980, *L'Analyse des Donnees,* v. 2, *L'Analyse des Correspondances*: Dunod, Paris, 628 pp.

Bock, H.-H., 1996, Probability models and hypotheses testing in partitioning cluster analysis, *in* Arabie, P., L.J. Hubert, and G. DeSoete [Eds.], *Clustering and Classification*: World Scientific Publ. Co., Singapore, p. 377–453.

Burt, C., 1937, Correlations between persons: *British Journal of Psychology,* General Section, v. 28, p. 59–96. *This early article discusses the relationship between*

R- and Q-mode factor analysis. The numerical example used to demonstrate the Eckart–Young theorem in this chapter is adapted from Burt's illustration.

Chung, H.M., and others, 1994, Source characteristics of marine oils as indicated by carbon isotope ratios of volatile hydrocarbons: *Bull. Am. Assoc. Petroleum Geologists,* v. 78, p. 396–408.

Clark, S.P., Jr. [Ed.], 1966, *Handbook of Physical Constants*: Geol. Soc. America, Memoir 97, 587 pp.

Cooley, W.W., and P.R. Lohnes, 1971, *Multivariate Data Analysis*: John Wiley & Sons, Inc., New York, 364 pp.

David, M., M. Dagbert, and Y. Beauchemin, 1977, Statistical analysis in geology: Correspondence analysis method: *Quart. Colorado Sch. Mines,* v. 72, no. 1, 60 pp.

Davis, J.C., 1970, Information contained in sediment-size analyses: *Mathematical Geology,* v. 2, no. 2, p. 105–112.

Draper, N.R., and H. Smith, 1998, *Applied Regression Analysis,* 3rd ed.: John Wiley & Sons, Inc., New York, 706 pp., diskette. *A very thorough treatment of all aspects of linear regression. The comparison of sequential techniques in chapter 15 is especially valuable.*

Dryden, I.L, and K.V. Mardia, 1998, *Statistical Shape Analysis*: John Wiley & Sons, Inc., New York, 347 pp.

Eckart, C., and B. Young, 1936, The approximation of one matrix by another of lower rank: *Psychometrika,* v. 1, no. 3, p. 211–218. *The fundamental reference on the relationship between R- and Q-mode eigenvectors, and the basis for singular value decomposition.*

Everitt, B.S., 1993, *Cluster Analysis,* 3rd ed.: Edward Arnold, Sevenoaks, U.K., 170 pp. *A compact discussion of similarity measures, hierarchical clustering, and other clustering procedures not considered in this text.*

Farnham, I.M., K.J. Stetzenbach, A.K. Singh, and K.H. Johannesson, 2000, Deciphering groundwater flow systems in Oasis Valley, Nevada, using trace element chemistry, multivariate statistics, and geographical information system: *Mathematical Geology,* v. 32, no. 8, p. 943–968.

Fisher, R.A., 1940, The precision of discriminant functions: *Annals of Eugenics,* v. 10, p. 422–429.

Folk, R.L., 1980, *Petrology of Sedimentary Rocks,* 4th ed.: Hemphill's, Austin, Texas, 184 pp.

Gabriel, K.R., 1971, The biplot display of matrices with application to principal component analysis: *Biometrika,* v. 58, p. 453–467.

Gnanadesikan, R., 1977, *Methods for Statistical Data Analysis of Multivariate Observations*: John Wiley & Sons, Inc., New York, 311 pp.

Golub, G.H., and C. Reinsch, 1971, Singular value decomposition and least squares solutions, *in* Wilkinson, J.H., and C. Reinsch [Eds.], *Linear Algebra: Computer Methods for Mathematical Computation,* v. 2: Springer-Verlag, Berlin, p. 134–151.

Gordon, A.D., 1999, *Classification,* 2nd ed.: Chapman & Hall/CRC, Boca Raton, Fla., 256 pp. *A survey of most aspects of cluster analysis and related classification procedures, written at an intermediate level of mathematical rigor.*

Gower, J.C., 1966, Some distance properties of latent root and vector methods used in multivariate analysis: *Biometrika,* v. 53, nos. 3, 4, p. 325–338.

Green, P.E., F.J. Carmone, Jr., and S.M. Smith, 1989, *Multidimensional Scaling: Concepts and Applications*: Allyn and Bacon, Needham Heights, Mass., 407 pp., 2 diskettes.

Greenacre, M.J., 1984, *Theory and Applications of Correspondence Analysis*: Academic Press, London, 364 pp.

Griffith, D.A., and C.G. Amrhein, 1997, *Multivariate Statistical Analysis for Geographers*: Prentice Hall, Inc., Upper Saddle River, N.J., 345 pp. *Written in a near-outline format, the book is heavily illustrated with SAS instructions and output.*

Griffiths, J.C., and C.W. Ondrick, 1969, Modelling the petrology of detrital sediments, *in* Merriam, D.F. [Ed.], *Computer Applications in the Earth Sciences*: Plenum Press, New York, p. 73–97. *The principal component analysis suggested in Exercise 6.8 is described on p. 86–88.*

Harman, H.H., 1967, *Modern Factor Analysis,* 2nd ed.: Univ. Chicago Press, Chicago, Ill., 474 pp. *Although most examples in this text are psychometric, the treatment is computationally oriented and relatively free of the jargon of psychology.*

Harris, R.J., 1985, *Primer of Multivariate Statistics,* 2nd ed.: Academic Press, Orlando, Fla., 576 pp. *Discussions of principal component and factor analysis, although written in the context of social and behavioral sciences, are equally applicable to geological data. Also covered are topics such as the effects of data transformation and factor rotation.*

Hartigan, J.A., 1975, *Clustering Algorithms*: John Wiley & Sons, Inc., New York, 351 pp. *A discussion of clustering using a strictly algorithmic approach. Contains numerous FORTRAN subroutines and test data sets, drawn from a variety of disciplines.*

Heckel, P.H., and J.F. Baesemann, 1975, Environmental interpretation of conodont distribution in Upper Pennsylvanian (Missourian) megacyclothems in eastern Kansas: *Bull. Am. Assoc. Petroleum Geologists,* v. 59, no. 3, p. 486–509.

Hill, M.O., 1974, Correspondence analysis: A neglected multivariate method: *Jour. Royal Statistical Soc.,* Ser. C, Appl. Stat., v. 23, no. 3, p. 340–354.

Hirschfeld, H.O., 1935, A connection between correlation and contingency: *Proc. Cambridge Philosophical Soc.,* v. 31, p. 520–524.

Illenberger, W.K., 1991, Pebble shape (and size!): *Jour. Sedimentary Petrology,* v. 61, p. 756–767.

Imbrie, J., and E.G. Purdy, 1962, Classification of modern Bahamian carbonate sediments, *in* AAPG, *Classification of Carbonate Rocks, a Symposium*: Memoir 1, Am. Assoc. Petroleum Geologists, Tulsa, Okla., p. 253–272. *The basic reference in the geological literature on principal coordinates analysis.*

Jackson, J.E., 1991, *A User's Guide to Principal Components*: John Wiley & Sons, Inc., New York, 569 pp. *A very readable discussion of all aspects of eigenvector analyses. Examples are mostly from industrial chemistry.*

Jardine, N., and R. Sibson, 1971, *Mathematical Taxonomy*: John Wiley & Sons, Ltd., London, 286 pp. *Clustering and classification discussed using set notation. The book contains a valuable glossary of the jargon of numerical taxonomy.*

Jobson, J.D., 1991, *Applied Multivariate Data Analysis*, v. I, *Regression and Experimental Design;* v. II, *Categorical and Multivariate Methods*: Springer-Verlag, New York, 621 pp. (v. I), 731 pp. (v. II), diskette. *This monumental work covers most of multivariate statistics from a business and industrial viewpoint. An appendix on singular value decomposition is especially helpful.*

Johnson, R.M., 1963, On a theorem stated by Eckart and Young: *Psychometrika*, v. 28, no. 3, p. 259–263. *A mathematical proof of the Eckart–Young theorem.*

Jöreskog, K.G., 1977, Factor analysis by least-squares and maximum-likelihood methods, *in* Enslein, K., and others [Eds.], *Statistical Methods for Digital Computers*, v. 3: John Wiley & Sons, Inc., New York, p. 125–153. *A complete discussion with flow charts and algorithms for maximum-likelihood factor analysis.*

Jöreskog, K.G., J.E. Klovan, and R.A. Reyment, 1976, *Geological Factor Analysis*: Elsevier Scientific Publ. Co., Amsterdam, 178 pp. *Essential reading for those interested in applying factor analysis to geological problems. The authors do not shrink from expressing opinions about the relative merits of different methods.*

Kaufman, L., and P.J. Rousseeuw, 1990, *Finding Groups in Data, an Introduction to Cluster Analysis*: John Wiley & Sons, Inc., New York, 342 pp.

Kendall, M.G., and A. Stuart, 1967, *Advanced Theory of Statistics*, v. 2, 2nd ed.: Charles Griffin & Co. Ltd., London, 690 pp.

Klovan, J.E., 1966, The use of factor analysis in determining depositional environments from grain-size distributions: *Jour. Sedimentary Petrology*, v. 36, p. 115–125. *A Q-mode analysis of sediments from Barataria Bay is included.*

Koch, G.S., Jr., and R.F. Link, 1980, *Statistical Analysis of Geological Data*: Dover Publications, Inc., New York, 850 pp.

Krumbein, W.C., and E. Aberdeen, 1937, The sediments of Barataria Bay: *Jour. Sedimentary Petrology*, v. 7, p. 3–17. *Data in file BARATARA.TXT were adapted from table 1 of this article.*

Krumbein, W.C., and R.L. Shreve, 1970, Some statistical properties of dendritic channel networks: *Office of Naval Research, Tech. Rept. 13, ONR Task No. 389-150*, 117 pp. [available from Documents Clearinghouse, Arlington, Va., as document AD 705 6251]. *Results of a study by geology and geography students on the drainage network of an area in eastern Kentucky. The classes were sufficiently large that extensive data on operator error could be gathered, as well as a large quantity of data on the subject itself. An excellent example of the use of statistics in the classroom. Data in file KENTUCKY.TXT are adapted from this study.*

Krzanowski, W.J., 1988, *Principles of Multivariate Analysis*: Oxford Univ. Press, New York, 563 pp.

Lawley, D.N., 1940, The estimation of factor loadings by the method of maximum likelihood: *Proc. Royal Soc. Edinburgh,* Ser. A60, p. 64–82.

Lawley, D.N., and A.E. Maxwell, 1971, *Factor Analysis as a Statistical Method,* 2nd ed.: Butterworth & Co., Ltd., London, 153 pp. *This short monograph describes factor analysis independently of its psychometric origins. Especially valuable are the worked examples with intermediate steps in the calculations.*

Lebart, L., A. Morineau, and K.M. Warwick, 1984, *Multivariate Descriptive Statistical Analysis*: John Wiley & Sons, Inc., New York, 231 pp. *Translation of a French text which discusses correspondence analysis as developed by Benzécri.*

Longley, J.W., 1967, An appraisal of least squares programs for the electronic computer from the point of view of the user: *Jour. American Statistical Assoc.,* v. 62, no. 319, p. 819–841.

Marascuilo, L.A., and J.R. Levin, 1983, *Multivariate Statistics in the Social Sciences: A Researcher's Guide*: Brooks/Cole Publ. Co., Monterey, Calif., 530 pp. *Clearly written discussions of multivariate techniques. The comparisons of popular multivariate statistics programs are dated but still useful.*

Maron, M.J., 1982, *Numerical Analysis: A Practical Approach*: MacMillan Publ. Co., Inc., New York, 471 pp. *Contains easy-to-understand descriptions of the steps in alternative methods of calculating eigenvalues and eigenvectors.*

Matalas, N.C., and B.J. Reiher, 1967, Some comments on the use of factor analysis: *Water Resources Research,* v. 3, p. 213–223.

McQueen, J., 1967, Some methods for classification and analysis of multivariate observations: 5th Berkeley Symposium on Mathematics, Statistics, and Probability, v. 1, p. 281–298.

Menke, W., 1989, *Geophysical Data Analysis: Discrete Inverse Theory,* revised ed.: Academic Press, Inc., San Diego, Calif., 260 pp.

Milligan, G.W., and M.C. Cooper, 1986, An examination of procedures for determining the number of clusters in a data set: *Multivariate Behavioral Research,* v. 21, p. 441–458.

Morrison, D.F., 1990, *Multivariate Statistical Methods,* 3rd ed.: McGraw-Hill, Inc., New York, 495 pp. *One of the most lucid of the multivariate statistics textbooks, containing a helpful section on the relationships between familiar univariate methods and their multivariate extensions. Also has an especially straightforward discussion of factor analysis.*

Murillo, F.A., 1991, An integrated development of correspondence analysis with applications to environmental data: Unpub. Ph.D. dissertation, Univ. Arizona, 201 pp.

Ondrick, C.W., and G.S. Srivastava, 1970, CORFAN–FORTRAN IV computer program for correlation, factor analysis (R- and Q-mode) and varimax rotation: Computer Contribution No. 42, Kansas Geol. Survey, Lawrence, Kansas, 92 pp.

Ostroff, A.G., 1967, Comparison of some formation water classification systems: *Bull. Am. Assoc. Petroleum Geologists,* v. 51, no. 3, p. 404–416.

Overall, J.E., and C.J. Klett, 1983, *Applied Multivariate Analysis*: R.E. Krieger Publ. Co., Malabar, Fla., 500 pp.

Potter, P.E., N.F. Shimp, and J. Witters, 1963, Trace elements in marine and fresh-water argillaceous sediments: *Geochimica et Cosmochimica Acta*, v. 27, p. 669–694.

Prather, B.E., 1992, Origin of dolostone reservoir rocks, Smackover Formation (Oxfordian), Northeastern Gulf Coast, U.S.A.: *Bull. Am. Assoc. Petroleum Geologists*, v. 76, p. 133–163.

Press, W.H., S.A. Teukolsky, W.T. Vetterling, and B.P. Flannery, 1992, *Numerical Recipes in FORTRAN*, 2nd ed.: Cambridge Univ. Press, New York, 963 pp.

Reyment, R.A., R.E. Blackith, and N.A. Campbell, 1984, *Multivariate Morphometrics*, 2nd ed.: Academic Press, London, 233 pp. *A thorough review of the application of multivariate statistical techniques to animal morphology, especially of fossils. Contains extensive citations to the literature and numerous summaries of published case studies.*

Reyment, R.A., and E. Savazzi, 1999, *Aspects of Multivariate Statistical Analysis in Geology*: Elsevier Science Publ. B.V., Amsterdam, 285 pp., CD. *This book is especially valuable for its insights into what can go wrong in geological data analysis, and its emphasis on the importance of checking sample distributions to ensure they conform to assumptions. Contains a program library and data sets.*

Rollinson, H., 1993, *Using Geochemical Data: Evaluation, Presentation, Interpretation*: Longman Scientific & Technical, Harlow, Essex, U.K., 352 pp. *Discusses the logical flaws and misleading nature of many plots, diagrams, and ratios used by geochemists and petrologists. Aitchison's logratio transformations are emphasized.*

Romesburg, H.C., 1984, *Cluster Analysis for Researchers*: Lifetime Learning Publ., Belmont, Calif., 334 pp.

Rowan, L.C., M.J. Pawlewicz, and O.D. Jones, 1992, Mapping thermal maturity in the Chainman Shale, near Eureka, Nevada, with Landsat Thematic Mapper images: *Bull. Am. Assoc. Petroleum Geologists*, v. 76, no. 7, p. 1008–1023.

Sherman, K.N., C.M. Bunker, and C.A. Bush, 1971, Correlation of uranium, thorium, and potassium with aeroradioactivity in the Berea area, Virginia: *Economic Geology*, v. 66, p. 302–308.

Sneath, P.H.A., and R.R. Sokal, 1973, *Numerical Taxonomy*: W.H. Freeman & Co., San Francisco, 573 pp. *The founding work on numerical taxonomy. Unfortunately, geologists may find the jargon confusing and the long arguments for a numerical phylogeny irrelevant to applications in other areas.*

Switzer, P., 1970, Numerical classification, *in* Merriam, D.F. [Ed.], *Geostatistics, a Colloquium*: Plenum Press, New York, p. 31–43.

Taylor, T.R., and C.H. Soule, 1993, Reservoir characterization and diagenesis of the Oligocene 64-Zone sandstone, North Belridge Field, Kern County, California: *Bull. Am. Assoc. Petroleum Geologists*, v. 77, no. 9, p. 1549–1566.

Teil, H., 1975, Correspondence factor analysis: An outline of its method: *Mathematical Geology*, v. 7, no. 1, p. 3–12.

Teil, H., and J.L. Cheminée, 1975, Application of correspondence factor analysis to the study of major and trace elements in the Erta Ale Chain (Afar, Ethiopia): *Mathematical Geology,* v. 7, no. 1, p. 13–30.

Temple, J.T., 1978, The use of factor analysis in geology: *Mathematical Geology,* v. 10, no. 4, p. 379–387. *A strong criticism of factor analysis, particularly as used in geology and related areas.*

Thurstone, L.L., 1947, *Multiple-factor Analysis*: Univ. Chicago Press, Chicago, Ill., 535 pp. *The classic work in factor analysis, in its original context as a psychometric tool. Written prior to the development of computers, much of the computational detail is now obsolete. However, the discussion of factor theory remains a basic reference.*

Wanke, H., and others, 1970, Major and trace elements and cosmic-ray produced radioisotopes in lunar samples: *Science,* v. 167, no. 3918, p. 523–525.

Ward, J.H., Jr., 1963, Hierarchical grouping to optimize an objective function: *Jour. American Statistical Assoc.,* v. 58, p. 236–244.

Wishart, D., 1969, An algorithm for hierarchical classification: *Biometrics,* v. 25, p. 165–170.

Zhou, D., T. Chang, and J.C. Davis, 1983, Dual extraction of R-mode and Q-mode factor solutions: *Mathematical Geology,* v. 15, no. 5, p. 581–606.

APPENDIX

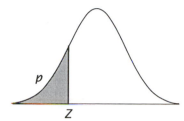

Table A.1. Cumulative probabilities for the standardized normal distribution. Z-scores are standard deviations from the mean. Probabilities are cumulative areas under the normal distribution. Especially useful critical values shown in bold italics.

Z	p	Z	p	Z	p	Z	p
−3.00	0.0013	−1.55	0.0606	0.05	0.5199	*1.64*	*0.9500*
−2.95	0.0016	−1.50	0.0668	0.10	0.5398	1.65	0.9505
−2.90	0.0019	−1.45	0.0735	0.15	0.5596	1.70	0.9554
−2.85	0.0022	−1.40	0.0808	0.20	0.5793	1.75	0.9599
−2.80	0.0026	−1.35	0.0885	0.25	0.5987	1.80	0.9641
−2.75	0.0030	−1.30	0.0968	0.30	0.6179	1.85	0.9678
−2.70	0.0035	*−1.28*	*0.1000*	0.35	0.6368	1.90	0.9713
−2.65	0.0040	−1.25	0.1056	0.40	0.6554	1.95	0.9744
−2.60	0.0047	−1.20	0.1151	0.45	0.6736	*1.96*	*0.9750*
−2.57	*0.0050*	−1.15	0.1251	0.50	0.6915	2.00	0.9772
−2.55	0.0054	−1.10	0.1357	0.55	0.7088	2.05	0.9798
−2.50	0.0062	−1.05	0.1469	0.60	0.7257	2.10	0.9821
−2.45	0.0071	−1.00	0.1587	0.65	0.7422	2.15	0.9842
−2.40	0.0082	−0.95	0.1711	0.70	0.7580	2.20	0.9861
−2.35	0.0094	−0.90	0.1841	0.75	0.7734	2.25	0.9878
−2.33	*0.0100*	−0.85	0.1977	0.80	0.7881	2.30	0.9893
−2.30	0.0107	−0.80	0.2119	0.85	0.8023	*2.33*	*0.9900*
−2.25	0.0122	−0.75	0.2266	0.90	0.8159	2.35	0.9906
−2.20	0.0139	−0.70	0.2420	0.95	0.8289	2.40	0.9918
−2.15	0.0158	−0.65	0.2578	1.00	0.8413	2.45	0.9929
−2.10	0.0179	−0.60	0.2743	1.05	0.8531	2.50	0.9938
−2.05	0.0202	−0.55	0.2912	1.10	0.8643	2.55	0.9946
−2.00	0.0228	−0.50	0.3085	1.15	0.8749	*2.57*	*0.9950*
−1.96	*0.0250*	−0.45	0.3264	1.20	0.8849	2.60	0.9953
−1.95	0.0256	−0.40	0.3446	1.25	0.8944	2.65	0.9960
−1.90	0.0287	−0.35	0.3632	*1.28*	*0.9000*	2.70	0.9965
−1.85	0.0322	−0.30	0.3821	1.30	0.9032	2.75	0.9970
−1.80	0.0359	−0.25	0.4013	1.35	0.9115	2.80	0.9974
−1.75	0.0401	−0.20	0.4207	1.40	0.9192	2.85	0.9978
−1.70	0.0446	−0.15	0.4404	1.45	0.9265	2.90	0.9981
−1.65	0.0495	−0.10	0.4602	1.50	0.9332	2.95	0.9984
−1.64	*0.0500*	−0.05	0.4801	1.55	0.9394	3.00	0.9987
−1.60	0.0548	0.00	0.5000	1.60	0.9452		

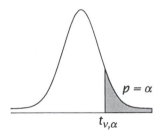

$p = \alpha$

$t_{\nu,\alpha}$

Table A.2. Critical values of t for ν degrees of freedom and selected levels of significance. For critical values in the left-hand tail, change the sign of the table value. Critical values are given for the right-hand tail.

	Significance Level, α, for:						
	One-tailed Test						
	.001	.005	.01	.025	.05	.1	.2
No. of Degrees	Two-tailed Test						
of Freedom, ν	.002	.010	.02	.05	.1	.2	.4
1	318.3088	63.6567	31.8205	12.7062	6.3138	3.0777	1.3764
2	22.3271	9.9248	6.9646	4.3027	2.9200	1.8856	1.0607
3	10.2145	5.8409	4.5407	3.1824	2.3534	1.6377	0.9785
4	7.1732	4.6041	3.7470	2.7764	2.1318	1.5332	0.9410
5	5.8934	4.0322	3.3649	2.5706	2.0150	1.4759	0.9195
6	5.2076	3.7074	3.1427	2.4469	1.9432	1.4398	0.9057
7	4.7853	3.4995	2.9980	2.3646	1.8946	1.4149	0.8960
8	4.5008	3.3554	2.8965	2.3060	1.8595	1.3968	0.8889
9	4.2968	3.2498	2.8214	2.2622	1.8331	1.3830	0.8834
10	4.1437	3.1693	2.7638	2.2281	1.8125	1.3722	0.8791
11	4.0247	3.1058	2.7181	2.2010	1.7959	1.3634	0.8755
12	3.9296	3.0545	2.6810	2.1788	1.7823	1.3562	0.8726
13	3.8520	3.0123	2.6503	2.1604	1.7709	1.3502	0.8702
14	3.7874	2.9768	2.6245	2.1448	1.7613	1.3450	0.8681
15	3.7328	2.9467	2.6025	2.1314	1.7531	1.3406	0.8662
16	3.6862	2.9208	2.5835	2.1199	1.7459	1.3368	0.8647
17	3.6458	2.8982	2.5669	2.1098	1.7396	1.3334	0.8633
18	3.6105	2.8784	2.5524	2.1009	1.7341	1.3304	0.8620
19	3.5794	2.8609	2.5395	2.0930	1.7291	1.3277	0.8610
20	3.5518	2.8453	2.5280	2.0860	1.7247	1.3253	0.8600
21	3.5272	2.8314	2.5176	2.0796	1.7207	1.3232	0.8591
22	3.5050	2.8188	2.5083	2.0739	1.7171	1.3212	0.8583
23	3.4850	2.8073	2.4999	2.0687	1.7139	1.3195	0.8575
24	3.4668	2.7969	2.4922	2.0639	1.7109	1.3178	0.8569
25	3.4502	2.7874	2.4851	2.0595	1.7081	1.3163	0.8562
26	3.4350	2.7787	2.4786	2.0555	1.7056	1.3150	0.8557

(Continued)

Table A.2. Concluded.

No. of Degrees of Freedom, ν	Significance Level, α, for: One-tailed Test						
	.001	.005	.01	.025	.05	.1	.2
	Two-tailed Test						
	.002	.010	.02	.05	.1	.2	.4
27	3.4210	2.7707	2.4727	2.0518	1.7033	1.3137	0.8551
28	3.4082	2.7633	2.4671	2.0484	1.7011	1.3125	0.8546
29	3.3962	2.7564	2.4620	2.0452	1.6991	1.3114	0.8542
30	3.3852	2.7500	2.4573	2.0423	1.6973	1.3104	0.8538
40	3.3069	2.7045	2.4233	2.0211	1.6839	1.3031	0.8507
50	3.2614	2.6778	2.4033	2.0086	1.6759	1.2987	0.8489
60	3.2317	2.6603	2.3901	2.0003	1.6706	1.2958	0.8477
70	3.2108	2.6479	2.3808	1.9944	1.6669	1.2938	0.8468
80	3.1953	2.6387	2.3739	1.9901	1.6641	1.2922	0.8461
90	3.1833	2.6316	2.3685	1.9867	1.6620	1.2910	0.8456
100	3.1737	2.6259	2.3642	1.9840	1.6602	1.2901	0.8452
110	3.1660	2.6213	2.3607	1.9818	1.6588	1.2893	0.8449
120	3.1595	2.6174	2.3578	1.9799	1.6577	1.2886	0.8446
130	3.1541	2.6142	2.3554	1.9784	1.6567	1.2881	0.8444
140	3.1495	2.6114	2.3533	1.9771	1.6558	1.2876	0.8442
150	3.1455	2.6090	2.3515	1.9759	1.6551	1.2872	0.8440
∞	3.0902	2.5758	2.2364	1.9600	1.6449	1.2816	0.8416

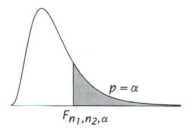

Table A.3a. Critical values of F for ν_1 and ν_2 degrees of freedom and 5% ($\alpha = 0.05$) level of significance.

df	1	2	3	4	5	6	7	8	9	10	15	20	25	∞
1	161.45	199.50	215.71	224.58	230.16	233.99	236.77	238.88	240.54	241.88	245.95	248.01	249.26	250.10
2	18.51	19	19.16	19.25	19.30	19.33	19.35	19.37	19.38	19.40	19.43	19.45	19.46	19.46
3	10.13	9.55	9.28	9.12	9.01	8.94	8.89	8.85	8.81	8.79	8.70	8.66	8.63	8.62
4	7.71	6.94	6.59	6.39	6.26	6.16	6.09	6.04	6.00	5.96	5.86	5.80	5.77	5.75
5	6.61	5.79	5.41	5.19	5.05	4.95	4.88	4.82	4.77	4.74	4.62	4.56	4.52	4.50
6	5.99	5.14	4.76	4.53	4.39	4.28	4.21	4.15	4.10	4.06	3.94	3.87	3.83	3.81

(Continued)

Table A.3a. Concluded.

df	1	2	3	4	5	6	7	8	9	10	15	20	25	∞
7	5.59	4.74	4.35	4.12	3.97	3.87	3.79	3.73	3.68	3.64	3.51	3.44	3.4	3.38
8	5.32	4.46	4.07	3.84	3.69	3.58	3.50	3.44	3.39	3.35	3.22	3.15	3.11	3.08
9	5.12	4.26	3.86	3.63	3.48	3.37	3.29	3.23	3.18	3.14	3.01	2.94	2.89	2.86
10	4.96	4.10	3.71	3.48	3.33	3.22	3.14	3.07	3.02	2.98	2.85	2.77	2.73	2.70
11	4.84	3.98	3.59	3.36	3.20	3.09	3.01	2.95	2.90	2.85	2.72	2.65	2.60	2.57
12	4.75	3.89	3.49	3.26	3.11	3.00	2.91	2.85	2.80	2.75	2.62	2.54	2.50	2.47
13	4.67	3.81	3.41	3.18	3.03	2.92	2.83	2.77	2.71	2.67	2.53	2.46	2.41	2.38
14	4.60	3.74	3.34	3.11	2.96	2.85	2.76	2.70	2.65	2.60	2.46	2.39	2.34	2.31
15	4.54	3.68	3.29	3.06	2.90	2.79	2.71	2.64	2.59	2.54	2.40	2.33	2.28	2.25
16	4.49	3.63	3.24	3.01	2.85	2.74	2.66	2.59	2.54	2.49	2.35	2.28	2.23	2.19
17	4.45	3.59	3.20	2.96	2.81	2.70	2.61	2.55	2.49	2.45	2.31	2.23	2.18	2.15
18	4.41	3.55	3.16	2.93	2.77	2.66	2.58	2.51	2.46	2.41	2.27	2.19	2.14	2.11
19	4.38	3.52	3.13	2.90	2.74	2.63	2.54	2.48	2.42	2.38	2.23	2.16	2.11	2.07
20	4.35	3.49	3.10	2.87	2.71	2.60	2.51	2.45	2.39	2.35	2.20	2.12	2.07	2.04
21	4.32	3.47	3.07	2.84	2.68	2.57	2.49	2.42	2.37	2.32	2.18	2.10	2.05	2.01
22	4.30	3.44	3.05	2.82	2.66	2.55	2.46	2.40	2.34	2.30	2.15	2.07	2.02	1.98
23	4.28	3.42	3.03	2.80	2.64	2.53	2.44	2.37	2.32	2.27	2.13	2.05	2.00	1.96
24	4.26	3.40	3.01	2.78	2.62	2.51	2.42	2.36	2.30	2.25	2.11	2.03	1.97	1.94
25	4.24	3.39	2.99	2.76	2.60	2.49	2.40	2.34	2.28	2.24	2.09	2.01	1.96	1.92
26	4.23	3.37	2.98	2.74	2.59	2.47	2.39	2.32	2.27	2.22	2.07	1.99	1.94	1.90
27	4.21	3.35	2.96	2.73	2.57	2.46	2.37	2.31	2.25	2.20	2.06	1.97	1.92	1.88
28	4.20	3.34	2.95	2.71	2.56	2.45	2.36	2.29	2.24	2.19	2.04	1.96	1.91	1.87
29	4.18	3.33	2.93	2.70	2.55	2.43	2.35	2.28	2.22	2.18	2.03	1.94	1.89	1.85
30	4.17	3.32	2.92	2.69	2.53	2.42	2.33	2.27	2.21	2.16	2.01	1.93	1.88	1.84
40	4.08	3.23	2.84	2.61	2.45	2.34	2.25	2.18	2.12	2.08	1.92	1.84	1.78	1.74
50	4.03	3.18	2.79	2.56	2.40	2.29	2.20	2.13	2.07	2.03	1.87	1.78	1.73	1.69
60	4.00	3.15	2.76	2.53	2.37	2.25	2.17	2.10	2.04	1.99	1.84	1.75	1.69	1.65
70	3.98	3.13	2.74	2.50	2.35	2.23	2.14	2.07	2.02	1.97	1.81	1.72	1.66	1.62
80	3.96	3.11	2.72	2.49	2.33	2.21	2.13	2.06	2.00	1.95	1.79	1.70	1.64	1.60
90	3.95	3.10	2.71	2.47	2.32	2.20	2.11	2.04	1.99	1.94	1.78	1.69	1.63	1.59
100	3.94	3.09	2.70	2.46	2.31	2.19	2.10	2.03	1.97	1.93	1.77	1.68	1.62	1.57
110	3.93	3.08	2.69	2.45	2.30	2.18	2.09	2.02	1.97	1.92	1.76	1.67	1.61	1.56
120	3.92	3.07	2.68	2.45	2.29	2.18	2.09	2.02	1.96	1.91	1.75	1.66	1.60	1.55
∞	3.85	3.00	2.61	2.38	2.22	2.11	2.02	1.95	1.89	1.84	1.68	1.58	1.52	1.47

Table A.3b. Critical values of F for ν_1 and ν_2 degrees of freedom and 2.5% ($\alpha = 0.025$) level of significance.

df	1	2	3	4	5	6	7	8	9	10	15	20	25	∞
1	647.79	799.50	864.16	899.58	921.85	937.11	948.22	956.66	963.28	968.63	984.87	993.10	998.08	1001.41
2	38.51	39.00	39.17	39.25	39.30	39.33	39.36	39.37	39.39	39.40	39.43	39.45	39.46	39.46
3	17.44	16.04	15.44	15.10	14.88	14.73	14.62	14.54	14.47	14.42	14.25	14.17	14.12	14.08
4	12.22	10.65	9.98	9.60	9.36	9.20	9.07	8.98	8.90	8.84	8.66	8.56	8.50	8.46
5	10.01	8.43	7.76	7.39	7.15	6.98	6.85	6.76	6.68	6.62	6.43	6.33	6.27	6.23
6	8.81	7.26	6.60	6.23	5.99	5.82	5.70	5.60	5.52	5.46	5.27	5.17	5.11	5.07
7	8.07	6.54	5.89	5.52	5.29	5.12	4.99	4.90	4.82	4.76	4.57	4.47	4.40	4.36
8	7.57	6.06	5.42	5.05	4.82	4.65	4.53	4.43	4.36	4.30	4.10	4.00	3.94	3.89
9	7.21	5.71	5.08	4.72	4.48	4.32	4.20	4.10	4.03	3.96	3.77	3.67	3.60	3.56
10	6.94	5.46	4.83	4.47	4.24	4.07	3.95	3.85	3.78	3.72	3.52	3.42	3.35	3.31
11	6.72	5.26	4.63	4.28	4.04	3.88	3.76	3.66	3.59	3.53	3.33	3.23	3.16	3.12
12	6.55	5.10	4.47	4.12	3.89	3.73	3.61	3.51	3.44	3.37	3.18	3.07	3.01	2.96
13	6.41	4.97	4.35	4.00	3.77	3.60	3.48	3.39	3.31	3.25	3.05	2.95	2.88	2.84
14	6.30	4.86	4.24	3.89	3.66	3.50	3.38	3.29	3.21	3.15	2.95	2.84	2.78	2.73
15	6.20	4.77	4.15	3.80	3.58	3.41	3.29	3.20	3.12	3.06	2.86	2.76	2.69	2.64
16	6.12	4.69	4.08	3.73	3.50	3.34	3.22	3.12	3.05	2.99	2.79	2.68	2.61	2.57
17	6.04	4.62	4.01	3.66	3.44	3.28	3.16	3.06	2.98	2.92	2.72	2.62	2.55	2.50
18	5.98	4.56	3.95	3.61	3.38	3.22	3.10	3.01	2.93	2.87	2.67	2.56	2.49	2.44
19	5.92	4.51	3.90	3.56	3.33	3.17	3.05	2.96	2.88	2.82	2.62	2.51	2.44	2.39
20	5.87	4.46	3.86	3.51	3.29	3.13	3.01	2.91	2.84	2.77	2.57	2.46	2.40	2.35
21	5.83	4.42	3.82	3.48	3.25	3.09	2.97	2.87	2.80	2.73	2.53	2.42	2.36	2.31
22	5.79	4.38	3.78	3.44	3.22	3.05	2.93	2.84	2.76	2.70	2.50	2.39	2.32	2.27
23	5.75	4.35	3.75	3.41	3.18	3.02	2.90	2.81	2.73	2.67	2.47	2.36	2.29	2.24
24	5.72	4.32	3.72	3.38	3.15	2.99	2.87	2.78	2.70	2.64	2.44	2.33	2.26	2.21
25	5.69	4.29	3.69	3.35	3.13	2.97	2.85	2.75	2.68	2.61	2.41	2.30	2.23	2.18
26	5.66	4.27	3.67	3.33	3.10	2.94	2.82	2.73	2.65	2.59	2.39	2.28	2.21	2.16
27	5.63	4.24	3.65	3.31	3.08	2.92	2.80	2.71	2.63	2.57	2.36	2.25	2.18	2.13
28	5.61	4.22	3.63	3.29	3.06	2.90	2.78	2.69	2.61	2.55	2.34	2.23	2.16	2.11
29	5.59	4.20	3.61	3.27	3.04	2.88	2.76	2.67	2.59	2.53	2.32	2.21	2.14	2.09
30	5.57	4.18	3.59	3.25	3.03	2.87	2.75	2.65	2.57	2.51	2.31	2.20	2.12	2.07
40	5.42	4.05	3.46	3.13	2.90	2.74	2.62	2.53	2.45	2.39	2.18	2.07	1.99	1.94
50	5.34	3.97	3.39	3.05	2.83	2.67	2.55	2.46	2.38	2.32	2.11	1.99	1.92	1.87
60	5.29	3.93	3.34	3.01	2.79	2.63	2.51	2.41	2.33	2.27	2.06	1.94	1.87	1.82
70	5.25	3.89	3.31	2.97	2.75	2.59	2.47	2.38	2.30	2.24	2.03	1.91	1.83	1.78
80	5.22	3.86	3.28	2.95	2.73	2.57	2.45	2.35	2.28	2.21	2.00	1.88	1.81	1.75
90	5.20	3.84	3.26	2.93	2.71	2.55	2.43	2.34	2.26	2.19	1.98	1.86	1.79	1.73
100	5.18	3.83	3.25	2.92	2.70	2.54	2.42	2.32	2.24	2.18	1.97	1.85	1.77	1.71
110	5.16	3.82	3.24	2.90	2.68	2.53	2.40	2.31	2.23	2.17	1.96	1.84	1.76	1.70
120	5.15	3.80	3.23	2.89	2.67	2.52	2.39	2.30	2.22	2.16	1.94	1.82	1.75	1.69
∞	5.04	3.70	3.13	2.80	2.58	2.42	2.30	2.20	2.13	2.06	1.85	1.72	1.64	1.58

Table A.3c. Critical values of F for ν_1 and ν_2 degrees of freedom and 1% ($\alpha = 0.01$) level of significance.

df	1	2	3	4	5	6	7	8	9	10	15	20	25	∞
1	4052.18	4999.50	5403.35	5624.58	5763.65	5858.99	5928.36	5981.07	6022.47	6055.85	6157.28	6208.73	6239.83	6260.65
2	98.50	99.00	99.17	99.25	99.30	99.33	99.36	99.37	99.39	99.40	99.43	99.45	99.46	99.47
3	34.12	30.82	29.46	28.71	28.24	27.91	27.67	27.49	27.35	27.23	26.87	26.69	26.58	26.50
4	21.20	18.00	16.69	15.98	15.52	15.21	14.98	14.80	14.66	14.55	14.20	14.02	13.91	13.84
5	16.26	13.27	12.06	11.39	10.97	10.67	10.46	10.29	10.16	10.05	9.72	9.55	9.45	9.38
6	13.75	10.92	9.78	9.15	8.75	8.47	8.26	8.10	7.98	7.87	7.56	7.40	7.30	7.23
7	12.25	9.55	8.45	7.85	7.46	7.19	6.99	6.84	6.72	6.62	6.31	6.16	6.06	5.99
8	11.26	8.65	7.59	7.01	6.63	6.37	6.18	6.03	5.91	5.81	5.52	5.36	5.26	5.20
9	10.56	8.02	6.99	6.42	6.06	5.80	5.61	5.47	5.35	5.26	4.96	4.81	4.71	4.65
10	10.04	7.56	6.55	5.99	5.64	5.39	5.20	5.06	4.94	4.85	4.56	4.41	4.31	4.25
11	9.65	7.21	6.22	5.67	5.32	5.07	4.89	4.74	4.63	4.54	4.25	4.10	4.01	3.94
12	9.33	6.93	5.95	5.41	5.06	4.82	4.64	4.50	4.39	4.30	4.01	3.86	3.76	3.70
13	9.07	6.70	5.74	5.21	4.86	4.62	4.44	4.30	4.19	4.10	3.82	3.66	3.57	3.51
14	8.86	6.51	5.56	5.04	4.69	4.46	4.28	4.14	4.03	3.94	3.66	3.51	3.41	3.35
15	8.68	6.36	5.42	4.89	4.56	4.32	4.14	4.00	3.89	3.80	3.52	3.37	3.28	3.21
16	8.53	6.23	5.29	4.77	4.44	4.20	4.03	3.89	3.78	3.69	3.41	3.26	3.16	3.10
17	8.40	6.11	5.18	4.67	4.34	4.10	3.93	3.79	3.68	3.59	3.31	3.16	3.07	3.00
18	8.29	6.01	5.09	4.58	4.25	4.01	3.84	3.71	3.60	3.51	3.23	3.08	2.98	2.92
19	8.18	5.93	5.01	4.50	4.17	3.94	3.77	3.63	3.52	3.43	3.15	3.00	2.91	2.84
20	8.10	5.85	4.94	4.43	4.10	3.87	3.70	3.56	3.46	3.37	3.09	2.94	2.84	2.78
21	8.02	5.78	4.87	4.37	4.04	3.81	3.64	3.51	3.40	3.31	3.03	2.88	2.79	2.72
22	7.95	5.72	4.82	4.31	3.99	3.76	3.59	3.45	3.35	3.26	2.98	2.83	2.73	2.67
23	7.88	5.66	4.76	4.26	3.94	3.71	3.54	3.41	3.30	3.21	2.93	2.78	2.69	2.62
24	7.82	5.61	4.72	4.22	3.90	3.67	3.50	3.36	3.26	3.17	2.89	2.74	2.64	2.58
25	7.77	5.57	4.68	4.18	3.85	3.63	3.46	3.32	3.22	3.13	2.85	2.70	2.60	2.54
26	7.72	5.53	4.64	4.14	3.82	3.59	3.42	3.29	3.18	3.09	2.81	2.66	2.57	2.50
27	7.68	5.49	4.60	4.11	3.78	3.56	3.39	3.26	3.15	3.06	2.78	2.63	2.54	2.47
28	7.64	5.45	4.57	4.07	3.75	3.53	3.36	3.23	3.12	3.03	2.75	2.60	2.51	2.44
29	7.60	5.42	4.54	4.04	3.73	3.50	3.33	3.20	3.09	3.00	2.73	2.57	2.48	2.41
30	7.56	5.39	4.51	4.02	3.70	3.47	3.30	3.17	3.07	2.98	2.70	2.55	2.45	2.39
40	7.31	5.18	4.31	3.83	3.51	3.29	3.12	2.99	2.89	2.80	2.70	2.55	2.45	2.39
50	7.17	5.06	4.20	3.72	3.41	3.19	3.02	2.89	2.78	2.70	2.52	2.37	2.27	2.20
60	7.08	4.98	4.13	3.65	3.34	3.12	2.95	2.82	2.72	2.63	2.42	2.27	2.17	2.10
70	7.01	4.92	4.07	3.60	3.29	3.07	2.91	2.78	2.67	2.59	2.35	2.20	2.10	2.03
80	6.96	4.88	4.04	3.56	3.26	3.04	2.87	2.74	2.64	2.55	2.31	2.15	2.05	1.98
90	6.93	4.85	4.01	3.53	3.23	3.01	2.84	2.72	2.61	2.52	2.27	2.12	2.01	1.94
100	6.90	4.82	3.98	3.51	3.21	2.99	2.82	2.69	2.59	2.50	2.24	2.09	1.99	1.92
110	6.87	4.80	3.96	3.49	3.19	2.97	2.81	2.68	2.57	2.49	2.22	2.07	1.97	1.89
120	6.85	4.79	3.95	3.48	3.17	2.96	2.79	2.66	2.56	2.47	2.21	2.05	1.95	1.88
∞	6.66	4.63	3.80	3.34	3.04	2.82	2.66	2.53	2.43	2.34	2.06	1.90	1.79	1.72

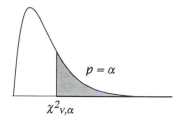

$p = \alpha$

$\chi^2_{v,\alpha}$

Table A.4. Critical values of χ^2 for v degrees of freedom and selected levels of significance.

No. of Degrees of Freedom, v	Significance Level, α				
	0.20	0.10	0.05	0.025	0.01
1	1.64	2.71	3.84	5.02	6.63
2	3.22	4.61	5.99	7.38	9.21
3	4.64	6.25	7.81	9.35	11.34
4	5.99	7.78	9.49	11.14	13.28
5	7.29	9.24	11.07	12.83	15.09
6	8.56	10.64	12.59	14.45	16.81
7	9.80	12.02	14.07	16.01	18.48
8	11.03	13.36	15.51	17.53	20.09
9	12.24	14.68	16.92	19.02	21.67
10	13.44	15.99	18.31	20.48	23.21
11	14.63	17.28	19.68	21.92	24.72
12	15.81	18.55	21.03	23.34	26.22
13	16.98	19.81	22.36	24.74	27.69
14	18.15	21.06	23.68	26.12	29.14
15	19.31	22.31	25.00	27.49	30.58
16	20.47	23.54	26.30	28.85	32.00
17	21.61	24.77	27.59	30.19	33.41
18	22.76	25.99	28.87	31.53	34.81
19	23.90	27.20	30.14	32.85	36.19
20	25.04	28.41	31.41	34.17	37.57
21	26.17	29.62	32.67	35.48	38.93
22	27.30	30.81	33.92	36.78	40.29
23	28.43	32.01	35.17	38.08	41.64
24	29.55	33.20	36.42	39.36	42.98
25	30.68	34.38	37.65	40.65	44.31
26	31.79	35.56	38.89	41.92	45.64
27	32.91	36.74	40.11	43.19	46.96
28	34.03	37.92	41.34	44.46	48.28
29	35.14	39.09	42.56	45.72	49.59
30	36.25	40.26	43.77	46.98	50.89
40	47.27	51.81	55.76	59.34	63.69
50	58.16	63.17	67.50	71.42	76.15
60	68.97	74.40	79.08	83.30	88.38

(Continued)

Table A.4. Concluded.

No. of Degrees of Freedom, ν	Significance Level, α				
	0.20	0.10	0.05	0.025	0.01
70	79.71	85.53	90.53	95.02	100.43
80	90.41	96.58	101.88	106.63	112.33
90	101.05	107.57	113.15	118.14	124.12
100	111.67	118.50	124.34	129.56	135.81
110	122.25	129.39	135.48	140.92	147.41
120	132.81	140.23	146.57	152.21	158.95

Table A.5. Probabilities of occurrence of specified values of the Mann–Whitney W_x statistic for testing the equality to two samples of size n and m, where $m \leq n \leq 8$. C_L is the lower critical value and C_U is the upper critical value.[1]

m = 3

C_L	3	C_U	4	C_U	5	C_U	6	C_U	7	C_U	8	C_U
6	.0500	15	.0286	18	.0179	21	.0119	24	.0083	27	.0061	30
7	.1000	14	.0571	17	.0357	20	.0238	23	.0167	26	.0121	29
8	.2000	13	.1143	16	.0714	19	.0476	22	.0333	25	.0242	28
9	.3500	12	.2000	15	.1250	18	.0833	21	.0583	24	.0424	27
10	.5000	11	.3143	14	.1964	17	.1310	20	.0917	23	.0667	26
11	.6500	10	.4286	13	.2857	16	.1905	19	.1333	22	.0970	25
12	.8000	9	.5714	12	.3929	15	.2738	18	.1917	21	.1394	24
13	.9000	8	.6857	11	.5000	14	.3571	17	.2583	20	.1879	23
14	.9500	7	.8000	10	.6071	13	.4524	16	.3333	19	.2485	22
15	1.0000	6	.8857	9	.7143	12	.5476	15	.4167	18	.3152	21
16			.942	8	.8036	11	.6429	14	.5000	17	.3879	20
17			.9714	7	.8750	10	.7262	13	.5833	16	.4606	19
18			1.0000	6	.9286	9	.8095	12	.6667	15	.5394	18
19					.9643	8	.8690	11	.7417	14	.6121	17
20					.9821	7	.9167	10	.8083	13	.6848	16
21					1.0000	6	.9524	9	.8667	12	.7515	15
22							.9762	8	.9083	11	.8121	14
23							.9881	7	.9417	10	.8606	13
24							1.0000	6	.9667	9	.9030	12

[1] Adapted from S. Siegel and N.J. Castellan, Jr., 1988,
Nonparametric Statistics for the Behavioral Sciences, 2ed.
Reproduced by permission of The McGraw-Hill Companies, New York.

(Continued)

Table A.5. Continued.

C_L	4	C_U	5	C_U	6	C_U	7	C_U	8	C_U
						n				

m = 4

C_L	4	C_U	5	C_U	6	C_U	7	C_U	8	C_U
10	.0143	26	.0079	30	.0048	34	.0030	38	.0020	42
11	.0286	25	.0159	29	.0095	33	.0061	37	.0040	41
12	.0571	24	.0317	28	.0190	32	.0121	36	.0081	40
13	.1000	23	.0556	27	.0333	31	.0212	35	.0141	39
14	.1714	22	.0952	26	.0571	30	.0364	34	.0242	38
15	.2429	21	.1429	25	.0857	29	.0545	33	.0364	37
16	.3429	20	.2063	24	.1286	28	.0818	32	.0545	36
17	.4429	19	.2778	23	.1762	27	.1152	31	.0768	35
18	.5571	18	.3651	22	.2381	26	.1576	30	.1071	34
19	.6571	17	.4524	21	.3048	25	.2061	29	.1414	33
20	.7571	16	.5476	20	.3810	24	.2636	28	.1838	32
21	.8286	15	.6349	19	.4571	23	.3242	27	.2303	31
22	.9000	14	.7222	18	.5429	22	.3939	26	.2848	30
23	.9429	13	.7937	17	.6190	21	.4636	25	.3414	29
24	.9714	12	.8571	16	.6952	20	.5364	24	.4040	28
25	.9857	11	.9048	15	.7619	19	.6061	23	.4667	27
26	1.0000	10	.9444	14	.8238	18	.6758	22	.5333	26
27			.9683	13	.8714	17	.7364	21	.5960	25
28			.9841	12	.9143	16	.7939	20	.6586	24
29			.9921	11	.9429	15	.8424	19	.7152	23
30			1.0000	10	.9667	14	.8848	18	.7697	22
31					.9810	13	.9182	17	.8162	21
32					.9905	12	.9455	16	.8586	20
33					.9952	11	.9636	15	.8929	19
34					1.0000	10	.9788	14	.9232	18

(Continued)

Table A.5. Continued.

C_L	5	C_U	6	C_U	7	C_U	8	C_U
m = 5				*n*				
15	.0040	40	.0022	45	.0013	50	.0008	55
16	.0079	39	.0043	44	.0025	49	.0016	54
17	.0159	38	.0087	43	.0051	48	.0031	53
18	.0278	37	.0152	42	.0088	47	.0054	52
19	.0476	36	.0260	41	.0152	46	.0093	51
20	.0754	35	.0411	40	.0240	45	.0148	50
21	.1111	34	.0628	39	.0366	44	.0225	49
22	.1548	33	.0887	38	.0530	43	.0326	48
23	.2103	32	.1234	37	.0745	42	.0466	47
24	.2738	31	.1645	36	.1010	41	.0637	46
25	.3452	30	.2143	35	.1338	40	.0855	45
26	.4206	29	.2684	34	.1717	39	.1111	44
27	.5000	28	.3312	33	.2159	38	.1422	43
28	.5794	27	.3961	32	.2652	37	.1772	42
29	.6548	26	.4654	31	.3194	36	.2176	41
30	.7262	25	.5346	30	.3775	35	.2618	40
31	.7897	24	.6039	29	.4381	34	.3108	39
32	.8452	23	.6688	28	.5000	33	.3621	38
33	.8889	22	.7316	27	.5619	32	.4165	37
34	.9246	21	.7857	26	.6225	31	.4716	36
35	.9524	20	.8355	25	.6806	30	.5284	35
36	.9722	19	.8766	24	.7348	29	.5835	34
37	.9841	18	.9113	23	.7841	28	.6379	33
38	.9921	17	.9372	22	.8283	27	.6892	32
39	.9960	16	.9589	21	.8662	26	.7382	31
40	1.0000	15	.9740	20	.8990	25	.7824	30

(Continued)

Table A.5. Continued.

C_L	$m = 6$					
			n			
	6	C_U	7	C_U	8	C_U
21	.0011	57	.0006	63	.0003	69
22	.0022	56	.0012	62	.0007	68
23	.0043	55	.0023	61	.0013	67
24	.0076	54	.0041	60	.0023	66
25	.0130	53	.0070	59	.0040	65
26	.0206	52	.0111	58	.0063	64
27	.0325	51	.0175	57	.0100	63
28	.0465	50	.0256	56	.0147	62
29	.0660	49	.0367	55	.0213	61
30	.0898	48	.0507	54	.0296	60
31	.1201	47	.0688	53	.0406	59
32	.1548	46	.0903	52	.0539	58
33	.1970	45	.1171	51	.0709	57
34	.2424	44	.1474	50	.0906	56
35	.2944	43	.1830	49	.1142	55
36	.3496	42	.2226	48	.1412	54
37	.4091	41	.2669	47	.1725	53
38	.4686	40	.3141	46	.2068	52
39	.5314	39	.3654	45	.2454	51
40	.5909	38	.4178	44	.2864	50
41	.6504	37	.4726	43	.3310	49
42	.7056	36	.5274	42	.3773	48
43	.7576	35	.5822	41	.4259	47
44	.8030	34	.6346	40	.4749	46
45	.8452	33	.6859	39	.5251	45
46	.8799	32	.7331	38	.5741	44
47	.9102	31	.7774	37	.6227	43
48	.9340	30	.8170	36	.6690	42
49	.9535	29	.8526	35	.7136	41
50	.9675	28	.8829	34	.7546	40
51	.9794	27	.9097	33	.7932	39

(Continued)

Table A.5. Concluded.

C_L	$m = 7$ 7	C_U	n 8	C_U	C_L	$m = 8$ n 8	C_U
28	.0003	77	.0002	84	36	.0001	100
29	.0006	76	.0003	83	37	.0002	99
30	.0012	75	.0006	82	38	.0003	98
31	.0020	74	.0011	81	39	.0005	97
32	.0035	73	.0019	80	40	.0009	96
33	.0055	72	.0030	79	41	.0015	95
34	.0087	71	.0047	78	42	.0023	94
35	.0131	70	.0070	77	43	.0035	93
36	.0189	69	.0103	76	44	.0052	92
37	.0265	68	.0145	75	45	.0074	91
38	.0364	67	.0200	74	46	.0103	90
39	.0487	66	.0270	73	47	.0141	89
40	.0641	65	.0361	72	48	.0190	88
41	.0825	64	.0469	71	49	.0249	87
42	.1043	63	.0603	70	50	.0325	86
43	.1297	62	.0760	69	51	.0415	85
44	.1588	61	.0946	68	52	.0524	84
45	.1914	60	.1159	67	53	.0652	83
46	.2279	59	.1405	66	54	.0803	82
47	.2675	58	.1678	65	55	.0974	81
48	.3100	57	.1984	64	56	.1172	80
49	.3552	56	.2317	63	57	.1393	79
50	.4024	55	.2679	62	58	.1641	78
51	.4508	54	.3063	61	59	.1911	77
52	.5000	53	.3472	60	60	.2209	76
53	.5492	52	.3894	59	61	.2527	75
54	.5976	51	.4333	58	62	.2869	74
55	.6448	50	.4775	57	63	.3227	73
56	.6900	49	.5225	56	64	.3605	72
57	.7325	48	.5667	55	65	.3992	71
58	.7721	47	.6106	54	66	.4392	70
59	.8086	46	.6528	53	67	.4796	69
60	.8412	45	.6937	52	68	.5204	68
61	.8703	44	.7321	51	69	.5608	67
62	.8957	43	.7683	50	70	.6008	66
63	.9175	42	.8016	49	71	.6395	65
					72	.6773	64
					73	.7131	63
					74	.7473	62
					75	.7791	61
					76	.8089	60

Table A.6. Critical values of Spearman's ρ for testing the significance of a rank correlation. Table gives upper critical value of Spearman's ρ for specified level of significance. Lower critical values are equal to $-\rho$.

	Significance, α, for One-tailed Test					
	.10	.05	.025	.01	.005	.001
	Significance, α, for Two-tailed Test					
	.20	.10	.05	.02	.01	.002
n						
4	.8000	.8000				
5	.7000	.8000	.9000	.9000		
6	.6000	.7714	.8286	.8857	.9429	
7	.5357	.6786	.7450	.8571	.8929	.9643
8	.5000	.6190	.7143	.8095	.8571	.9286
9	.4667	.5833	.6833	.7667	.8167	.9000
10	.4424	.5515	.6364	.7333	.7818	.8667
11	.4182	.5273	.6091	.7000	.7455	.8364
12	.3986	.4965	.5804	.6713	.7273	.8182
13	.3791	.4780	.5549	.6429	.6978	.7912
14	.3626	.4593	.5341	.6220	.6747	.7670
15	.3500	.4429	.5179	.6000	.6536	.7464
16	.3382	.4265	.5000	.5824	.6324	.7265
17	.3260	.4118	.4853	.5637	.6152	.7083
18	.3148	.3994	.4716	.5480	.5975	.6904
19	.3070	.3895	.4579	.5333	.5825	.6737
20	.2977	.3789	.4451	.5203	.5684	.6586
21	.2909	.3688	.4351	.5078	.5545	.6455
22	.2829	.3597	.4241	.4963	.5426	.6318
23	.2767	.3518	.4150	.4852	.5306	.6186
24	.2704	.3435	.4061	.4748	.5200	.6070
25	.2646	.3362	.3977	.4654	.5100	.5962
26	.2588	.3299	.3894	.4564	.5002	.5856
27	.2540	.3236	.3822	.4481	.4915	.5757
28	.2490	.3175	.3749	.4401	.4828	.5660
29	.2443	.3113	.3685	.4320	.4744	.5567
30	.2400	.3059	.3620	.4251	.4665	.5479

Table A.7. Critical values of D in the Kolmogorov–Smirnov goodness-of-fit test.

	Significance, α, for One-tailed Test					
	0.1	0.05	0.025	0.05	0.02	0.01
	Significance, α, for Two-tailed Test					
	0.2	0.1	0.05	0.025	0.01	0.005
n						
1	0.7275	0.8721	0.9950	0.9999	0.9999	0.9999
2	0.5551	0.6655	0.7592	0.8425	0.9413	0.9999
3	0.4671	0.5600	0.6389	0.7090	0.7922	0.8497
4	0.4114	0.4932	0.5627	0.6244	0.6977	0.7483
5	0.3720	0.4460	0.5088	0.5646	0.6309	0.6767
6	0.3422	0.4103	0.4681	0.5195	0.5804	0.6226
7	0.3187	0.3821	0.436	0.4838	0.5405	0.5798
8	0.2995	0.3591	0.4097	0.4546	0.5080	0.5448
9	0.2834	0.3398	0.3877	0.4302	0.4807	0.5156
10	0.2697	0.3234	0.3689	0.4094	0.4575	0.4907
11	0.2579	0.3092	0.3527	0.3914	0.4373	0.4691
12	0.2474	0.2967	0.3385	0.3756	0.4196	0.4501
13	0.2382	0.2856	0.3258	0.3616	0.4040	0.4333
14	0.2299	0.2757	0.3145	0.3490	0.3900	0.4183
15	0.2225	0.2668	0.3043	0.3377	0.3773	0.4048
16	0.2157	0.2586	0.2951	0.3274	0.3659	0.3924
17	0.2096	0.2513	0.2866	0.3181	0.3554	0.3812
18	0.2039	0.2444	0.2789	0.3095	0.3458	0.3709
19	0.1986	0.2382	0.2717	0.3015	0.3369	0.3614
20	0.1938	0.2323	0.2651	0.2942	0.3287	0.3526
21	0.1893	0.2270	0.2589	0.2873	0.3211	0.3443
22	0.1851	0.2219	0.2532	0.2810	0.3139	0.3367
23	0.1812	0.2172	0.2478	0.2750	0.3072	0.3296
24	0.1775	0.2128	0.2428	0.2694	0.3010	0.3228
25	0.1740	0.2086	0.2380	0.2641	0.2951	0.3166
26	0.1707	0.2047	0.2335	0.2591	0.2895	0.3106
27	0.1676	0.2010	0.2293	0.2544	0.2843	0.3049
28	0.1647	0.1975	0.2253	0.2500	0.2793	0.2996
29	0.1619	0.1941	0.2215	0.2458	0.2746	0.2946
30	0.1593	0.1910	0.2179	0.2418	0.2701	0.2897
31	0.1568	0.1880	0.2144	0.2379	0.2659	0.2852
32	0.1544	0.1851	0.2112	0.2343	0.2618	0.2808
33	0.1521	0.1823	0.2080	0.2308	0.2579	0.2766
34	0.1499	0.1797	0.2050	0.2275	0.2542	0.2726
35	0.1478	0.1772	0.2021	0.2243	0.2506	0.2688
36	0.1458	0.1748	0.1994	0.2212	0.2472	0.2652
37	0.1438	0.1725	0.1967	0.2183	0.2439	0.2616
38	0.1420	0.1702	0.1942	0.2155	0.2408	0.2582
39	0.1402	0.1681	0.1917	0.2128	0.2377	0.2550
40	0.1385	0.1660	0.1894	0.2102	0.2348	0.2519

(Continued)

Table A.7. Continued.

	Significance, α, for One-tailed Test					
	0.1	0.05	0.025	0.05	0.02	0.01
	Significance, α, for Two-tailed Test					
	0.2	0.1	0.05	0.025	0.01	0.005
41	0.1368	0.1640	0.1871	0.2076	0.2320	0.2489
42	0.1352	0.1621	0.1849	0.2052	0.2293	0.2460
43	0.1337	0.1602	0.1828	0.2029	0.2267	0.2431
44	0.1322	0.1585	0.1808	0.2006	0.2241	0.2404
45	0.1307	0.1567	0.1788	0.1984	0.2217	0.2378
46	0.1293	0.1551	0.1769	0.1963	0.2193	0.2353
47	0.1280	0.1534	0.1751	0.1942	0.2170	0.2328
48	0.1267	0.1519	0.1732	0.1923	0.2148	0.2304
49	0.1254	0.1503	0.1715	0.1903	0.2126	0.2281
50	0.1242	0.1488	0.1698	0.1885	0.2106	0.2259
51	0.1230	0.1474	0.1682	0.1866	0.2085	0.2237
52	0.1218	0.1460	0.1666	0.1849	0.2065	0.2216
53	0.1206	0.1447	0.1650	0.1831	0.2046	0.2195
54	0.1196	0.1433	0.1635	0.1815	0.2028	0.2175
55	0.1185	0.1421	0.1621	0.1798	0.2010	0.2155
56	0.1174	0.1408	0.1606	0.1783	0.1992	0.2137
57	0.1164	0.1396	0.1592	0.1767	0.1975	0.2118
58	0.1155	0.1384	0.1579	0.1753	0.1958	0.2100
59	0.1145	0.1373	0.1566	0.1738	0.1941	0.2082
60	0.1135	0.1361	0.1553	0.1723	0.1926	0.2065
61	0.1126	0.1350	0.1540	0.1709	0.1910	0.2049
62	0.1117	0.1340	0.1528	0.1696	0.1895	0.2033
63	0.1108	0.1329	0.1516	0.1683	0.1880	0.2017
64	0.1100	0.1319	0.1505	0.1670	0.1866	0.2001
65	0.1092	0.1309	0.1493	0.1657	0.1851	0.1986
66	0.1084	0.1299	0.1482	0.1645	0.1838	0.1971
67	0.1075	0.1290	0.1471	0.1633	0.1824	0.1957
68	0.1068	0.1280	0.1460	0.1621	0.1811	0.1942
69	0.1060	0.1271	0.1450	0.1609	0.1798	0.1928
70	0.1053	0.1262	0.1440	0.1598	0.1785	0.1915
71	0.1045	0.1253	0.1430	0.1587	0.1773	0.1901
72	0.1038	0.1245	0.1420	0.1576	0.1761	0.1888
73	0.1031	0.1236	0.1411	0.1565	0.1749	0.1876
74	0.1024	0.1228	0.1401	0.1555	0.1737	0.1863
75	0.1017	0.1220	0.1392	0.1544	0.1726	0.1851
76	0.1011	0.1212	0.1383	0.1534	0.1714	0.1839
77	0.1004	0.1204	0.1374	0.1525	0.1703	0.1827
78	0.0998	0.1197	0.1365	0.1515	0.1693	0.1816
79	0.0992	0.1189	0.1357	0.1506	0.1682	0.1805
80	0.0986	0.1182	0.1348	0.1496	0.1672	0.1793

(Continued)

Table A.7. Concluded.

	Significance, α, for One-tailed Test					
	0.1	0.05	0.025	0.05	0.02	0.01
	Significance, α, for Two-tailed Test					
	0.2	0.1	0.05	0.025	0.01	0.005
81	0.0980	0.1175	0.1340	0.1487	0.1662	0.1782
82	0.0974	0.1168	0.1332	0.1478	0.1652	0.1772
83	0.0968	0.1161	0.1324	0.1470	0.1642	0.1761
84	0.0962	0.1154	0.1316	0.1461	0.1632	0.1751
85	0.0957	0.1147	0.1309	0.1452	0.1623	0.1741
86	0.0951	0.1141	0.1301	0.1444	0.1613	0.1730
87	0.0946	0.1134	0.1294	0.1436	0.1604	0.1721
88	0.0941	0.1128	0.1287	0.1428	0.1595	0.1711
89	0.0935	0.1121	0.1279	0.1420	0.1586	0.1702
90	0.0930	0.1115	0.1272	0.1412	0.1578	0.1692
91	0.0925	0.1109	0.1265	0.1404	0.1569	0.1683
92	0.0920	0.1103	0.1259	0.1397	0.1561	0.1674
93	0.0915	0.1097	0.1252	0.1390	0.1552	0.1665
94	0.0911	0.1092	0.1245	0.1382	0.1544	0.1656
95	0.0906	0.1086	0.1239	0.1375	0.1536	0.1648
96	0.0901	0.1081	0.1233	0.1368	0.1528	0.1639
97	0.0897	0.1075	0.1226	0.1361	0.1520	0.1631
98	0.0892	0.1069	0.1220	0.1354	0.1513	0.1623
99	0.0888	0.1064	0.1214	0.1347	0.1505	0.1614
100	0.0883	0.1059	0.1208	0.1341	0.1498	0.1607

Table A.8. Critical values of the Lilliefors test statistic, T, for testing goodness-of-fit to a normal distribution.

Sample size, $n =$	Level of Significance, α				
	.20	.15	.10	.05	.01
4	.300	.319	.352	.381	.417
5	.285	.299	.315	.337	.405
6	.265	.277	.294	.319	.364
7	.247	.258	.276	.300	.348
8	.233	.244	.261	.285	.331
9	.223	.233	.249	.271	.311
10	.215	.224	.239	.258	.294
11	.206	.217	.230	.249	.284
12	.199	.212	.223	.242	.275
13	.190	.202	.214	.234	.268
14	.183	.194	.207	.227	.261
15	.177	.187	.201	.220	.257
16	.173	.182	.195	.213	.250
17	.169	.177	.189	.206	.245
18	.166	.173	.184	.200	.239
19	.163	.169	.179	.195	.235
20	.160	.166	.174	.190	.231
25	.142	.147	.158	.173	.200
30	.131	.136	.144	.161	.187
>30	$\dfrac{.736}{\sqrt{n}}$	$\dfrac{.768}{\sqrt{n}}$	$\dfrac{.805}{\sqrt{n}}$	$\dfrac{.886}{\sqrt{n}}$	$\dfrac{1.031}{\sqrt{n}}$

Table A.9. Maximum likelihood estimates of the concentration parameter κ for calculated values of $\overline{R}$ (adapted from Batschelet, 1965; and Gumbel, Greenwood, and Durand, 1953).

$\overline{R}$	κ	$\overline{R}$	κ	$\overline{R}$	κ
0.00	0.00000	0.35	0.74783	0.70	2.01363
.01	.02000	.36	.77241	.71	2.07685
.02	.04001	.37	.79730	.72	2.14359
.03	.06003	.38	.82253	.73	2.21425
.04	.08006	.39	.84812	.74	2.28930
.05	.10013	.40	.87408	.75	2.36930
.06	.12022	.41	.90043	.76	2.45490
.07	.14034	.42	.92720	.77	2.54686
.08	.16051	.43	.95440	.78	2.64613
.09	.18073	.44	.98207	.79	2.75382
.10	.20101	.45	1.01022	.80	2.87129
.11	.22134	.46	1.03889	.81	3.00020
.12	.24175	.47	1.06810	.82	3.14262
.13	.26223	.48	1.09788	.83	3.30114
.14	.28279	.49	1.12828	.84	3.47901
.15	.30344	.50	1.15932	.85	3.68041
.16	.32419	.51	1.19105	.86	3.91072
.17	.34503	.52	1.22350	.87	4.17703
.18	.36599	.53	1.25672	.88	4.48876
.19	.38707	.54	1.29077	.89	4.85871
.20	.40828	.55	1.32570	.90	5.3047
.21	.42962	.56	1.36156	.91	5.8522
.22	.45110	.57	1.39842	.92	6.5394
.23	.47273	.58	1.43635	.93	7.4257
.24	.49453	.59	1.47543	.94	8.6104
.25	.51649	.60	1.51574	.95	10.2716
.26	.53863	.61	1.55738	.96	12.7661
.27	.56097	.62	1.60044	.97	16.9266
.28	.58350	.63	1.64506	.98	25.2522
.29	.60625	.64	1.69134	.99	50.2421
.30	.62922	.65	1.73945	1.00	∞
.31	.65242	.66	1.78953		
.32	.67587	.67	1.84177		
.33	.69958	.68	1.89637		
.34	.72356	.69	1.95357		

Table A.10. Critical values of $\overline{R}$ for Rayleigh's test for the presence of a preferred trend. From Mardia (1972).

	Level of Significance, α			
	.10	.05	.025	.01
Sample size,				
$n =$ 4	0.768	0.847	0.905	0.960
5	.677	.754	.816	.879
6	.618	.690	.753	.825
7	.572	.642	.702	.771
8	.535	.602	.660	.725
9	.504	.569	.624	.687
10	.478	.540	.594	.655
11	.456	.516	.567	.627
12	.437	.494	.544	.602
13	.420	.475	.524	.580
14	.405	.458	.505	.560
15	.391	.443	.489	.542
16	.379	.429	.474	.525
17	.367	.417	.460	.510
18	.357	.405	.447	.496
19	.348	.394	.436	.484
20	.339	.385	.425	.472
21	.331	.375	.415	.461
22	.323	.367	.405	.451
23	.316	.359	.397	.441
24	.309	.351	.389	.432
25	.303	.344	.381	.423
30	.277	.315	.348	.387
35	.256	.292	.323	.359
40	.240	.273	.302	.336
45	.226	.257	.285	.318
50	.214	.244	.270	.301

Table A.11. Critical values of $\overline{R}$ for the test of uniformity of a spherical distribution.

	Level of Significance, α			
	.10	.05	.02	.01
Sample size,				
$n =$ 5	0.637	0.700	0.765	0.805
6	.583	.642	.707	.747
7	.541	.597	.659	.698
8	.506	.560	.619	.658
9	.478	.529	.586	.624
10	.454	.503	.558	.594
11	.433	.480	.533	.568
12	.415	.460	.512	.546
13	.398	.442	.492	.526
14	.384	.427	.475	.507
15	.371	.413	.460	.491
16	.359	.400	.446	.476
17	.349	.388	.443	.463
18	.339	.377	.421	.450
19	.330	.367	.410	.438
20	.322	.358	.399	.428
21	.314	.350	.390	.418
22	.307	.342	.382	.408
23	.300	.334	.374	.400
24	.294	.328	.366	.392
25	.288	.321	.359	.384
30	.26	.29	.33	.36
35	.24	.27	.31	.33
40	.23	.26	.29	.31
45	.22	.24	.27	.29
50	.20	.23	.26	.28
100	.14	.16	.18	.19

INDEX

A

Aberfan, Wales (UK) 284
ABOC.TXT 401
accuracy 26
added terms (curvilinear regression 210
additive rule of probability 21
A.E.C. (Atomic Energy Commission) 154
aerial photograph 445–7, 593
Africa 400–1
aggregated pattern of points 299
agricultural runoff 589
Agua Caliente Formation (Precambrian) 116
AGUACAL.TXT 116
airborne magnetometer survey 590
airborne radiometric measurement 570
Al_2O_3 590
Alabama (USA) 285, 444
Alaska (USA) 204–5
Alberta Basin (Canada) 371, 403–6
aliasing 274, 365
Allen's Creek, Indiana (USA) 284
alluvial fill 485
alluvial pediment 446
alternative hypothesis 61
ammonoid 502–3, 505–6
amphibolite 288
amplitude 267, 362
analcime 592
"analysis of associations" 552
analysis of variance (ANOVA) 78–92, 182, 196–
   200, 204–5, 210–2, 223–4, 288, 366–70, 407–12,
   464–5, 468–70, 487, 572
   – clustering 497
   – multiple regression 464–5, 468–70
   – nested 88–92, 367–9, 448
   – one-way 78–84, 117, 572, 589
   – regression 196–200, 204–5, 210–2, 223–4,
   288
   – segmenting 236
   – spatial analysis 366–70
   – trend surface 407–12
   – two-way 84–7, 116
ANDES.TXT 451
andesite 179, 202, 281
angle of strike 332
Anglo–Barren Oil Company 400–1
angular deviation 365
angular similarity (Q-mode factor analysis) 540
anhydrite 49, 154
anisotropy 264
Annapolis (Maryland) 251, 253
anomaly, magnetic 443
anorthite 153
anorthosite 116, 279, 312–3
ANOVA (See analysis of variance.)
Antarctica 77
anthropogenic origin 448

anticline 327–30
apatite 593
API gravity 244, 282, 591
apparent correlation, matrix of 492
apparent grain density 154
Appleby (UK) 117
aquifer 223, 431, 435–8, 485
AQUIFER.TXT 435
aragonite 495
arbitrary origin methods (classification) 488
Arbuckle Group (Ordovician) 301, 451
ARBUCKLE.TXT 301, 451
archaeology 357
Archie's equation 115
arcsine transformation 102
Arctic Ocean 119
area of closure 100, 104
area of object 356
area of rejection 63–4, 66
arenite 448
Argentine Limestone (Missourian) 557
arithmetic average 34
arithmetic averaging (in clustering) 490–1, 493,
   496–7
Arizona (USA) 444, 446–7
aromatics 564
arrowheads, shapes of 365–6
ARSENAL.TXT 250
ASO.TXT 178
Aso volcano (Japan) 178–80, 183, 281
asphaltics 564
association, coefficients of 490
astronomy 357
Atlantic Coastal Plain (USA) 322–3
Atokan (Pennsylvanian) 391
atoll 355, 364
auger sample 439
Australia 563
Austria 265, 287, 590
autocorrelation 161, 182–3, 214, 243–8, 278, 281,
   372–3, 388, 414, 590, 592
autocovariance 244, 258
autocovariogram 244
average linkage (clustering) 497–8
average rate of occurrence 179
axes of pebbles 46–7, 126
axes of oriented features 331
axial length 355–6
axial plane 327–8
azimuth 332–3, 338, 446
azurite 495

B

backward elimination 469–70
badlands topography 446
Bahia de Guasimas (Mexico) 589
balanced ANOVA 90, 367–8
Baltic Sea 140

BALTIC.TXT 140
Bangladesh 327–30
BANGLA.TXT 327, 329
BANKSAND.TXT 393
BANK.TXT 312–3
BARATARA.TXT 518, 523, 589
bar graph (*See also* histogram.) 517
Barataria Bay (Louisiana) 518, 520–5, 589
Bartlett's test 580–1
Basal Fish Scales (Cretaceous) 404–6
basalt 107–8, 286–7, 585–6, 593–4
"basket-of-eggs" topography 117
bathymetric profile 351
Bayes' theorem 23–4, 238
beach sand 439, 472, 474, 476
beam balance 113
bed thickness 211
Bellman's principle of optimality 237
Belmont, Virginia (USA) 284
BELRIDGE.TXT 593
beneficiation 287, 590
Benioff subduction zone 202–3, 449
BENIOFF.TXT 202, 204
bentonite 125, 178, 281–2, 404
Berea, Virginia (USA) 570–1
Bézier coefficient 378
BHTEMP.TXT 153
bias 29, 38, 196, 199, 220, 225–6, 414, 416
bicubic polynomial 378
Bighorn Basin 70
Billings County (North Dakota) 240
bimodal distribution 322, 325, 337
binary (presence–absence) variables 7–8, 490
binomial distribution 14–7, 25, 302–3
binomial, negative 17, 307–9
bioclast 448
biology 357, 488, 501
biotite 288
bitmap image 447
bivariate: 40–3, 447
  – data 191, 214–5, 221
  – ellipse 284
  – mean 220, 447
  – normal probability distribution 481
BIVARIAT.TXT 216
Black Hills (USA) 397
black-sand beach 439
black shale 281
Bladen County (North Carolina) 323
"blended" surface (gridding) 390
block data 507, 517, 531, 533, 536, 544, 550–1, 561–3, 569
block diagram 407
block kriging 437–43
blue galicia 146
Bolivia 451
Bonner Springs Shale (Missourian) 557–8
Bookstein coordinates 357–9, 447, 587–8

borehole televiewer 445
boron 160
bottomhole temperature 153
Bouger gravity 118–9
BOUGER.TXT 118
boundaries on maps 373, 394
boundary (segmenting sequences) 235
box-and-whisker plot 33
box counting 353
box data (*See* block data.)
BOXES.TXT 507–8, 517, 531, 535, 542, 560
brachiopod 45–6, 60, 62–4, 357–9, 447–8, 510, 517, 540, 587–8
Brancepeth colliery (County Durham, UK) 284
Brereton shale (Pennsylvanian) 366–8
BRERETON.TXT 367
brine 251, 575–6, 591
BRINE.TXT 575
brittlebush (*Encelia farinose*) 444–5
bryozoan 287–8
BRYOZOAN.TXT 287
buffer region (guard region) 391–2, 415
Buffon's problem 296
bulla (fossil skulls) 284

C

calcite 81, 495
calcium 486
Calcutta (India) 440
calibration 204–7
California (USA) 282, 406, 449, 593
Cambrian 586–7
Canada 278, 403, 405–6, 446
canonical:
  – correlation 577–84, 593–4
  – loading 579, 583
  – score 579, 581–3
  – variate 574, 577, 579, 581–2
canyon, submarine 283
CaO 590
Captain Creek Limestone (Missourian) 557
Carbon County (Wyoming) 446
carbon isotope ratio ($\delta^{13}C$) 591
Carbon-14 206
carbonate:
  – grains 114
  – marine 114, 591
  – mineral 79, 494–6
  – reef 403–6
  – rock 285, 403–6, 575–6
CARBONAT.TXT 494–5
Carboniferous 113, 115, 173
Caribbean Sea 446
Carlisle (UK) 117
"Carolina bays" (North Carolina) 322–3
CAROLINA.TXT 322
Cartesian coordinates 229, 331–4, 336, 358, 360, 362, 374, 436, 447, 449, 451, 587–8

Cathedral Bluffs Member (Eocene) 446
cation 486, 591
Cave Creek (Kentucky) 271–2
CAVECREK.TXT 271
cell (fractal analysis) 346–8, 350
cell (reservoir simulation) 439
Celtic Sea 114
CELTIC.TXT 114
cementation factor 115
center of gravity 360
centered logratio transformation 54, 523–4, 526, 547, 585
central limits theorem 58–60, 479
centroid 360–1, 448, 588
centroid (of cluster) 488
centroid method 497–9
cephelon (of trilobite) 587
cerussite 495
chabazite 592
Chainman Shale (Mississippian) 593
Chanute Shale (Missourian) 557–8
cheilostome bryozoan 287
chemical analysis 51–4, 146, 369, 543, 545, 575, 591–3
Chernobyl 33
chert pebbles 127
Chesapeake Bay (Maryland) 251, 253–4
Chile 451
China Sea 113
χ^2 distance 554–8
χ^2 distribution 92–6, 105, 171–2, 175, 178, 300–2, 304–7, 310, 326, 480, 485, 487, 539, 554, 581, 586
χ^2 similarity matrix 554–8
chlorite–actinolite schist 570
chromatogram 564
chromite 439
chromium 35, 38–9, 69–70
CICTUS Research Center, University of Sonora (Mexico) 589
circular:
  – data 316–30
  – distribution 316–30
  – histogram (*See* rose diagram.)
  – uniform distribution 322–3
  – variance 321–2
classification 471, 487–500, 545
clay 116, 243, 285, 520
closed data 48–50, 519–20, 523–6, 546–7, 549, 554, 560, 585, 594
closure (structural) 404
cluster analysis 238, 487–500, 526, 545, 548, 587–8
clustered pattern 299, 307–10, 312, 416
coal 160, 168–9, 172–4, 366–8, 440–3
Coal Measures (Carboniferous) 173
coastal lagoon 589
coastline (of Iceland) 345–7
cobalt 118

COBALT.TXT 118
coefficients of association 490
coefficient of variation 39
cofactors (evaluating determinant) 138
coin flipping 12–4, 25, 127, 185
cokriging 443
collapse feature 444
colliery spoil heap 284
COLLIERY.TXT 284
Colorado (USA) 31, 97, 101, 115, 250–1, 281, 348–9, 396–7
combinations 13–4, 20–1
common factors 527
communality 530–1, 534, 537, 543, 546
commutative matrices 153
compass (fractal dimension) 343–6
complete linkage 498
complex number 145, 276–7
Composita 45, 55, 60–4, 68
compositional data 48–53, 519–20, 523–6, 546–7, 591
compositional variation array 51–3
computer contouring 370–94
concentration parameter 322, 324–5, 330, 342
conditional probability 22–4, 169–71, 552–5
conditional relationship 22
conditional simulation 443
confidence interval 66–8, 72, 200–4, 206–7, 218, 225, 325–6, 342, 424–6, 428, 435, 437, 574
confounded 27, 79
conglomerate 397
conodont 364, 556–9
CONO.TXT 556, 559
constant-sum data 48–50, 519–20, 523–6, 546–7, 549, 554, 560, 585, 591, 594
continental shelf 287
contingency table 93–6, 552–4
continuous random variable 25–9
continuous spectrum 275–7
contouring density of points 341
contour map 294–5, 370–97, 417, 428, 449, 451
convex hull 391, 432
Cooper Basin (Australia) 563–5
COOPERBA.TXT 563–4
coordinates:
  – Bookstein 357–9, 447, 587–8
  – Cartesian 229, 331–4, 336, 358, 360, 362, 447, 449, 451, 587–8
  – Gauss–Krueger 369, 452
  – geographic 369, 398–400, 403, 412, 429, 436, 452
  – polar 332
  – principal 507, 548–52, 567–9
  – UTM 369, 435, 452
cophenetic correlation 493–500
copper 146, 439
core measurement 99, 285, 582, 584

correlation 43–8, 74, 105–7, 116–8, 147, 202, 219, 225, 406, 411, 415, 466–7, 494–7, 499-501, 509, 512, 515, 517, 584–6
- apparent, matrix of 492
- canonical 577–84, 593–4
- coefficient, Pearsonian 105, 116
- cophenetic 493–500
- cross- 161, 246, 248–54, 285–6
- geologic 162, 239, 254, 285–6
- induced negative 46–50, 54, 520
- lithostratigraphic 162, 239, 254, 285–6
- matrix 147, 466, 499–500, 509, 517, 528–30, 546, 571
    - reproduced 533, 537
    - residual 533, 537
- multiple (R) 195, 402
- partial (factor analysis) 527, 531
- serial 182, 245
- similarity measure 489–93, 554
- Spearman's rank 106–7, 116
- spurious negative 48–9
- stratigraphic 162–3, 254
correlogram 246
CORREL.TXT 43
correspondence analysis 507, 552–66
- axes 554, 557, 560, 562
- factor loadings 555, 558–9
cosine 267
cosine θ coefficient 540–4, 545–7
County Durham (UK) 284
covariance 40–3, 418–9, 480, 510, 512, 515
- directional 446
- matrix 147, 500, 514, 523, 536–7, 576, 586, 588
    - reproduced 531–3
covariogram 264–5, 417, 429, 433
COWURINE.TXT 118
Cramer's rule 139
creosote bush (*Larrea tridentata*) 444–5
Cretaceous 24, 31, 97, 281–2, 397, 401–2, 403–5, 446
Cretaceous–Tertiary boundary 287–8
critical region 63–6, 74, 76, 93, 170–2
Croatia 30, 33, 97, 101, 146
CROATRAD.TXT 30, 33
CROPB.TXT 97, 101
crossbed 331, 446
cross-correlation 161, 246, 248–54, 285–6
cross-correlogram 249–50, 254, 286
cross validation 390, 443
crystallographic axes 331
^{137}Cs 33
cubic polynomial 229
cumulative plot 18, 30–2
curvilinear regression 207–14
cycle 267
cyclicity 279
cyclostome bryozoan 287

cyclothem 160, 243

D

"Dansgaard–Oeschger events" 274
data 7–9, 93, 103, 106, 163, 452, 515
- bivariate 191, 214–5, 221
- block 507, 517, 531, 533, 536, 544, 550–1, 561–3, 569
- circular 316–30
- closed 48–50, 519–20, 523–6, 546–7, 549, 554, 560, 585, 594
- compositional 48–53, 519–20, 523–6, 546–7, 591
- constant-sum 48–50, 519–20, 523–6, 546–7, 549, 554, 560, 585, 591, 594
- dimensionality 523
- directional 316–42, 446
- interval 8, 159, 161, 393, 552, 560–1
- nominal 7, 93, 103, 161, 393, 549, 552
- ordinal 8, 93, 103, 106, 161, 549, 552, 560–3
- profile 592
- spherical 336–42
- stationary 183, 214, 256–8, 279, 447
- subsurface structural 380–1, 388–9, 391–4, 398–402, 404–7
- topographic 351–4, 370, 373–4, 378–80, 383, 386
decile 32
declination 446
decline curve 592
deep-sea core 116–7
Deep Sea Drilling Project (DSDP) 446, 593–4
deep-sea fan 283
DEEPSEA.TXT 594
degree of freedom 69, 75, 81, 87, 92, 94–5, 171, 178, 182, 197, 211, 244, 250, 288, 301, 304, 310, 326, 330, 368, 408–10, 414, 464, 469, 484–5, 487, 580–1, 588
Delaunay triangle 375–80
δ^{18}O record 273–4, 591
dendrogram 489, 491–4, 496, 499, 546
- distortion in 494, 499
density of points 294, 299–303, 308, 341
- contouring 341
density of rocks 288
density, well-log 239–400
Denver, Colorado (USA) 250–2
Denver–Julesburg Basin (Colorado) 31, 97, 101
dependent variable 194, 400, 462, 464, 577
depositional environment 518
derivatives of surface 372, 396–7
determinant 136–40, 481, 586
detrending 273, 276
Devonian 282, 371, 403–6
diabase 446, 548
diagenesis 592–3
diagonal matrix 124
- inverse 134

differentiated igneous body 543–5
diffusion-limited aggregation 349
diffusion profile 286
dihedral angle 446
DIHEDRAL.TXT 446
dimensionality, data 553
dimension, fractal 342–4
diorite 288, 548
dip 332–4, 338, 384–5, 404, 446, 450
dip projection 392–3
directional covariance 335–8, 446
directional data 316–42, 446
Dirichlet polygon 376
discontinuities in surface 372, 391–4
discovery well 102, 304
discrete power spectrum 270, 351
discrete probability 12
discrete variable 7–8, 12–24, 490
discriminant:
   – analysis 471–9, 484, 572–7, 590–1
     – multigroup 572–7, 592
   – axis 574
   – index, R_0 475
   – score 471, 475, 574, 577
disjunctive kriging 442
dispersion 319–21, 325, 334, 336–8, 341
dissimilarity 241, 489, 493, 498, 594
DISSIM.TXT 551
distance coefficient 493–9, 548, 567
distance-weighted averaging 382, 385–7, 389, 391
distributary channel 371–2
distribution [See type (χ^2, circular, F-, normal, t-, etc.)]
DJBASIN.TXT 97, 101
DJPOR.TXT 31
"D" and "J" sands (Cretaceous) 31
dolomite 279, 495, 449
dolomitization 591
DOLOMIT.TXT 591
double linear interpolation 396
dragon curve (fractal analysis) 343–4, 348–9
drainage basin 355, 357, 463–5, 468–9
drainage pattern 350–1
drape structure 404
drawdown 223–4
DRAWDOWN.TXT 223
drift 258–9, 261, 428–31, 433–8, 442
drilling mud 279
drillstem test 575
drumlin 117, 355
DRUMLIN.TXT 117
dune 351–5, 447
dye injection test 445
dynamic programming 237, 239, 241, 243

E
earthquake 178, 250–2, 449–51

Eastern Shelf area (of Permian Basin) 304 5
Eckart–Young theorem 502–4, 507, 541, 546, 552, 556, 566, 568, 570
Eden Valley (UK) 117
edge effect 391–2, 415–6
Edinburgh (Scotland) 108
Egypt 448
eigenvalue 141–52, 178, 334–8, 479, 487, 500–2, 505, 507, 512, 514–5, 517–8, 520, 524–5, 527–31, 539, 541–2, 546, 549–52, 554–8, 560–6, 568, 570–1, 573, 576–9, 581, 583–4
eigenvector 141, 152, 215, 217, 330, 334–8, 470, 500–3, 505, 507, 509, 511–2, 514–8, 520, 524–5, 527–31, 539, 541–2, 549–51, 554–8, 560–1, 564–9, 571–4, 576–7, 579
Eisenerz iron mine (Austria) 265, 287, 590
EISENERZ.TXT 287, 590
electron microprobe 286, 411
electron photomicrograph 446
elements, chemical 146, 584–5
elements of a matrix 123–4
elevation, topographic 118–9, 351–4, 373–4
Elk County (Kansas) 262
Ellenburger Dolomite (Cambro–Ordovician) 575–7
ellipse (search target) 296
ellipsoidal depression 322
elongation of drumlin 117
embedded Markov chain 173–4
empirical orthogonal function analysis 592
empirical survivor function 180–1
Encelia farinose (brittlebush) 445
end condition 230
end member 545
England 117, 235, 285, 406
enhanced recovery 114
ensemble 276, 417
environment 118, 211, 369, 591
Eocene 278, 285, 446
epicenter 450
equilibrium landscape 283
ergodicity 276, 417
erionite 592
error (petrographic and geochemical variates) 412
error sum of squares 80–2, 86–91, 195, 198–9, 218–9, 368
error variance (kriging) 418, 420, 424–7, 432–8, 442
Erzgebirge Mountains (Germany) 48, 117
Euclidean distance 236–8, 342–3, 477–8, 548–51, 567–9
Eudora Shale (Missourian) 557, 560
Europe 118
eutectic point 188
evaluating the determinant 136
evolutionary (time series) 214
exact interpolator (kriging) 418, 427
exinite 564

experimental error 27, 79
experimental psychology 500
experimental semivariogram 255, 260–1, 264, 285, 422, 452
exponential model 181, 221, 261–2
extracted organic material (EOM) 564
extrapolation 372, 432

F
factor:
   – analysis 237–9, 470, 479, 488, 500–3, 507–9, 514, 526–40, 538
     – maximum likelihood 528, 538–48
     – Q-mode 521, 540–71
     – R-mode 509, 526–40
   – axes 530–1, 535, 537
   – hypothesis 500, 527–8
   – loading 527, 529–31, 536–9, 541–2, 592
   – model 527–8
   – rotation 533–8, 545–7
     – Kaiser's varimax 533–8
     – oblique 537–40
   – score 535–7, 556
factorial 13, 303
FACTOR.TXT 528
fans, submarine 283
FANS.TXT 283
Farley Limestone (Missourian) 557, 560
Fast Fourier Transform (FFT) 276–7
fault 250, 340, 393–4, 373
fayalite 286
F-distribution 75–92
Fe (iron) 167, 265, 287, 411–2, 448, 486, 590–1
feldspar 188, 446, 490–1, 594
femic 545
FEOOID.TXT 448
ferruginous ooid 448
Festinger's test 105
Fick's second law (diffusion) 286
fiducial limits 206–7
filtering 273, 395–6, 405
finite element analysis 378
Finland 316–7, 325
FINLAND.TXT 316, 319
first-order Markov property 172
first-order stationarity (time series) 276
Fisher County (Texas) 304–5
Fisher distribution 341
Fisher, Sir Ronald 75
fit, lack of 198–200, 211, 228, 413–4
fixed-effects model (Model I) 83
fixed probability vector 170, 173
Florida (USA) 93–4, 96, 220, 285
fluid flow 349
fluoride 118
fold 327
formline structural map 396–7
forward selection 469

Fourier, Jean Baptiste 266
Fourier:
   – analysis 266–9, 276–7, 351–4, 359, 365, 447, 590
   – shape measurement 359–66
   – spectrum 270, 272–7, 353–4, 359, 361–2, 364–5, 447
   – transformation, circular 361–6
fractal analysis 342–54, 447
fractal dimension 342–4
fractional powers of matrices 131
fracture 340, 348–9, 445
France 254
Fremont County (Wyoming) 281–2
frequency 267
frequency analysis [*See* Fourier analysis.]
freshwater 251, 282, 589–91
Frisbee Limestone (Missourian) 557
Front Range (of Rocky Mountains, USA) 250
F-table 77, 482
F-test 76, 80, 197, 200, 211, 327, 330, 408–12, 414, 468–9, 478, 480–2, 484, 487, 573, 588

G
gabbro 548
gabbronorite 288
gambler's ruin 16, 21
gamma-ray log 49, 154, 243–5, 404, 581–4
Ganges River 327
Garden City (Kansas) 447, 351, 353–4
GARDENEW.TXT 447
GARDENNS.TXT 352, 447
garnet 166, 288
GARNETS.TXT 167
gas injection 445
Gauss–Krueger coordinates 369, 452
Gaussian semivariogram 256, 262, 442
generalized:
   – derivative (map) 396–7
   – distance 235, 574
   – variances, test of 484
General Linear Model (GLM) 369
Geochemical Map of the World (IUGS) 366
geochemical variable 4, 48, 51–4, 97, 101, 117–8, 366, 368, 412, 471, 590
geographic coordinates 369, 398–400, 403, 412, 429, 436, 452
geographic information system (GIS) 375
Geological Survey of Canada 366
geologic correlation 162, 239, 254, 285–6
geomagnetic field 331
geometric:
   – distribution 20
   – mean 34, 54, 98
   – probability 295–8
   – variance 99
geomorphic variable 463, 465, 468–9

geostatistics 254–6, 370, 390, 416–38, 442, 452
geothermal gradient 281–2
Germany 48, 117
girdle distribution 337
GIS (geographic information system) 375
GISP–2 ice core 272–4
glabella (of trilobite) 586–7
glacial striation 316–7, 325
glacial till 126–7
Glendon Limestone (Oligocene) 444
GLM (General Linear Model) 369
global zonation 236
GLOMAR.TXT 446
gneiss 288
gold 154, 278, 397
goniatite ammonoid (*Manticoceras*) 502–3
goodness of fit 93–6, 107, 184, 195, 220, 301, 326, 346, 402, 406–7, 467
Gosper island (fractal analysis) 343–4
Gower distance 490, 549
Graham County (Kansas) 395, 399, 406–7
GRAHAM.TXT 411
grain diameter 114, 472, 491
grain outline 359, 362–4
grain-size distribution 97, 116, 472, 518–23, 589, 592
granite 288, 364, 446
Grant, Louisiana (USA) 284
granulite 288
gravel 446, 570
Grayburg Dolomite (Permian) 445, 575–7
Great Basin region (USA) 586
Great Britain 114, 284
Greenland 77, 273–4
Green River Formation (Eocene) 279, 285–6
GREENRIV.TXT 278–9, 285–6
greenstone 593
greywacke 490–1
grid:
  – contouring 380–9, 391–6, 417, 449
  – fractal analysis 346–50, 447
  – node 380–6, 428
  – search 296–9
grid-to-grid operation 395
groundwater 91–2, 97, 110, 588
guard region 311, 391–2
Gulf of California (Mexico) 589
Gulf Coast (USA) 104, 472, 591
Gulf of Tonkin (Viet Nam) 113

H
halite 49, 154
harmonic number 268, 270, 272, 353
harmonic (spectral) analysis 266, 268–75, 361–4
Hausdorff dimension 343
heads or tails 12–5, 25–6, 127, 185–6
heavy metal 589–90

heavy oil 282
hemisphere 336, 338–41
Hermosillo (Mexico) 589
heteroscedasticity 214
heulandite 114, 592
HEULAND.TXT 114
hexagonal network 311–2
Hg (mercury) 220, 369–70, 448
HGCURVE.TXT 220
Hickory Creek Shale (Missourian) 557
hierarchical clustering 489–90, 498–9
hierarchical design (ANOVA) 88–90, 118, 366–70
High Plains aquifer (Kansas) 91–2, 260–1, 435, 437–8
Himalayas 327
histogram 29–31, 180, 304, 306, 309
  – circular 316–9, 446
Holocene 162, 191, 273
homogeneous series 276
homoscedasticity 214
honoring control points 388, 428
Hotelling's T^2 test 478, 481–2
hull, convex 432
"Humble Equation" 114
HUMBLE.TXT 115
Hunter-Shandaken, New York (USA) 284
Hutchinson Salt (Permian) 49, 154
hydrocarbon fraction (HC) 564
hydrocarbon source bed 119, 397, 565, 593
hydrogen index 119
hydrothermal origin 114, 592
hypergeometric distribution 20–1
hypersaline brine 591
hypersthene 548, 594
hypocenter, earthquake 451
hypothesis testing (*See* significance testing.)

I
ice core, GISP–2 272–4
ICECORE.TXT 272
Iceland 344–8
ice movement 325
Idaho (USA) 154
identity matrix (**I**) 124–5
Idria mercury mine (Slovenia) 369–70
igneous petrology 312–3
igneous rock analysis 543–8, 585–6, 593–4
IGNEOUS.TXT 543, 546–7
ill-conditioned matrix 140
Illinois (USA) 366–7
ilmenite 593
imaginary number 145, 276
immiscible fluids 349
inclination 332–3, 446
independent event 22
independent variable 194, 221, 246, 414, 469
India 440, 442–3

Indian subcontinent 327–30
indicator kriging 442
Indochinese peninsula 327
induced correlation 46–50, 140, 508, 520
industrial effluent 590
inertia, moment of 335–6
inertinite 564
inhomogeneity 412
initial saturated thickness 392
injection pressure, mercury 220
injection well 250–2
in situ pressure data 115
integer count 7–8, 92–3, 102
interaction 85, 468–9, 508
intergranular pores 490
interpolation 161, 163–8, 295, 396, 372
interval data 8, 159, 161, 393, 552, 560–1
intrusive 166
inverse distance weighting 386–7, 390
inverse matrix 132–6, 423
inverse regression 205–6, 217, 205
iodine 97
ion exchange 575
Ireton shale (Upper Devonian) 404
iron (Fe) 167, 265, 287, 411–2, 448, 486, 590–1
Island Creek Shale (Missourian) 557
isopach map 372, 395, 449
Istrian peninsula 30, 33, 97, 101, 146, 150
ISTRIA.TXT 146, 152
ITALNAVY.TXT 116–7
IUGS Geochemical Map of the World 366

J
Jaccard's coefficient 490
Japan 178
Java Sea 113
Jay Field (oil), Alabama–Florida, USA 285
joint probability 22–4, 169–71, 480, 553, 555, 562
joints 313
Jurassic 220, 285, 591

K
Kaiser's varimax (factor rotation) 533–8
KANSALT.TXT 154–5
Kansas (USA) 35, 39, 91, 110, 113, 118, 153–4, 223, 243–5, 260–2, 301–2, 350–4, 392–3, 395, 398–9, 406–7, 431, 435, 438–9, 447, 451, 485–6, 556–7, 581–3
karst 444
KENTUCKY.TXT 464, 470
Kentucky (USA) 271–2, 463–4
Kepler, Johannes 266
kerogen 282, 564
key landmark 357
kite diagram 319
k-means procedure 499-500

Kolmogorov–Smirnov statistic 107–8, 112, 184
kriging 255, 265, 295, 390, 416–43, 452
 – block 437–43
 – disjunctive 442
 – error variance 418, 420, 424–7, 432–8, 442
 – exact interpolator 418, 427
 – indicator 442
 – multigaussian 442
 – ordinary 420–30, 432, 437, 440, 452
 – punctual 437–40
 – simple 418–20, 430
 – universal 428–37, 443
Kruskal–Wallis test 105
K_2O (potassium) 48, 114, 202–3, 486, 570
kyanite 288
Kyushu (Japan) 178

L
La Chapelle bank (UK) 114
lack of fit 198–200, 211, 228, 413–4
lag 244, 248–9, 417
lagoon, coastal 589
Lagrange multiplier 420, 429–30, 432
Laguna Mountains (Arizona) 446–7
lake deposit 278, 285
Lambert projection (Schmidt net) 338–41, 446
Lamont sandstone (Mississippian) 380–1, 388
LAMONT.TXT 380–1, 388
landmark 357–8, 360, 447, 587
Landsat 327–30, 593
landscape, equilibrium 283
landslide 178
Lane Shale (Missourian) 557
Lansing–Kansas City Group (Pennsylvanian) 395, 398–9, 407
Laplace's problem 296
Laramie Range (Wyoming) 153, 279–80
large-sample statistics 68
Larrea tridentata (creosote bush) 445
latent factor (factor analysis) 527
latent value (*See* eigenvalue.)
latent vector (*See* eigenvector.)
"law of proportionate effect" 101
lead (Pb) 97, 101, 448
lease tract 303
least squares 191–227, 382, 385, 407, 462–3
 – piecewise linear 384
 – piecewise quadratic (gridding procedure) 384
Leduc Formation (Devonian) 403–4
LEDUC.TXT 371, 406, 411
level of significance 62–6
leveling (time series) 276
Ligonodina (conodont) 364
likelihood 12
Lilliefors procedure 109–10
limestone 88, 127, 160, 168, 172–3, 220, 243–5, 444, 581

line power spectrum 270, 275
lineament 313, 326–30
linear:
— drift 259, 429, 433–7
— interpolation 163–6
— projection (gridding) 385
— regression 199–200, 203–4, 273, 283, 288, 464
— semivariogram model 261, 435
lines:
— density of 314
— parallel survey 295–9, 443
— random pattern of 313–6
lithostratigraphic correlation 162, 239, 254, 285–6
loading 504, 514, 521, 527, 534–7, 551, 569
— diagram 525
"local boundary hunting" 235
local component 397–8, 412
locational analysis 299
Lodgepole Formation (Mississippian) 239–40
LODGEPOL.TXT 239
logarithmic distribution 307
logarithmic transformation 221–4, 226–8
LOGCORE.TXT 581
log empirical survivor function 181
logging tool 204, 583
log–log plot 222
lognormal distribution 97–101
"lognormal law" of geochemistry 97
logratio transformation 50–5, 117–8, 523–5, 585–6, 591
Lord Rayleigh 325
Louisiana (USA) 100, 104, 191, 518–21, 524, 589
LOUISMUD.TXT 191, 196, 198, 209–10
lunar basalt 116, 286, 482–3, 584–5
LUNARBAS.TXT 286

M

MAGELLAN.TXT 257
magnesite 495
magnesium (Mg) 163–5, 486, 590–1
magnetic anomaly 443
magnetic declination 154
magnetite 153, 279–80, 312–3
MAGNETIT.TXT 153
Mahalanobis' distance 478, 574, 588
major:
— axis (principal axis) 215
— diagonal (of matrix) 124
— oxide 51–4, 117, 547
— product matrix 130, 503, 566
manganese (Mn) 164, 590–1
Mann–Whitney test 103–5
MANOVA (multivariate analysis of variance) 487, 592
Manticoceras (goniatite ammonoid) 502–3

map 293–5, 300, 311, 338, 344–6, 354, 370–95, 405–6, 417, 442, 452, 593
— derivative 396–7
— drift 437–8
— error 425
— fault 393
— generalized derivative 396–7
— isopach 391–3, 395, 449
— kriging 417–8, 428, 435–8, 442–3
— standard error 425, 437
— trend residual 397, 399, 404–6, 412–4, 438, 451
— of water-table elevation 422, 427, 431, 437, 440
mapping, plane-table 374
marginal probability 170, 175, 553–4
MARINEOL.TXT 591
marine sediment 591
marine seismic survey 256–7, 263
Markov chain 161, 168–78
— embedded 173–4
Markov property, first-order 172
Maryland (USA) 251, 253
matrix:
— algebra 123–56, 194, 500–8
— cophenetic values 492
— correlation 147, 466, 492, 499–500, 509, 517
— covariance 474, 484, 500, 509–10, 519, 523, 568, 584
— diagonal 124
— distance 490, 493–5, 499, 548–52
— elements of 123–4
— filter 395
— identity ($\mathbf{I}$) 124–5
— ill-conditioned 140
— inverse 132–6, 423
— major product 130, 503, 506
— minor product 130, 503, 505, 566
— off-diagonal, elements of 124
— order of 124
— orthonormal, columns of 503, 507
— overdetermined 520
— pooled variance–covariance 473–4, 584, 588
— rank of 145, 505
— reproduced correlation 532–3, 537
— residual correlation 533, 537
— scalar 124
— similarity 488–90, 491–2, 499–500, 540–1
— singular 132, 139–40, 145, 152, 425, 502, 523
— sparse 136
— square 124
— standardized variance–covariance 531, 583
— symmetric 124
— transition 127, 168–70, 173–5
— tridiagonal 230

– unit 124
– variance–covariance 477, 482–5, 509–10, 515–20, 524–7, 529, 569, 578–9, 584, 586
– within-groups covariance 573, 576–7
maturity 564
maximum likelihood factor analysis 528, 538–48
MDS (multidimensional scaling) 560–66
– loadings 561, 564–5
mean 33–4, 61, 66, 72, 192, 202, 276, 306, 355
– deviation 35
– direction 319–20, 322, 326–7, 332, 341, 446
– rate of occurrence 179
– resultant 321–2, 325, 327, 330
– square 80, 369, 409–10, 197, 469
measurement 7
median 32–3, 103
median grain size 472, 521–2
Mediterranean Sea 116
megacyclothem 556–60
meltwater 77
mercury (Hg) 220, 369–70, 448
– displacement 115, 220, 285
Merriam Limestone (Missourian) 557
metamorphism 592
Mexico 116, 411, 589
Mg (magnesium) 163–5, 486, 590–1
Michigan Basin (USA) 397
microfossil 553
microlaterolog 583
microparticle 77
microprobe 411
Midland Basin (Texas) 445
MIDLAND.TXT 173
Midland Valley (Scotland) 173
Milankovitch cycle 274
mine 156, 265, 280, 287, 366–70, 437, 439, 590
mineralogy, normative 593–4
minor product matrix 130, 503, 505, 566
Miocene 282
misclassification ratio 476
Mississippian 153, 239–40, 380–1, 388, 593
Mississippi Delta 518
Mississippi River Valley 84
Missourian 556–9
mixed-effects model 83
Mn (manganese) 164, 590–1
mode 34
moisture 191–3, 198, 211
moment of inertia 335–6
Montana (USA) 281
Monterey Formation (Miocene) 282
montmorillonite 281
monzonite 548
MOONCRST.TXT 116
Mt. Gleason, California (USA) 284
moving average 246, 273, 383
Mowry Shale (Cretaceous) 190, 281–2

MOWRY.TXT 281
mud 191, 198
– drilling 279
mudstone 174
multidimensional scaling (MDS) 548, 552, 560–6
multigaussian kriging 442
multigroup discriminant analysis 572–7, 592
multinomial distribution 20
multiple correlation coefficient (R) 195, 402
multiple regression 400, 462–71, 479, 577
multiplicative model 223
multiplicative rule of probability 22
multivariate analysis of variance (MANOVA) 574, 592
multivariate morphometrics 357
multivariate normal 480, 483, 486, 584, 527, 590
Muncie Creek Shale (Missourian) 557, 560
Murray pluton (Canada) 446
mutually highest similarities 490, 493–4

N
Na (sodium) 48, 486, 591
Naga Hills 327
National Earthquake Information Center 449
National Geophysical Data Center 446
natural end condition 230
natural neighbor 377
nearest neighbor 310–5, 376–7, 387–9, 445, 449
negative binomial 17, 307–9
negative thickness 393
neighborhood 256, 258–9, 383–4, 388, 418–20, 428–9, 433–4
nested ANOVA 88–90, 118, 366–70, 448, 452
NESTED.TXT 88
neutron density 49, 214
Nevada Test Site 588
Nevada (USA) 588, 593
New Zealand 449–50
nickel 39
Noland County (Texas) 304–5
nominal data 7, 93, 103, 161, 393, 549, 552
nonnegative definite (semivariogram model) 261
nonparametric statistics 102–12
nonstationary 214, 246, 264, 428, 436
norite 548
normal distribution 27–8, 34, 36–7, 55–60, 69, 75, 92, 109, 111, 227, 246–8, 322, 341, 355, 412–3, 424–6, 435–6, 477, 479–80, 538
normal equation 194, 220, 224–5, 400, 418–21, 426, 429–30, 439, 462–3, 466
normalized eigenvector 152, 503
normative mineralogy 593–4
North America 449
North Belridge field (California) 593
North Carolina (USA) 323

North Dakota (USA) 239–40
North Slope (Alaska) 204
Norway 119
NOTREDAM.TXT 373–4, 378–9, 385, 390
Nubia Formation (Triassic) 448
nuclear device 588
nuclear waste 154
nugget effect 263, 285, 442
"nuisance factor" 543, 557
null hypothesis 61–4, 71–2, 76–8, 409, 481, 483–4
numerical taxonomy 471, 488–9, 492
Nyquist frequency 274–5

O

Oasis Valley (USA) 588
OASISVAL.TXT 588
oblique factor rotation 537–40
observation well 422, 425, 431, 435, 440
oceanic basalt 584–5, 593–4
oceanic trench 449
OCS.TXT 100
octant search 387–9, 436, 449
ODESSAN.TXT 445
ODESSANW.TXT 445
Odessa oil fields, Texas (USA) 445
ODESSAW.TXT 445
off-diagonal elements (of matrix) 124
offshore sand bar 371, 472, 474, 476
Ohio (USA) 380, 388
oil:
  – field 31, 36, 97–8, 100-1, 220, 285, 305, 327, 355, 393, 403–6, 412, 439, 445, 449, 451, 593
    – shape of 355
    – volume 100–1
  – gravity 282, 410
  – heavy 282
  – production decline curve 592
  – reservoir 239, 392
  – saturation 36, 392, 581, 583
  – shale 278, 285–6
  – well 14, 301–10, 385, 449, 589
Oklahoma (USA) 36, 99, 211–2, 225–6, 391–2, 406, 575–6
OKLA.TXT 211–2
Oligocene 284, 444, 593
olivine 153, 279, 594
one-tailed test 63–4, 71–2, 108–9, 187, 213
one-way analysis of variance (ANOVA) 78–84, 117, 572, 589
ONEOVA.TXT 79
Ontario (Canada) 364, 446
operational taxonomic unit, OTU 489
opisthoparian (trilobite) 587
ordered measurements 592
order of matrix 124
ordinal data 8, 93, 103, 106, 161, 549, 552, 560–3
ordinary kriging 420–30, 432, 437, 440, 452

ordinary regression 217, 284
ordination 239
ORDNALBX.TXT 562–3
Ordovician 84, 301, 357–8, 447, 451
oreodont 284
OREODONT.TXT 284
organic material 566, 591
orientation 316, 321–3, 329, 340–1
orthid brachiopod, *Resserella* sp. 357–9, 447, 587–8
ORTHID.TXT 359, 447, 587–8
orthogonal axes 150–2, 511, 515, 533
orthogonal regression 218–20
orthonormal (columns of matrix) 503, 507
orthoquartzite 84
ostracode 360
Ouachita Mountains (USA) 391
Outer Continental Shelf 100
outlier 116
overdetermined matrix 520
oxide, major 51–4, 117, 547
oxygen isotope ratio ($\delta^{18}O$) 273–4, 591
Ozark Dome (USA) 397

P

Pacific Ocean 364, 451, 584–5, 594
PAGELER.TXT 243
pair–group methods (clustering) 496
Paleocene 119
paleocurrent 326
paleoecology 553, 557
paleogeography 372
Paleolithic 366
Paleozoic 211–2
Paola Shale (Missourian) 557
parabola 402
parallel-line search 295–9, 443
partial correlation (factor analysis) 527, 531
partial regression coefficient 409–11, 463, 465–9
partitioning methods (classification) 488
pattern recognition 162, 271
PCA (*See* principal component.)
PCOORD.TXT 551
Pearce element ratio diagram 48
Pearsonian correlation coefficient 105, 116
pebbles 46, 75, 126–7, 490
pedicle valve 357–8
pegmatite 188
Pennsylvanian 35, 39, 70, 73, 113, 115, 243–5, 366, 391–2, 395, 398–9, 407, 581, 583
percentile 32–3
perimeter 355–6, 359–62, 448
period 267, 353
periodogram 161, 270–2, 274–5, 351–2
peristome 503
permeability 27, 84–5, 99, 115, 225–6, 331, 581, 583, 591
Permian 49, 154, 282, 445, 449

Permian Basin 304–7, 309–10
Perth Amboy, New Jersey (USA) 284
Peru 451
Petrified Forest, Arizona (USA) 284
petrofabric 331, 337–8, 341, 412, 591
petroleum 99–100, 113, 566, 591
  – exploration 414–6, 451
  – source-rock 564
petrophysical well log 102, 115, 154, 204, 583
"phantom black shale" 560
phase angle 267, 362
phi transformation 97
Phillippines 439
phillipsite 592
phosphate 118
Phosphoria shale (Permian) 282
photogeologic map 397, 593
photomicrograph 448
Piceance Basin 348–9
piecewise linear least squares 384
piecewise polynomial 229
piecewise quadratic least squares gridding procedure 384
pixel 348
plagioclase 279
plane-table mapping 374
Pleistocene 351–2, 447
plunge 340
pluton 545, 570
point density 294, 299–303, 308, 310, 341
point distribution 299–313
Poisson distribution 19, 102, 184–5, 302–10, 314, 368
polar coordinates 332, 359–63, 448
pole (on unit sphere) 340
polygon (triangulation) 376
polynomial 142, 207, 229
  – bicubic 378
  – drift 429
  – regression 207–14, 228, 268, 284, 288, 403, 410–2, 462
  – trend surface 403–7, 409–12, 415, 451
pooled estimate 73, 485
pooled variance–covariance matrix 473–4, 584, 588
population 28, 34, 61, 196
pores 88, 491
porosity 31, 70, 73, 76, 99, 113–6, 204, 206, 225–6, 285, 372, 439, 581, 583, 591, 593
porous medium 349
PORPERM.TXT 99, 225
positive definite (semivariogram) 261
postmultiplication 129
potassium (K_2O) 48, 114, 202–3, 486, 570
potassium-40 243
power 271, 352–5, 365, 415
  – plant 444
  – spectrum 270–4, 277, 362, 364, 447–8, 592
    – two-dimensional 354

  – transform 102
Precambrian 116, 279, 397
precision 26
premultiplication 129
primate 357
principal axis 215, 511–2, 537
principal component:
  – analysis (PCA) 235, 239, 470, 479, 507, 509–25, 527–8, 540, 566, 569, 577–8, 588–90, 592
    – loading 513–4, 517, 525
    – score 512–3, 519, 522, 526, 535, 556, 589
principal coordinates 507, 548–52, 567–9
principal diagonal 124
prism 277
pristane/n–C_{17} ratio 564
pristane/phytane ratio 564, 591
probabilistic similarity coefficient 490
probability 11–24, 127, 560
  – additive rule of 21
  – bivariate normal distribution 481
  – conditional 22–4, 169–71, 552–5
  – discrete 12
  – distribution, normal (*See* normal distribution.)
  – ellipses, bivariate 447
  – geometric 295–8
  – joint 22–4, 169–71, 480, 553, 555, 562
  – marginal 170, 175, 553–4
  – multiplicative rule of 22
Procrustes analysis 357
profile data 592
profile distance 560–1, 563–4
PROFILE.TXT 592
projection equation 338–9
proper value 141
proper vector 141
prospects (oil and gas) 104, 593
PROSPECT.TXT 154
provenance 364
Prudhoe Bay oil field (Alaska) 204–5
PRUDHOE.TXT 204
pseudo landmark 357
pseudopoint (triangulation) 380
P_2O_5 116
punctual kriging 437–40
pure error 198, 211
P-value 64
pygidium (of trilobite) 587
pyroxene 51, 54
Pythagorean theorem 320, 335, 383

Q
Q-mode:
  – analysis 500–1, 505–7
  – factor analysis 521, 540–71
  – loading 543–5, 560–3, 566, 568–9
  – score 504–6, 556, 568

QMODE.TXT 541–3
quadrant search 387–8
quadrat 300, 302–3, 445
quadratic equation 142, 209, 259, 287, 400
quantile 32
quartile 32–3
quartz 114, 116, 188, 446, 490–1, 594
 – diorite 548
 – monzonite 570
 – syenite 548
QUEBECAU.TXT 278
Quebec (Canada) 278
Quindaro Shale (Missourian) 557

R
radian 266, 325
radiation 30, 33, 444–5, 570
radioactivity 243, 366–8, 404
radiolarian 189–90
radionuclide 570–1
RADIO.TXT 570–1
random-effects model (Model II) 83
random:
 – error 196, 199, 227–8, 412–14, 462
 – function (geostatistics) 417–8
 – location 299, 302, 312–3
 – noise 246–7
 – order 515
 – sample 28–9, 408, 483
 – variable 25–9, 79, 196, 246–8, 516
 – walk 315
randomness, testing for 322–5, 341–2
range (geostatistics) 256–8, 433–6
Rangely oil field (USA) 115
RANGELY.TXT 115
Raniganj coal field (India) 440–3
RANIGANJ.TXT 440
rank of matrix 145, 505
rank of observations 8, 103–7, 514–5
Rappahannock, Virginia (USA) 284
rate of occurrence (of events) 184
ratio scale 8, 159, 161, 393, 552, 560–1
Raton Basin (Colorado) 396–7
Rayleigh's test 325–6
Raytown Limestone (Missourian) 557, 560
reaction rim 286
Recent (Holocene) 162, 191, 273
reciprocal matrix (*See* inverse matrix.)
rectangular integration 166–7
recursive procedure 237, 242
reduced major axis (RMA) 214–5, 217, 284
reef 371, 403–6, 449
REEF.TXT 449
regional dip 398, 404
regionalized variable 254–65, 295, 416–7, 420–1, 428–9, 433–6

regression 161, 191–227, 269, 284, 295, 346–7, 352–3, 397–417, 462–72, 590
 – "best possible" 468–70
 – curvilinear 207–14
 – generalized linear 417
 – inverse 204–7
 – ordinary 217, 284
 – orthogonal 218–20
 – through the origin 220–1
"reification" 517
relaxed end condition 230
remanent magnetism 446
remote sensing 444, 593
replicate 35, 78, 199, 413–4, 425
reproduced correlation matrix 532–3, 537
R.V. Glomar Challenger 446
RESENG.TXT 83
reserve estimates (coal) 441
reservoir, oil 239, 392
residual 226–7, 398, 401, 405–6, 408, 428, 433–4, 437–8
 – correlation matrix 533, 537
 – map 406
 – matrix (maximum likelihood) 539
 – stationary 428
 – trend map 451
resin 564
resistivity 114, 239–40
response surface 412
Resserella sp. (brachiopod) 357–9, 447, 587
RESSEREL.TXT 358, 447, 587–8
resultant 319–23, 325, 327, 329–30, 332–3, 341–2
rhodochrosite 495
rhyolitic volcanic ash 281
Rice County, Kansas (USA) 113, 154
Richardson's dimension 346–8
river 283, 463–5, 467, 469–70
RMA line 215–8
R-mode:
 – analysis 500, 504–6, 509–38, 566–71
 – correspondence axis 556
 – factor analysis 526–40, 542
 – loading 504–6, 556, 560, 562–3, 568–9
 – score 504, 506, 535–7, 566
R, multiple correlation coefficient 195, 402
rock analysis, igneous 543–4, 585–6, 593–4
Rock-Eval pyrolysis 119, 593
Rock Lake Shale (Missourian) 557
Rocky Mountain Arsenal (Colorado) 250–2
Rocky Mountains (USA) 189, 250–2, 278, 285, 403–4
rose diagram 316–9, 323, 329, 446
rotation, factor 533–8, 545–7
roughness (fractal analysis) 342, 353, 363
round-off error 209
roundness 106
R- and *Q*-mode analysis 501, 566–71
ruler method (fractal analysis) 343–46

runoff 271–2, 589
runs test 161, 185–90, 278

S

St. Peter Sandstone (Ordovician) 84
salinity 93–4, 96, 111, 251, 253–4
salt dome 100, 104
saltwater 575, 589
sample, definition of 28
sample, random 29, 486
sample size (trend-surface analysis) 415
sampling 6–7, 20, 28–9, 315, 368–9, 486
San Andres Limestone (Permian) 445
sand 116, 140, 351–2, 355, 359, 362–4, 371–2,
403, 446–7, 472, 520–3, 570, 572, 589
Sandford St. Martin (UK) 285
sandstone 78–9, 81, 106–7, 114–5, 119, 127, 160,
168, 172–4, 211–2, 348, 397, 593
SANDS.TXT 472, 474
San Jacinto County (Texas) 282
Santa Barbara Channel 282
Santa Maria basin (California) 282
satellite image 327, 444, 593
saturated thickness 393
saturates 564
scalar matrix 124
Schellerhau pluton 48, 117
SCHELLER.TXT 48, 117
Schmidt net (Lambert projection) 338–41, 446
Scotland 107, 173
sea level, changes in 557
search:
 – for control points 263, 383, 387–9, 394
 – nearest-neighbor 387–8, 449
 – octant 387–9, 449
 – pattern 294–9, 443
 – quadrant 387–8
 – systematic 294
seawater 251, 556
secondary dolomitization 591
second derivative 229, 396–7
second-order Markov (sequence) 172
second-order stationarity (time series) 276
sediment 114, 116–7, 283, 369, 404, 518–21, 589–
90
sedimentary zeolites 114, 592
sediment grain size 114, 116–7, 472, 518–23, 589,
592
sediment load 283
segmenting sequences 234
seismic reflection 256, 288, 296, 370, 380, 388,
390, 444, 449, 452
SEISMIC.TXT 449, 452
selenium 97
self-affine 342
self-similar 243, 342, 346
self-stationary 276
semiaxis 147–8, 152, 511

semimadogram 264
semivariance 254–64, 420, 422–4, 426, 431, 433–
4, 439–40
semivariogram 161, 255, 259, 287, 417, 422, 428–
9, 431, 433–4, 436, 452, 590
 – alternatives to 264–5
 – converting to covariogram 265
 – experimental 255, 260–1, 264, 285, 422,
452
 – Gaussian 262–3, 442
 – linear 261–2, 422–4, 431, 434–5
 – span of 285
 – spherical 261, 263, 436
 – theoretical 255–63, 419, 421
sequence 159
serial correlation 182, 245
seriation 161, 239
series of events 161, 178–85
serpentinite 593
shale 35, 38–9, 49, 69, 127, 154, 160, 168, 172–
4, 189–90, 243, 281, 366–8, 403, 563, 565, 591,
593
 – black 281, 366–8
 – oil 278–9, 285
 – "phantom black" 560
 – siliceous 90, 189, 281
shape 355–66, 448, 587–8
sharpening filter 395
shear stress 155, 284
shingle beach 46, 75
sialic rock 545
siderite 287, 495
Siegel–Tukey test 105
Sierpinski gasket (fractal analysis) 343–4
significance 64–7
 – level of 62–6
 – tests of 71–2, 74, 76, 82–3, 86–7, 89–90, 96,
106, 187, 197, 202, 210–2, 224–5, 307, 323–7,
342, 407–12, 465, 468–9, 477–9, 482, 484, 487,
580–1, 584
significant digits 466
silica (SiO_2) 486, 590, 594
siliceous shale 90, 189, 281
siliciclast 448
sill 256, 258, 261–3, 436, 442
sill, diabase 446
silt 116–7, 520
siltstone 119, 174, 211
similarities, mutually highest 490, 493–4
similarity:
 – cosine θ 540
 – mutually highest 493
 – within-cluster 498
 – fractal dimension 343
 – matrix 488–90, 500, 540, 554, 560, 562
simple kriging 418–20, 430, 437
simple matching coefficient 490
simple structure 531, 540
simplex 523–4

simulation, conditional 443
simultaneous equation 132, 194, 209, 400, 428, 470, 502
simultaneous R- and Q-mode analysis 566–71
sine wave 246–7, 268–9
single linkage clustering 496–9
singular matrix 132, 139–40, 145, 152, 425, 502, 523
singular value 503, 528–31, 541, 555–6, 568, 592
singular value decomposition (SVD) 136, 152, 502–3, 531, 556, 569, 573, 578
singular value 503, 528–31, 541, 555–6, 568, 592
sinkhole 444
sinusoidal (wave form) 268–9, 274–5
SiO_2 (silica) 486, 590, 594
64-Zone sandstone (Oligocene) 593
SLOFEPB.TXT 448, 452
slope 283, 384, 396, 449–50
slotting 239–43
Slovenia 369, 448
SLOVENIA.TXT 369, 448, 452
Smackover Formation (Jurassic) 285, 591
SMACKOVR.TXT 285
small-sample statistics 68
smithsonite 495
smoothing (filtering) 395
snow 273
social sciences 501, 552
sodium (Na) 48, 486, 591
soil 97, 101, 146, 235, 285, 351, 444, 448
solvent extraction 564
Solway Lowlands (UK) 117
sonic transit time 49, 154, 204, 206, 214, 285, 288, 581, 583
SONIC.TXT 288
Sonora area (Mexico) 589
SONORA.TXT 589
sorting, degree of 106, 521–2
source rock 566, 591
South Africa 397
South America 256–7, 263, 450
South Bend Limestone (Missourian) 557
span (semivariogram) 285
span (spline function) 229
sparse matrix 136
spatial covariance 417–22, 430, 433–4, 439, 443, 452
spatial domain 277, 396
Spearman's rank correlation 106–7, 116
specific gravity 113, 279–80
spectral:
  – analysis 266–77, 287, 351, 590
  – density 161, 272–5
  – method (fractals) 351–5, 447
  – window (filter) 273
spectrum, Fourier 270, 272–7, 353–4, 359, 361–2, 364–5, 447
sphalerite 411–2
SPHALRT.TXT 411

spherical:
  – angle 333
  – data 330–41, 446
  – model (semivariogram) 261, 435–6
  – variance 332, 334, 446
Spiro Sand (Pennsylvanian) 391–2
Spitzbergen Island 119
spline function 161, 228–34, 378
Spring Hill Limestone (Missourian) 557
SPTZBRGN.TXT 119
spurious negative correlation 48–9
squared Euclidean distance 548–9
square matrix 124
square network 312
square-root transformation 102
Sr 591
stagewise regression 469
standard deviation 35–9, 216, 465–6
standard error 59, 67, 201, 203, 218, 306, 314, 325–6, 364–5, 424–6, 435–8, 452
standardization 57, 61, 418, 477–9, 493, 495, 517, 519, 528
standardized variance–covariance matrix 531, 583
standard normal form 57, 466
stationarity, first-order (time series) 276
stationarity, second-order (time series) 418
stationarity, strong 276
stationary:
  – data 183, 214, 256–8, 279, 447
  – probability matrix 131
  – residual 428
statistics 29, 34–9, 479, 482
  – large- and small-sample 68
stepwise discriminant analysis 479
stepwise regression 469
stereographic projection (Wulff net) 338
stochastic 274–6, 342, 349
Stoner Limestone (Missourian) 557, 560
STPETER.TXT 84
Straits of Magellan 256–7, 263
stratigraphic correlation 162–3, 254
stratigraphic section 168, 239–41
stratovolcano 179
stream 163–5, 283, 351, 390, 464, 468
  – basin 283
  – profile 592
  – sediment analysis 590
stress 155–6
"stretchability" 147
striation 316–7, 325
strike and dip 331–3
strip mine 366
strong stationarity 276
strontianite 494–5
structural analysis 218, 259, 422, 429, 434–6
structural data 370–1, 394–5, 398, 405, 412–4
Student's t 68–74, 482
subcomposition 50, 591, 594

subduction surface 202–3, 449–51
submarine canyon 283
submarine fan 283, 593
subsurface structural data 380–1, 388–9, 391–4, 398–402, 404–7
sulfur 282, 591
SULFUR.TXT 282
sum of squares 37–8, 43, 45, 79–82, 86–91, 195, 197–9, 205, 210–3, 216–7, 236–7, 270–1, 401–2, 408–10, 465, 467–9, 473–4, 497
sunspot cycle 279
support (of regionalized variable) 255, 437
surface-wave dispersion curve 592
surveying 374
Sweden 590
SWEDEN.TXT 590–1
syenite 548
symmetric matrix 124
systematic error 79

T

TABLE612.TXT 510, 514–5
tally 168–9, 553–5
tangent plane 339–40
target 296–9
taxonomy, numerical 487–8
t-distribution 68–75, 480–4
tectonic plate 446
temperature 8, 281–2, 411, 446
  – bottomhole 153
TEMPER.TXT 282
Tensleep Sandstone (Pennsylvanian) 70, 72–3, 76
tensor 378
Tertiary 287–8, 348–9, 397
Tertiary basin (Wyoming) 397
tests of significance (*See* significance, tests of.)
Texas (USA) 100, 282, 304–5, 309, 406, 445, 449, 472, 575–6
textural maturity 106
Thames River valley (UK) 235
THEMATIC.TXT 593
theoretical semivariogram 255–63, 419, 421
"theory of breakage" 101
thermal maturity 281–2, 593
thermal radiation 445
THERMAL.TXT 444
thickness, negative 393
thickness, saturated 391–3
Thiessen polygon 376–7
thin section 88, 490–1
thorium 570
tidal cycle 251–3
time domain 277
time series 159, 178–83, 185, 213, 243–8, 250–4, 266–8, 275–8, 295, 417
TIN (triangulated irregular network) 375, 393–4

titanium 48
tolerance limit 219
Tonga–Kermadec Trench (New Zealand) 449–51
TONGA.TXT 449
topographic data 351–4, 370, 373–4, 378–80, 383, 386
topography, "basket-of-eggs" 117
topologic information 375
TOPSOIL.TXT 285
torus 311
total organic carbon (TOC) 119, 564
total regression coefficient 463
total sum of squares 80, 195
township (U.S. Public Land Survey system) 366
trace element 146, 588–9
trace of matrix 150, 512, 524, 551
tracts (containing discovery wells) 305, 309
transient 287
transition matrix 127, 168–70, 173–5
transition pair 178
transposition 126
trapezoidal approximation 360
tree diagram (*See* dendrogram.)
trend:
  – in observations) 161, 179, 182, 198, 213–4, 281
   – residual map 451
   – surface 195, 294–5, 378, 384, 397–416, 429, 449, 451, 462–3, 465, 479, 589
    – edge effects in 391, 415–6
triangular diagram 49
triangular network, Delaunay 375
triangulated irregular network (TIN) 375, 393–4
triangulation 374–80, 388, 449
Triassic 448
triaxial stress 284
tridiagonal matrix 230
trigonometric relationship 266–9
trilobite 586–7
TRILOBIT.TXT 586
truncation 44
T^2 test 474, 482–4, 487
t-test 70, 74, 116, 212, 250, 307, 480–4
Tukey–Hanning filter 273
two-tailed test 63, 108, 187, 213
two-way analysis of variance (ANOVA) 84–7, 116
two-way travel time 449, 452
TWOWAY.TXT 115
Tyne Gap (UK) 117
type I error 62
type II error 62, 415

U

ultimate production 98
ultrabasite 288
ultramafic 280

umbilicus 502–3
unbalanced design (ANOVA) 367–70
unbiased estimate 34, 192, 418, 429
unconditional 170
unconformity (stratigraphic) 388
underlying (latent) factor 527
uniform density 299–302
uniform distribution 301, 323, 337
unimodal vector 337
unique factor 527
unique variance 507, 536–9
U.S. Geological Survey 351, 366
U.S. Gulf Coast 191
unit matrix 124
unit vector 319–20
universal kriging 428–37, 443
Universal Transverse Mercator (UTM) projection 369, 435–7, 452
University of Sonora (Mexico) 589
unweighted average linkage 497–9
uranium 570
"Useful Heat Value" (UHV) 440–3
Utah (USA) 439, 586, 593
UTM coordinates 369, 435–7, 452

V

vanadium 39
variable:
    – continuous 8, 25–9
    – dependent 194, 400, 462, 464, 577
    – discrete 7–8, 12–24, 490
    – independent 194, 221, 246, 414, 469
    – random 25–9, 196, 254, 416
    – regionalized 254–65, 295, 416–7, 420–1, 428–9, 433–6
    – regressed 194
    – regressor 194
variance 35–9, 66–70, 69–70, 75–92, 94, 101, 195, 226, 270–2, 276, 306–7, 311, 355, 361–2, 398, 407
variance–covariance matrix 477, 482–5, 509–10, 515–20, 524–7, 529, 569, 578–9, 584, 586
variation, coefficient of 39
varimax factor score 547
varimax rotation 534–7
varve 273, 279, 285–6
vector 124
    – direction 316–341, 445–6
    – fixed probability 170, 173
    – latent (See eigenvector.)
    – proper 141
    – resultant 319–20, 445–6, 450
vegetation, distribution of 444
Vilas Shale (Missourian) 557–8
vineyard 146, 150
Viola Limestone (Ordovician) 575–7
Virginia (USA) 570–1
viscous fingering 349

vitrinite reflectance 564, 593
volcanic ash, rhyolitic 281
volcanic eruption 178–82, 202–3
von Mises distribution 322, 324–6, 341
Voronoi polygon 376

W

Wabaunsee County (Kansas) 244–5
Wales 284
Ward's method 238, 497–8
Wasatch Formation (Eocene) 348–9, 446
WASATCH.TXT 446
waste, injected 250–2
water:
    – connate 575
    – quality 485–6, 588
    – saturation 581, 583
    – table 91–110, 223–4, 260, 422, 424–7, 431–2, 434–40
water-flood breakthrough 445
WATER.TXT 91
"Waulsortian" (carbonate algal) mound 239
wavelength 267
wave number 267–8, 352, 362
weak stationarity (time series) 276
Weber Sandstone (Pennsylvanian) 115
Weichselian (Wisconsinan) glacial period 274
weighted
    – averaging 382–93, 395–6, 389–90, 449
    – pair–group linkage 493, 496–9, 546
    – projection 384–6, 390
    – regression 224–7
Wellington Formation (Permian) 49, 154
well-log analysis 227, 581–3
well-log density 239–400
well logs, automatic zoning of 234–40
well, oil 14, 301–10, 385, 449, 589
well, water 422–3, 485, 588
WELLWATR.TXT 486
West Lyons oil field (Kansas) 113
West Texas (USA) 36
WHITE.TXT 93
Whitewater Bay 93–4, 96, 111
whorl 502–3
Wilburton gas field (Oklahoma) 391–2
Wilcoxon test 105
Williston Basin (North Dakota) 239
Windfall Reef (Devonian) 406
Wind River Basin 73
Wisconsinan 274
Wishart's modification 498
witherite 494–5
within-cluster similarity 498
within-groups covariance matrix 573, 576–7
WLYONS.TXT 113
Wolf River (Kansas) 350–1
Woodford shale (Devonian) 282
Wulff net 338–40, 446

Wyoming (USA) 70, 72–3, 125, 153, 279–81, 286, 397, 406, 446

YUMA.PIC 447
YUMA.TIF 447

X
xenocryst 286–7
Xian province (China) 113
X-ray fluorescence 206

Y
Yellowcraigs (Scotland) 108
Yuma (Arizona) 446

Z
zeolites, sedimentary 114, 592
ZEOLITES.TXT 592
zero isopach problem 391–3, 449
zircon 102
zonation 234–41
z-score 57–8, 95, 110, 476–7
z-statistic 57–8, 61, 63–4, 66, 310

Colophon

All typesetting of this book was done by Jo Anne DeGraffenreid on a Macintosh PowerPC 8100 computer. The body of the text is 10-point Lucida Bright with Lucida Fax demibold roman and italic highlights. Chapter headings are 36- and 18-point Lucida Bright and first- and second-order headings are 17- and 14-point Computer Modern, sans serif. The manuscript was prepared using Microsoft Word v. 5.1 for Macintosh as text editor and set in type using Textures v. 2.0. The use of Textures, Blue Sky Software's Macintosh implementation of the TEX typesetting program, in combination with Plain TEX allowed enormous flexibility in the preparation of ANOVA and other tables. Equations created in Word using MathType v. 3.5 by Design Science were easily converted to TEX code and then modified by macros designed to suit the particular mathematical notation.

Illustrations were prepared by the author on a PowerPC Macintosh G3 computer using Canvas v. 3.5 and Adobe Illustrator v. 7.0 graphics software. Contour maps were made using SURFACE III v. 2.6 for Macintosh. Most statistical analyses were run using SAS Institute's JMP v. 3.2 interactive program for exploratory data analysis. Mathematical calculations were made using MathCad v. 3.1. Some Appendix tables were generated with e-HOP v. 1 running under VirtualPC emulation. Geostatistical calculations were made by R.A. Olea using the GSLIB (1st ed.) library running on a Sun SPARCstation 5. Singular value decompositions, Fourier analyses, some Appendix tables, and verification of all calculations were made by Geoff Bohling on a Sun Ultra 1 workstation using custom-programmed routines and S-Plus v. 4.

Woodcuts used for chapter heads were adapted from *Beschreibung: Alesfürnemisten Mineralischen Erzt und Berckwercksarten* by Lazarus Erckern (1574, Prague), *Geometriae Practicae Novae et Auctae* by M. Daniel Schwenter (1667, Nürnberg) and *De Re Metallica* by Georg Agricola (1556, Basle). Final camera-ready pages of the book were printed at 600 bpi on an Apple LaserWriter 1600/600.